Geophysical Monograph Series

Geophysical Monograph Series

186 **Amazonia and Global Change** *Michael Keller, Mercedes Bustamante, John Gash, and Pedro Silva Dias (Eds.)*
187 **Surface Ocean–Lower Atmosphere Processes** *Corinne Le Quèrè and Eric S. Saltzman (Eds.)*
188 **Diversity of Hydrothermal Systems on Slow Spreading Ocean Ridges** *Peter A. Rona, Colin W. Devey, Jérôme Dyment, and Bramley J. Murton (Eds.)*
189 **Climate Dynamics: Why Does Climate Vary?** *De-Zheng Sun and Frank Bryan (Eds.)*
190 **The Stratosphere: Dynamics, Transport, and Chemistry** *L. M. Polvani, A. H. Sobel, and D. W. Waugh (Eds.)*
191 **Rainfall: State of the Science** *Firat Y. Testik and Mekonnen Gebremichael (Eds.)*
192 **Antarctic Subglacial Aquatic Environments** *Martin J. Siegert, Mahlon C. Kennicut II, and Robert A. Bindschadler (Eds.)*
193 **Abrupt Climate Change: Mechanisms, Patterns, and Impacts** *HarunurRashid, LeonidPolyak, and EllenMosley- Thompson(Eds.)*
194 **Stream Restoration in Dynamic Fluvial Systems: Scientific Approaches, Analyses, and Tools** *Andrew Simon, Sean J. Bennett, and Janine M. Castro (Eds.)*
195 **Monitoring and Modeling the Deepwater Horizon Oil Spill: A Record-Breaking Enterprise** *Yonggang Liu, Amy MacFadyen, Zhen-Gang Ji, and Robert H. Weisberg (Eds.)*
196 **Extreme Events and Natural Hazards: The Complexity Perspective** *A. Surjalal Sharma, Armin Bunde, Vijay P. Dimri, and Daniel N. Baker (Eds.)*
197 **Auroral Phenomenology and Magnetospheric Processes: Earth and Other Planets** *Andreas Keiling, Eric Donovan, Fran Bagenal, and Tomas Karlsson (Eds.)*
198 **Climates, Landscapes, and Civilizations** *Liviu Giosan, Dorian Q. Fuller, Kathleen Nicoll, Rowan K. Flad, and Peter D. Clift (Eds.)*
199 **Dynamics of the Earth's Radiation Belts and Inner Magnetosphere** *Danny Summers, Ian R. Mann, Daniel N. Baker, and Michael Schulz (Eds.)*
200 **Lagrangian Modeling of the Atmosphere** *John Lin (Ed.)*
201 **Modeling the Ionosphere-Thermosphere** *Jospeh D. Huba, Robert W. Schunk, and George V. Khazanov (Eds.)*
202 **The Mediterranean Sea: Temporal Variability and Spatial Patterns** *Gian Luca Eusebi Borzelli, Miroslav Gacic, Piero Lionello, and Paola Malanotte-Rizzoli (Eds.)*
203 **Future Earth – Advancing Civic Understanding of the Anthropocene** *Diana Dalbotten, Gillian Roehrig, and Patrick Hamilton (Eds.)*
204 **The Galápagos: A Natural Laboratory for the Earth Sciences** *Karen S. Harpp, Eric Mittelstaedt, Noémi d'Ozouville, and David W. Graham (Eds.)*
205 **Modeling Atmospheric and Oceanic Flows: Insightsfrom Laboratory Experiments and Numerical Simulations** *Thomas von Larcher and Paul D. Williams (Eds.)*
206 **Remote Sensing of the Terrestrial Water Cycle** *Venkat Lakshmi (Ed.)*
207 **Magnetotails in the Solar System** *Andreas Keiling, Caitríona Jackman, and Peter Delamere (Eds.)*
208 **Hawaiian Volcanoes: From Source to Surface Rebecca** *Carey, Valerie Cayol, Michael Poland, and Dominique Weis (Eds.)*
209 **Sea Ice: Physics, Mechanics, and Remote Sensing** *Mohammed Shokr and Nirmal Sinha (Eds.)*
210 **Fluid Dynamics in Complex Fractured-Porous Systems** *Boris Faybishenko, Sally M. Benson, and John E. Gale (Eds.)*
211 **Subduction Dynamics: From Mantle Flow to Mega Disasters** *Gabriele Morra, David A. Yuen, Scott King, Sang Mook Lee, and Seth Stein (Eds.)*
212 **The Early Earth: Accretion and Differentiation** *James Badro and Michael Walter (Eds.)*
213 **Global Vegetation Dynamics: Concepts and Applications in the MC1 Model** *Dominique Bachelet and David Turner (Eds.)*
214 **Extreme Events: Observations, Modeling and Economics** *Mario Chavez, Michael Ghil, and Jaime Urrutia-Fucugauchi (Eds.)*
215 **Auroral Dynamics and Space Weather** *Yongliang Zhang and Larry Paxton (Eds.)*
216 **Low-Frequency Waves in Space Plasmas** *Andreas Keiling, Dong-Hun Lee, and Valery Nakariakov (Eds.)*
217 **Deep Earth: Physics and Chemistry of the Lower Mantle and Core** *Hidenori Terasaki and Rebecca A. Fischer (Eds.)*
218 **Integrated Imaging of the Earth: Theory and Applications** *Max Moorkamp, Peter G. Lelievre, Niklas Linde, and Amir Khan (Eds.)*
219 **Plate Boundaries and Natural Hazards** *Joao Duarte and Wouter Schellart (Eds.)*
220 **Ionospheric Space Weather: Longitude and Hemispheric Dependences and Lower Atmosphere Forcing Timothy Fuller-Rowell,** *Endawoke Yizengaw, Patricia H. Doherty, and Sunanda Basu (Eds.)*
221 **Terrestrial Water Cycle and Climate Change Natural and Human-Induced Impacts** *Qiuhong Tang and Taikan Oki (Eds.)*
222 **Magnetosphere-Ionosphere Coupling in the Solar System** *Charles R. Chappell, Robert W. Schunk, Peter M. Banks, James L. Burch, and Richard M. Thorne (Eds.)*
223 **Natural Hazard Uncertainty Assessment: Modeling and** *Decision Support Karin Riley, Peter Webley, and Matthew Thompson (Eds.)*
224 **Hydrodynamics of Time-Periodic Groundwater Flow: Diffusion Waves in Porous Media** *Joe S. Depner and Todd C. Rasmussen (Auth.)*
225 **Active Global Seismology** *Ibrahim Cemen and Yucel Yilmaz (Eds.)*
226 **Climate Extremes** *Simon Wang (Ed.)*
227 **Fault Zone Dynamic Processes** *Marion Thomas (Ed.)*
228 **Flood Damage Survey and Assessment: New Insights from Research and Practice** *Daniela Molinari, Scira Menoni, and Francesco Ballio (Eds.)*
229 **Water-Energy-Food Nexus – Principles and Practices** *P. Abdul Salam, Sangam Shrestha, Vishnu Prasad Pandey, and Anil K Anal (Eds.)*
230 **Dawn–Dusk Asymmetries in Planetary Plasma Environments** *Stein Haaland, Andrei Rounov, and Colin Forsyth (Eds.)*
231 **Bioenergy and Land Use Change** *Zhangcai Qin, Umakant Mishra, and Astley Hastings (Eds.)*
232 **Microstructural Geochronology: Planetary Records Down to Atom Scale** *Desmond Moser, Fernando Corfu, James Darling, Steven Reddy, and Kimberly Tait (Eds.)*
233 **Global Flood Hazard: Applications in Modeling, Mapping and Forecasting** *Guy Schumann, Paul D. Bates, Giuseppe T. Aronica, and Heiko Apel (Eds.)*
234 **Pre-Earthquake Processes: A Multidisciplinary Approach to Earthquake Prediction Studies** *Dimitar Ouzounov, Sergey Pulinets, Katsumi Hattori, and Patrick Taylor (Eds.)*
235 **Electric Currents in Geospace and Beyond** *Andreas Keiling, Octav Marghitu, and Michael Wheatland (Eds.)*
236 **Quantifying Uncertainty in Subsurface Systems** *Céline Scheidt, Lewis Li, and Jef Caers (Eds.)*
237 **Petroleum Engineering: Principles, Calculations, and Workflows** *Moshood Sanni (Ed.)*

Geophysical Monograph 238

Geological Carbon Storage

Subsurface Seals and Caprock Integrity

Stéphanie Vialle
Jonathan Ajo-Franklin
J. William Carey
Editors

AGU 100
ADVANCING EARTH
AND SPACE SCIENCE

WILEY

This Work is a co-publication between the American Geophysical Union and John Wiley & Sons, Inc.

This edition first published 2019 by John Wiley & Sons, Inc., 111 River Street, Hoboken, NJ 07030, USA and the American Geophysical Union, 2000 Florida Avenue, N.W., Washington, D.C. 20009

Published under the aegis of the AGU Publications Committee

Brooks Hanson, Executive Vice President, Science
Lisa Tauxe, Chair, Publications Committee

For details about the American Geophysical Union visit us at www.agu.org.

Wiley Global Headquarters
111 River Street, Hoboken, NJ 07030, USA

For details of our global editorial offices, customer services, and more information about Wiley products visit us at www.wiley.com.

Library of Congress Cataloging-in-Publication Data

ISBN: 9781119118640

Cover Design: James William Carey
Cover Images: Schematic diagram of geological phenomena in relation to the ability of caprock to permanently trap CO2 in the subsurface. Insets, from left to right, illustrate the role of matrix properties (FIB/SEM image), fracture-permeability relations (x-ray tomography), and fracture flow velocities in color during CO2-induced chemical reactions (x-ray tomography and simulation). Courtesy of Stéphanie Vialle, Jonathan Ajo-Franklin, and J. William Carey

Set in 10/12pt Times New Roman by SPi Global, Pondicherry, India

10 9 8 7 6 5 4 3 2 1

CONTENTS

CONTRIBUTORS

Jonathan Ajo-Franklin
Lawrence Berkeley National Laboratory
Berkeley, CA, USA

Lawrence M. Anovitz
Chemical Sciences Division
Oak Ridge National Laboratory
Oak Ridge, TN, USA

Jens T. Birkholzer
Earth and Environmental Sciences Area
Lawrence Berkeley National Laboratory
Berkeley, CA, USA

Robert C. Bissell
Carbon Fluids Limited
Bucks, UK

Bahman Bohloli
Norwegian Geotechnical Institute
Oslo, Norway

Etienne Brosse
IFP Energies nouvelles
Rueil-Malmaison, France

Andreas Busch
Heriot-Watt University
Lyell Centre
Edinburgh, UK

J. William Carey
Earth and Environmental Sciences
Los Alamos National Laboratory
Los Alamos, NM, USA

Cesar A. Castaneda-Herrera
Peter Cook Centre for CCS Research
The University of Melbourne
Melbourne, Australia;
Cooperative Centre for Greenhouse Gas
Technologies, CO2CRC
Carlton, Victoria, Australia

Li Chen
Earth and Environmental Sciences Division
Los Alamos National Laboratory
Los Alamos, NM, USA

M. Ben Clennell
CSIRO Energy
Perth, Australia

David R. Cole
School of Earth Sciences
The Ohio State University
Columbus, OH, USA

Baptiste Dafflon
Earth and Environmental Sciences Area
Lawrence Berkeley National Laboratory
Berkeley, CA, USA

Thomas M. Daley
Energy Geosciences Division
Lawrence Berkeley National Laboratory
University of California
Berkeley, CA, USA

Grant K. W. Dawson
School of Earth and Environmental Sciences
University of Queensland
Queensland, Australia

Claudio Delle Piane
CSIRO Energy
Perth, Australia

Russell L. Detwiler
Department of Civil and Environmental Engineering
University of California, Irvine
Irvine, CA, USA

Thomas Dewers
Geomechanics Department
Sandia National Laboratory
Albuquerque, NM, USA

David N. Dewhurst
CSIRO Energy
Perth, Australia

Lionel Esteban
CSIRO Energy
Perth, Australia

Alessandro Ferretti
TRE ALTAMIRA
Milan, Italy

Marc Fleury
IFP Energies nouvelles
Rueil-Malmaison, France

William Foxall
Energy Geosciences Division
Lawrence Berkeley National Laboratory
University of California
Berkeley, CA, USA

Luke P. Frash
Earth and Environmental Sciences
Los Alamos National Laboratory
Los Alamos, NM, USA

Bettina P. Goertz-Allmann
NORSAR
Kjeller, Norway

Joseph Grigg
Department of Earth and Environmental Sciences
New Mexico Tech
Socorro, NM, USA

Ralf R. Haese
Peter Cook Centre for CCS Research
The University of Melbourne
Melbourne, Australia;
Cooperative Centre for Greenhouse Gas Technologies, CO2CRC
Carlton, Victoria, Australia

Jeffrey De'Haven Hyman
Earth and Environmental Sciences Division
Los Alamos National Laboratory
Los Alamos, NM, USA;
Center for Nonlinear Studies
Theoretical Division
Los Alamos National Laboratory
Los Alamos, NM, USA

Jason Heath
Geomechanics Department
Sandia National Laboratory
Albuquerque, NM, USA

Helge Hellevang
Department of Geosciences
University of Oslo
Oslo, Norway;
The University Centre in Svalbard (UNIS)
Longyearbyen, Norway

Seyyed A. Hosseini
Bureau of Economic Geology
University of Texas at Austin
Austin, TX, USA

Susan Hovorka
Bureau of Economic Geology
University of Texas at Austin
Austin, TX, USA

Mathew Ingraham
Geomechanics Department
Sandia National Laboratory
Albuquerque, NM, USA

Matthew Josh
CSIRO Energy
Perth, Australia

Niko Kampman
Shell Technology Centre Amsterdam
The Netherlands

Qinjun Kang
Earth and Environmental Sciences Division
Los Alamos National Laboratory
Los Alamos, NM, USA

Zuleima Karpyn
Department of Energy and Mineral Engineering
Penn State University
State College, PA, USA

Valeri Korneev
Energy Geosciences Division
Lawrence Berkeley National Laboratory
University of California
Berkeley, CA, USA

Sanyog Kumar
Petroleum Engineering Department
Colorado School of Mines
Golden, CO, USA

Zhou Lei
Earth and Environmental Sciences Division
Los Alamos National Laboratory
Los Alamos, NM, USA

Ting Min
State Key Laboratory for Mechanical Behavior of Materials
Xi'an Jiaotong University
Xi'an, PR China

Rohaldin Miri
Department of Geosciences
University of Oslo
Oslo, Norway

Sergi Molins
Lawrence Berkeley National Laboratory
Berkeley, CA, USA

Joseph P. Morris
Atmospheric, Earth, and Energy Division
Lawrence Livermore National Laboratory
Livermore, CA, USA

Peter Mozley
Department of Earth and Environmental Sciences
New Mexico Tech
Socorro, NM, USA

Matthew Myers
CSIRO Energy
Kensington, New South Wales, Australia

Volker Oye
NORSAR
Kjeller, Norway

Julie K. Pearce
School of Earth and Environmental Sciences
University of Queensland
Queensland, Australia

Marina Pervukhina
CSIRO Energy
Perth, Australia

Ronny Pini
Department of Chemical Engineering
Imperial College
London, UK

Manika Prasad
Petroleum Engineering Department
Colorado School of Mines
Golden, CO, USA

Enrico Quintana
Non-Destructive Environments and Diagnostics Department
Sandia National Laboratories
Albuquerque, NM, USA

Abe Ramirez
Atmospheric, Earth, and Energy Division
Lawrence Livermore National Laboratory
Livermore, CA, USA

Matthew T. Reagan
Earth and Environmental Sciences Area
Lawrence Berkeley National Laboratory
Berkeley, CA, USA

Antonio Pio Rinaldi
Swiss Seismological Service
ETH, Zurich, Switzerland

Esteban Rougier
Earth and Environmental Sciences Division
Los Alamos National Laboratory
Los Alamos, NM, USA

Alessio Rucci
TRE ALTAMIRA
Milan, Italy

Jonny Rutqvist
Energy Geosciences Division
Lawrence Berkeley National Laboratory
University of California
Berkeley, CA, USA

Joel Sarout
CSIRO Energy
Perth, Australia

Mahmood Shakiba
Bureau of Economic Geology
University of Texas at Austin
Austin, TX, USA

Elin Skurtveit
Norwegian Geotechnical Institute (NGI)
Oslo, Norway;
Department of Geosciences
University of Oslo
Oslo, Norway

Geoffrey W. Stevens
Peter Cook Centre for CCS Research
The University of Melbourne
Melbourne, Australia;
Cooperative Centre for Greenhouse Gas Technologies, CO2CRC
Carlton, Victoria, Australia

Alexander Sun
Bureau of Economic Geology
University of Texas at Austin
Austin, TX, USA

Ruth M. Tinnacher
Earth and Environmental Sciences Area
Lawrence Berkeley National Laboratory
Berkeley, CA, USA;
Department of Chemistry & Biochemistry
California State University East Bay
Hayward, CA, USA

Robert C. Trautz
Electric Power Research Institute
Palo Alto, CA, USA

Charuleka Varadharajan
Earth and Environmental Sciences Area
Lawrence Berkeley National Laboratory
Berkeley, CA, USA

Donald W. Vasco
Energy Geosciences Division
Lawrence Berkeley National Laboratory
University of California
Berkeley, CA, USA

Hari Viswanathan
Earth and Environmental Sciences Division
Los Alamos National Laboratory
Los Alamos, NM, USA

Marco Voltolini
Lawrence Berkeley National Laboratory
Berkeley, CA, USA

Cameron White
CSIRO Energy
Kensington, New South Wales, Australia

Josh White
Atmospheric, Earth, and Energy Division
Lawrence Livermore National Laboratory
Livermore, CA, USA

Yuxin Wu
Earth and Environmental Sciences Area
Lawrence Berkeley National Laboratory
Berkeley, CA, USA

Li Yang
Lawrence Berkeley National Laboratory
Berkeley, CA, USA

Hongkyu Yoon
Geomechanics Department
Sandia National Laboratory
Albuquerque, NM, USA

Rui Zhang
School of Geosciences
University of Louisiana at Lafayette
Lafayette, LA, USA

Liange Zheng
Earth and Environmental Sciences Area
Lawrence Berkeley National Laboratory
Berkeley, CA, USA

PREFACE

Caprocks or sealing units are an essential component of subsurface hydrogeological systems, controlling the migration and trapping of hydrocarbons over geological timescales. With the current need to find safe storage sites for various energy-related waste streams (CO_2, the object of this monograph, but also nuclear waste), caprocks have recently received unprecedented attention, as an understanding of the integrity of such units, and their behavior over time, is crucial for the commercial deployment of storage technologies. Caprocks can be defined as a rock that prevents the flow of a given fluid at certain temperature, pressure, and chemical conditions; hence, the rock properties leading to sealing conditions may be distinct for different types of fluid. Caprocks encompass a broad range of rock types such as mudstones and shales (typically, clay mineral-rich rocks), evaporites such as anhydrite and halite, and tight carbonates (dolomite and marls). Although seals have been studied for decades by the oil and gas industry, a fundamental understanding of these units and of their evolution over time in the context of subsurface carbon storage is still lacking. Indeed new challenges have emerged with the case of CO_2 sequestration. First, sealing units contain minerals that are chemically reactive in the presence of CO_2-rich fluids, and hence, fluid-rock interactions such as dissolution, precipitation, and adsorption can result in changes to the pore space and resulting alteration of transport and mechanical properties. Second, injection of CO_2 into the storage reservoir changes the state of stress of both the reservoir and the adjacent sealing units: the caprock can then be mechanically damaged via reactivation of preexisting sealing faults and fractures or creation of new fracture systems. Third, hydrological, mechanical, and geochemical processes are intricately coupled: the fundamental understanding of the coupling between these different processes is still poor, and current models often fail at describing large-scale tests. These chemical and mechanical changes may compromise the caprock, allowing fluid migration out of the storage reservoir, potentially impacting groundwater in overlying aquifers.

The important role of caprock in successful CO_2 storage operations has inspired us to gather in a single volume a review of the state-of-the-art scientific research on the integrity of sealing units in the context of carbon storage. The monograph is organized into four parts: caprock characterization; coupled hydrological, geochemical, and geomechanical processes; monitoring techniques for caprock integrity; and environmental impacts of damaged caprock integrity alongside methods of remediating damaged caprock.

Part I of the monograph on caprock characterization begins with a review chapter presenting sample preservation methods and multiple mechanical and petrophysical characterization techniques (Dewhurst et al.). This is followed by a review and discussion paper on the transport properties of the caprock matrix at the core scale (Fleury and Brosse). A third chapter discusses characterization across pore-to-core-plug scales including both classical and recent cutting-edge technology aimed at development of constitutive laws that enable an unprecedented advance in the characterization of shale multiphysics (Dewers et al.). The last chapter of Part I focuses on a particular class of characterization techniques suitable for understanding the hierarchy of pore geometries down to the smallest scales, small and ultra small angle X-ray and neutron scattering methods (Anovitz and Cole).

Part II constitutes the core of the monograph and examines the coupling of geochemical, transport, and geochemical processes. Chapter 5 reviews processes associated with initiation, propagation, and reactivation of fractures and faults in caprock in light of changes to permeability and implications for developing an effective injection monitoring strategy (Detwiler and Morris). Chapter 6 considers the dynamic evolution of the effective permeability of matrix-fracture systems using 2D and 3D discrete fracture-matrix networks, combined finite discrete element methods to simulate fracture propagation processes, and a generalized Lattice Boltzmann model (Chen et al.). Chapter 7 is a review chapter on geochemical caprock reactivity covering observations from natural analogues worldwide, as well as experimental and modeling studies of caprock cores (Pearce and Dawson). Chapter 8 looks at the interdependency between fluid-rock interactions, fluid flow, and mechanical properties of seals, reviewing current theoretical models, experimental studies, and examples from sealing units in the North Sea (Skurtveit et al.). Chapter 9 describes a unique flow-through experiment conducted to image CO_2-induced fracture evolution in a dolomite sample with dynamic synchrotron X-ray micro-CT at appropriate stress states, providing a unique data set to probe and validate the next generation of fully coupled fracture-scale simulators (Ajo-Franklin et al.). The last chapter of Part II reviews the permeability of fractured shale, the potential for mitigation of CO_2 leakage by sorption to shale, and the detection by acoustic methods of CO_2 infiltration into shale (Carey et al.).

This leads to Part III of the monograph on monitoring techniques that can be used to assess CO_2 migration through caprocks. Chapter 11 reviews and discusses the most recent developments of in-zone and above-zone pressure monitoring techniques (Hosseini et al.). Chapter 12 describes the efforts made to monitor the fate of the injected carbon dioxide and the state of the caprock at the In Salah gas development project using a unique combination of wellbore, geophysical, and geochemical monitoring techniques (Vasco et al.). Chapter 13 presents a novel class of geochemical tracers, perfluorocarbon tracers, that provide a unique opportunity for monitoring caprock integrity for carbon storage and complements other geochemical and geophysical techniques (Myers and White).

Part IV of the monograph looks at environmental impacts and associated remediation techniques. Chapter 14 provides an overview of the physicochemical processes involved in fluid leakage from deep storage sites with a particular focus on the example of the Green River, Utah (Busch and Kampman). Chapter 15 reviews the literature on the primary concerns to groundwater quality from carbon sequestration and practices to mitigate or avoid impacts to water quality (Varadharajan et al.). Finally, Chapter 16 reviews the current research related to CO_2 leakage mitigation and remediation techniques (Castaneda-Herrera et al.).

The research presented in this monograph constitutes an inclusive survey of the wealth of recent scientific work on caprock integrity in the context of carbon storage. The research also highlights that caprock integrity is more than simply a measure of permeability and lateral continuity: caprock resilience to induced chemical and mechanical stress is a key quality of a robust storage system. The assessment of potential damage to caprock is complex, as transport, chemical, and mechanical processes are coupled and time dependent. Nonetheless, tremendous progress has occurred through the collection of experimental data on shale characterization (from the pore to the core scale) and on shale chemical reactivity and mechanical strength at in situ conditions and exposure to CO_2-rich fluids. This monograph also identifies knowledge gaps that need to be filled, some appearing to be within reach: how to link the different scales of observations? Can we derive a set of constitutive relationships (most likely rock specific) to feed predictive computational models using all the required couplings? Can we image or detect a leak in the subsurface before it reaches the surface?

Stephanie Vialle
Jonathan Ajo-Franklin
J. William Carey

Part I
Caprock Characterization

1

Microstructural, Geomechanical, and Petrophysical Characterization of Shale Caprocks

David N. Dewhurst, Claudio Delle Piane, Lionel Esteban, Joel Sarout, Matthew Josh, Marina Pervukhina, and M. Ben Clennell

ABSTRACT

Geological storage of carbon dioxide requires extensive characterization of potential selected sites in terms of injectivity, storage capacity, and containment integrity. The latter item on that list requires a multi-scale evaluation of all aspects of the subsurface geology that can trap CO_2 underground and keep it there long term. One part of a containment integrity strategy includes characterization of the caprock at a given site. Many selected sites have clay-rich shales as caprocks, and this contribution will concentrate on workflows and methods for characterizing such rocks in the laboratory. Shale preservation is the most critical step in the process as dehydration from the native in situ water content significantly affects shale properties. Various mineralogical, microscopical, petrophysical, and geomechanical properties and associated testing methods are discussed, and where possible, examples are shown of the impact of lack of preservation. Results are discussed in the context of interaction of CO_2 with caprocks and trapping mechanisms. Finally, the discussion looks at a number of the uncertainties associated with laboratory testing of shales in terms of both results obtained to date and our limited understanding as yet of the behavior and interaction of supercritical CO_2 with clay-rich caprocks.

1.1. INTRODUCTION

Geological storage of carbon dioxide (CO_2) has been mooted as a greenhouse gas mitigation strategy for over 20 years. The practical mechanics of such a strategy have been tested out at small scale at sites such as the Otway Basin in Australia [*Sharma et al.*, 2009] and Frio in Texas [*Doughty et al.*, 2008] and during industrial-scale projects, for example, at Sleipner [*Arts et al.*, 2008] and In Salah [*Ringrose et al.*, 2013]. Many years of effort have been put into defining the critical parameters for potential CO_2 storage sites [e.g., *IPCC*, 2005], and these include depth, storage capacity of the site, injectivity of the reservoir, and the containment integrity of the structure into which the CO_2 is injected. Containment integrity is usually thought of in similar terms as traps and seals in petroleum systems, and similar technologies can be used to evaluate the properties of the fault and/or top seals that provide the trapping mechanisms for keeping injected CO_2 in the deep subsurface. Fault seals usually result from the incorporation of material into the fault zone during fault movement, and this can comprise smearing out of ductile clay-rich units, abrasion of harder shales, cataclasis of rigid grains, and syn-/post-kinematic cementation of the fault rock products [e.g., *Lindsay et al.*, 1993; *Yielding et al.*, 1997; *Fisher and Knipe*, 1998; *Dewhurst et al.*, 2005]. Top seals are usually characterized in terms of their thickness (especially in relation to fault throw), areal extent, seal capacity (pore-scale capillary properties), and seal integrity (mechanical properties). There are multiple techniques for assessing the potential sealing capacity of faults [e.g., *Watts*, 1987; *Lindsay et al.*, 1993; *Yielding et al.*, 1997], and these will not be discussed further here. This paper will concentrate on methods that

CSIRO Energy, Perth, Australia

Geological Carbon Storage: Subsurface Seals and Caprock Integrity, Geophysical Monograph 238, First Edition.
Edited by Stéphanie Vialle, Jonathan Ajo-Franklin, and J. William Carey.

can be used to characterize caprocks in the laboratory and the relationship between these measurement techniques and the properties noted above. In this contribution, we will concentrate on shale-rich caprocks but acknowledge that other rocks such as anhydrites [e.g., *Hangx et al.*, 2010] are being evaluated as caprocks for CO_2 storage sites. However, it should be emphasized that any seal evaluation for a storage site or petroleum prospect should be fully integrated across both fault and top seals and for a wide range of scales.

Shale caprock properties are dependent on a number of factors, including depositional environment and resultant lithology, electrochemical conditions at deposition, mineralogy, the presence of organic matter, compaction, and diagenetic alteration. All of these processes have a significant impact on porosity and permeability, as well as the mechanical, capillary, and petrophysical properties of shales [e.g., *Bennett et al.*, 1991a,b; *Vernik and Liu*, 1997; *Dewhurst et al.*, 1998, 1999a, 1999b; *Clennell et al.*, 2006]. A number of these properties are also controlled by human intervention during and after the coring process, such as stress relief microfracture development and drying out and desiccation of recovered core, and care must be taken for certain properties that adequate sample preservation is undertaken [e.g., *Schmitt et al.*, 1994; *Dewhurst et al.*, 2012; *Ewy*, 2015]. This study will therefore review possible preservation methods and discuss multiple mechanical and petrophysical characterization techniques that can be used to either directly measure or estimate relevant properties required for shale caprocks.

1.2. SHALE PRESERVATION

The most critical stage for deriving high-quality laboratory results from shales is their immediate preservation on recovery. Loss of pore water from the in situ state can result in changing mechanical, physical, and petrophysical properties [*Schmitt et al.*, 1994] no matter whether the shale is soft, weak, and ductile or hard, strong, and brittle. Some pore fluid will always be lost from shales on recovery due to outgassing as cores are depressurized from the in situ conditions to the Earth's surface [*Schmitt et al.*, 1994]. However, most techniques that look to measure mechanical and rock physics properties, for example, would look to test the shale using a chemically compatible pore fluid under pressure, and this would generally drive any air into solution at fluid pressures >0.5 MPa. Running such tests under undrained conditions at low strain rates ($< 10^{-7}$ s^{-1}) allows monitoring of the pore pressure response either through Skempton B tests [*Skempton*, 1954] or during axial loading. Pore pressure increase under such conditions is indicative of full saturation. Hence, the slight loss of pore fluid during recovery can be alleviated for such tests. Equilibrating in relative humidity (RH) environments equivalent to shale native water activity can also mitigate this effect [e.g., *Steiger and Leung*, 1991; *Ewy*, 2014]. Other tests such as composition via X-ray diffraction (XRD), cation exchange capacity (CEC), or specific surface area (SSA) measurements generally would not be significantly affected by core preservation, although one should be careful to verify whether the presence of salts (e.g., halite, sylvite) or gypsum is real or artifacts of core storage [e.g., *Milliken and Land*, 1994].

Ideally, fully saturated core samples should be preserved under a nonpolar mineral spirit (e.g., Ondina 15 or Ondina 68) such that the fluid does not interact with the clays present and prevents native pore fluids escaping from the sample. Oil cannot intrude fully water saturated nanopores in shale at ambient pressures due to immiscibility and wettability issues. Other potential fluids that can be used for shale preservation include decane [e.g., *Ewy et al.*, 2008; *Ewy*, 2014]. Core plugs subsampled from recovered core should also be sealed in glass vials immersed in an appropriate preservative solution. Should such materials not be available, a short-term solution would be to coat cores or plugs in cling film, tin foil, and wax as a short-period (weeks to months) stopgap and kept cool in a fridge (but not frozen). However, it would be preferable to immerse in the fluids suggested above as soon as possible as wax is slightly permeable to air and samples will eventually begin to desiccate.

In order to avoid sample desiccation and concomitant alteration of rock properties (see examples below), a workflow has been developed to maximize high-quality results from preserved shale cores (Fig. 1.1). Initially, a whole core is X-ray CT scanned in order to look for fractures, limestone stringers, nodules, and the like. This allows the development of a coring plan (Fig. 1.2) directly linked to the workflow which avoids such features and means that when core plugs are taken, exposure to air is minimized. While conventional rotary coring is sometimes used for harder and more isotropic shales, in general a Murg diamond wire bandsaw is used to take core plugs, and these plugs are finished off on a cylindrical grinder. This allows significantly increased core plug recovery and better quality of plugs taken in these notoriously difficult-to-prepare rocks. Shales are at their weakest in tension parallel to bedding [e.g., *Fjær et al.*, 2008], and rotary core plugs often lead to biscuiting due to closely spaced fracture development parallel to the fabric anisotropy. The lack of stress induced by torque in the case of the diamond wire bandsaw means that plugs can be taken in more difficult rocks without imposing stresses on the intrinsic planes of weakness in the shale and better plug condition and recovery is the result.

This paper will discuss the methods and application of a number of the tests shown in the workflow with a view

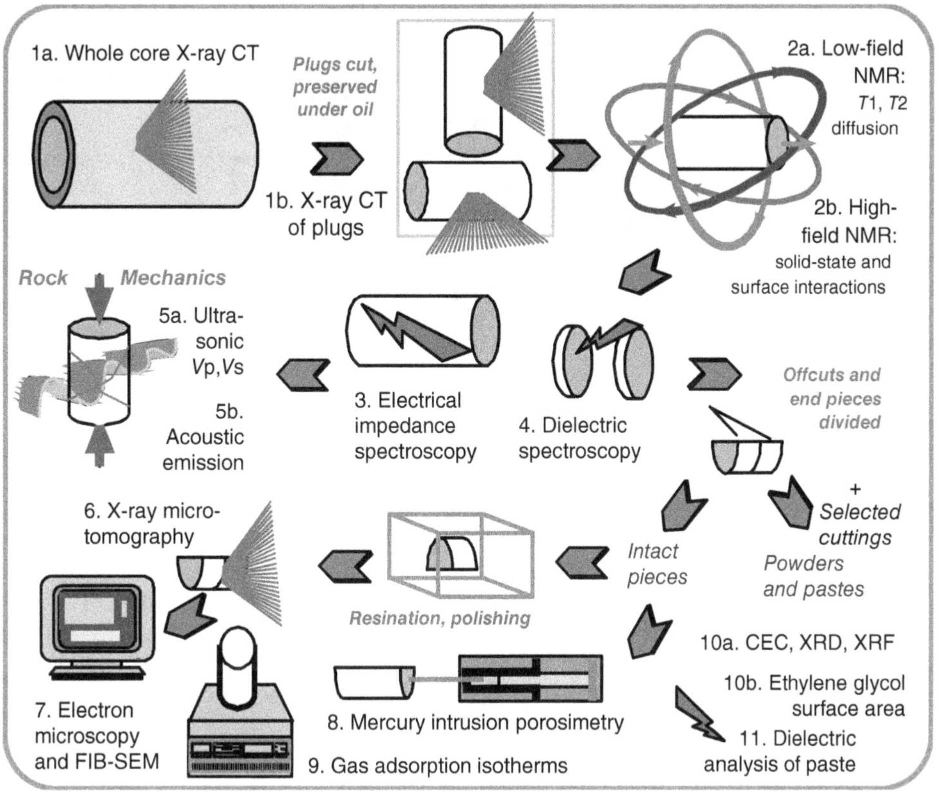

Figure 1.1 Schematic workflow for shale characterization. CT scanning of whole core and core plugs is followed by nondestructive petrophysical testing (NMR, electrical/dielectric properties) with plugs ending up in the rock mechanics laboratory for static and dynamic mechanical testing. Offcuts are sent for mineralogical, geochemical, and physicochemical properties testing, along with characterization of the shale microstructure and pore systems. Modified from *Clennell et al.* [2006]. Reprinted with permission of SPWLA.

to characterizing shale top seals for CO_2 storage and will also discuss the impact of poor preservation on results. Where possible, examples from CO_2 storage sites will be used; otherwise examples from top seals in petroleum systems will be shown.

1.3. MINERALOGY AND MICROSTRUCTURE

For a carbon capture and storage (CCS) project to be successful, it needs to guarantee the long-term trapping of the sequestered CO_2 underground. In this context, an important aspect to be considered is fluid-rock interaction which results from the contrast in physical properties between the injected CO_2 and the in situ pore fluids permeating the storage/sealing rocks and the potential reactivity between CO_2 as a free phase and as dissolved carbonic acid in the aqueous phase and the rock it comes in contact with. Through these physical and chemical interactions, the transport, elastic, and mechanical properties as well as the sealing effectiveness of shale caprocks could be changed, affecting the success of CO_2 geosequestration. A key factor to the understanding and prediction of shale behavior is the proper characterization of the various elements that collectively constitute their microstructure. Here we regard the microstructure of shales in the context of CO_2 geosequestration as subdivided into two components: (i) the minerals and (ii) the pore space; specific techniques to characterize these two components are discussed below.

A detailed characterization of the mineral component of shales is desirable as a number of common minerals have been identified as being reactive in the presence of sCO_2 and water at temperature and pressure conditions relevant for subsurface carbon dioxide storage. Numerous studies indicate that CO_2 trapping via mineral carbonation can occur by dissolution of albite ($NaAlSi_3O_8$) and the precipitation of dawsonite [$NaAlCO_3(OH)_2$; e.g., *Romanov et al.*, 2015] as well as the dissolution of Mg, Fe-silicates and sulfides, with precipitation of Mg, Fe-carbonates such as siderite and magnesite [e.g., *Kaszuba et al.*, 2005]. Moreover, recent studies showed that electrical and capillary interactions can occur between clay minerals and CO_2 in the liquid and supercritical state. CO_2 diffusion into the clay layered structure

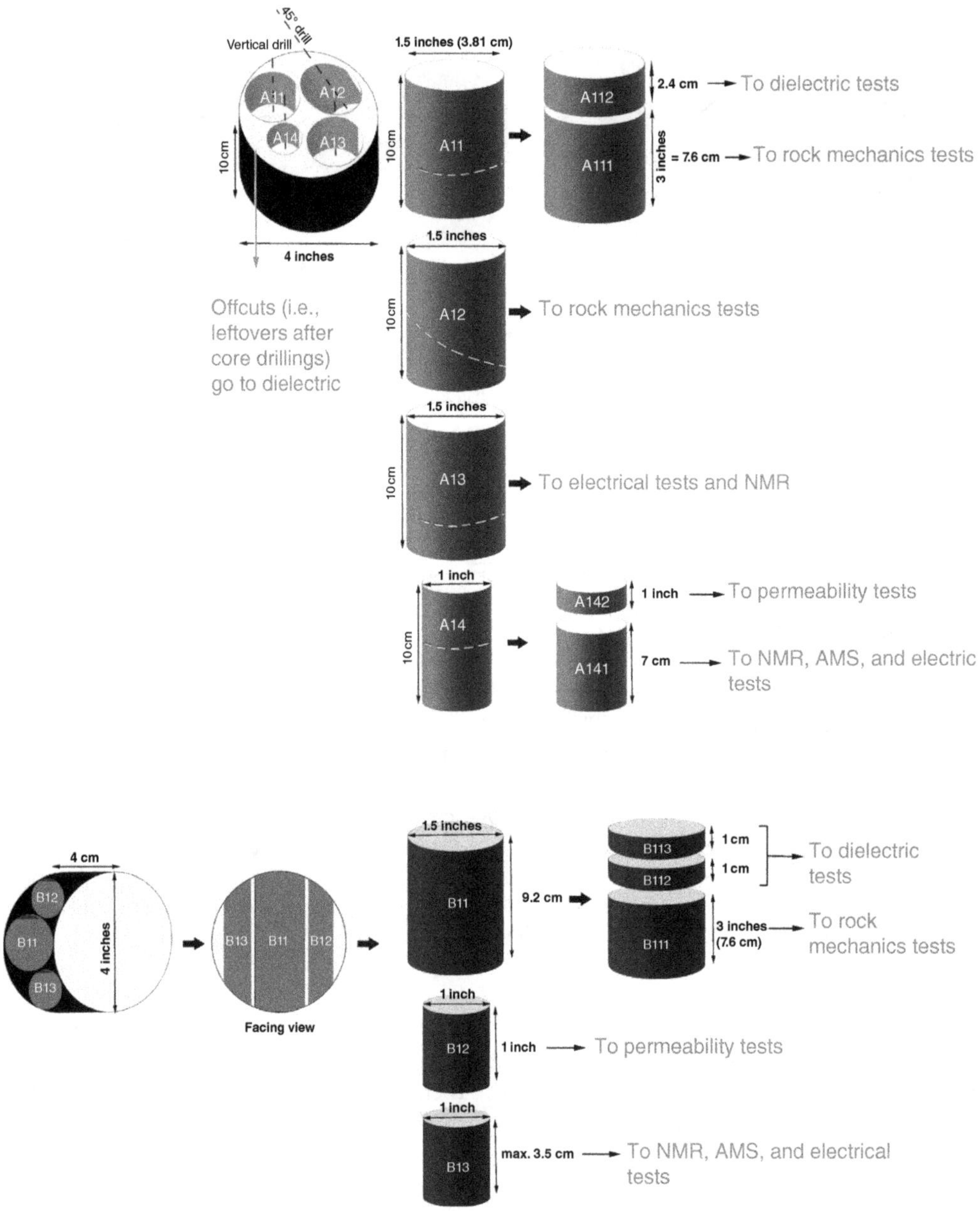

Figure 1.2 Example coring plan for subsampling whole cores normal to bedding (top) and parallel to bedding (bottom). Planning from medical CT scan images reduces exposure time to air, preventing further loss of shale pore fluids.

results in changes in the molecular clay-water chemistry leading to polarity changes in the internal electrical forces, eventually resulting in intraparticle/interparticle repulsion [e.g., *Espinoza and Santamarina*, 2012; *Berrezueta et al.*, 2013]. Electrical and capillary effects are modulated by the surface area of the particular clay minerals which in turn is known to be related to the type of clay [e.g., *Josh*, 2014]. Therefore, identification of the clay minerals in a shale caprock may help predict the potential fluid-rock interactions which could adversely affect the mechanical strength and seal capacity of the formation. The importance of clay types has been shown, for example, by *Busch et al.* [2008] who indicated that the CO_2 sorption capacity of Muderong Shale (a regional seal in offshore Western Australia) could be attributed entirely to its clay mineral constituents.

Identification of clay minerals is traditionally achieved on powdered samples by XRD [*Brindley*, 1952; *Moore and Reynolds*, 1997], a well-established technique that only requires a few grams of material for quantification. This, however, is a destructive procedure, meaning that all spatial information on the arrangement of the identified minerals is lost. Also, the technique is only sensitive to crystalline material such that any organic or amorphous matter will not be identified nor quantified; this could prove detrimental as organic matter is often found in shales and it is known to play a role in CO_2 sorption [*Krooss et al.*, 2002; *Busch et al.*, 2008]. Accurate determinations of grain size fractions (< 2 μm and <0.2 μm) are critical for determination of important clay parameters and composition. The clay fraction is defined as the wt% of material <2 μm in grain size and has been shown to be related to shale compaction and permeability [e.g., *Aplin et al.*, 1995; *Yang and Aplin*, 1998]. Accurate determination of wt% of clay minerals is also important as this defines the clay content parameter, which is directly linked to shale geomechanical and petrophysical properties [e.g., *Clennell et al.*, 2006; *Josh et al.*, 2012; *Dewhurst et al.*, 2015]. Finally, accurate determination of the composition and ordering of mixed layer clays, especially illite-smectites, requires analyses to be performed on the <0.2 μm fraction to allow clear distinction of peaks, especially in the low reflection angle region [*Moore and Reynolds*, 1997].

Complementary methods that can preserve the spatial information regarding the arrangement of different mineral phases at the microscale and allow for their identification and quantification are normally based on petrographic analysis of polished rock samples. As a result of the fine-grained nature of shales, the use of scanning electron microscopy (SEM) is necessary to resolve many of the single components of the rock matrix. SEM allows the visualization of microstructural features down to a resolution of few tens of nanometers and, at its best, can be used to visualize areas on the order of cm^2. Additionally, the interaction between the incident electron beam and the analyzed material gives rise to the emission of X-rays which with the use of appropriate detectors can be analyzed in terms of energy or wavelength to infer chemical information of the emitting substance. Examples of the use of SEM-based automated mineral analysis for the quantification of shale mineralogy include *Golab et al.* [2012] and *Timms et al.* [2015]. In geological materials, electron bombardment also stimulates the emission of light at relatively low energy by cathodoluminescence (CL); spectral analysis of the CL emission in the UV to IR region offers the ability to measure trace ionic species with relatively short acquisition times enabling large areas to be mapped with detection limits orders of magnitude below elemental detections levels acquired using X-rays. This technique can be useful, for example, in mapping the distribution of authigenic quartz cement in shale formations to understand their diagenetic history and help evaluate their mechanical and elastic properties [*Peltonen et al.*, 2009; *Thyberg et al.*, 2010; *MacRae et al.*, 2014; *Delle Piane et al.*, 2015].

All of the above techniques are applicable to study small samples in great detail and provide high-quality reference points along a depth profile. There may be cases though where continuous depth profiling on recovered cores is desirable, and this can be achieved, for example, via hyperspectral logging [e.g., *Haest et al.*, 2012]. The technique has been recently used to assess the variation in clay mineral content in cores extracted from the Harvey 1 well in the clay-rich, Late Triassic unit known as the Yalgorup Member of the Lesueur Sandstone [*Olierook et al.*, 2014], in order to evaluate potential containment stratigraphic horizons for CO_2 injection in the Perth Basin, Western Australia, as part of the South West Hub Australian Flagship Carbon Capture and Storage Project [*Stalker et al.*, 2013].

The second component of shale microstructure to be characterized in the context of CO_2 geosequestration is pore space. Research on this topic has increased dramatically in recent years, particularly since shale gas systems have become commercial hydrocarbon production targets [e.g., *Loucks et al.*, 2009; *Chalmers et al.*, 2012]. SEM imaging is certainly the most direct approach to investigate porosity, but it is generally limited by (i) the poor quality of the mechanically prepared surfaces and (ii) the relatively low resolution achievable in a traditional filament instrument with respect to shale pore sizes. The introduction of field emission gun (FEG) SEMs has allowed the attainment of much higher resolution with images being acquired with pixel sizes down to a few nanometers. Also, traditional sample preparation for SEM imaging include several steps of mechanical grinding and polishing of the surface under investigation via the use of fine grits and diamond suspensions. While effective for most geological materials, these methods tend to produce topographic irregularities due to the differential hardness of the fine-grained components of shales. *Loucks et al.* [2009] showed that these irregularities greatly exceed the size of nanopores typical of shales. To eliminate these conventional preparation limitations, argon-ion milling has been introduced as a technique to produce a much flatter surface where the minor topographic variations are unrelated to differences in hardness of the sample forming minerals but, rather, to slight variations in the path of the Ar-ion beam. Both focused ion beam (FIB) and broad ion beam (BIB) [e.g., *Holzer et al.*, 2004; *Desbois et al.*, 2009, respectively] preparations have been used to establish a pore classification scheme based on the morphology of pores from representative sample areas. *Desbois et al.* [2009] defined three main types of pore morphology occurring at the interfaces between mineral particles: (i) *type I* (elongated pores between similarly oriented clay sheets), (ii) *type II* (crescent-shaped pores in saddle reefs of folded sheet of clay, and (iii) *type III* (large jagged pores surrounding clast grains). *Type III* pores are typically larger than 1 μm, *type II* pores are between 1 μm and 100 nm, and *type I* pores are <100 nm. It should be noted that both FIB-SEM and BIB-SEM have relatively small areas/volumes of investigation, with the former on the scale of a few square microns and the latter around 1 mm^2.

Loucks et al. [2009], on the other hand, focused on the porosity observed within the organic matter defined as intraparticle organic nanopores and later introduced a simple classification scheme for the spectrum of pore types normally found in mudrocks, dividing them into mineral matrix pores (intraparticles and interparticles) and organic matter pores [*Loucks et al.*, 2012]. The classification follows the schemes normally adopted to describe pore space in sandstones and carbonates with the additional variable of pores contained within the organic matter. The classification is based on a ternary diagram populated via point counting of pores from SEM images. The authors concluded that interparticle and organic matter pores have generally better connectivity than intraparticle pores. The latter pores will provide storage and contribute to some permeability but will not have the same level of connectivity as the other pore types [*Loucks et al.*, 2012].

The 2D assessment of pore morphology has been complemented by the recent development of field emission microscopes coupled with ion milling (FIB) tools. This advance has allowed the production of in situ high-quality polished cross sections suitable for high-resolution SEM imaging of pores down to the nanoscale and the backstripping of the sample surface, in at best 5 nm steps, to visualize three-dimensional (3D) volumes of the specimen. Previous studies based on mercury injection highlighted that pore sizes of preserved shales are often well below the micron scale [e.g., *Dewhurst et al.*, 1998, 1999a, 1999b, 2002; *Yang and Aplin*, 1998]. FIB nanotomography is therefore ideal to describe porous networks in detail as it enables the 3D reconstruction of microstructural features on the 5–100 nm scale and can serve as a basis for quantitative microstructural analysis. *Keller et al.* [2011] used this approach on samples of Opalinus Clay considered as a candidate host rock formation for the disposal of radioactive waste. Their image acquisition and analysis workflow was adapted to the study of pores as small as about 10 nm, revealing preferential alignment of the pore paths along the bedding planes of the rock. *Heath et al.* [2011] investigated the properties of pore types and networks from a variety of geologic environments using core samples from continental and marine mudstones. They recognized seven dominant pore types distinguished by geometry and connectivity based on the quantitative and qualitative 3D observations. In particular, a dominant planar pore type was recognized in all investigated mudstones generally characterized by a number of neighboring connected pores. The authors argued that connected networks of pores of this type likely control most matrix transport in these mudstones. Each sample was defined using cubes of 3D pore network reconstructions of 101.5 μm^3 obtained through the interpolation of 2D slices collected at a spacing of 25 nm. *Thyberg et al.* [2010] also noted planar geometrical bodies of diagenetic quartz cement in mudstones which may have precipitated in such pores and would serve to further impede fluid transport in materials already at nano-Darcy level permeability.

FIB nanotomography is a very powerful technique that allows imaging of porosity in situ; that is, the spatial relationships between pores and surrounding mineral matrix are preserved and constitute part of the visualization results. However, the technique has strong limitations in terms of sample volume and requires some image manipulation procedures in order to obtain quantitative information on the pore space. A novel approach is to complement the information provided by local high-resolution imaging with bulk measurements of pore size distribution (size ranging from nanometers to micrometers) obtained from small and ultra small angle neutron scattering (SANS/USANS) on cm-scale shale specimens [*Gu et al.*, 2015; *King et al.*, 2015] or small angle X-ray scattering (SAXS) [e.g., *Mildner et al.*, 1986; *North et al.*, 1990] to gain a more complete representation of the pore architecture and connectivity in shales (see below).

In terms of material preservation, mineralogical composition, CEC, and SSA are not significantly affected by desiccation, and indeed, generally the samples are dried out and powdered before such tests are made. However, microstructural imaging and pore size distribution measurements can be affected by lack of preservation. In soft and stiff clays, particle orientation can change on desiccation as can clay mineral shape, with resultant impacts on pore size distribution [*Diamond*, 1970; *Delage et al.*, 1982; *Griffiths and Joshi*, 1989, 1990]. Capillary threshold pressure and resultant seal capacity calculations for shales using mercury porosimetry can also be affected by drying method. *Dewhurst et al.* [2002] found that freeze-drying produced the most consistent seal capacity calculations, whereas air-drying and vacuum-drying results were more scattered.

1.4. GEOMECHANICAL PROPERTIES

Seal integrity for both hydrocarbons and CCS is usually assessed from geomechanical properties of the rocks in question. Laboratory properties of shales can be used to inform geomechanical models used to investigate the impact of injection on a prospective CCS site [e.g., *Rinaldi et al.*, 2015; *Zhang et al.*, 2015a] or as inputs for geomechanical techniques such as slip tendency [*Morris et al.*,

1996], critically stressed fractures [*Barton et al.*, 1995], or fracture stability [*Mildren et al.*, 2005]. These latter approaches have been used to assess potential CCS sites [e.g., *Vidal-Gilbert et al.*, 2010; *Rasouli et al.*, 2013; *Tenthorey et al.*, 2013, 2014; *Zhang et al.*, 2015b] in combination with magnitude and orientation of the situ stress field and pore pressure measurements. However, these field-scale approaches are beyond the scope of this paper which is reviewing laboratory measurements, and as such, we will concentrate on the latter in this section.

Standard triaxial testing and multistage triaxial testing are the best methods to obtain shale geomechanical properties [e.g., *Marsden et al.*, 1992; *Horsrud et al.*, 1998; *Dewhurst and Hennig*, 2003; *Nygard et al.*, 2004; *Dewhurst et al.*, 2015; *Skurtveit et al.*, 2015]. While unconfined compressive strength (UCS) testing and Brazilian testing are also standard tests, they are not recommended for shales as samples can dry out during testing and thus properties can be altered (see below). UCS can be calculated from standard triaxial tests anyway [e.g., *Fjær et al.*, 2008]. Standard triaxial testing involves placing samples in a cell surrounded by hydraulic oil at different confining pressures and loading axially until failure occurs and residual strength is reached. Best practice multistage triaxial testing involves axially loading samples to 80–90% of their failure strength at a given confining pressure, unloading to zero axial load, increasing confining pressure, allowing consolidation, and repeating axial loading [e.g., *Fjær et al.*, 2008; *Dewhurst et al.*, 2015]. To judge where individual axial loading stages should be terminated, assessment of proximity to failure can be determined through deviations from linearity in the load-displacement curve (or stress-strain curve) or by monitoring volumetric strain and the onset of dilatancy [e.g., *Fjær et al.*, 2008; *Youn and Tonon*, 2010; *Dewhurst et al.*, 2015]. This train of events is then repeated multiple times, and the final stage is taken through to failure and residual strength. See *Dewhurst et al.* [2015] for a full description of the multistage methodology and associated limitations, plus *Dewhurst and Siggins* [2006] for definitions of geomechanical terminology used in this paper.

Geomechanical properties were required to evaluate the fault and top seal for the CO2CRC Otway Project [*Vidal-Gilbert et al.*, 2010], a pilot-scale CO_2 storage site in Victoria, Australia. Standard triaxial testing was performed at varying confining pressures on seven 50 mm long × 25 mm diameter core plugs of preserved Belfast Mudstone plugged parallel to bedding. This orientation was necessary due to closely spaced stress relief fractures parallel to bedding which did not allow the recovery of long enough plugs normal to bedding. Two plugs failed early, splitting axially along the bedding, and were excluded from the analysis of rock strength, so results from five plugs are shown here (Fig. 1.3). The Belfast Mudstone is shown to be a weak rock with a cohesive strength of ~2 MPa and a friction coefficient <0.5 (Fig. 1.4). It should be remembered that these properties are measured parallel to bedding and that shales are significantly anisotropic in regard to strength and elastic properties [e.g., *Niandou et al.*, 1997; *Ajalloeian and Lashkaripour*, 2000]. In general, shales are stronger and stiffer normal and parallel to bedding and weaker at 30–60° to bedding,

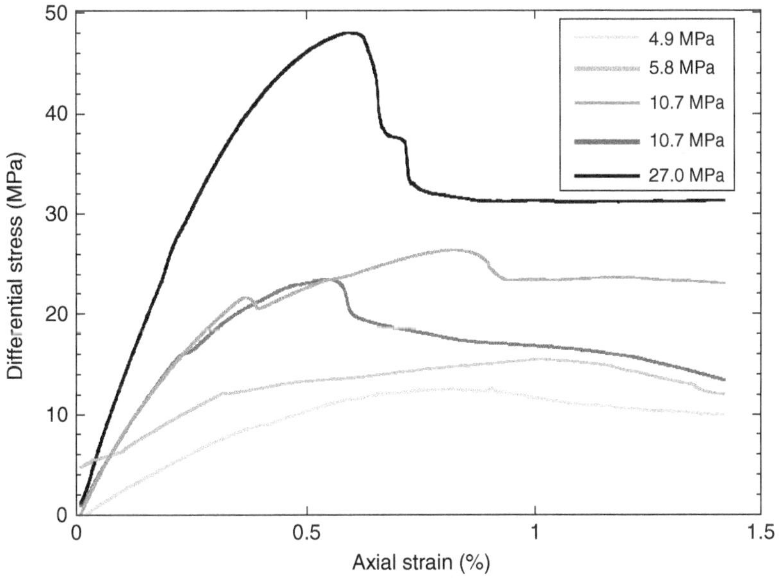

Figure 1.3 Stress-strain curves for Belfast Mudstone under increasing confining pressures with deformation parallel to bedding. The samples seem to become more brittle with increasing confining pressure, with the largest stress drop observed at the highest confining pressure used.

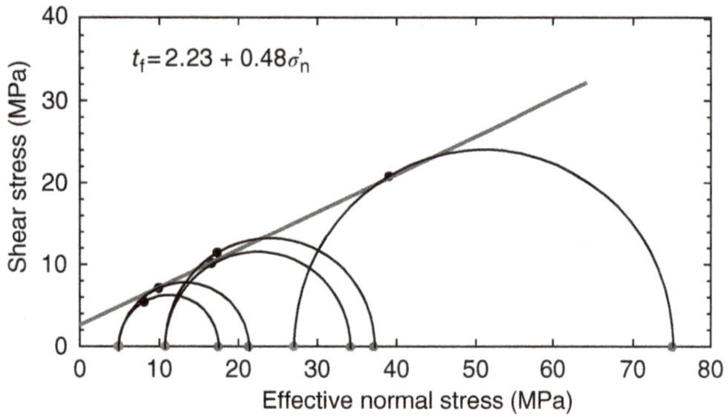

Figure 1.4 The Mohr circles derived from the peak strength takes from the stress-strain curves in Figure 1.3. The failure envelope appears to be linear with low cohesive strength and a friction coefficient lower than the standardly used values for non-shaly rocks.

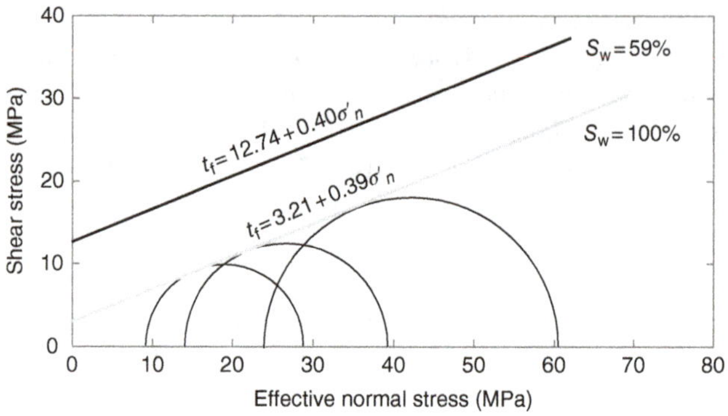

Figure 1.5 Failure envelopes for Pierre Shale deformed when fully water saturated (grey line) and partially water saturated (black line). S_w is the percentage of water-saturated porosity before samples were confined and axially loaded. The Mohr circles for the upper failure envelope have been omitted for clarity. Pierre Shale becomes substantially stronger as water saturation decreases although the friction coefficient does not change much.

usually weakest at ~45° [e.g., *Valès et al.*, 2004; *Skurtveit et al.*, 2015]. However, it is not clear whether shales are stronger normal to bedding or parallel to bedding or if the strengths in these orientations are similar as various past research shows contrasting results [e.g., *Niandou et al.*, 1997; *Ajalloeian and Lashkaripour*, 2000; *Valès et al.*, 2004; *Skurtveit et al.*, 2015]. The anisotropy of shale properties will be returned to in the discussion section below. These results were used in field-scale geomechanical assessments of the Otway site and the nearby Iona gas storage site by *Tenthorey et al.* [2013] and *Vidal-Gilbert et al.* [2010].

As noted above, preservation of shales for geomechanical testing is critical in terms of measurement of rock properties that would be applicable to in situ scenarios. Loss of in situ pore fluids from shales can completely change the rock properties being measured, and this is particularly the case in terms of parameters derived from triaxial testing such as strength and static/dynamic elastic properties such as Young's modulus or Poisson's ratio [e.g., *Valès et al.*, 2004; *Dewhurst et al.*, 2012; *Pervukhina et al.*, 2015]. An example of the change of properties on loss of water is shown in Figure 1.5. Here a preserved set of samples of Pierre Shale have been deformed at room temperature under different confining pressures to determine a failure envelope, which shows a cohesive strength of ~3 MPa and a friction coefficient of 0.39. A second set of Pierre Shale samples were equilibrated in a desiccator in a lower RH environment, resulting in a water saturation of 59%. Hence, these samples are no longer fully preserved and have lost some of their in situ water content. It should be noted that the 59% water saturation quoted is the value measured before shearing and that saturation will change during deformation due to porosity loss. These samples were deformed under the same effective stress and temperature conditions as the preserved samples. However, these partially saturated samples are significantly stronger than the preserved samples, with a cohesive strength of >12 MPa. The friction coefficient remains similar though for both sets of samples. Other examples of the effects of changing water saturation in shales can be found in *Pervukhina et al.* [2015]. Again, this shows the significant impact the lack of preservation can have on the results of rock properties tests on shales and why it is critical to preserve samples as soon as a shale core is recovered to the surface.

1.5. POROSITY AND PERMEABILITY

Transport properties and sealing capacity are two fundamental aspects for shale barriers when evaluating a site for CO_2 geosequestration. These parameters will control the volume and displacement velocity of fluid(s) able to move through the sealing layer. Other factors also affect flow properties, namely, (i) pore network tortuosity, (ii) pore network geometry (e.g., equant/tubular pores, planar microcracks) and grain surface, (iii) pore network connectivity, and (iv) wettability. Most of the characteristics of the pore network are affected by stress (nature and magnitude) to variable extents; for instance, microcracks are more prone to deformation and closure under compressive stress than equant pores [e.g., *Paterson*, 1977]. In addition, these characteristics of the pore network are not independent from each other, and in fact, a complex interaction between them is usually observed; as such, microcrack closure under increasing compressive stress affects pore network connectivity and hence fluid transport properties.

In clay-rich shales, porosity and permeability properties are linked by scale from nanometers to kilometers as illustrated by *Neuzil* [1994] and *Sondergeld et al.* [2010]. Over this wide range of scale, the notions of *free fluid* and *bound fluid* are fundamental as they affect the storage/sealing capacity and flow properties in different manner and intensity. At the reservoir scale, fractures can play a major role in the leakage risk and can be defined as the *free fluid* component of the shale formation. The fractures can have various origins such as overpressure, tectonic or induced during drilling and/or injection. In most cases, fractures, if preexisting, can be identified with logging tools and seismic exploration, and their scale and connectivity will govern the overall permeability (integrity) of the sealing barrier. The large-scale effects are beyond the scope of the paper however and will not be discussed further here [see, e.g., *Neuzil*, 1994; *Ingram and Urai*, 1999; *Dewhurst and Hennig*, 2003; *Lash*, 2006].

At the pore scale, the pore sizes generally range from sub-nanometer to micron scale in clay-bearing shale formations, making the notion of *free fluid* and effective porosity less applicable in the very small pores. Bound water is related to the interactions of the fluid(s) with minerals, mostly the clay minerals due to their strong affinity for water. As a result of the extreme surface area of clays, the water binds to clays over a large volume that is in most cases the dominant contribution to the total water content. Some clays, for example, smectites, can swell significantly by incorporating water (and ions) into interlayers in the mineral structure, which can complicate porosity determination [*Brown and Ransom*, 1996]. The various different types of water (free, clay-bound, capillary, and interlayer) make porosity determination in these materials rather more complex than it first appears [e.g., *Pearson*, 1999].

1.5.1. Porosity and Seal Capacity Determination

The recent interests in *shales* by the petroleum industry pushed development of new and existing technologies to investigate microscale and even nanoscale porosity in these rocks. Beyond the knowledge of porosity to estimate volumes and some fluid transport aspects, the small porosity scales occurring in shales have been found to have an impact on seismic-scale phenomena and rock physics modeling, for instance. Thorough porosity analysis in shales requires combined techniques to overlap all scales of investigations. The first rule to apply when investigating porosity in shales is to have preserved samples (i.e., original saturated water content) and handle samples with an extreme care to avoid dehydration and crack generation. Porosity can be assessed by various laboratory approaches: (i) 2D and 3D images, (ii) gas methods, and (iii) liquid methods. All of these approaches and methods have pros and cons.

1.5.1.1. Image Approaches

While assessment of nanopore scales and above have become easier and more accurate with the recent technologies such as SEM or X-ray micro-CT images, porosity determination remains challenging at the nanometer scale. Sample preparation can dry the samples that will change texture and potentially form some microcracks, although sometimes this can be mitigated by freeze-drying or critical point drying [*Chiou et al.*, 1991; *Dewhurst et al.*, 1998; *Holzer et al.*, 2010]. However, such images are still useful to detect clay locations that could affect the flow properties (e.g., pore throat clogging, grain coating, pore/crack bridging) and will give essential insights about diagenetic processes. In all cases, image processing is not trivial, particularly the segmentation to separate pores from the various minerals composing the rock. Establishing representative elementary volumes for material property determination from images is also a nontrivial task [e.g., *Keller et al.*, 2013].

1.5.1.2. Gas Approaches

These methods are usually used on *unpreserved* samples with all the inherent risks on result quality (e.g., clay shrinkage, cracks, residual pore fluid), although again, such methods could use freeze-dried or critical point dried samples. Nitrogen or helium can be injected into solid plugs. The principle of gas methods is to invade the pore network with a high-diffusivity inert gas. Gas expansion is monitored, and the pore volume (i.e., porosity) of the rock sample is calculated from Boyle's law:

$$V_1 = \frac{P_2 \cdot V_2}{P_1}$$

where V_1 is the volume of gas permeating the rock sample, P_2 and V_2 are the pressure and the calibrated volume of gas before being released into the sample, and P_1 is the pressure of gas after sample infiltration. Overburden stress can be applied to close any potential microcracks. The low permeability of shales makes such measurements very time consuming to allow gas to permeate through the entire volume of the sample. Variations in temperature during the experiment will also affect the results. As a result of these limitations, such a method is rarely accurate and repeatable [e.g., *Sinha et al.*, 2013; *Wang et al.*, 2013; *Heller et al.*, 2014].

Helium pycnometry on granulated samples and powders is a good method to measure bulk and grain density in order to compute the total porosity. The best procedures involve granulation of preserved samples into 0.5–1 mm diameter size and immediately perform accurate mass and volume measurement by gas pycnometry to compute bulk density. The measurement is then repeated after drying the granulated samples in oven at 105°C for up to 2 weeks until mass stabilization occurs.

The bulk density and grain density of the sample are therefore accurately measured under the same conditions, and total porosity can be calculated (assuming negligible amounts of residual salts from pore fluid evaporation) from

$$\phi = \frac{(\rho_{\text{grain}} - \rho_{\text{bulk}})}{(\rho_{\text{grain}} - \rho_{\text{fluid}})} \times 100$$

where ϕ is the porosity in percent and densities (ρ) are in g/cm^3. Gas pycnometry uses the same gas expansion principle described above but works at very low gas pressure (usually around 20 psi) to measure the sample volume with an accuracy usually around 0.0001 cm^3. Most of the issues with grain density computations in laboratory come from the drying process. Many experiments have demonstrated that solid chunks/plugs cannot be fully dehydrated at 105°C due to isolated water-filled pores and clay-bound water that require time to dehydrate by diffusion transport, and this can result in lower than expected grain density measurements.

The molecular size of the gas used for the porosity measurement in the laboratory (N_2, 0.421 nm equivalent spherical van der Waals radius; He, 0.356 nm) compared to the actual size of carbon dioxide (CO_2, 0.454 nm) can affect the gas measurement results [*Sondergeld et al.*, 2010]. Helium is often used for porosity measurements, although its size is significantly smaller than carbon dioxide. As pore throats approach the molecular diameter of the gas used for testing, helium can pass into adjacent pores where CO_2 will not be able to pass. Layers of bound ions or water make the constriction filtration effect even more prominent. Therefore, helium determined porosity is potentially greater than the effective porosity to CO_2. The magnitude of these porosity discrepancies is controlled by pore size distributions, which are poorly constrained and can reach up to a factor of 2.

1.5.1.3. Liquid Approaches

Several methods exist to access the porosity using injection of liquid (brine or mercury). Some of the techniques try to use the original water content without affecting the preservation state of the samples, and others push liquid through the pore network using dried samples (freeze-dried, critical point dried, or unpreserved).

Water immersion porosity (WIP) consists of saturating a solid plug or solid chunk of shale with water to invade all the pores. The wettability, fluid type, fluid composition, and pore pressure will affect the water mass intake of the sample during saturation, leading to considerable inconsistency. The saturation process can take a very long time as all fluids move at a diffusive speed and will only access the connected pores with a capillary force less or equal to the water injection pressure. In other words, it is virtually impossible to reach nanopores that require a large capillary force to invade, especially as the nanopores are water wet. However, it is fundamental to understand in the case of injecting water that any variation in the salinity and/or salt composition of the injected water from that in equilibrium with the shale will generate osmotic forces. Clay-water reactions and capillary forces in the case of partial saturation can quickly destroy the shale structure during water injection and re-saturation (Fig. 1.6) and is particularly significant in smectite-rich shales that can swell or shrink. However, the presence of organic matter and organic matter coatings can result in different wettability, and this may affect how clay minerals and shales as a whole respond to water injection.

If the sample is well preserved (i.e., approximately fully saturated), the sample can be dried under 105°C until mass stabilization and the difference in saturated and dry mass corresponds to the total water content. Assuming or knowing a pore fluid density allows the equivalent porosity to be computed from the sample volume [e.g.,

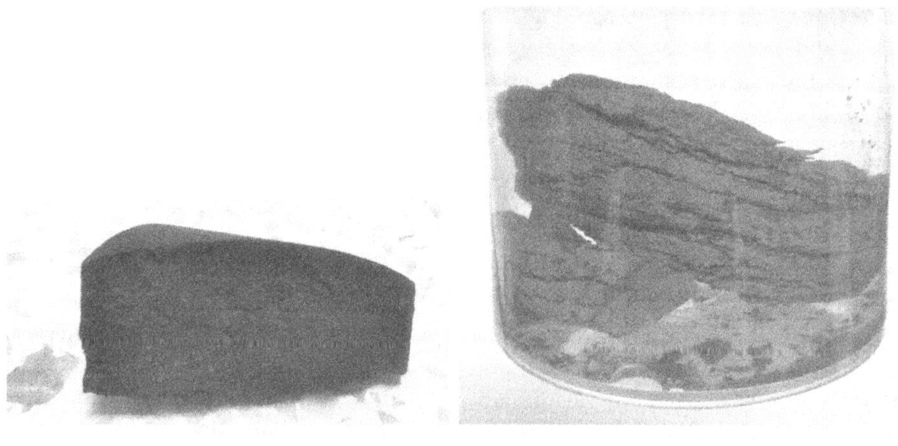

Figure 1.6 Opalinus Clay preserved (left) and the same sample after a few seconds in a non-appropriate brine solution (right): note the massive structural deterioration.

Head, 1980]. This is a relatively standard method and easy to use for well-preserved samples, and porosity calculated this way is directly related to the strength of overburden shales [*Dewhurst et al.*, 2015].

Mercury intrusion porosimetry (MIP) consists of recording the volume of injected non-wettable mercury at increasing injection pressures through a dry solid resin-coated shale cylindrical plug or, in a worst case scenario, cuttings [e.g., *Vavra et al.*, 1992; *Dewhurst et al.*, 2002; *Esteban et al.*, 2006]. Assuming cylindrical pores, the Laplace-Washburn equation [*Washburn*, 1921] can relate the pressure of mercury injection to an equivalent pore throat size. The cumulative pore throat volumes or total volume of injected mercury corresponds to the connected porosity. The typical maximum pressure of injection is ~400 MPa which is equivalent to a cylindrical pore throat diameter of ~4 nm. However, MIP is always lower than water or gas-derived porosity as mercury cannot access all pores. Some authors also demonstrated that pore orientations in shales can be anisotropic [e.g., *Keller et al.*, 2013], such that there is a strong orientation of pores in the foliation plane of shales resulting from grain shape and compaction, although MIP measurements performed parallel and normal to bedding in shales often show little difference in intrusion spectra [e.g., *Dewhurst et al.*, 2002; *Sarout et al.*, 2014].

MIP has also be used to estimate the capillary seal capacity of shales, that is, the height of a hydrocarbon or CO_2 column that can be retained by a given seal under in situ conditions before the non-wetting phase begins to move into and potentially breach the seal, assuming an idealized cylindrical tube pore morphology. Essentially, a capillary threshold pressure (P_{th}) for the air-Hg system can be calculated from a plot of pressure against cumulative intruded volume of mercury (or mercury saturation). This is considered to be the point where a continuous filament of non-wetting phase traverses the seal. There are a number of ways to pick P_{th}, including (i) the inflection point on the pressure-cumulative volume curve [e.g., *Dewhurst et al.*, 2002]; (ii) a percentage of mercury saturation, usually between 5 and 10% [e.g., *Schowalter*, 1979]; or (iii) by using the entry (or displacement) pressure where mercury first intrudes the seal sample [e.g., *Vavra et al.*, 1992]. None of these techniques has much rigor around them, and in the main, they are based on experimental results from *Schowalter* [1979] or from *Katz and Thompson* [1986] who evaluated mercury penetration through sandstones using MIP and electrical properties. Once an air-Hg threshold pressure has been calculated, this can be converted to an in situ capillary pressure for a given CO_2 column height through the use of air-Hg and brine-CO_2 contact angles and interfacial tensions; finally, a column height can be calculated from the seal and reservoir threshold pressures and the difference in brine and CO_2 fluid densities. There are multiple publications that cover these methods in great detail [e.g., *Schowalter*, 1979; *Watts*, 1987; *Vavra et al.*, 1992]. It should be noted that estimated MIP threshold pressures for CO_2 do not often match CO_2 breakthrough pressures from laboratory experiments on shales [*Hildenbrand et al.*, 2002]. In addition, while it has long been known that surface conditions of samples need to be corrected for [so-called conformance; e.g., *Vavra et al.*, 1992], more recently, it has been suggested that for tight samples such as shales, parts of the MIP curve represent sample damage or compressibility and not intrusion of mercury [e.g., *Clarkson et al.*, 2012], and there is some concern that in tight shales, mercury might not even be entering the pores at all [*Hildenbrand and Urai*, 2003; *Q. Fisher, pers. comm.*, 2015]. As a final word of warning, MIP assumes that the shale system is water wet and there is some uncertainty around wettability in CO_2-mineral systems, although there is minimal research as yet on clay minerals which usually form the bulk (the load- and pore-bearing framework) of shaly caprocks [*Iglauer et al.*, 2015]. Capillary pressure techniques for measuring mudrock properties have been extensively examined by *Busch and Amann-Hildenbrand* [2013] and further discussed by *Amann-Hildenbrand et al.* [2013] in the context of CO_2 storage.

Ferrofluid injection combined with magnetic susceptibility consists of injection of a non-wettable ferrofluid into the pore network under high pressure. Ferrofluid is a liquid made of low-viscosity isoparaffin saturated with nanoparticles of magnetite. Such nanoparticles will be able to invade the small pores and will return a strong magnetic signal correlated to their amount [*Pfleiderer and Halls*, 1994]. Magnetic susceptometers can measure the intensity of the magnetic signal of the ferrofluid-saturated shales. The amount of ferrofluid in the shales (i.e., porosity, assuming full saturation) can be calibrated against the magnetic signal from a specific amount of ferrofluid. The use of an MIP apparatus with mercury replaced by ferrofluid was successfully applied on shales and provided information about porosity and pore shape/pore connectivity [*Esteban et al.*, 2006, 2007].

Low-field nuclear magnetic resonance (NMR) is probably the most promising tool for porosity and pore size distribution in shales. NMR is a nondestructive method that records the processing of protons occurring in liquid water for shales [*Martinez and Davis*, 2000; *Dunn et al.*, 2002]. The magnitude and time relationship of the magnetization decay signal after applying an external magnetic field (the so-called relaxation time) is directly proportional to the amount of protons and their interactions with the pore volumes and mineral surfaces. The water content measured by NMR can then be converted into equivalent porosity (assuming that water fully saturates the pore network) knowing the sample volume.

Beyond porosity measurements, NMR can provide details of the pore fluid distribution such as the clay-bound water, capillary water, and potential movable water in cracks. Models can be used to tentatively compute permeability [*Coates et al.*, 1991; *Martinez and Davis*, 2000; *Hidajat et al.*, 2003; *Arns et al.*, 2004; *Minh and Sundararaman*, 2006; *Josh et al.*, 2012; *Rezaee et al.*, 2012]. However, extreme care must be taken due to NMR machine resolution as a 2 MHz spectrometer (equivalent to NMR logging tool frequency) will not be able to resolve pores smaller than 10 nm [*Nicot et al.*, 2006]. Therefore, the total NMR porosity usually underestimates the real porosity by missing the part of the water signal from clay-bound and interlayer water. To overcome this limitation, a 20 MHz NMR spectrometer is more adapted to resolve nanopores, but the macroporosity will not be detected. A combination of 2 and 20 MHz NMR instruments is a good compromise to fully assess the pore size distribution and porosity of shales. In regard to the NMR technique and sample preservation, two spectra are shown in Figure 1.7, illustrating the impact of drying on NMR response in Opalinus Clay. The solid line shows the NMR spectrum for a preserved sample and clearly has a large peak at relaxation times between 0.4 and ~3 μs, which corresponds to capillary and clay-bound water. On dehydration, this water is mostly lost, and all that remains is water with relaxation times below 0.4 μs which is likely tightly bound and interlayer water in mixed layer illite-smectite.

Taken individually, each method will generate different results for porosity measurements [*Howard*, 1991]. Method combinations are necessary to overlap the different scales accessible by each method [*Al Hinai et al.*, 2014] and for quality control in the overlap regions (Fig. 1.8). Repeatability of the measurements is the major laboratory issue due to difficult pore fluid accessibility related to the different sample treatment (grinding, sieving, drying, and partial re-saturation, same environmental conditions). Therefore, the best measurements are always achieved on preserved shales, particularly given the issues with crushing samples, limited pore access using MIP, or the clay-water reactions that can result from liquid re-saturation [*Kuila*, 2013]. Consistent sample treatment and environmental conditions (temperature, pressure, and RH) for each method, and ideally a combination of methods, can help overcome this repeatability limitation.

1.5.1.4. Radiation-based Approaches

In recent years, with the advent of gas shales, radiation-based approaches such as SAXS and SANS/USANS have come back into fashion after initial application to shales back in the 1980s [e.g., *Hall et al.*, 1986], and neutron-based methods have recently been applied to CO_2 storage research by *Busch et al.* [2014]. These methods use an incident beam of thermal neutrons or X-rays of known intensity and fixed wavelength impacting on the sample and then recording of the angle and intensity of scattering [e.g., *Radlinski et al.*, 2004]. These

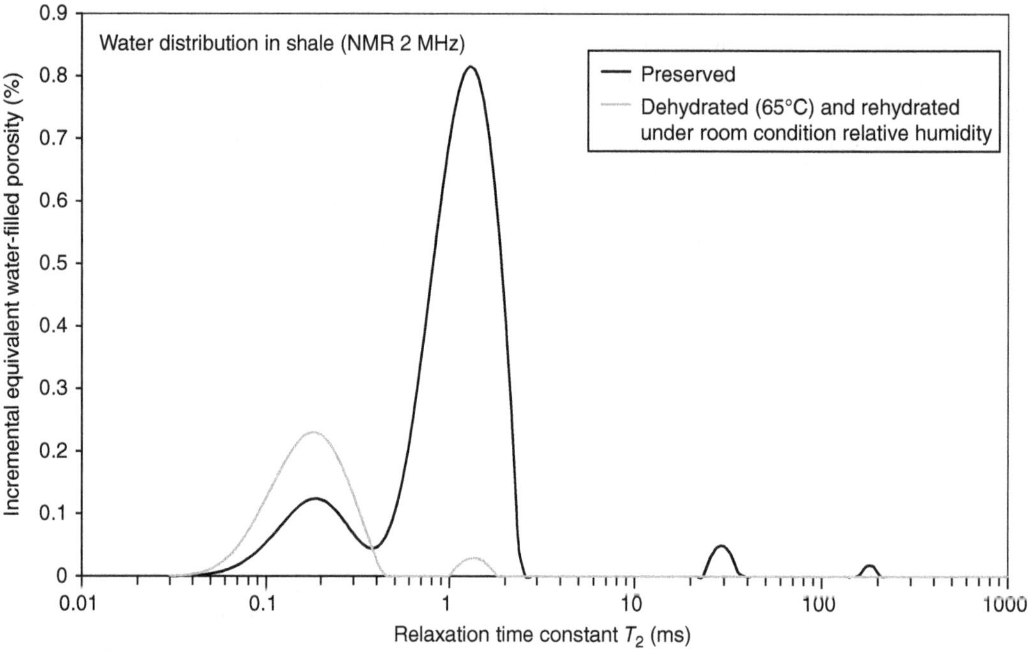

Figure 1.7 Nuclear magnetic resonance spectra for preserved and dehydrated Opalinus Clay showing significant differences in response due to dehydration. A clay-bound water peak is visible in the preserved sample but is almost gone in the dried-out sample.

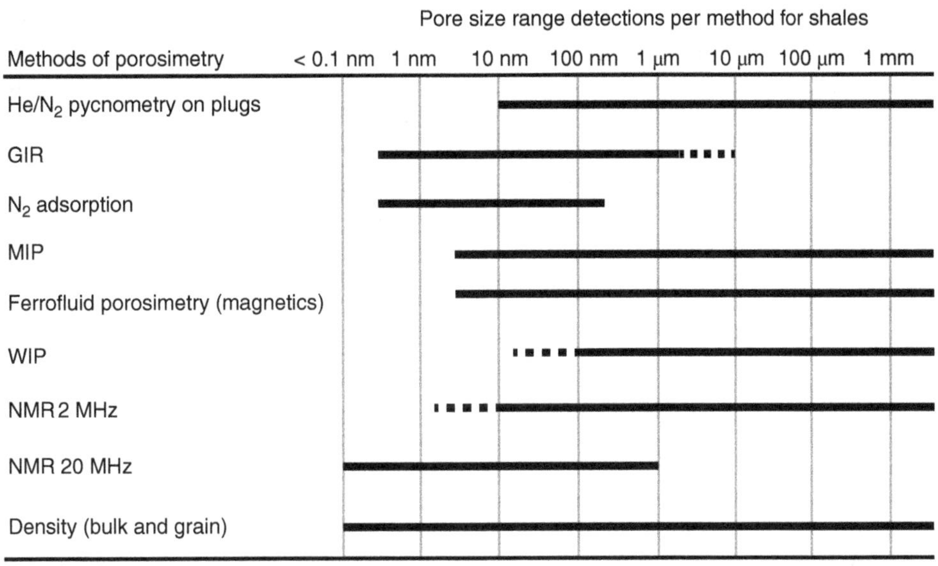

Figure 1.8 Overview of the pore size ranges detected by different methods of porosimetry for clay-bearing shales.

parameters are dependent on the geometry of the pore-matrix interface where scattering occurs [e.g., *Hall et al.*, 1986; *North and Dore*, 1993; *Radlinski et al.*, 2004]. Data obtained from such experiments can give information on total porosity, SSA, and pore size distribution (over a range from ~0.5 nm to ~10 μm) as long as certain reasonable assumptions are made (e.g., spherical pore models). However, results can show significant anisotropy on sections normal to bedding due to preferred orientation of particles (especially clays), although isotropy is usually noted in the bedding plane [e.g., *Hall et al.*, 1986; *Gu et al.*, 2015]. Essentially, preferred orientation of particles resulting from compaction and diagenesis leads to preferred orientation and shape of pores, with strongly oriented slit and tube-shaped pores causing the observed anisotropy. This is observed in overburden shales as well as gas shales [e.g., *Keller et al.*, 2011]. This anisotropy gives different porosity and SSA in different orientations although this can be corrected for [e.g., *Gu et al.*, 2015].

Various studies have also compared the results of SAXS and SANS/USANS to pore size distribution and porosity determined through MICP and N_2 adsorption, for example. Differences are often noted between the results of these tests as they measure different properties, for example, total porosity and pore size distribution for the radiation-based methods and connected porosity and pore size distribution from MICP and N_2 adsorption [*Hall et al.*, 1986; *Clarkson et al.*, 2013; *Gu et al.*, 2015]. The pore size distributions calculated from these methods also differ in that some measure pore bodies (N_2) and others pore throats (MICP), while the radiation-based methods do not distinguish between throats and bodies [e.g., *Clarkson et al.*, 2013]. SAXS and SANS/USANS usually give higher porosities than N_2 adsorption which in turn is higher than MICP due to pore accessibility issues for the larger molecules, as well as different assumptions between scattering models, adsorption models, and MICP models used to estimate porosity and SSA [*Gu et al.*, 2015].

SAXS and SANS/USANS have different sensitivities to water-filled pores, especially when pore sizes are <2 nm and such pores are expected to hold water under ambient conditions [*Hall et al.*, 1986]. For SANS/USANS, the contrast between hydrocarbon- or water-filled pores is strong compared to empty pores, but for SAXS the contrast is much less [e.g., *Hall et al.*, 1986]. While SANS/USANS, for example, nominally evaluates total porosity, *Clarkson et al.* [2013] used deuterated methane (CD_4) to look at connected porosity in gas shales with SANS/USANS and noted that not all micropores (< 2 nm) and surprisingly not all large pores (> 0.3 μm) were accessible to CD_4.

Overall, radiation-based methods provide a further string to the bow in analyzing and evaluating shale properties, especially those that are porosity and pore size distribution based. In combination with other techniques which analyze different parts of the porosity spectrum, they can provide additional information useful in fully characterizing shale behavior at the nanometer to micron scale.

1.5.2. Permeability Techniques

Clay-rich caprocks are common in sites evaluated for geological storage of CO_2 due to high capillary seal capacity and often nano-Darcy level permeability. Additionally, it is known that CO_2 can be adsorbed onto clay minerals; as such, if capillary leakage occurred, this could result in extra storage capacity [e.g., *Busch et al.*,

2008]. Once capillary seal leakage occurs, then fluid transport properties such as water and effective CO_2 permeability become important [*Hildenbrand et al.*, 2002]. There are a number of methods for measuring water permeability of materials such as constant head tests, falling head tests, steady state flow, and transient pulse decay methods. The first two are generally unsuitable for clay-bearing materials, especially softer materials due to high hydraulic gradients which are imposed that can lead to sample deformation [Olsen *et al.*, 1985]. Pulse decay methods work well and have been used by *Brace et al.* [1968] and *Kwon et al.* [2004], for example, for testing of low-permeability materials as the technique is often held to be the most rapid of all the low-permeability testing techniques. However, in general, steady state methods have been used for the assessment of CO_2-related caprocks in that this method can assure full saturation of a given caprock sample with brine under pressure as well as measuring permeability to said brine before CO_2 flooding through a sample is attempted [e.g., *Hildenbrand et al.*, 2002; *Wollenweber et al.*, 2010].

An example instrument for steady state flow testing by water injection and measurement of permeability is shown in Figure 1.9, along with some results on a cemented shale. Nominally, a disc of 25.4 mm in diameter and 10–15 mm in length is confined between two platens inside a Viton jacket and surrounded by pressurized hydraulic oil in a small steel cell. During a test, temperature and confining pressure are maintained at constant values, typically 22°C and 20 MPa, respectively. Saturation with an appropriate brine composition (determined from petrophysical measurements) occurs through injection at the upstream side of the sample (bottom end) for several days while monitoring the downstream fluid pressure (top end) with the valve closed (Fig. 1.9). The downstream fluid circuit is initially flooded with brine at atmospheric pressure and then closed. During the saturation process, it shows an increase in pressure up to the desired value, equaling the pressure imposed at the upstream end (typically 10 MPa). The injection procedure helps us ensure that the specimen is saturated prior to the permeability test, although 100% water saturation is still an assumption, given there are almost no methods that can assess this properly in tight shales. The relatively high pore fluid pressure imposed (10 MPa) allows any residual air within the pore network to be dissolved in the brine. Following saturation, the downstream pressure is regulated at a specific pressure, typically 10 MPa. The upstream pressure is suddenly increased from 10 MPa to a higher pressure (15 MPa, for instance) and regulated at that value using a high-accuracy volumetric pump. The resolution of the upstream pump volume monitoring is 10^{-6} ml/min. The fluid pressure difference between the upstream and downstream circuits is then held constant at $\Delta P = 5$ MPa throughout the test. Once a steady flow is achieved (straight line in Fig. 1.9), the constant fluid flow imposed by the pump is used to compute permeability using Darcy's law. Once this steady state flow regime is

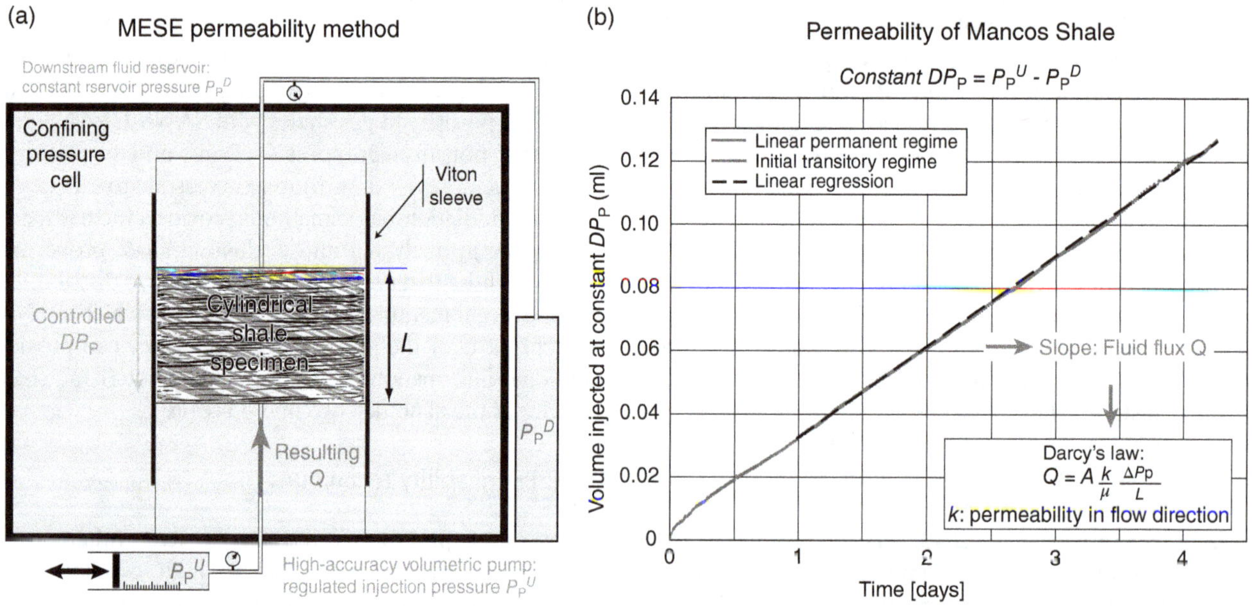

Figure 1.9 (a) Schematics of an experimental setup for permeability measurement of a cylindrical shale specimen. (b) Example of a permeability measurement for Mancos Shale, parallel to bedding, showing the fit of displaced water volume versus time during the permeability test.

established, the average pore pressure within the shale sample is expected to be 12.5 MPa so that the sample experiences a mean effective stress of 7.5 MPa.

Figure 1.9 shows the evolution of the volume of water displaced by the injection pump in order to maintain the constant pressure difference across the specimen. Steady state is achieved after about a day from the beginning of the permeability test for this Mancos Shale sample cut parallel to bedding. The interval between 1 and 4 days is chosen for the determination of the permeability of this shale as a linear fit with a good correlation factor can be obtained. From Darcy's law,

$$Q = A\frac{k}{\mu}\frac{\Delta P}{L}$$

where Q is the fluid flow rate in (m^3/s), k is the permeability in (m^2), μ is the fluid dynamic viscosity (equal to 10^{-3} Pa.s for water at room temperature), ΔP is the difference in fluid pressure across the specimen in (Pa), A is the cross-sectional area of the cylindrical specimen in (m^2), and L is its length in (m). In this case, the permeability of the Mancos Shale parallel to bedding was 1.6 nD (L = 12.9 mm and Q = 1.82 10^{-5} ml/min for ΔP = 4.98 MPa), and that of the Mancos Shale normal to bedding was 1.1 nD (L = 15.1 mm and Q = 1.08 10^{-5} ml/min for ΔP = 4.77 MPa). Anisotropy is a topic that will be returned to in the discussion. During these tests, the flow through the specimens reaches a steady value that is consistent with time if left over longer time periods. The small length of the specimens used for the permeability tests helps to reduce the duration of the test to few days. Note that very small flow rates are achieved during the test; however, thanks to the high accuracy of the pump volume monitoring, this flow rate is resolvable.

Following full brine saturation of a shale sample and measurement of water permeability, CO_2 breakthrough experiments are then performed [e.g., *Hildenbrand et al.*, 2002; *Angeli et al.*, 2009; *Wollenweber et al.*, 2010]. Essentially, the pressure of CO_2 is raised significantly on one side of the specimen and breakthrough monitored on the opposite side through pressure transducers, geochemical sampling, and/or measurements of P-wave velocity. Both drainage and imbibition-type experiments have been used to attempt to define a seal capacity to CO_2 for such experiments, although there does not seem to be good agreement between these methods nor with estimations of seal capacity from mercury porosimetry tests [e.g., *Hildenbrand et al.*, 2002; *Wollenweber et al.*, 2010], with potentially all methods having issues that are not yet fully resolved in terms of getting consistent results. These inconsistencies could be caused by the various different gas injection methods that are used, methods of interpreting seal capacity, and issues mentioned with mercury porosimetry earlier. However, what appears certain is that effective permeability to CO_2 is at least an order of magnitude lower than the brine permeability in the measured rocks and, in most cases, the brine permeability was already at the nano-Darcy level. Such tests also noted increases in both diffusivity and permeability with repeated cycling of pressure, some of which was ascribed to mineral reactions or adsorption of CO_2 on mineral surfaces [e.g., *Chiquet et al.*, 2007; *Busch et al.*, 2008; *Shah et al.*, 2008]. However, *Wollenweber et al.* [2010] noted no significant alteration of samples during such tests although they raised the possibility that redistribution of phases could be occurring, that is, dissolution and re-precipitation on the experimental timescale, which would not be detectable by traditional mineralogical analysis methods.

It is not clear as yet as to the specific mechanisms behind CO_2 breakthrough in such rocks. Traditional seal capacity analysis would mean that when capillary breakthrough occurs, the non-wetting fluid flows through the pore structure, which remains unaltered by either the fluid pressure (or effective stress change) or chemical reactions, that is, essentially a passive process. However, experiments where sample dimensions and P-wave velocity have been monitored during CO_2 injection [e.g., *Angeli et al.*, 2009; *Skurtveit et al.*, 2012, 2015] seem to suggest that the process is far from passive and that either pre-existing microcracks are dilated or new ones are forcibly formed. This is also consistent with suggested methods of gas flow through argillaceous rocks in the nuclear waste disposal industry [e.g., *Harrington and Horseman*, 1999; *Harrington et al.*, 2009]. This topic will be expanded upon in the discussion.

1.6. DIELECTRIC METHODS

Most mudrocks contain abundant charge carriers in the pore fluids and on surfaces that are mobilized and/or become reoriented in an electrical field, in a process termed polarization. The degree of polarization can be measured in terms of the dielectric permittivity of a given material (also known as the dielectric constant, ε'). The anomalously high conductivity (low resistivity) of clay-bearing rocks containing pore fluid of a given salinity is well known in petrophysics and can be related to electrical conduction on and close to clay mineral surfaces. The charges on and around clay particles become increasingly polarized by space-charge mechanisms as frequency decreases so that a clay or mudrock can have a dielectric permittivity that easily exceeds 100 at a frequency of 10 MHz. The electrical properties of clays and shales exhibit a complex frequency dependence controlled by (i) water content and its salinity, (ii) surface area and surface charge density that are essentially controlled by clay content and clay type, and (iii) other microstructural characteristics including pore and grain shape and pore space tortuosity and interconnectedness. In general, dielectric properties of clay-rich shales are dominated by water content at high frequency ($\gtrsim$ 300 MHz) and by CEC at frequencies $\lesssim$50 MHz [e.g., *Myers*, 1991; *Josh*, 2014]. These aspects make dielectric properties appealing in clay-bearing shales as CEC and water content are directly related to strength [*Dewhurst et al.*, 2015] and brittleness/ductility [*Marsden et al.*, 1992; *Petley*, 1999; *Dewhurst and Hennig*, 2003] in overburden shales.

The dielectric spectrum of preserved shales can be measured with appropriate instruments in the laboratory, including (i) parallel-plate-based methods with an impedance analyzer and (ii) coaxial or end-terminated probe measurements using a vector network analyzer [see *Josh et al.*, 2012; *Josh*, 2014 for details]. Cuttings measurements can also be useful to characterize mudrock sequences [e.g., *Leung and Steiger*, 1992]; much useful information about mineralogy and surface activity of shales can be derived from studies of pulverized cuttings using a simple end-terminated dielectric probe without using complex preparation steps found in previous literature [*Josh*, 2014]. An example of dielectric and conductivity response with frequency using the end-loaded probe method [*Josh et al.*, 2012; *Josh*, 2014] on standard clay mineral *dry* powders and pastes (Fig. 1.10) shows larger differences at low frequency (kHz) than at higher frequency (GHz). In the case of *dry* powders, small amounts of moisture are adsorbed from air, and this occurs to a larger degree for the minerals with a higher SSA. In the case of Figure 1.10 and SSA, smectite SSA > illite > kaolinite > quartz, and as such, at high frequency, the smectite contains more water than illite, kaolinite, and quartz and has a higher dielectric constant. The same result is visible for the mineral pastes, except that the illite and kaolinite seem to have similar responses. However, in this particular case, the illite used is Ward Illite, a clay mineral standard, which is around 50% illite and 50% quartz. Given illite usually has about twice the SSA of kaolinite [e.g., *van Olphen*, 1977] and quartz has negligible SSA, it is consistent that the results are similar for these two minerals in the hydrated state. Such results are interesting from a point of view of mineral typing in shales, with larger dispersion (frequency dependence) of the dielectric constant likely to relate to increased smectite content, which would also likely equate to increasing ductility of a given caprock and less likelihood of significant fracturing. In terms of pure minerals, the order of smectite-illite-kaolinite-quartz is the order of increasing strength of those minerals as well as decreasing CEC/SSA. Hence, in addition to the trends linking shale strength and ductility to water content and CEC [*Dewhurst et al.*, 2015], there are sound mineralogical and theoretical reasons to tie dielectric response to these parameters [see also *Steiger and Leung*, 1988; *Leung and Steiger*, 1992]. Strength and ductility are important parameters to understand in any evaluation of caprock integrity for CO_2 storage.

As the dielectric response of shales is strongly correlated to water content, it should come as no surprise that poor sample preservation will significantly affect dielectric results. Figure 1.11 shows the dielectric results obtained on multiple samples of the same shale kept in different RH atmospheres using a parallel plate capacitance cell [*Josh et al.*, 2012; *Josh*, 2014]. This method covers a wider range of frequencies than the end-loaded probe used in generating the results in Figure 1.10. The sample at 97% RH would have the highest water saturation, and as the RH decreases, so does water saturation. The figure shows that the dielectric constant at high frequency increases slightly with increasing RH as would be expected as there is more brine in the samples. However, as the frequency decreases, the difference between the degrees of dispersion is seen to significantly increase between samples at different RH and is highest in the 97% RH sample. This is likely to do with clay-fluid charged surface interactions increasing with increasing fluid saturation allowing increasing ion mobility.

Overall, the techniques outlined above provide a brief outline of various potential shale characterization techniques and show some example workflows, procedures, and results. There are a number of aspects of shale behavior that have not been covered or only lightly touched on in the above descriptions, and we will use the discussion to expand on some of these issues further.

1.7. DISCUSSION

A number of experimental techniques for characterizing shale caprocks for CO_2 storage and containment have been discussed here. The review is intended to be in the area of the authors' expertise and as such is not fully comprehensive. For example, considerable effort has been invested in CO_2 breakthrough, permeability, and diffusion experiments on mudrocks [e.g., *Hildenbrand et al.*, 2002, 2004; *Amann-Hildenbrand et al.*, 2013; *Busch et al.*, 2008, 2009; *Wollenweber et al.*, 2010; *Skurtveit et al.*, 2012, 2015], and the reader is referred to these papers and references therein for full details. Most of these papers show that tight mudrocks have high CO_2 breakthrough pressures or do not break through at all under simulated in situ pressure and temperature conditions. In addition, in general, little mineral alteration is noted in these intact siliciclastic mudrocks as any CO_2 penetration is along discrete flow paths which only exposes a small amount of mineral surface to flowing sCO_2 [*Amann-Hildenbrand et al.*, 2015]. *Harrington et al.* [2009] measured compressibility, permeability, and its anisotropy, plus specific storage on

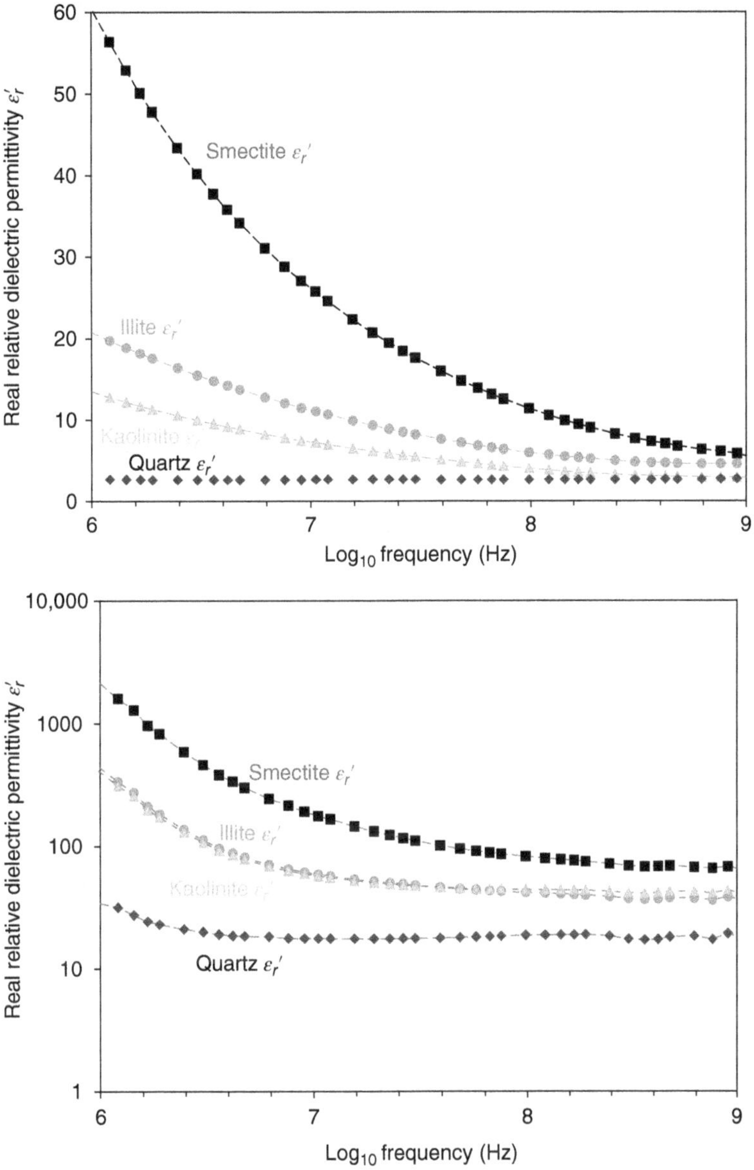

Figure 1.10 Dielectric response of *dry* powders of pure clay minerals and quartz (top) and pastes made from the same materials mixed with water (bottom). The dry powders are room dry and so contain moisture adsorbed from the air. The dielectric constant increases with decreasing frequency in both cases, the degree of dispersion related directly to the CEC of the minerals (smectite > illite > kaolinite > quartz). At high frequency, the dielectric constant is generally related to the water content alone and is much higher in the pastes than in the room dry materials. From *Josh et al.* [2012] and *Josh* [2014].

water-saturated Nordland Shale from the North Sea. Permeability was at the micro-Darcy level with an anisotropy of ~2.5 to water. Nitrogen gas breakthrough pressures were ~3 MPa. Modeling indicated that gas would have a greater anisotropy than water although the experimental data could not be matched. However, it appeared that standard concepts of two-phase flow through an unchanged pore system were not adequate to describe gas flow through this shale. It appeared that pressure-induced discrete pathways were developing rather than passive Darcy flow [e.g., *Harrington and Horseman*, 1999]. *Skurtveit et al.* [2012, 2015] also looked at CO_2 breakthrough during triaxial testing of shales and noted both physical sample dilation on radial gauges during CO_2 injection associated and a drop in *P*-wave and *S*-wave velocity in Draupne Shale from the North Sea. The velocity drops could not be accounted for by saturation changes as these were very low due to flow along discrete

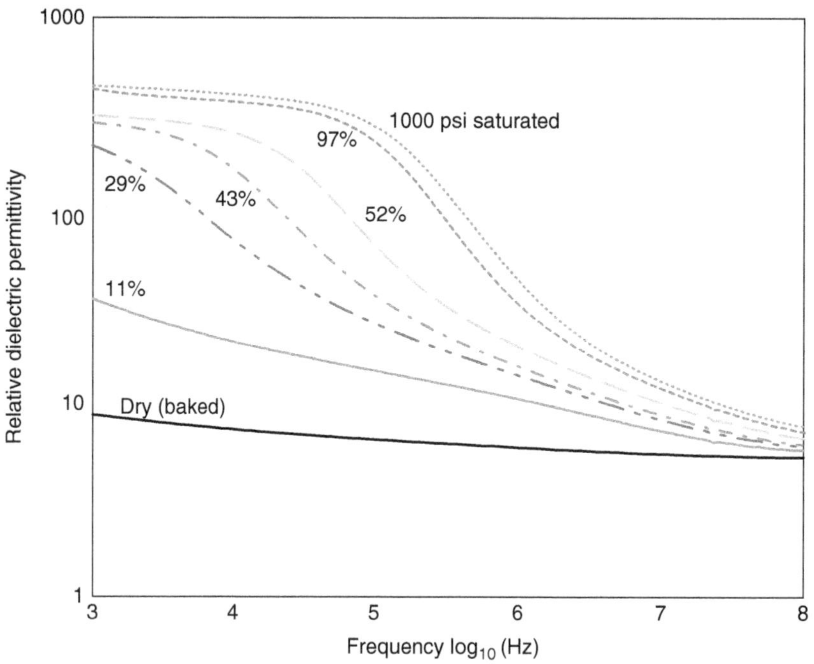

Figure 1.11 Dielectric response of multiple samples of the same shale re-saturated under different relative humidity (RH) environments. The 97% RH sample has the highest water saturation, with lower water saturations commensurating with relative humidity levels. Decreased saturation as a result of poor preservation would therefore completely change dielectric response.

flow paths as noted above. Once CO_2 flow was established, there was no change in velocity with changes in differential pressure across the sample. They noted that the CO_2 breakthrough pressure depended on both confining pressure and effective stress but not the gas pressure difference induced across the sample. At the highest confining pressures (~20 MPa), no CO_2 breakthrough was observed. This suggested that in addition to capillary displacement, there was a pressure-induced microfracture opening to accommodate the CO_2 transport through the shale and noted that the flow through the sample was consistent with fracture flow models, similar to the pressure-induced gas flow pathways noted by *Harrington and Horseman* [1999] and *Harrington et al.* [2009]. The effective CO_2 permeability was more sensitive to volumetric deformation than the water permeability and fitted a power law dependency which is typical of flow in a fractured medium. *Skurtveit et al.* [2012, 2015] also noted similarities between the CO_2 breakthrough pressure in the triaxial rig and that estimated from MIP tests, although this is not always the case [e.g., *Hildenbrand et al.*, 2002].

One aspect only briefly discussed so far is that of anisotropy. In general, shales are thought of as transversely isotropic materials, with similar rock properties within the bedding or clay foliation plane and different properties orthogonal or at angles to that plane. Strength anisotropy is a critical parameter both for drilling wells and for fracture/fault reactivation [e.g., *Fjær et al.*, 2008]. In general, shales are weaker at 30–60° to bedding and often weakest at 45° as compared to deformation normal to or parallel to bedding. However, although elastic deformation shows that static and dynamic properties in shales are almost always stiffer parallel to bedding [e.g., *Dewhurst et al.*, 2011], it is unclear in terms of failure properties which orientation is weaker. Some studies suggest both orientations have similar strength [e.g., *Niandou et al.*, 1997; *Valès et al.*, 2004], while others note that parallel to bedding is in fact stronger than normal to bedding in shales [*Ajalloeian and Lashkaripour*, 2000; *Fjær and Nes*, 2014]. *Clennell et al.* [1999] investigated the development of anisotropic permeability in clays undergoing consolidation and found that the intrinsic anisotropy is only modest when there is no compositional layering, that is, where fabric anisotropy comes from clay particle alignment alone. The most extreme cases were found in pure clays and may reach a factor of 3–4. However, in natural shales, anisotropy of permeability, and other physical properties, can be enhanced by orders of magnitude where shales contain silty laminations [e.g., *Armitage et al.*, 2012]. Note that many studies claiming high values of permeability anisotropy in mudrocks are either looking at laminated rocks, containing silt layers, or else are obviously affected by microcracks, which close at higher confining stress. There should be some general correspondence between electrical/dielectric anisotropy and permeability anisotropy, as both depend on particle shape and pore space tortuosity [e.g., *Clennell*, 1997]. Anisotropy of *P*-wave and *S*-wave velocity (and other rock physics properties such as attenuation) are well documented [e.g., *Banik*, 1984; *Johnston and Christensen*, 1994, 1995; *Vernik and Liu*, 1997; *Dewhurst and Siggins*, 2006; *Sarout and Guéguen*, 2008; *Delle Piane et al.*, 2011, 2014] and in experiments on shales can range up to 30% for *P* waves and 70% for *S* waves in extreme cases. However, of the petrophysical properties, electrical properties (resistivity and conductivity, including dielectric methods) usually show the most extreme anisotropy, in some cases up to a factor of 6–8 [*Clavaud*, 2008; *Josh and Clennell*, 2015]. As such, dependent on the properties required and local conditions, anisotropy should not be neglected when evaluating caprock shales for CO_2 storage purposes.

A significant amount of research has looked at the chemical interaction between sCO_2 and both the reservoir and caprocks. It seems that local conditions and mineralogy strongly govern whether significant alteration to mineralogy and pore structure does occur. Studies in siliciclastic reservoir rocks which contain natural CO_2 accumulations report minor to significant mineralogical reactions in such systems. For example, *Major et al.* [2014] examined a natural CO_2 reservoir-seal system in sandstones and cemented siltstones. They noted that the reservoir sands that had hosted CO_2 were weaker than those that had not, but the sealing siltstones showed little change in properties measured. The weakness was suggested as the result of the dissolution of iron-rich cements in the framework and their re-precipitation as secondary carbonates and clays. *Hangx et al.* [2015] noted that in general, unreacted rock strength was similar to sCO_2-saturated rock strength in reservoir sandstones, except for where anhydrite dissolution had occurred. *Watson et al.* [2004] compared two adjacent gas fields with low and high concentrations of CO_2 to study potential mineral reactions in siliciclastic reservoirs. In the high CO_2 case, most of the reactive minerals have altered or dissolved, with kaolinite, quartz, and less soluble carbonates being precipitated. *Haese and Watson* [2014] noted that the CO_2 mineral trapping potential in two siliciclastic reservoirs was low compared to volcaniclastic formations with abundant reactive minerals. *Higgs et al.* [2015] note alteration of chlorite to kaolinite and iron-bearing carbonates at moderate to high CO_2 concentrations, but no effect where CO_2 concentrations were low. While these studies were performed in reservoir rocks, it seems that in clay-rich caprocks at least, chemical reactions tend to have much lesser effects, although they can be observed in microfabric analysis or measured through changes in pore structure and water chemistry. *Liu et al.* [2012] noted minor dissolution of feldspar and anhydrite in Eau Claire shale on exposure to supercritical CO_2, associated with precipitation of pore-bridging illite and smectite, as well as minor precipitation of siderite adjacent to pyrite. Their review of the literature around caprock mineral interaction with sCO_2 showed that in silicate-rich systems, reactivity is low and generally confined to the reservoir-caprock interface. Reaction rates were temperature controlled unsurprisingly, with little occurring at temperatures of 50–80°C. sCO_2 can react with organic matter, potentially resulting in alteration of pore structure [*Busch et al.*, 2008]. *Lu et al.* [2009] investigated a natural CO_2 field sealed by Kimmeridge Clay and noted through geochemical examination that there was ~12 m zone of interaction in the caprock, involving dissolution of carbonates, decarboxylation of organic matter, and infiltration of reservoir CO_2. *Garrido et al.* [2013] noted that exposure to sCO_2 resulted in increases in pore volume in pore sizes <10 nm in a Tournemire shale sample that contained a significant amount of calcite. This change in pore structure was attributed to calcite dissolution, although minor aluminosilicate dissolution was also indicated by pore water chemistry. *Wollenweber et al.* [2010] evaluated CO_2 transport in a clay-rich shale and a marl under single-phase and multiphase flow conditions. Cyclic applications of gas pressure led to a reduction in capillary threshold pressure and decreased effective diffusion coefficients. Water permeability also increased after each cycle, although there was little change in composition (by XRD), SSA (BET), or pore size distribution (MIP). Overall, it appears that mineral reactions and changes in properties due to sCO_2 exposure are limited in larger intact volumes of clay-rich shales. Extensive reviews of caprock-CO_2 interaction have been performed by *Liu et al.* [2012], *Griffith et al.* [2011], and *Shukla et al.* [2010].

One area of research which is not yet well understood is the adsorption of sCO_2 onto clay surfaces and into clay interlayers. It has been previously noted [e.g., *Busch et al.*, 2008] that shales of all mineralogies (i.e., with any or a mixture of the common clays such as kaolinite, illite, illite-smectite, and smectite) can strongly adsorb sCO_2, although it should be noted that many of such tests have been done on powdered samples which would significantly increase surface area able to be contacted by CO_2 compared to the case should sCO_2 enter a caprock along discrete pressure-driven microfracture pathways as discussed above. While there is a vast literature on the incorporation of water into smectite interlayers, there is little published about the incorporation of sCO_2 into such interlayers, either in pure dry form or containing dissolved water. *Loring et al.* [2014] worked on smectitic clay standards at realistic in situ pressures and temperatures. They noted that sCO_2 can intercalate into the smectite interlayers but generally only if levels of water intercalation were low. *Michels et al.* [2015] noted that smectitic clays can capture CO_2 in the interlayer positions and that the degree of capture was dependent upon the interlayer cation type (e.g., Na^+, Li^+, etc.). Although the conditions they used to test the samples were very different to those encountered in situ. *De Jong et al.* [2014] noted swelling strains of up to ~2.5% in sCO_2 atmospheres at 45°C and CO_2 pressures up to 15 MPa in smectitic clay standards. Such swelling if occurring in situ would likely close pores and microfractures. Mechanisms of CO_2 adsorption however are not well understood. Little swelling is noted in *dry* clays, for example [e.g., *Giesting et al.*, 2012], and dehydration can occur when there are more than two water interlayers in smectite, dependent on the interlayer cation [*Schaef et al.*, 2012]. *Jin and Firoozabadi* [2014] use molecular dynamics models of montmorillonite to show that the presence of water strongly reduces sCO_2 adsorption. Under water-saturated conditions and in pore sizes >2 nm, water forms the first layer on the clay surface, and CO_2 forms the second through interaction between the CO_2 and water molecules. When smectite has an RH representative of discrete states, so 0 (dry), 1, or 2 water layers, no or little swelling is observed. When the states are in between these discrete states, then significant swelling is measured [*Busch et al.*, 2016].

However, it is not just smectitic clays which can adsorb CO_2. Illite and kaolinite have also been observed to adsorb CO_2, which therefore must be a mineral surface phenomenon rather than an interlayer one [*Busch et al.*, 2008]. Most work on smectite swelling looks at a smear of clay paste and examines swelling through the change in d-spacings. The *de Jong et al.* [2014] study used larger samples (1 mm cubes) in order to investigate a more *bulk rock* scale sample size to try to draw conclusions regarding bulk scale parameters such as geomechanical properties. In their tests, they noted that the amount of swelling strain was highest in the highest hydration state, although these hydration states lay between fully dehydrated and one interlayer of water. CO_2 contact with a smectitic caprock is likely to be limited to the reservoir-caprock interface and small volumes around preexisting fractures [*de Jong et al.*, 2014]. Any swelling that occurred in smectites with lower amounts of interlayer water (< 1) would likely lead to the closure of pores and small fractures, resulting in lower permeability, and would thus be a self-limiting process. However, where smectites are highly hydrated, with two or more water interlayers, dehydration may occur in contact with sCO_2, which may result in shrinkage, and this would be an issue in regard to the potential for fracture development [*de Jong et al.*, 2014]. There is currently little data under realistic in situ conditions for such smectites. It is also unclear as yet as to how CO_2 swelling would affect clay-rich fault gouges and whether this would impact on fault stability [*de Jong et al.*, 2014], although natural leakage of CO_2 along faults in uplifted terrains is observed [e.g., *Shipton et al.*, 2004]. However, all these results above looking at d-spacings or small cubes were all undertaken under unconfined conditions. When pressure-driven sCO_2 injection is performed on smectitic shales under a simulated in situ confining pressure, it appears that such shales dilate and small microfractures are generated that allow sCO_2 flow through the rock [e.g., *Harrington et al.*, 2009; *Skurtveit et al.*, 2012, 2015]. It may be possible that confining pressure suppresses any swelling that might occur and that the injection of sCO_2 results in a pressure-driven flow that opens dynamic and localized microscale fluid pathways [*Harrington and Horseman*, 1999; *Hildenbrand and Krooss*, 2003]. After CO_2 breakthrough and establishment of a steady flow, reduction in fluid pressure resulted in decreasing CO_2 flow and eventually complete cessation at ~1 MPa due to capillary re-imbibition of water.

There is a large uncertainty around wettability in clay-bearing caprocks as few measurements have been made on shales [*Iglauer et al.*, 2015]. No measurements of contact angles for clay-water-sCO_2 systems have been performed on pure clays due to their fine grain size, although *Borysenko et al.* [2009] made measurements on shales in brine-hydrocarbon systems. Micas such as muscovite and biotite are usually used as substitutes for illite, assuming that, for example, the chemical similarity of muscovite and illite would automatically mean a similarity in wettability. This has not been verified, and it is likely that illite, for example, has much higher surface charge density than muscovite and certainly the SSA and CEC of illite is much higher than that of muscovite. *Amann-Hildenbrand et al.* [2013] note that the assumption of a water-wet caprock may not always be correct and there have been studies in quartz, calcite, and micas which suggest water wetness may not be endemic [see review by *Heath et al.*, 2012]. However, once again, this latter review found no papers investigating wettability of clay minerals in a CO_2 storage context. The vastly different surface characteristics of clays compared to the minerals above and their well-known strongly hydrophilic behavior [e.g., *van Olphen*, 1977] should be noted, in combination with experimental breakthrough pressures and transport rates of CO_2 through shales under pressure which approach molecular diffusion slowness [e.g., *Amann-Hildenbrand et al.*, 2013]. However, natural CO_2 gas accumulations are trapped in nature over long time periods [*Watson et al.*, 2004; *Haszeldine et al.*, 2005; *Major et al.*, 2014], suggesting that shale caprock wettability or its alteration may not be too significant on the timescale required for CO_2 injection. However, long-term exposure to organic molecules may potentially change caprock wettability [e.g., *Heath et al.*, 2012; *Iglauer et al.*, 2015], although bitumen coatings have been noted to preserve otherwise reactive minerals [*Hangx et al.*, 2015].

It is fair to say that the wettability of clays is currently a known unknown in terms of shale seals for CO_2 storage.

The discussion above highlights a number of uncertainties around testing of shales for CO_2 storage, and it is difficult to determine the relative importance and priorities of such research. However, as noted in the main section of the manuscript, there are many techniques available for shale characterization for CO_2 storage, many of which have long-term provenance in the evaluation of hydrocarbon trapping in the oil and gas industry. It should be remembered however that although these multiple techniques are useful for application to CO_2 containment integrity, experimental procedures such as these should not be used as stand-alone techniques but should form part of a microscale to macroscale evaluation of seal integrity and containment running from seismic scale through a full structural fault and top seal analysis down to pore-scale evaluation of capillary and flow properties. Full-scale integration of all available measurements and data, modeling, and calibration to larger scales are critical in order to fully evaluate containment integrity for CO_2 storage.

1.8. SUMMARY

In this contribution, we have described a workflow and a number of methods for characterizing caprocks for geological CO_2 storage sites. These methods are predicated on high-quality samples being obtained and preserved at the point of recovery, so they can be tested with minimized loss of in situ pore fluids. Dehydration of clay-rich shales causes significant changes to their mechanical, physical, and petrophysical properties. Ideally, properties that are critical to assess include mineralogy, microstructure, porosity, seal capacity, permeability (both water and CO_2), diffusivity, electrical response, and geomechanics. Such properties can then feed into larger-scale integrated site analysis and models of subsurface behavior. There are still significant uncertainties around the interaction of supercritical CO_2 and clay-rich caprocks. These uncertainties include chemical reactivity of supercritical CO_2 with caprock minerals, although in general it appears that such reactions would only occur close to the reservoir-caprock interface and reactive surface area would be small due to the discrete flow paths taken by sCO_2 should it penetrate into the caprock. Swelling of smectitic clays through incorporation of sCO_2 into interlayers can be significant, but in general such tests have been run under unconfined conditions, and it is not clear whether such a response would occur in situ. Uncertainty also surrounds the wettability in the shale-brine-sCO_2 system and whether the assumption of water wetness is a good one. There are no data at all on wettability of individual clay minerals, so this remains a fertile area for research. However, even with the uncertainties around reactivity, clay swelling, and wettability, what can be stated is that there are numerous natural CO_2 accumulations that have survived for hundreds of thousands to millions of years, long in excess of the timescales currently being investigated for geological storage of CO_2, and some of those have seals which seem considerably more permeable than typical shale caprocks discussed here. As a final point, it should be noted that experimental characterization of shales such as that discussed here would be one part of a much larger strategy to evaluate containment integrity which should cover an integrated fault and top seal methodology across a range of scales from kilometer to nanometer, integrating regional seismic, wireline log and well data with laboratory approaches and modeling.

ACKNOWLEDGMENTS

Much of the work carried out for this paper would not have been possible without the expertise and efforts of Bruce Maney, Stephen Firns, Shane Kager, Leigh Kiewiet, David Nguyen, Ian Penny, and Neil Sturrock. We thank Andreas Busch and an anonymous reviewer for their sterling efforts to improve the original manuscript.

REFERENCES

Ajalloeian, R. and G. R. Lashkaripour, (2000), Strength anisotropies in mudrocks, *Bulletin of Engineering Geology and the Environment*, *59*, 195–199.

Al Hinai, A., R. Rezaee, L. Esteban, and M. Labani, (2014), Comparisons of pore size distribution: a case from the western Australian gas shale formations, *J. Unconv. Oil Gas Res.*, *8*, 1–13. DOI:10.1016/j.juogr.2014.06.002.

Amann-Hildenbrand, A., P. Bertier, A. Busch, and B. M. Krooss, (2013), Experimental investigation of the sealing capacity of generic clay-rich caprocks, *International Journal of Greenhouse Gas Control*, *19*, 620–641. DOI:10.1016/j.ijggc.2012.12.040.

Amann-Hildenbrand, A., B. M. Krooss, P. Bertier, and A. Busch, (2015), Laboratory testing procedure for CO_2 capillary entry pressures on caprocks. In *Carbon Dioxide Capture for Storage in Deep Geological Formations*, edited by Gerdes, K.F., vol. *4*, pp. 355–384, CPL Press and BPCNAI.

Angeli, M., M. Soldal, E. Skurtveit, and E. Aker, (2009), Experimental percolation of supercritical CO_2 through a caprock, *Energy Procedia*, *1*, 3351–3358. DOI:10.1016/j.egypro.2009.02.123.

Aplin, A.C., Y. Yang, and S. Hansen, (1995), Assessment of β, the compression coefficient of mudstones and its relationship with detailed lithology, *Marine and Petroleum Geology*, *12*, 955–963. DOI:10.1016/0264-8172(95)98858-3.

Armitage, P.J. D.R. Faulkner, R.H. Worden, A.C. Aplin, A.R. Butcher, and J. Iliffe, (2012), Experimental measurement of, and controls on, permeability and permeability anisotropy of caprocks from the CO_2 storage project at the Krechba Field, Algeria, *Journal of Geophysical Research*, *116*, B12208. DOI:10.1029/2011JB008385.

Arns, H., H. Averdunk, F. Bauget, A. Sakellariou, T. J. Senden, A. P. Sheppard, R. M. Sok, W. V. Pinczewski, and M. A. Knackstedt, (2004), Digital core laboratory: reservoir core analysis from 3D images, In: Gulf Rocks 2004, the 6th North America Rock Mechanics Symposium (NARMS), ARMA Paper 04–498, Houston, Texas.

Arts, R., A. Chadwick, O. Eiken, S. Thibeau, and S. L. Nooner, (2008), Ten Years' experience of monitoring CO_2 injection in the Utsira sand at Sleipner, Offshore Norway, *First Break*, *26*, 65–72.

Banik, N.C., (1984), Velocity anisotropy of shales and depth estimation in the North Sea basin, *Geophysics*, *49*, 1411–19. DOI:10.1190/1.1441770.

Barton, C. A., M. D. Zoback, and D. Moos, (1995), Fluid flow along potentially active faults in crystalline rock, *Geology*, *23*, 683–686. DOI:10.1130/0091-7613.

Bennett, R. H., W. R. Bryant, and M. H. Hulbert, (1991a), *Microstructures of Fine Grained Sediments, From Mud to Shale*, Springer, New York, 582 pp. DOI:10.1007/978-1-4612-4428-8.

Bennett, R. H., N. R. O'Brien, and M. H. Hulbert, (1991b), Determinants of clay and shale microfabric signatures: processes and mechanisms. In *Microstructures of Fine Grained Sediments*, edited by Bennett, R.H., W. R. Bryant, and M. H. Hulbert, pp. 5–32, Springer-Verlag, New York.

Berrezueta, E., L. González-Menéndez, D. Breitner, and L. Luquot, (2013), Pore system changes during experimental CO_2 injection into Detritic rocks: studies of potential storage rocks from some sedimentary basins of Spain, *International Journal of Greenhouse Gas Control*, *17*, 411–22. DOI:10.1016/j.ijggc.2013.05.023.

Borysenko, A., B. Clennell, R. Sedev, I. Burgar, J. Ralston, M. Raven, D. Dewhurst, and K. Liu, (2009), Experimental investigations of the wettability of clays and shales, *J. Geophys. Res.*, *114*, B07202. DOI:10.1029/2008JB005928.

Brace, W. F., J. B. Walsh, and W. T. Frangos, (1968), Permeability of granite under high pressure, *J. Geophys. Res.*, *73*, 2225–2236. DOI:10.1029/JB073i006p02225.

Brindley, G.W., (1952), Identification of clay minerals by X-ray diffraction analysis, *Clays and Clay Minerals*, *1*(1), 119–129. DOI:10.1346/CCMN.1952.0010116.

Brown, K. M., and B. Ransom, (1996), Porosity corrections for smectite-rich sediments: impact on studies of compaction, fluid generation, and tectonic history, *Geology*, *24*, 843–846. DOI:10.1130/0091-7613(1996)024.

Busch, A. and A. Amann-Hildenbrand, (2013), Predicting capillarity of mudrocks, *Marine and Petroleum Geology*, *45*, 208–223. DOI:10.1016/j.marpetgeo.2013.05.005.

Busch, A., P. Bertier, Y. Gensterblum, G. Rother, C. J. Spiers, M. Zhang, and H. M. Wentinck, (2016), On sorption and swelling of CO_2 in clays, *Geomechanics and Geophysics for Geo-Energy and Geo-Resources*, *2*(2), 111–130. DOI:10.1007/s40948-016-0024-4.

Busch, A., N. Kampman, S. J. Hangx, J. Snippe, M. Bickle, P. Bertier, H. Chapman, C. J. Spiers, R. Pijnenburg, J. Samuelson, J. P. Evans, A. Maskell, J. Nicholl, V. Pipich, Z. Di, G. Rother, and M. Schaller, (2014), The Green River natural analogue as a field laboratory to study the Long-term fate of CO_2 in the subsurface, *Energy Procedia*, *63*, 2821–2830. DOI:10.1016/j.egypro.2014.11.304.

Busch, A., S. Alles, B. M. Krooss, H. Stanjek, and D. Dewhurst, (2009), Effects of physical sorption and chemical reactions of CO_2 in shaly caprocks, *Energy Procedia*, *1*, 3229–3235. DOI:10.1016/j.egypro.2009.02.107.

Busch, A., S. Alles, Y. Gensterblum, D. Prinz, D. N. Dewhurst, M. D. Raven, and B. M. Krooss, (2008), Carbon dioxide storage potential of shales, *International Journal of Greenhouse Gas Control*, *2*, 297–308. DOI:10.1016/j.ijggc.2008.03.003.

Chalmers, G. R., R. M. Bustin, and I. M. Power, (2012), Characterization of gas shale pore systems by porosimetry, pycnometry, surface area, and field emission scanning electron microscopy/ transmission electron microscopy image analyses: examples from the Barnett, Woodford, Haynesville, Marcellus, and Doig units, *AAPG Bulletin*, *96*, 1099–1119. DOI:10.1306/10171111052

Chiou, W. A., W. R. Bryant, and R. H. Bennett, (1991), Clay fabric of gassy submarine sediments. In *Microstructures of Fine Grained Sediments*, edited by Bennett, R. H., W. R. Bryant, and M. H. Hulbert, pp. 333–52, Springer-Verlag, New York.

Chiquet, P., D. Broseta, and S. Thibeau, (2007), Wettability alteration of caprock minerals by carbon dioxide, *Geofluids*, *7*, 1–11. DOI:10.1111/j.1468-8123.2007.00168.x.

Clarkson, C. R., J. L. Jensen, and S. Chipperfield, (2012), Unconventional gas reservoir evaluation: what do we have to consider, *Journal of Natural Gas Science and Engineering*, *8*, 9–33. DOI:10.1016/j.jngse.2012.01.001.

Clarkson, C. R., N. Solano, R. M. Bustin, A. M. M. Bustin, G. R. L. Chalmers, L. He, Y. B. Melnichenko, A. P. Radlinski, and T. P. Blach, (2013), Pore structure characterization of north American shale gas reservoirs using USANS/SANS, gas adsorption, and mercury intrusion, *Fuel*, *103*, 606–616. DOI:10.1016/j.fuel.2012.06.119.

Clavaud, J.-B., (2008), Intrinsic electrical anisotropy of shales: the effect of compaction, *Petrophysics*, *49*, 243–260.

Clennell, M. B., (1997), Tortuosity: a guide through the maze. In *Development in Petrophysics*, edited by Lovell, M. A. and P. K. Harvey, Geological Society London Special Publications, *122*, 299–344. 10.1144/GSL.SP.1997.122.01.18.

Clennell, M. B., D. N. Dewhurst, and M. Raven, (2006), Shale petrophysics: electrical, dielectric and nuclear magnetic resonance studies of shales and clays. Proceedings of the 47th SPWLA Annual Logging Symposium, Veracruz, Mexico, 4–7 June, 13 pp.

Clennell, M. B., D. N. Dewhurst, K. M. Brown, and G. K. Westbrook, (1999), Permeability anisotropy of consolidated clays. In *Muds and Mudstones: Physical and Fluid Flow Properties*, edited by Aplin, A. C., A. J. Fleet, and J. H. S. Macquaker, Geological Society of London Special Publication, *158*, pp. 79–96. 10.1144/GSL.SP.1999.158.01.07.

Coates, G. R., R. C. A. Peveraro, and D. Roberts, (1991), The magnetic resonance imaging log characterized by comparison with petrophysical properties and laboratory core data. SPE Paper 22723, 66th Annual Technical Conference and Exhibition of the SPE, Dallas, Texas, 6–9 October.

de Jong, S. M., C. J. Spiers, and A. Busch, (2014), Development of swelling strain in smectite clays through exposure to carbon dioxide, *International Journal of Greenhouse Gas Control*, *24*, 149–161. DOI:10.1016/j.ijggc.2014.03.010.

Delage, P., D. Tessier, and M. Marcel-Audiguier, (1982), Use of the Cryoscan apparatus for observation of freeze-fractured planes of a sensitive Quebec clay in scanning electron microscopy, *Canadian Geotechnical Journal*, *19*, 111–114.

Delle Piane, C., B. S. G. Almqvist, C. M. MacRae, A. Torpy, A. J. Mory, and D. N. Dewhurst, (2015), Texture and diagenesis of ordovician shale from the canning basin, western Australia: implications for elastic anisotropy and geomechanical properties, *Marine and Petroleum Geology*, *59*, 56–71. DOI:10.1016/j.marpetgeo.2014.07.017.

Delle Piane, C., C. Madonna, J. Sarout, E.H. Saenger, D.N. Dewhurst, and M.D. Raven, (2014), Frequency dependent seismic attenuation in shales: experimental results and theoretical analysis, *Geophysical Journal International*, 1–12. DOI:10.1093/gji/ggu148.

Delle Piane, C., D. N. Dewhurst, A. F. Siggins, and M. D. Raven, (2011), Stress-induced anisotropy in saturated shale, *Geophysical Journal International*, *184*, 897–906. DOI:10.1111/ j.1365-246X.2010.04885.x, 10 pp.

Desbois, G., J. L. Urai, and P. A. Kukla, (2009), Morphology of the pore space in claystones – evidence from BIB/FIB ion beam sectioning and cryo-SEM observations, *Earth Discussions*, *4*, 1–19. DOI:10.5194/eed-4-1-2009.

Dewhurst, D. N. and A. Hennig, (2003), Geomechanical properties related to top seal leakage in the Carnarvon Basin, Northwest Shelf, Australia, *Petroleum Geoscience*, *9*, 255–263. DOI:10.1144/1354-079302-557.

Dewhurst, D. N. and A. F. Siggins, (2006), Impact of fabric, microcracks and stress field on shale anisotropy, *Geophysical Journal International*, *165*, 135–148. DOI:10.1111/j.1365-246X.2006.02834.x

Dewhurst, D. N., A. C. Aplin, J.-P. Sarda and Y. Yang, (1998), Compaction-driven evolution of porosity and permeability in natural mudstones: an experimental study, *Journal of Geophysical Research, Solid Earth*, *103*(B1), 651–661. DOI:10.1029/97JB02540.

Dewhurst, D. N., Y. Yang, and A. C. Aplin, (1999a), Fluid flow through natural mudstones. In *Muds and Mudstones: Physical and Fluid Flow Properties*, edited by Aplin, A. C., A. J. Fleet, and J. H. S. Macquaker, Geological Society of London Special Publication, *158*, pp. 23–43. DOI:10.1144/GSL.SP.1999.158.01.03.

Dewhurst, D. N., A. C. Aplin, and J.-P. Sarda, (1999b), Influence of clay fraction on pore-scale properties and hydraulic conductivity of experimentally compacted mudstones, *Journal of Geophysical Research, Solid Earth*, *104*(B12), 29261–29274. DOI:10.1029/1999JB900276.

Dewhurst, D. N., P. J. Boult, R. M. Jones, and S. A. Barclay, (2005), Fault healing and fault sealing in impure sandstones. In *Evaluating Fault and Cap Rock Seals*, edited by Boult, P. J. and J. Kaldi, AAPG Hedberg Series, no. *2*, pp. 37–56. DOI:10.1306/1060755H23161.

Dewhurst, D. N., R. M. Jones, and M. D. Raven, (2002), Microstructural and petrophysical characterisation of Muderong shale: application to top seal risking, *Petroleum Geoscience*, *8*, 371–383. DOI:10.1144/petgeo.8.4.371.

Dewhurst, D. N., B. Maney, B. Clennell, C. Delle Piane, C. Madonna, E. Saenger, and N. Tisato, (2012), Impact of saturation change on shale properties. Extended abstract for 3rd EAGE Shale Conference, Barcelona, 4 pp. DOI:10.3997/2214-4609.20143919.

Dewhurst, D. N., J. Sarout, C. Delle Piane, A. F. Siggins, and M. D. Raven, (2015), Empirical strength prediction for preserved shales, *Marine and Petroleum Geology*, *67*, 512–525. DOI:10.1016/j.marpetgeo.2015.06.004.

Dewhurst, D. N., A.F. Siggins, J. Sarout, M. Raven, and H. M. Nordgard-Bolas, (2011), Geomechanical and ultrasonic characterisation of a Norwegian Sea shale, *Geophysics*, *76*(3), WA101–WA111, DOI:10.1190/1.3569599.

Diamond, S., (1970), Pore size distribution in clays, *Clays and Clay Minerals*, *18*, 7–23. DOI:10.1346/CCMN.1970.0180103.

Doughty, C., B. Freifeld, and R. Trautz, (2008), Site characterization for CO_2 geologic storage and vice versa: the Frio brine pilot, Texas USA as a case study, *Environ. Geol.*, *54*, 1635–1656. DOI:10.1007/s00254-007-0942-0.

Dunn, K. J., D. J. Bergmann, and G. A. Latorraca, (2002), *Nuclear Magnetic Resonance; Petro-Physical and Logging Applications*. Handbook of Geophysical Exploration, Amsterdam, Pergamon.

Espinoza, D. N., and J. C. Santamarina, (2012), Clay interaction with liquid and supercritical CO_2: the relevance of electrical and capillary forces, *International Journal of Greenhouse Gas Control*, *10*, 351–362. DOI:10.1016/j.ijggc.2012.06.020.

Esteban, L., J. L. Bouchez, and A. Trouiller, (2006), The Callovo-Oxfordian argillites from eastern Paris basin: magnetic data and petrofabrics, *Comptes Rendus Geosciences*, *338*, 867–881. 10.1016/j.crte.2006.03.011.

Esteban, L., Y. Géraud, and J. L. Bouchez, (2007), Pore network connectivity anisotropy in Jurassic argillite specimens from eastern Paris Basin (France), *Physics and Chemistry of the Earth*, *32*, 161–169. DOI:10.1016/j.pce.2005.11.001.

Ewy, R. T., (2015), Shale/claystone response to air and liquid exposure, and implications for handling, sampling and testing, *International Journal of Rock Mechanics and Mining Sciences*, *80*, 388–401. DOI:10.1016/j.ijrmms.2015.10.009.

Ewy, R. T., (2014), Shale swelling/shrinkage and water content change due to imposed suction and due to direct brine contact, *Acta Geotechnica*, *9*, 869–886. DOI:10.1007/s11440-013-0297-5.

Ewy, R. T., C. A. Bovberg, and R. J. Stankovich, (2008), Shale Triaxial strength alteration due to brine exposure. Proceedings of the 42nd US Rock Mechanics Symposium, ARMA 08-304, San Francisco, CA, 29 June–2 July, 9 pp.

Fisher, Q. J. and R. J. Knipe, (1998), Fault sealing processes in siliciclastic sediments. In *Faulting, Fault Sealing and Fluid Flow in Hydrocarbon Reservoirs*, edited by Jones, G., Q. J. Fisher, and R. J. Knipe, Geological Society of London, Special Publication, 147, pp. 117–134. DOI:10.1144/GSL.SP.1998.147.01.08.

Fjær, E. and O.-M. Nes, (2014), The impact of heterogeneity on the anisotropic strength of an outcrop shale, *Rock Mechanics and Rock Engineering*, *47*, 1603–1611. DOI:10.1007/ s00603-014-0598-5.

Fjær, E., R. M. Holt, P. Horsrud, A. M. Raaen, and R. Risnes, (2008), *Petroleum Related Rock Mechanics. Developments in Petroleum Science*, *33*, Elsevier, Amsterdam, 338 pp.

Garrido, D. R. R., S. Lafortunea, H. Soulib, and P. Dubujet, (2013), Impact of supercritical CO_2/water interaction on the caprock nanoporous structure, *Procedia Earth and Planetary Science*, *7*, 738–741. DOI:10.1016/j.proeps.2013.03.123.

Giesting, P., S. Guggenheim, A. F. K. van Groos, A. Busch, (2012), Interaction of carbon dioxide with Na-exchanged montmorillonite at pressures to 640 bars: implications for CO_2 sequestration, *International Journal of Greenhouse Gas Control*, *8*, 73–81. DOI:10.1016/ j.ijggc.2012.01.011.

Golab, A. R. Romeyn, H. Averdunk, M. Knackstedt, and T. J. Senden, (2012), 3D characterisation of potential CO_2 reservoir and seal rocks, *Australian Journal of Earth Sciences*, *59*, 1–13. DOI:10.1080/08120099.2012.675889.

Griffith, C., D. Dzombak, and G. Lowry, (2011), Physical and chemical characteristics of potential seal strata in regions considered for demonstrating geological saline CO_2 sequestration, *Environmental Earth Sciences*, *64*, 925–948. DOI:10.1007/s12665-011-0911-5.

Griffiths, F. J. and R. C. Joshi, (1989), Change in pore size distribution due to consolidation of clays, *Geotechnique*, *39*, 159–167.

Griffiths, F. J. and R. C. Joshi, (1990), Clay fabric response to consolidation, *Applied Clay Science*, *5*, 37–66. DOI:10.1016/0169-1317(90)90005-A.

Gu, X., D.R. Cole, G. Rother, D. F. R. Mildner, and S. L. Brantley, (2015), Pores in Marcellus shale: A neutron scattering and FIB-SEM study, *Energy and Fuels*, *29*, 1295–1308. DOI:10.1021/acs.energyfuels.5b00033.

Haese, R. R. and M. Watson, (2014), Comparison of the mineral trapping capacity in three reservoirs with variable mineral compositions under CO_2 saturated conditions, *Energy Procedia*, *63*, 5479–5482. DOI:10.1016/j.egypro.2014.11.579

Haest, M., T. Cudahy, C. Laukamp, and S. Gregory, (2012), Quantitative mineralogy from infrared spectroscopic data. I. Validation of mineral abundance and composition scripts at the Rocklea channel iron deposit in Western Australia, *Economic Geology*, *107*, 209–228. DOI:10.2113/econgeo.107.2.209.

Hall, P. L., D. F. R. Mildner, and R. L. Borst, (1986), Small-angle scattering studies of the pore spaces of Shaly rocks, *Journal of Geophysical Research*, *91*(B2), 2183–2192. DOI:10.1029/ JB091iB02p02183.

Hangx, S. J. T., C. J. Spiers, and C. J. Peach, (2010), Mechanical behaviour of anhydrite caprock and implications for CO_2 sealing capacity, *J Geophys Res.*, *115*, B07402. DOI:10.1029/2009JB006954

Hangx, S., E. Bakker, P. Bertier, G. Nover, and A. Busch, (2015), Chemical-mechanical coupling observed for depleted oil reservoirs subjected to long-term CO_2 exposure – a case study of the Werkendam natural CO_2 analogue field, *Earth and Planetary Science Letters*, *428*, 230–242. DOI:10.1016/j.epsl.2015.07.044.

Harrington, J. F., D. J. Noy, S. T. Horseman, D. J. Birchall, and R. A. Chadwick, (2009), Laboratory study of gas and water flow in the Nordland Shale, Sleipner, North Sea. In *Carbon Dioxide Sequestration in Geological Media – State of the Science*, edited by Grobe, M, J. C. Pashin, and R. L. Dodge, AAPG Studies in Geology, *59*, pp. 521–543. DOI:10.1306/13171259St593394.

Harrington, J. F. and S. T. Horseman, (1999), Gas transport properties of clays and mudrocks. In *Muds and Mudrocks: Physical and Fluid-Flow Properties*, edited by Aplin, A. C., A. J. Fleet, and J. H. S. Macquaker, Geological Society of London Special Publications, 158, pp. 107–124. DOI:10.1144/GSL.SP.1999.158.01.09.

Haszeldine, R. S., O. Quinn, G. England, M. Wilkinson, Z. K. Shipton, J. P. Evans, J. Heath, L. Crossey, C. J. Ballentine, and C. M. Graham, (2005), Natural geochemical analogues for carbon dioxide storage in deep geological porous reservoirs, a United Kingdom perspective, *Oil & Gas Science and Technology*, *60*, 33–49. DOI:10.2516/ogst:2005004.

Head, K. H., (1980), *Manual of Soil Laboratory Testing. Volume 1: Soil Classification and Compaction Tests*. Pentech Press Ltd, Plymouth, 339 pp.

Heath, J. E., T. A. Dewers, B. J. O. L. McPherson, M. B. Nemer, and P. G. Kotula, (2012), Pore-lining phases and capillary breakthrough pressure of mudstone caprocks: sealing efficiency of geologic CO_2 storage sites, *International Journal of Greenhouse Gas Control*, *11*, 204–220. DOI:10.1016/j.ijggc.2012.08.001.

Heath, J. E., T. A. Dewers, B. J. O. L. McPherson, R. Petrusak, T. C. Chidsey, A. J. Rinehart, and P. S. Mozley, (2011), Pore networks in continental and marine mudstones: characteristics and controls on sealing behavior, *Geosphere*, *7*, 429–454. DOI:10.1130/GES00619.1

Heller, R., J. Vernylen, and M. Zoback, (2014), Experimental investigation of matrix permeability of gas shales, *AAPG Bull.*, *98*(5), 975–995. DOI:10.1306/09231313023.

Hidajat, I., M. Singh, and K. K. Mohanty, (2003), NMR response of porous media by random walk algorithm: a parallel implementation, *Chem. Eng. Commun. 190*, 1661–1680. DOI:10.1080/00986440390225233.

Higgs, K. E., R. Haese, S. D. Golding, U. Schacht, and M. N. Watson, (2015), The Pretty Hill Formation as a natural analogue for CO_2 storage; An investigation of mineralogical and isotopic changes associated with sandstones exposed to low, intermediate and high CO_2 concentrations over geological time, *Chemical Geology*, *399*, 36–64. DOI:10.1016/ j.chemgeo.2014.10.019.

Hildenbrand, A. and B. Krooss, (2003), CO_2 migration processes in argillaceous rocks: pressure-driven volume flow and diffusion, *Journal of Geochemical Exploration*, *78*, 169–172. DOI:10.1016/S0375-6742(03)00077-3.

Hildenbrand, A., and J. L. Urai, (2003), Investigation of the morphology of pore space in mudstones – first results, *Marine and Petroleum Geology*, *20*, 1185–1200. DOI:10.1016/j.marpetgeo.2003.07.001.

Hildenbrand, A., S. Schlömer, B. Krooss, and R. Littke, (2004), Gas breakthrough experiments on pelitic rocks: comparative study with N_2, CO_2 and CH_4, *Geofluids*, *4*, 61–80. DOI:10.1111/j.1468-8123.2004.00073.x.

Hildenbrand, A., S. Schlömer, and B. M. Krooss, (2002), Gas breakthrough experiments on fine-grained sedimentary rocks, *Geofluids*, *2*, 3–23. DOI:10.1046/j.1468-8123.2002.00031.x.

Holzer, L., F. Indutnyi, P. Gasser, B. Münch, and M. Wegmann, (2004), Three-dimensional analysis of porous $BaTiO_3$ ceramics using FIB nanotomography, *Journal of Microscopy*, *216*, 84–95. DOI:10.1111/j.0022-2720.2004.01397.x.

Holzer, L., B. Münch, M. Rizzi, R. Wepf, P. Marschall, and T. Graule, (2010), 3D microstructure analysis of hydrated bentonite with cryo-stabilized pore water, *Applied Clay Science*, *47*, 330–342. DOI:10.1016/j.clay.2009.11.045.

Horsrud, P., E. F. Sønstebø, and R. Bøe, (1998), Mechanical and petrophysical properties of North Sea shales, *Int. J. Rock Mech. Min. Sci.*, *35*, 1009–1020. DOI:10.1016/S0148-9062(98)00162-4.

Howard, J. J., (1991), Porosimetry measurement of shale fabric and its relationship to illite/smectite diagenesis, *Clays and Clay Minerals*, *39*, 355–361. DOI:10.1346/CCMN.1991.0390403

Iglauer, S., C. H. Pentland, and A. Busch, (2015), CO_2 wettability of seal and reservoir rocks and the implications for carbon geo-sequestration, *Water Resources Research*, *50*, 1–46. DOI:10.1002/2014 WR015553.

Ingram, G. M. and J. L. Urai, (1999), Top seal leakage through faults and fractures: the role of mudstone properties. In *Muds and Mudstones: Physical and Fluid Flow Properties*, edited by Aplin, A. C., A. J. Fleet, and J. H. S. Macquaker, Geological Society of London Special Publication, 158, pp. 125–135. DOI:10.1144/GSL.SP.1999.158.01.10

IPCC, (2005), Special Report on Carbon Dioxide Capture and Storage: Prepared by Working Group III of the Intergovernmental Panel on Climate Change. Cambridge University Press, Cambridge and New York, 442 pp.

Jin, Z. and A. Firoozabadi, (2014), Effect of water on methane and carbon dioxide sorption in clay minerals by Monte Carlo simulations, *Fluid Phase Equilibria*, *382*, 10–20. DOI:10.1016/j.fluid.2014.07.035.

Johnston, J. E. and N. I. Christensen, (1994), Elastic constants and velocity surfaces of indurated anisotropic shales, *Surveys in Geophysics*, *15*, 481–494. DOI:10.1007/BF00 690171.

Johnston, J. E. and N. I. Christensen, (1995), Seismic anisotropy of shales, *Journal of Geophysical Research*, *100*, 5991–6003. DOI:10.1029/95JB00031.

Josh, M., (2014), Dielectric permittivity: A petrophysical parameter for shales, *Petrophysics*, *55*, 319–332.

Josh, M. and B. Clennell, (2015), Broadband electrical properties of clays and shales: comparative investigations of remolded and preserved samples, *Geophysics*, *80*, D129–D143. DOI:10.1190/GEO2013-0458.1.

Josh, M., L. Esteban, C. Delle Piane, J. Sarout, D.N. Dewhurst, and M.B. Clennell, (2012), Laboratory characterisation of shale properties, *Journal of Petroleum Science and Engineering*, *88–89*, 107–124. DOI:10.1016/ j.petrol.2012.01.023.

Kaszuba, J. P., D. R. Janecky, and M. G. Snow, (2005), Experimental evaluation of mixed fluid reactions between supercritical carbon dioxide and NaCl brine: relevance to the integrity of a geologic carbon repository, *Chemical Geology*, *217*, 277–293. DOI:10.1016/j.chemgeo.2004.12.014.

Katz, A. J. and Thompson, A. H., (1986), Quantitative prediction of permeability in porous rock, *Physical Reviews*, *B34*, 8179–8181. DOI:10.1103/PhysRevB.34.8179.

Keller, L. M., L. Holzer, P. Schuetz, and P. Gasser, (2013), Pore space relevant for gas permeability in Opalinus clay: statistical analysis of homogeneity, percolation, and representative volume element, *Journal of Geophysical Research Solid Earth*, *118*, 2799–2812. DOI:10.1002/jgrb.50228.

Keller, L. M., L. Holzer, R. Wepf, and P. Gasser, (2011), 3D geometry and topology of pore pathways in Opalinus clay: implications for mass transport, *Applied Clay Science*, *52*, 85–95. DOI:10.1016/j.clay.2011.02.003.

King, H. E., A. P. R. Eberle, C. C. Walters, C. E. Kliewer, D. Ertas, and C. Huynh, (2015), Pore architecture and connectivity in gas shale, *Energy and Fuels*, *29*, 1375–1390. DOI:10.1021/ef502402e.

Krooss, B. M, F. van Bergen, Y. Gensterblum, N. Siemons, H. J. M. Pagnier, and P. David, (2002), High-pressure methane and carbon dioxide adsorption on dry and moisture-equilibrated Pennsylvanian coals, *International Journal of Coal Geology*, *51*, 69–92. DOI:10.1016/ S0166-5162(02)00078-2.

Kuila, U., (2013), Measurement and interpretation of porosity and pore-size distribution in mudrocks, PhD Thesis, Colorado School of Mines, 269 pp.

Kwon, O., A. K. Kronenberg, A. F. Gangi, B. Johnson, and B. E. Herbert, (2004), Permeability of illite-bearing shale: 1. Anisotropy and effects of clay content and loading, *Journal of Geophysical Research*, *109*, B10205. DOI:10.1029/2004JB003052.

Lash, G., (2006), Top seal development in the shale-dominated upper Devonian Catskill Delta complex, western New York State, *Marine and Petroleum Geology*, *23*(3), 317–335. DOI:10.1016/j.marpetgeo.2006.02.001.

Leung, P. K. and R.P. Steiger, (1992), Dielectric constant measurements: a new, rapid method to characterize shale at the wellsite. IADC/SPE Drilling Conference, Paper IADC/SPE 23887, pp. 401–408. DOI:10.2118/23887-MS.

Lindsay, N. G., F. C. Murphy, J. J. Walsh, and J. Watterson, (1993), Outcrop studies of shale smears on fault surfaces. In *The Geological Modelling of Hydrocarbon Reservoirs and Outcrop Analogues*, edited by Flint, S. S. and I. D. Bryant, International Association of Sedimentologists Special Publication, 15, 113–123. DOI:10.1002/9781444303957.ch6.

Liu, F., P. Lua, C. Griffith, S. W. Hedges, Y. Soong, H. Hellevang, and C. Zhu, (2012), CO_2–brine–caprock interaction: reactivity experiments on Eau Claire shale and a review of relevant literature, *International Journal of Greenhouse Gas Control*, *7*, 153–167. DOI:10.1016/ j.ijggc.2012.01.012.

Loring, J. S., E.S. Ilton, J. Chen, C. J. Thompson, P. F. Martin, P. Bénézeth, K. M. Rosso, A. R. Felmy, and H. T. Schaef, (2014), In situ study of CO_2 and H_2O partitioning between Na−montmorillonite and variably wet supercritical carbon dioxide, *Langmuir*, *30*, 6120–6128, DOI:10.1021/la500682t.

Loucks, R. G., R. M. Reed, S. C. Ruppel, and U. Hammes, (2012), Spectrum of pore types and networks in mudrocks and a descriptive classification for matrix-related mudrock pores, *AAPG Bulletin*, *96*, 1071–1098. DOI:10.1306/08171111061.

Loucks, R. G., R. M. Reed, S. C. Ruppel, and D. M. Jarvie, (2009), Morphology, genesis, and distribution of nanometer-scale pores in siliceous mudstones of the Mississippian Barnett shale, *Journal of Sedimentary Research*, *79*, 848–861. DOI:10.2110/jsr.2009.092.

Lu, J., M. Wilkinson, R. S. Haszeldine, and A. E. Fallick, (2009), Long-term performance of a mudrock seal in natural CO_2 storage, *Geology*, *37*, 35–38. DOI:10.1130/G25412A.1.

MacRae, C. M., N. C. Wilson, A. Torpy, S. Kidder, Z. Li, C. Delle Piane, and D. N. Dewhurst, (2014), Quartz overgrowths in shales and sandstones studied by EPMA and SIMS. *Microscopy and Microanalysis*, *20*, 914–915. DOI:10.1017/S1431927614006291.

Major, J. R., P. Eichhubl, T. A. Dewers, A. S. Urquhart, J. E. Olson, and J. Holder, (2014), The effect of CO_2-related diagenesis on geomechanical failure parameters: fracture testing of CO_2-altered reservoir and seal rocks from a natural analog at crystal geyser, Utah. Proceedings of the 48th US Rock Mechanics/Geomechanics Symposium, Minneapolis, MN, 1–4 June, ARMA 14-7463, 5 pp.

Marsden, J. R., R. M. Holt, S. J. Nakken, and A. M. Raaen, (1992), Mechanical and petrophysical characterisation of highly stressed mudstone. In *Rock Characterisation*, edited by Hudson, J. R.. Proceedings of the ISRM Conference, Eurock, 92, pp. 51–56. DOI:10.1680/rc.35621.0010.

Martinez, G. A. and L. A. Davis, (2000), Petrophysical measurements on shales using NMR. Paper presented at SPE/AAPG Western Regional Meeting, SPE Paper number 62851-MS, Long Beach, CA. DOI:10.2118/62851-MS.

Michels, L., J. O. Fossum, Z. Rozynek, H. Hemmen, K. Rustenberg, P. A. Sobas, G. N. Kalantzopoulos, K. D. Knudsen, M. Janek, T. S. Plivelic, and G. J. da Silva, (2015), Intercalation and retention of carbon dioxide in a Smectite clay promoted by interlayer cations. Scientific Reports, 5, 8775. DOI:10.1038/srep08775.

Mildner, D. F. R., R. Rezvani, P. L. Hall, and R. L. Borst, (1986), Small-angle scattering of shaley rocks with fractal pore interfaces, *Applied Physics Letters*, *48*, 1314–1316. DOI:10.1063/1.96964

Mildren, S. D., R. R. Hillis, D. N. Dewhurst, P. J. Lyon, J. J. Meyer, and P. J. Boult, (2005), FAST: a new technique for Geomechanical assessment of the risk of reactivation-related breach of fault seals. In *Evaluating Fault and Cap Rock Seals*, edited by Boult, P. and J. Kaldi, AAPG Hedberg Series, *2*, pp. 73–85. DOI:10.1306/1060757H23163.

Milliken, K. and L. S. Land, (1994), Evidence of fluid flow in microfractures in geopressured shales: discussion, *AAPG Bulletin*, *78*, 1637–1640.

Minh, C. C. and P. Sundararaman, (2006), NMR petrophysics in thin sand/shale laminations. SPE Paper 102435, 81st SPE Annual Technical Conference and Exhibition, San Antonio, TX. DOI:10.2118/102435-MS.

Moore, D. M. and R. C. Reynolds, (1997), *X-Ray Diffraction and the Identification and Analysis of Clay Minerals*. Oxford: Oxford University Press, 378 pp.

Morris, A., D. A. Ferrill, and D. B. Henderson, (1996), Slip tendency analysis and fault reactivation, *Geology*, *24*, 275–278. DOI:10.1130/0091-7613(1996)024

Myers, M. T., (1991), A Saturation Model for the Dielectric Constant of Shaly Sands. Paper 9118, Preprints, SCA Annual Conference, Dallas, TX, 21–22 August.

Neuzil, C. E., (1994), How permeable are clays and shales? *Water Resources Research*, *30*, 145–150. DOI:10.1029/93WR02930.

Niandou, H., J. F. Shao, J. P. Henry, and D. Fourmaintreaux, (1997), Laboratory investigation of the mechanical behaviour of Tournemire shale, *International Journal of Rock Mechanics and Mining Sciences*, *34*, 3–16. DOI:10.1016/S0148-9062(96)00053-8.

Nicot, B., S. Gautier, M. Fleury, and S. Durucan, (2006), Pore Structure Analysis of Coals Using Low Field NMR Measurements and Thermogravimetric Analysis. SCA paper 2006-27, International Symposium of the Society of Core Analysts, Trondheim, Norway, 12 pp.

North, A. N. and J. C. Dore, (1993), Characterisation of porous solids using small-angle x-ray and neutron scattering. In *Multifunctional Mesoporous Inorganic Solids*, edited by Sequeira, C. A. C. and M. J. Hudson, NATO Advanced Science Institutes Series, Series C, Mathematical and Physical Sciences, 400, Kluwer Academic Publishers, Dordrecht/Boston, 379–400.

North, A. N., J. C. Dore, A. R. Mackie, A. M. Howe, and J. Harries, (1990), Ultrasmall-angle X-ray scattering studies of heterogeneous systems using synchrotron radiation techniques, *Nuclear Instruments and Methods in Physics Research Section B: Beam Interactions with Materials and Atoms*, *47*, 283–290. DOI:10.1016/0168-583X(90)90758-M.

Nygard, R., M. Gutierrez, R. Gautam, and K. Høeg, (2004), Compaction behaviour of argillaceous sediments as function of diagenesis, *Marine and Petroleum Geology*, *21*, 349–362. DOI:10.1016/j.marpetgeo.2004.01.002.

Olierook, H. K. H., C. Delle Piane, N. E. Timms, L. Esteban, R. Rezaee, A.J. Mory, and L. Hancock, (2014), Facies-based rock properties characterization for CO_2 sequestration: GSWA Harvey 1 well, Western Australia, *Marine and Petroleum Geology*, *50*, 83–102. DOI:10.1016/j.marpetgeo.2013.11.002.

Olsen, H. W., R. W. Nichols, and T. L. Rice, (1985), Low gradient permeability measurements in a triaxial system, *Geotechnique*, *35*, 145–157.

Paterson, M. S., (1977), *Experimental Rock Deformation – The Brittle Field*. Springer-Verlag, New York, 254 pp.

Pearson, F. J., (1999), What is the porosity of a mudrock? In *Muds and Mudstones: Physical and Fluid Flow Properties*, edited by Aplin, A. C., A. J. Fleet, and J. H. S. MacQuaker, Geological Society of London, Special Publication, 158, 9–21. DOI:10.1144/GSL.SP. 1999.158.01.02

Peltonen, C., Ø. Marcussen, K. Bjørlykke, and J. Jahren, (2009), Clay mineral diagenesis and quartz cementation in mudstones: the effects of smectite to illite reaction on rock properties, *Marine and Petroleum Geology*, *26*, 887–898. DOI:10.1016/j.marpetgeo.2008.01.021.

Pervukhina, M., B. Gurevich, D. N. Dewhurst, P. Golodoniuc, and M. Lebedev, (2015), Rock physics analysis of shale reservoirs. In *Fundamentals of Gas Shale Reservoirs*, edited by Rezaee, R., John Wiley & Sons, Inc., Hoboken, NJ, pp. 191 205.

Petley, D. N., (1999), Failure envelopes of mudrocks at high confining pressures. In *Muds and Mudstones: Physical and Fluid Flow Properties*, edited by Aplin, A. C., A. J. Fleet, and J. H. S. Macquaker, Geological Society of London Special Publication, 158, 61–71. DOI:10.1144/GSL.SP.1999.158.01.05

Pfleiderer, S. and H. C. Halls, (1994), Magnetic pore fabric analysis: a rapid method for estimating permeability anisotropy, *Geophys. J. Int.*, *116*, 39–45. DOI:10.1111/j.1365-246X.1994.tb02125.x

Radlinski, A. P., M. Mastalerz, A. L. Hinde, M. Hainbuchner, H. Rauch, M. Baron, J. S. Lin, L. Fan, and P. Thiyagarajan, (2004), Application of SAXS and SANS in evaluation of porosity, pore size distribution and surface area of coal, *International Journal of Coal Geology*, *59*, 245–271. DOI:10.1016/j.coal.2004.03.002.

Rasouli, V., M. Pervukhina, T. M. Müller, and R. Pevzner, (2013), In-situ stresses in the southern Perth Basin at the GSWA Harvey-1 well site, *Exploration Geophysics*, *44*, 289–298. DOI:10.1071/EG13046.

Rezaee, R., A. Saeedi, and B. Clennell, (2012), Tight gas sands permeability estimation from mercury injection capillary pressure and nuclear magnetic resonance data, *J. Pet. Sci. Eng.*, *88–89*, 92–99. DOI:10.1016/j.petrol.2011.12.014.

Rinaldi, A. P., V. Vilarrasa, J. Rutqvist, and F. Cappa, (2015), Fault reactivation during CO_2 sequestration: effects of well orientation on seismicity and leakage, *Greenhouse Gases-Science and Technology*, *5*, 645–656. DOI:10.1002/ghg.1511.

Ringrose, P. S., A. S. Mathieson, I. W. Wright, F. Selama, O. Hansen, R. Bissell, N. Saoula, J. Midgley, (2013), The In Salah CO_2 storage project: lessons learned and knowledge transfer, *Energy Procedia*, *37*, 6226–6236. DOI:10.1016/j.egypro.2013.06.551.

Romanov, V., Y. Soong, C. Carney, G. E. Rush, B. Nielsen, and W. O'Connor, (2015), Mineralization of carbon dioxide: a literature review, *ChemBioEng Reviews*. DOI:10. 1002/cben.201500002.

Sarout, J. and Y. Guéguen, (2008), Anisotropy of elastic wave velocities in deformed shales: part 1 – experimental results, *Geophysics*, *73*(5), D75–D89. DOI:10.1190/1.2952744.

Sarout, J., L. Esteban, C. Delle Piane, B. Maney, and D. N. Dewhurst, (2014), Elastic anisotropy of Opalinus clay under variable saturation and triaxial stress, *Geophysical Journal International*, *198*, 1662–1682. DOI:10.1093/gji/ggu231.

Schaef, H. T., E. S. Ilton, O. Qafoku, P. F. Martin, A. R. Felmy, K. M. Rosso, (2012), In situ XRD study of Ca^{2+} saturated montmorillonite (STX-1) exposed to anhydrous and wet supercritical carbon dioxide, *International Journal of Greenhouse Gas Control*, *6*, 220–229. DOI:10.1016/j.ijggc.2011.11.001.

Schmitt, L., T. Forsans, and F. J. Santarelli, (1994), Shale testing and capillary phenomena, *Int. J. Rock Mech. M. Sci. and Geomech. Abstr.*, *31*, 411–427. DOI:10.1016/0148-9062 (94)90145-7

Schowalter, T. T., (1979), Mechanisms of secondary hydrocarbon migration and entrapment, *Bulletin of the American Association of Petroleum Geologists*, *63*, 723–760.

Shah, V., D. Broseta, and G. Mouronval, (2008), Capillary alteration of caprocks by acid gases. SPE Paper 113353. SPE/DOE Improved Oil Recovery Symposium, Tulsa, OK. DOI:10.2118/113353-MS

Sharma, S., P. Cook, T. Berly, and M. Lees, (2009), The CO2CRC Otway project: overcoming challenges from planning to execution of Australia's first CCS project, *Energy Procedia*, *1*, 1965–1972. DOI:10.1016/j.egypro.2009.01.256.

Shipton, Z. K., J. P. Evans, D. Kirschner, P. T. Kolesar, A. P. Williams, and J. Heath, (2004), Analysis of CO_2 leakage through 'low-permeability' faults from natural reservoirs in the Colorado plateau, east-central Utah. In *Geological Storage of Carbon Dioxide*, edited by Baines, S. J. and R. H. Worden, Geological Society of London Special Publications, 223, 43–58. DOI:10.1144/GSL.SP.2004.233.01.05

Shukla, R., P. Ranjith, A. Haque, and X. Choi, (2010), A review of studies on CO_2 sequestration and caprock integrity, *Fuel*, *89*, 2651–2664. DOI:10.1016/j.fuel.2010.05.012.

Sinha, S., E. M. Braun, M. D. Determan, Q. R. Passey, S. A. Leonardi, J. A. Boros, A. C. Wood, T. Zirkle, and R. A. Kudva, (2013), Steady-state permeability measurements on intact shale samples at reservoir conditions – effect of stress, temperature, pressure, and type of gas. SPE 164263, 15 pp. DOI:10.2118/164263-MS

Skempton, A. W., (1954), The pore pressure coefficients A and B, *Geotechnique*, *4*, 143–147.

Skurtveit, E., E. Aker, M. Soldal, M. Angeli, and Z. Wang, (2012), Experimental investigation of CO_2 breakthrough and flow mechanisms in shale, *Petroleum Geoscience*, *18*, 3–15. DOI:10.1144/1354-079311-016

Skurtveit, E., L. Grande, O. Y. Ogebule, R. H. Gabrielsen, J. I. Faleide, N. H. Mondol, R. Maurer, and P. Horsrud, (2015), Mechanical testing and sealing capacity of the Upper Jurassic Draupne Formation, North Sea. Proceedings of the 49th US Rock Mechanics/Geomechanics Symposium, San Francisco, CA, 28 June–1 July, ARMA 15-331, 8 pp.

Sondergeld, C. H., K. E. Newsham, J. T. Comisky, M. C. Rice, and C. S. Rai, (2010), Petrophysical considerations in evaluating and producing shale gas resources. SPE Unconventional Conference, Pittsburg, Pennsylvania. SPE Paper 131768. DOI:10.2118/131768-MS.

Stalker, L., S. Varma, D. van Gent, J. Haworth, and S. Sharma, (2013), South west hub: a carbon capture and storage project, *Australian Journal of Earth Sciences*, *60*, 45–58. DOI:10.1080/08120099.2013.756830.

Steiger, R. P. and P. K. Leung, (1988), Quantitative determination of the mechanical properties of shales. SPE Paper 18024, 8 pp. DOI:10.2118/18024-PA

Steiger, R. P. and P. K. Leung, (1991), Consolidated undrained triaxial test procedure for shales. Proceedings 32nd U.S. Symposium on Rock Mech., ARMA Paper 91-637, Norman, OK, Balkema, Rotterdam, pp. 637–646.

Tenthorey, E., T. Dance, Y. Cinar, J. Ennis-King, and J. Strand, (2014), Fault modelling and geomechanical integrity associated with the CO2CRC Otway 2C injection experiment, *Int. J. Greenhouse Gas Control*, *30*, 72–85. DOI:10.1016/j.ijggc.2014.08.021.

Tenthorey, E., S. Vidal-Gilbert, G. Backé, R. Puspitasari, Z. J. Pallikathekathil, B. Maney, and D. N. Dewhurst, (2013). Modelling the geomechanics of gas storage: a case study from the Iona gas field, Australia, *Int. J. Greenhouse Gas Control*, *13*, 138–148. DOI:10.1016/ j.ijggc.2012.12.009

Thyberg, B., J. Jahren, T. Winje, K. Bjørlykke, J. I. Faleide, and Ø. Marcussen, (2010), Quartz cementation in late cretaceous mudstones, northern North Sea: changes in rock properties due to dissolution of smectite and precipitation of micro-quartz crystals, *Marine and Petroleum Geology*, *27*, 1752–1764. DOI:10.1016/j.marpetgeo.2009.07.005.

Timms, N. E, H. K. H. Olierook, M. E. J. Wilson, C. Delle Piane, P. J. Hamilton, P. Cope, and L. Stütenbecker, (2015), Sedimentary facies analysis, mineralogy and diagenesis of the mesozoic aquifers of the Central Perth Basin, Western Australia, *Marine and Petroleum Geology*, *60*, 54–78. DOI:10.1016/j.marpetgeo.2014.08.024.

Valès, F., D. Nguyen Minh, H. Gharbi, and A. Rejeb, (2004), Experimental study of the influence of the degree of saturation on physical and mechanical properties in Tournemire shale (France), *Applied Clay Science*, *26*, 197–207. DOI:10.1016/j.clay.2003.12.032.

Van Olphen, H., (1977), *An Introduction to Clay Colloid Chemistry*, 2nd edition, John Wiley, New York.

Vavra, C. L., J. G. Kaldi, and R. M. Sneider, (1992), Geological applications of capillary-pressure – a review, *Bulletin of the American Association of Petroleum Geologists*, *76*, 840–850.

Vernik, L. and X. Liu, (1997), Velocity anisotropy in shales; a petrophysical study, *Geophysics*, *62*, 521–532. DOI:10.1190/1.1444162.

Vidal-Gilbert, S., E. Tenthorey, D. Dewhurst, J. Ennis-King, P. van Ruth, and R. Hillis, (2010), Geomechanical analysis of the Naylor field, Otway Basin, Australia: implications for CO_2 injection and storage, *International Journal of Greenhouse Gas Control*, *4*, 827–839. DOI:10.1016/j.ijggc.2010.06.001.

Wang, R., N.-S. Zhang, X.-J. Liu, X.-M. Wu, and J. Yan, (2013), Apparent permeability of shale considering the adsorption and diffusion of gas and the effects of temperature and pressure on it. Research progress of mechanism of adsorption and desorption of gas in shale, Science Technology and Engineering, 2013-19.

Washburn, E. W., (1921), A note on the method of determining the distribution of pore sizes in a porous material, *Proceedings of the National Academy*, *7*, 115–116.

Watson, M. N., N. Zwingmann, and N. M. Lemon, (2004), The Ladbroke Grove–Katnook carbon dioxide natural laboratory: a recent CO_2 accumulation in a lithic sandstone reservoir, *Energy*, *29*, 1457–1466. DOI:10.1016/j.energy.2004.03.079.

Watts, N., (1987), Theoretical aspects of cap-rock and fault seals for single- and two-phase hydrocarbon columns, *Marine and Petroleum Geology*, *4*, 274–307. DOI:10.1016/0264-8172(87)90008-0.

Wollenweber, J., S. Alles, A. Busch, B. M. Krooss, H. Stanjek, and R. Littke, (2010), Experimental investigation of the CO_2 sealing efficiency of caprocks, *International Journal of Greenhouse Gas Control*, *4*, 231–241. DOI:10.1016/j.ijggc.2010.01.003.

Yang, Y. and A. C. Aplin, (1998), Influence of lithology and compaction on the pore size distribution and modelled permeability of some mudstones from the Norwegian Margin, *Marine and Petroleum Geology*, *15*, 163–175. DOI:10.1016/S0264-8172(98)00008-7.

Yielding, G., B. Freeman, and T. D. Needham, (1997), Quantitative fault seal prediction, *AAPG Bulletin*, *81*, 897–917.

Youn, H. and F. Tonon, (2010), Multi-stage triaxial test on brittle rock, *International Journal of Rock Mechanics and Mining Sciences*, *47*, 678–684. DOI:10.1016/j.ijrmms.2009.12.017.

Zhang, Y., L. Langhi, C. Delle Piane, P. Schaubs, D. Dewhurst, L. Stalker, and K. Michael, (2015a), Geomechanical stability of CO_2 containment at the southwest hub, Western Australia: a coupled geomechanical-fluid flow modelling approach, *International Journal of Greenhouse Gas Control*, *37*, 12–23. DOI:10.1016/j.ijggc.2015.03.003.

Zhang, Y., L. Langhi, C. Delle Piane, P. Schaubs, D. Dewhurst, L. Stalker, and K. Michael, (2015b), Geomechanical evaluation of fault reactivation potential and uplift at the South West Hub geological CO_2 storage site, Western Australia. Paper 15-127, 49th ARMA Conference, San Francisco, CA, 8 pp.

2

Transport in Tight Rocks

Marc Fleury and Etienne Brosse

ABSTRACT

We describe in this overview the transport properties of tight rocks at the core scale representing the matrix between fractures (if any) at larger scale. Techniques for measuring single-phase transport due to diffusion (induced by a concentration gradient) and convection (induced by a pressure gradient) are critically reviewed. The possible link between these properties and some characteristics of the pore space (porosity, pore size distribution, clay content) is listed. In a two-phase situation, the gas entry pressure is discussed as a key aspect.

2.1. INTRODUCTION

Transport properties of tight rocks are of primary interest to prove the long-term sealing efficiency of caprock formations for CO_2 storage. In this document, we gather all available information on low-permeability formation in various domains and useful for the subject. For example, natural gas storage, although short term, has a critical parameter in common with CO_2 storage, the gas entry pressure. For underground nuclear waste storage, transport properties are also key issues. In the oil and gas industry, progress in terms of measurement techniques and understanding has been obtained recently for unconventional oil and gas production (*shale gas and shale oil*).

For all above applications, the definition of tight rocks can vary; in hydrogeology, three levels are traditionally used from low to high permeability: aquiclude $1 \cdot 10^{-13}$ m/s (≈ 10 nD; see below the difference between hydraulic conductivity and intrinsic permeability), aquitard $1 \cdot 10^{-13}$ m/s to $1 \cdot 10^{-8}$ m/s (≈ 10 nD to ≈1 mD), and aquifer. In the oil and gas industry, the limit between producible and non-producible reservoir using standard technique may be around 0.1–1 mD. Below 0.1 mD for tight gas reservoir or shales, oil and gas production is often not economical without artificial fracturing.

IFP Energies nouvelles, Rueil-Malmaison, France

We will consider in this chapter the transport properties of the matrix at the laboratory scale. The difficult issue of upscaling these properties for transport at field scales or the comparison with well testing is not treated here.

2.2. POROSITY AND DIFFUSION PROPERTIES

2.2.1. Porosity

Porosity is the most basic petrophysical parameter, and yet it is not a trivial measurement due to the existence of small or very small pore sizes ranging down to the nanometer scale. Porosity is the primary parameter driving diffusion properties, and this is the reason why these two notions are described together in this chapter.

Porosity can be defined as the ratio of the volume of voids to the total volume, or equivalently, the volume not occupied by solids to the total volume. Several techniques exist for measuring porosity:

- Mass difference between dry and saturated states
- Determination of the solid volume in the dry state by gas expansion using helium (called helium porosity)
- Determination of the volume invaded by mercury at the highest pressure in experiments called mercury injection capillary pressure (MICP) tests
- Determination of the volume of water by nuclear magnetic resonance (NMR) usually at low magnetic field (<0.5 T)

Geological Carbon Storage: Subsurface Seals and Caprock Integrity, Geophysical Monograph 238, First Edition.
Edited by Stéphanie Vialle, Jonathan Ajo-Franklin, and J. William Carey.

These techniques allow determining either the solid volume or the volume of voids. The total volume is determined by diameter and length measurements in the case of a simple well-defined geometry or by pycnometry.

The first three techniques abovementioned are described in standard textbooks [*Dullien*, 1992] and are performed on a routine basis. They require however a dry sample. Since pore size can reach nanometer sizes, the comparison between these different techniques revealed severe inconsistencies (50%) when using standard protocols applied for reservoir [*Lalanne et al.*, 2014]. The accuracy of porosity measurements is of course crucial in the context of gas shale production, whereas it is less for caprock characterization and may only introduce additional scatter in porosity-permeability relationships (if any). The key aspect is the sample drying to remove water from the smallest pores, including interlayer water in some swelling clays, requiring elevated temperature up to 150°C [*Lalanne et al.*, 2014], and the capacity of such samples to quickly adsorb water vapor contained in the air, requiring a careful storage at the different preparation steps. The drying temperature varies from one laboratory to another, and the value of 105°C seems to be the most popular [*Busch et al.*, 2017]. On the opposite, the NMR technique requires a fully saturated state, and this technique may be probably the best for the determination of porosity in undisturbed conditions. A detailed study performed on clay powders reveal that the interlayer water in smectites can be detected with the appropriate NMR instrument and that this water is in fast diffusive exchange with intraparticle water [*Fleury et al.*, 2013b]. Recent carefully conducted comparisons [*Busch et al.*, 2017] reveal that water and helium porosity gives the same results even in the presence of nanometer pore sizes. This is expected since van der Waals molecular diameters of helium and water molecules are 0.28 and 0.34 nm, respectively; hence, size exclusion effect is insignificant, but connectivity of the pore network may play a role. As mentioned, preparation protocols are a key aspect. Finally, helium is often considered as a reference measurement, but a standard error calculation reveals that it could be in error by about 20–50% at very low values of porosity (1–2 porosity units); indeed, since the helium technique determines the solid volume, the difference between the solid and total volume becomes increasingly uncertain as porosity decreases and errors accumulate.

The MICP experiment [*Purcell*, 1949] is commonly used to determine the pore size distribution but is destructive. It is usually performed on trim ends or smaller pieces of rocks that cannot be used or are too small for permeability measurements. Since mercury is non-wetting for most minerals, it will not penetrate spontaneously into the pore network. In the MICP experiment, the volume V of mercury penetrating the porous media is recorded as a function of pressure P which is increased gradually by small steps up to 4000 bar. Using Laplace's law, the pressure can be linked to a pore throat radius assuming a cylindrical geometry according to

$$P = \frac{2\gamma\cos(\theta)}{r} \tag{2.1}$$

where γ is the surface tension and θ is the air-mercury contact angle. Considering each couple of points (V,P) at capillary equilibrium, the derivative of the experimental curve $S(P)$ is exploited to obtain the pore throat size distribution $g(r)$ plotted in logarithmic scale according to

$$g(r) = P\frac{dS}{dP} \tag{2.2}$$

where $S = V/V_{max}$ is the saturation. The MICP derived distribution only reflects the volume accessible through the throats at a given pressure. At a pressure of 4000 bar, the smallest measurable diameter is about 4 nm; see, for example, Figure 2.1.

Tight rocks do not have necessarily a low porosity and one must understand the depositional context. The most common tight rocks are made of fine-grained sediments called with a variety of names (mudstones, claystone, shale, siltstone, argillite, etc.) expressing a very large variety of composition and compaction state. Hence, without considering any compaction or diagenesis effect, the grain size is governing permeability but not porosity. Indeed, theoretically, if we take a simple system of grain packs of uniform grain size, permeability depends typically on the

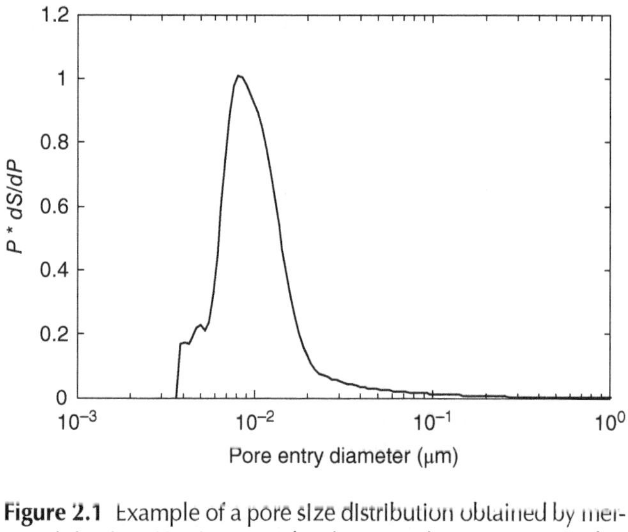

Figure 2.1 Example of a pore size distribution obtained by mercury injection, Ketzin caprock [*Fleury et al.*, 2013a]. Note that the resolution is limited to 4 nm and that smaller pores may exist. Fleury, Gautier, Gland, Boulin, Norden, and Schmidt-Hattenberger. https://ogst.ifpenergiesnouvelles.fr/articles/ogst/pdf/first/ogst120213.pdf. Licensed under CC BY 4.0.

square of the grain size, whereas porosity may be constant. As proposed by *Pearson* [1999], one could define porosity as a function of the transport process; this is indeed true when considering ions for which only a fraction of the water porosity is accessible due to anion exclusion.

2.2.2. Diffusion

Compared to convection and to the timescale of injection in a storage site, molecular diffusion is a very slow process. However, diffusion of CO_2 in the pore water may affect the pore structure at the base of the caprock due to geochemical reaction. In the caprock, we consider the diffusion coefficient of dissolved CO_2 which has similar properties as the diffusion of water. Indeed, for bulk solution, diffusion of dissolved CO_2 is lower but close to the self-diffusion of water [*Hayduk and Laudie*, 1974]. CO_2 may however adsorb onto natural solid surfaces [*Busch et al.*, 2008], a small but not negligible effect. Hence, considering the diffusion of water only will give upper limits for the diffusion of dissolved CO_2, and we can take advantage of the much larger corresponding bibliography.

A confusing terminology exists about diffusion in porous media: effective, apparent, or pore diffusion. If a flux J of a tracer is measured through a porous media along a single dimension, then an effective diffusion coefficient D_e must be considered according to Fick's law:

$$J = -D_e \frac{\partial c}{\partial z} \tag{2.3}$$

However, following *Berne et al.* [2009], the diffusive transport is finally calculated from the following equation in which porosity Φ is present:

$$\frac{\partial c}{\partial t} = D_p \frac{\partial^2 c}{\partial z^2} \quad \text{where} \quad D_p = \frac{D_e}{\Phi} \tag{2.4}$$

The pore diffusion coefficient D_p includes porosity and is therefore larger than the effective diffusion coefficient. When comparing two porous media, the pore diffusion should be used and not the effective diffusion. The ratio D_m/D_p is also used to define tortuosity τ, where D_m is the molecular diffusivity of bulk water. Finally, an apparent diffusivity is usually defined for species interacting with the solid surface (e.g., ions):

$$D_{app} = \frac{D_e}{R} = \frac{D_p}{\Phi} \tag{2.5}$$

where R is commonly called the retardation factor. Diffusion can also be measured along several directions, usually across and along the bedding.

For nuclear waste storage, diffusion has been studied extensively for a long time using tracers [*Shackelford*, 1991]. For water, the most common technique is the through-diffusion setup in which a sample is placed between two reservoirs; one of the reservoirs is then spiked with a tracer (e.g., tritium) and monitored in the other as a function of time. For CO_2 diffusion, the sensitive devices used for gas breakthrough measurements can also be adapted to measure diffusion [*Wollenweber et al.*, 2010]; CO_2 is placed upstream of a sample saturated with water and monitored downstream, ensuring the absence of pressure gradient. Pulsed field gradient nuclear magnetic resonance (PFG-NMR) is another technique for measuring water diffusion coefficient without the need for tracer [*Callaghan*, 2011]. However, in natural tight porous media, the lifetime of the magnetization needed to follow the molecular motions is too short to permit such measurements. Using deuterium as a tracer, NMR can be used to measure water diffusion by immersing a sample into deuterium and monitoring the amount of water inside the sample as a function of time [*Berne et al.*, 2009; *Fleury et al.*, 2009].

Water or dissolved CO_2 diffusion coefficient D_p is typically in a range $5 \cdot 10^{-11}$ to $5 \cdot 10^{-10}$ m²/s even though permeability may be below 10^{-18} m² for a large variety of sedimentary formations spanning from limestones to clay-rich mudstones [*Andra*, 2005; *Motellier et al.*, 2007; *Busch et al.*, 2008; *Fleury et al.*, 2009; *Wollenweber et al.*, 2010; *Berthe et al.*, 2011; *Berthe*, 2012; *Fleury et al.*, 2013a; *Fleury and Romero-Sarmiento*, 2016]. Indeed, permeability and diffusion are two separate processes not sensitive to the same parameters of the pore network. Using the analogy between electrical and diffusion properties [*Berne et al.*, 2009], the primary parameter governing diffusion is porosity according to

$$\frac{D_p}{D_m} = \Phi^{m-1} \tag{2.6}$$

where m is the standard cementation exponent used in conductivity measurements [*Dullien*, 1992]. The exponent m has a default value of 2 and can typically vary between 1.7 and 2.5. Data taken from the above-mentioned references and plotted in Figure 2.2 indicate that the cementation exponent varies around 2.5. In the standard interpretation of the cementation exponent [*Glover*, 2009], values larger than 2 correspond to a less connected pore space, and such values are therefore expected. The above model may only give a rough estimate.

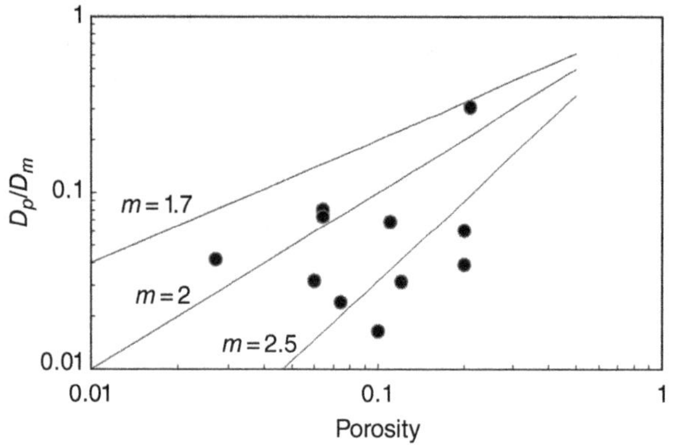

Figure 2.2 Compilation of diffusion coefficients found in literature. Permeability is in the range of 1–100 nD for most of the samples. D_m is the molecular diffusion of water taken at the temperature of the experiment.

2.3. PERMEABILITY

Permeability K is the second most important petrophysical parameter. It is a coefficient relating the flow rate to the pressure drop in Darcy's law valid for incompressible fluids:

$$q\left[\mathrm{m^3s^{-1}}\right]=\frac{K\left[\mathrm{m^2}\right]}{\mu\left[\mathrm{Pa.s}\right]}\frac{\Delta P\left[\mathrm{Pa}\right]}{L\left[\mathrm{m}\right]}S\left[\mathrm{m^2}\right] \quad (2.7)$$

where q is the flow rate, μ is the viscosity of the fluid flowing through the porous media, ΔP is the pressure gradient over the length L, and S is the cross-sectional area. This definition shows that permeability is an intrinsic property of a porous media and has units of m^2. A very common unit is also the Darcy $(D) = 0.987 \cdot 10^{-12}\,m^2$. In hydrogeology or geology and many related publications, the hydraulic conductivity K' is used instead of permeability (but also called sometimes permeability) with units of m/s. It is a coefficient relating the flow rate to a pressure drop generated by a water column of height H and length L according to

$$q\left[\mathrm{m^3s^{-1}}\right]=K'\left[\mathrm{ms^{-1}}\right]\frac{H\left[\mathrm{m}\right]}{L\left[\mathrm{m}\right]}S\left[\mathrm{m^2}\right] \quad (2.8)$$

There is no unique correspondence between K and K', but for storage condition, $K = \alpha K'$ with $\alpha = 3\text{–}6 \cdot 10^{-7}$ when taking appropriate values for water viscosity, function of storage temperature.

As noted by *Dewhurst et al.* [1999], permeability data for tight rocks were virtually lacking before 2000. For the oil and gas industry, the need to understand or predict overpressure zones resulted in some studies in the past [*Thomas et al.*, 1968]. Besides the fact that caprocks were generally not cored, permeability measurements were also time consuming because most experimental devices were designed to deal with permeabilities at least three orders of magnitude larger. In the last 15 years, the situation completely changed due to the need of characterization for various applications not only for CO_2 storage but also for nuclear waste storage issues and for oil and gas shale production. Hence, experimental techniques were revisited and improved.

2.3.1. Direct Measurements of Intrinsic Permeability

There exist a large variety of techniques for measuring permeability. They can be classified in several groups and subgroups:

• Steady state (SS) techniques in which a steady flow is generated; from Eq. (2.7), one can either impose a pressure gradient and measure a flow rate or impose a flow rate and measure the upstream and downstream pressures.

• Unsteady state (USS) techniques first proposed by *Brace et al.* [1968]: a pulsed pressure is imposed at one end of the sample, and the pressure response is recorded at the other end; similarly, an oscillating pressure [*Kranz et al.*, 1990] can also be applied.

Within these two techniques, water or gas can be used, or a combination of both. If we include measurement techniques using cuttings and small chips that are less commonly used [*Lenormand et al.*, 2012; *Egermann et al.*, 2013], the variety of techniques is extremely large, and the resulting values can vary by several orders of magnitude. Hence, the subject has been quite controversial in the recent years. It is necessary to distinguish several effects:

• The measurement technique itself, including experimental difficulties

• Type of fluid and confining stress associated with a given technique

• Heterogeneity related to the size of the sample

For the latter, it has been clearly demonstrated on artificial drill cuttings made from a plug with a known permeability that the observed variability (about a factor of 5) is due to the pore structure itself [*Egermann et al.*, 2013]. Hence, this variability can and must be compensated by a large number of measurements. It has been shown in a systematic study that permeability from crushed rocks is much higher than permeability measured on preserved samples [*Ghanizadeh et al.*, 2015a].

If Darcy's law is applicable, SS and USS techniques should give the same intrinsic permeability. Based on the work of *Brace et al.* [1968], USS techniques have been preferred recently by several authors because they are believed to be faster. Using water, a detailed comparison has been performed by *Boulin et al.* [2012]; USS is indeed very fast and simple in terms of interpretation (an exponential curve must be analyzed) assuming an instantaneous pressure equilibrium in the sample; however, this hypothesis can be questioned for low or very low (nD) permeable media due to the rock and fluid compressibility. These effects are summarized in a coefficient called the specific storage S_s defined as

$$\frac{S_s}{\rho_w g}\frac{\partial P_w}{\partial t} = -\nabla\left(\frac{K_w}{\mu_w}\cdot\nabla P_w\right) \tag{2.9}$$

The specific storage, of the order of 10^{-6} to 10^{-5} m^{-1}, corresponds to the amount of water that can be released by the formation while remaining fully saturated, per unit volume, per unit pressure gradient. It can be related directly to Young's modulus and Poisson's ratio [*Marsily*, 1986]. During a pulse decay, both the permeability and storage factor must be adjusted, yielding some possible ambiguity about the fitted values because multiple solutions of K_w and S_s can yield the same curve (compensation effect). This ambiguity can be cleared up by measuring K_w using an SS experiment and fitting only S_s when interpreting the USS experiment. It was also shown that both USS and SS techniques may take similar duration, providing flow rates are measured with a high sensitivity, of the order of 1 µl/min. Besides these effects, there exist also numerous experimental difficulties (leaks, accurate calibration of volume measurements) yielding a difference between USS and SS techniques as discussed in *Amann-Hildenbrand et al.* [2013]. These issues may become critical when reaching very low permeability (Fig. 2.3).

Although caprock formation is saturated with water or brine, available samples may be already dried after several years of storage, and it may be more convenient to consider gas permeability. For gas flow, the comparison of USS and SS is more complex. First, the apparent gas permeability K_{app} is deduced from an SS experiment using

$$K_{app} = 2.\frac{\mu.L.Q.P}{s.\left(P_u^2 - P_d^2\right)} \tag{2.10}$$

which takes into account gas compressibility. The determination of the intrinsic permeability K when using gas requires multiple experiments at different mean gas pressure in order to extrapolate at infinite pressure, as illustrated in Figure 2.4. This is the well-known Klinkenberg effect, also called slippage effect and described by the relation

$$K_{app} = K\left(\frac{\beta}{P_m}+1\right) \tag{2.11}$$

where β is the so-called Klinkenberg coefficient. Indeed, for a non-dense gas flowing in a tight porous media, the mean free path of molecules may be of the same order of magnitude as the pore size, or equivalently, the number of collisions with the solid walls and with other molecules may be similar. Hence, the energy dissipation is not solely due to intermolecular interaction as in regular flow but also to interactions with solids; the Klinkenberg regime can be considered as a combination of the Knudsen and

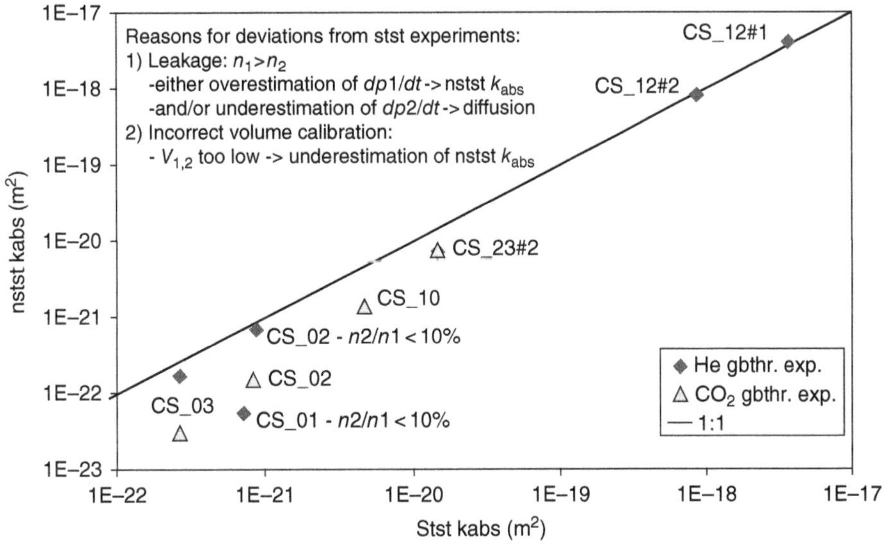

Figure 2.3 Comparison of water permeability using steady state (stst) and unsteady state (nstst) methods [*Amann-Hildenbrand et al.*, 2013]. Reprinted with permission from Elsevier.

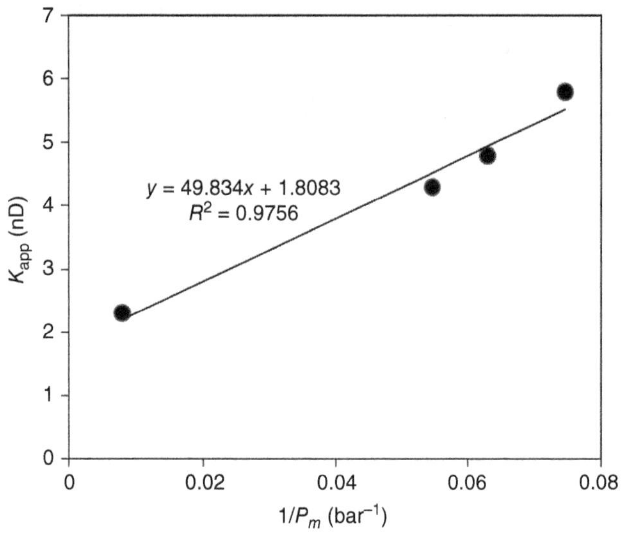

Figure 2.4 Typical example of measurement of the apparent gas permeability versus mean gas pressure on a shale. The mean pressure was varied between 125 and 13 bar. The Klinkenberg coefficient is β = 27.6 bar^{-1} in this example; the apparent permeability at atmospheric pressure is about 51 nD, more than one order of magnitude larger than the intrinsic permeability of 1.8 nD.

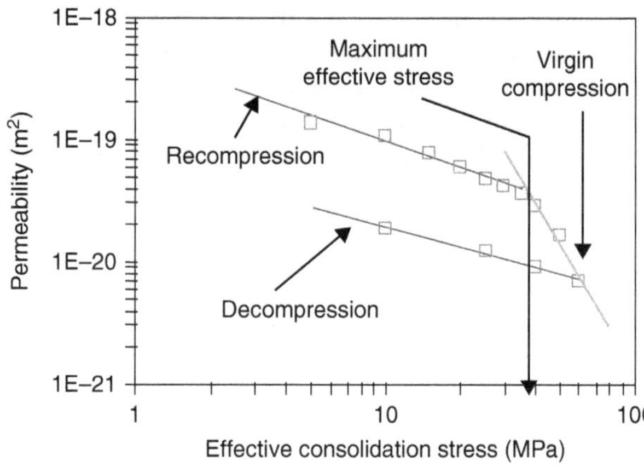

Figure 2.5 Illustration of stress effect on permeability [*Yang and Aplin*, 2007]. A variation of 1 order of magnitude can be observed. Reprinted with permission from Wiley.

the Poiseuille regime. For the same pressure drop, the flow is *faster* as the mean pressure is decreased; the apparent permeability is larger than K. The validity of the Klinkenberg regime for low permeable media is not obvious and has been a long-standing problem. The value of β varies from one pore structure to another, but also on the confining stress for the same sample, as well as the gas. However, for the latter, a correlation relating the most common gases (He, N_2, CO_2, air, H_2) can be used [*Civan*, 2010]. Hence, β must be measured for each sample in well-defined conditions. While the determination of K and β is straightforward for the SS technique using Eqs. (2.10) and (2.11) and a minimum number of experiments, the design and analysis of gas USS techniques has been the object of recent new development [*Jannot et al.*, 2008; *Jannot and Lasseux*, 2012; *Lasseux et al.*, 2012]. A single pressure decay experiment does not contain enough information and may not suffice for the determination of K and β. Hence, these authors recommend to perform not only one experiment but a series of experiments realized in an optimum way.

2.3.1.1. Stress Effects

Besides compressibility effect or Klinkenberg correction, the effect of stress is major in mudstones with high clay content, as shown in the example of Figure 2.5 taken from *Yang and Aplin* [2007]. Well coring inevitably involves a reduction of stress when bringing samples to the surface, potentially generating microcracks. This is also a well-known issue for borehole stability [*Horsrud et al.*, 1998] in relation with the choice of appropriate drilling fluids. In the nuclear industry, specific protocols have been tested based on trial and error approach; besides avoiding drying by careful protection of the cores, samples collected on site are put back under axial stress as soon as possible after extraction from coring tubes [*Andra*, 2005]. In the laboratory, it is therefore necessary to reproduce the in situ stress and verify the absence of cracks in order to have representative measurements of matrix permeability [*Dewhurst and Siggins*, 2006].

2.3.2. Examples

Together with the testing of measurement methods, many permeability data are available in literature, and we will give here only a few examples. As the measurement techniques improved and clarified in the last 15 years, it is necessary to examine the method used before using the data for practical applications. Covering a large range of mudstones, two systematic and remarkable studies have been performed by *Hildenbrand et al.* [2004] and *Yang and Aplin* [2007]. For the latter, the formations are not specified, but the data set is well documented with clay content (Fig. 2.6), pore structure information, and in situ effective stress. These studies cover permeabilities ranging from 10^{-21} up to 10^{-15} m^2 and porosities up to 25%. More closely related to specific sites, data are available for the Jurassic Draupne Formation [*Angeli et al.*, 2009], for the caprock above the pilot site in Ketzin [*Fleury et al.*, 2013a], for the Nordland Shale above the Sleipner Utsira Formation [*Harrington et al.*, 2009], for the caprock above the Midale Formation at the Weyburn site [*Li et al.*, 2005], and for the Montney Formation [*Ghanizadeh et al.*, 2015b, 2015c]. Related to nuclear waste storage, two formations have been extensively studied: the Opalinus Clay [*Marschall et al.*, 2005] and the Callovo-Oxfordian Clay [*Andra*, 2005]. See Table 2.1 for an overview of the range of permeabilities covered in all these studies.

In clay-rich mudstones, the issue of anisotropy at small scale cannot be avoided. Indeed, due to the sedimentation process yielding mudstones, clay particles may be oriented at the millimeter or centimeter scale. It is usually considered that permeability perpendicular to bedding is two to three times smaller than parallel to bedding [*Harrington et al.*, 2009; *Fleury*, 2013, unpublished data for the Callovo-Oxfordian Clay]. However, permeability ratio up to 8 has also been measured [*Yang and Aplin*, 2007].

2.3.3. Permeability from Pore Structure Information

It is of great practical interest to try to relate pore structure information to permeability. As can be seen in Figure 2.6, general permeability-porosity relationships are very difficult to evidence, especially in the range of 10^{-19} to 10^{-21} m^2. A very common characterization of the pore structure is the mercury injection experiment already discussed. Many models have been developed, essentially relying on the idea that a porous medium can be viewed as a collection of tortuous tubes. For one tortuous tube of diameter d, permeability can be calculated as

$$k = \frac{\Phi d^2}{32\tau^2} \tag{2.12}$$

where Φ is porosity and τ is tortuosity. This simple relation indicates the important parameters but does not express at all the difficulty of deriving with enough accuracy permeability from pore size distribution. The pore size is a key and sensitive indicator of permeability. For common permeable media, a typical pore throat size can be 10 μm with a permeability value of 100 mD (10^{-13} m^2); hence, the permeability in tight rocks for pore sizes around 10 nm can be shifted by 6 orders of magnitude down to 10^{-19} m^2 or less, while tortuosity may be similar. In the presence of a distribution of pore size, the main problem is to choose a representative one or calculate an average size using an appropriate method. Hence, there exist in literature a large number of correlations in which a different choice is made. In the famous Kozeny-Carman relationship, permeability is proportional to the square of the specific surface. In other correlations, a specific point in the MICP curve is chosen [*Swanson*, 1981; *Katz and Thompson*, 1986]. An interesting comparison of various correlations performed in tight gas sands can be found in *Comisky et al.* [2007], in the range of 0.1 μD–0.2 mD. In a range further down (down to 1 nD), a model is proposed for mudstones [*Yang and Aplin*, 2007] taking into account pore size (squared dependence), pore alignment due to stress to distinguish vertical from horizontal permeability, and clay content.

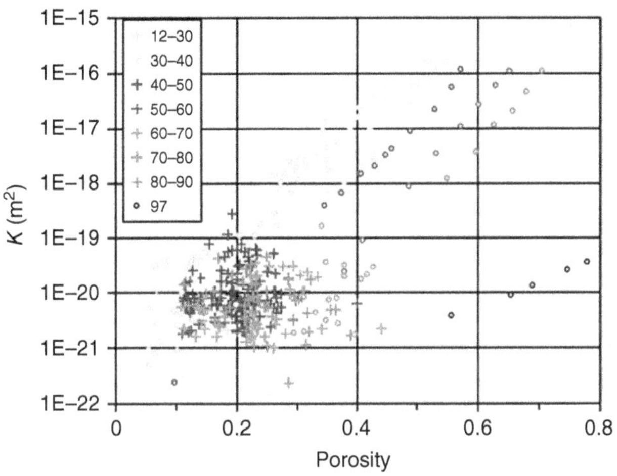

Figure 2.6 Porosity and permeability for a series of 376 mudstones with variable amount of clay (mass fraction indicated in the inset). From *Yang and Aplin* [2007]. Reprinted with permission from Wiley.

Table 2.1 Range of Intrinsic Permeability for Various Formations.

Reference	Location or formation	Permeability range (nD or 10^{-21} m^2)
Andra [2005]	Callovo-Oxfordian	1–100
Angeli et al. [2009]	Jurassic Draupne	60
Fleury et al. [2013a]	Ketzin	27
Ghanizadeh et al. [2015b]	Montney	1–1·10^5
Harrington et al. [2009]	Nordland Shale, Utsira	130–900
Hildenbrand et al. [2004]	Paleocene, Early Jurassic Opalinus, Permian, Oligocene (Boom Clay), Rhetien	0.01–1000
Li et al. [2005]	Weyburn	6–2000
Marschall et al. [2005]	Opalinus	0.5–8
Yang and Aplin [2007]	North Sea, Gulf of Mexico	0.5–0.5·10^6

Note that an important consequence of Eq. (2.12) is that only a small fraction of pores, the largest ones, contribute to permeability.

Using the same principle as in MICP, NMR measurement could also be used to estimate permeability in tight formations. This is of particular interest because NMR is now more often acquired in well logging programs in tight formations, although specific acquisition protocols must be used [*Hook et al.*, 2011]. The result of NMR measurements is a relaxation time distribution (T_2) that can be interpreted in certain conditions as a pore size distribution [*Dunn et al.*, 2002]. In the oil and gas industry, NMR is often interpreted as a permeability indicator using the following relationship [*Dunn et al.*, 2002]:

$$K = C\,\Phi^a\,T_{2m}^b \tag{2.13}$$

where T_{2m} is an appropriate relaxation time average of the measured distribution (often taken as the log average). The constant C and the exponents a and b should ideally be adjusted to a given data set in order to interpret the NMR log into a permeability log. In a similar way to mercury and the parallel tube model, default values for a and b are 1 and 2, respectively. Whether or not such relationship can be used for tight formations has yet to be demonstrated.

As for permeable formations, the use of porosity alone as an indicator of permeability often does not yield convincing results. An interesting and successful attempt to include another indicator has been performed by *Yang and Aplin* [2010] for a large suite of samples (376 data points) in which the clay content has been measured. These authors derived the following relationship:

$$\begin{aligned} &\ln(K) = a_1 + a_2 e + a_3\sqrt{e} \\ &\text{with } a_i = c_{i1} + c_{i2}CF + c_{i3}\sqrt{CF} \quad \text{and} \quad e = \frac{\Phi}{1-\Phi} \end{aligned} \tag{2.14}$$

where CF is the clay content defined as the mass fraction of particles smaller than 2 μm. The scatter in the data set presented in Figure 2.6 could be reduced considerably, down to a factor of 3 compared to several orders of magnitude at a single porosity value (correlation coefficient of 0.93). Although 9 fitting coefficients (c_{ij}) have been used, this successful correlation clearly indicates the potential important role of the clay content.

2.4. ENTRY PRESSURE AND TWO-PHASE PROPERTIES

2.4.1. Breakthrough Pressure

In the context of CO_2 storage (but also for methane storage), a key property is the breakthrough pressure P_{BT} of the gas into the caprock. Other terminologies also exist: entry pressure and displacement pressure expressing sometimes different physical mechanisms. Here we chose the term *breakthrough* to emphasize that the gas can flow steadily after reaching P_{BT}, at the scale of a plug (a few centimeters). The term *entry* may be confusing because it could mean that the gas can penetrate the porous media but not percolate. The breakthrough pressure defines the maximum height h of the gas column under the caprock according to

$$h = \frac{P_{BT}}{\left(\rho_b - \rho_{CO_2}\right)} \tag{2.15}$$

where ρ_b and ρ_{CO_2} are the specific density of brine and CO_2 at storage conditions, respectively. Taking a high value $\rho_{CO_2} = 500\,\text{kg/m}^3$, an entry pressure of 1 bar is typically required for $h = 20\,\text{m}$. Therefore, a minimum of a few bars is required in general, not taking into account a safety coefficient. P_{BT} depends on gas-water interfacial tension and contact angle θ according to Laplace's law in a tube of radius r [*Dullien,* 1992]:

$$P_{BT} = \frac{2\gamma\cos(\theta)}{r} \tag{2.16}$$

Interfacial tension γ has been measured at different pressures and brine salinities [*Chiquet et al.*, 2007b; *Chalbaud et al.*, 2009]; a typical value is 27 mN/m; for a compilation of data with different gases, see *Heath et al.* [2012]. When the gas is non-wetting, $\theta = 0$ and the entry pressure is maximum for a given pore structure. However, the possibility of CO_2 to partially wet solid minerals has been a great concern, with some alarming values as large as 80° ($\cos(\theta) = 0.174$), reducing considerably the breakthrough pressure [*Chiquet et al.*, 2007a]. Reviewed and improved experimental protocols [*Broseta et al.*, 2012] indicate smaller value closer to 30–40° depending on the mineral ($\cos(\theta) > 0.77$), implying little reduction compared to the perfectly non-wetting state. Since breakthrough pressure measurements are often performed with nitrogen ($\theta = 0$), the conversion to the brine-CO_2 system should also take into account this small effect. The comparison of breakthrough pressures measured on the same sample but with different gases such as N_2, CH_4, and CO_2 [*Hildenbrand et al.*, 2004] confirms that a contact angle correction should be made.

The accurate evaluation of P_{BT} values in low-permeability materials can be a very long process. The main technique used to determine P_{BT} in low-permeability media is the step-by-step approach initially proposed by Thomas *et al.* [1968]. Gas is placed in contact with the upstream surface of a water-saturated sample (Fig. 2.7). Initially, the gas pressure is equal to the pore pressure and is then increased by steps. Choices of pressure steps and duration depend on sample permeability and on the accuracy required for P_{BT} estimates. When the capillary pressure (gas pressure minus pore pressure) is higher than P_{BT}, water is displaced out of the sample. A continuous flow of water results, ultimately followed by a free gas phase.

Besides the standard method described above, other methods or concepts exist:

2.4.1.1. Residual Pressure Difference [*Hildenbrand et al.*, 2002, 2004]

In this method, gas is present both upstream and downstream in closed vessels in which the pressure is recorded. Gas pressure is initially set at a higher value than expected; as gas is flowing, upstream and downstream pressure vessels equilibrate (as in gas pulse decay experiments), but after stabilization, a pressure difference is found due to gas snap-off (the gas phase is then discontinuous). This method gives a minimum value of P_{BT}, underestimated by about a factor of two.

2.4.1.2. The Dynamic Approach [*Egermann et al.*, 2006]

The dynamic method introduced later is another technique for measuring P_{BT}, combined with a measurement of water permeability. Prior to testing, the sample is fully saturated. Gas is injected upstream at a constant pressure P_g. On the upstream reservoir, the gas displaces the water until the gas is in contact with the sample surface. Downstream pressure is maintained constant at P_w throughout the experiment. The order of magnitude of P_{BT} should be known as the gas pressure should be high enough to allow the gas to penetrate the sample. Two different flow rates are observed: before (Q_1) and after (Q_2) gas entry. Before gas entry, water is produced due to a pressure gradient within the sample equal to $\Delta P_1 = P_g - P_w$. When gas enters the sample, owing to capillary forces, pore pressure drops. Upstream pressure drops from P_g to $P_g - P_{BT}$. However, $P_g - P_{BT}$ should be higher than P_w; otherwise, the water stops moving and no flow will be observed. The new pressure gradient is thus $\Delta P_2 = P_g - P_{BT} - P_w$. The flow rate is measured continuously, and since P_g, P_w are known values, using Darcy's law, P_{BT} can be estimated from the second observed flow rate Q_2:

$$Q_2 = -\frac{K_w}{\mu_w}\frac{S}{L}\Delta P_2 \qquad (2.17)$$

where K_w is the sample permeability (m^2), S is the cross-sectional area (m^2), L is the sample length (m), and μ_w is the water viscosity (Pa.s). K_w can be estimated by Darcy's law applied to the first flow rate Q_1 if the pressure gradient ΔP_1 is known.

2.4.1.3. Pressure-Induced Preferential Pathways [*Harrington and Horseman*, 1999]

In the context of nuclear waste storage, these authors suggest that gas breakthrough pressures can be lower than measured in standard experiments. The gas flow does not occur through the existing porosity but through pathways that are created under the effect of the imposed pressure gradient during the test. This concept is evidenced by long-term laboratory experiments (1 year) in which gas flow is finally observed without water production. These gas pathways were later evidenced on different formations [*Harrington et al.*, 2009, 2012]. This disturbing concept may apply in systems with high clay content and high breakthrough pressure (50–100 bar). Obviously, it cannot be described by Darcy's laws.

Whatever the technique, a sensitive detection of the volumes is required, together with stable temperature. Typically, visual observations using a burette (Fig. 2.7) can be made with an accuracy of 0.02 ml, and the time required may reach 1 month. With more advanced setups or techniques, experimental duration may be reduced down to a few days.

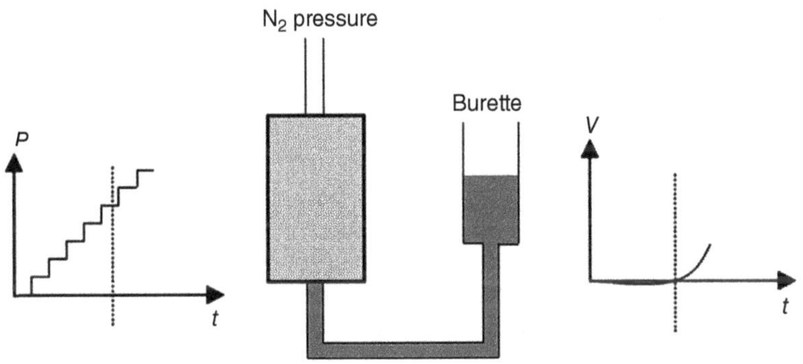

Figure 2.7 Schematic description of the simplest setup for breakthrough pressure measurement by classical approach (*P*, pressure; *V*, water accumulation). The dashed line indicates the pressure step at which a water flow is observed.

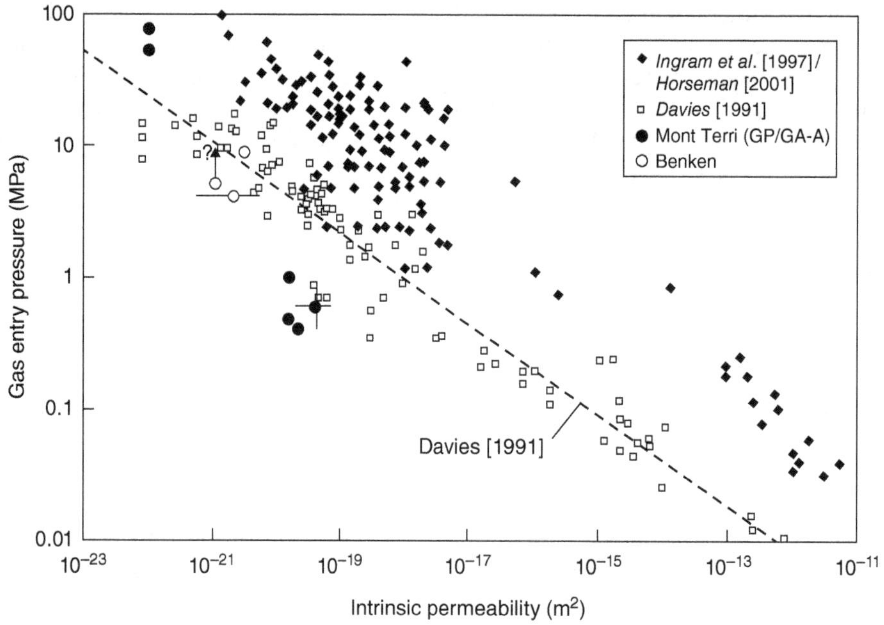

Figure 2.8 Gas entry pressure versus intrinsic permeability. After *Marschall et al.* [2005]. https://ogst.ifpenergiesnouvelles.fr/articles/ogst/abs/2005/01/marchal_vol60n1/marchal_vol60n1.html. Licensed under CC BY 4.0.

In a similar way to permeability, sample preservation and quality are important aspects to obtain representative measurements. The effect of stress may have a similar impact, but with a smaller sensitivity (P_{BT} is expected to depend linearly on pore size, whereas a square dependence is expected for permeability). For a fractured carbonated system [*Tonnet et al.*, 2010], it was found that P_{BT} depends critically on the confining pressure. P_{BT} values can also be evaluated roughly using correlations such as the relationship between cation exchange capacity (CEC) and P_{BT} [*Al-Bazali et al.*, 2005] or the relationship between permeability and P_{BT} (Thomas *et al.*, 1968). For the latter, it was found that P_{BT} is proportional to $K^{-0.45}$ close to the expected square root dependence, Eq. (2.12). A compilation of measured values in the context of nuclear waste storage is presented in Figure 2.8. The use of MICP curves may be an alternate way of estimating P_{BT}, but in a very similar way to permeability, an empirical choice of a pore throat size must be made, and this is an important source of uncertainty.

2.4.2. Two-Phase Flow Properties

Once the gas has penetrated the porous medium, two-phase flow properties should be used, usually described with a set of relative permeability curves for the gas and water in the context of generalized Darcy's equations. Of interest in a caprock are the curves corresponding to first drainage; these curves are well documented for permeable media but poorly known for tight formations. In general, for the first drainage, the water relative permeability decreases strongly by a factor of 10 at a water saturation of about 80%. Compared to gas, the water mobility is much lower and will therefore determine the kinetics of the desaturation. Hence, if this is still valid for tight formations, once the breakthrough pressure is reached, the desaturation will occur very slowly because the water flow will be also 10 times slower than in single-phase flow. Furthermore, the mobility of the gas phase even at high pressure is much larger than the water phase due to the low gas viscosity compared to water. Hence, the mobility ratio is very unfavorable, and a uniform displacement of water may not occur.

Effective gas permeability is often measured after breakthrough [*Hildenbrand et al.*, 2002, 2004]. They are smaller at least by one order of magnitude than water permeability (Fig. 2.9) as expected, but no saturation has been given for this particular case. Strictly speaking, the measurement of the gas flow after breakthrough may not be representative of two-phase flow properties if the gas is flowing only in desaturated preferential pathways; this would be the case for high water saturation (~90%). Rigorously, one should also determine the Klinkenberg effect if the pore pressure is low. Relative permeability curves for both water and gas have been determined for a caprock of the Alberta basin, two Colorado shales, and one anhydrite sample [*Bennion and Bachu*, 2008]. Despite that these experiments and their interpretation may be

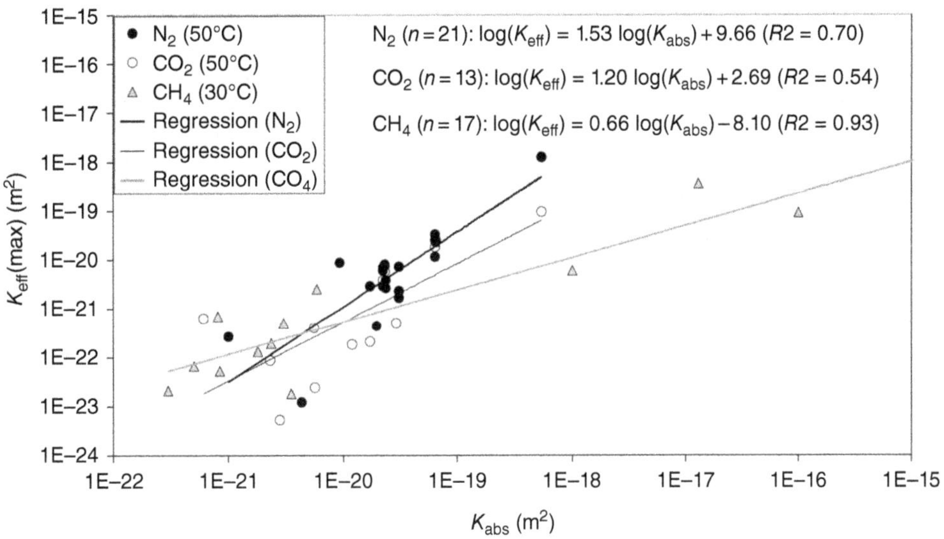

Figure 2.9 Effective gas permeability after breakthrough as a function of water permeability [*Hildenbrand et al.*, 2004]. Reprinted with permission from Wiley.

criticized, the results indicate that water relative permeability curve is decreasing sharply by a factor of 10 at a water saturation of 80%.

2.5. CONCLUSION

At the core scale, the two main mechanisms of transport in tight rocks, diffusion and convection, are reviewed. For diffusion, we consider essentially the diffusion properties of water confined in tight rocks which can be considered as an upper limit for the diffusion of dissolved CO_2 in these rocks. Measurement techniques and results are described, originating essentially from the nuclear industry. We suggest that a rough estimate of diffusion coefficients as a function of porosity is given by an Archie relationship in which the exponent m is about 2.5. For convection, the important difficulty of measuring very low permeability using water or gas and various techniques is discussed; Darcy's law is still applicable for tight rocks, but for gas, the measured permeability strongly depends on pressure (Klinkenberg effect). Then, the possibility of predicting permeability from the knowledge of porosity, pore size distribution, and clay content is listed from existing literature and by extending existing relationships valid for high-permeability formation to tight formation. Establishing relationships between various pore structure parameters and permeability is a challenging task for the future. Finally, in two-phase situation, the key aspect of the gas entry pressure is mainly considered, a parameter driving the presence of gas in the caprock. After breakthrough, the effective gas permeability is at least 10 times smaller than water permeability.

REFERENCES

Al-Bazali, T. M., J. Zhang, M. E. Chenevert, and M. M. Sharma (2005), Measurement of the sealing capacity of shale caprocks, *SPE Annual Technical Conference and Exhibition*(SPE-96100), doi:10.2118/96100-MS.

Amann-Hildenbrand, A., P. Bertier, A. Busch, and B. M. Krooss (2013), Experimental investigation of the sealing capacity of generic clay-rich caprocks, *International Journal of Greenhouse Gas Control*, *19*, 620–641, doi:10.1016/j.ijggc.2013.01.040.

Andra (2005), Argile: Référentiel du site de Meuse/haute-Marne, Dossier Argiles, Andra, Chatenay-Malabry, France.

Angeli, M., M. Soldal, E. Skurtveit, and E. Aker (2009), Experimental percolation of supercritical CO_2 through a caprock, *Energy Procedia*, *1*(1), 3351–3358, doi:10.1016/j.egypro.2009.02.123.

Bennion, B., and S. Bachu (2008), Drainage and imbibition relative permeability relationships for supercritical CO_2 / brine and H_2S / brine systems in intergranular sandstone, carbonate, shale, and anhydrite rocks, *SPE Reservoir Evaluation & Engineering* (*June*), 487–496.

Berne, P., P. Bachaud, and M. Fleury (2009), Diffusion properties of carbonated caprocks from the Paris basin, *Oil & Gas Science and Technology – Revue de l'Institut Français du Pétrole*, *65*(3), 473–484, doi:10.2516/ogst/2009072.

Berthe, G. (2012), Évolution des propriétés de confinement des roches-couvertures type argilite soumises à des fluides enrichis en CO_2: impact des discontinuités naturelles et artificielles, Université Paris Sud - Paris XI.

Berthe, G., S. Savoye, C. Wittebroodt, and J.-L. Michelot (2011), Changes in containment properties of claystone caprocks induced by dissolved CO_2 seepage, *Energy Procedia*, *4*, 5314–5319, doi:10.1016/j.egypro.2011.02.512.

Boulin, P. F., P. Bretonnier, N. Gland, and J. M. Lombard (2012), Contribution of the steady state method to water permeability measurement in very low permeability porous media, *Oil & Gas Science and Technology – Revue d'IFP Energies nouvelles*, *67*(3), 387–401, doi:10.2516/ogst/2011169.

Brace, W. F., J. B. Walsh, and W. T. Frangos (1968), Permeability of granite under high pressure, *Journal of Geophysical Research*, *73*(6), 2225–2236, doi:10.1029/JB073i006p02225.

Broseta, D., N. Tonnet, and V. Shah (2012), Are rocks still water-wet in the presence of dense CO_2 or H_2S?, *Geofluids*, *12*(4), 280–294, doi:10.1111/j.1468-8123.2012.00369.x.

Busch, A., S. Alles, Y. Gensterblum, D. Prinz, D. N. Dewhurst, M. Raven, H. Stanjek, and B. Kroos (2008), Carbon dioxide storage potential of shales, *International Journal of Greenhouse Gas Control*, *2*(3), 297–308, doi:10.1016/j.ijggc.2008.03.003.

Busch, A., K. Schweinar, N. Kampman, A. Coorn, V. Pipich, A. Feoktystov, L. Leu, A. Amann-Hildenbrand, and P. Bertier (2017), Determining the porosity of mudrocks using methodological pluralism, *Geological Society, London, Special Publications*, *SP454.1*, doi:10.1144/SP454.1.

Callaghan, P. T. (2011), Translational dynamics and magnetic resonance: principles of pulsed gradient spin echo NMR, 1 online resource (XVII, 547 S.), Oxford University Press, Oxford.

Chalbaud, C., M. Robin, J.-M. Lombard, F. Martin, P. Egermann, and H. Bertin (2009), Interfacial tension measurements and wettability evaluation for geological CO_2 storage, *Advances in Water Resources*, *32*(1), 98–109, doi:10.1016/j.advwatres.2008.10.012.

Chiquet, P., D. Broseta, and S. Thibeau (2007a), Wettability alteration of caprock minerals by carbon dioxide, *Geofluids*, *7*(2), 112–122, doi:10.1111/j.1468-8123.2007.00168.x.

Chiquet, P., J.-L. Daridon, D. Broseta, and S. Thibeau (2007b), CO_2/water interfacial tensions under pressure and temperature conditions of CO_2 geological storage, *Energy Conversion and Management*, *48*(3), 736–744, doi:10.1016/j.enconman.2006.09.011.

Civan, F. (2010), Effective correlation of apparent gas permeability in tight porous media, *Transport in Porous Media*, *82*, 375–384.

Comisky, J. T., K. E. Newsham, J. A. Rushing, and T. A. Blasingame (2007), A comparative study of capillary-pressure-based empirical models for estimating absolute permeability in tight gas sands: SPE 110050, *Proceedings of SPE Annual Technical Conference and Exhibition*, 11–14 November, Anaheim, CA.

Davies, P. B. (1991), Evaluation of the role of threshold pressure in controlling flow of waste-generated gas into bedded salt at the Waste Isolation Pilot Plant (WIPP). *Sandia Rep. SAND 90-3246*, Sandia National Laboratories, Albuquerque, NM.

Dewhurst, D. N., and A. F. Siggins (2006), Impact of fabric, microcracks and stress field on shale anisotropy, *Geophysical Journal International*, *165*(1), 135–148, doi:10.1111/j.1365-246X.2006.02834.x.

Dewhurst, D. N., Y. Yang, and A. C. Aplin (1999), Permeability and fluid flow in natural mudstones, *Geological Society, London, Special Publications*, *158*(1), 23–43, doi:10.1144/GSL.SP.1999.158.01.03.

Dullien, F. A. L. (1992), *Porous Media: Fluid Transport and Pore Structure*, 2nd ed., 574 pages, Academic Press, San Diego, CA.

Dunn, K.-J., D. J. Bergman, and G. A. Latorraca (2002), *Nuclear Magnetic Resonance: Petrophysical and Logging Applications*, 1st ed., 293 pages, *Seismic exploration*, v. *32*, Pergamon, Amsterdam and New York.

Egermann, P., J.-M. Lombard, and P. Bretonnier (2006), A fast and accurate method to measure threshold capillary pressure of caprocks under representative conditions, *International Symposium of the Society of Core Analysts*.

Egermann, P., R. Lenormand, D. G. Longeron, and C. Zarcone (2013), A fast and direct method of permeability measurements on drill cuttings, *SPE Reservoir Evaluation & Engineering*, *8*(4), 269–275, doi:10.2118/77563-PA.

Fleury, M., and M. Romero-Sarmiento (2016), Characterization of shales using T1–T2 NMR maps, *Journal of Petroleum Science and Engineering*, *137*, 55–62, doi:10.1016/j.petrol.2015.11.006.

Fleury, M., P. Berne, and P. Bachaud (2009), Diffusion of dissolved CO_2 in caprock, *Energy Procedia*, *1*(1), 3461–3468, doi:10.1016/j.egypro.2009.02.137.

Fleury, M., S. Gautier, N. Gland, P. Boulin, B. Norden, and C. Schmidt-Hattenberger (2013a), Advanced and integrated petrophysical characterization for CO_2 storage: application to the Ketzin site, *Oil & Gas Science and Technology*, *68*(3), 557–576.

Fleury, M., E. Kohler, F. Norrant, S. Gautier, J. M'Hamdi, and L. Barré (2013b), Characterization and quantification of water in smectites with low-field NMR, *Journal of Physical Chemistry C*, *117*, 4551–4560.

Ghanizadeh, A., S. Bhowmik, O. Haeri-Ardakani, H. Sanei, and C. R. Clarkson (2015a), A comparison of shale permeability coefficients derived using multiple non-steady-state measurement techniques: examples from the Duvernay formation, Alberta (Canada), *Fuel*, *140*, 371–387, doi:10.1016/j.fuel.2014.09.073.

Ghanizadeh, A., C. R. Clarkson, S. Aquino, O. H. Ardakani, and H. Sanei (2015b), Petrophysical and geomechanical characteristics of Canadian tight oil and liquid-rich gas reservoirs: I. Pore network and permeability characterization, *Fuel*, *153*, 664–681, doi:10.1016/j.fuel.2015.03.020.

Ghanizadeh, A., C. R. Clarkson, S. Aquino, O. H. Ardakani, and H. Sanei (2015c), Petrophysical and geomechanical characteristics of Canadian tight oil and liquid-rich gas reservoirs: II. Geomechanical property estimation, *Fuel*, *153*, 682–691, doi:10.1016/j.fuel.2015.02.113.

Glover, P. (2009), What is the cementation exponent? A new interpretation, *The Leading Edge*, *28*(1), 82–85, doi:10.1190/1.3064150.

Harrington, J. F., and S. T. Horseman (1999), Gas transport properties of clays and mudrocks, *Geological Society, London, Special Publications*, *158*(1), 107–124, doi:10.1144/GSL.SP.1999.158.01.09.

Harrington, J. F., D. J. Noy, S. T. Horseman, D. J. Birchall, and R. A. Chadwick (2009), Laboratory study of gas and water flow in the Nordland shale, Sleipner, North Sea, *AAPG Studies in Geology*, *59*, 521–543, doi:10.1306/13171259St593394.

Harrington, J. F., de la Vaissière, R., D. J. Noy, R. J. Cuss, and J. Talandier (2012), Gas flow in Callovo-Oxfordian claystone (COx): results from laboratory and field-scale measurements,

Mineralogical Magazine, *76*(8), 3303–3318, doi:10.1180/minmag.2012.076.8.43.

Hayduk, W., and H. Laudie (1974), Prediction of diffusion coefficients for nonelectrolytes in dilute aqueous solutions, *AIChE Journal*, *20*(3), 611–615, doi:10.1002/aic.690200329.

Heath, J. E., T. A. Dewers, B. J. McPherson, M. B. Nemer, and P. G. Kotula (2012), Pore-lining phases and capillary breakthrough pressure of mudstone caprocks: sealing efficiency of geologic CO_2 storage sites, *International Journal of Greenhouse Gas Control*, *11*, 204–220, doi:10.1016/j.ijggc.2012.08.001.

Hildenbrand, A., S. Schlömer, B. M. Krooss, and R. Littke (2002), Gas breakthrough experiments on fine-grained sedimentary rocks, *Geofluids*, *2*, 3–23.

Hildenbrand, A., S. Schlömer, B. M. Krooss, and R. Littke (2004), Gas breakthrough experiments on pelitic rocks: comparative study with N_2, CO_2 and CH_4, *Geofluids*, *4*, 61–80.

Hook, P., D. L. Fairhurst, E. I. Rylander, R. Badry, H. Bachman, S. F. Crary, K. Chatawanich, and T. D. Taylor (2011), Improved precision magnetic resonance acquisition: application to shale evaluation: SPE 146883, *Proceedings of SPE Annual Technical Conference and Exhibition*, doi:10.2118/146883-MS.

Horseman, S. T. (2001), Gas migration through indurated clays (mudrocks). In: *Gas Generation and Migration in Radioactive Waste Disposal. Proceedings of the NEA Workshop in Reims 2000*, France.

Horsrud, P., E. F. Sonstebo, and R. Boe (1998), Mechanical and petrophysical properties of North Sea shales, *International Journal of Rock Mechanics and Mining Sciences*, *35*(8), 1009–1020, doi:10.1016/S0148-9062(98)00162-4.

Ingram, G. M., J. L. Urai, and M. A. Naylor (1997), Sealing processes and top seal assessment. In: *Hydrocarbon Seals: Importance for Petroleum Exploration and Production* (eds. P. Molle-Pedersen and A. G. Koesler) Norwegian Petroleum Society, Special Publication 7, Elsevier, Amsterdam, 165–174.

Jannot, Y., and D. Lasseux (2012), A new quasi-steady method to measure gas permeability of weakly permeable porous media, *The Review of Scientific Instruments*, *83*(1), 015113, doi:10.1063/1.3677846.

Jannot, Y., D. Lasseux, L. Delottier, and G. Hamon (2008), A simultaneous determination of permeability and Klinkenberg coefficient from an unsteady-state pulse-decay experiment, *Proceeding of the International Symposium of the Society of Core Analysts*, 29 October–2 November 2008, Abu Dhabi, UAE.

Katz, A. J., and A. H. Thompson (1986), Quantitative prediction of permeability in porous rock, *Physical Review B*, *34*(11), 8179–8181, doi:10.1103/PhysRevB.34.8179.

Kranz, R. L., J. S. Saltzman, and J. D. Blacic (1990), Hydraulic diffusivity measurements on laboratory rock samples using an oscillating pore pressure method, *International Journal of Rock Mechanics and Mining Science and Geomechanics Abstracts*, *27*(5), 345–352, doi:10.1016/0148-9062(90)92709-N.

Lalanne, B., A. Le-Bihan, E. Elias, R. Poyol, and L. Martinez (2014), How to cope with some of the challenges associated with laboratory measurements on gas shale core samples: SPE 167709, *SPE/EAGE European Unconventional Conference and Exhibition*.

Lasseux, D., Y. Jannot, S. Profice, and M. Mallet (2012), The "step decay": a new transient method for the simultaneous determination of intrinsic permeability, Klinkenberg coefficient and porosity on very tight rocks, *Proceeding of the International Symposium of the Society of Core Analysts*, 27–30 August 2012, Aberdeen, Scotland.

Lenormand, R., F. Bauget, and G. Ringot (2012), Permeability measurement on small rock samples, *Proceeding of the International Symposium of the Society of Core Analysts*, 4–7 October 2010, Halifax, Nova Scotia, Canada.

Li, S., M. Dong, Z. Li, S. Huang, H. Qing, and E. Nickel (2005), Gas breakthrough pressure for hydrocarbon reservoir seal rocks: implications for the security of long-term CO_2 storage in the Weyburn field, *Geofluids*, *5*, 326–334.

Marschall, P., S. Horseman, and T. Gimmi (2005), Characterisation of gas transport properties of the Opalinus clay, a potential host rock formation for radioactive waste disposal, *Oil & Gas Science and Technology – Revue de l'IFP*, *60*(1), 121–139, doi:10.2516/ogst:2005008.

Marsily, G. d. (1986), *Quantitative Hydrogeology: Groundwater Hydrology for Engineers*, xix, 440, Academic Press, Orlando, FL.

Motellier, S., I. Devol-Brown, S. Savoye, D. Thoby, and J.-C. Alberto (2007), Evaluation of tritiated water diffusion through the Toarcian clayey formation of the Tournemire experimental site (France), *Journal of Contaminant Hydrology*, *94*(1–2), 99–108, doi:10.1016/j.jconhyd.2007.05.012.

Pearson, F. J. (1999), What is the porosity of a mudrock?, *Geological Society, London, Special Publications*, *158*(1), 9–21, doi:10.1144/GSL.SP.1999.158.01.02.

Purcell, W. R. (1949), Capillary pressures – their measurement using mercury and the calculation of permeability therefrom, *Transactions of AIME*, *186*(2), 39–48, doi:10.2118/949039-G.

Shackelford, C. D. (1991), Laboratory diffusion testing for waste disposal – a review, *Journal of Contaminant Hydrology*, *7*(3), 177–217, doi:10.1016/0169-7722(91)90028-Y.

Swanson, B. F. (1981), A simple correlation between permeabilities and mercury capillary pressures, *Journal of Petroleum Technology* (*December*), 2498–2504.

Thomas, L. K., D. L. Katz, and M. R. Tek (1968), Threshold pressure phenomena in porous media, *Society of Petroleum Engineers Journal*, *243*, doi:10.2118/1816-PA.

Tonnet, N., D. F. Broseta, and G. Mouronval (2010), Evaluation of petrophysical properties of a carbonate-rich caprock for CO_2 geological storage purposes, *Proceeding of the SPE EUROPEC/EAGE Annual Conference and Exhibition*, Barcelona, Spain, doi:10.2118/131525-MS.

Wollenweber, J., S. Alles, A. Busch, B. M. Krooss, H. Stanjek, and R. Littke (2010), Experimental investigation of the CO_2 sealing efficiency of caprocks, *International Journal of Greenhouse Gas Control*, *4*(2), 231–241, doi:10.1016/j.ijggc.2010.01.003.

Yang, Y., and A. C. Aplin (2007), Permeability and petrophysical properties of 30 natural mudstones, *Journal of Geophysical Research*, *112*(B3), doi:10.1029/2005JB004243.

Yang, Y., and A. C. Aplin (2010), A permeability – porosity relationship for mudstones, *Marine and Petroleum Geology*, *27*(8), 1692–1697, doi:10.1016/j.marpetgeo.2009.07.001.

3

Pore-to-Core Characterization of Shale Multiphysics

Thomas Dewers[1], Jason Heath[1], Hongkyu Yoon[1], Mathew Ingraham[1], Joseph Grigg[2], Peter Mozley[2], Enrico Quintana[3], and Zuleima Karpyn[4]

ABSTRACT

Shale, mudstone, and claystone are fine-grained rocks that are heterogeneous at all scales and possess pore sizes ranging from nanometer to supra-micron scale. Such small sizes are not amenable to optical investigation, which has until recently prevented pore-scale investigations into shale behavior, particularly under in situ conditions. Macroscopic thermal, hydrological, mechanical, and chemical constitutive behaviors (e.g., swelling, dry-out, yield and failure, osmotic flux) are all macroscopic manifestations of pore scale, that is, nano-behavior, with nanopores containing fluid phases that differ in properties from the bulk. Characterization is not just a description of *what*, but needs to include an assessment of the *how* and *why* of constitutive behaviors. Recent technological advances are enabling an unprecedented advancement in the characterization of shale multiphysics. This chapter examines new and ongoing efforts in shale characterization, with special attention given to in situ imaging methods. These methods are yielding unprecedented understanding of shale behavior in the subsurface and attendant responses to engineering perturbations of fluid injection, waste storage, and resource extraction including geologic carbon storage.

3.1. INTRODUCTION

Mudstone and similar rock formations are receiving unprecedented attention due to the rise of unconventional resources and renewed interest in geologic storage. Many studies of mudstone use grain size for classification and nomenclature assignment [*Hart et al.*, 2013]. The term *mudstone* refers to a consolidated rock that consists of more than 50% grains of a diameter 62.5 μm or less [*Potter et al.*, 2005; *Lazar et al.*, 2015]. The term *shale* is generally used in proper names of formations (e.g., Mancos Shale), but we use it here as a general descriptor of mudrocks as is common in the literature. Shale is the dominant rock volumetrically in the sedimentary record [*Blatt*, 1982], forms the seals for hydrocarbon reservoirs, is held to be the sealing formation for geologic carbon storage, and is under renewed interest as a repository for domestic nuclear waste. As such, shale arguably is the most relevant rock type for US domestic energy security. Resources in shales are predicted by some to enable US energy independence and export, providing substantial amounts of domestic energy. A myriad of depositional environments have produced a huge variation in mud-sourced sediments across geologic time that have transformed, via biological, chemical, and mechanical means (i.e., diagenesis), into mudstone. This imparts shale rock with heterogeneity at all scales, with nanometer- to micron-sized pores and predominantly clay, silica, organic, and carbonate phases forming complex microlithofacies. Mudstones remain, of all rock types, the least studied and understood quantitatively

[1]Geomechanics Department, Sandia National Laboratory, Albuquerque, NM, USA
[2]Department of Earth and Environmental Sciences, New Mexico Tech, Socorro, NM, USA
[3]Non-Destructive Environments and Diagnostics Department, Sandia National Laboratories, Albuquerque, NM, USA
[4]Department of Energy and Mineral Engineering, Penn State University, State College, PA, USA

Geological Carbon Storage: Subsurface Seals and Caprock Integrity, Geophysical Monograph 238, First Edition.
Edited by Stéphanie Vialle, Jonathan Ajo-Franklin, and J. William Carey.

from a geomechanical, hydrological, and geochemical standpoint [*Olgaard et al.*, 1997; *Dewhurst et al.*, 2002; *Lyon et al.*, 2005; *Aplin et al.*, 2006; *Dewhurst and Siggins*, 2006].

Fundamental understanding of shale for subsurface science and engineering applications such as these is however lacking. There is no industry consensus as to factors underlying observed rapid declines in gas production in shale gas fields. Observed interactions of subsurface carbon dioxide (CO_2) plumes with mudstone caprock have not followed model prediction [e.g., *Bickle et al.*, 2007]. Current coupled hydro-thermo-mechanical models do a poor job describing large-scale tests in clay formations in European repository underground laboratories [e.g., *Sanchez et al.*, 2011]. Essentially the response of anisotropic heterogeneous shale lithologies to engineered three-dimensional (3D) thermal, fluid flow, and stress perturbations is not currently quantifiable to a degree where prediction is viable. One can argue that this is largely because developments of applicable constitutive laws describing the coupled multiphysics and incorporating heterogeneity have not kept pace with advancements in high-performance computing. Science-based solutions to energy needs will require both a means to handle extreme heterogeneity and advancements in coupled thermal-hydrological-mechanical-chemical (THMC) constitutive behavior of shale.

In this chapter, we focus on methods applicable to advancing understanding of shale multiphysics. We discuss characterization across pore-to-core-plug scales including electron tomography with aberration-corrected transmission electron microscopy (AC-TEM), dual focused ion beam/scanning electron microscopy (FIB-SEM), multi-beam scanning electron microscopy (mSEM), energy dispersive compositional mapping (MAPS-EDS, a product of FEI™), laser scanning confocal microscopy (LSCM), and epifluorescence microscopy. Understanding of coupled constitutive behavior is being advanced by linking imaging deformation under in situ conditions. Toward this, we examine the more common techniques of nanoindentation and differential image correlation, but augment our treatment with new in situ imaging techniques that permit measurement of 3D fracture and microlithofacies and pore compression. These include oedometric small angle neutron scattering (SANS), radiographical and micro-computed tomography (μCT) imaging of fracture propagation and morphology under stress, and differential volume correlation with X-ray μCT, which uses positions of pyrite grains or iodine-labeled organic material to measure micro deformations. All of these methods are applied to conceptually answer the following scientific questions: What is the appropriate representative elementary volume (REV, defined as the volume over which statistical fluctuations in pore-scale properties are minimized, usually in terms of the standard deviation of the property divided by the mean of the property [*Yoon and Dewers*, 2013; *Kelly et al.*, 2016]) to apply for this heterogeneous geomaterial, and is it the same for different physics [*Nitao and Bear*, 1996; *Kelly et al.*, 2016]? What are poromechanical behaviors of mudstone microlithofacies at the REV level? How does deformation partition between organic-rich and clay-rich portions of shale microlithofacies? How does REV response scale up in common lithofacies to something approaching a rock core plug, that is, measurable by accessible experimental means or observable in a thin section, and what does this imply for constitutive response at that scale? What are the pore network topological factors controlling dominant force-flux relations for material transport? What are the most important deviations from simple isotropic soil mechanics constitutive laws that attend shale deformation under true in situ stress and pore pressure conditions? Our treatment is by no means a comprehensive review of characterization or experimental techniques applied to shale but rather reflects our interests and research developments in answering these questions. Relevant topics in pore-to-core characterization such as permeability and wettability have been discussed elsewhere [*Odusina et al.*, 2011; *Sulucarnain et al.*, 2012; *Tinni et al.*, 2014; *Ashish et al.*, 2016; *Abdelmalek et al.*, 2017].

3.2. BACKGROUND

This chapter aims to summarize efforts toward improvement of constitutive laws (either derived from first principles or often phenomenological, such laws couple forces and fluxes such as Fickian diffusion or Darcian fluid flux; these can be linear, that is, Hooke's law for linearly elastic solids, or nonlinear, such as rate and state variable friction applied to earthquake mechanics) for multiphysics modeling applications for subsurface engineering. There have been incredible advancements made in high-performance computing, that is, current THMC models utilize massively parallel processing, efficient solvers, unprecedented meshing capabilities, and so on [*Mikelic' et al.*, 2014; *Gens et al.*, 2011; *Yin et al.*, 2011; *Edwards*, 2006]. However, constitutive laws within those models are still reliant on simplified geometries (e.g., penny-shaped cracks, spherical pores, cubic law fracture permeability, volume fraction representation of solids, and porosity in reactive transport laws), linearized couplings, and assumptions of homogeneity and isotropy. For shale geomaterials in particular, it is not known if there is a separation of scales between what is an REV for continuum mechanics and fluctuations in texture associated with depositional or other heterogeneity. It is clear that next-generation constitutive models are needed for shale science and engineering that can include all the textural complexity and required couplings [e.g., *Sanchez et al.*, 2008] and that much advancement will come from the combination of nanoscale imaging, digital representations of textural elements, new upscaling statistical methods, and experimental validation.

Shale characterization from the pore-to-core and larger scales is a subject of much recent attention. From a geological standpoint, *Schieber and Zimmerle* [1998] trumpet the need for understanding shale heterogeneity from a depositional and/or sequence stratigraphy standpoint, and this is a valuable concept for both understanding the nature of geomechanical heterogeneity [*Slatt and Abousleiman*, 2011] and pore-scale fabrics [*Macquaker et al.*, 2007; *Scheiber*, 2013]. Using the organizing principles of sequence stratigraphy, *Slatt et al.* [2011] describe a *workflow for characterization* which purports to integrate sedimentology, petrography, high-resolution SEM, lithostratigraphy, geochemistry, biostratigraphy, well logs (including borehole image logs), and 3D seismic processing and interpretation. Additionally, the electron microscopy company FEI markets a multi-scale shale imaging workflow for characterization and interpretation from the pore-to-core scale aimed mostly at improving recovery for the shale gas industry [e.g., *Goergen et al.*, 2013].

In this chapter, we introduce new methods that, by themselves or integrated with these types of workflows, can be used to further understanding of shale science focusing on pore-scale textural attributes that impact measurable continuum properties. We begin by examining extensions of traditional two-dimensional (2D) optical microscopy, used to identify and segment textural characteristics that at once may prove useful for both sampling at finer scales and upscaling pore-scale results to core and larger scales. 3D X-ray μCT methods have resolutions that are generally too large for pore analysis of shale samples of reasonable size but can identify microfacies variability and microfracture and macrofracture morphologies and may prove useful for some in situ testing methods. Beam methods have long been used in pore characterization, and we introduce extensions in rapid 2D data acquisition over larger areas and in 3D characterization of nanosized pore topologies. Neutron methods are gaining attention for the ability to characterize pore distributions over orders of magnitude of scale, and we discuss new methods for examining the partitioning of pore volume changes attending deformation and water or CO_2 intercalation and swelling in clays. These pore-scale imaging methods can be useful guides in modeling both mechanics and flow at scales relevant to the physics but require experimental validation. We discuss methods for experimental testing at the REV and slightly larger scale for shale coupled multiphysics model validation and then potential routes for upscaling to continuum-level constitutive law development. These imaging and testing methods probe into pore geometries and topologies contained within shale microlithofacies that are the basis for macroscopic continuum behavior manifest in measurable petrophysical and geomechanical properties at the core-plug scale. We feel these efforts will prove worthy in developing improved constitutive laws for THMC models that honor realistic shale textures and heterogeneity, in the spirit of *digital rock physics* [*Bosl et al.*, 1998; *Keehm et al.*, 2001; *Walls and Sinclair*, 2011].

3.3. OPTICAL, EPIFLUORESCENCE, AND LASER CONFOCAL MICROSCOPY

Optical petrography at the thin section scale is useful in characterizing microlithofacies variability in 2D although typical optical resolutions are not amenable for discriminating micron and less grain and pore sizes. Figure 3.1a shows microlithofacies interpretations from a typical thin section of Mancos Shale (Cretaceous, Utah, USA, from *Grigg*, 2016). Different gray levels in the right-hand portion show lithofacies interpretations ranging from fine mudstone to muddy sandstone. The textural differences are separated on the basis of observable grain size variations, clay content and amount, and types of visible cement as well as interpretations of bedding, sedimentary structures, and bioturbation. These features are useful to the petrographer in shale classification schemes [*Shepard*, 1954; Folk, 1980; *Lewan*, 1978; *Macquaker and Adams*, 2003; *Ulmer-Scholle et al.*, 2014; *Lazar et al.*, 2015] and predicting qualitative mechanical behavior [e.g., *brittle* or *ductile*; *Ulmer-Scholle et al.*, 2014] but are limiting in scale and in describing 3D variability.

3.3.1. Imaging Heterogeneity

The additional technique of epifluorescence can be applied to shale for further types of characterization and in particular has long been applied to recognition of bitumen and thermal maturity of kerogen organic material. Figure 3.1b shows an example of a muddy sandstone (Pennsylvanian Morrow B from West Texas) that has been impregnated with a low-viscosity fluorochrome-dyed epoxy. The gray-scale image in the upper left shows regular halogen light illumination, but the lower left shows the same image but excited via a 546 nm filter. The red fluorescence, here shown in gray scale, shows connected pore pathways in the clay matrix in between the sandier portions of the sample, which would otherwise not be visible. The upper and lower right panels of the figure show UV and fluorescein (FITC) fluorescence showing calcite cement (light gray) and live oil within the sample. FITC itself is used commonly as a tracer for flow but can be used to detect some organics and to distinguish textural elements within a sample.

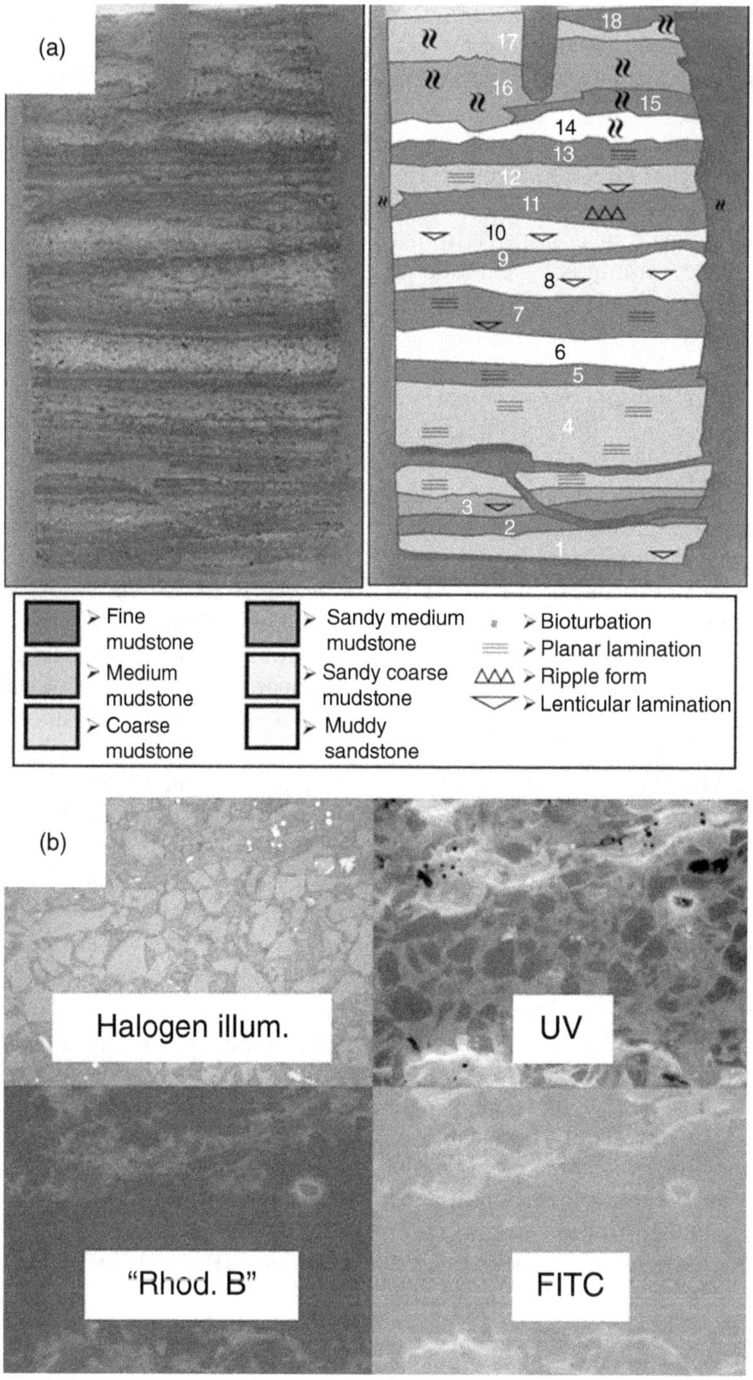

Figure 3.1 (a) Thin section of Mancos Shale [from *Grigg*, 2016] and associated interpretations of microlithofacies (gray-level bands) and sedimentary structures (symbols). (b) Examples of epifluorescence petrography of sandy mudstone using vacuum impregnation with rhodamine-B-dyed epoxy. UV (300–375 nm) excitation with 395 nm beam split and 420 nm LP filter for emission; *rhodamine B* filter set with 546 nm excitation, 580 nm beam split, and 590 nm LP filter for emission; *FITC* filter set with 450–490 nm BP filter excitation, 510 nm beam split, and 515–565 nm BP emission detection.

Another means to distinguish textural variations is LSCM, which has found recent application in the geosciences in geomicrobiology, paleontology, fission track dating, fluid inclusion analysis, hydrocarbon and source rock characterization, fracture characterizations, and imaging contamination in soils and in modeling pore networks and transport properties [*Petford and Miller*, 1990, 1992, 1993; *Fredrich et al.*, 1993, 1995; O'Connor and

Fredrich, 1999; *Petford et al.*, 1999; *Stasiuk*, 1999; *Aplin et al.*, 2000; *Ribes et al.*, 2006]. The main advantage of the confocal technique is that the confocal pinhole excludes light from optical planes other than the focal plane. This results in the confining of illumination and detection to a single location on and within a sample and the elimination of scattering from other planes. This enhances resolution in the *x-y* optical plane (resolution to 200 nm is achievable using this method), and optical sections can be resolved (and thus digitized) by changing the depth of the focal plane. For coarse-grained rocks and fractures in mudrocks, this enables a nondestructive reconstruction of 3D structure [e.g., *Fredrich et al.*, 1995].

For fine-grained rocks, similar to epifluorescence microscopy, LSCM can be useful for mapping variations in porosity across a sample within which the pore spaces are occupied by a fluorescent agent, even if the pore spaces are too small to be detected optically (< ~ 1 μm). In Figure 3.2a (left image), we show a 2D cross section of a 2.54 cm diameter plug of Selma Chalk (Cretaceous, Gulf Coast, USA), a carbonate mud with pore spaces too small to discern by optical methods [*Yoon and Dewers*, 2013]. Pore spaces have been impregnated with a fluorochrome-dyed low-viscosity epoxy and fluoresced with a 543 nm helium-neon laser. The brightness in gray level correlates to increasing pore volume fraction as is seen by comparing the region of interest (ROI) 1 consisting of chalk pores with the ROI 2 containing clay seams or horsetail structures.

The small white spots in Figure 3.2a correspond to macrofossils completely cemented with calcite and devoid of pore space. A map of fluorescence in the ROI 2 obtained using a HeNe laser at 543 nm excitation is shown to the right, and a 3D reconstruction of the clay microstructure 96 μm deep with optical sections 1 μm in width is shown also at right. Pore spaces occupying the

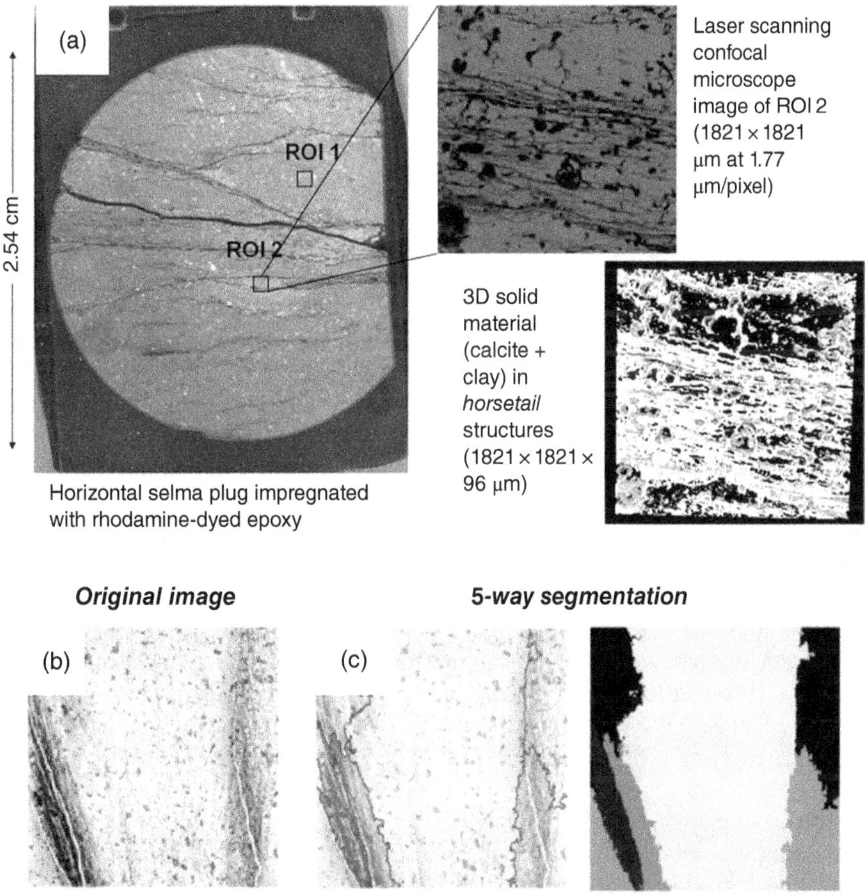

Figure 3.2 (a) Selma Chalk plug impregnated with fluorochrome-dyed epoxy (left panel). Two regions of interest are delineated; the ROI 2 is a 1.8 by 1.8 mm region shown in the upper right panel, excited with 543 nm laser radiation. The lower right panel shows a 3D rendering from serial sectioning with LSCM showing the distribution of clay and pore within ROI 2. (b) ROI 1 consists of a 1.8 mm by 1.8 mm region of chalk containing clay-bearing regions darker in gray level. (c) Textural segmentation example using ROI 1 of (B) with five principal components.

ROI 1 are similar to those shown at nm resolution using FIB-SEM methods (discussed below) by *Yoon and Dewers* [2013].

3.3.2. Texture Segmentation

In addition to 3D rendering of microstructures, the fluorescence mapping can be useful in discerning different textural types for further sample selection by FIB-SEM or other electron methods. In Figure 3.2b and c, as an example of this, we apply a spectral segmentation algorithm *FN-CUT* developed by *Kim et al.* [2013] to the pores in ROI 2, which serves as a principal component analysis (PCA) delineating microlithofacies or other texturally distinct regions on the basis of fluorescence. From right to left, we see the original image (Fig. 3.2a), the segmented image, and a false color representation of the textural segmentation (both in Fig. 3.2b). This particular example delineates how fluorescence mapping can be used to down-select regions for statistically robust sampling for FIB-SEM or other beam methods and as a potential method for upscaling pore-scale models, such as the lattice Boltzmann results obtained for ROI 1 by *Yoon and Dewers* [2013] to the core scale. This is discussed below in the last section.

3.4. X-RAY COMPUTED TOMOGRAPHY: IMAGING MULTIPHYSICS IN SITU

X-ray computed tomography (CT) is an established 3D imaging technique, and new advances are enabling improvements in resolution and in situ measurement of shale rock physics [*Karpyn et al.*, 2009; Wildenschild and Sheppard, 2013; *Rodriguez et al.*, 2014]. X-ray imaging relies on variations in X-ray transmittance, which correlates to density and material composition. For core-sized samples, resolution from millimeters in medical-type scanners to tens of microns or less in μCT scanners can yield useful information on microlithofacies distributions and microfracture morphology but is too large to image pore structures in clastic or organic components. *Carey et al.* [2015] have utilized a triaxial core flood system with direct shear-like capability that enables simultaneous rendering of fracturing shale core at 1″ diameter at 25 μm resolution.

In Figure 3.3, we show examples of in situ CT imaging of shale deformation using a similar system to that of *Carey et al.* [2015]. A typical stress-strain curve (right in Fig. 3.3a) shows standard regions of elastic behavior (I–II as noted on the curve), initial yielding (circa point II), and post-failure and residual strength during sliding on produced localized fractures (III). 2D scans of the shale sample at initial, yielding, and residual strength post-failure are shown to the right of the stress-strain curve in Figure 3.3a. The five panels in Figure 3.3b are radiographs taken every 0.1 s imaging the shale sample just prior to and during fracture propagation near peak stress. The fracture system is seen to have barely initiated at the lower right end of the left-hand initial figure and propagates from both top and bottom to join as a network 0.4 s later. Figure 3.3c shows an example of μCT imaging of fractures in the Mancos including a high-resolution 3D reconstruction of the produced fracture network.

Advanced synchrotron methods [e.g., X-ray nano-CT; *Harrison et al.*, 2016] are being used to visualize effects of chemical floods in small shale samples down to 30 nm resolution, small enough to image some kerogen pore networks and discrete pyrite framboids, and changes in surface properties attending reactive flows. 4D synchrotron techniques have been applied to imaging hydrocarbon generation and migration in fractures generated by heating millimeter-sized cores at about 5 μm resolution [*Panahi et al.*, 2013; *Kobchenko et al.*, 2011]. Interestingly, these authors apply a differential image correlation technique to observed surface deformation with heating and fracturing. We investigate this technique in simple Brazilian fracture tests in a later section. We propose that abundant pyrite could act as strain markers for 3D digital volume correlation (DVC) [*Jin and Lu*, 2013] during in situ triaxial deformation of shale cores, and this is discussed below. It is possible that the variations in pyrite seen in the Mancos in Figure 3.3d, for example, could enable partitioning of bulk deformation into different strains attending the variations in mineralogy, kerogen, cement content, and porosity composing the microlithofacies within each discrete lamina. These and other developments show that X-ray CT has a place in understanding shale multiphysics and will undoubtedly continue as an important characterization technology.

3.5. ELECTRON BEAM METHODS

Here we briefly review electron beam methods useful for characterizing 2D and 3D properties of shale, including pores in kerogen, networks of nanosized pores in clay aggregates, micropores and macropores of various types, microfractures, and mineral types. Scanning electron microscopy (SEM) utilizes focused electron beams that are scanned across a surface of a specimen. Interactions of the beam with the specimen occur as a function of the specimen surface and sub-volume properties and cause emission of secondary electrons (SE) (at low energy), backscattered electrons (BSE) (at higher energy), photons (cathodoluminescence), and X-rays. Both SE images of rough 3D shale surfaces and BSE images of polished or ion-milled surfaces have both been used to examine 2D structures and pore types [*Loucks et al.*, 2009; *Curtis et al.*, 2011; *Spain and McLin*, 2013;

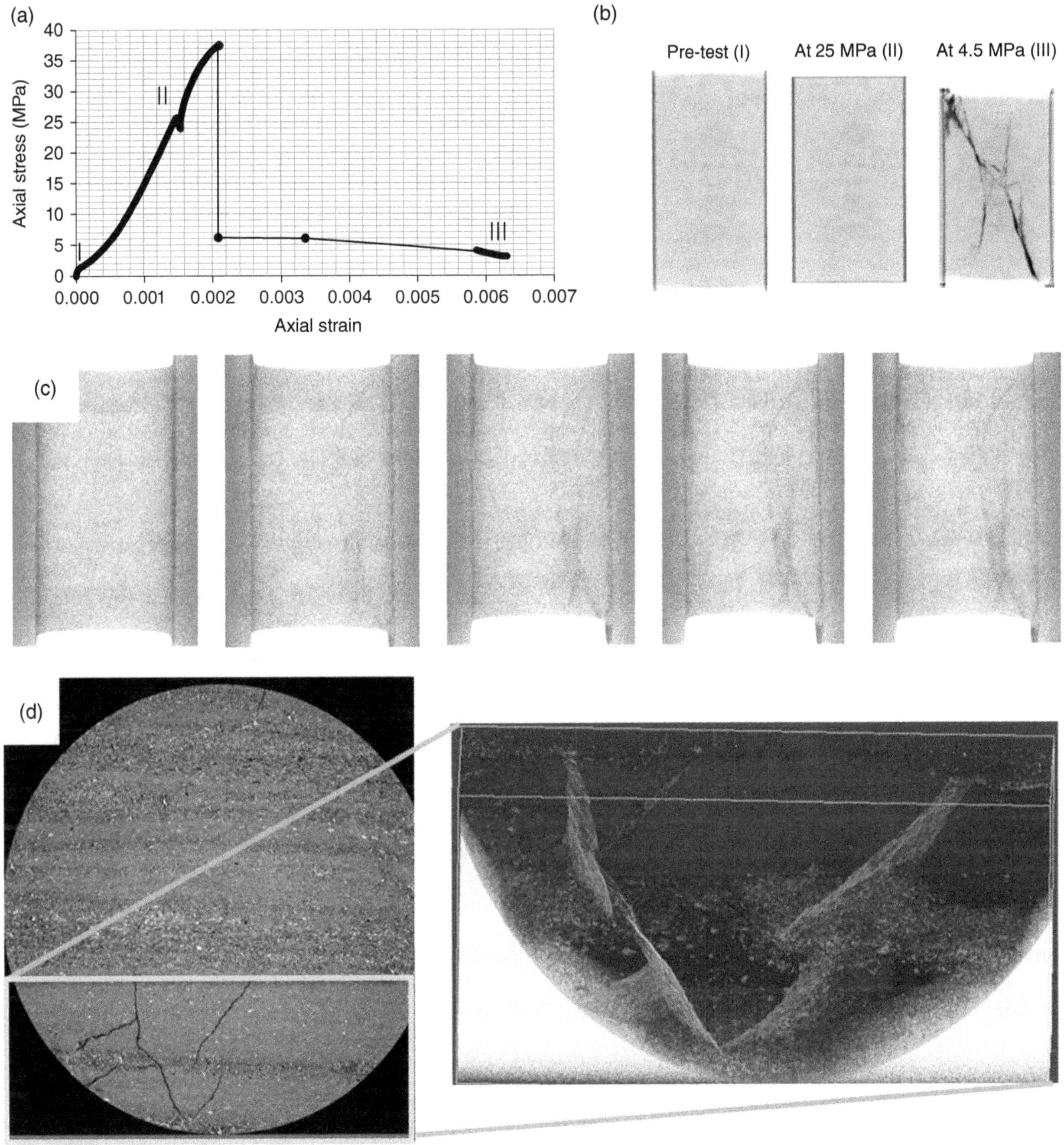

Figure 3.3 Use of X-ray CT in imaging fracture development in Mancos Shale. (a) Stress-strain curve and CT reconstructions (shown in 2D) including (b) the initial shale at 25 MPa axial stress post-yield and pre-fracture development and at 4.5 MPa post fracture (residual strength) showing conjugate shear bands and interference with strong horizontal laminae at sample center. (c) Raw radiographic images of fracture propagation in Mancos Shale, taken 0.1 s apart. (d) 2D image of Mancos Shale (same as in Figure 3.1, 17 μm resolution) showing lithologic variation as horizontal laminae visible both by changes in layer gray level and in abundance and distribution of bright high-Z pyrite. To the right is seen a 3D X-ray CT reconstruction of fractures seen at left.

Scheiber, 2013; *Milliken and Curtis*, 2016]. A recent synopsis of the field of electron microscopy as applied to shale hydrocarbon reservoirs is found in the recent memoir edited by *Camp et al.* [2013]. For our purposes here, we discuss focused ion beam (FIB) methods to retrieve 3D pore topological information in shale lithologies useful for flow modeling in shales, broad ion beam and mSEM methods that permit unprecedented throughput rates of nm-resolution imaging over large (potentially up to cm-scale) regions in 2D, and a new 3D electron tomographic technique that holds promise for collecting information on pore topologies at the scale of a single nm.

3.5.1. Dual Focused Ion Beam/SEM

Dual focused ion beam/scanning electron microscopy methods combine ion and electron beams in one instrument. The ion beam mills a sample to produce finely polished surfaces, which are imaged by an SEM. The SEM imaging modes commonly include BSE, SE, and energy dispersive spectroscopy (EDS). The ions are typically from a liquid metal ion source such as Ga^{3+} but may include other elements such as Ar or Xn. A sequential slicing and imaging through a material at a fixed interval produces serial sections that are amenable to digital reconstruction. This technique is finding wide use in mudrock science and engineering [*Camp et al.*, 2013; *Bennett et al.*, 2015; *Gu et al.*, 2015; *Hemes et al.*, 2015; *Peng et al.*, 2015; *Saraji and Piri*, 2015; *Vega et al.*, 2015]. Our efforts at FIB-SEM in mudrocks have been used to determine REVs and pore architecture for flow modeling [*Dewers et al.*, 2012; *Yoon and Dewers*, 2013], examined pore types for caprock integrity of continental and marine mudstones [*Heath et al.*, 2011], and examined pore-lining phases addressing wettability in shales [*Heath et al.*, 2012]. We have achieved voxel resolutions of about 7 nm, which is about the limit of pore sizes that contain bulk water and thus about the cutoff for Darcy flow behavior in shales. Typical problems with FIB-SEM reconstructions involve charging (absorption of SE at corners and edges of pores), so-called *curtaining* or nonplanar ion milling, segmentation or thresholding of pores that extend within a greater depth than one slice and thus appear in multiple images, and irregular illumination or shading issues in images. In most FIB-SEM applications, the SEM imaging is done at an angle to the planar surface cut parallel to the ion beam direction, which needs to take into account aligning sequential images. Newer techniques not yet widely applied in shale science involve plasma FIB dual beam technology and integrating cryo-stages and cryogenic transfer to image cryogenically frozen samples. The latter (see, e.g., *Desbois et al.*, [2009] for application of this technique in 2D) could be useful in minimizing vacuum-produced damage and even in determining pore solution composition with EDS analysis of frozen crystallites.

An example of this technique, examining 3D pore topology in kerogen, is shown in Figure 3.4. A sample of Tuscaloosa Shale from the US Gulf Coast was taken from a fresh core obtained as part of the Southeast Regional Carbon Sequestration Partnership project [*Koperna et al.*, 2012]. Samples were gold-palladium coated and painted with silver dag to mitigate specimen charging. In some cases, a piece of copper tape applied to the sides of the sample near the ion-milled trench improves charging effects. The sample assembly itself is then electrically grounded. Serial sectioning and imaging was performed with FEI Company's Helios NanoLab DualBeam instrument and semiautomated Slice and View software. Minimization of curtaining effects was done by depositing a layer of platinum over the area of sample to be milled using the FIB deposition system. During milling, the Ga ion beam was normal to the sample surface, and the electron beam was at an angle of 52° from the ion beam. Acceleration voltage for Ga^+ was 30 kV, and the beam current was 2.8 nA. BSE imaging mode with a through-the-lens detector allowed a discrimination of mineral types and kerogen on the basis of mean atomic number.

Figure 3.4a shows a typical FIB-SEM image of the Marine Tuscaloosa cored at 7925 ft of burial (see animation 3, or figure 7, in *Heath et al.* [2011]). Evident in the figure is a dark gray stringer of kerogen intercalated with the lighter clastic matrix, composed of mostly clay minerals. Pore spaces, seen as dark holes on the surface of ion-milled surface in Figure 3.4a, are evident in both the kerogen and clay masses, showing that both water- and oil-wetting pore spaces occur in organic-rich shales. Figure 3.4b and c show 2D and 3D reconstructions 4.66 μm on a side of the kerogen mass, and Figure 3.4d shows a 3D reconstruction of the pore spaces within the kerogen. It is thought that the pore spaces in kerogen are associated with the generation of oil and gas [*Tiwari et al.*, 2013] and, in this sample, appear to be arranged in elongate tubules within the kerogen with a preferred orientation (Fig. 3.4d). A computational fluid dynamics simulation of gas flow within one of the tubules (Fig. 3.4d) with streamlines is shown as an example of pore-scale modeling of gas flow in actual pore geometries in kerogen.

3.5.2. Multi-beam Scanning Electron Microscopy

As pore sizes in mudstones can vary from nm to μm scales, investigating REV and scale-dependent connectivity of pore networks is difficult. Conventional TEM, FE-SEM, and FIB-SEM systems achieve the high resolution needed to examine mudstone pores, but these techniques are limited in throughput, may be practical only on less-than-REV scales, and are expensive to implement. Conventional methods for multi-scale imaging and assessment involve high-resolution imaging of limited areas or volumes with a single electron beam, the data of which is mapped to methods, such as X-ray CT or thin section petrography, which may have lower resolution

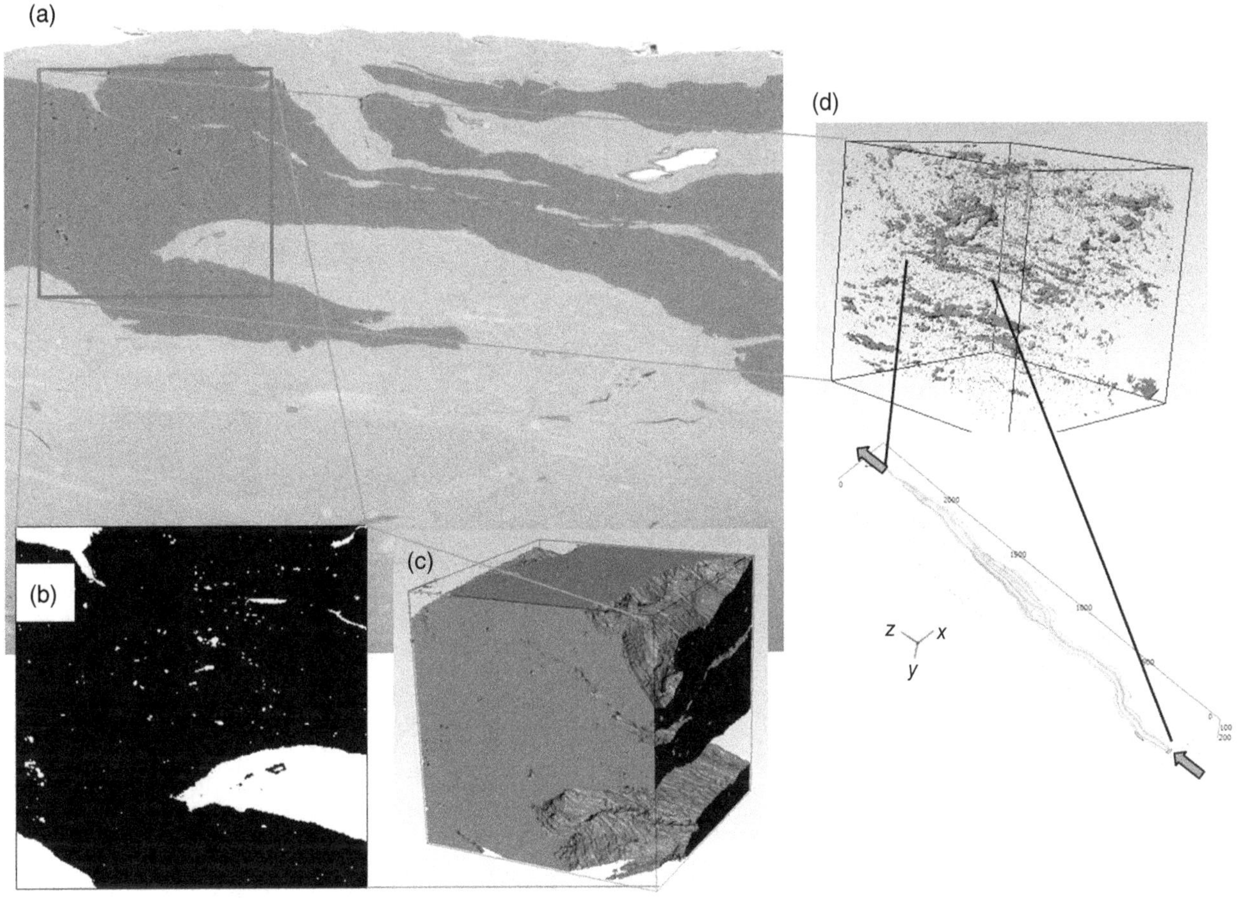

Figure 3.4 Dual focused ion beam image and reconstruction of FIB-SEM data of Marine Tuscaloosa Shale. (a) Backscattered electron image of ion-milled surface, 16 μm on the horizontal. Gray-level variations are associated with phases of varying *Z*, with kerogen appearing as dark gray and silicate and carbonate minerals appearing as lighter gray. (b) 2D segmented image of kerogen. (c) 3D rendering of kerogen mass. (d) 3D rendering of pores in kerogen mass which appear as elongate tubules with preferred orientation and CFD simulation of gas flow in one of the tubules.

while imaging a larger region. Ideally imaging techniques would collect data at high resolution (at the scale of the pores) over several orders of magnitude to allow studies of hierarchical multi-scale diagenetic or sedimentary structures and determination of REVs or scale separation as a function of lithological variability.

Materials science and subfields such as connectomics are advancing methods to automate collection of large-area electron microscopy data [*Lichtman et al.*, 2014]. mSEM involves simultaneous electron beams in a single column and a dedicated detector for each beam [*Eberle et al.*, 2015; *Michael et al.*, 2015]. Each of the primary electron beams of the mSEM thus creates beams of SE that are collected in parallel (Fig. 3.5a). Interest in large-scale imaging of biological samples is driven by the need to understand hierarchical organization and connectivity across many scales. Connectomics is pushing the frontiers of brain imaging at high resolution over large areas and volumes [*Lichtman et al.*, 2014], including extension to three dimensions using automated microtomes or other serial techniques in conjunction with systems such as the mSEM. Such rapid data collection by the mSEM will enable nanometer- to centimeter-scale examination of rock samples with SE images.

As a proof of concept for application of mSEM methods to rocks, we examined a siliceous-argillaceous mudstone provided by Kitty Milliken of the Bureau of Economic Geology, University of Texas (Fig. 3.5). We used the Zeiss multi-beam SEM 505, which projects 61 beams with a maximum throughput of 1.2 billion pixels/second [*Michael et al.*, 2015] with maximum resolution of 4 nm/pixel. The imaging mode is SE only (i.e., no BSE nor X-ray imaging).

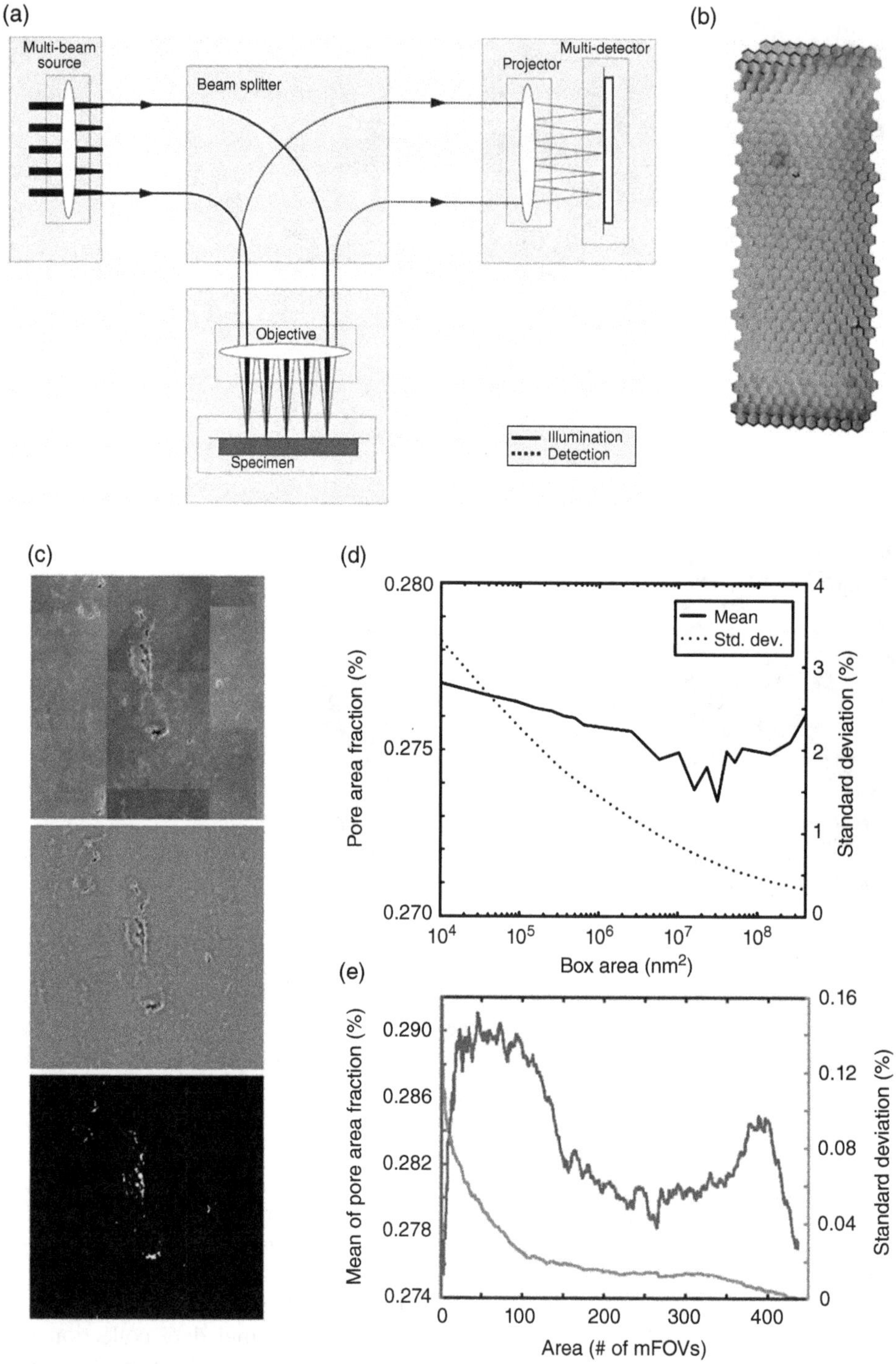

Figure 3.5 Multi-beam SEM schematic and images. (a) Schematic of multiple beams in parallel [from *Eberle et al.*, 2015]). (b) 3.1 mm × 0.9 mm mosaic of 26,657 images at 4 nm resolution of a siliceous-argillaceous mudrock. (c) Raw, histogram-adjusted and aligned, and thresholded images. Each image is 25 μm on a side. (d) and (e) Pore area fraction versus area for combined 61 images (single mFOV, (d)) and larger scales (e).

One challenge for sample preparation is obtaining a flat sample on the order of <10 nm for the optics of the mSEM system. Broad ion beam methods [*Klaver et al.*, 2015; *Scheiber*, 2013] hold promise in preparing mm-sized undamaged polished surfaces for mSEM or regular SEM investigation. The siliceous-argillaceous sample in Figure 3.5 was argon-ion polished. We collected 26,657 images at individual image size of 12.5 μm × 10.9 μm over a

3.1 mm × 0.9 mm area at 4 nm/pixel resolution in 16 minutes (Fig. 3.5b; note that the hexagons represent packing of 61 images).

The application of continuum methods to modeling mechanical, hydrologic, and chemical processes in mudstones relies on assumptions of REVs and scale separation for parameters of interest. Testing continuum assumptions is difficult as it requires investigating progressively larger measurements of a parameter of interest over a wide range of scales. Using the images of the siliceous-argillaceous sample, we post-processed the raw images to obtain histogram-adjusted and aligned images, which were then thresholded to identify pores (Fig. 3.5c). Sets of 61 images, or mFOVs for *multiple fields of view*, were combined into a single large image, which was then divided into many smaller, non-overlapping areas of increasingly larger sizes. The mean and standard deviation of pore area fraction was then calculated as a function of area of these regions across the entire data set, as given in Figure 3.5d. The mean pore area fraction does not vary greatly; however, the standard deviation drops from ~3% to <0.5% as a function of the area and appears to be asymptotically approaching a value. The representative elementary area (as a proxy for REV) thus may exist at a length scale of greater than 0.02 mm, as based on the greatest area size of 4×10^8 nm^2 for the sub-mFOV scale analysis (Fig. 3.5d). The pore area fraction was also calculated as a function of the groups of 61 images or mFOVs (each mFOV is 0.000137 mm^2). The mFOV scale fluctuations vary from ~0.275 to 0.292, which is small for pore area fractions, and thus heterogeneity of this sample is low. The multi-scale mFOV scale fluctuations (Fig. 3.5d) illustrate capturing multi-scale variation in porosity.

3.5.3. Electron Tomography for Nanopore Characterization

Results from SANS [*Navarre-Sitchler et al.*, 2013] and mercury porosimetry [e.g., *Heath et al.*, 2011; *Dewers et al.*, 2012] along with FIB-SEM results [*Heath et al.*, 2011, 2012; *Dewers et al.*, 2012] reveal that a large pore volume fraction (perhaps up to 50% by volume) in gas shales such as in the Haynesville Formation (Jurassic, Louisiana, USA) resides in the <10 nm fraction, at or below typical FIB-SEM resolution. Fluids in pores of this size will deviate from those of the bulk [*Ho et al.*, 2016], which can influence non-Darcy flow and mechanical response. The flow of gases in these pore networks can result in molecular sieving and resulting elemental fractionation. How, then, to quantify pore structure and topology in this important class of shale pores?

Here we present a state-of-the art technique for visualizing 3D connectivity and structure of the <10 nm pore population in shales: tomography using aberration-corrected scanning transmission electron microscopy (AC-STEM) [*Susan et al.*, 2014; *MacLaren and Ramasse*, 2014]. This rapidly evolving method can characterize connectivity and structure of a large fraction of otherwise inaccessible mudstone pore volumes, for example, for transport properties. Z-contrast imaging, generated by high-angle annular dark-field (HAADF) scanning transmission electron microscopy (STEM), can be used for tomographic reconstruction of nanopores in clay aggregates, authigenic calcite, kerogen, and pyrite framboids.

As an example, microanalysis was performed on a sample of Haynesville Shale with an FEI Company Titan G2 80-200 operated at 200 kV and equipped with a high-brightness electron source, spherical aberration corrector on the probe-forming optics, and an array of silicon X-ray drift detectors with a combined solid angle of 0.7 sr. The same instrument was used to collect tomograms in STEM mode at 2° intervals over a range from −60 to +60°. Both bright-field (BF) and annular dark-field STEM images (Fig. 3.6) were acquired at each projection. The Inspect3D software package from FEI Company was used to reconstruct the projection data into a 3D model (Fig. 3.7) with simultaneous iterative reconstruction tomography (SIRT method) and two iterations.

HAADF images show minimal diffraction, and their intensity goes as ~Z^2, ideal for tomographic reconstruction as it generates strong contrast that is fully monotonic with thickness, minimizing problems with overlap. While still in preliminary stages, this type of nanotomography of clay aggregates is being used to develop better transport and poroelastic-plastic constitutive models. Current work is developing rigorous thresholding methods to isolate better resolved nanopores, image analysis methods combining BF and HAADF image information, and pore-scale CFD models for Knudsen gas and solute reactive transport that are dependent on pore characteristics and pore-lining phases.

3.6. NEUTRON SCATTERING METHODS

Carbon capture, utilization, and storage (CCUS) of CO_2, production of hydrocarbon from shale, and use of shale as repositories for nuclear waste spur research on in situ evolution of pore structure of mudstones as a function of pore pressure, non-hydrostatic stress states, temperature, and fluid composition and phase behavior. Pore structure evolution affects key parameters of porosity, permeability, and mechanical behavior. Previous work has focused on pore pressure, temperature, and fluid composition, but not realistic non-hydrostatic stress states. The CCUS community has identified shrink-swell properties of clay-rich caprock as a risk for creation and/

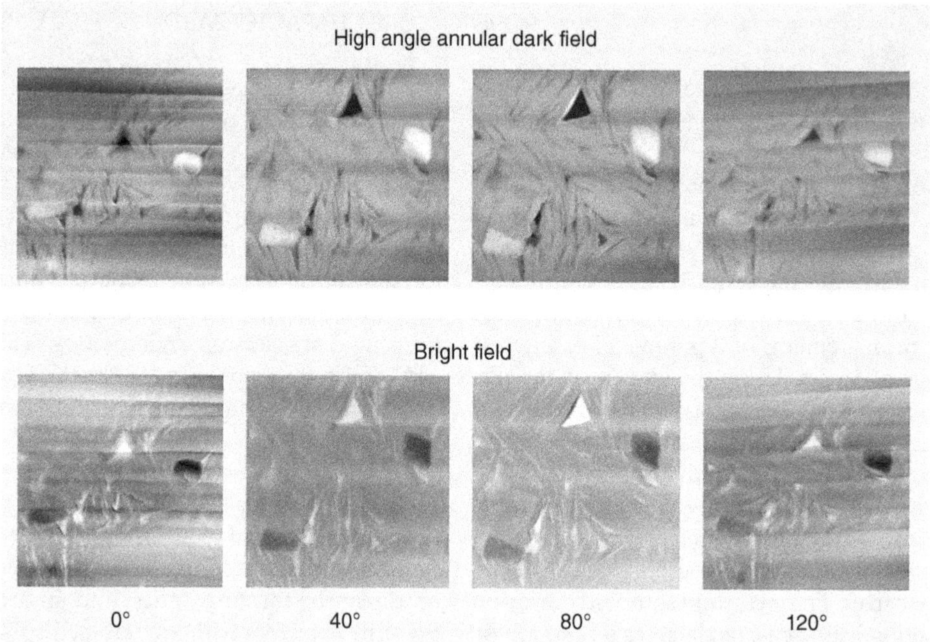

Figure 3.6 Examples of HAADF and BF images collected at every 2° over a 120° angle, used for tomographic reconstruction. The AC-STEM enables high spatial resolution/high sensitivity analyses of *Z* differences and compositional information routinely, potentially to angstrom sensitivity. Horizontal axis is 1 μm sampled at 1 nm resolution.

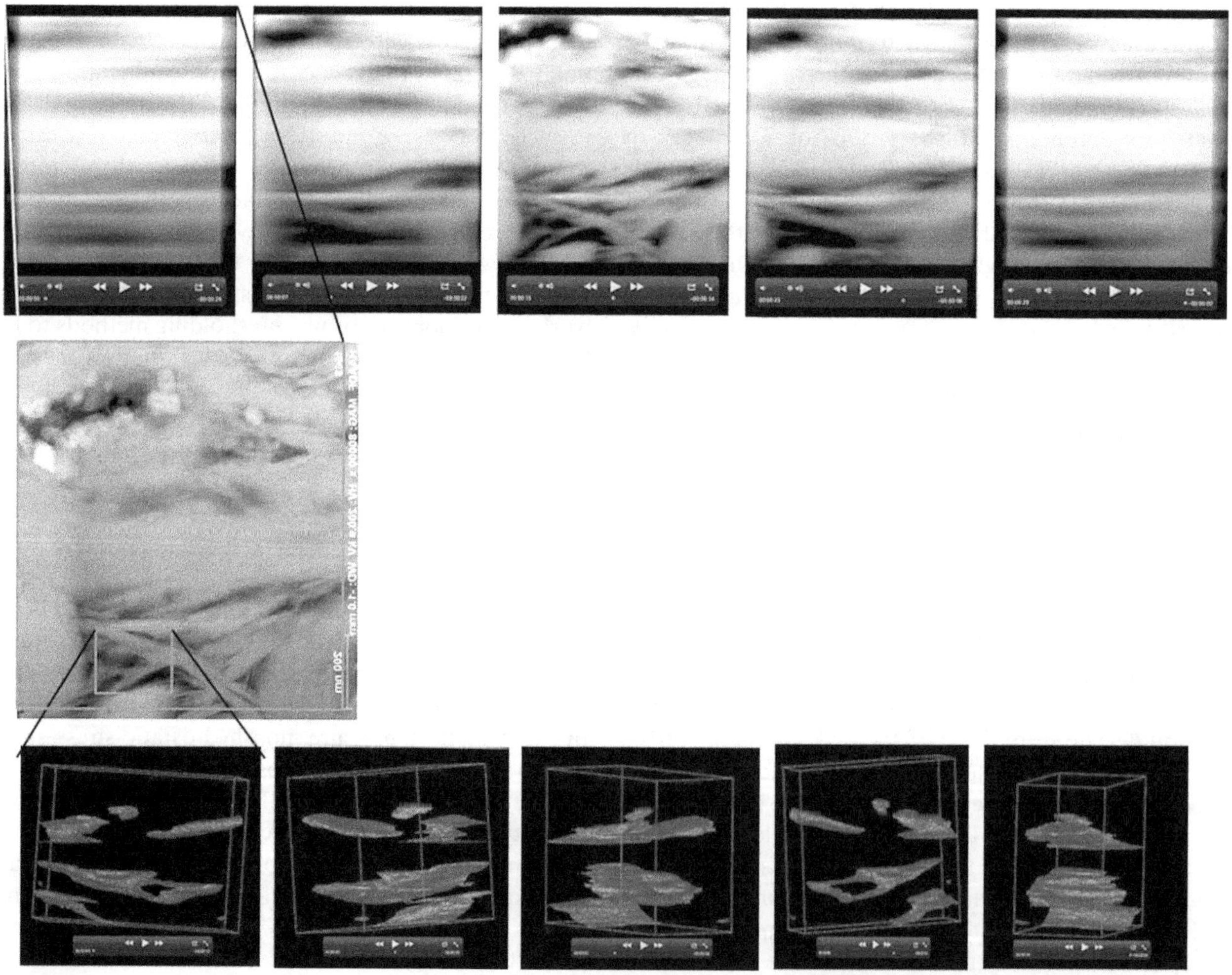

Figure 3.7 HAADF and electron tomographic reconstructions (ranging over 120°) of pores in clay aggregates of Haynesville Shale. Horizontal dimension is 1 μm, sampled at 1 nm resolution.

or exacerbation of leakage pathways at subsurface CO_2 storage sites [*Gaus*, 2010]. Injected, anhydrous CO_2 will dissolve water to solubility limits such that subsequent dry-out of clays by CO_2 near wellbores may occur. *Ilton et al.* [2012] and *Schaef et al.* [2012], using in situ hydrostatic high-pressure X-ray diffraction (XRD), observed changes in smectite clay interlayer spacing upon exposure to CO_2 at variable water content, which suggests dehydration and potential shrinkage of caprock. In contrast, using Monte Carlo and molecular dynamics simulations, *Botan et al.* [2010] investigated interlayer spacing of smectite under varying amounts of water and CO_2, but neither shrinkage nor swelling was observed. For the conditions Botan *et al.* considered, the chemical potential likely remained nearer that of pure water as compared to the work by *Ilton et al.* [2012] and *Schaef et al.* [2012].

Clay swelling against an applied pressure or under constant volume conditions and water imbibition can produce swelling pressures up to, for example, ~10 MPa (1450 psi) [*Schanz and Tripathy*, 2009]. A recent study examines CO_2 intercalation and shrink-swell under conditions of non-hydrostatic stress [*de Jong et al.*, 2014], but without water. In the context of the previous work, there is a need to understand, under realistic non-hydrostatic stress states typical of the sequestration environment, how deformation partitions between bulk volume strains, pore volume strains, swelling, and H_2O and/or CO_2 intercalation. To examine the partitioning of deformation of clays subject to multiple fluid types, we are developing novel oedometric SANS in collaboration with Los Alamos National Laboratory's Neutron Science Center (LANL-LANSCE) [*Los Alamos Science*, 2006]. Oedometric SANS involves a non-hydrostatic pressure vessel system that controls an axial load and shortening, pore pressure, and pore fluid composition (Fig. 3.8).

We have performed test-of-concept experiments using two designs of oedometric SANS pressure vessels at the Low-Q Diffractometer (LQD) at LANSCE. The LQD examines length scales from 1 to 100 nm. We note that pore structures can vary over much greater length scales, which can be accessed with other versions of scattering techniques such as ultra small angle neutron scattering (USANS) [*Anovitz and Cole*, 2015]. Our intent here is to advance scattering methods under non-hydrostatic stress states and pore pressures with reactive fluids (H_2O and CO_2) while identifying and overcoming if possible any issues that may complicate data interpretation. The first design was built out of titanium: titanium (i.e., Ti-6Al-4V) has low neutron attenuation, experiences minor activation under the neutron beam, and is resistant to chemical corrosion. The sample chamber of the Ti oedometer is 25.4 mm (1 in.) diameter, which proved to be overly large for SANS in terms of causing multiple scattering (as discussed below); however, the Ti oedometer sample chamber may still be suited for other neutron scattering instruments at LANSCE (e.g., the High-Pressure-Preferred Orientation (HIPPO) time-of-flight diffractometer) or other long-term consolidation or shrink-swell testing in the corrosion-resistant vessel. The Ti oedometer was designed with a factor of safety of 4× for maximum allowable working pressure (MAWP) within the sample chamber of 27.6 MPa (4000 psi).

Preliminary attempts at interpreting the SANS data indicate multiple neutron scattering: neutrons path scattered more than once while traveling through the clay pellet. Multiple scattering can distort scattering curves and, depending on how severe it is, make data interpretation difficult to impossible in terms of accurately estimating the size and number of scattering objects [*Mang and Hjelm*, 2013]. Effects of multiple scattering and how it affects scattering patterns are generally predictable, but difficult to correct [*Anovitz and Cole*, 2015], and thus should be avoided by designing sample and vessel thicknesses that preclude or minimize multiple scattering.

To counter multiple scattering, a newer stainless steel and aluminum oedometer (Fig. 3.8) was designed and built to be better optimized for the optics of SANS. The major change is a decrease in sample thickness in a slit geometry to ~3.2 mm normal to the neutron beam. The oedometer was tested at air and gaseous to liquid CO_2 at dry and wet (with D_2O) and under different axial loads with pore pressure up to ~5.5 MPa (Fig. 3.9). Multiple scattering still occurs as indicated by some transmission data for the sample divided by the transmission of the empty vessel versus time of flight that are below 0.75%. However, preliminary analysis indicates that the data also do not fall off at greater than Q^4 (Fig. 3.9), which would be the case for severe multiple scattering, and thus ~25% error in average pore size in single scattering approach is probable [*J. Mang*, pers. commun., 2017]. The SANS I(Q) versus Q curves were converted to pore size distributions (Fig. 3.9a–g) using the Irena package in Igor Pro with the *size distribution* and *unified fit* options [Ilavsky and Jemian, 2009] and assuming spherical scatters (i.e., spherical pores). Preliminary interpretation of oedometric SANS shows that initial compaction shifts to slightly smaller pores as expected. Release of pore pressure for dry CO_2 seems reversible with pore size distributions falling back near the initial compaction curve. Release to lower pressure of wet CO_2 shows a change in pore sizes and thus irreversible strain. Oedometric SANS has distinct results for the clay sample at different stress conditions and for different fluids. Each step in changing conditions produces measurable changes in the fractal dimension and pore size distribution of bentonite pore networks. In situ neutron methods thus are a promising tool to reveal pore structure information across a range of scales during deformation and exposure to fluids of different chemistries.

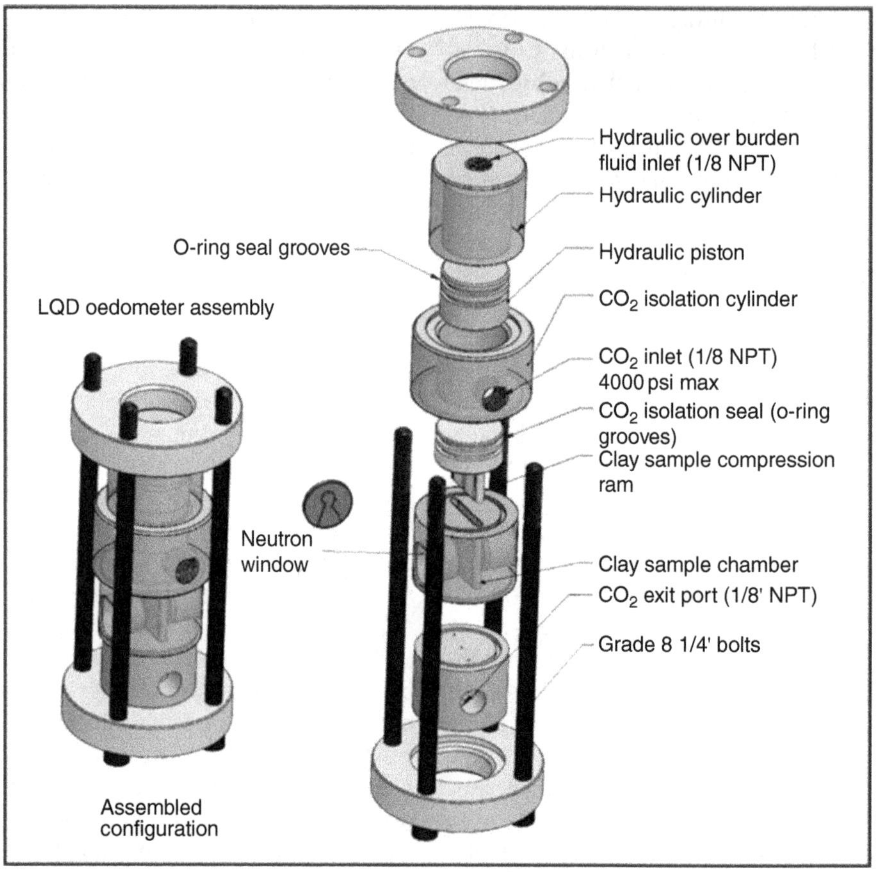

Figure 3.8 Schematic diagram of second oedometer with annotation indicated by arrows. Note penny for scale. Figure courtesy of Mark Taylor, Los Alamos National Laboratories.

3.7. MECHANICAL PROPERTIES OF REPRESENTATIVE VOLUME ELEMENTS

Shales are heterogeneous at all scales, and this poses a problem for determining mechanical constitutive behavior in typical rock mechanics testing configurations involving 1 in. – diameter or larger core plugs, such as the Mancos Shale shown in Figures 3.1 and 3.3. Mudstone mechanical testing is often limited by poor core recovery and sample size, preservation, and preparation issues, which can lead to sampling bias, damage, and time-dependent effects. Nanoindentation is often used to determine mechanical properties such as Young's modulus or some material strength parameter [*Abousleiman et al.*, 2007; *Shukla et al.*, 2013], but there is a clear need for additional methods to examine shale mechanics at the scale of an REV.

3.7.1. Micropillar Compression

A micropillar compression technique, originally developed by *Uchic et al.* [2004], is being applied to elastoplastic deformation of small volumes of mudstone, as part of a study investigating size-scale and anisotropy effects on mechanical response [*Dewers et al.*, 2010]. Figure 3.10 shows pillars of pristine silicon and deformed samples of Gothic Shale for comparison (the basal unit of the Ismay zone of the Pennsylvanian Paradox Formation, Utah, USA, a potential shale gas play; *Heath et al.*, 2017). Focused Ga^{+} ion milling and SEM imaging with an FEI NanoLab including pillar machining and 50 nm slice-and-view examined a mudstone sample from 5390 ft (1643 m) below ground surface (FIB-SEM results are given in *Heath et al.* [2011]). Micropillar compression was performed with a Hysitron Performech Triboscan nanoindenter and flat diamond indenter. Figure 3.10b shows a newly milled pillar composed of silicon, Figure 3.10b shows locations on the surface of a shale sample, and Figures 3.10c and d show before and after images of Pillar II. Pillar II is likely composed of clay-organic matrix, Pillar IV is likely composed of detrital clay-sized quartz grain (and behaves similarly to single quartz crystal indentation experiments, e.g., *Dubach et al.*, 2009; *Maeder et al.*, 2010), and Pillar III is composed of both.

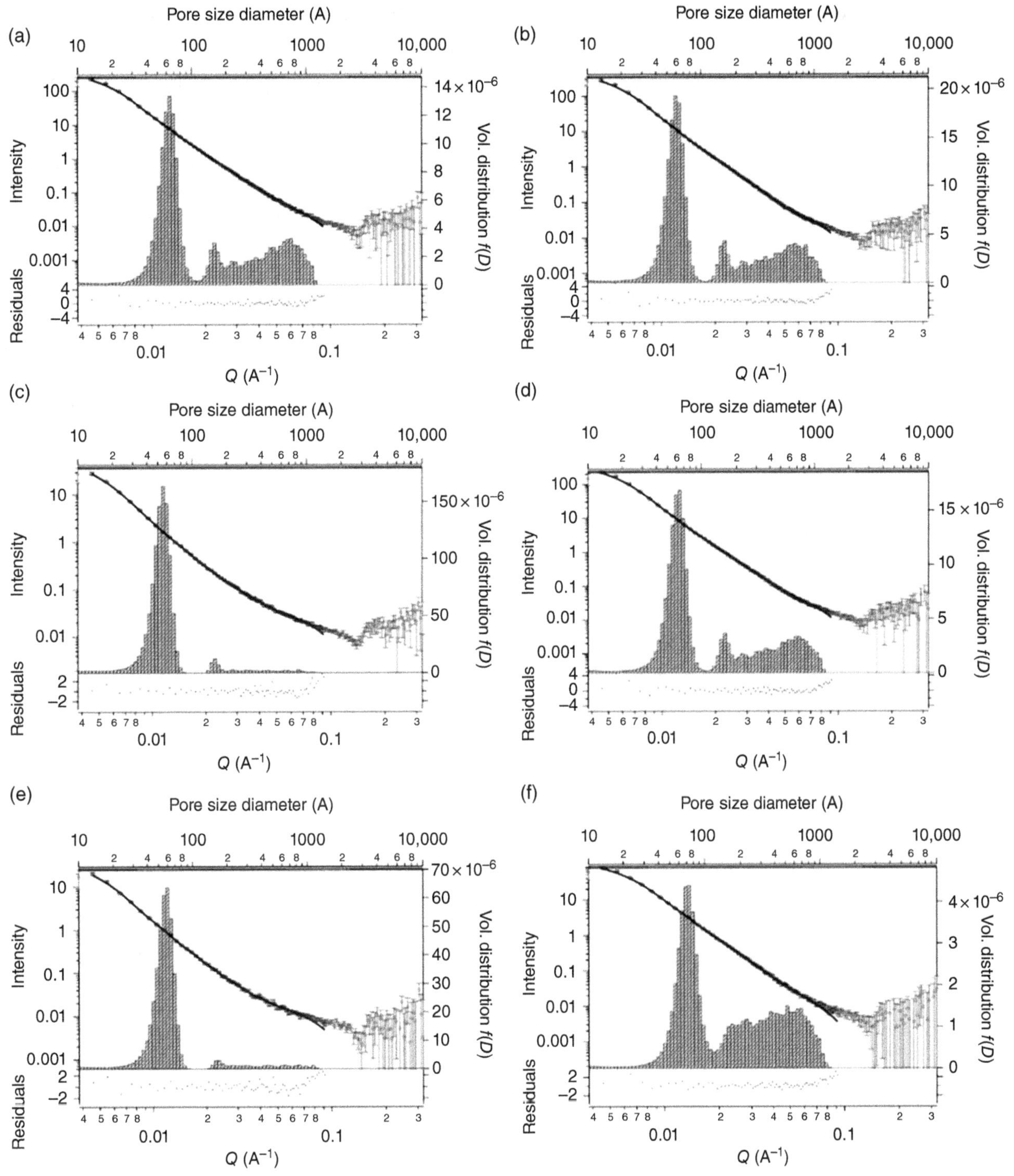

Figure 3.9 Intensity $I(Q)$ versus scattering vector (Q) for the various sample states. (a) Initial dry compaction (fluid port open to atmosphere) to estimated void ratio and axial strain of 1.012 and 0.001, respectively. (b) Second dry compression (fluid port open to atmosphere) to estimated void ratio and axial strain of 0.738 and 0.126. (c) Dry CO_2 injected via fluid port to hydrostatic conditions with estimated void ratio and axial strain of 0.878 and 0.110. (d) Bleed-off of CO_2 to axial load of part b, with resultant void ratio and axial strain of 0.741 and 0.168, respectively. (e) Injection of wet CO_2 (with D_2O) to hydrostatic conditions with estimated void ratio and axial strain of 0.734 and 0.171, respectively. (f) Bleed off of wet CO_2 to atmosphere but residual stress of 89 psig remained on piston.

Pillar II deformed plastically via localized shear and shows a shear band (Fig. 3.10d). At 5 μm diameter by 10 μm length, this may be on the order of certain mudstone REVs although this remains to be rigorously tested for the Gothic. The unconfined compressive strength (UCS) for Pillars II, III, and IV are 1.1, 3.8, and 6.3 GPa, respectively, whereas a 1 in. diameter by 2 in. long core of Gothic Shale has a UCS of ~88 MPa. Similarly, Young's modulus for Pillars II through IV are 29, 37, and 44 MPa, respectively, whereas Young's modulus for the 1 in. core is 18 MPa. Thus the Gothic Shale UCS and Young's modulus (E) is scale dependent over several orders of magnitude and likely have a first-order dependence on size and orientation of flaws (i.e., bedding planes, alignment of kerogen, microcracks, etc.). This method shows promise for advancing understanding of shale anisotropy, and scale effects on strength, but still suffers from occasional fabrication damage and alignment issues.

3.7.2. Digital Image Correlation, Mineralogical Mapping, and Phase Field Modeling

Imaging both in 2D and 3D has become an invaluable tool in understanding rock deformation. New optical measurement techniques, specifically *full field optical techniques* such as 2D and 3D digital image correlation (DIC), are becoming increasingly useful tools for mechanics experimentation [*Sutton et al.*, 2009; *Bay et al.*, 1999; *Louis et al.*, 2007; *Hall et al.*, 2009, 2010] and evaluating dependence on heterogeneity [*Lenoir et al.*, 2007]. We show examples utilizing 2D DIC in understanding and modeling mode I fracture tests in so-called Brazilian tests of Mancos Shale (Fig. 3.11). Four core plugs of Mancos Shale, measuring 1.0 in. in diameter and 0.5 in. thick, were prepared and tested according to ASTM standards in a 22 kip load frame at a rate of 2×10^{-6} mm/s with failure typically occurring approximately 2.5 min after contact. Prior to testing, specimens

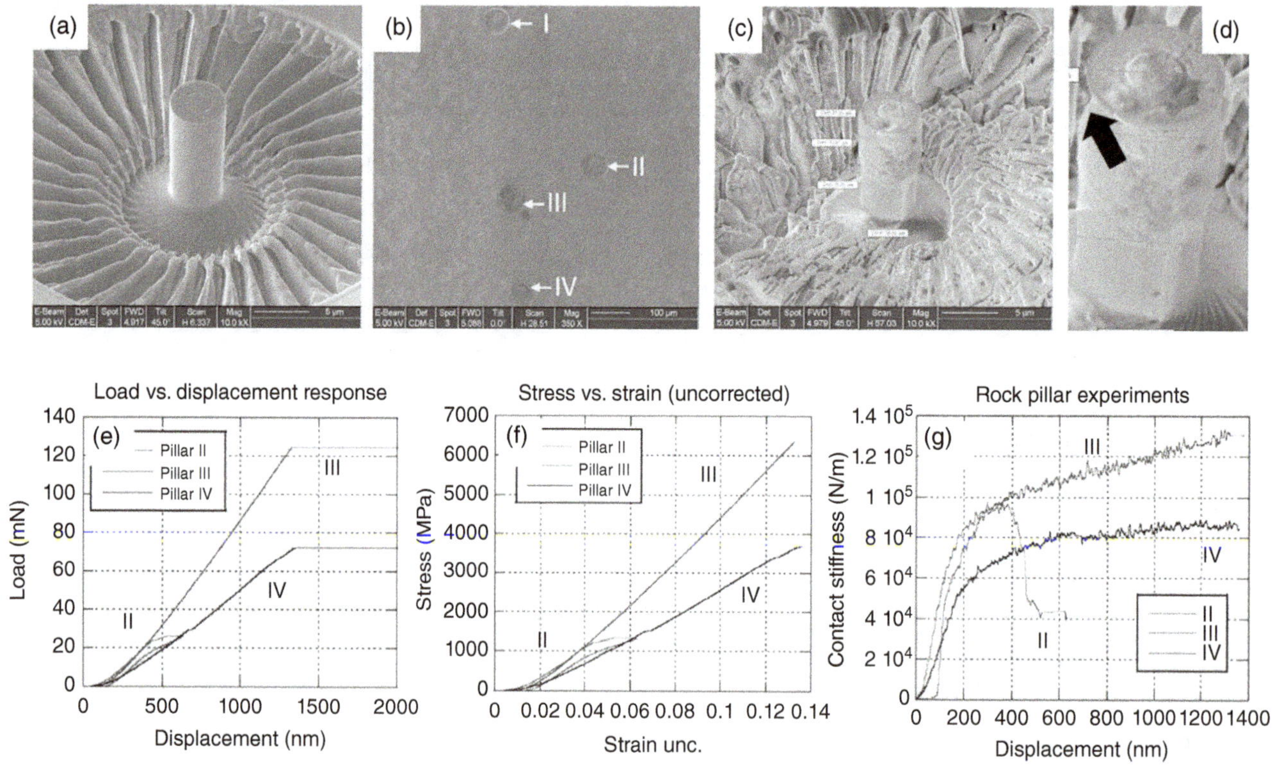

Figure 3.10 Images and results of micropillar compression tests on Gothic Shale. (a) Example pillar composed of silicon. The radial markings surrounding the pillar are from the ion beam essentially acting as a lathe in cutting the pillar. (b) Map view of locations on the shale sample surface. (c and d) Before- and after-pillar deformation by a flat nanoindenter of Pillar II. Pillar II (d) survived the deformation, and the black arrow shows a shear band beneath the platinum cap. (e and f). Load versus displacement and stress versus strain for three pillar compression measurements. (g) Contact stiffness and local slopes of load versus displacement curves.

were placed in a constant humidity chamber at 70%, non-condensing, for at least 2 weeks. These were painted with flat black spray paint and speckled with white spray paint for the purposes of performing DIC during the testing [*Na et al.*, 2017]. The DIC system consisted of two Point Grey Research 5.0 Megapixel Monochromatic Grasshopper cameras. These were equipped with a pair of Edmund Optics 35 mm machine vision lenses with a 5 mm extension tube. Data was collected with Correlated Solutions VIC-Snap software at a rate of four frames per second and processed with Correlated Solutions VIC-3D 7. Subsets and step size were optimized to minimize non-correlated points.

Following testing, one sample (Fig. 3.11, row 3) was polished by ion milling and then scanned with BSE scanning and MAPS Mineralogy (Fig. 3.12a–c), a new FEI analysis tool useful for high-resolution images over large areas with automated compositional analysis. For this sample, loading was applied at ~45° to bedding. The fracture in Figure 3.12 corresponds to the inclined crack or deformed zone in DIC mapping (e.g., Fig. 3.11, row 3, c, and d), and the horizontal *jog* shows preferential propagation along a bedding plane until the crack propagates vertically. Initiation and propagation of cracks are influenced by many factors including a fraction of stiff and soft layers, thickness of the layer, the presence of weak materials (e.g., defect, organics, existing cracks), and mechanical properties of each component [*Na et al.*, 2017]. Nonetheless, the features of cracks at a local scale are predominantly governed by local mineralogical and mechanical heterogeneity.

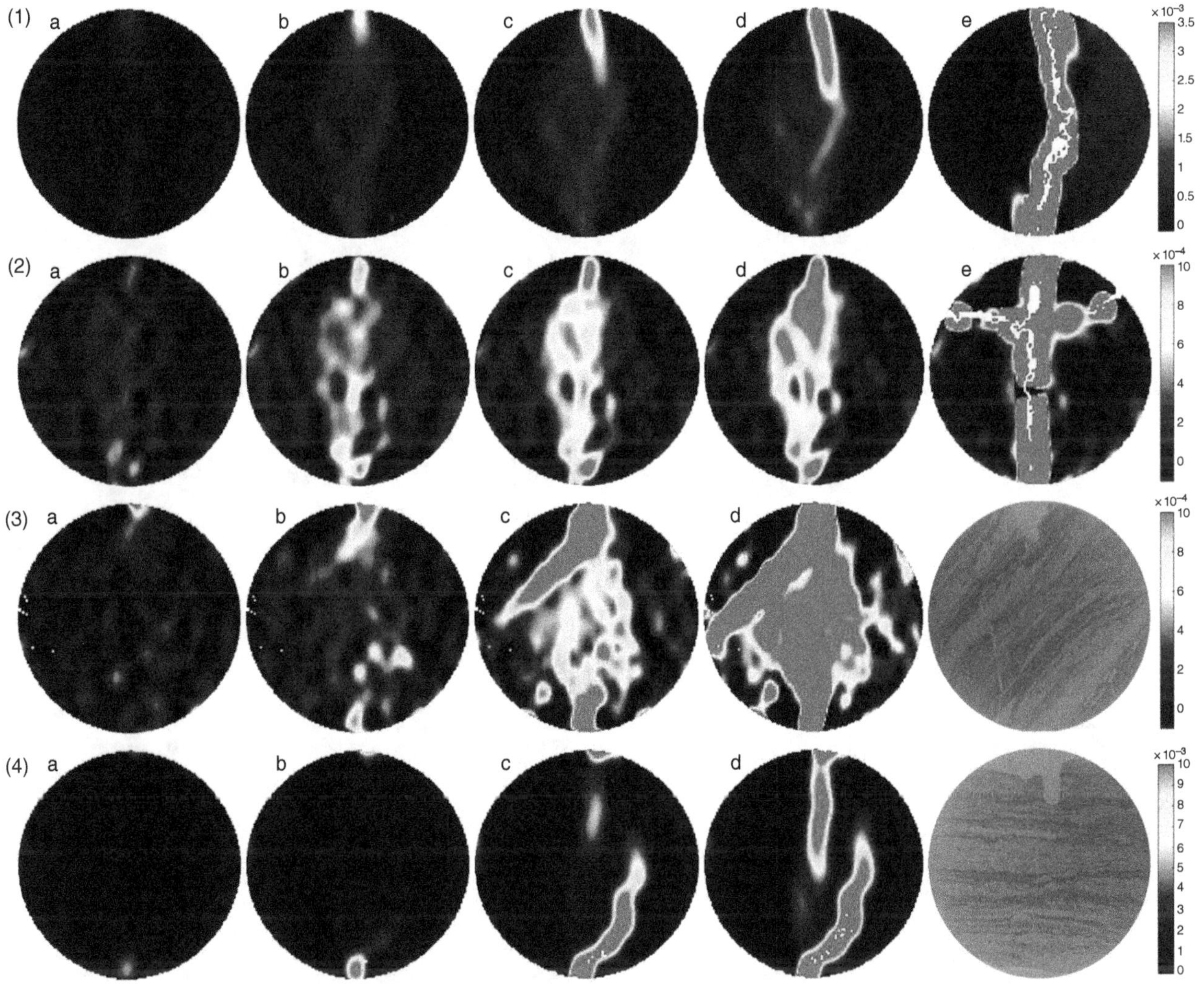

Figure 3.11 Brazilian tests and DIC analysis (horizontal displacement, mm) of four Mancos Shale samples. Rows 1 and 2 are results where bedding is parallel to the loading direction. Row 3 is at a 45° angle to the loading direction, and row 4 is perpendicular to the loading direction. Note that the light gray surrounded within white color corresponds to the highest value of displacement in gray scale. Modified from *Na et al.* [2017], with snapshots at different times. Reprinted with permission from Wiley.

As a preliminary work, nanoindentation tests on another Mancos Shale sample were performed at 16 × 16 regular grid with 20 μm spacing using a Hysitron TriboScan (Fig. 3.12d–g). A Berkovich tip at a constant strain rate of $0.1\,s^{-1}$ was used with maximum load of 10 mN. The BSE and mineralogical distribution before testing and optical image of indentation markers and calculated elastic modulus are shown in Figure 3.12d–g. Based on the mineralogy distribution, mechanical properties of shale can be directly linked to compositional and textural heterogeneity at a fine resolution including textural features.

Compositional and mechanical heterogeneity can be directly incorporated into numerical modeling frameworks such as phase field models to examine shale mechanics at the scale of an REV. This direct correlation can overcome manual and statistical interpretation of indentation results [*Kumar et al.*, 2012; *Akono and Kabir*, 2016]. Figure 3.13 shows modeling results of the Brazilian tests as a function of load angle with respect to bedding, based on experimental conditions in Figure 3.11. A phase field model was used to evaluate the impact of layer orientation on crack initiation and propagation in anisotropic heterogeneous rocks where shale heterogeneity conceptually consists of two main constituents including stiff and soft materials. The phase field model treats the mechanical discontinuity (e.g., cracks) using a diffusive transition zone, and the diffuse interface field is obtained via the minimization of a free energy function [*Moelans et al.*, 2008]. This diffuse interface model can overcome some of the difficulties inherent to discrete fracture approaches that need sharp interfaces with complex evolving topologies and explicit

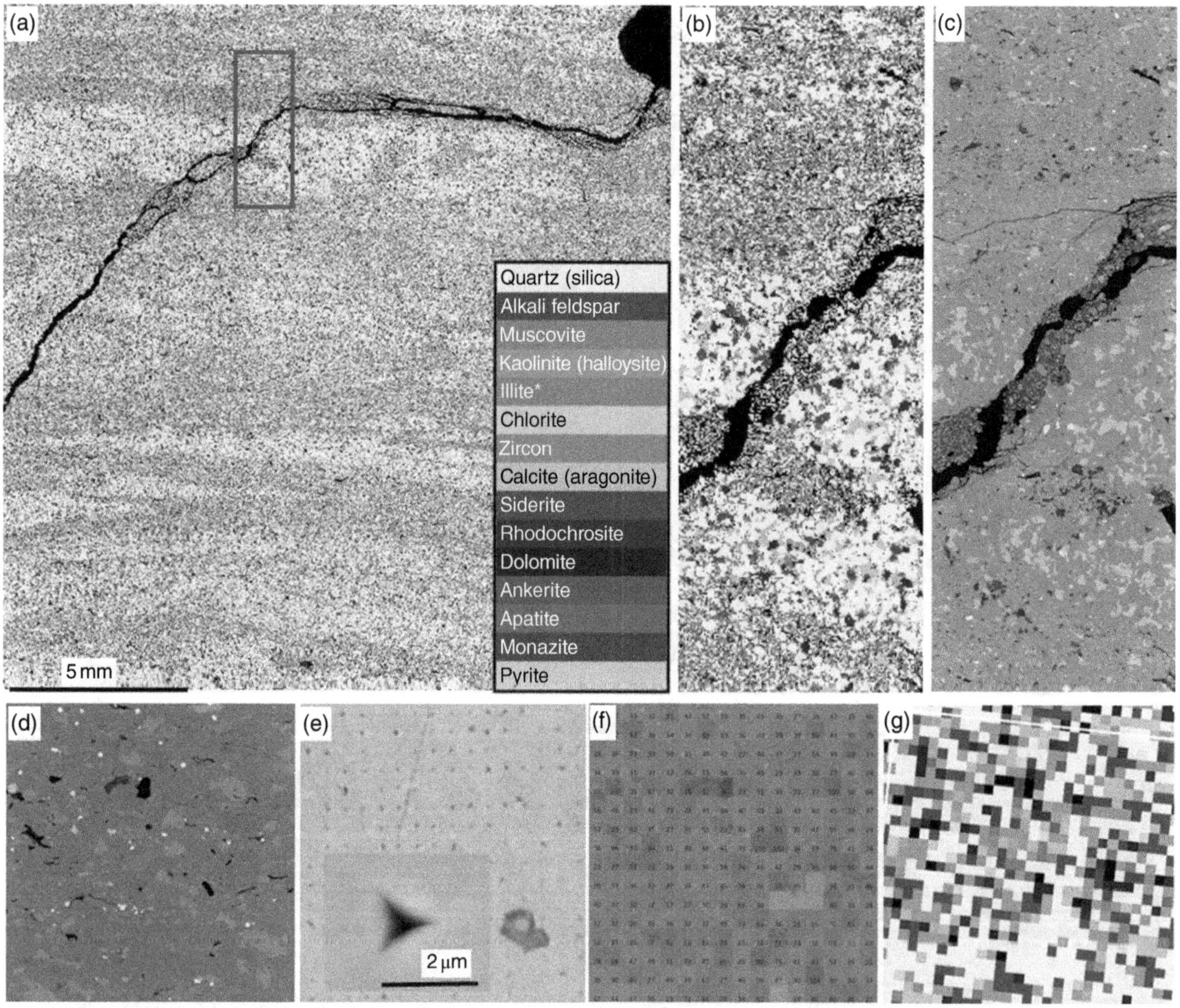

Figure 3.12 Mineralogy mapping at 2 μm resolution using FEI MAPS Mineralogy. Scale bar is 5 mm. Two panels on the right side (corresponding to the box) show the mineralogy distribution at 2 μm and BSE images at 0.2 μm resolution, respectively. Cracks were formed after the Brazilian test shown in Figure 3.11 (row 3).

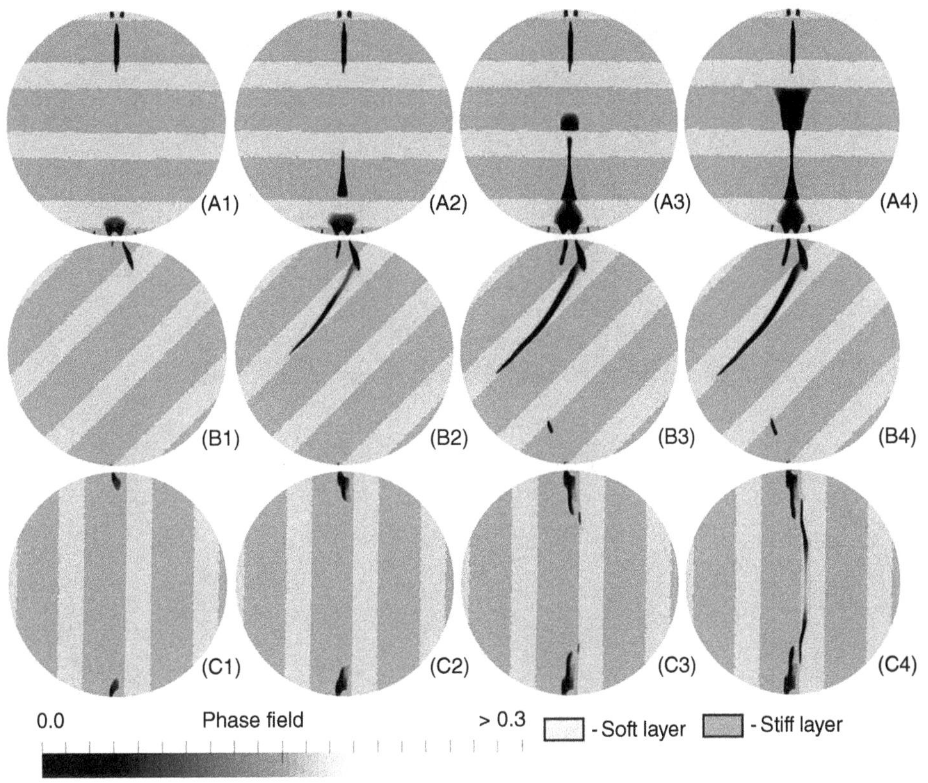

Figure 3.13 Phase field modeling of fracture propagation during the Brazilian test at different bedding angles along the vertical loading direction in the layered system mimicking Mancos Shale. Modified from *Na et al.* [2017] with snapshots at different times. The layered system consists of the stiff (62.5%) and soft (37.5%) materials.

heuristics to govern their propagation and interaction. In particular, an adaptive mesh refinement scheme is employed to account for crack initiation and propagation at fine scale [*Heister et al.*, 2015] using the open source finite element library deal.II [*Bangerth et al.*, 2007, 2013].

Modeling results show cracks initiate from top and bottom parts of loading area, and crack propagation is strongly influenced by the layer orientation. For the inclined layer case, modeling results capture the main crack propagation along the soft layer as seen in the experiment (Fig. 3.11). For the vertical layer, the main crack develops along the vertical line of the domain, and partial cracks are barely identified. This example highlights the importance of elastic heterogeneity and anisotropy on fracture propagation which has been reported in detail in *Na et al.* [2017]. Phase field modeling is proving to be an effective tool to capture crack initiation and localized propagation processes in brittle fracture.

3.7.2.1. Digital Volume Correlation (DVC)

Digital volume correlation (DVC), first developed by *Bay et al.* [1999], is an extension of the 2D DIC technique that can be used not only with 3D imaging techniques discussed herein including μCT and confocal microscopy but also with magnetic resonance imaging (MRI) and positron emission tomography (PET) [*Sutton et al.*, 2009]. In Figure 3.14, we demonstrate how DVC can be utilized on shales to measure true 3D strains varying by lithofacies under in situ loading conditions. This takes advantage of high concentrations of pyrite that distribute differently among microlithofacies. Pyrite can act as natural tracer particles with CT imaging due to the high contrast in CT number, compared to other silicates and carbonate minerals. This is achieved by loading a specimen while in an X-ray μCT scanner utilizing a special loading system designed for X-ray transparency, seen in Figure 3.14a. Pyrite contrasts within the material are used to generate a 3D image (14B). By comparing subsequent 3D images with DVC algorithms, strains can be calculated, provided there is enough local density variation in pyrite grains in the sample. Results show that despite the sparse distribution of particles, highly accurate results of specimen displacement heterogeneity can be successfully tracked using the DVC technique. These results can then be used to determine the mechanical properties based on full field measurements. Figure 3.14c shows the results of the DVC analysis: the left and middle panels show displacements in the x and y direction, while the right panel shows

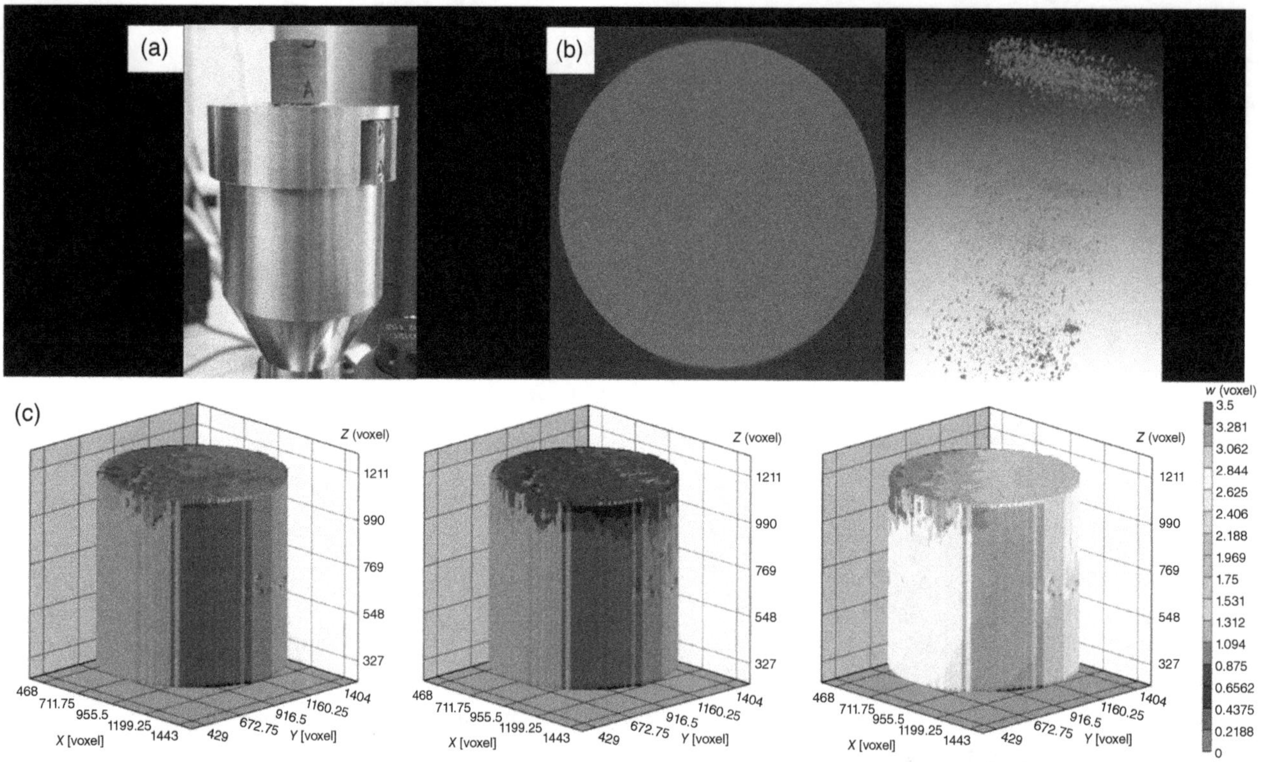

Figure 3.14 DVC analysis of a sample of Woodford Shale from Oklahoma. The sample is 0.5 in. in diameter and ~1 in. long. (a) Image of the loading system used in the μCT scanner to apply load. Shale specimen is shown on top of the loading platform marked with an *A*. (b) The left image is one slice of the specimen as raw output from the scanner. The right image is a 3D rendering of the high-density regions (pyrite grains) within the sample. (c) Images showing the *X* (left), *Y* (center), and *Z* (right) displacements measured by DVC of pyrite grain displacements.

displacements in the *z* direction. Note that there is essentially no motion in the *x* and *y* directions (any small amount that exists can be attributed to noise in the μCT scans) while there is significant motion in the *z* direction, which is the direction of actuation for the loading system. In-depth analysis of the DVC signals, how these relate to microlithofacies, and comparison with other strain-measuring techniques are topics of future work. Recent works on CT contrast agents utilizing iodine *fixing* or staining of organic matter in shales [*Fogden et al.*, 2014] are additional options for applying the DVC techniques which could prove useful in understanding differences in deformational styles of the organic versus clastic fractions.

3.8. PUTTING IT BACK TOGETHER: UPSCALING TO CORE SCALE

Currently, much work is being done on various THMC processes by the larger shale research community, but community consensus and understanding of coupled processes requires much further integration [*Heath and Ilgen*, 2015; *Ilgen et al.*, 2017]. Key to the applicability and fit for purpose of the techniques and methods discussed herein is the application and suitability of upscaling techniques. The broad research community lacks conceptual mechanisms for ready integration of the data and understanding by very different disciplines that study shale. Materials and biological sciences are currently applying abstract mathematics (e.g., category theory) in designing materials from the bottom up [*Buehler*, 2013; *Brommer et al.*, 2016] and could be applied to upscaling issues for shale science and engineering. These methods use novel knowledge representation schemes that incorporate many characteristics that we care about, such as upscaling and multiphysics or coupled processes, and prediction [*Spivak and Kent*, 2012; *Wong et al.*, 2012; *Ehresmann and Gomez-Ramirez*, 2015; *Buehler and Genin*, 2016]. One promising technique is the ontology log or *olog*, which, for example, has been used to build spatially hierarchical representations of silk fiber production starting at the level of proteins [*Spivak and Kent*, 2012; *Wong et al.*, 2012] and may be quite applicable for shale.

Given a typical sample size of TEM and FIB-SEM analysis (~10–30 μm in each dimension), 3D characterization of nanopores with advanced imaging techniques described in this chapter requires tremendous image data collection at the scale of a thin section (2D) or core (3D). Hence, it is strongly necessary to identify key features (e.g., microlithofacies, pore structure, and/or phase distribution) critical to processes of interest which can be reconstructed using advanced multi-scale image analysis for texture classification [e.g., *Cour et al.*, 2005; *Mallat*, 2012; *Kim et al.*, 2013]. With primary pore textures honored, 3D digital pore networks can be reconstructed using several stochastic approaches such as multiple-point simulation [*Remy et al.*, 2009], directional cluster functions for anisotropic inclusion [*Jiao and Chawla*, 2014], and/or multiscale pore network models [e.g., *Blunt et al.*, 2013]. To overcome a highly expensive computational cost of generating and utilizing stochastic ensemble members of 3D reconstructed pore structures, model dimension reduction approaches based on a support vector machine scheme and group similarity in PCA space can be adopted to develop representative ensemble members (e.g., ~10–20 members), resulting in a computationally intensive outcome. For example, *Niezgoda et al.* [2013] developed a novel microstructure quantification framework by applying PCA to construct the reduced-order representation (e.g., 8 class members instead of 100–200 randomly generated members). Similarly, *Samudrala et al.* [2014] developed a data dimensionality reduction framework based on graph analysis, which was applied for cluster analysis of chemical crystallography. These recent advancements in multi-scale imaging techniques coupled with 3D reconstruction algorithms permit development of databases [e.g., *Prodanovic et al.*, 2015] which are being used to evaluate the impact of rock heterogeneities such as anisotropy, pore structures, fluid distribution, and microfracture networks on static and dynamic poromechanical and multiphase flow properties.

3.9. CONCLUSIONS AND FUTURE DIRECTIONS

The past decade has seen an unprecedented advancement in technologies and methodologies for shale characterization. This chapter has discussed but a few advancements, and there are plenty of methods not discussed herein that are driving shale characterization forward (electrical streaming potential, microfluidics, and ultrasonics, to name a few). Shales possess macroporosity (> 1 μm), microporosity (< 1 μm and>10 nm), and nanoporosity (< 10 nm). Each type appears to have distinct scaling. Variable water properties in these pores impart lots of dynamic behavior specific to these rocks (such as osmotic membrane behavior, swelling, or intercalation). The nanopore distribution contributes perhaps 20–50% of total porosity and at least 80% of the total surface area. Mineralogy of (micro-) pore-lining phases does not correspond well to bulk XRD mineralogy.

Geologists do not seem to be close to reaching consensus on how to classify shale as evidenced by the continued creation and lively debate of classification schemes [*Macquaker and Adams*, 2003; *Milliken*, 2014; *Lazar et al.*, 2015; Camp et al., 2013; *Milliken*, 2016]. Assigning macroscopic properties, such as *brittleness* or *ductility* on the basis of bulk XRD measurement, while being a simple and practical consideration, may be an oversimplification and a wealth of newer technologies (e.g., Fourier transform infrared spectroscopy) can provide information to supplement and supplant older XRD-based interpretations.

The increase in availability and use of high-resolution imaging techniques for shale (e.g., ion milling and SEM techniques) seems to have actually made classification more difficult [*Milliken*, 2014] as the massive amounts of small-scale data show greater diversity in grain types and at least the full range of post-depositional effects as compared to conventional sandstone and limestone reservoirs. Despite the technological advancements then, relating geologic controls in qualitative and quantitative ways, such as primary depositional and diagenetic alterations, to constitutive behavior in natural and engineered systems is a fruitful area of future research.

ACKNOWLEDGMENTS

We acknowledge helpful suggestions from two anonymous reviewers which greatly improved the quality of this chapter. We thank Bill Carey, Anastasia Ilgen, Pat Brady, Carlos Jove-Colon, and Yifeng Wang for useful discussions. Joe Michael, Paul Kotula, Michael Rye, and Gary Bryant contributed greatly to the beam sections of the chapter, and Rex Helm, Mark Taylor, Mei Deng, and Alexis Navarre-Sitchler were invaluable contributors to our understanding of neutron methods. We thank Russ Ewy and Luca Duranti from Chevron Energy Technology Company for guidance, motivation, and Haynesville Shale samples. We thank Eric Shields of Sandia National Laboratories for assistance in post-processing and analyzing multi-beam scanning electron microscopy data sets and Michelle Williams and Carl Jacques for help with μCT imaging and interpretation. Bill Mook from Sandia's Center for Integrated Nanotechnologies provided much guidance and assistance for the nanoindentation results. This work was funded by the US Department of Energy, Office of Science, Basic Energy Sciences under Award Number DE-SC0006883. This work benefited from the use of the small angle neutron scattering instrument LQD at the Manuel Lujan Jr. Neutron Scattering Center of the Los Alamos National Laboratory supported by the US Department of Energy at Los Alamos National Laboratory operated by Los Alamos National Security LLC under contract number DE-AC52-06NA25396.

Sandia National Laboratories is a multimission laboratory managed and operated by National Technology and Engineering Solutions of Sandia, LLC, a wholly owned subsidiary of Honeywell International, Inc., for the US Department of Energy's National Nuclear Security Administration under contract DE-NA0003525.

REFERENCES

Abdelmalek, B., Z. Karpyn, S. Liu, H. Yoon, and T. Dewers, (2017), Gas permeability measurements from pressure-pulse decay laboratory data using pseudo-pressure and pseudo-time transformations. *Journal of Petroleum Exploration and Production Technology*, doi:10.1007/s13202-017-0376-5.

Abousleiman, Y., M. Tran, S. Hoang, C. Bobko, A. Ortega, and F.J. Ulm, (2007), Geomechanics field and lab characterization of Woodford shale: the next gas play. SPE paper 110120, SPE Annual Technical Conference Exhibition, Anaheim, CA, 11–14 November.

Akono, A.T., and P. Kabir, (2016), Nano-scale characterization of organic-rich shale via indentation methods. In (Jin C., Cusatis, G. eds.) *New Frontiers in Oil and Gas Exploration*. Cham, Springer International Publishing, 209–233.

Anovitz, L.M., and D.R. Cole, (2015), Characterization and analysis of porosity and pore structures. *Reviews in Mineralogy and Geochemistry 80*, 61–164.

Aplin, A., S. Larter, M. Ashley Bigge, G. Macleod, R. Swarbrick, and D. Grunberger, (2000), Confocal microscopy of fluid inclusions reveals fluid-pressure histories of sediments and an unexpected origin of gas condensate. *Geology 28*, 1047–1050.

Aplin, A.C., I.F. Matenaar, D. McCarty, and B.A. van der Pluijm, (2006), Influence of mechanical compaction and clay mineral diagenesis on the microfabric and pore-scale properties of deep-water Gulf of Mexico mudstone. *Clays and Clay Minerals 54*, 501–515.

Ashish, M., C.H. Sondergeld, and C.S. Rai, (2016), Comparison of steady-state and transient methods for measuring shale permeability, SPE 180259.

Bangerth, W., R. Hartmann, and G. Kanschat, (2007), Deal. IIa: general-purpose object-oriented finite element library. *ACM Transactions on Mathematical Software. 33*, 24.

Bangerth, W., T. Heister, L. Heltai, G. Kanschat, M. Kronbichler, M. Maier, B. Turcksin, and T. Young, (2013), The Deal.II Library, version 8, 1–5.

Bay B.K., T.S. Smith, D.P. Fyhrie, and M. Saad, (1999), Digital volume correlation: three-dimensional strain mapping using X-ray tomography. *Experimental Mechanics 39*, 217–226.

Bennett, K.C., L.A. Berla, W.D. Nix, and R.I. Borja, (2015), Instrumented nanoindentation and 3D mechanistic modeling of a shale at multiple scales. *Acta Geotechnica 10*, 1–14.

Bickle, M., A. Chadwick, H.E. Huppert, M. Hallworth, and S. Lyle, (2007), Modelling carbon dioxide accumulation at Sleipner: implications for underground carbon storage. *Earth and Planetary Science Letters 255*, 164–176.

Blatt, H., (1982), *Sedimentary Petrology*. Freeman Press, San Francisco, 564 p.

Blunt, M.J., B. Bijeljic, H. Dong, O. Gharbi, S. Iglauer, P. Mostaghimi, A. Paluszny, and C. Pentland, (2013), Pore-scale imaging and modelling. *Advances in Water Resources, 51*, 197–216.

Bosl, W., J. Dvorkin, and A. Nur, (1998), A study of porosity and permeability using a lattice Boltzmann simulation. *Geophysics Research Letters 25*, 1475–1478.

Botan, A., B. Rotenberg, V. Marry, P. Turq, and B. Noetinger, (2010), Carbon dioxide in Montmorillonite clay hydrates: thermodynamics, structure, and transport from molecular simulation. *Journal of Physical Chemistry C 114*, 14962–14969.

Brommer, D.B., T. Giesa, D.I. Spivak, and M.J. Buehler, (2016), Categorical prototyping: incorporating molecular mechanisms into 3D printing. *Nanotechnology 27*, doi:10.1088/0957-4484/27/2/024002.

Buehler, M.J., (2013), Materials by design: a perspective from atoms to structures. *Materials Research Bulletin 38*, 169–176.

Buehler, M.J., and G.M. Genin, (2016), Integrated multiscale biomaterials experiment and modelling: a perspective. *Interface Focus 6*, doi:10.1098/rsfs.2015.0098.

Camp, W., E. Diaz, and B. Wawak, (2013), *Electron Microscopy of Shale Hydrocarbon Reservoirs*. AAPG Memoir 102. Tulsa, OK, American Association of Petroleum Geologists, 177 p.

Carey, J.W., Z. Lei, E. Rougier, H. Mori, and H.S. Viswanathan, (2015), Fracture permeability behavior of shale. *Journal of Unconventional Oil and Gas Resources, 11*, 27–43.

Cour, T., F. Benezit, and J. Shi, (2005), Spectral segmentation with multiscale graph decomposition. *Computer Vision and Pattern Recognition, CVPR, IEEE Computer Society Conference*, Vol. *2*, 1124–1131.

Curtis, M.E., R. Ambrose, C. Sondergeld, and C. Rai, (2011), Transmission and scanning electron microscopy investigation of pore connectivity of gas shales on the nanoscale. *North American Unconventional Gas Conference and Exhibition*, The Woodlands, TX, Society of Petroleum Engineers, SPE 144391.

Desbois, G., J.L. Urai, and P.A. Kukla, (2009), Morphology of the pore space in claystones – evidence from BIB/FIB ion beam sectioning and cryo-SEM observations. *eEarth 4*, 15–22.

Dewers, T.A., B. Boyce, T. Buchheit, J.E. Heath, T. Chidsey, and J. Michael, (2010), Micropillar compression technique applied to micron scale mudstone elasto-plastic deformation. *Abstract MR11B-1882 presented at 2010 Fall Meeting*. American Geophysical Union, San Francisco, CA, 13–17 December.

Dewers, T., J.E. Heath, R. Ewy, and L. Duranti, (2012), Three-dimensional pore networks and transport properties of a shale gas formation determined from focused ion beam serial imaging. *International Journal of Oil, Gas and Coal Technology 5*, 229–248.

Dewhurst, D., and A. Siggins, (2006), Impact of fabric, micro-cracks and stress field on shale anisotropy. *Geophysical Journal International 165*, 135–148.

Dewhurst, D., R. Jones, and M. Raven, (2002), Microstructural and petrophysical characterization of Muderong shale: application to top seal risking. *Petroleum Geoscience 8*, 371–383.

Dubach, A., R. Raghavan, J.F. Loffler, J. Michler, and U. Ramamurty, (2009), Micropillar compression studies on a bulk metallic glass in different structural states. *Scripta Materialia 60*, 567–570.

Eberle, A.L., S. Mikula, R. Schalek, L. Lichtman, M.L. Knothe Tate, and D. Zeidler, (2015), High-resolution, high-throughput imaging with a multibeam scanning electron microscope. *Journal of Microscopy 259*, 114–120.

Edwards, H.C., (2006), Managing complexity in massively parallel, adaptive, multiphysics applications. *Engineering with Computers 22*, 135–156.

Ehresmann, A.C., and J. Gomez-Ramirez, (2015), Conciliating neuroscience and phenomenology via category theory. *Progress in Biophysics & Molecular biology 119*, 347–359.

Fogden, A., Latham, S., McKay, T., Marathe, R., Turner, M., Kingston, A., and T. Senden, (2014), Micro-CT analysis of pores and organics in unconventionals using novel contrast strategies. *Unconventional Resources Technology Conference*, Denver, CO, August (2014), doi:10.15530/urtec-2014-1922195.

Folk, R.L., (1980), *Petrology of Sedimentary Rocks*. Hemphill Publ. Co.: Austin, TX, *184* p.

Fredrich, J.T., K.H. Greaves, and J.W. Martin, (1993), Pore geometry and transport properties of Fontainebleau sandstone. *International Journal of Rock Mechanics and Mining Science and Geomechanics Abstracts 30*, 691–697.

Fredrich, J., B. Menendez, and T.-F. Wong, (1995), Imaging the pore structure of geomaterials. *Science 268*, 276–279.

Gaus, I., (2010), Role and impact of CO_2-rock interactions during CO_2 storage in sedimentary rocks. *International Journal of Greenhouse Gas Control 4*, 73–89.

Gens, A., B. Vallejan, M. Sanchez, and H-Ch. Cho, (2011), Hydromechanical behavior of a heterogeneous compacted soil: experimental observations and modelling. *Geotechnique 61*, 367–386.

Goergen, E., K.H. Skinner, H. Lemmens, and A. Benidictus, (2013), From core to pore: multi-scale, multi-dimensional characterization of fine-grained reservoir rocks. *Abstract in CSPG/CSEG/CWLS Geoconvention (2013)*. Calgary, Alberta, 6–12 May.

Grigg, J., (2016), Macroscopic and microscopic controls on mechanical properties of mudstones. Unpublished Master Thesis, New Mexico Tech, 98 p.

Gu, X., D.R. Cole, G. Rother, D.F.R. Mildner, and S.L. Brantley, (2015), Pores in Marcellus shale: a neutron scattering and FIB-SEM study. *Energy & Fuels 29*, 1295–1308.

Hall, S., N. Lenoir, G. Viggian, J. Desrues, and P. Bésuelle, (2009), Strain localisation in sand under triaxial loading: characterisation by x-ray micro tomography and 3D digital image correlation. *Proceedings of the 1st International Symposium on Computational Geomechanics* 26, 239–247. Juan-les-Pins, Cote d'Azur, France, April 29–May 1, (2009).

Hall, S., M. Bornert, J. Desrues, Y. Pannier, N. Lenoir, G. Viggiani, and P. Bésuelle, (2010), Discrete and continuum analysis of localized deformation in sand using X-ray μCT and volumetric digital image correlation. *Geotechnique 60*, 315–322

Harrison, A.L., A. Jew, M. Dustin, A. Kiss, A. Kohli, D. Thomas, C. Joe-Wong, Y. Liu, J.-H. Lim, G. Brown, K. Maher, and J. Bargar, (2016), Physical and chemical alteration of shales during hydraulic fracturing. *Abstract Presented at Geochemical Society 26th Goldschmidt Conference*, Yokohama, Japan, 16 June–1 July.

Hart, B.S., J.H.S. Macquaker, and K.G. Taylor, (2013), Mudstone (shale) depositional and diagenetic processes: implications for seismic analyses of source-rock reservoirs, *Interpretation 1*, B7–B26.

Heath, J.E., and A. Ilgen, (2015), Understanding multiscale coupled processes in shale. *Eos 96*, doi:10.1029/2015EO036447.

Heath, J., T. Dewers, B. McPherson, R. Petrusak, T. Chidsey, and A. Rinehart, (2011), Pore networks in continental and marine mudstones: characteristics and controls on sealing behavior. *Geosphere 7*, 429–454.

Heath, J., T. Dewers, B. McPherson, M. Nemer, and P. Kotula, (2012), Pore-lining phases and capillary breakthrough pressure of mudstone caprocks: sealing efficiency of geologic storage sites. *International Journal of Greenhouse Gas Control 11*, 204–220.

Heath, J., Dewers, T., Chidsey, T., Carnery, S., and Bereskin, S., (2017), The gothic shale of the Pennsylvanian paradox formation, greater Aneth field (Aneth unit), southeastern Utah: seal of hydrocarbons and carbon dioxide. Utah Geological Survey Miscellaneous Publication 17-1, 31 p.

Heister, T., M.F. Wheeler, and T. Wick, (2015), A primal-dual active set method and predictor-corrector mesh adaptivity for computing fracture propagation using a phase field approach. *Computer Methods in Applied Mechanics and Engineering 290*, 466–495.

Hemes, S., G. Desbois, J.L. Urai, B. Schroeppel, and J.-O. Schwarz, (2015), Multi-scale characterization of porosity in boom clay (HADES-level, Mol, Belgium) using a combination of X-ray mu-CT, 2D BIB-SEM and FIB-SEM tomography. *Microporous and Mesoporous Materials 208*, 1–20.

Ho, T.A., Criscenti, L., and Wang, Y., (2016), Nanostructural control of methane release in kerogen and its implications to wellbore production decline. *Scientific Reports 6*, 28053, doi:10.1038/srep28053.

Ilavsky, J., and P.R. Jemian, (2009), Irena: tool suite for modeling and analysis of small-angle scattering. *Journal of Applied Crystallography 42*, 347–353.

Ilgen, A., J.E. Heath, I.Y. Akkutlu, L.T. Bryndzia, D. Cole, Y. Kharaka, T. Kneafsey, K. Milliken, L. Pyrak-Nolte, and R. Suarez-Rivera, (2017), Shales at all scales: exploring coupled processes in mudrocks, *Earth-Science Reviews 166*, 132–152.

Ilton, E.S., H.T. Schaef, O. Qafoku, K.M. Rosso, and A.R. Felmy, (2012), In situ X-ray diffraction study of Na+ saturated montmorillonite exposed to variably wet super critical CO_2. *Environmental Science & Technology 46*, 4241–4248.

Jiao, Y., and N. Chawla, (2014), Modeling and characterizing anisotropic inclusion orientation in heterogeneous material via directional cluster functions and stochastic microstructure reconstruction. *Journal of Applied Physics*, *115*, 093511, doi:10.1063/1.4867611.

Jin, H., and W.-Y. Lu, (2013), 3D Deformation Field throughout the Interior of Materials. Sandia National Laboratories SAND Report 2013–7824, Livermore, CA, 25p.

de Jong, S.M., C.J. Spiers, and A. Busch, (2014), Development of swelling strain in smectite clays through exposure to carbon dioxide. *International Journal of Greenhouse Gas Control 24*, 149–161.

Karpyn, Z.T., P.M. Halleck, and S. Grader, (2009), An experimental study of spontaneous imbibition in fractured sandstone with contrasting sedimentary layers. *Journal of Petroleum Science and Engineering 67*, 48–56.

Keehm, Y., T. Mukerji, and A. Nur, (2001), Computational rock physics at the pore scale: transport properties and diagenesis in realistic pore geometries. *The Leading Edge 20*, 180–183.

Kelly, S., El-Sobky, H., Torres-Verdin, C., and M.T. Balhoff, (2016), Assessing the utility of FIB-SEM images for shale digital rock physics, *Advances in Water Resources*, *95*, 302–316, doi:10.1016/j.advwatres.2015.06.010.

Kim, T.H., K.M. Lee, and S.U. Lee, (2013), Learning full pairwise affinities for spectral segmentation, *IEEE Transactions on Pattern Analysis and Machine Intelligence 35*, 1690–1703.

Klaver, J., S. Hemes, M. Houben, G. Desbois, Z. Radi, and J. Urai, (2015), The connectivity of pore space in mudstones: insights from high-pressure Wood's metal injection, BIB-SEMimaging, and mercury intrusion porosimetry. *Geofluids 15*, 577–591.

Kobchenko, M., H. Panahi, F. Renard, D.K. Dysthe, A. Malthe-Sorenssen, A. Mazzini, J. Scheibert, B. Jamtveit, and P. Meakin, (2011), 4D imaging of fracturing in organic-rich shales during heating. *Journal of Geophysical Research 116*, B12201, doi:10.1029/2011JB008565.

Koperna, G., D. Riestenberg, V. Kuuskraa, R. Rhudy, R. Trautz, G. Hill, and R. Esposito, (2012), The SEACARB Anthropogenic Test: a US integrated CO_2 capture, transportation, and storage test. *International Journal of Clean Coal and Energy 1*, 13–26.

Kumar, V., C.H. Sondergeld, and C.S. Rai, (2012), Nano- to macro- mechanical characterization of shale, Society of Petroleum Engineers SPE 159804. Presented at SPE Annual Technical Conference and Exhibition, San Antonio, TX, October 2012.

Lazar, O.R., K.M. Bohacs, J.H.S. Macquaker, J. Schieber, and T.M. Demko, (2015), Capturing key attributes of fine-grained sedimentary rocks in outcrops, cores, and thin sections; nomenclature and description guidelines. *Journal of Sedimentary Research 85*, 230–246.

Lenoir, N., M. Bornert, J. Desrues, P. Bésuelle, and G. Viggiani, (2007), Volumetric digital image correlation applied to X-ray microtomography images from triaxial compression tests on argillaceous rock. *Strain 43*, 193–205

Lewan, M.D., (1978), Laboratory classification of very fine grained sedimentary rocks, *Geology 6*, 745–748.

Lichtman, J.W., H. Pfister, and N. Shavit, (2014), The big data challenges of connectomics. *Nature Neuroscience 17*, 1448–1454.

Los Alamos Science, (2006), LANSCE into the future, Los Alamos Science, Number 30, Los Alamos National Laboratory, Los Alamos, NM.

Loucks, R.G., R.M. Reed, S.C. Ruppel, and D.M. Jarvie, (2009), Morphology, genesis, and distribution of nanometer-scale pores in siliceous mudstones of the Mississippian Barnett shale. *Journal of Sedimentary Research 70*, 848–861.

Louis, L., T.-F. Wong, and P. Baud, (2007), Imaging strain localization by X-ray radiography and digital image correlation: deformation bands in Rothbach sandstone. *Journal of Structural Geology 29*, 129–140.

Lyon, P., P. Boult, R. Hillis, and S. Mildren, (2005), Sealing by shale gouge and subsequent seal breach by reactivation: a case study of the Zema Prospect, Otway Basin. In (P. Boult, Kaldi, J. eds.), *Evaluating Fault and Cap Rock Seals*. AAPG Hedberg Series 2. Tulsa, OK, American Association of Petroleum Geologists, 179–197.

MacLaren, I., and Q.M. Ramasse, (2014), Aberration-corrected scanning transmission electron microscopy for atomic-resolution studies of functional oxides. *International Materials Reviews 59*, 115–131.

Macquaker, J.H.S., and A.E. Adams, (2003), Maximizing information from fine-grained sedimentary rocks: an inclusive nomenclature for mudstones. *Journal of Sedimentary Research 73*, 735–744.

Macquaker, J, K. Taylor, and R. Gawthorpe, (2007), High-resolution facies analyses of mudstones: implications for paleoenvironmental and sequence stratigraphic interpretations of offshore ancient mud-dominated successions. *Journal of Sedimentary Research 77*, 324–339.

Maeder, X., R. Ghisleni, and J. Michler, (2010), Plastic deformation of quartz at room temperature by SEM in situ micropillar compression. *Abstract MR41A-1975 Presented at 2010 Fall Meeting*. American Geophysical Union, San Francisco, CA, 13–17 December.

Mallat, S., (2012), Group invariant scattering. *Communications on Pure and Applied Mathematics 65*, 1331–1398.

Mang, J.T., and R.P. Hjelm, (2013), Fractal networks of intergranular voids in pressed TATB. *Propellants, Explosives, Pyrotechnics 35*, 1–10.

Michael, J.R., C. Nakakura, T. Garbowski, A.L. Eberle, T. Kemen, and D. Zeidler, (2015), *Applications of High-Throughput Multibeam SEM in Materials Science*. SAND2015-6762C, Sandia National Laboratories, Albuquerque, NM.

Mikelic', A., B. Wang, and M. Wheeler, (2014), Numerical convergence study of iterative coupling for coupled flow and geomechanics. *Computational Geosciences 18*, 325–341.

Milliken, K., (2014), A compositional classification for grain assemblages in fine-grained sediments and sedimentary rocks. *Journal of Sedimentary Research 84*, 1185–1199.

Milliken, K.L., (2016), A compositional classification for grain assemblages in fine-grained sediments and sedimentary rocks-reply. *Journal of Sedimentary Research 86*, 6–10.

Milliken, K.L., and M.E. Curtis, (2016), Imaging pores in sedimentary rocks: foundation of porosity prediction: *Marine and Petroleum Geology 73*, 590–608.

Moelans, N., B. Blanpain, and P. Wollants, (2008), An introduction to phase field modeling of microstructure evolution. *Calphad 32*, 268–294.

Na, S.H., W.C. Sun, M.D. Ingraham, and H. Yoon, (2017), Effects of spatial heterogeneity and material anisotropy on the fracture pattern and macroscopic effective toughness of Mancos shale in Brazilian tests, *Journal of Geophysical Research - Solid Earth 122*, doi:10.1002/2016JB013374.

Navarre-Sitchler, A.K., K. Mouzakis, G. Rother, T. Dewers, J. Heath, and J.E. McCray, (2013), Nano-to-micro-meter scale characterization of pore networks in fine-grained rocks using electron microscopy and small angle neutron scattering. *Microscopy and Microanalysis 18*, 1954–1955

Niezgoda, S.R., A.K. Kanjarla, and S.R. Kalidindi, (2013), Novel Microstructure Quantification Framework for Databasing, Visualization, and Analysis of Microstructure Data. *Integrating Materials and Manufacturing Innovation 2*, 1–27.

Nitao, J.J., and J. Bear, (1996), Potentials and their role in transport in porous media. *Water Resources Research 32*, 225–250.

O'Connor, R.M., and J.T. Fredrich, (1999), Microscale flow modeling in geologic materials. *Physics and Chemistry of the Earth 24*, 611–616.

Odusina, E., C.H. Sondergeld, and C.S. Rai, (2011), An NMR study on shale wettability, CSUG/SPE 147371 presented at the Canadian Unconventional Resources Conference held in Calgary, Alberta, Canada, 15–17 November.

Olgaard, D., J. Urai, L. Dell'Angelo, R. Nuesch, and G. Ingram, (1997), The influence of swelling clays on the deformation of mudrocks. *International Journal of Rock Mechanics and Mining Sciences 34*, 3–4.

Panahi, H., M. Kobchenko, F. Renard, A. Mazzini, J. Scheibert, D.K. Dysthe, B. Jamveit, A. Malthe-Sorenssen, and P. Meakin, (2013), A 4D synchrotron X-ray tomography study of the formation of hydrocarbon migration pathways in heated organic-rich shale. *Journal of Society of Petroleum Engineers 18*, 366–377.

Peng, S., J. Yang, X. Xiao, B. Loucks, S.C. Ruppel, and T. Zhang, (2015), An integrated method for upscaling pore-network characterization and permeability estimation: example from the Mississippian Barnett shale. *Transport in Porous Media 109*, 359–376.

Petford, N., and J. Miller, (1990), Scanning laser microscopy: an improved way of viewing fission tracks. *Journal of Geological Society of London 147*, 217–218.

Petford, N., and J. Miller, (1992), Three-dimensional imaging of fission tracts using confocal scanning laser microscopy. *American Minerologist 77*, 529–533.

Petford, N., and Miller, J., (1993), The study of fission tracks and other crystalline defects using confocal scanning laser microscopy. *Journal of Microscopy 170*, 201–212.

Petford, N., G. Davidson, and J. Miller, (1999), Pore structure determination using confocal scanning laser microscopy. *Physics and Chemistry of the Earth 24*, 563–567.

Potter, P.E., J. Maynard, and P.J. Depetris, (2005), *Mud and Mudstones: Introduction and Overview*, Springer, New York. 581p.

Prodanovic, M., M. Esteva, M. Hanlon, G. Nanda, and P. Agarwal, (2015), Digital Rocks Portal: a repository for porous media images, doi:10.17612/P7CC7K.

Remy, N., A. Boucher, and J. Wu, (2009), *Applied Geostatistics with SGeMS*, Cambridge University Press, Cambridge, NY.

Ribes, A., F. Rack, G. Tsintzouras, S. Damaskinos, and A. Dixon, (2006), Applications of confocal macroscope-microscope luminescence imaging to sediment cores. *Geological Society, London, Special Publications 267*, 141–150.

Rodriguez, R., D. Crandall, X. Song, C. Verba, and D. Soeder, (2014), Imaging techniques for analyzing shale pores and minerals. NETL-TRS-6-2014; NETL Technical Report Series; U.S. Department of Energy, National Energy Technology Laboratory: Morgantown, WV, 40 p.

Samudrala, S.K., P.V. Balachandran, J. Zola, K. Rajan, and B. Ganapathysubramanian, (2014), A software framework for data dimensionality reduction: application to chemical crystallography. *Integrating Materials and Manufacturing Innovation 3*, 17.

Sanchez, M., A. Gens, and S. Olivella, (2008), Implementation algorithm of a generalized plasticity model for swelling clays. *Computers and Geotechnics*, doi:10.1016/j.compgeo.2008.08.004.

Sanchez, M., A. Gens, and S. Olivella, (2011), Thermo-hydro-mechanical analysis of a large scale heating test incorporating material fabric changes. *International Journal for Numerical and Analytical Methods in Geomechanics*, doi:10.1002/nag.1011.

Saraji, S., and M. Piri, (2015), The representative sample size in shale oil rocks and nano-scale characterization of transport properties, *International Journal of Coal Geology 146*, 42–54.

Schaef, H.T., E.S. Ilton, O. Qafoku, P.F. Martin, A.R. Felmy, and K.M. Rosso, (2012), In situ XRD study of Ca^{2+} saturated montmorillonite (STX-1) exposed to anhydrous and wet supercritical carbon dioxide. *International Journal of Greenhouse Gas Control 6*, 220–229.

Schanz, T., and S. Tripathy, (2009), Swelling pressure of a divalent-rich bentonite: diffuse double-layer theory revisited, *Water Resources Research 45*, doi:10.1029/2007WR006495.

Scheiber, J., (2013), SEM observations on ion-milled samples of Devonian black Shales from Indiana and New York. In (Camp, W., Diaz, E., and Wawak, eds.) *Electron Microscopy of Shale Hydrocarbon Reservoirs*, AAPG Memoir 102. Tulsa, OK, American Association of Petroleum Geologists, 153–172.

Schieber, J., and W. Zimmerle, (1998), The history and promise of shale research. In (J. Schieber, W. Zimmerle, and P. Sethi, eds.), *Shales and Mudstones (vol. 1): Basin Studies, Sedimentology and Paleontology*, Stuttgart, Schweizerbart'sche Verlagsbuchhandlung, 1–10.

Shepard, F.P., (1954), Nomenclature based on sand-silt-clay ratios. *Journal of Sedimentary Petrology 24*, 151–158.

Shukla, P., V. Kumar, M. Curtis, C. Sondergeld, and C.S. Rai, (2013), Nanoindentation studies in shales, American rock mechanics association ARMA 13-578. Presented at the 47th US Rock Mechanics/Geomechanics Symposium, San Francisco, CA, June 2013.

Slatt, R.M., and Y. Abousleiman, (2011), Merging sequence stratigraphy and geomechanics for unconventional gas shales. *The Leading Edge 30*, 274–282.

Slatt, R.M., P. Philp, Y. Abousleiman, P. Singh, R. Perez, R. Portas, K.J. Marfurt, S. Madrid-Arroyo, N. O'Brien, E. Eslinger, and E.T. Baruch, (2011), Pore-to-regional-scale integrated characterization workflow for unconventional gas shales. In (J. Breyer, ed.) *Shale Reservoirs: Giant Resources for the 21st Century*, AAPG Memoir 97. Tulsa, OK, American Association of Petroleum Geologists, 127–150.

Spain, D.R., and R. McLin, (2013), SEM characterization of shale gas reservoirs using combined secondary and backscatter electron methods: an example from the Haynesville shale, Texas and Louisiana. In (Camp, W., E. Diaz, and B.

Wawak, eds.) *Electron Microscopy of Shale Hydrocarbon Reservoirs*, AAPG Memoir 102. Tulsa, OK, American Association of Petroleum Geologists, 45–52.

Spivak, D.I., and R.E. Kent, (2012), Ologs: a categorical framework for knowledge representation. *Plos One 7*, doi:10.1371/journal.pone.0024274.

Stasiuk, L.D., (1999), Confocal laser scanning fluorescence microscopy of Botryococcus alginite from boghead oil shale, Boltysk, Ukraine: selective preservation of various microalgal components. *Organic Geochemistry 30*, 1021–1026.

Sulucarnain, I., C.H. Sondergeld, and C.S. Rai, (2012), An NMR study of shale wettability and effective surface relaxivity. *Canadian Unconventional Resource Conference*, SPE 162236, 30 October–1 November.

Susan, D., Z. Ghanbari, P.G. Kotlua, J.R. Michael, and M.A. Rodriguez, (2014), Characterization of continuous and discontinuous precipitation phases in Pd-rich precious metal alloys. *Metallurgical and Materials Transactions 45A*, 3755–3766.

Sutton, M.A., J-J. Orteu, and H.W. Schreier, (2009), *Image Correlation for Shape, Motion and Deformation Measurements: Basic Concepts, Theory, and Applications*. Springer, New York. 320p.

Tinni, A., E. Odusina, I. Sulucarnian, C. Sondergeld, and C. Rai, (2014), Nuclear-magnetic-resonance response of brine, oil and methane in organic rich shales. *SPE Reservoir Evaluation & Engineering 18* (3), 400–406.

Tiwari, P., Deo, M., Lin, C.L., and Miller, J.D., (2013), Characterization of oil shale pore structure before and after pyrolysis by using X-ray micro CT. *Fuel 107*, 547–554.

Uchic, M.D., D.M. Dimiduk, J.N. Florando, and W.D. Nix, (2004), Sample dimensions influence strength and crystal plasticity. *Science 305*, 986–988.

Ulmer-Scholle, D., P.A. Scholle, J. Schieber, and R.J. Raine, (2014), *A Color Guide to the Petrography of Sandstones, Siltstones, Shales and Associated Rocks*, AAPG Memoir 109. Tulsa, OK, American Association of Petroleum Geologists, 509p.

Vega, B., C.M. Ross, and A.R. Kovscek, (2015), Imaging-based characterization of calcite-filled fractures and porosity in shales. *SPE Journal 20*, 810–823.

Walls, J.D., and S.W. Sinclair, (2011), Eagle ford shale reservoir properties from digital rock physics. *European Association of Geoscientists and Engineers 29*, 97–101.

Wildenschild, D., and A.P. Sheppard, (2013), X-ray imaging and analysis techniques for quantifying pore-scale structure and processes in subsurface porous medium systems, *Advances in Water Resources 51*, 217–246.

Wong, J.Y., J. McDonald, M. Taylor-Pinney, D.I. Spivak, D.L. Kaplan, and M.J. Buehler, (2012), Materials by design: merging proteins and music. *Nano Today 7*, 488–495.

Yin, S., M. Dusseault, and L. Rothenburg, (2011), Coupled THMC modeling of CO_2 injection by finite element methods. *Journal of Petroleum Science and Engineering 80*, 53–60.

Yoon, H., and T. Dewers, (2013), Nanopore structures, statistically representative elementary volumes, and transport properties of chalk, *Geophysical Research Letters 40*, 4294–4298.

4

Analysis of the Pore Structures of Shale Using Neutron and X-Ray Small Angle Scattering

Lawrence M. Anovitz[1] and David R. Cole[2]

ABSTRACT

Analysis of the multi-scale pore structures of shale is of critical importance to understanding their role as cap-rocks and oil/gas reservoirs, as well as their contribution to global elemental cycles because of weathering. While several techniques have been employed to understand these structures, small angle neutron and X-ray scattering, which can quantify several features of the pore structure from the nanometer scale to tens of microns, provides a unique perspective not found in more standard petrophysical approaches. Although large-scale regional studies using this approach are not yet available, over 30 years of work have led to a number of important conclusions. Among the results obtained are data suggesting that multi-scale pores structures, even of less fissile shales, are highly asymmetric. However, this is both scale and maturity dependent. Pore surfaces commonly exhibit highly rough surface fractals, which often smooth with diagenetic alteration and weathering. Mass fractal behavior of larger-scale interpore structures has also been observed. In many cases, pore structures are bimodal, reflecting interlayer spacing in clays at the smaller scales and larger spacing between clay packets or between clay folia and detrital grains at larger scales. Non-clay porosity may also be observed, including significant nanoporosity associated with organic matter (OM). Data obtained as a function of thermal maturity suggest that while the overall porosity does decrease with increasing maturity, the change is not linear with R_o (percent vitrinite reflectance), but occurs primarily in the early stages of maturation. Pore size or volume distributions obtained from scattering experiments often differ considerably from those obtained from nitrogen or carbon dioxide absorption or mercury intrusion porosimetry, which may reflect the overall accessibility of various pore sizes to external fluids. Despite these results, however, a significant amount of work and a number of basic questions remain to be answered that can be addressed either uniquely by scattering methods or in a complementary fashion using these and other approaches.

4.1. INTRODUCTION

From a global perspective, shales are an extremely important rock type. Weathering of shale plays an important role in landscape formation and global element cycles, and shale can form seals or caprocks for deep geologic reservoirs exploited for oil and gas or for formations that could store anthropogenic CO_2. More recently, with the advent of advanced drilling technologies, vast reserves of oil and gas have been realized in tight formations including shale, which has transformed the international energy production landscape [*Arthur and Cole*, 2014].

Shale is a fine-grained, clastic sedimentary rock. It is formed from fine-grained mud and is compositionally, therefore, best referred to as a mudstone. It is composed of a mixture of silt-sized flakes of clays and other minerals, most commonly quartz and calcite, as well as variable

[1] *Chemical Sciences Division, Oak Ridge National Laboratory, Oak Ridge, TN, USA*

[2] *School of Earth Sciences, The Ohio State University, Columbus, OH, USA*

Geological Carbon Storage: Subsurface Seals and Caprock Integrity, Geophysical Monograph 238, First Edition.
Edited by Stéphanie Vialle, Jonathan Ajo-Franklin, and J. William Carey.

amounts of organic matter (OM). The ratios between these components may be quite variable, and some materials referred to as shales may be more accurately described as marls. Texturally, shales are characterized by parallel layering or bedding caused by alignment of the clays and variable levels of fissile parting [*Blatt and Tracy*, 1996].

As noted by *Anovitz et al.* [2015a], several interrelated properties control the extraction of resources from the Earth's crust and the economics of the process including temperature gradients, permeability, physical rock properties, seismicity, local stress regime, and indigenous stored fluids. Many of these factors are, in turn, controlled by the porosity and pore structure of the rock in question. While such porosity is commonly highly multi-scalar, for shales in particular, the nanoscale (≤100 nm) pore structure is both highly important and poorly understood.

Caprocks are overlying strata composed of relatively impermeable rocks that restrict the buoyant migration of lower-density fluids to the surface (Fig. 4.1). Understanding the spatial extent and integrity of caprocks is clearly essential for determining reservoir storage capacities [*Bachu et al.*, 2007]. Common caprock lithologies are shales and evaporites with thicknesses from tens to hundreds of meters [*Grunau*, 1987]. The sealing properties of caprocks are derived in large part from their clay content and sedimentological setting [*IPCC*, 2005; *Benson and Cole*, 2008]. The main clay components identified in shales overlying some CO_2 sequestration sites (Sleipner, Cranfield, Muderong, Krechba, and SMB-Paris Basin) are illite, smectite, and kaolinite [*Hildenbrand et al.*, 2004; *Gaus et al.*, 2005; *Busch et al.*, 2008; *Kohler et al.*, 2009; *Armitage et al.*, 2010; *Lu et al.*, 2012], but it is the depositional history and subsequent diagenetic evolution of these materials during burial that determine the effectiveness of the seal. The integrity of caprock sealing can be degraded by geochemical and geomechanical processes, depending on the caprock formation and imposed chemical perturbations and physical stresses [*Li et al.*, 2005; *Rutqvist et al.*, 2007; *Wollenweber et al.*, 2009; *Ellis et al.*, 2011].

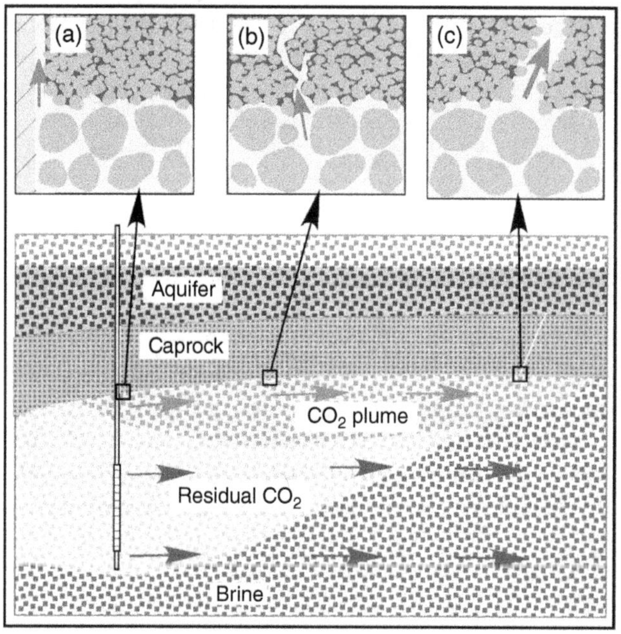

Figure 4.1 Possible advective CO_2 pathways through caprocks. (a) Wellbore seal defects. (b) Capillary breakthrough. (c) Faults and fractures [*T. K. Tokunaga*, pers. comm.].

Given the large scale of heterogeneous reservoir structures [*IPCC*, 2005; *Bachu et al.*, 2007], numerical modeling has become the primary approach for investigating various sequestration scenarios, although analytical solutions remain useful for simpler systems. To date, many field-scale simulations have modeled caprocks as perfect (impermeable) seals [*Doughty*, 2010; *Juanes et al.*, 2010; *Liu and Zhang*, 2011]. Others have assumed that hydraulic relations developed originally to describe unsaturated flow of water through soils can be applied to variably saturated flow of brine-scCO_2 [*Gherardi et al.*, 2007; *Alkan et al.*, 2010; *Bildstein et al.*, 2010; *Vilarrasa et al.*, 2011]. Faults, fractures, and wellbores have been treated as idealized discrete leakage pathways [*Suekane et al.*, 2005; *Nordbotten et al.*, 2009].

While the key properties of caprocks, permeability and capillary breakthrough pressure, are macroscopic, many of the underlying processes determining these properties are nanoscale because pore sizes in these systems are generally in the nanometer range. In addition, spatially rare but very important exceptions to the dominance of nanoscale controls in caprocks are found in leakage through highly transmissive macroscopic fractures and faults and through crack defects in wellbore seals. Even in these more macroscopic defects, however, interactions between caprocks and leaking fluids may ultimately involve transport and reactions at solid-fluid interfaces. Thus, studies of the nanoscale properties of caprocks have the potential to improve substantially our understanding of the integrity of shales as seals.

Shales have also become increasingly important because of their role in a new source of hydrocarbons, tight oil/gas formations. The USDOE notes that "Natural gas production from hydrocarbon rich shale formations, known as 'shale gas' is one of the most rapidly expanding trends in onshore domestic oil and gas exploration and production today" [*USDOE*, 2009]. US reserves are very large (2009 est. 1744 tcf, 2013 est. 1864 tcf), and "total annual production volumes of 3–4 tcf may be sustainable for decades" [*USDOE*, 2009, 2013]. Gas shale formations are located in a number of the lower 48 states. As of 2009 [*USDOE*, 2009], the most active in terms of exploration were the Barnett Shale (Texas), the Haynesville/Bossier

Shale (Texas and Louisiana), the Antrim Shale (Michigan), the Fayetteville Shale (Arkansas), the Marcellus Shale (Pennsylvania, New York, Ohio, Virginia), and the New Albany Shale (Illinois). Significant exploration is also ongoing in the Eagle Ford Shale (Texas) (both oil and natural gas), the Utica/Pt. Pleasant (Ohio), and the Niobrara (Kansas, Nebraska, and Colorado), among others. By 2013, most were producing significant gas (the Barnett had generated 14 tcf), and the list included the Woodford Shale (Oklahoma), Pearsall Shale (Texas), Mancos Shale (several Rocky Mountain basins), and Lewis Shale (Rocky Mountains as well) [*USDOE*, 2013]. It has also been suggested that depleted shale gas formations may make suitable locations for CO_2 sequestration [*Godec et al.*, 2013; *Liu et al.*, 2013; *Tao and Clarens*, 2013; *Edwards et al.*, 2015]. While tight gas shales may contain enormous energy resources (30–300 tcf recoverable in the Marcellus; *Lee et al.*, 2014), the rate of gas flow through natural porosity is almost negligible. As this resource cannot be utilized until it is extracted, therefore, a critical factor in making it economically viable has been advances in hydraulic fracturing. As rocks with at least some connected permeability through fractures or pore spaces are more likely to result in a connected circulation system after stimulation, analysis of pore structure, which plays an important role in controlling permeability, rock physical properties (e.g., fracture mechanics), the stress regime, the presence of other indigenous fluids, and susceptibility to seismicity [e.g., *Ingebritsen and Sanford*, 1999; *Ague*, 2004], is critical. These factors, taken together, not only control the physical processes of extracting gas but also play a major role in determining the economics of energy production and are critical to understanding both the storage and recovery of this important resource.

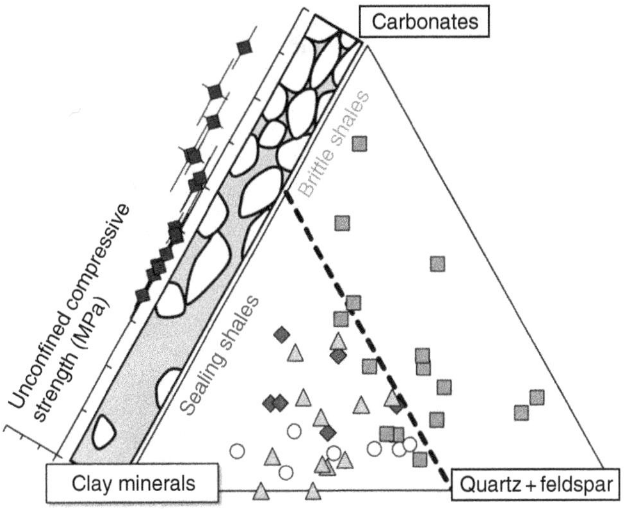

Figure 4.2 Ternary diagram showing the three main types of minerals present in shale: phyllosilicates (clay minerals), tectosilicates (quartz and feldspar), and other minerals (primarily carbonates). Only well-characterized shales and mudstones for which data on rock strengths were available were included. Symbols indicate the current or proposed uses of each shale: square, hydrocarbon extraction; triangle, carbon capture and storage; diamond, radioactive waste storage; circle, other. The dashed line indicates rocks with X_{clay} = 1/3. This threshold coincides with the predictions of the simple conceptual model shown to the left of the microstructure of sedimentary rocks and is reflected in the uses of shale and mudstone formations for CCS, radioactive waste storage, and shale gas extraction. Unconfined compressive strengths range from near zero to approximately 200 MPa Reprinted with permission from *Bourg* [2015]. Copyright 2015 American Chemical Society.

4.2. SHALE CHARACTERISTICS RELEVANT TO SEAL INTEGRITY

In general, shales are fine-grained sediments, typically with mean grain diameters <4 μm. This classification, however, groups together rocks with a wide variety of microstructural, morphological, chemical, mineralogical, and physical properties [*Aplin et al.*, 1999]. *Bourg* [2015] showed that if the mineralogy of shales is plotted in a ternary diagram (Fig. 4.2) representing their clay, carbonate, and quartz-feldspar contents, shales can be separated into brittle (clay-poor) and sealing (clay-rich) types. He suggested that the separation between these fields is controlled by whether the larger quartz, feldspar and carbonate grains, or the clay matrix are load bearing, which is, in turn, controlled by the clay content.

Critically, shale micro-architectures are intrinsically multi-scalar with structures and physical properties stemming from organization of clay minerals (10^{-9} m), porous composites of clay minerals (10^{-7}–10^{-6} m), porous-clay-silt-sized inclusion composites (e.g., detrital grains) and microfractures (10^{-5}–10^{-4} m), and larger-scale fractures and bedding features (> 10^{-3} m; *Ortega et al.*, 2009). For compacted clays and shales, high-pressure mercury injection capillary pressure (MICP) porosimetry suggests that most porosity consists of pores finer than 100 nm [*Katsube et al.*, 1992]. Newer 3D imaging approaches with resolutions closer to the nanometer scale (e.g., focused ion beam-scanning electron microscopy (FIB-SEM), transmission electron microscopy (TEM) tomography, and nano-X-ray computed tomography (CT)) have led to more detailed characterization of the morphology, connectivity, and phase relationships of shale mesopores [*Loucks et al.*, 2009, 2012], as shown in Figure 4.3. As we will show in more detail, a similar expansion of our understanding of the pore structure of shales has been provided by small angle scattering (SAS) techniques [cf., *Anovitz and Cole*, 2015].

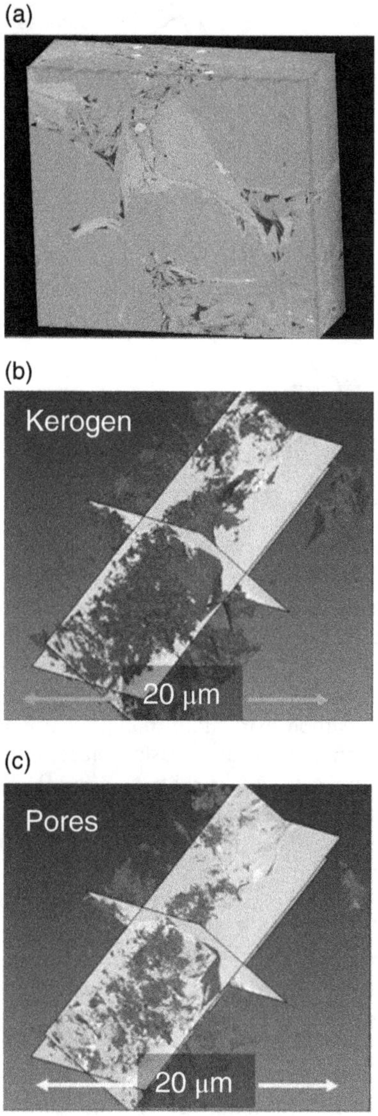

Figure 4.3 (a) Dual-beam FIB-SEM reconstructed volume of a Utica shale sample [*Anovitz and Cole*, 2015]. (b) 3D reconstructed images of kerogen (red) and (c) pore (blue) distributions. The 3D reconstructed volumes have dimensions of 20 × 17 × 6.5 mm [*Arthur and Cole*, 2014; *Anovitz and Cole*, 2015]. (*See electronic version for color representation of the figure.*)

One relevant example is the marine middle Tuscaloosa mudstone (ca. 3100 m depth) that forms the primary confining unit at the Cranfield CO_2 sequestration pilot site [Cranfield, MS; *Lu et al.*, 2012]. This formation is located directly above the lower Tuscaloosa Formation into which over one million tons of $scCO_2$ was injected over a roughly 2-year period. Powder XRD reveals significant kaolinite and illite, with the remainder composed largely of quartz and K-feldspar, although the formation is quite heterogeneous with eight microfacies over the interval examined. Permeability was below 1 microdarcy ($< 10^{-18}$ m^2) for all samples analyzed, and porosity measured by MICP was <4%. Analysis of pore size distributions (PSDs) by FIB-SEM and MICP revealed modal pore sizes <20 nm with estimated sealing capacities of between 49 and 237 m of CO_2 (column height) at 100% water saturation. Within the middle Tuscaloosa, some regions with high calcite fractions showed very low porosity and permeability values despite containing relatively coarse silt fractions. These regions would likely be susceptible to reactive permeability enhancement. *Lu et al.* [2012] concluded that the middle Tuscaloosa was an excellent seal, with the major source of uncertainty in storage reliability being fault transmissibility.

Although shale permeabilities are on the low end of the permeability spectrum, they cover a wide range of values. *Neuzil* [1994] reported porosities ranging from 10 to 40% in well-characterized samples with single-phase permeabilities ranging over six orders of magnitude from 10^{-17} to 10^{-23} m^2 (from hundreds of microdarcies to sub-nanodarcy). *Hildenbrand et al.* [2004] reported permeabilities of various finer-grain sedimentary rocks that ranged from 6×10^{-22} m^2 to 5.5×10^{-19} m^2 and porosities ranging from 0.7 to 30%. The minimum and average pore radii estimated from nitrogen gas breakthrough tests were 8 and 42 nm, respectively. In many shales, permeability seems to be controlled primarily by grain size distribution on the core scale [*Dewhurst et al.*, 1999], with secondary control by fabric alignment, compaction history, and diagenetic overprint. Microfractures can generate permeability anisotropy ratios >10, while shales of the same porosity can exhibit permeabilities differing by up to three orders of magnitude [*Dewhurst et al.*, 1999; *Bolton et al.*, 2000]. *Bourg et al.* [2015] compiled experimental data and model predictions of the core-scale permeability and porosity of shales and mudstones and found that reported porosity values ranged from ~ 5 to 40% and permeabilities from ~10^{-16} to 10^{-21} m^2. They found that for weakly consolidated sediments with porosities greater than about 20%, permeability roughly decreased with decreasing porosity (ϕ) as a function of ϕ^{10}. They interpreted this as consistent with the bimodal distribution of pores in many shales [cf., *Swift et al.*, 2014a; *Anovitz et al.*, 2015a] with larger, hydrodynamically active pores compressing before nanopores. At porosities from 10 to 20%, permeability was essentially independent of porosity, suggesting the disappearance of hydrodynamically active pores. They noted that this is consistent with the concept of a critical depth of burial (~ 2.5–3.2 km), below which the permeability of clay-rich rocks becomes approximately constant [*Best and Katsube*, 1995]. Below 10% porosity, however, the data of *Bourg et al.* [2015] suggest that permeability may again increase.

A major challenge to using these data to study mass transfer in shales is that it is strongly coupled with mineral-water interfacial phenomena because of the significant contribution of nanometer-scale pores to overall porosity [*Kim et al.*, 1999; *Loucks et al.*, 2009; *Anovitz et al.*, 2015a]. These couplings remain poorly characterized, except for a few studies of clayey shales [*Fritz*, 1986; *Sherwood*, 1995; *Jougnot et al.*, 2009], but they have been extensively studied for engineered clay barriers, where the nanopore and mesopore space may be similar [*Jullien et al.*, 2005]. For example, ion adsorption in clay interlayer nanopores is known to influence strongly osmotic effects [*Shackelford and Lee*, 2003], swelling mechanics [*Gajo and Loret*, 2004], and ion diffusion coefficients in engineered clay barriers [*Bourg and Sposito*, 2010; *Bourg and Tournassat*, 2015]. Water confined in smectite interlayer nanopores behaves as a *subaqueous phase* that has a greater affinity for hydrophobic solutes than bulk liquid water [*Teppen and Aggarwal*, 2007]. Among several unexpected phenomena suggested by these previous studies are (i) a possible enhancement of CO_2 solubility in water-filled nanopores relative to bulk liquid water and (ii) a possible influence of the osmotic pressure of caprock pore water on the CO_2 capillary entry pressure and fracturing pressure of caprocks.

The importance of pore structure from the nanoscale up to our understanding of shale permeability is undeniable, and our ability to quantify porosity in such complex heterogeneous matrices has evolved from the application of standard petrophysical methods to the use of sophisticated techniques such as X-ray and neutron scattering [*Anovitz and Cole*, 2015]. However, despite the availability of these methods and the importance of shale as a key lithology, only a limited body of scattering data (neutron, X-ray, or light) is available. Early work [*Hall and Mildner*, 1983; *Hall et al.*, 1986; *Mildner et al.*, 1986; *Wong et al.*, 1986; *North et al.*, 1988, 1990] showed that the pore surfaces in shales had a fractal character due to the presence of clay and that scattering was asymmetrical for samples cut perpendicular to bedding. *Knudsen et al.* [2003, 2004] studied the effects of pressure on the pore structure of a synthetic smectite (fluorohectorite) and showed that the asymmetric scattering patterns were due to platelet orientation in the direction of sedimentation and that absorption of water consisted of both intercalation of water molecules between platelets and filling of mesopores. *Radlinski et al.* [1996] showed data on maturing oil shales that *Anovitz et al.* [2015a] interpreted as indicative of the presence of non-fractal *fuzzy* boundaries [*Ruland*, 1971; *Schmidt*, 1991; *Schmidt et al.*, 1991; *Kim*, 2004]. Such boundaries are formed when the scattering length density (SLD) changes gradually, rather than abruptly, at the mineral/pore boundary, which can occur when pore surfaces become coated with another substance. *Itakura et al.* [2005] examined kaolinitic clays on the nanoscale and their response to uniaxial compression using small and ultra small angle neutron scattering ((U)SANS). Their data show alignment of clay particles and compaction. Pressures of 400 kPa led to significant sample restructuring. Increase of pressure to 800 kPa did not change this structure much. *De Stefanis et al.* [2007] used small angle neutron scattering (SANS) and sorption to study smectite clays and measure surface roughness and fractal dimensions. *Jin et al.* [2011] investigated the effect of weathering on nanoporosity development in Rose Hill shale from the Susquehanna/Shale Hills Observatory (SSHO) using ultra small angle neutron scattering (USANS). They found that the initial shale had 5–6% porosity once the carbonate in deeper material dissolved. The onset of feldspar dissolution increased both porosity and surface area, and these increased again near the surface with the initiation of chlorite and illite dissolution. Pore geometries also changed during weathering. A number of other studies [e.g., *Ma et al.*, 1999a, 1999b; *Qi et al.*, 2002; *Clarkson et al.*, 2012, 2013; *Mastalerz et al.*, 2012, 2013; *Jin et al.*, 2013; *Ruppert et al.*, 2013; *Anovitz et al.*, 2014, 2015a; *Bahadur et al.*, 2014, 2015; *Lee et al.*, 2014; *Swift et al.*, 2014a, 2014b; *Gu et al.*, 2015] have also used SAS to describe the pore structures of shales. The goal of this paper is to describe this work, to summarize the basic conclusions, and to suggest where further studies are needed.

4.3. SCATTERING METHODS FOR ANALYSIS OF PORE STRUCTURES

While many techniques have been used to examine and quantify the pore structures of shales, it is not the purpose of this paper to review them all [cf., *Anovitz and Cole*, 2015]. Instead, this paper will only review those using neutron and X-ray scattering. Where such data were obtained on identical samples, however, we will also compare SAS results to those obtained by other techniques. Although SAS techniques have been used to study the pore structures of shales since the inaugural work of *Hall et al.* [1983], these approaches (which include X-ray, neutron, and light) are, perhaps, less well known to the geological community than more classical petrophysical approaches. This paper, therefore, hopes to summarize the results obtained in this manner, as well as point out some important future research directions. We will provide a short description of the general technique (following *Anovitz et al.*, 2011), but the reader is directed to the recent review by *Anovitz and Cole* [2015] for a more thorough coverage, including descriptions of instrumentation, sample preparation, and data analysis.

The pore structure of geological materials is an inherently multi-scale system. However, limited attention has been paid to the texture and overall volume of pores at the submicron scale because they have been difficult to characterize in a statistically meaningful manner. This is due to the wide variation in pore features (i.e., length scales, pore shapes, connectivity, etc.) involved. Even though electron microscopy can provide detailed two-dimensional (2D) nanoscale images, and this can be extended to 3D using focused ion beam techniques [*Goldstein et al.*, 2003; *Heath et al.*, 2011; *Zhang and Klimentidis*, 2011; *Curtis et al.*, 2012; *Landrot et al.*, 2012; *Silin and Kneafsey*, 2012; *Huang et al.*, 2013; *Arthur and Cole*, 2014; *Anovitz and Cole*, 2015], the high magnifications mean the total volume imaged is quite small. In fact, *Howard and Reed* [2005] calculated that all the material that had ever been studied in all the transmission electron microscopes in the world amounts to a total of $<1\,cm^3$ in volume. Although the statistical utility of high-resolution imaging is limited, a combination of small and ultra small angle neutron or X-ray scattering (SANS and USANS or SAXS and USAXS) is well suited to a statistical characterization of pores from the nanoscale to the microscale and, with the addition of imaging techniques, enables a comprehensive structural characterization from nm to cm scales.

The intensity of scattering is a function of the volume of scatterers (pores) in a sample and the contrast between scatterers and their matrix. Scattering contrast in rock samples arises primarily from the difference in the coherent SLDs between the rock and the pores within it, because the SLDs of different minerals are often similar, making their individual scattering contributions negligible. Thus, scattering analysis of a rock sample provides a direct characterization of the pore structure. For X-rays, the SLD is a function of the average atomic number of the scatterer. For neutrons, which primarily scatter from the nucleus rather than the electron cloud, this is not the case (Fig. 4.4). Thus, the two approaches can be used in a complementary manner to better understand pore structure. Unlike with X-rays, the total neutron scattering cross section for hydrogen is large, allowing studies of water-rock interactions at nanoscales to microscales (and at various timescales with inelastic/quasielastic scattering techniques). In addition, neutron beam cross sections are typically several cm^2, thus providing averaged results from a relatively large rock volumes (typically approximately $30\,mm^3$; *Anovitz et al.*, 2009, 2013b). Small angle X-ray systems, on the other hand, often have much smaller beam sizes and higher intensity and can be used to map small-scale changes within a sample [*Cheshire et al.*, 2017].

Because both neutron and X-ray beams are relatively penetrating and access all interfaces within a given sample (see *Anovitz et al.* [2009] and *Anovitz and Cole* [2015] for a description of sample sizes and preparation), scattering sees both pores that are connected to the overall network and those that are not. Because the neutron SLD is a nuclear characteristic, it depends on the isotope, as well as the atomic number of a given atom. It is quite helpful, therefore, that the SLD of H_2O (-0.561×10^{-6} Å^2) is significantly different from that of D_2O (6.335×10^{-6} Å^2). These values span the range of SDL values for many minerals (quartz has an SLD of 4.18×10^{-6} Å^2). If a rock is saturated with a H_2O/D_2O (or CH_4/CD_4) mixture whose SLD has been matched to that of the surrounding matrix scattering from connected porosity disappears, leaving only that from unconnected porosity. This allows neutron scattering to differentiate between connected and unconnected porosity without significantly changing the chemical properties of the pore fluid.

Figure 4.5 shows a schematic scattering curve plotting the log of the scattering intensity (I, units of cm^{-1}) as a function of the log of the scattering vector (q, units of

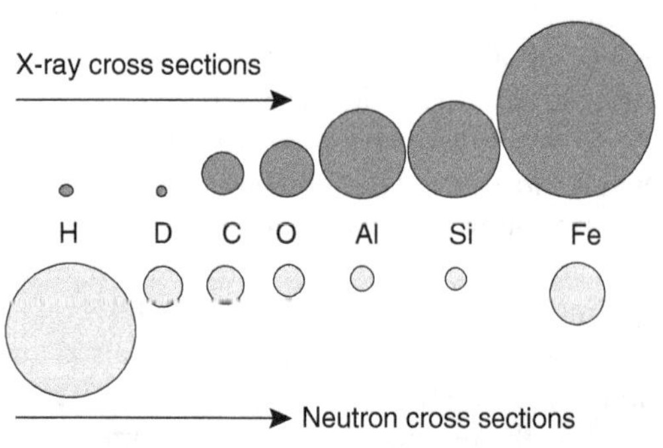

Figure 4.4 Comparative scattering cross sections of X-rays and neutrons for several common elements [*Anovitz et al.*, 2011].

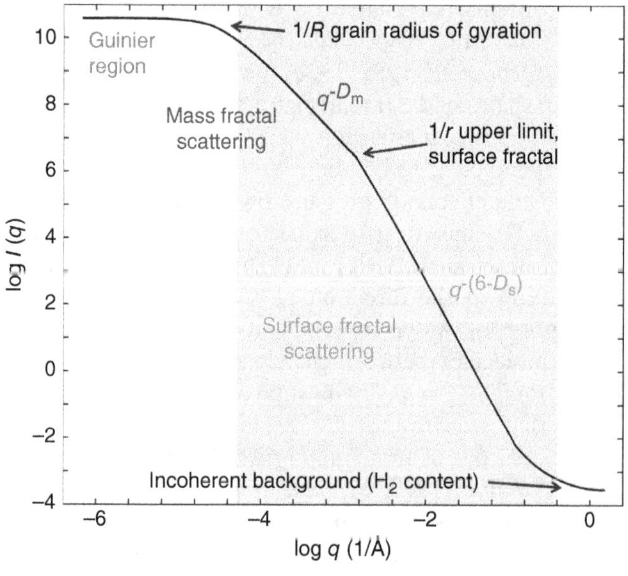

Figure 4.5 Schematic of typical scattering curve. The gray area shows the q range typically reached by (U)SANS measurements. From *Swift et al.* [2014a], AAPG© [2014], reprinted by permission of the AAPG whose permission is required for further use.

Å^{-1}). In many (U)SANS experiments scattering intensity is constant at high q. In this range, coherent (structural) scattering is overwhelmed by a q-independent incoherent background, which is primarily a function of the hydrogen content of the sample. The slopes of the remaining parts of the SAS curve are determined by the surface from which scattering occurs. Slopes between −2 and −3 are characteristic of mass fractal systems (in which density scales with total volume), those between −3 and −4 are of surface fractal systems (where surface area scales with ruler length), and those between −4 and −5 are of non-fractal *fuzzy* interfaces in which the SLD varies monotonically between two phases. Smooth interfaces give rise to scattering with a power law slope of −4, referred to as Porod scattering. In some samples, there may be a point, known as the surface fractal correlation length r, which forms the upper scaling limit of surface fractal behavior. At q values below this point (larger scales), the scaling exponent is dominated by mass fractal behavior, with a dimension (D_m) equal to the slope, which describes the pore distribution within the sample. At yet lower q values corresponding to length scales greater than the largest aggregates in a sample, the slope of I versus q should flatten. This *Guinier region* is seldom observed in rock samples, implying that the maximum size of features covered by the (U)SANS measurements is smaller than the largest grain size of the rock. From Bragg's law, one can derive that (pore diameter) = $2\pi/q$. *Radlinski et al.* [2000b] and *Clarkson et al.* [2013] suggest, however, that for polydisperse porous media, a more appropriate relationship may be $r \sim 2.5/q$ ($d = 5/q$), where r is the pore radius.

Figure 4.6 shows a schematic illustration of what the pore structure of a shale is assumed to look like [*Gu et al.*, 2015]. Several different types of pores occur in shales [*Loucks et al.*, 2012], and features of these pores appear in the scattering data [*Anovitz et al.*, 2015; *Gu et al.*, 2015]. *Loucks et al.* [2012] described the pores within mudrock as falling within three structural types: (i) pores within organic matter (OM pores), (ii) pores within crystalline particles (intraP pores), and (iii) pores between grains (interP pores). *Gu et al.* [2015] suggested that "As a first approximation, the sizes of the pores decrease in the order interP pores > OM pores ≈ intraP pores. Compaction is the most significant process in controlling the interP pore size and shape. During compaction, interP pores around rigid grains (e.g., quartz, calcite, and pyrite) become elongated but remain relatively randomly distributed, while interP pores between soft and ductile grains (e.g., clay, kerogen) are elongated roughly parallel to the bedding. The shapes of intraP pores in mechanically strong minerals such as quartz are mainly controlled by their genesis and are less affected by compaction. For instance, intraP pores in pyrite framboids reflect the structure of the pyrite crystals within the framboid, while intraP pores in clays are sheetlike and roughly parallel to the bedding. Pores in kerogen are bubblelike or irregular in shape, generating symmetric scattering. Overall, the asymmetric scattering in perpendicular bedding results largely from the pores related to clay minerals." This asymmetry leads to a distinctive asymmetry in the 2D scattering data for samples cut perpendicular to bedding. As might be expected, this is not generally observed for samples cut parallel to bedding, suggesting that the 3D scattering curve is an ellipsoid of rotation. In this case, all the data obtainable from two sections, one cut parallel and one perpendicular to bedding, should be obtainable from sector analysis along the major and minor axes of the ellipsoidal data obtained from the perpendicularly cut sections.

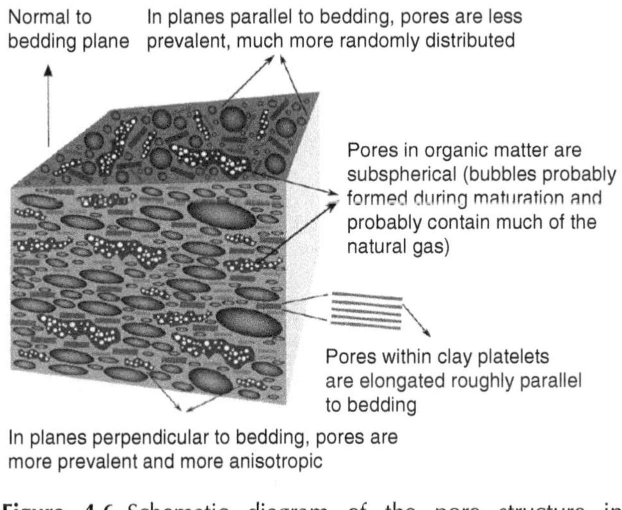

Figure 4.6 Schematic diagram of the pore structure in Marcellus shale. Reprinted with permission from *Gu et al.* [2015]. Copyright 2015 American Chemical Society.

It would clearly also be useful to obtain PSDs from scattering data. With certain limiting assumptions, several authors have suggested methods of doing so. For example, *Hinde* [2004] and *Radlinski and coworkers* [2004] developed the polydisperse spherical pore (PDSP) model in which fractal pore shapes are treated as an assemblage of spherical pores. Maximum entropy techniques [e.g., *Potton et al.*, 1988a, 1988b; *Morrison et al.*, 1992; *Jemian et al.*, 1999], Fourier calculations [e.g., *Muller et al.*, 1995], total nonnegative least squares (TNNLS) [*Merrit and Zhang*, 2004], Monte Carlo approaches [e.g., *Martelli and Di Nunzio*, 2002], and regularization [cf., *Ilavsky and Jemian*, 2009] have also been applied. Many of these approaches, however, require assumptions about pore shapes, an issue made more complicated by the heterogeneous, commonly fractal shapes of rock pores.

4.4. PUBLISHED WORK

4.4.1. Early Work

To our knowledge, the first use of SAS to analyze the pore structure of any rock material was the study of several shales, including the Woodford, Burbank, and New Scotland shales, New York red slate, East Kentucky and Texas oil shales, and Utah mudstone by *Hall et al.* [1983]. They introduced the method as an alternative to more traditional techniques such as TEM, gas sorption, and mercury intrusion porosimetry and observed many of the features discussed in greater detail in later studies. Both powdered and sliced samples were examined, the latter used to observe the effects of anisotropy caused by bedding. Both SANS and SAXS measurements were performed, the latter on powdered samples. They observed a broad distribution of pores from 750 to at least 20 Å and demonstrated that samples cut perpendicular to bedding showed ellipse-like anisotropic scattering (Fig. 4.7), whereas those cut parallel did not. They found that the Debye model [*Debye and Beuche*, 1949; *Debye et al.*, 1957], with a log (I) versus log (q) slope of −4 at high q, only fitted the data over a short q range and, therefore, expanded it [*Hall and Mildner*, 1983] to cover anisotropic data. *Hall et al.* [1986] reported additional analyses of many of the same samples and suggested that the extent of asymmetry corresponded to the clay content, although this was not universally true. The theoretical origin of this asymmetry was shown by *Summerfield and Mildner* [1983].

Hall et al. [1983] also showed that porosities obtained from SANS are often smaller than those derived from density data, suggesting that porosity outside the range measured is probably important as well. In addition, they found significant differences between cumulative porosities derived from nitrogen adsorption, MICP, and SANS (Fig. 4.8). Using the model of *Vonk* [1976], they calculated the PSD and showed that it was bimodal (Fig. 4.9), a result repeated in later studies [e.g., *Swift et al.*, 2014a, 2014b], and suggested that when compared to nitrogen sorption data, SANS results compared better to adsorption than desorption data. *Hall et al.* [1986] concluded that where the nitrogen adsorption isotherm showed little hysteresis, the results tended to correspond with those of SANS, but not with mercury porosimetry results. Where hysteresis was present in one case, the MICP data agreed either with the nitrogen adsorption branch at smaller pore sizes and the desorption branch at larger pore sizes or with neither, while the SANS results followed the adsorption isotherm at all scales.

Hall et al. [1983] also observed that the differences in SLD for X-rays and neutrons can be exploited to better understand the location of OM in a pore system. SAXS

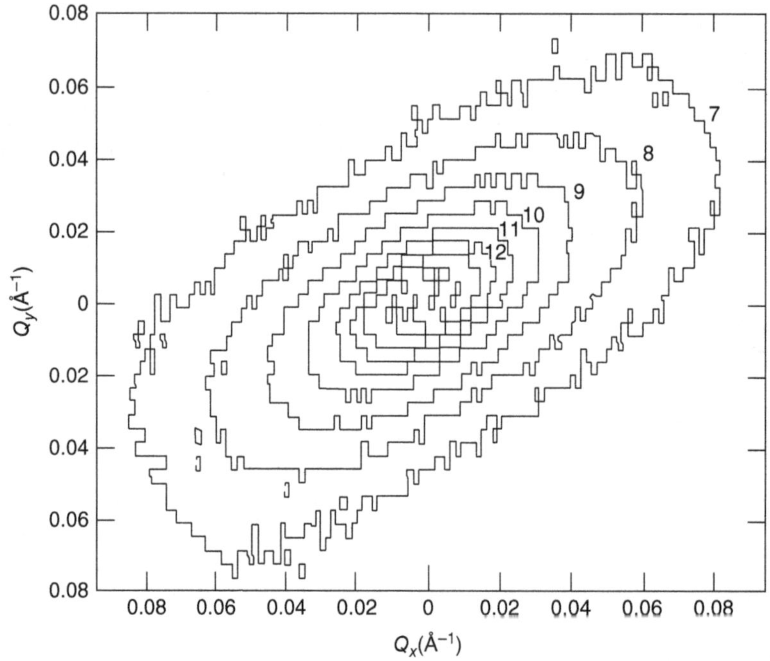

Figure 4.7 Contours of constant intensity of SANS data obtained from a vertical cut (perpendicular to bedding) of a Woodford Shale core taken at a depth of 6190 ft. The contour index refers to the binary power of the intensity so that each additional contour represents a 50% reduction in intensity. Note the asymmetry of the scattering. Reprinted from *Hall et al.* [1983] with the permission of AIP Publishing.

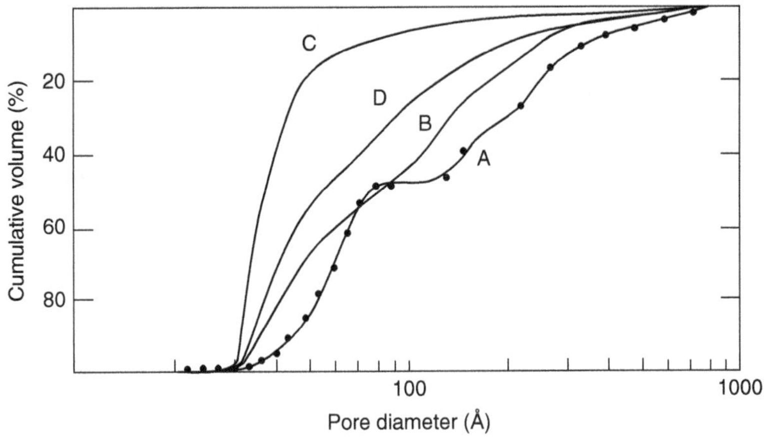

Figure 4.8 (A) Cumulative pore size distribution calculated by the Vonk method from the SANS data obtained from a horizontal cut of East Kentucky oil shale (outcrop shale). These data are compared with data taken from mercury intrusion porosimetry (D) and from nitrogen adsorption (B) and desorption (C) isotherms. Note both the bimodal distribution of the results from the scattering data (weakly suggested in the nitrogen adsorption and mercury intrusion data) and the significant differences between results from the four methods. Reprinted from *Hall and Mildner* [1983] with the permission of AIP Publishing.

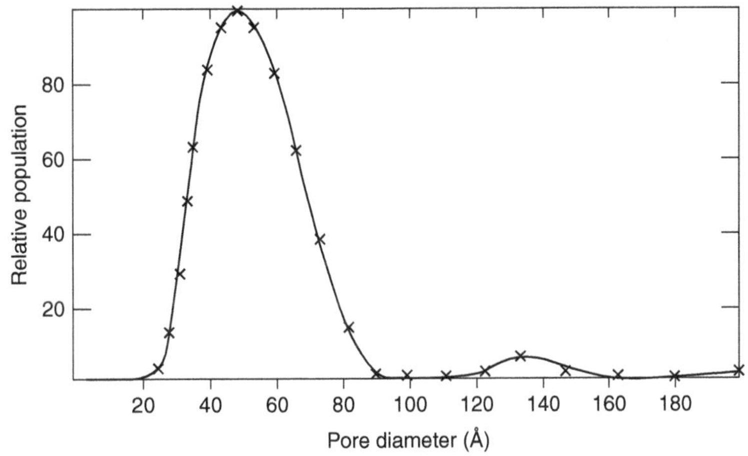

Figure 4.9 Pore size distribution calculated by the Vonk method from the SANS data obtained from a horizontal cut of a Utah mudstone core taken at a depth of 15,666 ft. Note the bimodal distribution of the pores and the large relative proportion of the smaller pore sizes. Reprinted from *Hall and Mildner* [1983] with the permission of AIP Publishing.

data were obtained, but not reported [*Mildner*, pers. comm.]. The comparison was explored later by *Mildner et al.* [1986] and *Hall et al.* [1986]. They found that the power law slopes observed by SANS were different from those observed by SAXS (Fig. 4.10). Noting that for SANS, water (H_2O)-filled or organic-filled pores would have SLDs similar to those of empty pores (this would not be true, or course, for D_2O), whereas for SAXS they would not. They suggested that SANS data are relatively insensitive to the presence of a fluid, while SAXS measurement senses a diffuse boundary between the mineral matrix and empty pores, increasing the slope of the scattering curve (cf., *Anovitz et al.* [2009] discussion of *fuzzy* interfaces). *Hall et al.* [1986] found that SANS analyses showed a broad range of pores covering the entire accessible pore range, while SAXS showed no porosity below ~2 nm, a result they attributed to water remaining in the smaller pores. They also found that the experimental incoherent cross sections were significantly higher than those suggested by the bulk chemistry of the samples and suggested that this could be due to diffuse scattering caused by microheterogeneity. Later work, however, has suggested that Bragg peaks from the clays may also play a role in increasing background values.

Following results by *Schmidt* [1982], *Pfeifer and Avnir* [1983], and *Bale and Schmidt* [1984], *Mildner et al.* [1986]

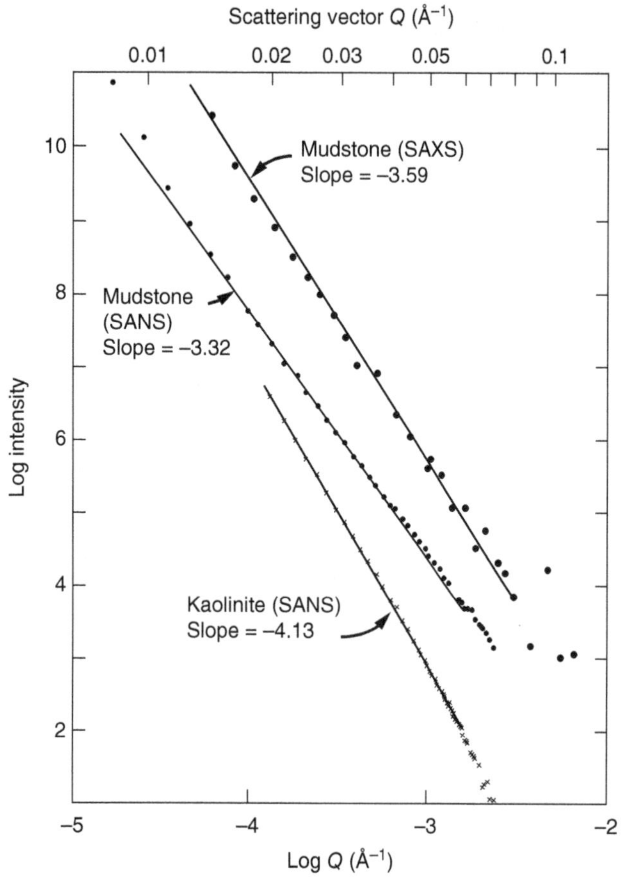

Figure 4.10 Porod plots of the SANS and SAXS data for the Utah mudstone. Reprinted from *Mildner et al.* [1986] with the permission of AIP Publishing. Units for scattering intensity are not given on the original figure. Note the difference in slope between the SAXS and SANS data on the same sample. The kaolinite data were shown for comparison.

also suggested that the nonintegral power law slopes observed by *Hall et al.* [1983] could be attributed to pore systems with fractal boundary interfaces. They concluded that "the surfaces are so irregular that the geometric concept of surface area is meaningless" and showed how the surface fractal dimension could be obtained from the scattering data. The connection between the slopes of the scattering curves and fractal dimensions was not, however, discussed by *Hall et al.* [1986]. *Wong and Bray* [1988] clarified the fractal interpretation of scattering data, such that on a Porod plot (log (I) as a function of log (q)), slopes between 2 and 3 represent mass fractals where the fractal dimension is the negative of the slope and slopes between 3 and 4 represent surface fractals where the slope (s) gives

$$s = -(6 - D_s) \tag{4.1}$$

At $D_s = 3$ for a surface fractal $s = -3$. This is thus identical to a $D_m = 3$ mass fractal.

Hall et al. [1986] also investigated changes in pore structure after extraction of soluble organics with methylene chloride (dichloromethane). Using the asymmetric Debye model of *Hall and Mildner* [1983], they calculated the characteristic pore width and volume before and after extraction and found that low organic content or high-maturity samples change little with extraction, while the changes in lower-maturity, organic-rich samples were significant. The effect of the fractal analysis approach of *Mildner et al.* [1986] on these results is, however, uncertain.

A second set of analyses of the pore structure of shales using SANS was performed by *Wong et al.* [1986]. They examined 26 rocks, four of them shales, and showed that limestones and dolostones tended to have smooth surfaces at scales above 50 Å but were rougher at smaller scales. The extent to which the latter was affected by background is, however, unclear because background tends to cut off scattering at high q (small sizes) and may affect the slope of the scattering curve near the intersection (Fig. 4.5). For sandstones and shales, *Wong et al.* [1986] found that the surfaces were fractal at all scales observed (5–500 Å), but the fractal dimension was not consistent from sample to sample (i.e., the fractal dimension was non-universal). No evidence of mass fractal scattering was observed (but see below wrt. mass fractal scattering in USANS data). They further suggested that fractal behavior and scattering intensity are functions of clay content, which was essentially zero in the carbonates. *Wong et al.* [1986] then related these results to an Ising model of porous solids described earlier [*Wong*, 1985] and suggested that during carbonate precipitation from *clean* water, roughness is "determined by the competition between thermal fluctuation and surface tension … analogous to the roughening of domain walls in a pure Ising model." This implies that there exists a roughening transition temperature T_r, above which the domain wall is rough. This temperature is reduced by the presence of impurities [cf., *Grinstein and Ma*, 1983], possibly explaining the rough surfaces they and later authors [e.g., *Anovitz et al.*, 2009] observed at smaller scales. For shales and sandstones, however, they suggested that clays prefer to maximize surface area on all possible length scales, implying a negative surface tension [*Cohen and Anderson*, 1985] caused by counterbalancing charge defects due to substitutions (and other defects) in the lattice with hydrated surface cations that limit surface growth. Where the energies of these counterion bonds are dominant, the lowest energy state of the system will maximize surface area. This implies that the reactive surface in shales and carbonates will be, at least partly, controlled by the composition of the reacting fluid.

As part of a study of the analysis of heterogeneous systems using SAS techniques, *North et al.* [1988] examined a sample of the Bakken shale using both SAXS and SANS. This yielded a linear relationship of ln $I(q)$ versus ln q over the whole range measured, with a slope of −3.55 (D_s = 2.45) for SAXS and −3.67 (D_s = 2.33) for SANS. They noted that this agreed with the results of *Mildner et al.* [1986], indicating that X-rays and neutrons were being differently affected by hydrocarbons adsorbed at the pore interface. However, in the samples examined by *Mildner et al.* [1986], the slope of the SAXS data was generally steeper than that of the SANS data, while *North et al.* [1988] observed the opposite effect. The range of q investigated was not large enough to determine whether a mass fractal scattering region was present.

North et al. [1990] extended the study of *North et al.* [1988] by examining USAXS data. They also added a sample of *Utah clay*. The fractal dimension determined from the Bakken shale decreased to 2.27 in the USAXS range, while the Utah clay yielded D_s = 2.55 for SAXS and 2.56 for USAXS. In neither case was a mass fractal region detected. They suggested that the long linear dependence of ln $I(q)$ on ln (q) could be due to either a fractal surface or a very wide distribution of pore sizes as suggested by *Bale and Schmidt* [1984]. A more recent study by *Leu et al.* [2016] explored the use of SAXS and WAXS in describing the pore system of three samples: the Opalinus shale, an oil shale from Jordan, and a schist from France. This study focused on the nuances of how to properly apply these two methods considering variable spot size and sample thickness.

4.4.2. Organic-Rich Mudstones

Perhaps not surprisingly given their economic impact, most of the more recent studies of shales using SAS approaches have concentrated on organic-rich mudstones. While the studies described above were concerned with performing the first examinations of the scattering properties of rocks in general, including shales, many of the latter have concentrated on specific formations, either as an example of a specific rock type or as a means of understanding the properties and evolution of a particular formation. Some, however, have again examined samples of material from a range of shales. The next section of this paper will summarize the results obtained for individual shale formations in more or less chronological order. It is important to reiterate that while these are the source of economic quantities of gas and/or oil, several could act as caprocks for CO_2 storage. This is particularly true for organic-rich shales located in the mid-continent of the United States.

4.4.2.1. Velkerri Formation

The Velkerri Formation was examined using SAS techniques by Radlinski and coworkers [*Radlinski et al.*, 2000a, 2000b]. The Velkerri [*Donnelly and Crick*, 1988; *Warren et al.*, 1998; *Jackson and Raiswell*, 1991] is part of the Roper Group in the McArthur Basin in the Northern Territory, Australia. It is a mid-Proterozoic (~ 1.43 Ga) shale consisting of alternating organic-rich black shales with up to 8.7% total organic carbon (TOC) alternating with gray-green organic-lean layers (TOC < 2%). It has been a target for oil exploration since the late 1970s, but some parts have relatively low permeability and porosity, in part because of blocked pore throats due to clay diagenesis. *Radlinski et al.* [1996] used SANS and SAXS to analyze mudstones and siltstones containing organic-rich intervals at seven depths where an oil-generating window was indicated. While they recognized the asymmetry expected for samples cut perpendicular to bedding, only parallel-cut samples were examined. They also created synthetic organic-rich rocks by annealing crushed shale in air at 600°C for 24 h to remove the natural organics and then replaced these with hydrogenated or deuterated eicosane in dichloromethane. The effects of this heating on the shale mineralogy were not examined.

Radlinski et al. [1996] noted that because the SLDs for SANS are different from those for SAXS, the number of phases that must be considered also differs (Fig. 4.11). The SANS values for macerals and shales are similar, as are those for void, oil and water, so a two-phase approximation is appropriate. For X-rays, however, shale, void, macerals, and oil/water had to be considered separately. Nonetheless, the shapes of both the SAXS and SANS curves were similar, although the SANS curves had a higher background due to hydrogen incoherent scattering and small-scale inhomogeneities at high q. Both SAXS and SANS spectra were described by power law relations in log q versus log $I(q)$. Lower-maturity samples showed a single power law with a slope of −3.5 ± 0.1 and an estimated surface area of 31 m²/g. This slope is indicative of a surface fractal geometry, but was not affected by the presence or absence of OM. Higher-maturity samples, however, showed two slopes. Between 50 and 600 Å, the slope increases to −2.8 ± 0.1. This is in the mass fractal region, and *Radlinski et al.* [1996] interpreted it as due to a network of microcracks formed by pressure increases created by petroleum generation. At larger scales (600–2000 Å), the slope approaches −4, a smooth Porod slope with an estimated specific surface area (SSA) of only 1.55 m²/g. They interpreted this smoothing as due to either melting or redistribution of initially highly dispersed or newly formed macerals. *Radlinski et al.* [2000a] expanded the size range investigated by performing USANS measurements. This was then extended to larger scales using SEM imaging. The latter was analyzed by counting the number of *features* that intersect randomly positioned straight lines [*Katz and Thompson*, 1985] rather than

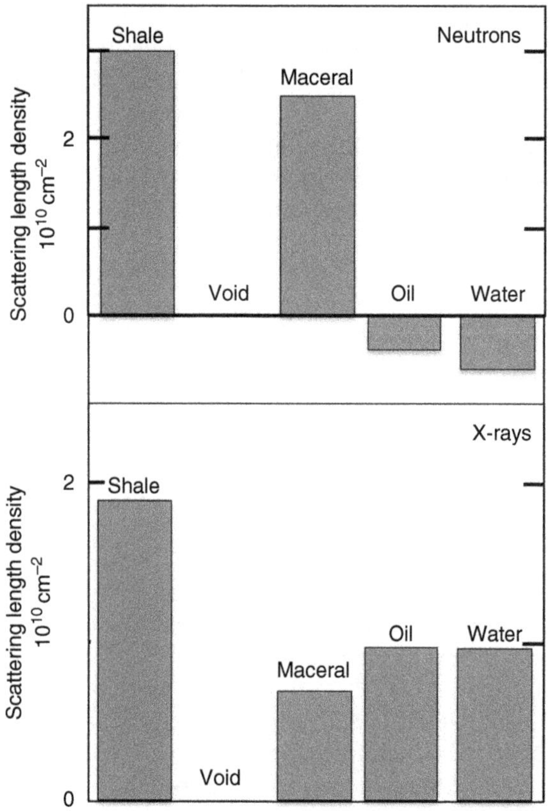

Figure 4.11 Neutron (ρ_n) and electron ($l_e\rho_e$) scattering length densities for the five phases present in BMR Urapunga 4 source rocks calculated from the known chemical composition. Reprinted figure with permission from *Radlinski et al.* [1996] by the American Physical Society.

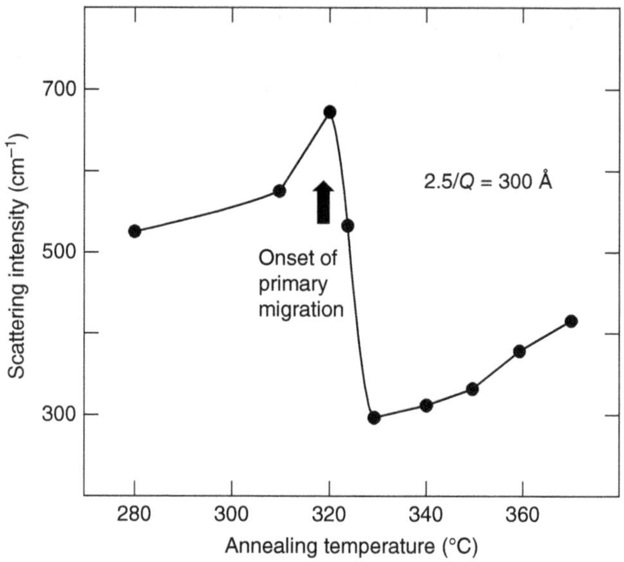

Figure 4.12 The variation of SANS intensity at a single q value versus the annealing temperature for an immature hydrocarbon source rock. Data correspond to the average pore size of 300 Å ± 50%. The marked decrease of the scattering intensity constitutes the SANS fingerprint of the hydrocarbon generation and primary migration. The rising trend is caused by increased hydrogen depletion as the annealing temperature increases. Redrawn after *Radlinski et al.* [2000b] and *Radlinski and Hinde* [2002] (Fig. 4.6).

two-point autocorrelations [cf., Anovitz *et al.*, 2013a, 2015b; *Wang et al.*, 2013]. The results suggested that fractal behavior was limited to scales below ~4 μm.

Radlinski et al. [2000b] further explored the effects of maturation by comparing SANS analyses of natural and experimentally matured shales. For the latter, samples were pyrolyzed at temperatures from 310 to 370°C for 48 h. These were measured on SANS instruments at both ILL (14 Å neutrons, 7 mm thick samples) and ORNL/HFIR (4.75 Å neutrons, 0.7 mm thick samples), and the authors suggest that multiple scattering may have been significant in the former. The scattering curves consisted, first, of a relatively flat high-q background. Intermediate-range scattering followed a power law with an exponent of -3.5 ± 0.05, suggesting a fractal surface with $D_s = 2.5 \pm 0.05$. With increasing temperature, the scattering intensity in this range decreased between 320 and 330°C correlated with the onset of oil generation, followed by a continuous but smaller increase with further maturation related to hydrogen depletion (Fig. 4.12). This was interpreted as due to a decrease in pore-rock contrast caused by filling pores with the hydrocarbons generated. Natural samples showed a similar decrease at a vitrinite reflectance R_o of ~0.8%. These decreases were strongest at small length scales (large q), suggestive of filtering of migrating bitumen through small pores, leaving behind the heavier components to fill the smaller pores. This was explained using the calculated differences in SLD of OM, which they showed varies from -0.4×10^{10} cm^{-2} for n-alkanes up to 2.0×10^{10} cm^{-2} for asphaltenes. A plot showing these results (Fig. 4.13), and comparing them to other relevant values, was presented by *Radlinski and Hinde* [2002]. Invasion of larger pores appeared to require the formation of microcracks, explaining the changes in slope at higher maturity [cf., *Radlinski et al.*, 1996]. Both *Radlinski et al.* [2000b] and *Radlinski and Hinde* [2002] interpreted their results in terms of a polydisperse assemblage of spherical pores, a model later incorporated as the PDSP model in PRINSAS [*Hinde*, 2004]. Using this, *Radlinski and Hinde* [2002] showed that there was a decrease in SSA with increasing maturity (measured as vitrinite reflectance R_o) but that this change occurred primarily in the earlier stages of maturation. One can conjecture that similar effects might be observed during leaching of OM by supercritical CO_2 during geologic carbon sequestration (GCS), although there have, as yet, been no scattering data obtained to test this idea.

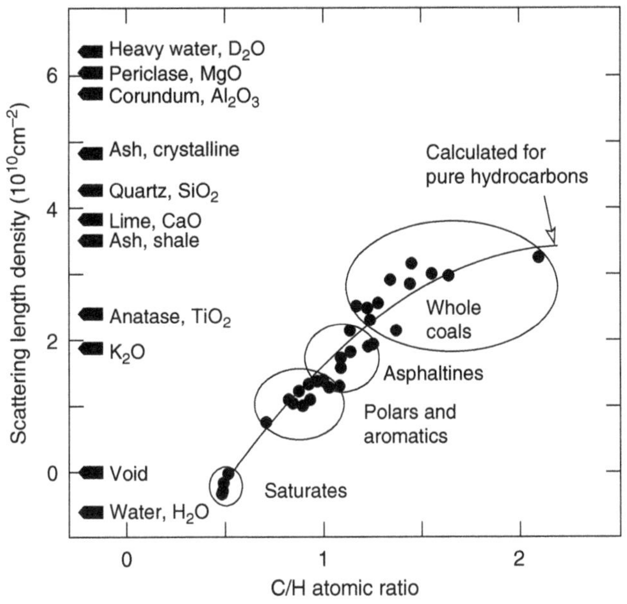

Figure 4.13 Neutron scattering length density for major minerals and OM types present in sedimentary rocks. The solid curve was calculated for saturated hydrocarbons (C_nH_{2n+2}). The scatter reflects the departure from this formula for saturates, polars, aromatics, asphaltenes, and macerals in whole coals and kerogens. Redrawn from *Radlinski and Hinde* [2002]. Reprinted with permission from Taylor & Francis Ltd (Figure 4.2.)

4.4.2.2. Frio Shale

The Frio shale [*Hower et al.*, 1976; *Galloway et al.*, 1982; *Lynch*, 1997] is an Oligocene formation in southeastern Texas, deposited between about 33.5 and 25 my bp. It is in the same area but lies significantly above the Cretaceous Eagle Ford formation discussed below. The Frio Formation also contains significant sandstone layers and has been considered as a potential location for geological CO_2 sequestration [*Xu et al.*, 2003, 2005; *Hovorka et al.*, 2006; *Hovorka*, 2009]. SAS was used to investigate the pore structure of the Frio shale by *Ma et al.* [1999a, 1999b] and *Qi et al.* [2002]. These studies were primarily concerned with comparing SAS results with those obtained from adsorption studies.

Ma et al. [1999a] compared the surface area obtained for two samples of the Frio shale by *Wong et al.* [1986] with results from nitrogen adsorption measurements. The SANS studies yielded fractal dimensions of 2.83 and 2.75; a difference the authors argued was outside of experimental uncertainties. They found that two different approaches to analyzing the BET adsorption isotherm yielded fractal dimensions of 2.54 and 2.43 and 2.63 and 2.50, respectively, both significantly lower than obtained from the SANS data. Applying a thick-film model to the sorption data based on the capillary condensation data, which provided a minimum scale of 12 Å, yielded even poorer agreement, with D_s values of 2.39 and 2.24 and 2.34 and 2.31. Thus, in all cases, the fractal dimensions from the nitrogen adsorption experiments were lower than those obtained from neutron scattering. This analysis was expanded by *Ma et al.* [1999b] for three samples (H2, H4, and H8) of the Frio shale, including the two analyzed by Ma *et al.* [1999a, H2 and H4]. Considering the limitations of the Kelvin equation to relate the minimum radius of curvature to pressure in the adsorption experiment, they concluded that the difference is primarily due to how adsorption isotherms and SAS probe the nature of the surface. Scattering is a direct probe of the interfacial roughness, although it may be affected by short-range correlations [see, e.g., *Anovitz et al.*, 2015a]. In adsorption studies, however, the liquid-vapor interface is smoothed by surface tension. Further analysis of this effect was carried out by *Qi et al.* [2002], who used a power law distribution of uncoupled spherical pores to represent a fractal surface for a numerical simulation of the adsorption isotherm. They found that adsorption measurements always fall in a crossover region as a function of pore radius between regions dominated by van der Waals forces and those dominated by capillary condensation. This implied that adsorption isotherms could not be used to accurately determine pore surface fractal dimensions.

4.4.2.3. New Albany Formation

Following the work of *Qi et al.* [2002] on the Frio shale, there was a 10-year gap in neutron or X-ray scattering studies of shales. With the rise of gas shale as an important energy resource, and perhaps following the use by *Anovitz et al.* [2009] of (U)SANS to analyze changes in the pore structures of geologic systems, rather than individual samples, several studies were performed. The first of these was the work of Mastalerz and colleagues [*Mastalerz et al.*, 2012, 2013; *Bahadur et al.*, 2015] on the pore structure of the New Albany Shale.

The New Albany Shale is an Upper Devonian-Mississippian shale from the Illinois Basin [*Campbell*, 1946]. It is part of a depositional environment that included the Marcellus and Chattanooga shales further east and south. It is a significant potential source of gas and has been considered for geological CO_2 sequestration [Strapoć *et al.*, 2010; *Lahann et al.*, 2013; *Liu et al.*, 2013]. *Mastalerz et al.* [2012] examined two shale samples, one each from the Clegg Creek and Camp Run members of the New Albany, to compare the pore structure of organic-rich shales to that of coals, as well as the pore structure obtained from (U)SANS studies to that obtained from gas absorption (N_2 and CO_2). Two samples of the shale were examined: MM1 and MM3. The first was organic rich (15.8 vol.%) with a vitrinite R_o of 0.55%. The second contained less OM and had a slightly higher R_o

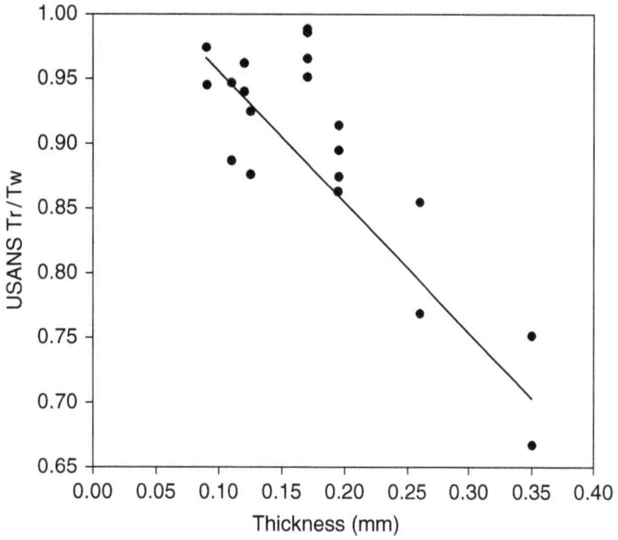

Figure 4.14 USANS transmission data given as a ratio of the rocking curve and wide transmissions for a series of clay and carbonate-rich samples from the Eagle Ford Shale, TX. While there is significant scatter, it is apparent that transmissions reach values near 0.9 at thicknesses near 150 μm. Smaller values imply significant multiple scattering, suggesting that this is a reasonable maximum thickness for USANS analyses of shales [*Anovitz and Cole*, 2015].

value (0.62%). Both are predominantly composed of illite. (U)SANS measurements were made on 0.5–1 mm grains loaded into a thin-walled aluminum holder with an internal thickness of 1 mm. As pointed out by *Anovitz and Cole* [2015] (Fig. 4.14), however, this thickness is probably too great to avoid multiple scattering in the USANS measurements, and their low-q data are, therefore, probably somewhat distorted. This approach also does not permit analysis of the asymmetry of the scattering. Analysis of mesoporosity by nitrogen absorption showed no maximum in the mesopore range, just a near linear increase in pore volume from 2 to 50 nm. In addition, the unclosed hysteresis loop observed in MM1 suggested swelling or micropore adsorption not observed in MM2. Nitrogen surface areas for MM1 and MM2 were 2.4 and 0.2 m²/g, respectively. CO_2 adsorption, however, yielded 12.8 and 10.9 m²/g and suggested the presence of subnanometer pores. In both cases, the larger surface areas correlated with greater amounts of OM. (U)SANS analysis of MM1 and MM3 yielded identical slopes of -3.2 ± 0.1 for both samples. Specific surface areas based on the PDSP analysis [*Hinde*, 2004] and a probe (ruler) size of 4 Å yielded surface areas (14.8, 23.3 m²/g) significantly larger than those obtained by either nitrogen or carbon dioxide adsorption, suggesting considerable inaccessible porosity. Attempts to measure accessible versus inaccessible porosity by injecting CD_4 at various pressures yielded very little change, however, reflecting very low permeability.

The work of *Mastalerz et al.* [2012] was extended by *Mastalerz et al.* [2013] and *Bahadur et al.* [2015] to five samples covering a range of maturities (R_o = 0.35–1.41). The first study concentrated on gas adsorption, and the latter on (U)SANS. The number of samples examined was rather small, however, and stochastic variation could, therefore, not be adequately documented, a fact the authors recognized. *Mastalerz et al.* [2013] argued that results from gas adsorption (He, CO_2, and N_2) and MICP were consistent, both in terms of pore volume and PSDs. Total OM content did not necessarily correlate with maturity, but two early-mature (low R_o) samples suggested a strong relationship between TOC and pore features. At this maturity, TOC did not affect total pore volume, but did increase the contribution of micropores to overall porosity. This did not, however, extend to the entire sample set. Total porosity increased with clay and quartz content and decreased with carbonate content. Much of the former was associated with microporosity from clays. Increasing maturity was associated with reduction in total porosity except in the most mature sample (IL-1, R_o = 1.41%) in which new pores appeared to form. This sample was, however, from a much shallower environment than the next most mature sample, which may have affected the pore structure. For the remaining samples, maturation was associated with an increase in microporosity and tended to eliminate nanoporosity. *Mastalerz et al.* [2013] argued that porosity generation was related to hydrocarbon production, while its decline at high maturity was due to pore filling by oil and bitumen.

Bahadur et al. [2015] used (U)SANS to examine the samples analyzed by *Mastalerz et al.* [2013]. The sample preparation was the same as that described by *Mastalerz et al.* [2012], and thus the scattering curve in the USANS region is likely distorted due to multiple scattering. *Bahadur et al.* [2015] also used an SLD for the organic components estimated for coal on the high end of the values predicted by *Radlinski and Hinde* [2002]. Its applicability to these samples, therefore, remains uncertain. Plots of ln $I(q)$ versus ln (q) showed significantly high, flat background intensities at high q, especially for the lower-maturity samples. Using the PDSP model from PRINSAS yielded extremely high and unrealistic porosities (a pore fraction of 0.828 for one sample) when this range was considered. Truncation of the high-q data, however, yielded porosities significantly below those observed by He adsorption. The extent to which these differences were due to the SLD value assumed for the OM, however, was not investigated. The lower-q data were approximately linear and suggested surface fractal behavior, although nonlinear details are apparent that might represent a more complex pore structure [cf., *Anovitz et al.*, 2015a, 2015b]. Analysis of changes in pore structure

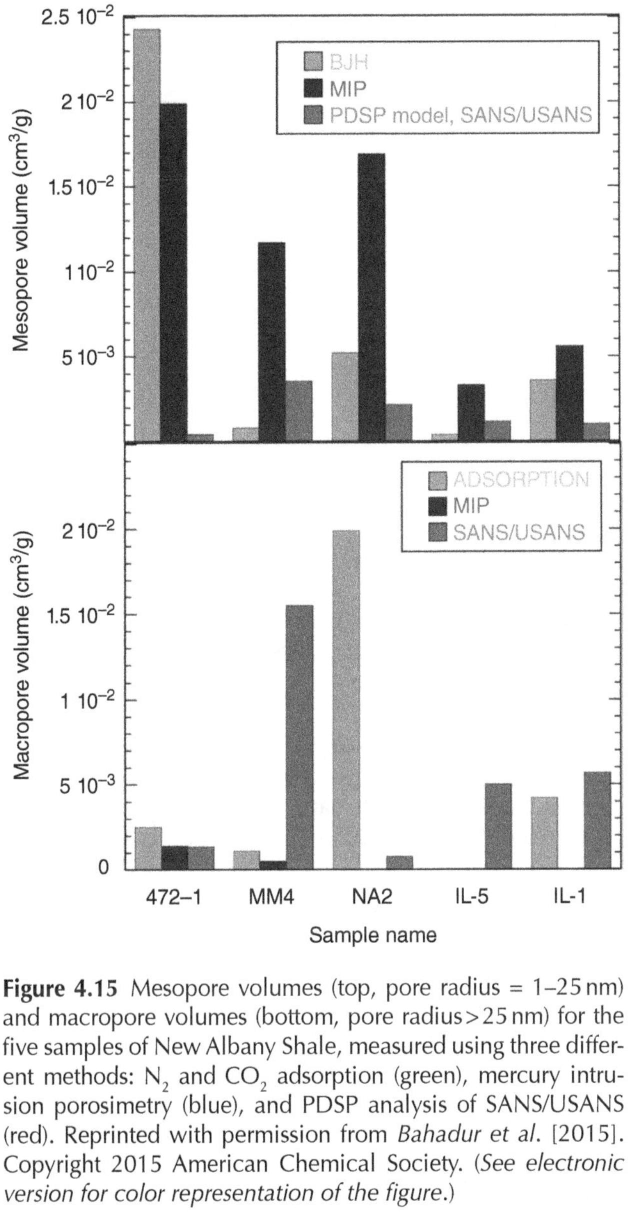

Figure 4.15 Mesopore volumes (top, pore radius = 1–25 nm) and macropore volumes (bottom, pore radius > 25 nm) for the five samples of New Albany Shale, measured using three different methods: N_2 and CO_2 adsorption (green), mercury intrusion porosimetry (blue), and PDSP analysis of SANS/USANS (red). Reprinted with permission from *Bahadur et al.* [2015]. Copyright 2015 American Chemical Society. (*See electronic version for color representation of the figure.*)

suggested a significant decrease in total porosity with maturity, but not in the truncated porosity range. Much of this, however, was observed at the lowest maturity levels, as was observed in the Eagle Ford Shale by *Anovitz et al.* [2014, 2015a]. Comparisons of porosity estimated from nitrogen and carbon dioxide adsorption and (U)SANS (Fig. 4.15) showed that in the mesopore range (pore radius = 1–25 nm), porosity estimated from the SANS data was consistently lower than that of the other techniques. This was expected because the (U)SANS porosity was estimated using the high-q truncated data, thus missing finer-scale porosity. In the macropore range (pore radius > 25 nm), the results were less consistent, with good agreement for two samples, but (U)SANS giving a much higher value in one sample and much lower in the other than the other two approaches.

Bahadur et al. [2015] also attempted to explain the observed high background values. They first calculated the hydrogen content of the rock from the TOC values. Although the details of this calculation for OM of changing maturity were not provided, more detailed data were provided by *Mastalerz et al.* [2013]. For three or four of the samples (those with the lowest background values), however, background may have been difficult to estimate accurately because a high-q cutoff caused by the limited acceptance angle of the sample holders used limited the data set to q values where the background was still decreasing. They found that the incoherent background calculated from this hydrogen content was significantly smaller than that observed. However, this did not account for hydrogen in the clays, which comprised from 2 to 42 wt% of the rocks. They found a poor correlation between TOC and the background levels, although it was primarily one sample, the least mature, for which this correlation was extremely poor. They then showed that the difference between the He adsorption porosity and that from the truncated scattering data (termed excess porosity) correlated much better with the high-q background minus that calculated for incoherent scattering and estimated that this was due to nanoporosity *hidden* in the background region. As would be expected from the total porosity, the magnitude of the nanoporosity was largest for the lowest-maturity sample.

4.4.2.4. Barnett Shale

The Mississippian Barnett Shale is another example of a tight oil/gas reservoir, located in the Bend Arch-Fort Worth Basin near Fort Worth, Texas. It was deposited in a deep, low-oxygen marine basin, and parts of the basin may have reached 100–180°C during diagenesis [*Gutschick and Sandberg*, 1983; *Jarvie et al.*, 2007; *Loucks and Ruppel*, 2007; *Ruppel and Loucks*, 2008; *Rowe et al.*, 2008; *Loucks et al.*, 2009]. It is one of the largest shale gas sources in the United States [*EIA*, 2018]. Two studies [*Clarkson et al.*, 2013; *Ruppert et al.*, 2013] have used SAS to investigate the pore structures of samples from this formation. Between them, however, only three samples of the Barnett were examined.

Clarkson et al. [2013] used USANS and SANS along with nitrogen and carbon dioxide adsorption and Hg intrusion porosimetry to investigate the pore structure of samples from several shale formations. Approximately half of their analyses were for the sample from the Barnett, which was run at 80°C under a range of CD_4 pressures (see below for discussion for results for the Milk River, Duvernay, Eagle Ford, Haynesville, and Marcellus). (U)SANS samples were cut to 1 mm thickness, again probably leading to significant multiple scattering in the USANS measurements. The samples were cut parallel to bedding, so the anisotropy was not investigated. They were also preheated to 60°C under vacuum for at least 24 h to dry them. As pointed out by *Lee et al.* [2014], however, such drying is likely to modify the pore structure of shales. Combined nitrogen and carbon dioxide adsorption suggested that porosity in the Barnett is bimodal, with peaks near 10 and 1000 Å. The smaller-scale peak may represent interlayer spacing in the clays, while the latter may represent spacing between clay packets. Hg intrusion data do not, however, show the larger pores. *Clarkson et al.* [2013] suggested that this was either due to damage caused by Hg intrusion or to the fact that Hg intrusion measures pore throat sizes, not pore body distributions. The slope of the scattering curve, −3.3, suggests a surface fractal dimension D_s = 2.7. Analysis of pore distributions from the (U)SANS data using PRINSAS/PDSP yielded a power law distribution of the log of pore concentrations as a function of the log of pore radius.

The multi-pressure (U)SANS analyses of the Barnett sample by *Clarkson et al.* [2013] were performed to determine the percent of total porosity accessible to methane. As summarized by *Anovitz and Cole* [2015], if pores are filled with a material whose SLD matches that of the surrounding fluid, those pores *disappear* during a scattering experiment. As the SLD is a function of density, changing the pressure of the CD_4 injected into the rock changes the scattering contrast. Inaccessible pores, therefore, yield residual scattering. This may be a function of q if the fraction of inaccessible porosity is pore size dependent. In the Barnett sample, the fraction of accessible porosity was reasonably constant at about 85% for pore radii from 400 to 3000 Å but decreased from 400 to 60 Å and at pore radii larger than 3000 Å.

Ruppert et al. [2013] investigated two samples of the Barnett (72 and 152) using SANS and USANS. The samples were obtained from similar depths (2239 and 2288 m, respectively), and vitrinite reflectance R_o values (1.89 and 1.86%, respectively) suggesting maturities near the upper end of the wet gas zone [*Tissot and Welte*, 1984]. Pressurized CD_4 and H_2O/D_2O mixtures were used to investigate connectivity. Samples were ~0.8 mm thick, which may have led to significant multiple scattering and deformation of the scattering curve at low q. The two samples showed similar PSDs and total porosities. Slopes near −3 are shallower than observed in the Velkerri, New Albany, and Frio, possibly reflecting a more mature sample. Pressurization with CD_4 caused intensity to drop at low q (larger sizes) and increase at high q (smaller sizes). The origin of the latter change remains uncertain. For larger pore sizes (> 250 nm), >85% of the pores were accessible to both CD_4 and H_2O/D_2O. In sample 72, the fraction of accessible pores was similar at all scales, but in sample 152 the accessibility declined to ~65% at higher q. H_2O/D_2O mixtures appeared to penetrate fine pores more easily than did CD_4 (Fig. 4.16). This may reflect relative penetrability of pore throats. SLDs calculated from the pressurized CD_4 experiments yielded values significantly below those estimated from the overall shale composition, consistent with pores primarily in pyrite, illite, or OM. SEM analyses showed significant porosity only in the organic samples, but the SLD values required (3.7×10^{-6} Å^{-2}) are much larger than calculated from R_o (or estimated by *Radlinski and Hinde*, 2002). The suggestion that pores were not primarily present in the illite also differed from the results of *Clarkson et al.* [2013]. SLD values from H_2O/D_2O mixtures (4.6×10^{-6} Å^{-2}) were also higher than those from the CD_4 experiments and larger than that of the quartz, which makes up the bulk of the mineralogy (4.2×10^{-6} Å^{-2}). *Ruppert et al.* [2013] suggested that this could be due to exchange of deuterium with OH groups in illite.

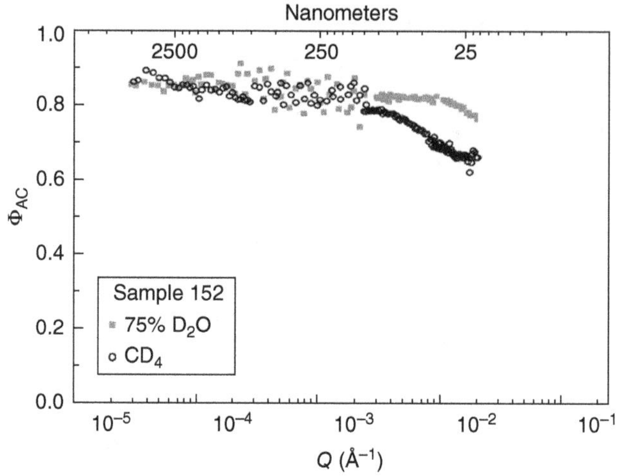

Figure 4.16 Comparison of the fraction of accessible pores in sample 152 by D_2O and CD_4. Reprinted with permission from *Ruppert et al.* [2013]. Copyright 2013 American Chemical Society. D_2O and CD_4 are approximately equally accessible to pores in the USANS range (larger pores), but CD_4 accessibility decreases to about 70% in the SANS range, while any decrease for D_2O is much more minor.

4.4.2.5. Eagle Ford Shale

Anovitz et al. [2014, 2015a] and *Clarkson et al.* [2013] reported scattering results from samples of the Eagle Ford Shale, which is a Late Cretaceous unit underlying significant portions of southeastern Texas and adjacent parts of Mexico. Some portions are relatively clay rich, while others contain as much as 89% calcite and are essentially dirty limestones. The Eagle Ford is hydrocarbon rich and increases in depth to the southeast from surface exposures that occur in an arc from north of Austin to San Antonio and Kinney County. The Eagle Ford is also an active target for tight oil/gas shale exploration and production.

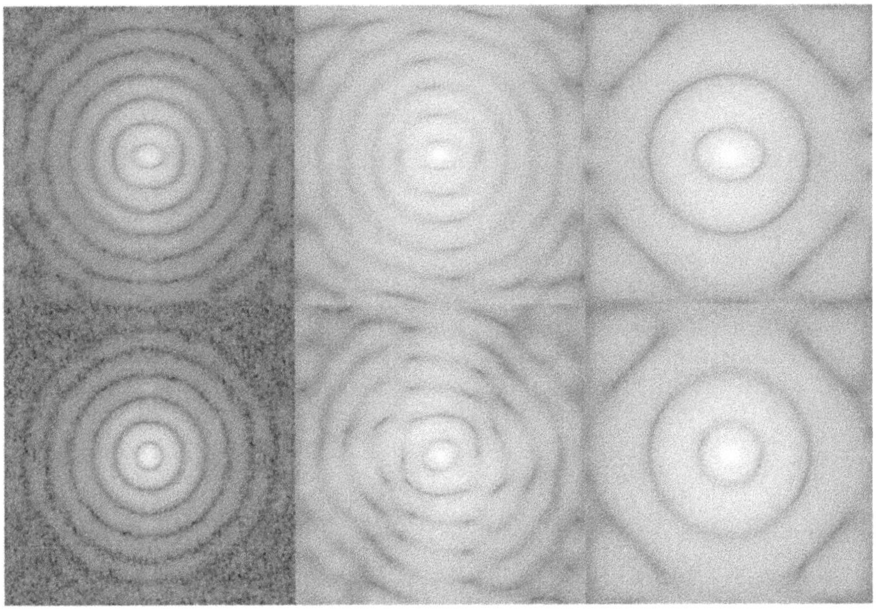

Figure 4.17 (Top) Autocorrelation analysis of the lowest- and (bottom) highest-maturity carbonate-rich samples of the Eagle Ford Shale. The three images for each maturity were calculated from scattering patterns obtained at detector distances of 1, 4, and 13 m (left to right) on NG3-SANS instrument at NIST/NCNR. Focusing on the innermost ring, note the loss of pattern asymmetry with increasing maturity. From *Anovitz et al.* [2015a]. AAPG© [2015], reprinted by permission of the AAPG whose permission is required for further use.

Clarkson et al. [2013] analyzed one sample from the Eagle Ford formation. It was reported as coming from a fresh core, but the origin of that core was not reported. It had a TOC of 2.63 wt%, a R_o of ~1.65%, and a relatively high calcite content near 41 wt%. It had the largest nitrogen adsorption pore volume of the shales investigated (~ 0.05 cm^3/g) but an intermediate surface area (~9 m^2/g). Micropore analysis by CO_2 adsorption again yielded an intermediate value (~0.0025 cm^3/g). Hg intrusion showed a maximum in pore throat dimensions near 30 Å, but no evidence of bimodality. (U)SANS analysis showed a power law slope of −3.2 (D_s = 2.8), a pore fraction of 0.0683, and an SSA of 4.59×10^{-5} cm^2/cm^3 for a calculated SLD of 4.35×10^{-6} Å^{-2} and a probe size of 6.31 Å.

Anovitz *et al.* [2015a] reported preliminary results for a suite of 10 samples from the Eagle Ford. These were in pairs, with one carbonate-rich and one clay-rich sample, each obtained from drill cores from several different counties. R_o maturities ranged from 0.58 to 1.57%. Samples were cut perpendicular to bedding, to a thickness of 0.15 mm to limit multiple scattering [cf., *Anovitz et al.*, 2009; *Anovitz and Cole*, 2015] (Fig. 4.14). The elliptical scattering patterns were analyzed in segments around the major and minor axes to quantify changes in asymmetry with maturity. A bend in the scattering curve near 4×10^{-4} Å^{-1} suggested a shift from surface fractal (pore surface roughness) scattering at smaller length scales to mass fractal (pore aggregate) scattering at larger scales. *Anovitz et al.* [2015a] also showed that by converting the 2D scattering patterns into 2D autocorrelation patterns, the asymmetry could be evaluated in real space, at least for the SANS data where 2D patterns are collected (Figs. 4.17 and 4.18). They found that in carbonate-rich samples, increasing maturity eliminated the scattering asymmetry at all scales. In clay-rich samples, however, the finer-scale asymmetry was eliminated, but larger-scale asymmetry was not. *Anovitz et al.* [2015a] offered the caveat, however, that these apparent directional differences are due to projections of 3D structures onto the 2D plane of the detector.

Anovitz et al. [2015a] also questioned the applicability of the two-phase assumption [cf., *Radlinski*, 2006]. That is, scattering patterns can be analyzed as resulting solely from scattering between empty pores and a matrix with a single SLD. They noted significant variability in the SLD values for the minerals typically present in shales (e.g., carbonates vs. framework silicates such as feldspars and quartz or sulfides such as pyrite) and that this was complicated by the variable and frequently unknown values for any OM present. They found that mineral-mineral scattering might be significant and that the treatment of the OM made a large difference in the calculated porosity. Calculating porosity assuming empty pores yielded very high values. These were corrected to some extent using the measured TOC values. This suggested not only the change in asymmetry with maturation discussed above,

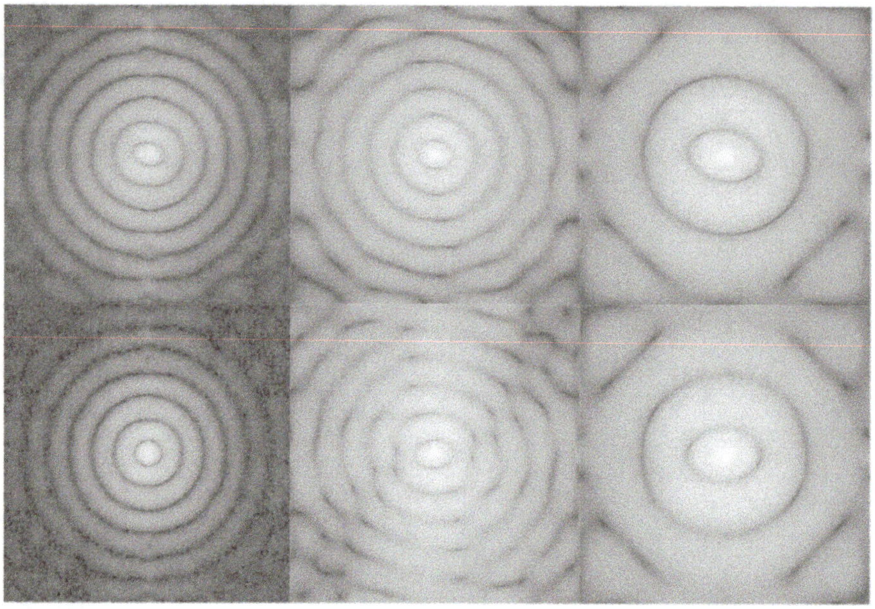

Figure 4.18 (Top) Autocorrelation analysis of the lowest- and (bottom) highest-maturity clay-rich samples of the Eagle Ford Shale. The three images for each were calculated from scattering patterns obtained at detector distances of 1, 4, and 13 m on NG3-SANS instrument at NIST/NCNR. As with the carbonate-rich samples (Fig. 4.17), increasing maturity reduces the asymmetry. In this case, however, the effect is incomplete even at the largest scales (13 m) and highest maturities, as can be seen by focusing on the innermost ring, especially of the highest-maturity sample at 13 m. From *Anovitz et al.* [2015a]. AAPG© [2015], reprinted by permission of the AAPG whose permission is required for further use.

but also that the porosity decrease with maturation occurs in the earliest stages, at low R_o values, although there may be some local variation (Fig. 4.19). This is a very similar interpretation to that suggested by *Radlinski et al.* [2001] and *Radlinski and Hinde* [2002] for coals from the Illinois and Appalachian basins and by *Bahadur et al.* [2015] in the New Albany formation.

4.4.2.6. Marcellus Formation

The Marcellus Formation is a Middle Devonian unit in the Hamilton Group, a thick sequence of marine rocks regionally distributed within the Appalachian Basin of the eastern United States. It is host to proven large reserves of natural gas that have been exploited via hydraulic fracturing. An organic-rich black shale dominates the Marcellus Formation, but it also contains low organic content shale units and interbedded limestones that resulted from variations in sea level during deposition [*Harper et al.*, 2004]. The black shale was deposited in relatively deep water devoid of oxygen and is only sparsely fossiliferous [*Sageman et al.*, 2003].

The Marcellus has been the subject of several studies focused on stratigraphy, structure, geochemistry, mineralogy, and pore features [*Lash and Blood*, 2014; *Bruner et al.*, 2015; *Wendt et al.*, 2015] but has been interrogated with neutron scattering in only a few. To put these studies

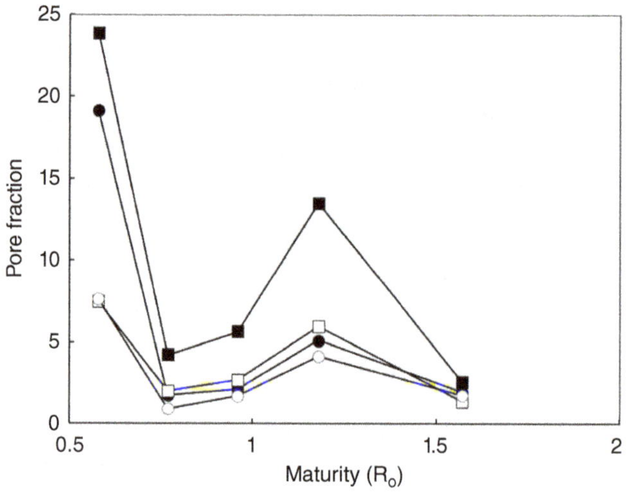

Figure 4.19 The remaining porosity (scattering porosity minus estimated organic pore volume) as a function of maturity (R_o) for samples from the Eagle Ford Shale was calculated assuming an organic density of 1 g/cm^3. Solid squares, clay rich, parallel to the bedding; open squares, clay rich, perpendicular to the bedding; solid circles, carbonate rich, parallel to the bedding; and open circles, carbonate rich, perpendicular to the bedding. Note that most of the decrease in porosity appears to occur at low R_o values. From *Anovitz et al.* [2015a]. AAPG© [2015], reprinted by permission of the AAPG whose permission is required for further use.

in perspective, we should first consider the work of *Loucks et al.* [2012] who provided a detailed summary of pore types and their networks for a variety of mudstones. They classified the pore type in the Marcellus as OM dominant with minor contribution from intraparticle pores. *Clarkson et al.* [2013] reported SANS/USANS results on one sample of the Marcellus whose location was not reported. This was low in total organic carbon (1.57 wt%) with a R_o value of ~1.2%. The slope of the log $I(q)$ versus log q curve was −3.3 corresponding to a surface fractal D_s = 2.7, that is, a reasonably rough pore-matrix interface similar to other shales described above. Assuming a polydisperse spherical model (PRINSAS software), *Clarkson et al.* [2013] estimated a total porosity of 3.55% and an SSA of 0.34×10^5 cm²/cm³. They did not compare the Marcellus scattering data against results obtained from BET or mercury porosimetry, nor did they provide pore volume or pore size distribution results derived from the scattering data.

A more comprehensive study of the Marcellus shale was performed by *Gu et al.* [2015], who investigated the porosity of its Union Springs (Shamokin) Member from a core drilled in Centre County, PA, USA, using SANS, USANS, FIB-SEM, and nitrogen gas adsorption and examined the relationship between anisotropy, porosity generation, and host mineral environment. For polished thin sections cut in the plane of bedding (~150–200 μm thick rock slices), the scattering pattern was isotropic, while for thin sections cut perpendicular to the bedding, the scattering pattern was anisotropic (Fig. 4.20). FIB-SEM observations demonstrated that the anisotropy could be attributed to elongated pores predominantly associated with clay. The apparent porosities calculated from the bedding plane sections were lower than those calculated from sections cut perpendicular to bedding. *Gu et al.* [2015] presented a preliminary method for estimating total porosity from measurements made on the two orientations. Results from this method were in good agreement with nitrogen BET adsorption for both porosity and SSA. Contrast matching using H_2O/D_2O mixtures combined with FIB-SEM revealed that the dominant nanosized pores in organic-poor, clay-rich shale samples are water-accessible sheetlike pores within clay aggregates. In contrast, bubble-like organophilic pores in kerogen dominate organic-rich samples (Fig. 4.6).

In more recent efforts, *Gu et al.* [2016] and *Gu and Mildner* [2016] investigated the topic of organic porosity and wettability and developed an improved method to assess azimuthal asymmetry in the Marcellus, respectively. In the former study, Gu *et al.*, used contrast matching of D_2O/H_2O mixtures in SANS and USANS experiments on cm-sized samples of the Marcellus shale to quantify the total and water-accessible porosity over the length scale range of nm to μm. Scattering was also conducted on samples that were combusted at 450°C. In this case of the Marcellus, 24–47% of the total porosity is contained in OM for both organic-rich and organic-poor samples. Further, nearly 30% of the volume of the OM is composed of void space. More interestingly, despite conventional wisdom that OM porosity is organophilic and not water wetting, Gu *et al.*, demonstrate that nontrivial water accessibility is possible for OM pores with diameters of greater than 20 nm.

The work of *Gu and Mildner* [2016] addresses the issue of how to best connect SANS results obtained from a pinhole geometry SANS detector with USANS data derived from a double-crystal diffractometer. The challenge is to convert data obtained from the latter to the former, known as desmearing. When scattering for a given shale, or any rock for that matter, exhibits azimuthal symmetry (i.e., circular contours on a 2D detector for samples cut parallel to bedding), this conversion is reasonably straightforward. The real challenge comes from performing this conversion for azimuthally asymmetric samples, which we encounter with scattering from shale samples cut perpendicular to bedding. These types of samples commonly show elliptically dependent intensity contours with the long axis corresponding to the normal to the bedding plane (see Fig. 4.20). They proposed that this issue of connecting SANS and USANS scattering data may be improved if the SANS elliptical contours have the same aspect ratio for all measurable values of the scattering vector. Their method used the ratio of the long and short axes of the elliptical contours observed in the SANS data and applies this result to modify the raw USANS intensity data. The outcome is a quasi-symmetric data set that can then be desmeared in the usual way by concatenating the USANS data with the corresponding SANS data obtained from the pinhole geometry. The interested reader is encouraged to refer to their study for the mathematical details of their approach.

The impact of weathering on the mineralogy, elemental geochemistry, and associated pore features in the Marcellus was described by *Jin et al.* [2013]. This study focused on alteration of the Oatka Creek Member of the Marcellus, which is characterized by its black color, high pyrite content, organic-rich nature, enrichment of trace metals, and lack of fossils [*Potter et al.*, 1980; *Roen*, 1983a, 1983b, *Obermajer et al.*, 1997, *Shultz*, 1999; *Sageman et al.*, 2003]. SANS and USANS data revealed that nanoporosity (pores <100 nm) first decreases upward from the shale bedrock interface to 80 cm depth and then increases above 20 cm. This trend cannot be explained by OM because the SLD for organic-filled pores was similar to that of air-filled pores. *Jin et al.* [2013] proposed that dissolution of pyrite and plagioclase increased porosity as deep as 100 cm, but secondary minerals such as kaolinite and Fe oxyhydroxides precipitate within the shale,

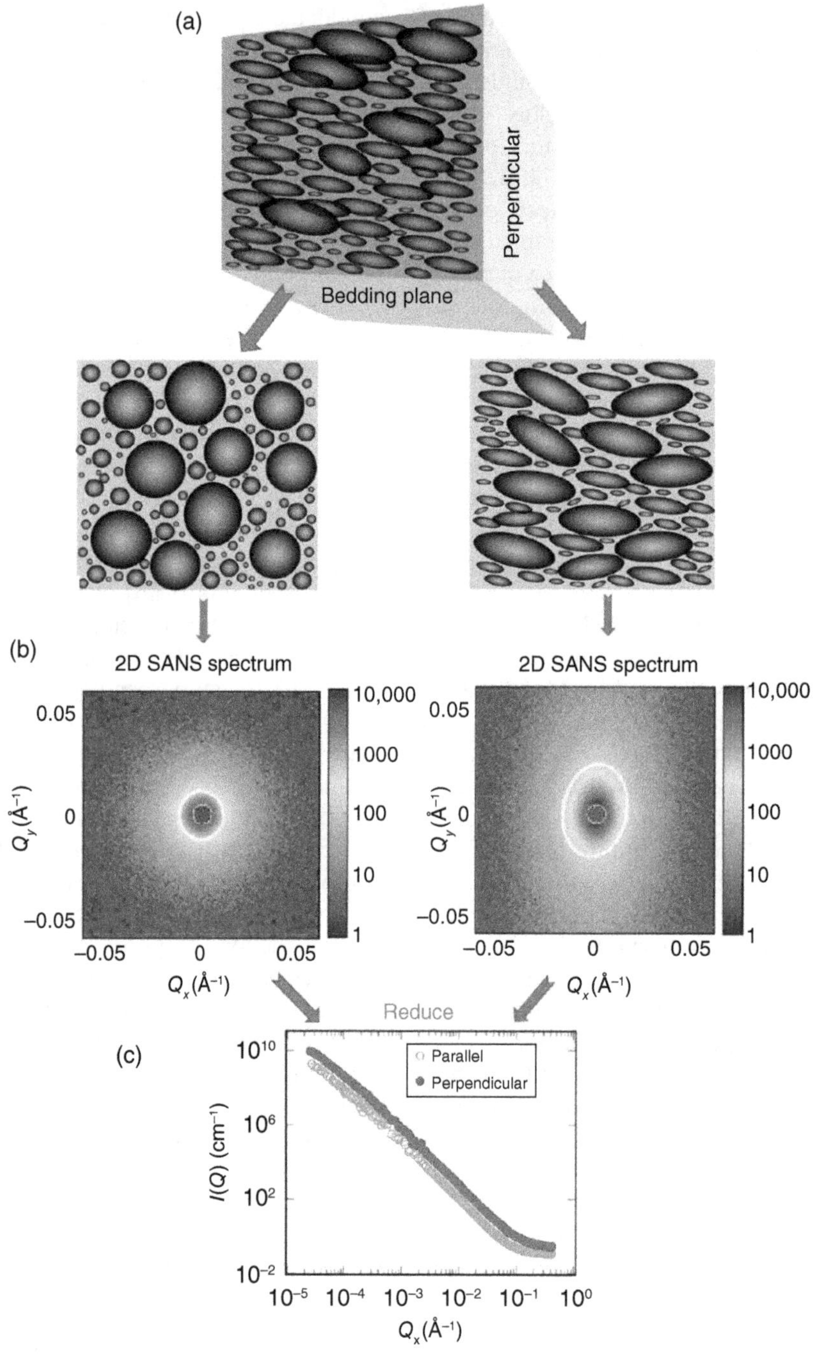

Figure 4.20 (a) Schematics showing the shapes of mineral grains in shale samples cut parallel or perpendicular to the bedding. (b) 2D SANS spectra for the BE850 sample cut parallel (left) and perpendicular (right): for the bedding-parallel sample, the scattering pattern is isotropic, and that for the bedding-perpendicular sample is anisotropic. The preferred direction in reciprocal space (the direction of the long axis of the elliptical contours shown in b) is orthogonal to the preferred direction in real space (the direction of grain alignment). (c) SANS and USANS neutron scattering intensities for a sample cut parallel and perpendicular to the bedding, plotted as functions of the scattering vector Q. Error bars are too small relative to the data symbols to be apparent. Reprinted with permission from *Gu et al.* [2015]. Copyright 2015 American Chemical Society.

maintaining a near-constant nanoporosity from 80 to 20 cm depth. The unweathered Marcellus shale has a surface fractal dimension near 3. Except for the two samples at about 30 cm depth, the fractal dimension generally decreases as weathering intensity increases. Thus, the mineral-grain interface becomes smoother and less space filling with weathering. The largest change in fractal dimension occurs between 120 and 80 cm depth, consistent with the depth where pyrite, OM, and feldspar may have been removed, and again at 10 cm, consistent with the zone of clay dissolution and kaolinite precipitation.

4.4.2.7. Utica and Pt. Pleasant Formations

The Utica and Pt. Pleasant Formations (commonly referred to as the Utica Shale) are an organic and clay-rich, calcareous black mudstone sequence deposited about 440–460 million years ago during the Late Ordovician in the Appalachian Basin of North America. The Utica Shale is thicker than the Marcellus, is more geographically extensive, and supports large-scale commercial production of gas, condensates, and oil. The Utica Formation lies above the Pt. Pleasant and is much richer in clay but lower in TOC. Current exploration and production in Ohio, Pennsylvania, and West Virginia target the gas-rich Pt. Pleasant, which is an organic-rich marl. Like the Marcellus, it has the potential to become an enormous natural gas resource.

Swift et al. [2014b] used detailed SEM imaging and micro X-ray CT, coupled with SANS and USANS experiments, to assess the relationships among rock fabric, mineralogy, and the extent of anisotropy for the carbonate-rich Pt. Pleasant Formation. Samples (~150 μm thick) were cut parallel and normal to bedding or laminations. Parallel sections consist of mostly irregularly shaped calcite, with lesser amounts of dolomite and quartz that tend to mechanically polish well and form *plateaus*, interrupted by lower-lying dendritic patches containing phyllosilicates (mostly illitic clay) along with clay-sized albite, quartz, pyrite, carbonate, and organic material. Minor large grains of calcium phosphate and celestite are also observed in all samples. Phyllosilicates are not only oriented with folia parallel to bedding but are also observed at an angle or vertical, especially where adjacent to non-clay mineral grains. In the third dimension, the clays often wrap around the other minerals. Because these phyllosilicates distributed at an angle to bedding also change in azimuthal angle, it was hypothesized that any anisotropic shape effect was effectively eliminated. Sections cut perpendicular to bedding show more obvious textural laminations that impart an anisotropic shape factor to the fabric, especially for more clay-rich samples. As expected, folia are mostly parallel to bedding, and OM is commonly elongated parallel to bedding planes. This does not, however, describe textural zones where phyllosilicates form grain boundaries with very large, irregularly shaped grains (often fossils). Figure 4.21 shows that changes in azimuthal angle are controlled by the shape of boundaries defining the larger grains. Therefore, *Swift et al.* [2014b] hypothesized that samples composed of contrastingly different textural zones might reduce the observed anisotropy.

The scattering intensity of neutrons travelling normal to the sample surface was azimuthally symmetric for samples cut parallel to bedding and asymmetric for samples cut perpendicular. This indicates that defining the *z* axis as being perpendicular to bedding, (a) scattering objects appear similar to neutrons passing parallel to *z*, regardless of sample rotation around the *z* axis, and (b) they change in apparent density or shape depending on sample rotation around the *x* or *y* axis [*X. Gu*, pers. comm.]. Therefore, textural orientation matters for a sample cut perpendicular to bedding, as shown in

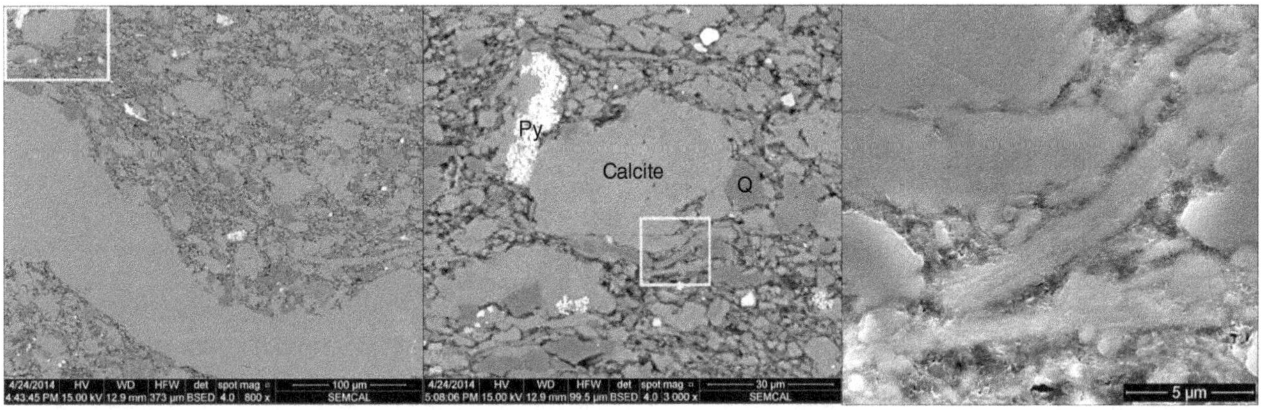

Figure 4.21 Backscattered (left and center panels) and secondary electron (c) SEM images of a carbonate-rich fossiliferous sample (Utica Formation), similar in texture to Utica 19. (center and right panels) are close-up views of the region shown in the white box within the image to its left. Q = quartz, Py = pyrite. From *Swift et al.* [2014b]. (*See electronic version for color representation of the figure.*)

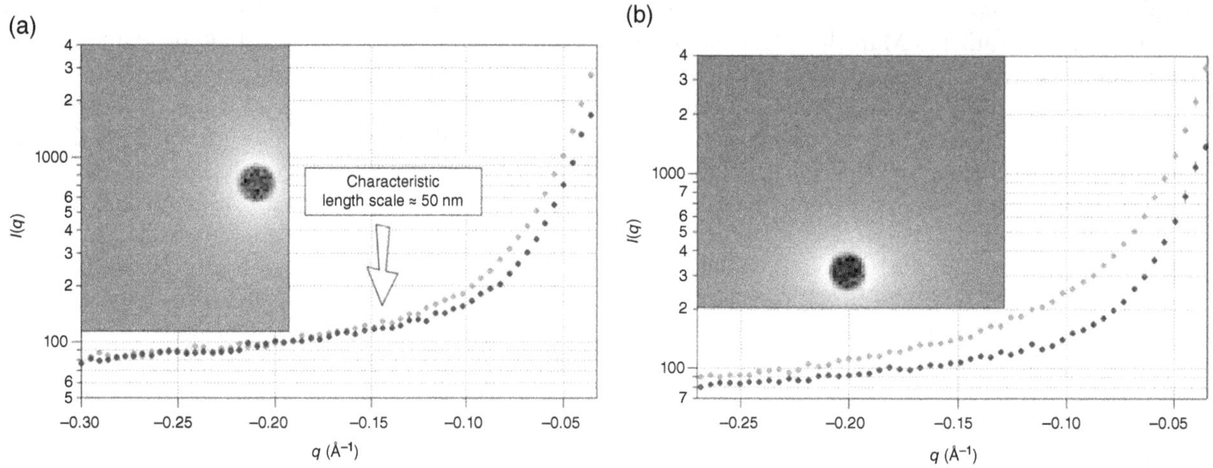

Figure 4.22 Neutron scattering intensity curves and actual detector responses for Pt. Pleasant samples (a (left) calcite rich) and (b (right) clay rich) as a function of scattering vector (q), which can be converted to a characteristic scale of scattering object (d) through Bragg's equation: $d = 2\pi/q$ (note that q is always positive). Therefore, the intensity plotted at a $-q$ value of −0.15 is most sensitive to scattering objects ~40–50 nm in diameter. From *Swift et al.* [2014b].

Figure 4.22 left and center panel. In these figures, relative anisotropy may be visually estimated by observing the distance between the two curves at each q value.

Figure 4.22 left panel shows data for a sample with a R_o value of 1.06% composed of calcite > illite > quartz, whereas Figure 4.22 center panel shows data for a more mature Pt. Pleasant sample (R_o of 2.02%) composed of illite > quartz > calcite. The calcite-rich sample was anisotropic at characteristic length scales of 50 nm and larger and isotropic at smaller length scales ($\gtrsim$5 nm). The clay-rich sample, however, was anisotropic at every scale of measurement. The findings for the calcite-rich sample, therefore, run counter to the expected results: slit-shaped clay flakes that tend to dominate the smaller length scales, yielding greater anisotropy at the nanoscale. Clasts and fossils at length scales from microns to a millimeter or more dictate the orientation of juxtaposed clay packets. This local perturbation of the azimuthal angle of bedding is scale dependent. SEM images shown in Figure 4.21 explain this behavior. We see that multiple scales of objects impact the orientation of nearby clays. Therefore, at smaller scales, local geometries are more randomized. The clay foliations are slit shaped, but in a sufficiently fossiliferous (or bioturbated or clastic) mudstone, they lack a strong preferred orientation.

Zhao et al. [2017] used a combination of SANS with low-pressure N_2 physisorption (BET) and MICP to study the pore structure of calcite (51.5%)-muscovite (20.6%)-quartz (15.3%)-rich Utica (1719.7 m depth, Coshocton Co., Ohio) and three other *oil* shale samples (Niobrara, Wolfcamp, Bakken). SANS-derived porosity for the Utica was 1.48% with a slightly rough pore-matrix interface (D = 2.8–2.9). This study highlighted a number of features of this particular Utica sample worth noting: (i) the high clay content may have caused the lower volume of mercury injection, (ii) the relatively large difference of PSD between SANS and BET suggests a poorly connected pore system which is consistent with the low porosity and cumulative pore volume determined from MICP, and (iii) over half of the measured pore volume in the pore size range of 10–50 nm is inaccessible, and ~90% of the measured pores are inaccessible for pores 5–10 nm in size. The main take-home message from the *Zhao et al.* [2017] study was the combination of different methods including SANS revealed multi-scale inaccessible porosity, in this case, five different pore-diameter intervals for the Utica as well as the Niobrara, Wolfcamp, and the Bakken mudstones.

4.4.3. Other Organic-Rich Mudstones

In addition to the formations discussed above, a combination of (U)SANS and other methods such as mercury, He, and/or BET-based porosimetry has been used to quantify pore features for a few samples of Triassic and Cretaceous organic-rich shales. *Bahadur et al.* [2014] investigated porosity as a function of mineralogy in the Late Cretaceous Second White Specks River and Belle Fourche Formations from the Western Canadian basin in Alberta. Their samples came from a rather narrow depth interval (1818.8–1826.6 m) and exhibited very similar TOC (2.04–2.6 wt%) and vitrinite reflectance (0.78–0.85%) values. Mineralogy, however, ranged from carbonate rich (calcite and dolomite) and clay rich (illite and kaolinite) in the shallowest sample to quartz-albite-clay rich in the deepest sample. Differences in the scattering results were attributed to variations in mineralogy and not changes in OM. Closed porosity (total scattering porosity minus He porosity) increased as quartz, albite, and, to a lesser degree, illite increased relative to the carbonates. Scattering experiments also revealed the presence of fractal and non-fractal pores. For all samples, there was a pore size regime between ~40 and 150 nm where pores were relatively smooth (non-fractal), but surface fractal behavior occurred above and below this scale range. Both porosity and SSA were dominated by contribution from mesopores and micropores. In a non-scattering porosity and PSD study, *Furmann et al.* [2016] concluded that (i) OM contributed to the overall microporosity with additional contribution from kaolinite and chlorite, (ii) the increase in total porosity seemed to correlate with the increase in quartz content, and (iii) in-core variations may be attributable to greater abundances of kaolinite and illite, which contribute to the mesopore volumes.

Clarkson et al. [2013] examined one sample from the Late Cretaceous Milk River Formation from central Alberta with a TOC of 1.28 wt% and a vitrinite reflectance value of 0.7%. The mineralogy was composed of montmorillonite (~45%), quartz (~28%), muscovite (~10%), albite (~8%), kaolinite (~7%), and pyrite (1–2%). While this is only one sample, it is instructive to take a closer look at the scattering results because of the uncharacteristic abundance in smectitic clays and other phyllosilicates (total ≈ 63%) compared to many typical gas shales, which tend to contain <40% clay composed generally of illite and/or chlorite. *Clarkson et al.* [2013] claimed that a log $I(q)$ versus log q slope of −3.2, corrected for background, was indicative of polydispersed 3D objects rather than 2D sheets. A significant fraction of the pore sizes and pore throat dimensions should, however, be approximately the same in clay-rich rocks, that is, slit-shaped pores, which makes the application of the PDSP model perhaps less effective in analyzing the pore features. A more suitable approach would be to assess scattering in terms of a dual-geometry model that considers both slit and spherical pores [*Beckinham et al.*, 2017]. Setting this issue aside, the scattering results from this sample suggest multiple scales of porosity: (i) pores with diameters between ~2 and 3.5 nm possibly related to the interlayer repeat spacing plus layer thicknesses known for montmorillonites and (ii) pores between roughly 120 and 800 nm representing pore space between coarser grains of quartz and/or albite wrapped by clay folia.

In contrast to this high clay content sample, *Clarkson et al.* [2012] examined Lower Triassic tight gas siltstones (depth range 2197.7–2213 m) from the Montney Formation using (U)SANS and petrophysical methods. In tight gas reservoirs, permeability is typically controlled by a combination of very fine laminae, extensive cementation, and moderate clay content. The mineralogy is dominated by quartz (29.8–41.4%), dolomite (20–29.8%), and feldspar (orthoclase > albite) (14.3–23.7%) with somewhat less abundant phyllosilicates (muscovite > chlorite > illite) (10.5–17.2%). U(SANS) experiments conducted at ambient pressure and temperature on lithologically diverse subsamples of three core plugs found broad PSDs. The pore structure of the three samples was fairly uniform, with small differences in the small pore range (< 200 nm), possibly related to differences in degree of cementation and mineralogy, in particular the clay content. Total porosity was similar to (but systematically higher than) He porosities measured on the whole core plug. As expected, porosity determined by scattering exceeded that measured by He porosimetry as the former accesses all pores and the later only accesses connected porosity. Combined, these yielded an estimate of the percentage of open porosity as a ratio of connected porosity, established from He adsorption, to total porosity, estimated from scattering. *Clarkson et al.* [2012] found that the greatest open porosity correlated with the highest permeability as determined by both pressure (probe) and pulse-delay permeametry. Surface area calculated from low-pressure nitrogen and carbon dioxide adsorption was significantly less than estimates from U(SANS). This is due, in part, to limited accessibility of the gases to all pores. To facilitate more direct comparison of the (U)SANS PSD and the sorption isotherms, scattering from one sample was converted to a pore volume distribution (Fig. 4.23). The resulting PSDs show a remarkable consistency in trends, particularly for pore radii from 4 to 100 nm.

King et al. [2015] used SANS coupled with helium ion microscopy (HIM) and conventional petrophysics to assess the PSD and pore architecture in gas shales from

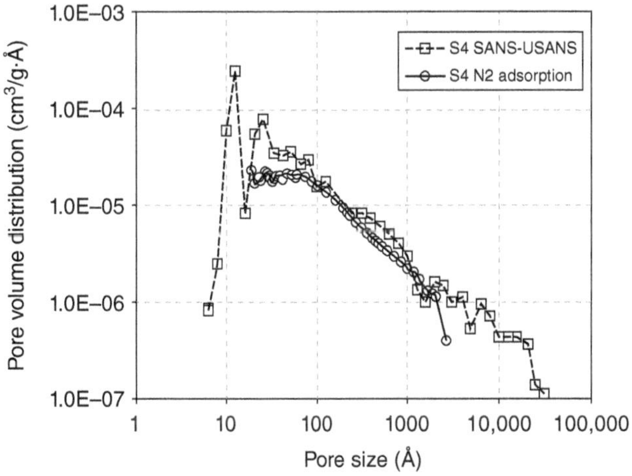

Figure 4.23 Comparison of pore volume distribution (PSD) obtained from (U)SANS and nitrogen adsorption data for one example of the Montney Formation. From *Clarkson et al.* [2012]. Reprinted with permission of Elsevier.

the Powder River (Late Cretaceous Mowry Formation) and Horn River basins (Middle to Late Devonian Middle Otter Creek Park and Evie Members of the Horn River Formation). The focus was on finer length scale features and pores associated with OM. SANS results demonstrated that the PSD did not adhere strictly to the standard power law distribution across many length scales typically found in sedimentary rocks [but see *Anovitz et al.*, 2015b], but exhibited an anomalous population of ~2 nm pores. These were generally hosted in the OM (Fig. 4.24) and comprised roughly one third of total porosity. The power law distribution was associated with inter/intragrain mineral porosity. *King et al.* [2015] drew upon insights from well-established models of kerogen diagenesis to propose a model of *foamy porosity* at small length scales that reflects the diagenetic evolution of the OM (kerogen). The concept they described identifies the cross-linking of kerogen combined with phase separation of either gas or oil, leading to arrested coarsening at a size determined by the cross-link density, yielding a diameter of about 2 nm. The HIM images supported the SANS size distribution, confirming the small pore size associated with OM. Combined SANS and MICP analyses indicated that pore/pore throat ratios in shale are nearly independent of pore size, quite distinct from typical rocks, and unexpected from simple sphere packing. They proposed that interpore throats are the result of microfractures generated during kerogen diagenesis when the phase separated oil/gas internal pressure exceeds the surrounding matrix yield strength. A similar idea was suggested by *Radlinski et al.* [1996, 2000b] in the Velkerri Formation.

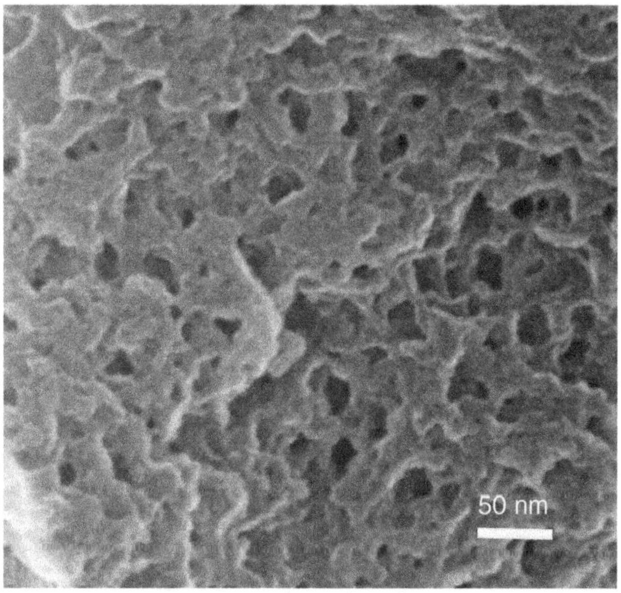

Figure 4.24 Helium ion microscopy image of organic porosity in the Mowry Formation shale from the Powder River Basin, Wyoming. Reprinted (adapted) with permission from *King et al.* [2015]. Copyright (2015) American Chemical Society.

There have been recent reports on the use of either SANS or SAXS to interrogate the pore features of organic-bearing mudstones in China. Specifically, *Sun et al.* [2017a] used SAXS to assess the pore system in the Paleogene-age Shahejie Formation in Dongying Sag, Bohai Bay Basin; *Yang et al.* [2017] employed SANS and SEM to examine nanoscale pore structure in the Longmaxi Formation, Sichuan Basin; and similarly, *Sun et al.* [2017b, 2018] used SANS in concert with MICP and He pycnometry to characterize the Lower Silurian Longmaxi Formation, northwest of Guizhou, China, and the Lower Cambrian Niutitang Formation, southeast Chongqing, China. In general, the ultimate goal of these studies is to better understand the gas storage capacity and production from hydraulic fracturing.

The *Sun et al.* [2017a] study of the Shahejie Formation is noteworthy because they used solvent extraction methods (e.g., dichloromethane, methanol, NaOCl) to remove both the bitumen (as low-maturity, high-wax compounds) and the OM. In doing so, they were able to compare SAXS results from before and after the extraction process, which allowed them to quantify the change in volume percent of the mesopores, thus revealing the oil content. The relative content of fine mesopores (2–10 nm) increased significantly after extraction, reflecting shale oil storage space that contains high-wax crude oil, thus providing important evidence for the lower limit of occurrence of high-wax crude oil in Dongying Sag. The negative correlation between the diameter of the maximum peak point and the clay mineral content in samples after extraction reveals that the clay mineral content affects the PSD of shale oil reservoir space.

The Longmaxi Formation has been the major gas shale target in the Sichuan Basin since about 2014. *Yang et al.* [2017] analyzed that four distinct shale lithologies with SANS along with MICP and He pycnometry: one silt rich, two siliceous rich, and one very carbonaceous rich. Quartz and TOC increase with depth, whereas the carbonate and feldspar content tend to decrease. Curiously, *Yang et al.* [2017] report that all four samples exhibit mass fractal behavior, which seems counter to many other shale scattering studies that have revealed both surface and mass fractal behavior. The Porod invariant method and the PDSP model yielded data on PSD, SSA, and porosity that compared favorably with MICP results. The scattering intensity decreased by over five orders of magnitude with increasing depth from 2371 to 2402 m, likely reflecting a change in porosity and perhaps pore structure. *Yang et al.* [2017] report that the fraction of closed pores also increased with increasing depth and was attributed to an increase in TOC and silica content.

A similar study of the Longmaxi Formation was conducted by *Sun et al.* [2017b] using SANS, He pycnometry, and low-pressure N_2 and CO_2 adsorption. Nine samples were investigated from two wells (XY1, 610, 643.2 m; TY1, 6673, 674 m). They found a positive correlation between TOC and the porosity obtained from both He pycnometry and SANS (using the PDSP approach). The percentage of closed porosity estimated from the difference between the SANS and gas adsorption techniques ranged from ~6.7 to 42.6%, but this varied strongly between samples. They also found a positive correlation between the mass fractal dimension, which ranged from 2.79 to 3, and the fraction of closed porosity.

A range of petrophysical methods such as MICP, BET, and He pycnometry were used to complement a SANS study of the quartz-clay-rich Niutitang Formation [*Sun et al.*, 2018]. A comparison of MICP and SANS data allowed for an estimate of closed pores for each of the four samples analyzed. The mean pore throat size was <20 nm, whereas the average pore size determined from SANS was estimated to be about 3 nm. The trend of closed pores approximately correlated with the clay to quartz ratio wherein roughly 35% of the available pores space was closed at a clay/quartz ratio of ~0.8 versus less than a few percent closed pores for a clay/quartz ratio of 0.33. Despite this interesting correlation, *Sun et al.* [2018] claim that most of the closed pores are associated with OM and not the clay.

4.4.4. Organic-Poor Mudstones

While the economic impact of organic-rich shales has clearly led to a preponderance of SAS studies on these materials, this is clearly not the only importance of shales. They are, for example, important features of many landscapes, and their weathering behavior plays a key role in landscape development. In addition, many shales are important caprocks. This is especially true with respect to proposed geological sequestration of carbon dioxide. Studies have, therefore, also been completed to evaluate the properties of shales in these environments.

4.4.4.1. Eau Claire Formation

The Middle to Upper Cambrian-age Eau Claire Formation, which consists of a fine to very fine grained quartz-rich sandstone interbedded with variably colored shales and mudstones, is a regionally extensive aquitard that impedes the exchange of groundwater between the Mt. Simon Sandstone below and overlying aquifers in parts of Wisconsin, Illinois, Indiana, and Ohio [*Neufelder et al.*, 2012]. The Eau Claire has been recognized as a good caprock for the Illinois Basin-Decatur GCS demonstration project in the Mt. Simon Sandstone, Illinois [*Barnes et al.*, 2009; *Carroll et al.*, 2013; *Freiburg et al.*, 2014; *Mehnert et al.*, 2014]. The Eau Claire has also been considered for other CO_2 sequestration projects such as FutureGen [e.g., *Bonneville et al.*, 2013; *Gilmore et al.*, 2014].

Swift et al. [2014a] focused on three representative lithofacies of the Eau Claire for characterization by (U)SANS, mercury porosimetry, BET, gas permeametry, polarized light microscopy (PLM), and XRD: an illite-rich shale (2710.1 ft. depth), a fine- to medium-grained carbonate-rich mudstone (2846 ft. depth), and a fine- to medium-grained glauconite-rich mudstone (3102.8 ft. depth). The neutron scattering results obtained by *Swift et al.* [2014a] are shown in Figure 4.25(a–d). Figure 4.25a shows the scattering curves for the three samples. Figure 4.25b shows the Porod transform of the same data. The Porod transform plots log q^4I as a function of log q and normalizes the slope of the scattering data to the value (−4) expected for scattering from a smooth interface. In such a plot, "fuzzy" boundaries will have a negative slope, smooth interfaces will have a slope of zero, surface fractals will have a slope between 0 and 1, mass fractals will have a slope between 1 and 2, and the background should have a slope of 4. By removing much of the log scaling from the intensity, this transform allows more subtle details in the scattering pattern and hence pore features to be observed [cf., *Anovitz et al.*, 2013a]. At low q, the scattering intensity from the glauconite-rich mudstone was larger than that of the other two samples. Both the illitic shale and the glauconite-rich mudstone show slight increases in scattering intensity (small humps) at high q, approximately at $q = 0.1$ (~5.5 nm) and $q = 0.05$ (~12.5 nm), respectively. If such a peak exists in the carbonate-rich mudstone, it is at the very high q limit of the data and mostly hidden in the background. While the high-q peaks make definition of fractal slopes more difficult, all three appear to have high-q background slopes near −3. These flatten at lower q ($<10^{-3}$ Å^{-1}) in the Porod transform, and the slope of the scattering curve approaches the −4 value expected for smooth surfaces. The scale at which this occurs increases with depth from the illitic shale to the glauconite-rich mudstone. At larger scales (q values near 10^{-4} Å^{-1}), the illitic shale (and possibly the glauconite-rich mudstone) shows evidence of mass fractal behavior as the slope of the scattering curve drops below −3.

Using a method described by *Anovitz et al.* [2013a], *Swift et al.* [2014a] transformed the scattering data to yield cumulative porosity curves. These (Fig. 4.25c) showed both similarities and dramatic differences among the three lithofacies. All three showed bimodal porosity, with inflection points between 1 and 10 μm (1.9, 5.9, and 4.7 μm for the illitic shale, the carbonate-rich mudstone, and the glauconite-rich mudstone, respectively). The fractions of the total porosity below this point, however, differ. For the illitic shale, the most clay rich, 86.6% of

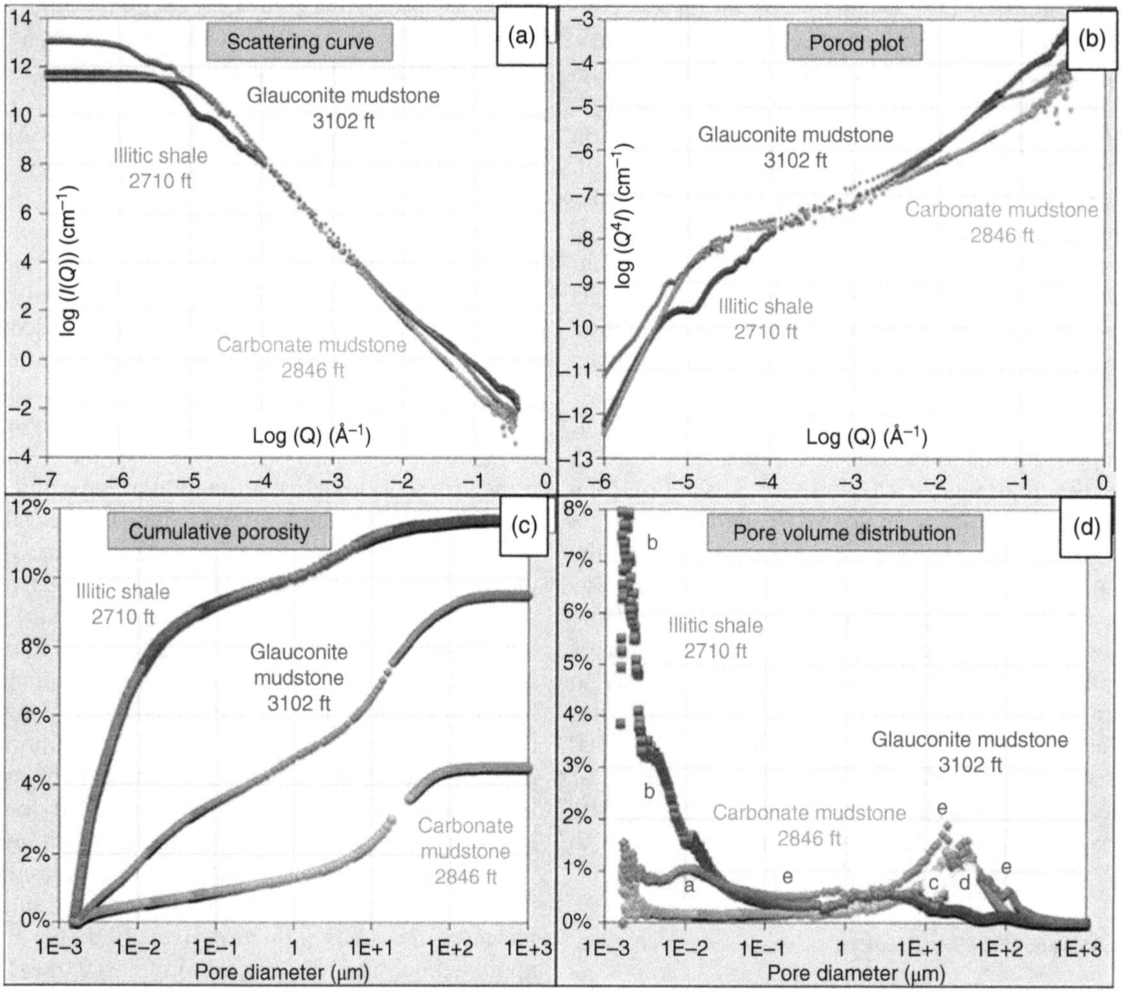

Figure 4.25 Neutron-scattering (NS) results on the Eau Claire Formation. Lithofacies and associated depths are: illitic shale (2710.1 ft/826.0 m), carbonate mudstone (2846 ft/867.5 m), and glauconitic mudstone (3102.8 ft/945.7 m). (a) Scattering curves. (b) Porod plots for the same samples. (c) Cumulative porosity versus pore size for the same samples. (d) Differential pore volume distribution versus pore diameter, defined as change in percent porosity at each pore diameter, divided by the difference in spherical pore diameters, multiplied by the average pore diameter. The letters correspond to SEM images shown in Swift et al (2014a) and indicate the length scales at which each mineral-related pore type contributes to total porosity. (From *Swift et al.* 2014a, Relationship between mineralogy and porosity in seals relevant to geologic CO_2 sequestration. Environ Geosci 21(2): 48; AAPG© [2014], reprinted by permission of the AAPG whose permission is required for further use.)

total porosity is at smaller scales, and fully 70% is ≲50 nm. This contrasts with the carbonate-rich mudstone, which has a somewhat larger inflection point, a smaller fraction of porosity below the inflection point (43.5%), and only 17.2% below about 50 nm. Results for the glauconite-rich mudstone are intermediate between the other two, with 59% of the total porosity below the inflection point and 30.3% below 50 nm.

Once the cumulative pore volumes had been obtained, *Swift et al.* [2014a] calculated PSDs. They urged caution, however, as this requires an assumption about pore shape. *Anovitz et al.* [2009, 2011] documented extremely variable pore shapes using TEM down to the tens of nm scale. In addition, the fractal pore/solid boundary further complicates such assumptions. However, stepwise subtraction of the cumulative porosity curves yields a pore volume distribution without pore shape assumptions (Fig. 4.25d). Pore scales fell into several groups, not all of which are present in each material. Two pore regimes occur near 25 and 135 Å, and a broad, larger-scale regime is centered around 10–20 μm in both mudstones. The illitic shale is dominated by the first of the two nanoscale distributions. Microscale pores form only a small fraction of the total in this sample and are polydispersed, with a broad hump around 2 μm. While nanoscale pores appear to be present in both mudstones, only the glauconitic mudstone has a significant peak near 10 nm, mirroring the weaker pore size cluster at that scale in the shale. These reflect the high-q *humps* described above. In the microscale regime, the glauconite-rich mudstone has a larger peak at 20 μm, which may have a shoulder around 115 μm. The carbonate mudstone has a narrower peak at 30 μm that may correspond to part of the wider distribution observed in the glauconitic sample.

Swift et al. [2014a] observed a bimodal porosity distribution as a function of both pore diameter and pore throat diameter for the shale and mudstones, but the importance of pores at each scale to total and connected porosity differed between lithologies. Pore/pore throat ratios at these two scales diverge markedly, being almost unity at the nanoscale regime dominated by illitic clay and micas, and one and a half orders of magnitude at the microscale in a clastic mudstone. Individual minerals, primarily illite and glauconite, have unmistakable pore and pore throat signatures and contribute disproportionately to connected reactive surface area. The pore types created during diagenesis influence the mix of pore types at the microscale and mediate profound differences between bulk and pore network-accessible mineralogies in the mudstones.

4.4.4.2. Rose Hill Shale

Jin et al. [2011] used (U)SANS in concert with geochemical and mineralogical data to characterize the evolution of nanoscale features in a deep weathering profile in Rose Hill shale within the SSHO. The goals of this effort were to determine (i) porosity evolution as a function of depth in the shale, (ii) the relation between changes in chemistry and porosity, (iii) the mineral-water interfacial area in the unaltered rock and how it changes with reaction progress, (iv) the fractal nature of the rock-pore interface, and (v) what proportions of pores are connected and thus accessible for water-rock interaction.

The study site was a V-shaped catchment basin underlain by the Rose Hill Formation in central Pennsylvania. The Silurian-age shale is olive-pink to grayish-buff with a few interbedded limestones and is locally characterized by fracture cleavage that has produced pencil-like fragments up to 30 cm long. Shale chips were recovered from several shallow holes (maximum depth 20 m) augered or drilled at different locations within the catchment. Collectively, the holes covered from the regolith zone through weathered rock (saprolite) to relatively unaltered shale.

Overall, absolute scattering intensities from the shallow regolith are greater than those for samples from deeper regolith and weathered rock, especially in the USANS region; that is, scattering intensity increased with increased weathering. Samples from relatively shallow depths of the drilled core (30–170 cm depth) scattered neutrons more intensely than the deeper samples (520–2000 cm depth) over the entire q range interrogated. At low q, shale chips from shallow depths in the regolith also scattered neutrons more strongly than those from deeper depths. However, the trend was reversed at high q: chips from deeper regolith scatter more strongly than those from shallow depths. Porod transforms yielded information about surface and mass fractal behavior as a function of depth (Fig. 4.26a). The deepest samples from the two regolith profiles and all those from the drilled core were characterized by Porod exponents (n, the negative of the slope of the log $I(q)$ vs. log (q) curve) values between 3 and 2 (mass fractal dimensions D_m between 3 and 2). In contrast, the chips from the shallowest regolith samples have Porod exponents between 4 and 3 (surface fractal dimensions D_s between 2 and 3).

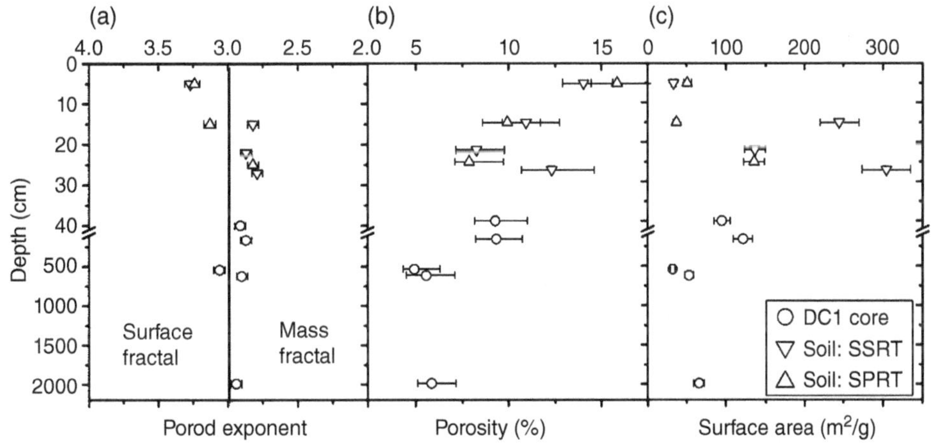

Figure 4.26 Variations of power law Porod exponent (a), porosity (b), and specific surface area (c) as a function of depth for Rose Hill shales that have been weathered to different extents. The Porod exponent, n, indicates the type of fractal ($2<n<3$, mass fractal, $3<n<4$, surface fractal) [*Jin et al.*, 2011].

Jin et al. [2011] took advantage of the fact that scattering intensities depend on the density of the interfacial features that scatter neutrons. They calculated porosity using a random two-phase approximation (mineral matrix-pore space). The shale chips from the drilled saprolite comprised two distinct groups: an uppermost group (< 5 m) with porosity as high as 9.3% and a deeper group (> 5 m) with an average porosity of 5–6% (Fig. 4.26b). This increase continued above the saprolite-regolith interface to the ground surface, reaching ~8% at 30 cm and 14–16% at 10 cm. These data indicated that more and more pores were produced in the shallower samples, especially larger pores. The porosity and fractal dimension of the two regolith localities (depths 0–30 cm) were very similar.

To obtain SSA, *Jin et al.* [2011] used the PRINSAS program. SSA varied significantly with depth (Fig. 4.26c). The least weathered samples (depths 520–2000 cm) exhibited low SSA (between 30 and 60 m^2/g) well within the range of SSA determined by gas adsorption for typical shales from depths of 500–3500 m [*Radlinski et al.*, 1999]. As weathering progressed, SSA increased gradually to 100 m^2/g (depths 30–170 cm) and eventually reached 300 m^2/g 30 cm below the surface (20–30 cm depth). SSA then decreased to 40 m^2/g at 5 cm (0–10 cm depth), although porosity continued to increase.

Jin et al. [2011] also conducted scattering experiments on shale samples soaked with a D_2O/H_2O mixture with an SLD similar to that of the average rock matrix. They compared the results to scattering from dry samples to assess pore connectivity versus extent of weathering. Even with the soaked samples, the trend of increased porosity with increased weathering was preserved. The largest connected porosity was observed at 0–30 cm (5.0% of total rock volume), followed by samples from depths of 25–29 cm (4.1%), 30–40 cm (2.5%), and 610–630 cm (0%). Unconnected porosities, however, were constant for the relatively unweathered samples (around 2%). Intriguingly, more unconnected pores were observed in the chip derived from the shallowest regolith, 0–10 cm depth, than in the deeper samples. Figure 4.27 emphasizes this point: that is, unconnected porosity decreased with weathering only until the very surface, while connected porosity increased with weathering in all samples.

The changes in pore evolution revealed by scattering were correlated to variations in regolith and saprolite mineralogy. At ~20 m, dissolution was inferred to have depleted the bedrock of ankerite. All the chips investigated with (U)SANS were from above the ankerite dissolution zone. Neutron scattering documented that 5–6% of the ankerite-free rock volume was composed of isolated intraparticle pores. At 5 m depth, an abrupt increase in porosity and surface area corresponded to the onset of feldspar dissolution in the saprolite. This was attributed mainly to periglacial processes from 15 ka. At tens of centimeters below the saprolite-regolith interface, porosity and surface area increased markedly as chlorite and illite began to dissolve. These clay reactions contributed to the transformation of saprolite to regolith. Throughout the regolith, intraparticle pores connected to form larger interparticle pores, and scattering changed from a mass fractal at depth to a surface fractal near the surface. In the most weathered regolith, kaolinite and Fe oxyhydroxides precipitated, blocking some connected pores. These precipitates, coupled with exposure of more quartz by clay weathering, contributed to the decreased mineral-pore interfacial area in the uppermost samples.

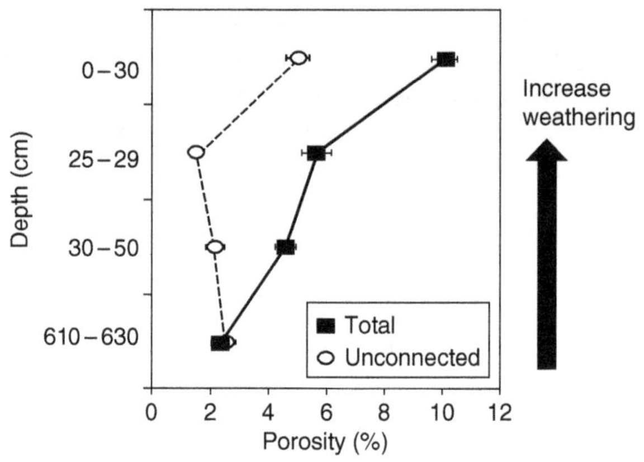

Figure 4.27 Estimates of total and unconnected porosity as a function of depth derived from a comparison of scattering from dry Rose Hill shale versus samples soaked with a D_2O/H_2O mixture [*Jin et al.*, 2011].

4.5. OM POROSITY EVOLUTION DUE TO SHALE REACTIONS

Scattering studies have also been effective at delineating the impact of dehydration, maturation, or dissolution of OM (kerogen) and reaction with supercritical CO_2 on pore features. Recently, *Lee et al.* [2014] investigated how pore size, porosity, and fractality changed because of dehydration of Silurian black shale from the Baltic Basin, Poland. They performed ultra small angle X-ray scattering (USAXS) measurements on shale from a wide range of depths along a burial diagenetic sequence. They point out that USAXS covers the range from 2.5 to 1000 nm, making it a good match for the range of pore sizes responsible for much of the total porosity and fluid transport in shales and mudrocks. Samples were selected for variation in OM content and thermal maturity, ranging from immature to overmature, providing the opportunity to track changes in pore structure with burial depth and temperature within the same geologic strata. The samples were gently crushed, sieved to ≤425 μm, and split into two groups: one air-dried at a relative humidity of 50% and the other heated under vacuum at 200°C for 24 h.

Two major trends were observed: porosity values (i) decreased with depth as expected and (ii) increased upon sample dehydration. Total USAXS porosity decreased with depth from 10–16% at <2000 m to 4–9% at >2500 m. Dehydration removed capillary water and water bound to clay surfaces. Measured porosity values for the dehydrated samples matched those from the same samples measured by kerosene immersion porosimetry. The increase in total porosity upon dehydration was primarily controlled by an increase in the proportion of pores between 100 and 1000 nm in diameter. This may be an important contributor to the structural deformation of pores. Additionally, *Lee et al.* [2014] quantified the nature of the microstructural pore-shale interfaces, noting that characterization of the pore-rock interface is essential to estimate how irregularity affects diffusive transport or chemical interaction phenomena. Surface fractal dimensions decreased in the smallest pores, trending to more mass fractal-like behavior. Surface roughness increased because of the removal of capillary water and turbostratic stacking of clay minerals upon dehydration of the interlayers. The larger pores showed little change in fractal dimension.

To correctly interpret and normalize scattering data, the SLD of the phases in the material is required. For shale, this means knowing values for each mineral, the OM (kerogen), and pores. As discussed above, *Hall et al.* [1983] and *Radlinski et al.* [1996, 2000b] explored the use of the varying contrasts afforded by SAXS and SANS to study the relative locations of organic phases. *Radlinski and Hinde* [2002] calculated SLDs for the major organic components of sedimentary rocks. *Bahadur et al.* [2015] used SLD values derived from coal, but this yielded unreasonably high porosities. *Thomas et al.* [2014] point out that the SLD of many phases can be estimated with reasonable accuracy from compositional data but that it is not clear this is the case for kerogen, as its SLD is expected to vary considerably with OM type and thermal maturity. To address this issue, they measured the SLD of pure kerogen samples separated from the shale by acid demineralization. By varying the kerogen type (I (immature Green River Formation, Colorado), II (overmature Marcellus Formation, Pennsylvania; immature Woodford Formation, Oklahoma), IIS (high-sulfur Monterey Formation, California), or III (Pond Creek bituminous coal)) and the associated thermal maturity, samples with a broad range of hydrogen to carbon atomic ratios ($0.5 < H/C < 1.4$) were obtained. The SANS measurements used deuterated methanol to vary the SLD of the pore space in contact with the kerogen, allowing the SLD of the kerogen to be determined. This was also calculated from the elemental composition and density.

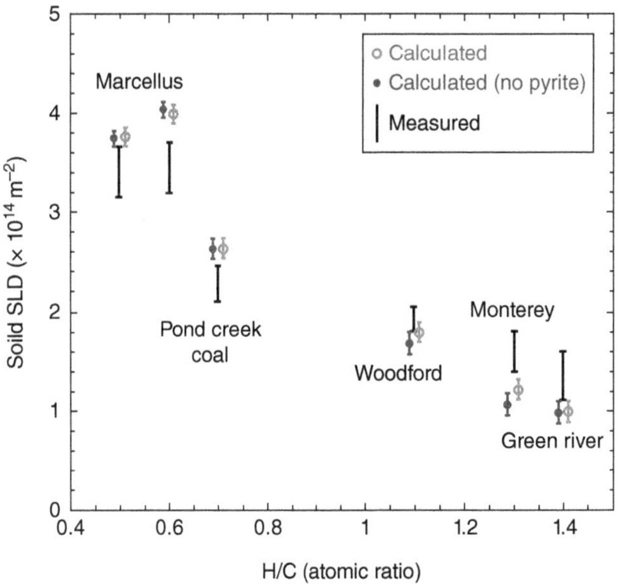

Figure 4.28 Comparison of calculated and measured values of the solid SLD for four shale formations and one coal. The vertical extent of the measured values reflects the variation observed over the range in Q, which is greater than other sources of uncertainty. Uncertainties on the calculated values of SLD are calculated from the reported standard deviations for the H and C elemental compositions and for the mass density. From *Thomas et al.* [2014]. Reprinted with permission from Elsevier.

They observed that the SLD of kerogen varies over a rather large range, from 1 to $4 \times 10^{14}\,m^{-2}$, increasing significantly with decreasing H/C, and that the measured and calculated values are in reasonable agreement (Fig. 4.28). A key result from their study is that for shales containing immature kerogen with an SLD close to 1, the scattering between the kerogen and mineral phases may be significant, such that the shale cannot be treated as a two-phase material, a result similar to that reported by *Radlinski et al.* [1996]. Moreover, the surface area of extracted kerogen does not increase with thermal maturity. This result contrasts with the contemporary literature which clearly indicates that catagenesis leaves nanopores in kerogen. Therefore, they provided a cautionary note that the acid demineralization procedure alters the kerogen structure.

To quantify how native *extractable* OM is distributed with respect to pore size, *DiStafano et al.* [2016] investigated the relationship between hydrocarbon chemistry and pore structure in two shales, the Eagle Ford (four samples) and the Marcellus (one sample). These shales were subjected to different extraction methods: pyrolysis gas chromatography (GC), thermogravimetric analysis, and series of organic solvent extractions (acetone, dichloromethane [DCM], hexane, dodecane, methanol, and toluene). In general, the amount of native OM

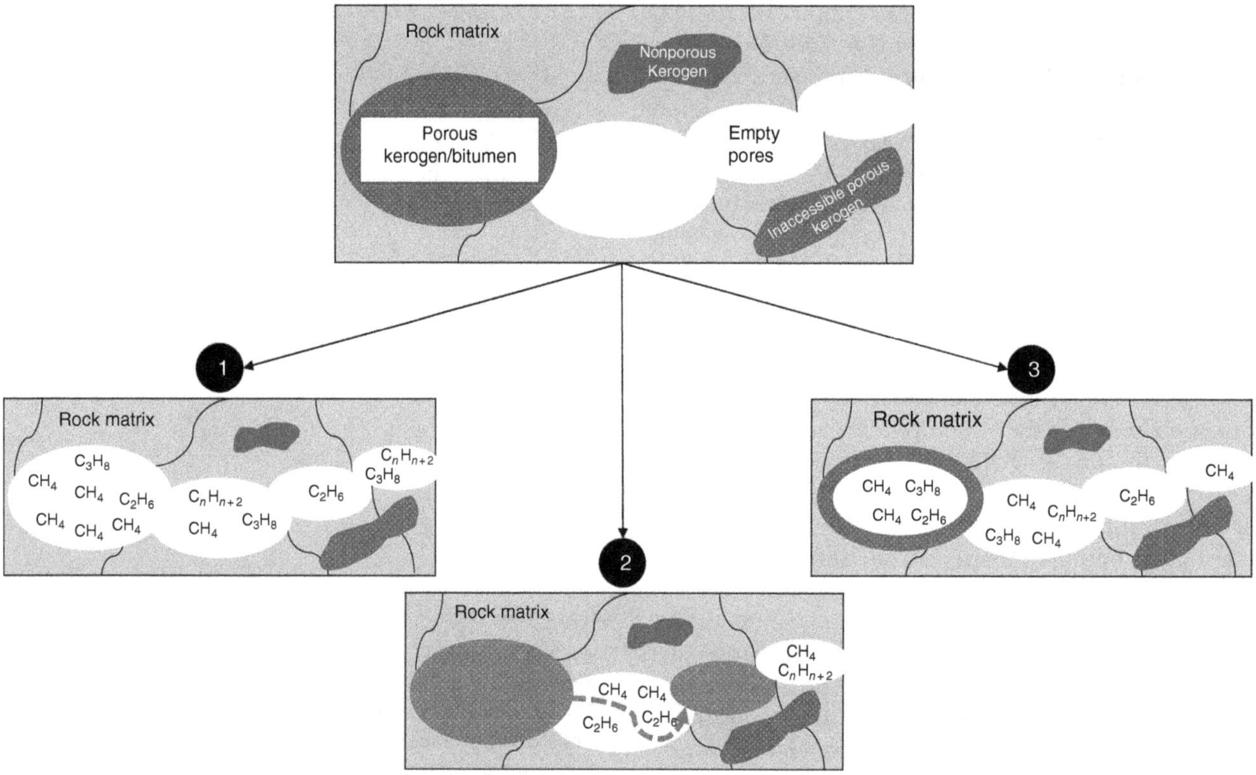

Figure 4.29 Possible mechanisms of porosity development due to extraction by solvents that could be observed by SANS. Porous kerogen and bitumen in the rock matrix (composed of minerals and nonporous or inaccessible porous kerogen) could (1) completely break down into smaller molecules that can be extracted, (2) partially dissolve and migrate into smaller pores, or (3) partially break down, leaving behind smaller pores. From *DiStafano et al.* [2016]. Reprinted with permission from Elsevier.

extracted directly correlated with the percentage of clay in the shale. OM extraction efficiency was best for toluene, DCM, and hexane because they appear to be more chemically compatible to the paraffinic hydrocarbons present in the shales. SANS was used to determine the porosity and PSD before and after solvent extraction. They interpreted that the changes in porosity were due to either the extraction or breakdown of high molecular weight bitumen with high C/H ratios (e.g., asphaltenes and resins). Indeed, pyrolysis GC examination did reveal the presence of longer-chain molecules (C30–C40), whereas GC-MS analysis failed either because the OM was too heavy to be extracted by conventional GC-MS methods or not effectively dissolved by the solvent. Despite this limitation, DiStefano *et al.*, proposed three possible mechanisms for the extraction of some of the higher molecular weight compounds (Fig. 4.29). mechanism 1 (complete breakdown of asphaltenes and resins), mechanism 2 (partial dissolution and migration of asphaltenes and resins which become trapped in select portions of the pore matrix), and mechanism 3 (partial breakdown of refractory OM, creating smaller pores). Interestingly, porosity did generally increase for the carbonate-rich Eagle Ford (high and low maturity) and Marcellus samples in contrast to the clay-rich Eagle Ford samples which exhibited a general trend of porosity reduction especially for the low-maturity sample. This decrease in porosity was attributed to swelling of the pore matrix during uptake of solvent or migration of asphaltenes and/or resins. In contrast to porosity changes, there was no observed shift in the PSD in any of the samples treated with solvents. *DiStafano et al.*, [2016] concluded from these observations that mechanisms 1 and 2 tend to dominate over mechanism 3.

As noted above, the permanence of CO_2 storage at injection sites may depend on the distribution and connectivity of pores in caprocks and potential changes induced by reactions with supercritical CO_2. To explore this topic, *Mouzakis et al.* [2016] experimentally reacted either synthetic brine or synthetic brine + CO_2 with two different caprocks, the Gothic Shale, Utah (carbonate-rich) and the Marine Tuscaloosa Shale, MS (more siliciclastic rich) at 160°C and 15 MPa for ~45 days (brine + CO_2) or ~35 days (brine only; Ar as the pressure medium). The synthetic brine chemistries were based on Geochemist's Workbench modeling of either sampled brines plus original mineralogy (as in the case of the Tuscaloosa Marine Shale) or mineralogy only (as for the Gothic Shale). Samples, in the form of rock chips and powder, were interrogated with SEM, BET, and SANS prior to and after reaction. Further samples were soaked with a contrast mixture of 75% D_2O/25% H_2O to assess the connected versus unconnected porosity using SANS.

Both shales exhibited mineral dissolution, which was observed to be greater in the presence of the supercritical CO_2. The total porosity and pore connectivity increased in the Gothic Shale, whereas total porosity increased but connectivity decreased in the Marine Tuscaloosa. *Mouzakis et al.* [2016] concluded that greater dissolution of carbonate minerals compared to the silicates in the Gothic Shale in the presence of scCO_2 produced the increased porosity and connectivity for pores less than about 200 nm. Conversely, silicate phases in the Marine Tuscaloosa dissolved more than carbonates, resulting in increased porosity at all measurable length scales particularly for coarser pores greater than about 1 μm. A decrease in pore connectivity by almost 50% compare to the unreacted rock, however, was observed for the Marine Tuscaloosa reacted with scCO_2. Speculation pointed to either precipitation in pore throats or hydration of the abundant clay assemblage.

The study by *Pearce et al.* [2018] explored a similar theme wherein they reacted two different types of caprock mudstones from the Evergreen Formation, Australia, with impure CO_2 (+SO_2 and O_2) and low-salinity brine. One was a clay- and organic-rich mudstone, and the other was a more organic-lean feldspar-rich fine-grained sandstone. MICP indicated that the majority of pores in both cores had pore throat radii ~5–150 nm with porosities of 5.5–8.4%. After reaction with impure CO_2-brine, the measured pore throats decreased in the clay-rich mudstone core. Dissolution and precipitation of carbonate and silicate minerals were observed during impure CO_2 reaction of both cores via changes in water chemistry. SEM identified macroporosity in clays, mica, and amorphous silica cements. After impure CO_2-brine reaction, precipitation of barite, Fe oxides, clays, and gypsum was observed. SANS measured the fraction of total and non-accessible pores (~10–150 nm radii pores) before and after reaction. The fraction of pores that was accessible in both virgin caprocks had a decreasing trend to smaller pore size. The clay-rich caprock had a higher fraction of accessible pores (~0.9) at the smallest SANS measured pore size than the feldspar-rich fine-grained sandstone (~0.75). Both core samples showed a decrease in SANS accessible pores after impure CO_2-water reaction at CO_2 storage conditions. The clay-rich caprock showed a more pronounced decrease. After impure CO_2-brine reaction, the fraction of accessible pores at the smallest pore size was ~0.85 in the clay-rich caprock and ~0.75 in the feldspar-rich fine-grained sandstone.

4.6. SUMMARY AND OUTLOOK

The studies described above highlight how both neutron and X-ray small angle scattering techniques have been used for over 30 years to study the pore structures of shales. These results are summarized in Table 4.1. This began with the first analyses of rocks using SAS [*Hall et al.*, 1983] and has continued through more recent studies, many of which have focused on shales associated with hydrocarbons such as gas shales. There have been several key outcomes from these studies that impact our fundamental understanding of the multi-scale pore structures of shales relevant to caprocks for CO_2 sequestration, barriers for nuclear waste isolation, hydrocarbon generation, hosting and migration, and regional as well global weathering.

4.6.1. Key Outcomes of Research to Date

• Although the pore structure of shales is highly asymmetric, there are regularities to the pattern. Asymmetry is not observed in samples cut parallel to bedding, but is observed, and commonly yields elliptical scattering patterns, for samples cut perpendicular to bedding. The extent of asymmetry decreases with increasing maturity, but this depends on both scale and the overall mineralogy of the specific shale sample [cf., *Hall et al.*, 1983; *Swift et al.*, 2014b; *Anovitz et al.*, 2015a; *Gu et al.*, 2015, 2016; *Gu and Mildner*, 2016].

• Scattering patterns commonly show power law variations in the log of the scattering intensity as a function of the log of the scattering vector q. This is in keeping with scattering from a matrix-pore interface that is a surface fractal at smaller scales and, in some cases, a mass fractal at larger scales. The values vary between different shales and are not universal, although a pattern of increasing surface fractal dimension, followed by a shift to mass fractal behavior with depth, has been observed during weathering [*Jin et al.*, 2011].

• Changes in intensity during maturation may reflect pore invasion by generated hydrocarbons. In general, an increase in scattering intensity can be correlated with an increase in porosity.

• Mass fractal scattering in some shales may be related to formation of microfractures formed by local pressure increases during petroleum and gas generation [*Radlinski et al.*, 2000b; *Radlinski and Hinde*, 2002; *King et al.*, 2015], although the effects of depressurization and microfracturing during sample recovery also need to be considered.

Table 4.1 Summary of Results from SAS Papers on Shales.

Formation	Location	Age	Author [date]	Observations	Fractal dimensions	Comments
Miscellaneous shales			*Hall and Mildner* [1983] *Hall et al.* [1983] *Hall et al.* [1986] *Mildner et al.* [1986]	Anisotropic scattering for cuts perpendicular to bedding; symmetric scattering for cuts parallel to bedding; exploited combined SANS/SAXS; scattering porosities different from those from density, nitrogen BET, and MICP; bimodal pore distributions; SANS but no SAXS porosity below 2 nm; incoherent cross sections higher than suggested by bulk chemistry; fractal boundary interfaces; high-maturity samples change little with DCM extraction, and low-maturity samples change significantly	Differed between SAXS and SANS measurements due to pore filling	Earliest studies of small angle scattering of rocks. Observed most of the primary characteristics. Did not apply further to a general, regional, or local geological problem
Miscellaneous shales			*Wong et al.* [1986]	Carbonate surfaces: smooth >50 Å, rough <50 Å; sandstones and shales fractal at all observed scales; suggested roughness due to an Ising-like model with roughening transition dependent on fluid composition	Not universal; functions of clay content	26 samples, 4 shales
Bakken shale,	Williston Basin, North Dakota, Montana	Late Devonian/early Mississippian	*North et al.* [1988, 1990]	Difference between SAXS and SANS opposite *Mildner et al.* [1988]	(Bakken) $D_s = 2.45$ SAXS, 2.33 SANS, 2.27 USAXS	One sample of each
Utah clay	unknown	unknown			(Utah clay) $D_s = 2.55$ SAXS, 2.56 USAXS	
Velkerri Formation	McArthur Basin, NW Territory, Australia	Mid-Proterozoic (~1.43 Ga)	*Radlinski et al.* [1996] *Radlinski et al.* [2000a, 2000b] *Radlinski* [2006]	SAXS and SANS, also synthetic organic-rich shales; number of phases that must be considered for SAXS and SANS different. Smoothing with maturity due to melting or redistribution of macerals. Pyrolyzation experiments: scattering intensity decrease with onset of oil generation caused by filling of pores strongest at small length scales	Low maturity $D_s = 2.5$, SA = 31 m²/g Higher maturity 50–600 Å, $D_m = 2.8$; network of microcracks, larger scales smooth SA = 1.55 m²/g	Only parallel-cut samples
Frio shale	SE Texas	Oligocene	*Wong et al.* [1986] *Ma et al.* [1999a, 1999b] *Qi et al.* [2002]	SANS and adsorption experiments yield different fractal dimensions. Scattering a direct probe of interface; adsorption smoothed by surface tension	D_s (SANS) = 2.75, 2.83; D_s (BET) 2.43, 2.54 or 2.50, 2.63 D_s (thin film) = 2.24, 2.39 or 2.31, 2.34	Potential GCS caprock. Total of three samples

New Albany Formation	Illinois Basin	Upper Devonian/ Mississippian	*Mastalerz et al.* [2012] *Mastalerz et al.* [2013] *Bahadur et al.* [2015]	Initially one each from the Clegg Creek and Camp Run members SA (nitrogen) = 2.4, 0.2 m^2/g, (CO_2) = 12.8, 10.9 m^2/g. Difference suggests subnanometer pores. Larger surface area correlates with more OM. SA (SANS) = 14.8, 23.3 m^2/g, suggesting considerable inaccessible porosity. Increased TOC increases fraction of micropores at low maturity. Increasing maturity increases microporosity and reduces total porosity except at the highest R_o values	D_s = 2.8 for both initial samples.	Gas source and potential GCS caprock. Note 10-year gap from previous studies. Total of five samples. Thick samples for USANS measurements likely led to multiple scattering. Grain samples do not allow analysis of asymmetry. SLD used for OM estimated from coal may be incorrect. Did not account for H in clays when calculating backgrounds
Barnett Shale	Bend Arch-Fort Worth Basin, Texas	Mississippian	*Clarkson et al.* [2013] *Ruppert et al.* [2013]	***Clarkson et al.* [2013]:** (U)SANS/adsorption/ MICP comparison. One Barnett sample at 80°C under a range of CD_4 pressures. Porosity bimodal (clay interlayers and between clay packets). 85% accessible porosity from 400 to 3000 Å, less for smaller and larger pores ***Ruppert et al.* [2013]:** two samples from similar depths with similar maturities. >85% of pores accessible for sizes >250 nm, declined to ~65% for smaller pores in one sample, SLD matching suggested pores primarily in pyrite, illite, or OM	***Clarkson et al.* [2013]:** D_s = 2.7 ***Ruppert et al.* [2013]:** D_s = 3	***Clarkson et al.* [2013]:** 1 mm thick samples likely led to multiple scattering distortion, at least in the USANS range. Cut parallel to bedding; anisotropy not observable. Dried at 60°C under vacuum before analysis; effect unknown **Rupert *et al.* [2013]:** –0.8 mm samples also likely led to significant multiple scattering
Eagle Ford Shale	SE Texas and NE Mexico	Late Cretaceous	*Clarkson et al.* [2013] *Anovitz et al.* [2014] *Anovitz et al.* [2015a]	***Clarkson et al.* [2013]:** one sample. Largest N_2 adsorption pore volume (~ 0.05 cm^3/g), intermediate surface area (~9 m^2/g). CO_2 adsorption intermediate (~0.0025 cm^3/g). Hg intrusion maximum pore throat dimensions ~ 30 Å, no bimodality. (U)SANS, pore fraction = 0.0683, SA = 4.59×10^{-5} cm^2/cm^3 ***Anovitz et al.* [2014, 2015a]:** 10 samples cut perpendicular to bedding covering a range of depths, compositions (clay and carbonate rich) and R_o. Both surface and mass fractal behavior observed. Asymmetry reduced in clay-rich samples and eliminated in carbonate-rich samples with maturity. Primary loss of total porosity during early maturation	***Clarkson et al.* [2013]:** D_s = 2.8	***Anovitz et al.* [2015a]** questioned the applicability of the two-phase approximation for shales [cf., *Radlinski*, 2006]. Samples thin enough to eliminate most multiple scattering

(*Continued*)

Table 4.1 (Continued)

Formation	Location	Age	Author [date]	Observations	Fractal dimensions	Comments
Marcellus Fm.	Appalachian Basin, USA	Middle Devonian	*Clarkson et al.* [2013] *Jin et al.* [2013] *Gu et al.* [2015] *Gu et al.* [2016] *Gu and Mildner* [2016]	***Clarkson et al.* [2013]:** one sample, TOC = 1.57 wt%, R_o ~ 1.2%. (U)SANS total porosity = 3.55% SSA = 0.34×10^5 cm^2/cm^3 ***Jin et al.* [2013]:** weathering study. Oatka Creek Member, (U)SANS. Nanoporosity (pores <100 nm) decreases upward from bedrock to 80 cm depth and then increases above 20 cm ***Gu et al.* [2015]:** (U)SANS/FIB-SEM. Isotropic scattering for samples parallel to bedding and anisotropic for samples cut perpendicular to bedding. Anisotropy associated with elongated pores associate with clay. Method of combining directions yielded porosities similar to N_2 BET. Dominant nanopores in organic-pore, clay-rich shales are water-accessible sheetlike pores within clay aggregates. Bubble-like organophilic pores in kerogen dominate organic-rich samples ***Gu et al.* [2016]:** H_2O/D_2O contrast matching (U)SANS. 24–47% of total porosity is in OM. Nearly 30% of OM volume composed of void space. Nontrivial water accessibility is possible for OM pores with diameters > than 20 nm ***Gu and Mildner* [2016]:** how to connect 2D SANS results with 1D USANS measurements for asymmetric scattering. Assumed SANS elliptical contours have same ratio at all q	***Clarkson et al.* [2013]:** D_s = 2.7 ***Jin et al.* [2013]:** D_s ~ 3 (unweathered); decreases with increase weathering	***Clarkson et al.* [2013]:** location not reported ***Gu et al.* [2015]:** sample thickness was thin enough to eliminate most multiple scattering

Utica and Pt. Pleasant Fms.	Appalachian Basin, USA	Late Ordovician	*Swift et al.* [2014b]	(U)SANS, XCT, SEM. Sections parallel and perpendicular to bedding. Anisotropy reduced when clays wrap around other grains. Calcite-rich sample anisotropic at scales >50 nm and isotropic at smaller length scales (≳5 nm) despite expected dominance of these scales by clay because clasts and fossils dictate clay orientation. This local perturbation is scale dependent. At smaller scales, local geometries are more randomized. Clay-rich sample is anisotropic at all scales. All samples cut perpendicular to bedding showed isotropic scattering		Sample thickness was thin enough to eliminate most multiple scattering
			Zhao et al. [2017]	SANS, low-pressure N_2 physisorption, MICP. High clay content may have caused the lower MICP volume; difference of pore size distribution between SANS and BET suggests a poorly-connected pore system consistent with the low porosity; over half of the 10–50 nm pores and ~90% of the 5–10 nm pores are inaccessible. The combination of different methods revealed multi-scale inaccessible porosity.	$D_s = 2.8–2.9$	*Zhao et al.* also looked at samples from the Niobrara, Wolfcamp, Bakken
Second White Specks River and Belle Fourche Formations	Western Canadian Basin, Alberta	Late Cretaceous	*Bahadur et al.* [2014] *Furmann et al.* [2016]	Closed porosity increased with quartz, albite, and illite content relative to carbonates. Both fractal and non-fractal pores observed (smooth at ~40–150 nm), surface fractal behavior for larger and smaller pores. Mesopores and micropores controlled porosity and SSA. OM (+kaolinite, chlorite) contributed to overall microporosity. Increase in total porosity correlated with quartz content. In-core variations possibly due to kaolinite and illite mesopores		Samples from narrow depth interval and had similar TOC and R_o. Variable mineralogy controlled scattering
Milk River Formation	Central Alberta	Late Cretaceous	*Clarkson et al.* [2013]	Fractal dimension suggested due to polydisperse 3D objects. Pores with between ~2 and 3.5 nm possibly related to montmorillonites. Pores between ~120 and 800 nm represent grains of quartz and/or albite wrapped by clay folia	$D_s = 2.8$	One sample; large amount of smectite and other phyllosilicates. Slit-like pores suggest PDSP model may not be appropriate
Montney Formation	Western Canadian Sedimentary Basin, British Columbia and Alberta	Lower Triass c	*Clarkson et al.* [2012]	(U)SANS. The pore structure is fairly uniform; small differences <200 nm possibly related to cementation and clay content. Total porosity like (but systematically higher than) He porosities. The greatest open porosity correlated with the highest permeability. Surface area calculated from low-pressure N_2 and CO_2 adsorption significantly less than from U(SANS)		Tight gas siltstone. Quartz, dolomite, and feldspar dominated. Three core plugs

(Continued)

Table 4.1 (Continued)

Formation	Location	Age	Author [date]	Observations	Fractal dimensions	Comments
Mowry Formation and Otter Creek Park and Evie Members of the Horn River Formation	Powder River Basin, Montana and Wyoming and Horn River Basin British Columbia	Powder River: Late Cretaceous Mowry Formation Horn River: Middle to Late Devonian Middle Otter Creek Park and Evie Members of the Horn River Formation	*King et al.* [2015]	SANS and helium ion microscopy. Pore size distribution did not adhere to power law distribution across many length scales. An anomalous population of ~2 nm pores generally hosted in OM comprised roughly one third of total porosity. Power law distribution associated with inter/intragrain porosity. Proposed *foamy porosity* at small length scales reflecting diagenetic evolution of OM. Cross-linking of kerogen combined with oil/gas phase separation leading to arrested coarsening at a size determined by cross-link density ~2 nm. Combined SANS and MICP indicated pore/pore throat ratios ~ independent of pore size. Suggest interpore throats are the result of microfractures generated during kerogen diagenesis		Focus on finer length scale features and pores associated with OM
Eau Claire Formation	Parts of Wisconsin, Illinois, Indiana, and Ohio	Middle to Upper Cambrian	*Swift et al.* [2014a]	At low q $I(q)$ for the glauconite-rich mudstone is > the other two samples. The illitic shale and the glauconite-rich mudstone show small humps at ~q = 0.1 (~5.5 nm) and q = 0.05 (~12.5 nm), respectively. All showed bimodal porosity (inflection points 1.9, 5.9, and 4.7 μm, with 86.6, 43.5, and 59% of porosity below this point for the illitic shale, carbonate-rich mudstone, and glauconite-rich mudstone, respectively). They urged caution about calculating pore size distributions because of variable pore shape. Calculated pore volumes: two at ~25 and 135 Å, one at 10–20 μm in both mudstones. The smallest dominates the illitic shale with a second at ~2 μm. The glauconitic mudstone has peaks ~10 nm and 20 μm. The carbonate mudstone has a narrower peak at 30 μm. Pore/pore throat ratios ~1 at the nanoscale dominated by illite and one and a half orders of magnitude at the microscale in a clastic mudstone. Illite and glauconite contribute disproportionately to connected reactive surface area	High-q background near D_s = 3 for all three samples. At lower q (< 10^{-3} Å^{-1}), D_s approaches −4. The scale at which this occurs increases with depth. At ~10^{-4} Å^{-1}), the illitic shale (and possibly the glauconite-rich mudstone) shows evidence of mass fractal behavior	Three samples were analyzed, one each of three lithofacies: an illitic clay-dominated shale (2710.1 ft. depth), a fine- to medium-grained carbonate-rich mudstone (2846 ft. depth), and a fine- to medium-grained glauconite-rich mudstone (3102.8 ft. depth)

Rose Hill Shale	Pennsylvania	Silurian	*Jin et al.* [2011]	Shallow regolith $I(q)$>for deeper regolith and weathered rock, that is, $I(q)$ increased with weathering. Saprolite shales comprised an uppermost group (< 5 m) with porosity as high as 9.3% and a deeper group (> 5 m) with an average porosity of 5–6%. This increase continued in the regolith to the ground surface: ~8% at 30 cm and 14–16% at 10 cm. Least weathered samples (depths 520–2000 cm) exhibited low SSA of 30–60 m^2/g. SSA increased with weathering to 300 m^2/g 30 cm below the surface and then decreased to 40 m^2/g at 5 cm (0–10 cm depth), although porosity continued to increase. Largest connected porosity at 0–30 cm (5.0% total rock volume), followed by 25–29 cm (4.1%), 30–40 cm (2.5%), and 610–630 cm (0%). Unconnected porosities constant for unweathered samples (~2%). At ~20 m, dissolution depleted the bedrock of ankerite. At 5 m, an increase in porosity and SSA corresponded to the onset of feldspar dissolution. Below the saprolite-regolith interface, porosity and surface area increased as chlorite and illite dissolved. Throughout the regolith intraparticle pores connected to form larger interparticle pores and scattering changed from mass fractal to surface fractal. In the most weathered regolith, kaolinite and Fe oxyhydroxides precipitated, blocking some pores	Deepest samples: D_m = 3–2 Shallowest regolith samples: D_s = 2–3	Characterizes the evolution of nanoscale features in a deep weathering profile. Samples from augered or drilled holes
Unnamed black shale	Baltic Basin, Poland	Silurian	*Lee et al.* [2014]	How pore size, porosity, and fractality change with dehydration. Porosity values decreased with depth and increased upon dehydration. Dehydration effects primarily controlled by pores between 100 and 1000 nm	On dehydration, D_s decreased in the smallest pores, trending to more mass fractal-like behavior. Larger pores showed little change in fractal dimension	
Green River, Monterey, Marcellus and Woodford shales and Pond Creek seam coal	Colorado, California, Pennsylvania, Oklahoma, Eastern Kentucky	Eocene, Miccene, Middle Devonian, Late Devonian/ early Mississippian, Pennsylvanian	*Thomas et al.* [2014]	Kerogen SLD is expected to vary considerably with OM type and thermal maturity. SANS experiments with H_2O/D_2O matching. SLD of kerogen varies from 1 to 4×10^{14} m^{-2}, increasing with decreasing H/C. Measured and calculated values are in reasonable agreement. For shales containing immature kerogen with an SLD close to 1, scattering between kerogen and mineral phases may be significant. OM-rich shale cannot be treated as a two-phase material		

(*Continued*)

Table 4.1 (Continued)

Formation	Location	Age	Author [date]	Observations	Fractal dimensions	Comments
Eagle Ford Shale Marcellus shale	SE Texas and NE Mexico Appalachian Basin, USA	Late Cretaceous Middle Devonian	*DiStafano et al.* [2016]	To quantify how native *extractable* OM is distributed with respect to pore size, shales were subjected to extraction by pyrolysis gas chromatography, thermogravimetric analysis, acetone, dichloromethane, hexane, dodecane, methanol, and toluene. OM extraction correlated with % clay and was greatest for toluene, DCM, and hexane. Changes in porosity were due to extraction or breakdown of high molecular weight bitumen with high C/H ratios. Proposed three mechanisms for extraction of higher molecular weight compounds: (i) complete breakdown of asphaltenes and resins, (ii) partial dissolution and migration of asphaltenes and resins which become trapped in select portions of the pore matrix, and (iii) partial breakdown of refractory OM, creating smaller pores. Porosity increased for carbonate-rich samples, generally decreased for clay-rich samples. Decrease attributed to swelling of the pore matrix during solvent uptake or migration of asphaltenes and/or resins. There was no observed shift in the pore size distribution, suggesting mechanisms 1 and 2 dominate over mechanism 3		
Gothic Shale Marine Tuscaloosa Shale	Paradox Basin, Utah, Colorado Mississippi and Louisiana	Pennsylvanian (Desmoinesian) Cretaceous	*Mouzakis et al.* [2016]	Distribution and connectivity of pores in shale caprocks and potential changes induced by reactions with supercritical CO_2. Reacted shale with brine or brine + CO_2. Assessed connectivity with H_2O/D_2O mixture. Both shales exhibited mineral dissolution, greater with supercritical CO_2. Total porosity and connectivity increased in the Gothic Shale. Total porosity increased but connectivity decreased in the Marine Tuscaloosa. Concluded due to greater dissolution of carbonates compared to the silicates in the Gothic Shale with $scCO_2$ for pores ~200 nm. Silicates in the Marine Tuscaloosa dissolved more than carbonates, resulting in increased porosity at all measurable length scales. Either precipitation in pore throats or hydration of the abundant clay assemblage may have caused decreased connectivity		Other sources refer to the second source as the Tuscaloosa Marine Shale

Miscellaneous shales and schists	Opalinus shale, oil shale from Jordan, schist from France		*Leu et al.* [2016]	Nuances of how to properly apply SAXS and WAXS considering variable spot size and sample thickness.	
Shahejie Formation	Dongying Sag, Bohai Bay Basin, China	Paleogene	*Sun et al.* [2017a]	Used solvent extraction to remove both bitumen and organic matter. Compared SAXS results before and after extraction. Quantified volume change of mesopores, revealing oil content. Relative fraction of fine mesopores (2–10 nm) increased after extraction, indicating these contain the high wax crude oil and suggesting the lower limit of high wax crude oil in Dongying sag. Negative correlation between the diameter of the maximum peak and clay content after extraction reveals that clay mineral content affects the pore size distribution of shale oil reservoir space.	
Longmaxi Formation	Upper Yangtze Platform, Sichuan Basin, China	Lower Silurian	*Yang et al.* [2017]	SANS, MICP, He pycnometry. Four shale lithologies were analyzed: one silt rich, two silica-rich and one carbonaceous. Quartz and TOC increase, carbonate and feldspar decrease with depth. All four samples exhibit mass fractal behavior, contrary to many other shales with both surface and mass fractal behavior. Porosity distributions from Porod invariant and PDSP compare favorably with MICP. Scattering intensity decreased by over five orders of magnitude with increasing depth from 2371 m to 2402 m, Fraction of closed pores increased with increasing depth and was attributed to an increase in TOC and silica.	$D_m = 2.8–3$
			Sun et al. [2017b]	SANS, He pycnometry, low pressure N_2 and CO_2 adsorption. Nine samples from two wells. Positive correlation between TOC and porosity. Percentage of closed porosity ranged from ~6.7 to 42.6 percent, but varied strongly between samples. Positive correlation between mass fractal dimension and the fraction of closed porosity.	$D_m = 2.79–3$

(*Continued*)

Table 4.1 (Continued)

Formation	Location	Age	Author [date]	Observations	Fractal dimensions	Comments
Niutitang Formation	Southeast Chongqing, China	Lower Cambrian	*Sun et al.* [2018]	MICP, BET and He pycnometry, and SANS. Four samples. Comparison of MICP and SANS allowed an estimate of closed pores. The mean pore throat size was <20 nm, average SANS pore size ~3 nm. Closed pores approximately correlated with the clay to quartz ratio. ~35% of pores space was closed at a clay/quartz ratio of ~0.8, less than a few percent of pores were closed for a clay/quartz ratio of 0.33. Despite this correlation, *Sun et al.* [2018] claim most of the closed pores are associated with organic matter not clay.		
Evergreen Formation	Surat Basin, Queensland, Australia	Early Jurassic	*Pearce et al.* [2018]	Reacted two types of caprock mudstones (clay and organic-rich mudstone, organic-lean feldspar-rich fine-grained sandstone) with impure CO_2 (+SO_2 and O_2) and low salinity brine. MICP indicated the majority of pores in both had pore throat radii ~5–150 nm with porosities of 5.5–8.4%. After reaction measured pore throats decreased in the clay-rich mudstone core. SEM identified macroporosity in clays, mica and amorphous silica cements. Precipitation of barite, Fe-oxides, clays and gypsum was observed after reaction. The SANS fraction of accessible pores before reaction in both decreased at smaller pore sizes. The clay-rich caprock had a higher fraction of accessible pores (~0.9) at the smallest pore size than the sandstone (~0.75). After reaction the fraction of the smallest accessible pores was ~0.85 in the clay-rich caprock and ~0.75 in the sandstone.		

• There are significant differences between total porosities and pore volume distributions obtained by different techniques including (U)SANS/(U)SAXS, nitrogen and carbon dioxide adsorption, mercury intrusion porosimetry, and density [cf., *Hall et al.*, 1983; *Ma et al.*, 1999a, 1999b; *Mastalerz et al.*, 2012, 2013; *Clarkson et al.*, 2012; *Swift et al.*, 2014a; *Bahadur et al.*, 2015]. The observation by *Hall et al.* [1983] that differences between (U)SANS and (U)SAXS can be used to understand the distribution of OM within the pore system has, as yet, not been well exploited [but see *Radlinski et al.*, 1996].

• The background values at high q appear to exhibit more scattering intensity than can be explained by incoherent scattering from hydrogen [*Radlinski et al.*, 1996, 2000b; *Anovitz et al.*, 2015a; *Bahadur et al.*, 2015].

• The decrease in porosity with increasing maturity (measured as R_o) is not linear. Instead, it appears to occur primarily at lower maturity values. This tends to be associated primarily with a decrease in nanoporosity [*Anovitz et al.*, 2014, 2015a; *Bahadur et al.*, 2015].

• In some cases, shale porosity appears to be bimodal [*Clarkson et al.*, 2013; *Anovitz et al.*, 2014, 2015a; *Swift et al.*, 2014a, 2014b]. This may reflect a contribution from the interlayer spacing in clays and/or OM-bearing nanopores at the smaller scales, compared to scattering between clay packets.

• A large fraction (up to 85%) of the pores in the shales investigated [*Clarkson et al.*, 2013; *Ruppert et al.*, 2013] appear to be accessible to methane. In some cases, however, a smaller fraction may be accessible to water. In both cases, this may be pore size dependent.

• A significant fraction of nanoporosity can be associated with OM especially if it has been diagenetically altered to yield gas, gas condensates, or oil. The burial process leads to the cross-linking of kerogen combined with phase separation of either gas or oil, resulting in arrested coarsening at a size determined by the cross-link density, yielding a pore diameter of about 2 nm. The helium ion microscope images supported by the SANS size distribution confirm the small pore size associated with the OM [e.g., *King et al.*, 2015].

• As expected in organic-poor, clay-rich shales, most of the nanoporosity occurs within clay bundles or between clays that wrap around very small detrital grains such as quartz and feldspar [e.g., *Swift et al.*, 2014a, 2014b].

• The comparison between scattering and the MICP method tends to improve at the smaller length scales dominated by porosity in clays where pores and pore throats are of the same approximate shape and size [*Clarkson et al.*, 2013; *Swift et al.*, 2014a].

• In low-organic shales, closed or unconnected porosity tends to increase with an increase in clay content as well as feldspar and quartz. Weathering or diagenetic reactions that produce clays also tend to result in reduced connectivity [*Jin et al.*, 2011; *Badahur et al.*, 2104].

• Diagenetic alteration and weathering can also lead to smoother pore-mineral interfaces, that is, loss of surface fractal behavior, and reduce or eliminate anisotropy [*Jin et al.*, 2011, 2013; *Gu et al.*, 2015].

4.6.2. Key Issues, Outstanding Questions, Knowledge Gaps, and Future Directions

Although the work to date, as was summarized above, has outlined several important features of the porosity of shales, there remains a great deal to learn about the pore structure of shales using neutron and X-ray small angle scattering, either alone or in combination with other complementary techniques. Below, we identify some of the key outstanding challenge questions:

• While a few individual samples have been studied, and a few recent studies [*Lee et al.*, 2014; *Anovitz et al.*, 2015a; *Gu et al.*, 2015] have enlarged the number of samples included in a single study, there exists no truly extensive systematic regional study showing how shale pore structures vary vertically and laterally within a shale horizon or how such variations correlate with known changes in the regional geology (faulting, folding, facies changes, etc.). It is clear, however, from the limited data that exist that such shales do show tremendous variability and that this may have a dramatic effect both on the suitability of a given caprock horizon and on oil and gas recovery. Larger studies are needed to examine the extent of variability in each formation, providing a baseline against which apparent changes due to CO_2 interaction or thermal maturation can be assessed.

• How can the often-discrepant results between nitrogen and carbon dioxide adsorption, Hg intrusion porosimetry, and SAS studies be used to more effectively to quantify multi-scale porosity?

• How do wettability phenomena scale down to mesopores and nanopores and scale up to the sediment cores? What are the causes and extents of wetting alteration over longer times for different types of caprocks and minerals in brine-scCO_2 systems?

• How might SAS be used effectively to quantitatively assess the impact of scCO_2 flowing through microfractures in reactive caprocks that leads to either a self-inhibiting (precipitation-controlled) or self-enhancing (dissolution-controlled) process in typical shales?

• How can the flat background due to incoherent scattering be separated from scattering due to nanoscale porosity?

• What is the origin of the observed mass and surface fractality of shales, and how does this vary with OM type and maturity or with shale mineralogy? Can the results obtained be explained in terms of the surface roughening model proposed by *Wong et al.* [1986]?

• To what extent does the PDSP model developed in PRINSAS reflect actual pore shapes?

• What is the correlation between the development of nanoporosity/microporosity and both the mineralogy and timing of the diagenesis of clays in shales?

• What is the relationship between the magnitude, structure, and connectivity of nanoporosity/microporosity and fracture mechanisms in shales?

• The need for wider realization that some USANS data sets are far less accurate than they could be because sample thickness was greater than necessary, leading to complications of multiple scattering.

• Scattering studies have traditionally assumed a two-phase approximation, that is, pore-matrix. Analysis to date of organic-rich shales suggests that this may be insufficient. Mineral-mineral scattering may also play a role, as may variations in the mineral phases present on the pore interface and changes in the maturity and pore filling of the organic phases.

• To what extent can scattering patterns from shales be considered uniaxial so that a single section, cut perpendicular to bedding, contains all the significant data on the anisotropy of the pore structure?

• How do mineral dissolution/precipitation reactions affect the multi-scale pore structure, transport properties, and mechanical integrity of shales? How do these vary with shale composition, burial history, and diagenesis?

• How is porosity modified by maturation and transport of organic materials during burial?

• How can a combination of data from X-ray and/or neutron small and ultra small angle scattering, TEM, adsorption isotherms, Hg intrusion porosimetry, and CT be used to build 3D structures suitable to aid the interpretation of experimental evidence on mobility, permeability, adsorption, dissolution/precipitation, and fracture?

• Based on conventional imaging analysis, *Loucks et al.* [2012] defined several pore types in shales. Is this and related studies sufficient to capture the pore groups (bimodal distributions, connected vs. unconnected porosity, mineral association) observable by SAS?

• From an experimental perspective, the intrinsically low permeabilities of caprocks make these materials extremely challenging to study because reactions are strongly transport limited. Unusually high pressure gradients are often employed in laboratory experiments to circumvent mass transfer limitations imposed by very low permeabilities. Whether such extreme experimental procedures cause significant artifacts needs to be considered.

ACKNOWLEDGMENTS

This work was supported as part of the Center for Nanoscale Controls on Geologic CO2 (NCGC), an Energy Frontier Research Center funded by the US Department of Energy, Office of Science, Basic Energy Sciences, under Award DE-AC02-05CH11231.

REFERENCES

Ague, J. J. (2004), Fluid flow in the deep crust. In. *Treatise of Geochemistry, Vol. 3 The Crust*, RL Ridnick (ed.), 195–228, Elsevier.

Alkan, H., Y. Cinar, and E. B. Ulker (2010), Impact of capillary pressure, salinity and in situ conditions on CO_2 injection into saline aquifers. *Transport in Porous Media*, *84*, 799–819.

Anovitz, L. M., and D. R. Cole (2015), Characterization and analysis of porosity and pore structures. *Reviews in Mineralogy and Geochemistry*, *80*, 61–164.

Anovitz, L. M., G. W. Lynn, D. R. Cole, G. Rother, L. F. Allard, W. A. Hamilton, L. Porcar, and M. H. Kim (2009), A new approach to quantification of metamorphism using ultra-small and small angle neutron scattering. *Geochimica et Cosmochimica Acta*, *73*(24), 7303–7324.

Anovitz, L. M., G. Rother, and D. R. Cole (2011), Characterization of rock pore features in geothermal systems using small angle neutron scattering (SANS). *Proceedings of the 36th Workshop on Geothermal Reservoir Engineering*, no. SGP-TR-191, 571-11. https://pangea.stanford.edu/ERE/pdf/IGAstandard/SGW/2011/anovitz.pdf.

Anovitz, L. M., D. R. Cole, G. Rother, L. F. Allard, A. Jackson, and K. C. Littrell (2013a), Diagenetic changes in macro-to nano-scale porosity in the St. Peter Sandstone: An (Ultra) small angle neutron scattering and backscattered electron imaging analysis. *Geochimica et Cosmochimica Acta*, *102*, 280–305.

Anovitz L. M., H.-W. Wang, D. R. Cole, J. Sheets, G. Rother, D. D. Faulder, and M. Walters (2013b), Analysis of Multiscale Porosity at the Coso Geothermal Field. *Proceedings of the 38th Workshop on Geothermal Resource Engineering*, SPG-TR-198, Stanford, California, USA, February 11–13, 2013.

Anovitz L. M., D. R. Cole, A. Swift, J. Sheets, H. Elston, S. Welch, S. J. Chipera, K. C. Littrell, D. F. R. Mildner, and M. J. Wasbrough (2014), Multiscale (nano to mm) porosity in the Eagle Ford Shale: Changes as a function of maturity. *Unconventional Resources Technology Conference Proceedings*, American Association of Petroleum Geologists Publication URTeC: 1923519, Denver, Colorado, August 25–27, 2014.

Anovitz, L. M., D. R. Cole, J. Sheets, A. Swift, H. W. Elston, S. Welch, S. J. Chipera, K. C. Littrell, D. F. R. Mildner, and M. J. Wasbrough (2015a), Effects of maturation on multiscale (nanometer to millimeter) porosity in the Eagle Ford Shale. *Interpretation*, *3*, 1–11.

Anovitz, L. M., D. R. Cole, A. J. Jackson, G. Rother, K. C. Littrell, L. F. Allard, A. D. Pollington, and D. J. Wesolowski (2015b), Effect of quartz overgrowth precipitation on the multiscale porosity of sandstone: A (U)SANS and imaging analysis. *Geochimica et Cosmochimica Acta*, *158*, 199–222.

Aplin, A. C., A. J. Fleet, and J. H. S. Macquaker (1999), Muds and mudstones: physical and fluid-flow properties. *Geological Society, London, Special Publication*, *158*, 1–8.

Armitage, P. J., R. H. Worden, D. R. Faulkner, A. C. Aplin, A. R. Butcher, and J. Iliffe (2010), Diagenetic and sedimentary controls on porosity in Lower Carboniferous fine-grained lithologies, Krechba field, Algeria: A petrological study of a caprock to a carbon capture site. *Marine and Petroleum Geology*, *27*(7), 1395–1410.

Arthur M, and D. R. Cole (2014), Unconventional hydrocarbon resources: prospects and problems. *Elements*, *10*, 257–264.

Bachu, S., D. Bonijoly, J. Bradshaw, R. Burruss, S. Holloway, N. P. Christensen, and O. M. Mathiassen (2007), CO_2 storage capacity estimation: Methods and gaps. *International Journal of Greenhouse Gas Control*, *1*, 430–443.

Bahadur J, Y. B. Melnichenko, M. Mastalerz, A. Furmann, and C. R. Clarkson (2014), Hierarchical pore morphology of Cretaceous shale: a small-angle neutron scattering and ultrasmall-angle neutron scattering study. *Energy and Fuels*, *28*, 6336–6344.

Bahadur, J., A. P. Radlinski, Y. B. Melnichenko, M. Mastalerz, and A. Schimmelmann (2015), Small-angle and ultrasmall-angle neutron scattering (SANS/USANS) study of New Albany shale: A treatise on microporosity. *Energy and Fuels*, *29*, 567–576.

Bale, H. D., and P. W. Schmidt (1984), Small-angle X-ray-scattering investigation of submicroscopic porosity with fractal properties. *Physics Review Letters*, *53*(6), 596–599.

Barnes, D. A., D. H. Bacon, and S. R. Kelley (2009), Geological sequestration of carbon dioxide in the Cambria Mount Simon Sandstone: Regional storage capacity, site characterization, and large-scale injection feasibility, Michigan Basin. *American Association of Petroleum Geologists*, *16*(3), 163–183.

Beckingham, L. E., C. I. Steefel, A. M. Swift, M. Voltolini, L. Yang, L. Anovitz, J. M. Sheets, D. R. Cole, T. J. Kneafsey, E. H. Mitnick, S. Zhang, G. Landrot, J. Ajo-Franklin, D. DePaolo, S. Mito, and Z. Xue (2017), Evaluation of accessible mineral surface areas for improved prediction of mineral reaction rates in porous media. *Geochimica et Cosmochimica Acta*, *205*, 31–49.

Benson, S. M., and D. R. Cole (2008), CO_2 sequestration in deep sedimentary formations. *Elements*, *4*, 325–331.

Best, M. E., and T. J. Katsube (1995), Shale permeability and its significance in hydrocarbon exploration. *The Leading Edge*, *14*, 165 170.

Bildstein, O., C. Kervevan, V. Lagneau, P. Delaplace, A. Credoz, P. Audigane, E. Perfetti, N. Jacquemet, and M. Jullien (2010), Integrative modeling of caprock integrity in the context of CO_2 storage: Evolution of transport and geochemical properties and impact on performance and safety assessment. *Oil and Gas Science and Technology*, *65*(3), 485–502.

Blatt, H., and R. J. Tracy (1996), *Petrology: Igneous, Sedimentary and Metamorphic*, 2nd ed., New York, Freeman.

Bolton, A. J., A. J. Maltman, and Q. Fisher (2000), Anisotropic permeability and bimodal pore-size distributions on fine-grained marine sediments. *Marine and Petroleum Geology*, *17*, 657–672.

Bonneville, A., T. Gilmore, C. Sullivan, V. Vermeul, M. Kelley, S. White, D. Appriou, B. Bjornstad, J. Gerst, N. Gupta, J. Horner, C. McNeil, M. Moody, W. Rike, F. Spane, P. Thorne, E. Zeller, F. Zhang, J. Hoffmann, and K. Humphreys (2013), Evaluating the suitability for CO_2 storage at the FutureGen 2.0 site, Morgan County, Illinois, USA, *Energy Procedia*, *37*, 6125–6132.

Bourg, I. C. (2015), Sealing shales versus brittle shales: a sharp threshold in the material properties and energy technology uses of fine-grained sedimentary rocks. *Environmental Science and Technology Letters*, *2*(10), 255–259.

Bourg, I. C., and G. Sposito (2010), Connecting the molecular scale to the continuum scale for diffusion processes in smectite-rich porous media. *Environmental Science and Technology*, *44*(6), 2085–2091.

Bourg, I. C., and C. Tournassat (2015), Self-diffusion of water and ions in clay barriers. In: *Developments in Clay Science, Natural and Engineered Clay Barriers*. Ed. by C. Tournassat, C. I. Steefel, I. C. Bourg and F. Bergaya. Vol. *6*, 189–226, Elsevier.

Bourg, I. C., L. E. Beckingham, and D. J. DePaolo (2015), The nanoscale basis of CO_2 trapping for geologic storage. *Environmental Science and Technology*, *49*, 10265–10284.

Bruner, K. R., M. Walker-Milani, and R. Smosna (2015), Lithofacies of the Devonian Marcellus shale in the eastern Appalachian Basin, U.S.A.. *Journal of Sedimentary Research*, *85*, 937–954.

Busch, A., S. Alles, Y. Gensterblum, D. Prinz, D. N. Dewhurst, M. D. Raven, H. Stanjek, and B. M. Krooss (2008), Carbon dioxide storage potential of shales. *International Journal of Greenhouse Gas Control*, *2*(3), 297–308.

Campbell, G. (1946), New Albany shale. *Bull. Geol. Soc. Amer.*, *57*, 829–908.

Carroll, S.A., W.W. McNab, Z. Dai, and S.C. Torres (2013), Reactivity of Mount Simon sandstone and the Eau Claire shale under CO_2 storage conditions. *Environmental Science and Technology*, *47*, 252–261.

Cheshire, M. C., A. G. Stack, J. W. Carey, L. M. Anovitz, T. R. Prisk, and J. Ilavsky (2017), Wellbore Cement Porosity Evolution in Response to Mineral Alteration during CO_2 Flooding. *Environ Sci Technol.*, *51*(1), 692–698.

Clarkson, C. R., M. Freeman, L. He, M. Agamalian, Y. B. Melnichenko, M. Mastalerz, R. M. Bustin, A. P. Radlinski, and T. P. Blach (2012), Characterization of tight gas reservoir pore structure using USANS/SANS and gas adsorption analysis. *Fuel*, *95*, 371–385.

Clarkson, C. R., N. Solano, R. M. Bustin, A. M. M. Bustin, G. R. L. Chalmers, L. He, Y. B. Melnichenko, A. P. Radlinski, and T. P. Blach (2013), Pore structure characterization of North American shale gas reservoirs using USANS/SANS, gas adsorption, and mercury intrusion. *Fuel*, *103*, 606–616.

Curtis M. E., C. H. Sondergeld, R. J. Ambrose, and C. S. Rai (2012), Microstructural investigation of gas shales in two and three dimensions using nanometer-scale resolution imaging. *American Association of Petroleum Geologists Bulletin*, *96*, 665–677.

De Stefanis, A., A. A. G. Tomlinson, T. A. Steriotis, G. C. Charalambopoulou, and U. Keiderling (2007), Study of structural irregularities of smectite clay systems by small-angle neutron scattering and adsorption. *Applied Surface Science*, *253*(13), 5633–5639.

Debye P., and A. M. Beuche (1949), Scattering by an inhomogeneous solid. *Journal of Applied Physics*, *20*(6), 518–525.

Debye, P., H. R. Anderson Jr., and H. Brumberger (1957), Scattering by an inhomogeneous solid. II. The correlation function and its application. *Journal of Applied Physics*, *28*, 679–683.

Dewhurst, D. N., Y. L. Yang, and A. C. Aplin (1999), Permeability and fluid flow in natural mudstones. In Aplin, A. C., A. J. Fleet, and J. H. S. Macquaker, Eds. *Muds and Mudstones: Physical and Fluid Flow Properties*, Geological Society, London: London, *158*, 23–43.

DiStafano, V. H., J. McFarlane, L. M. Anovitz, A. G. Stack, A. D. Gordon, K. C. Littrell, S. J. Chipera, R. D. Hunt, S. A. Lewis, Sr., R. E. Hale, and E. Perfect (2016), Extraction of organic compounds from representative shales and the effect on porosity. *Journal of Natural Gas Science and Engineering*, *35*, 646–660.

Donnelly, T. H., and I. H. Crick (1988), Depositional environment of the Middle Proterozoic Velkerri Formation in northern Australia: Geochemical evidence. *Precambrian Research*, *42*(1/2), 165–172.

Doughty, C. (2010), Investigation of CO_2 plume behavior for a large-scale pilot test of geologic carbon storage in a saline formation. *Transport in Porous Media*, *82*, 49–76.

Edwards, R. W. J., M. A. Celia, K. W. Bandilla, F. Doster, and C. M. Kanno (2015), A model to estimate carbon dioxide injectivity and storage capacity for geological sequestration in shale gas wells. *Environmental Science and Technology*, *49*, 9222–9229.

EIA (Energy Information Administration) (2018), U.S. crude oil and natural gas reserves, Year-end 2016. EIA. https://www.eia.gov/naturalgas/crudeoilreserves/.

Ellis, B. R., G. S. Bromhal, D. L. McIntyre, and C. A. Peters (2011), Changes in caprock integrity due to vertical migration of CO_2 -enriched brine. *Energy Procedia*, *4*, 5327–5334.

Freiburg, J. T., D. G. Morse, H. E. Letaru, R. P. Ross, and Q. Yan (2014), A depositional and diagenetic characterization of the Mt. Simon Sandstone at the Illinois Basin – Decatur project carbon capture and storage site, Decatur, Illinois, USA. *Illinois State Geological Survey*, *Prairie Research Institute*, Circular 583, 73 p.

Fritz, S. J. (1986), Ideality of clay membranes in osmotic processes: A review. *Clays and Clay Minerals*, *34* (2), 214–223.

Furmann, A., M. Mastalerz, D. Bish, A. Schimmelmann, and P. K. Pedersen (2016), Porosity and pore size distribution in mudrocks from the Belle Fourche and Second White Specks Formations in Alberta, Canada. *American Association of Petroleum Geologists*, *100*(8), 1265–1288.

Gajo, A., and B. Loret (2004), Transient analysis of ionic replacements in elastic-plastic expansive clays. *International Journal of Solids and Structures*, *41*(26), 7493–7531.

Galloway, W. E., D. K. Hobday, and K. Magara (1982), Frio Formation of Texas Gulf coastal plain: Depositional systems, structural framework, and hydrocarbon distribution. *AAPG Bull.*, *6*, 649–688.

Gaus, I., M. Azaroual, and I. Czernichowski-Lauriol (2005), Reactive transport modelling of the impact of CO_2 injection on the clayey cap rock at Sleipner (North Sea). *Chemical Geology*, *217*(3–4), 319–337.

Gherardi, F., T. Xu, and K. Pruess (2007), Numerical modeling of self-limiting and self-enhancing caprock alteration induced by CO_2 storage in a depleted gas reservoir. *Chemical Geology*, *244*, 103–129.

Gilmore, T. J., A. Bonneville, V. Vermeul, F. Spane, M. E. Kelley, C. Sullivan, and J. Hoffmann (2014), Overview of the CO_2 geological storage site for the FutureGen project in Morgan County Illinois, USA, *Energy Procedia*, *63*, 6361–6367.

Godec, M., G. Koperna, R. Petrusak, and A. Oudinot (2013), Potential for enhanced gas recovery and CO_2 storage in the Marcellus Shale in the Eastern United States. *International Journal of Coal Geology*, *118*, 95–104.

Goldstein J., D. E. Newbury, D. C. Joy, C. E. Lyman, P. Echlin, E. Lifshin, L. Sawyer, and J. R. Michael (2003), *Scanning electron microscopy and X-ray microanalysis*, 3rd ed.: New York, Springer, 695 p.

Grinstein, G., and S.-K. Ma (1983), Surface tension, roughening, and lower critical dimension in the random-field Ising model. *Physical Review B* *28*(5), 2588–2601.

Grunau, H. R. (1987), A worldwide look at the cap-rock problem. *Journal of Petroleum Geology*, *10*(3), 245–265.

Gu, X., and D. F. R. Mildner (2016), Ultra-small-angle neutron scattering with azimuthal asymmetry. *Journal of Applied Crystallography*, *49*, 934–943.

Gu, X., D. R. Cole, G. Rother, D. F. R. Mildner, and S. L. Brantley (2015), Pores in Marcellus Shale: A neutron scattering and FIB-SEM Study. *Energy and Fuels*, *29*, 1295–1308.

Gu, X., D. F. R. Mildner, D. R. Cole, G. Rother, R. Slingerland, and S. L. Brantley (2016), Quantification of organic porosity and water accessibility in Marcellus Shale using neutron scattering. *Energy and Fuels*, *30*, 4438–4449.

Gutschick, R., and C. Sandberg (1983), Mississippian continental margins on the conterminous United States, in Stanley, D. J., and Moore, G. T., eds., *The Shelf Break: Critical Interface on Continental Margins*, Society of Economic Paleontologists and Mineralogists, Special Publication 33, 79–96. DOI: doi.org/10.2110/pec.83.06.0025

Hall, P. L., and Mildner, D. F. R. (1983), On the analysis of small-angle scattering with elliptical azimuthal symmetry. *Journal of Applied Physics*, *1*, 427–428.

Hall, P., D. Mildner, and R. Borst (1983), Pore-size distribution of shaly rock by small-angle neutron scattering, *Applied Physics Letters*, *43*, 252–254.

Hall, P. L., D. F. Mildner, and R. L. Borst (1986), Small angle scattering studies of the pore spaces of shaly rocks. *J. Geophys. Res.: Solid Earth*, *91*, 2183–2192.

Harper, J. A., C. D. Laughrey, J. Kostelnik, D. P. Gold, and A. G. Doden (2004), Trenton and Black River Carbonates in the Union Furnace Area of Blair and Huntingdon Counties, Pennsylvania: Introduction. Field Trip Guidebook for the Eastern Section AAPG Annual Meeting, 10 September 2003 and the PAPG Spring Field Trip, 26 May 2004. Pennsylvania Geological Survey.

Heath J. E, Dewers, T. A., McPherson, B. J. O. L., Petrusak, R., Chidsey, T. C., Rinehart, A. J., and Mozley, P. S. (2011), Pore networks in continental and marine mudstones: Characteristics and controls on sealing behavior. *Geosphere*, *7*(2), 429–454.

Hildenbrand, A., S. Schlomer, B. M. Krooss, and R. Littke (2004), Gas breakthrough experiments on pelitic rocks: comparative study with N_2, CO_2 and CH_4. *Geofluids*, *4*(1), 61–80.

Hinde, A. L. (2004), PRINSAS – a Windows-based computer program for processing and interpretation of small-angle scattering data tailored to the analysis of sedimentary rocks. *Journal of Applied Crystallography*, *37*, 1020–1024.

Hovorka, S. D. (2009), Frio brine pilot: the first U.S. sequestration test. *Southwest Hydrology*, *8*(5), 26–31.

Hovorka, S. D., S. M. Benson, C. Doughty, B. M. Freifeld, S. Sakurai, T. M. Daley, Y. K. Kharaka, M. H. Holtz, R. C. Trautz, H. S. Nance, L. R. Myer, and K. G. Knauss (2006), Measuring permanence of CO_2 storage in saline formations: The Frio experiment. *Environmental. Geosciences*, *13*(2), 105–121.

Howard, C. V., and M. G. Reed (2005), Unbiased stereology: Garland Science/BIOS Scientific.

Hower, J., E. Eslinger, M. E. Hower, and E. A. Perry (1976), Mechanism of burial metamorphism of argillaceous sediment: 1. Mineralogical and chemical evidence. *Geol Soc. Am. Bull.*, *87*, 725–737.

Huang, J., T. Cavanaugh, and B. Nur (2013), An introduction to SEM operational principles and geologic applications for shale hydrocarbon reservoirs. *Am. American Association of Petroleum Geologists Memoirs*, *102*, 1–6.

Ilavsky, J., and P. R. Jemian (2009), Irena: tool suite for modeling and analysis of small-angle scattering. *Journal of Applied Crystallography*, *42*(2), 347–353.

Ingebritsen, S. E., and W. E. Sanford (1999), *Groundwater in Geologic Processes*. Cambridge, Cambridge University Press, 341p.

IPCC (2005), Underground geological storage. In *IPCC Special Report on Carbon Dioxide Capture and Storage*, Cambridge University Press: Cambridge, 195–276.

Itakura, T., W. K. Bertram, and R. B. Knott (2005), The nanoscale structural response of a natural kaolinitic clayey soil subjected to uniaxial compression. *Applied Clay Science*, *29*(1), 1–14.

Jackson, M. J., and R. Raiswell (1991), Sedimentology and carbon-sulphur geochemistry of the Velkerri Formation, a mid-Proterozoic potential oil source in northern. *Australia Precambrian Research*, *54*(1), 81–108.

Jarvie, D. M., R. J. Hill, T. E. Ruble, and R. M. Pollastro (2007), Unconventional shale-gas systems: the Mississippian Barnett Shale of north central Texas as one model for thermogenic shale-gas assessment, in Hill, R. J., and D. M. Jarvie eds., Special Issues: Barnett Shale, *American Association of Petroleum Geologists Bulletin*, *91*, 475–499.

Jemian, P., G. Long, F. Lofaj, and S. Wiederhorn (1999), Anomalous ultra-small-angle x-ray scattering from evolving microstructures during tensile creep. Meeting of the Materials Research Society, Vol. 590, edited by S. Mini, D. Perry, and S. Stock, Boston, MA (Materials Research Society), p. 131–136.

Jin, L. X., G. Rother, D. R. Cole, D. F. R. Mildner, C. J. Duffy, and S. L. Brantley (2011), Characterization of deep weathering and nanoporosity development in shale-A neutron study. *American Mineralogist*, *96*(4), 498–512.

Jin, L., R. Mathur, G. Rother, D. R. Cole, E. Bazilevskaya, J. Williams, A. Carone, and S. L. Brantley (2013), Evolution of porosity and geochemistry in Marcellus Formation black shale during weathering. *Chemical Geology*, *356*, 50–63.

Jougnot, D., A. Revil, and P. Leroy (2009), Diffusion of ionic tracers in the Callovo-Oxfordian clay-rock using the Donnan equilibrium model and the formation factor. *Geochimica et Cosmochimica Acta*, *73*(10), 2712–2726.

Juanes, R., C. W. MacMinn, and M. L. Szulczewski (2010), The footprint of the CO_2 plume during carbon dioxide storage in saline aquifers: Storage efficiency for capillary trapping at the basin scale. *Transport in Porous Media*, *82*(1), 19–30.

Jullien, M., J. Raynal, E. Kohler, and O. Bildstein (2005), Physicochemical reactivity in clay-rich materials: Tools for safety assessment. *Oil and Gas Science and Technology*, *60* (1), 107–120.

Katsube, T. J., M. Williamson, and M. E. Best (1992), Shale pore structure evolution and its effect on permeability. 33rd Annual Symposium of the Society of Professional Well Log Analysts (SPWLA), Society of Core Analysts, SCA-9214, pp. 1–22.

Katz, A. J., and A. H. Thompson (1985), Fractal sandstone pores: implications for conductivity and pore formation. *Physical Review Letters*, *54*, 1325–1328.

Kim, M. H. (2004), Modified Porod's law estimate of the transition-layer thickness between two phases: test of triangular smoothing function. *Journal of Applied Crystallography*, *37*, 643–651.

Kim, J.-W., W. R. Bryant, J. S. Watkins, and T. T. Tieh (1999), Electron microscopic observations of shale diagenesis, offshore Louisiana, USA, Gulf of Mexico. *Geo-Marine Letters*, *18* (3), 234–240.

King, H. E., A. P. R. Eberle, C. C. Walters, C. E. Kliewer, D. Ertas, and C. Huynh (2015), Pore architecture and connectivity in gas shale. *Energy and Fuels*, *29*, 1375–1390.

Knudsen, K. D., J. O. Fossum, and G. Helgesen, and Bergaplass, V. (2003), Pore characteristics and water absorption in a synthetic smectite clay. *Journal of Applied Crystallography*, *36*, 587–591.

Knudsen, K. D., J. O. Fossum, G. Helgesen, and M. W. Haakestad (2004), Small-angle neutron scattering from a nano-layered synthetic silicate. *Physica B-Condensed Matter*, *352*(1–4), 247–258.

Kohler, E., T. Parra, and O. Vidal (2009), Clayey cap-rock behavior in H_2O-CO_2 media at low pressure and temperature conditions: An experimental approach. *Clays and Clay Minerals*, *57*(5), 616–637.

Lahann, R., M. Mastalerz, J. A. Rupp, and A. Drobniak (2013), Influence of CO_2 on New Albany Shale composition and pore structure. *International Journal of Coal Geology*, *108*, 2–9.

Landrot, G., J. B. Ajo-Franklin, L. Yang, S. Cabrini, and C. I. Steefel (2012), Measurement of accessible reactive surface area in a sandstone, with applications to CO_2 mineralization. *Chemical Geology*, *318–319*, 113–125.

Lash, G. G., and D. R. Blood (2014), Organic matter accumulation, redox, and diagenetic history of the Marcellus Formation, southwestern Pennsylvania, Appalachian Basin. *Marine and Petroleum Geology*, *57*, 244–263.

Lee, S, T. B. Fischer, M. R. Stokes, R. J. Klingler, J. Ilavsky, D. K. McCarty, M. O. Wigand, A. Derkowski, and R. E. Winans (2014), Dehydration effect on pore size, porosity, and fractal parameters of shale rocks: USAXS study. *Energy and Fuels*, *28*(11), 6772–6779.

Leu, L., A. Georgiadis, M. J. Blunt, A. Busch, P. Bertier, K. Schweinar, M. Liebi, A. Menzel, and H. Ott (2016), Multiscale description of shale pore systems by scanning SAXS and WAXS microscopy. *Energy and Fuels*, *30*, 10282–10297.

Li, S. W., M. Dong, Z. Li, S. Huang, H. Qing, and E. Nickel (2005), Gas breakthrough pressure for hydrocarbon reservoir seal rocks: implications for the security of long-term CO_2 storage in the Weyburn field. *Geofluids*, *5*, 326–334.

Liu, B. Z., and Y. Zhang (2011), CO_2 modeling in a deep saline aquifer: A predictive uncertainty analysis using design of experiment. *Environmental Science and Technology*, *45*(8), 3504–3510.

Liu, F. Y., K. Ellett, Y. Xiao, and J. A. Rupp (2013), Assessing the feasibility of CO_2 storage in the New Albany Shale (Devonian-Mississippian) with potential enhanced gas recovery using reservoir simulation. *International Journal of Greenhouse Gas Control*, *17*, 111–126.

Loucks, R. G., and S. C. Ruppel (2007), Mississippian Barnett Shale: lithofacies and depositional setting of a deep-water shale-gas succession in the Fort Worth Basin, Texas. *American Association of Petroleum Geologists Bulletin 91*, 579–601.

Loucks, R. G., R. M. Reed, S. C. Ruppel, and D. M. Jarvie (2009), Morphology, genesis, and distribution of nanometer-scale pores in siliceous mudstones of the Missippian Barnett Shale. *Journal of Sedimentary Research*, *79*, 848–861.

Loucks, R. G., R. M. Reed, S. C. Ruppel, and U. Hammes (2012), Spectrum of pore types and networks in mudrocks and a descriptive classification for matrix-related mudrock pores. *American Association of Petroleum Geologists Bulletin*, *96*(6), 1071–1098.

Lu, J. M., K. Milliken, R. M. Reed, and S. Hovorka (2012), Diagenesis and sealing capacity of the middle Tuscaloosa mudstone at the Cranfield carbon dioxide injection site, Mississippi, U.S.A. *Environmental Geosciences*, *18*(1), 35–53.

Lynch, F. L. (1997), Frio shale mineralogy and the stoichiometry of the smectite-to-illite reaction: the most important reaction in clastic sedimentary diagenesis. *Clays and Clay Minerals*, *45*(5), 618–631.

Ma, J., H. Qi, and P. Z. Wong (1999a), Probing the fractal character of pore surfaces in shale with adsorption technique. Dynamics in Small Confining Systems IV, ed. by Drake, J. M., G. S. Grest, J. Klafter, and R. Kopelman, Materials Research Society Symposium Proceedings 543, pp. 243–248.

Ma, J., H. Qi, and P. Z. Wong (1999b), Experimental study of multilayer adsorption on fractal surfaces in porous media. *Physical Review E*, *59*, 2049–2059.

Martelli, S., and P. E. Di Nunzio (2002), Particle size distribution of nanospheres by Monte Carlo fitting of small angle X-ray scattering curves. *Particle and Particle Systems Characterization*, *19*, 247–255.

Mastalerz, M., L. He, Y. B. Melnichenko, and J. A. Rupp (2012), Porosity of coal and shale: Insights from gas adsorption and SANS/USANS techniques. *Energy and Fuels*, *26*, 5109–5120.

Mastalerz, M., A. Schimmelmann, A. Drobniak, and Y. Y. Chen (2013), Porosity of Devonian and Mississippian New Albany Shale across a maturation gradient: Insights from organic petrology, gas adsorption, and mercury intrusion. *American Association of Petroleum Geologists Bulletin*, *97*(10), 1621–1643.

Mehnert, E., J. Damico, S. Frailey, H. Leetaru, R. Okwen, B. Storsved, and A. Valocchi (2014), Basin-scale modeling for CO_2 sequestration in the basal sandstone reservoir of the Illinois Basin – Improving the geologic model. *Energy Procedia*, *63*, 2949–2960.

Merrit, M., and Y. Zhang (2004), An interior-point gradient method for large-scale totally nonnegative least squares problems, Technical report TR04-08, Department of Computational and Applied Mathematics Rice University, Houston, TX (May 2004).

Mildner, D., R. Rezvani, P. Hall, and R. Borst (1986), Small-angle scattering of shaly rocks with fractal pore interfaces. *Applied Physics Letters*, *48*, 1314–1316.

Morrison, J. D., J. D. Corcoran, and K. E. Lewis (1992), The determination of particle-size distributions in small-angle scattering using the maximum-entropy method. *Journal of Applied Crystallography*, *25*, 504–513.

Mouzakis, K. M., A. K. Navaare-Sitchler, G. Rother, J. L. Banuelos, X. Wang, J. P. Kaszuba, J. E. Heath, Q. R. S. Miller, V. Alvarado, and J. E. McCray (2016), Experimental study of porosity change in shale caprocks exposed to CO_2-saturated brines I: Evolution of mineralogy, pore connectivity, pore size distribution, and surface area. *Environmental Engineering Science*, *33*(10), 725–735.

Muller, J. J., S. Hansen, G. Lukowski, and K. Gast (1995), Multimodal particle-size distribution or fractal surface of acrylic acid copolymer nanoparticles: A small-angle X-ray scattering study using direct Fourier and indirect maximum-entropy methods. *Journal of Applied Crystallography*, *28*, 774–781.

Neufelder, R. J., B. B. Bowen, R. W. Lahann, and J. A. Rupp (2012), Lithologic, mineralogical, and petrophysical characteristics of the Eau Claire Formation: Complexities of a carbon storage system seal. *Environmental Geosciences*, *19*(3), 81–23.

Neuzil, C. E. (1994), How permeable are clays and shales? *Water Resources Research*, *30*(2), 145–150.

Nordbotten, J. M., D. Kavetski, M. A. Celia, and S. Bachu (2009), Model for CO_2 leakage including multiple geological layers and multiple leaky wells. *Environmental Science and Technology*, *43*(3), 743–749.

North, A. N., J. C. Dore, R. M. Keenan, A. R. Mackie, A. M. Howe, B. H. Robinson, and C. Nave (1988), Small-angle X-ray scattering studies of heterogeneous systems using synchrotron radiation techniques. *Nuclear Instruments and Methods in Physics Research*, *B34*, 188–202.

North, A. N., J. C. Dore, A. R. Mackie, A. M. Howe, and J. Harries (1990), Ultrasmall-angle X-ray scattering studies of heterogeneous systems using synchrotron radiation techniques. *Nuclear Instruments and Methods in Physics Research, B47*, 283–290.

Obermajer, M., M. G. Fowler, F. Goodarzi, and L. R. Snowdon (1997), Organic petrology and organic geochemistry of Devonian black shales in southwestern Ontario, Canada. *Organic Geochemistry, 26*(3–4), 229–246.

Ortega, J. A., F. J. Ulm, and Y. Abousleiman (2009), The nanogranular acoustic signature of shale. *Geophysics, 74*(3), D65–D84.

Pearce, J. K., Dawson, G. K., Blach, T. P., Bahadur, J., Melnichenko, Y. B., and Golding, S. D. (2018), Impure CO_2 reaction of feldspar, clay, and organic matter rich cap-rocks: Decreases in the fraction of accessible mesopores measured by SANS. *International Journal of Coal Geology, 185*, 79–90.

Pfeifer, P., and D. Avnir (1983), Chemistry in noninteger dimensions between two and three. I. Fractal theory of heterogeneous surfaces. *Journal of Chemical Physics, 79*, 3562–3565.

Potter, P. E., J. B. Maynard, and W. A. Pryor (1980), Final report of special geological, geochemical, and petrological studies of the Devonian shales in the Appalachian Basin. Final Report prepared for US Department of Energy Eastern Gas Shales Project, Contract DE-AC21-76MCO5201.

Potton, J. A., G. J. Daniell, and D. Melville (1988a), A new method for the determination of particle-size distributions from small-angle neutron-scattering measurements. *Journal of Applied Crystallography, 21*, 891–897.

Potton, J. A., G. J. Daniell, and B. D. Rainford (1988b), Particle size distributions from SANS data using the maximum entropy method. *Journal of Applied Crystallography, 21*, 663–668.

Qi, H., J. Ma, and P. Z. Wong (2002), Adsorption isotherms of fractal surfaces. *Colloids and Surfaces A, 206*, 401–407.

Radlinski, A. P. (2006), Small-angle neutron scattering and rock microstructure. In: Neutron Scattering in Earth Sciences, ed. by H.-R. Wenk, *Reviews in Mineralogy and Geochemistry, 63*, 363–397.

Radlinski, A. P., and Hinde, A. L. (2002), Small angle neutron scattering and petroleum geology. *Neutron News, 13*(2), 10–14.

Radlinski, A. P., C. J. Boreham, G. D. Wignall, and J.-S. Lin (1996), Microstructural evolution of source rocks during hydrocarbon generation: a small-angle-scattering study. *Physical Review B, 53*, 14152–14160.

Radlinski, A. P., E. Z. Radlinska, M. Agamalian, G. D. Wignall, P. Lindner, and O. G. Randl (1999), Fractal geometry of rocks. *Physical Review Letters, 82*, 3078–3081.

Radlinski, A. P., E. Z. Radlinska, M. Agamalian, G. D. Wignall, P. Lindner, and O. G. Randl (2000a), The fractal microstructure of ancient sedimentary rocks. *Journal of Applied Crystallography, 33*, 860–862.

Radlinski, A. P., C. J. Boreham, P. Lindner, O. Randl, G. D. Wignall, A. Hinde, and J. M. Hope (2000b), Small angle neutron scattering signature of oil generation in artificially and naturally matured hydrocarbon source rocks. *Organic Geochemistry, 31*, 1–14.

Radlinski, A. P., M. Mastalerz, A. L. Hinde, M. Hainbuchner, H. Rauch, M. Baron, J.-S. Lin, L. Fan, and P. Thiyagarajan (2001), Non-invasive measurements of pore size distribution in coal pellets using X-ray and neutron techniques. International Coalbed Methane Symposium, Proceedings, 14–18 May, Tuscaloosa, AL, pp. 163–175.

Radlinski, A. P., M. A. Ioannidis, A. L. Hinde, M. Hainbuchner, M. Baron, H. Rauch, and S. R. Kline (2004), Angstrom-to-millimeter characterization of sedimentary rock microstructure. *Journal of Colloid and Interface Science, 274*, 607–612.

Roen, J. B. (1983a), Geology of the Devonian Black Shales of the Appalachian Basin. *The American Chemical Society, 185*, 45.

Roen, J. B. (1983b), Geology of the Devonian Black Shales of the Appalachian Basin. *Organic Geochemistry, 5*, 241–254.

Rowe, H. D., R. G. Loucks, S. C. Ruppel, and S. M. Rimmer (2008), Mississippian Barnett Formation, Fort Worth Basin, Texas: bulk geochemical inferences and Mo-TOC constraints on the severity of hydrographic restriction. *Chemical Geology, 2571-2*, 16–25.

Ruland, W. (1971), Small-angle scattering of 2-phase systems – determination and significance of systematic deviations from Porod's law. *Journal of Applied Crystallography, 4*, 70–73.

Ruppel, S. C., and R. G. Loucks (2008), Black mudrocks: lessons and questions from the Mississippian Barnett Shale in the southern Midcontinent. *The Sedimentary Record, 6*(2),ij

Ruppert L. F., R. Sakurovs, T. P. Blach, L. He, Y. B. Melnichenko, D. F. R. Mildner, and L. Alcantar-Lopez (2013), A USANS/SANS study of the accessibility of pores in the Barnett shale to methane and water. *Energy and Fuels, 27*, 772–779.

Rutqvist, J., J. T. Birkholzer, and C.-F. Tsang (2007), Coupled reservoir–geomechanical analysis of the potential for tensile and shear failure associated with CO_2 injection in multilayered reservoir–caprock systems. *International Journal of Rock Mechanics and Mining Sciences, 45*(2), 132–143.

Sageman, B. B., A. E. Murphy, J. P. Werne, C. A. Ver Straeten, D. J. Hollander, and T. W. Lyons (2003), A tale of shales: The relative roles of production, decomposition, and dilution in the accumulation of organic-rich strata, Middle-Upper Devonian, Appalachian Basin. *Chemical Geology, 195*, 229–273.

Schmidt, P. W. (1982), Interpretation of small-angle scattering curves proportional to a negative power of the scattering vector. *Journal of Applied Crystallography, 15*, 567–569.

Schmidt, P. W. (1991), Small-angle scattering studies of disordered, porous and fractal systems. *Journal of Applied Crystallography, 24*, 414–435.

Schmidt, P. W., D. Avnir, D. Levy, A. Hohr, M. Steiner, and A. Roll (1991), Small-angle X-ray-scattering from the surfaces of reversed-phase silicas – power-law scattering exponents of magnitudes greater-than-4. *Journal of Chemical Physics, 94*(2), 1474–1479.

Shackelford, C. D., and J.-M. Lee (2003), The destructive role of diffusion on clay membrane behavior. *Clays and Clay Minerals, 51*(2), 186–196.

Sherwood, J. D. (1995), Ionic transport in swelling shale. *Advances in Colloid Interface Science, 61*, 51–64.

Shultz, C. H. (1999), *The Geology of Pennsylvania*. Harrisburg, PA: Pennsylvania Geological Survey.140–145.

Silin, D., and T. Kneafsey (2012), Shale gas: nanometer-scale observations and well modelling. *Journal of Canadian Petroleum Technology*, *46*, 464–475.

Strapoć, D., M. Mastalerz, A. Schimmelmann, A. Drobniak, and N. R. Hasenmueller (2010), Geochemical constraints on the origin and volume of gas in the New Albany Shale (Devonian–Mississippian), eastern Illinois Basin. *American Association of Petroleum Geologists Bulletin*, *94*(11), 1713–1740.

Suekane, T., S. Soukawa, S. Iwatani, S. Tsushima, and S. Hirai (2005), Behavior of supercritical CO2 injected into porous media containing water. *Energy*, *30*, 2370–2382.

Summerfield, G. C., and D. F. R. Mildner (1983), Small angle scattering with azimuthal asymmetry. *Journal of Applied Crystallography*, *16*, 384–389.

Sun, C., Yao, S., Li, J., Liu, B., Liu, H., and Xie, Z. (2017a) Characteristics of pore structure and effectiveness of shale oil reservoir space in Dongying Sag, Jiyang Depression, Bohai Bay Basin. *Journal of Nanoscience and Nanotechnology*, *17*, 6781–6790.

Sun, M., Yu, B., Hu, Q., Zhang, Y., Li, B., Yang, R., Melnichenko, Y. B., and Cheng, G. (2017b), Pore characteristics of Longmaxi shale gas reservoir in the Northwest of Guizhou, China: Investigations using small-angle neutron scattering (SANS), helium pycnometry, and gas sorption isotherm. *International Journal of Coal Geology*, *171*, 61–68.

Sun, M., Yu, B., Hu, Q. Y., Yang, R., Zhang, Y., Li, B., Melnichenko, Y. B., and Cheng, G. (2018), Pore structure characterization of organic-rich Niutitang shale from China: Small angle neutron scattering (SANS) study. *International Journal of Coal Geology*, *186*, 115–125.

Swift, A. M., L. M. Anovitz, J. M. Sheets, D. R. Cole, S. A. Welch, and G. Rother (2014a), Relationship between mineralogy and porosity in seals relevant to geologic CO_2 sequestration. *Environmental Geosciences*, *21*(2), 39–57.

Swift, A., J. M. Sheets, D. R. Cole, L. M. Anovitz, S. Welch, X. Gu, D. Mildner, S. Chipera, E. Buchwalter, and A. Cook (2014b), Nano- to microscale pore characterization of the Utica shale. *Unconventional Resources Technology Conference (URTeC) Proceedings*, AAPG Publication, URTeC: 1923519, Denver, Colorado, August 25–27, 2014.

Tao, Z., and A. Clarens (2013), Estimating the carbon sequestration capacity of shale formations using methane production rates. *Environmental Science and Technology*, *47*(19), 11318–11325.

Teppen, B. J., and V. Aggarwal (2007), Thermodynamics of organic cation exchange selectivity in smectites. *Clays and Clay Minerals*, *55*(2), 119–130.

Thomas, J. J., and J. J. Valenza II, P. R. Craddock, K. D. Bake, and A. E. Pomeranz (2014), The neutron scattering length density of kerogen and coal as determined by CH_3OH/CD_3OH exchange. *Fuel*, *117*, 801–808.

Tissot, B. P., and D. H. Welte (1984), *Petroleum Formation and Occurrence*, 2nd ed., New York, Springer-Verlag, 699 p.

USDOE (2009), *Modern Shale Gas Development in the United States: A Primer*, Washington, DC, Department of Energy.

USDOE (2013), *Modern Shale Gas Development in the United States: An Update*, Washington, DC, Department of Energy.

Vilarrasa, V., S. Olivella, and J. Carrera (2011), Geomechanical stability of the caprock during CO_2 sequestration in deep saline aquifers. *Energy Procedia*, *4*, 5306–5313.

Vonk, C. G. (1976), On two methods of determination of particle size distribution functions by means of small-angle X-ray scattering. *Journal of Applied Crystallography*, *9*(6), 433–440.

Wang, H.-W., L. M. Anovitz, A. Burg, D. R. Cole, L. F. Allard, A. J. Jackson, A. G. Stack, and G. Rother (2013), Multi-scale characterization of pore evolution in a combustion metamorphic complex, Hatrurim basin, Israel: Combining (ultra) small-angle neutron scattering image analysis. *Geochimica et Cosmochimica Acta*, *121*, 339–362.

Warren, J. K., S. C. George, P. J. Hamilton, and P. Tingate (1998), Proterozoic source rocks: Sedimentology and organic characteristics of the Velkerri Formation, Northern Territory, Australia. *American Association of Petroleum Geologists Bulletin*, *82*(3), 442–463.

Wendt, A. K., Arthur, M. A., Slingerland, R., Kohl, D., Bracht, R., and Engelder, T. (2015), Geochemistry and depositional history of the Union Springs Member, Marcellus Formation in central Pennsylvania. *Interpretation*, August, SV17-SV33.

Wollenweber, J., S. Alles, A. Kronimus, A. Busch, H. Stanjek, and B. M. Krooss (2009), Caprock and overburden processes in geological CO_2 storage: An experimental study on sealing efficiency and mineral alterations. *Energy Procedia*, *1*, 3469–3476.

Wong, P.-Z. (1985), Scattering by inhomogeneous systems with rough internal surfaces: porous solids and random-field Ising systems. *Physical Review B*, *32*(11), 7417–7424.

Wong, P.-Z., and A. J. Bray (1988), Porod scattering from fractal surfaces. *Physical Review Letters*, *60*(13), 1344.

Wong, P.-Z., J. Howard, and J.-S. Lin (1986), Surface roughening and the fractal nature of rocks. *Physical Review Letters*, *57*(5), 637–640.

Xu, T., J. A. Apps, and K. Pruess (2003), Reactive geochemical transport simulation to study mineral trapping for CO_2 disposal in deep arenaceous formations. *Journal of Geophysical Research*, *108*(B2), 2071.

Xu, T., J. A. Apps, and K. Pruess (2005), Mineral sequestration of carbon dioxide in a sandstone–shale system. *Chemical Geology*, *217*, 295–318.

Yang, R., He, S., Hu, Q., Sun, M., Hu, D., and Yi, J. (2017), Applying SANS technique to characterize nano-scale pore structure of Longmaxi shale, Sichuan Basin (China). *Fuel*, *197*, 91–99.

Zhang S., and R. E. Klimentidis (2011), Porosity and permeability analysis on nanoscale FIB_SEM 3D imaging of shale rock. Presentation at the International Symposium of the Society of Core Analysts held in Austin, TX, 18–21 September 2011. http://www.scaweb.org/assets/papers/2011_papers/SCA2011-30.pdf.

Zhao, J., Jin, Z., Hu, Q., Jin, Z., Barber, T. J., Zhang, Y., and Bleuel, M. (2017), Integrating SANS and fluid invasion methods to characterize pore structure of typical American shale oil reservoirs. *Nature Scientific Reports*, *7*, 15413.

Part II
Fracture Generation, Permeability, and Geochemical Reactions in Damaged Shale

5

Fracture Initiation, Propagation, and Permeability Evolution

Russell L. Detwiler[1] and Joseph P. Morris[2]

ABSTRACT

Injection into deep aquifers or depleted oil and gas reservoirs provides a promising long-term approach for isolating waste fluids such as CO_2. The viability of this approach depends upon the ability of overlying caprocks to minimize leakage of injected fluids to shallow aquifers and the atmosphere. Leakage through low-permeability caprocks requires the presence of fractures, which typically have permeabilities many orders of magnitude greater than the rock matrix. In this paper, we review processes associated with the initiation and propagation of fractures in caprocks and reactivation of faults, both of which can create leakage pathways. In addition, we explore the processes that control the permeability of existing or induced fractures. This includes geomechanical stresses and geochemical perturbations, which can both lead to fracture permeability alteration. Finally, we discuss the implications of these different mechanisms on developing an effective strategy for monitoring injection facilities to ensure safe long-term isolation of injected fluids.

5.1. INTRODUCTION

There is significant concern over the potential for leakage (of either CO_2, leached minerals, or other contaminated fluids) to shallow aquifers and the atmosphere during subsurface injection associated with CO_2 storage [*Siirila et al.*, 2012; *Plampin et al.*, 2014]. Because the permeable formations targeted for injection are typically overlain by thick layers of low-permeability rock, if leakage to shallow aquifers and the atmosphere occurs, it is most likely through fractures in the caprock or along interfaces within the wellbore system [*Carroll et al.*, 2014].

Fractures can provide fast flow paths through otherwise low-permeability rock. For this reason, creating and controlling fractures and fracture networks is of key importance in managing the performance of systems where flow and transport is intended to be either maximized or minimized. It is often the case that applications involve both enhancing permeability in one region and minimizing it in another. For example, in enhanced geothermal systems, it is not desirable to induce significant fracturing outside of the target geothermal resource or create short circuits of the geothermal system that minimize fluid contact with the energy resource [*Fu et al.*, 2015]. Similarly, hydraulic fracturing is performed to improve production from low-permeability petroleum reservoirs. Extension of hydraulic fractures outside of the target formation has been studied in detail [*Fisher and Warpinski*, 2012] because it is regarded as a potential loss of revenue or potential leakage pathway from the target formation. Finally, for CO_2 storage, it is clear that injection should not reactivate existing fractures or introduce new fractures in the caprock because they may provide leakage pathways out of the target storage formation.

Leakage of injected fluids through layers of caprock requires both interconnected fractures with permeabilities significantly larger than the surrounding low-permeability rock and a vertical pressure gradient sufficient to drive flow through the fractures. Fracture permeability is often treated as a static property of a rock mass and estimated by analyzing geophysical data and pressure histories for

[1]Department of Civil and Environmental Engineering, University of California, Irvine, Irvine, CA, USA
[2]Atmospheric, Earth, and Energy Division, Lawrence Livermore National Laboratory, Livermore, CA, USA

Geological Carbon Storage: Subsurface Seals and Caprock Integrity, Geophysical Monograph 238, First Edition.
Edited by Stéphanie Vialle, Jonathan Ajo-Franklin, and J. William Carey.

input into static reservoir models [e.g., *Pamukcu et al.*, 2011]. These estimated fracture permeabilities may then be calibrated to improve agreement between simulations and observations. However, fracture apertures will change due to both changing effective stresses in the caprock and possible chemical alteration induced by fluids that are out of equilibrium with minerals on fracture surfaces. Changes in fluid pressure induced by injection or production of fluids from a reservoir will alter the stress state of the rock and may dilate preexisting fractures, leading to fracture permeability evolution [*Pyrak-Nolte and Morris*, 2000]. Additionally, the injected supercritical CO_2 will be at significantly lower temperature than the reservoir, and the induced cooling can lead to significant thermal stresses and strains that can both dilate existing fractures [*Rutqvist et al.*, 2008] and create new ones [*Bissell et al.*, 2011]. After injection ceases, pore pressures will decrease as the plume dissipates, causing apertures to decrease. Fluids migrating upward from the injection formation through overlying caprocks will likely be out of chemical equilibrium with minerals along fracture surfaces. This leads to mineral dissolution or precipitation and alteration of fracture permeabilities over time [*Johnson et al.*, 2005]. Furthermore, injection of fluids such as CO_2 will significantly alter the chemical composition of reservoir fluids and potentially enhance both the rate and magnitude of these permeability changes.

In this paper, we review the processes that control the initiation and propagation of fractures in caprocks (Section 5.2), reactivation of existing fractures and faults (Section 5.3), and the subsequent evolution of fracture permeability (Section 5.4). We then discuss the implications of these processes on developing effective monitoring strategies and identify mechanical, fluid, and chemical monitoring technologies that aid assessment of the role of fractures in storage performance (Section 5.5).

5.2. CONTAINMENT OF FLUID-DRIVEN FRACTURES

A number of analytical and numerical approaches have been designed to predict the propagation and extent of fluid-driven fractures, largely in support of hydraulic fracture design. The propagation and extent of fluid-driven fractures are controlled by competition among several mechanisms including the following [*Garagash et al.*, 2011]:

1. Fracture toughness, tensile strength, and shear strength
2. Fluid viscosity and flow rate
3. Confining stress and fluid pressure
4. Fluid leak-off into the formation
5. Interaction with preexisting fractures/faults

Fluid viscosity and flow rate can control whether a fluid-driven fracture crosses a preexisting fracture [*Chuprakov et al.*, 2013]. Conservatively, it is appropriate to assume that the fracture toughness and tensile strength are negligible and that interfaces between lithological units are insufficient to arrest the fracture. Consequently, when considering fluid-driven fractures with the specific intent of containment, the overriding property considered is the variations in stress within the formation. For a vertically propagating fracture, it is the minimum horizontal stress σ_{hmin} that controls propagation, and when a fracture reaches a horizon where the stress contrast is greater than the net pressure in the fracture at that point, it will arrest [*Simonson et al.*, 1978]. This phenomenon has subsequently been confirmed through modeling, laboratory experiment [*Jeffrey and Bunger*, 2009], and field-scale mineback [*Warpinski and Teufel*, 1987]. Depending upon the quality of the data in the model, the robustness and reliability of the stress barrier may be marginal. Consequently, it is important to be able to quantify the stress state of the reservoir *and* overburden as accurately as possible.

It is possible to relate the stress state of the formation to the elastic properties of the formation and the tectonic strain. The basic concept can be demonstrated for an isotropic medium subjected to a prescribed vertical overburden and laterally constrained, where it can be shown [*Eaton*, 1969] that

$$\sigma_{hmin} = \frac{\nu}{1-\nu}(\sigma_v - \alpha p) + \alpha p \tag{5.1}$$

where ν is the Poisson ratio, σ_v is the overburden stress, α is the Biot coefficient, and p is the pore pressure. This simple equation predicts that lateral stresses can increase or decrease with depth through a layered medium (e.g., sedimentary basin) due to variation in elastic properties, and therefore the lateral stress can be higher in a caprock than in a reservoir, which is good for fracture containment. This simple equation demonstrates that for a layered medium (such as most underburden, reservoir, and overburden combinations), it can be expected that similar vertical/overburden stress will induce contrasting lateral stresses (σ_{hmin}). *Dunphy and Campagna* [2011] utilized this approach to predict containment of fractures within the Horn River Basin. A good candidate caprock should have contrasting properties that encourage the development of high lateral stress in order to provide vertical containment.

In practice, more general equations than (5.1) have been developed that take into account specified far-field tectonic strains [e.g., *Plumb et al.*, 2000].

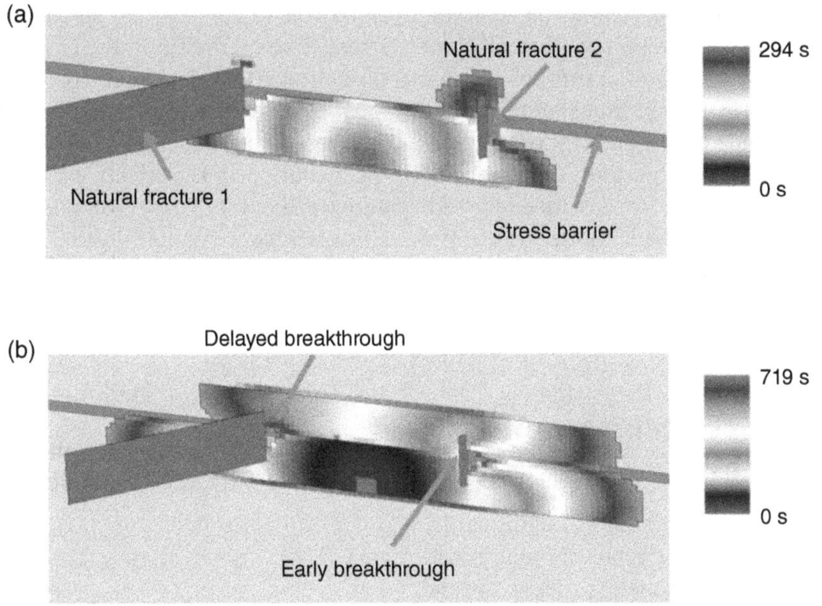

Figure 5.1 *Morris et al.* [2016] investigated the interactions between natural fractures and stress barriers. This side view of hydraulic fracture development at (a) 300 and (b) 720 s (colors reflect the time at which that portion of the fracture was opened) suggest that height growth may be encouraged if the hydraulic fracture passes near the tip of a natural fracture (NF2). If the hydraulic fracture passes through the midpoint of the natural fracture, the height growth may even be delayed. (*See electronic version for color representation of the figure.*)

Using these approaches, formation elastic properties (obtained via seismic, downhole acoustic, or core measurement) can be used to obtain an estimate of stresses. In addition, fully 3D finite element modeling has also been performed to help constrain the stress profile [*Qiuguo et al.*, 2013]. These results are typically supplemented by a much smaller number of direct stress measurements such as diagnostic fracture injection tests and wellbore failure data [*Chiaramonte et al.*, 2008].

There is evidence that preexisting natural fractures may influence the effectiveness of stress barriers. For example, *Morris et al.* [2016] investigated the interactions between natural fractures and stress barriers (see Fig. 5.1). Their analysis suggested that height growth may be encouraged if the hydraulic fracture passes near the tip of a natural fracture and impeded if the hydraulic fracture passes through the midpoint of the natural fracture.

Different storage site geologies and operation parameters place different demands upon the storage system. Although there are few large-scale storage projects to date, they include a range of target formations from tight to high permeability, and consequently a range of geomechanical responses (both desirable and undesirable) have been observed. *Verdon et al.* [2013] compared the mechanical deformation associated with several megaton-scale storage projects and observed that lower-permeability formations experienced more significant pressure changes and associated geomechanical deformation. For low-permeability storage targets, it may in fact be advantageous to fracture the reservoir to improve injectivity. Even if the fluid storage scenario is designed with no intention of fracturing the reservoir, it is possible to unintentionally fracture the reservoir due to combined uncertainties around downhole thermal and stress conditions. For example, *Bissell et al.* [2011] reported that well injection data indicated that the Krechba reservoir was likely fractured by combinations of pressure and thermal stresses. Specifically, they found that depending upon the details of the wellbore flow model, cooling of the formation could reduce the fracture pressure by 1.5–5 MPa. However, they concluded that the fracturing did not propagate into the caprock. Later analysis by *White et al.* [2014] also indicated that no significant hydraulic fracturing of the caprock had occurred.

Vilarrasa and Laloui [2015] investigated the influence of cooling upon potential fracturing of caprocks in a range of storage scenarios and argued that in the instance of normal faulting, the contraction within the reservoir tended to induce compressive stresses in the adjacent caprock that slightly locally stabilize the caprock against tensile fracturing. It is worth noting that there are relevant case studies from outside CO_2 sequestration applications where the potential creation and extension of fractures due to thermomechanical stresses are addressed. *Khan et al.* [2011] coupled a reservoir model to a finite element geomechanical model to predict the influence of injection-induced changes in temperature and pressure upon caprock integrity and develop a safe injection procedure for a specific site.

5.3. REACTIVATION OF EXISTING FRACTURES AND FAULTS

Permeability normal to faults is often significantly lower than the intact formation due to the combined influence of deformation and diagenetic processes in the immediate vicinity of faults [*Ahmadov et al.*, 2007]. However, reactivation of faults is likely to cause damage to the adjacent thin regions of reduced permeability, which may increase the permeability in these regions [*Uehara and Shimamoto*, 2004]. For example, *Guglielmi et al.* [2015] directly measured a 20-fold increase of permeability during fault slip resulting from fluid injection into a natural fault. It is well established that fractures and faults that are favorably oriented for slip under the prevailing in situ stress regime (so-called critically stressed fractures) tend to provide conduits for fluid flow [*Barton et al.*, 1995]. The propensity for slip on a fracture or fault can be quantified using the concept of slip tendency [*Morris et al.*, 1996]:

$$T_s = \frac{\sigma_s}{\sigma'_n} \tag{5.2}$$

where σ_s is the magnitude of the *current local* shear stress onto the plane of the fracture/fault and σ'_n is the so-called effective stress:

$$\sigma'_n = \sigma_n - \alpha p \tag{5.3}$$

where σ_n is the normal component of the total stress, $\boldsymbol{\sigma}$, and p is the pore pressure. σ_n quantifies the component of normal stress that is being carried by the fracture contacts. It is typically assumed that as the slip tendency approaches the value of 0.6, which is commonly adopted for the coefficient of friction, the fault fails in shear and becomes a conduit for flow [*Streit and Hillis*, 2004].

As we have seen, it is the stress state of the rock that controls and limits the progress of hydraulic fractures. Similarly, fault reactivation is heavily influenced by the in situ stress state along with the fault geometry. Equation (5.2) assumes perturbations to the effective stress state on the fault due to injection trigger reactivation; however this equation requires that the current local perturbed stress state be quantified.

Several approaches have been developed for estimating the evolution in the effective stress state during injection and production of reservoirs. If we assume that the total stress field, $\boldsymbol{\sigma}$, remains constant during injection, then the change in effective normal stress upon the fracture/fault is entirely due to the influence of pore pressure upon the effective stress:

$$\Delta\sigma'_n = -\alpha\Delta p \tag{5.4}$$

Consequently, in this instance, the critical change in pore pressure required to induce shear slip is

$$\Delta p_c = \frac{\sigma_n^0 - \sigma_s}{\mu_f} \tag{5.5}$$

where σ_n^0 is the original total normal stress and μ_f is the Coulomb friction coefficient [*Wiprut and Zoback*, 2002].

Equation (5.5) assumes that the only effect of the injected fluid is to change the pore pressure. In reality, the formation will deform in response, modifying the stress within the reservoir. For example, *Segall and Fitzgerald* [1996] considered the case of vertically oriented fractures/faults in the limit of an isotropic, porous elastic reservoir with static total overburden stress and zero lateral strain. This limit is approached for reservoirs with large lateral extent (exceeding 10 times the thickness) where the vertical strains dominate over the lateral strains. For this case, it can be shown that an increase in pore pressure induces (using geomechanics sign convention)

$$\Delta\sigma'_h = \Delta\sigma'_h = -\alpha\Delta p \frac{\nu}{1-\nu} \tag{5.6}$$

We observe that for physically meaningful values of ν, $\nu/(1-\nu)$ is less than one and the change in effective normal stress is reduced relative to the previously considered limit (5.4). As a result, the influence of pore pressure increase on the *slip tendency* of vertical fractures is weaker than the simpler theory would predict. Another way to consider this result is to rearrange (5.6)

$$\Delta\sigma'_h = -\alpha\Delta p + \alpha\frac{1-2\nu}{1-\nu}\Delta p \tag{5.7}$$

and compare with (5.4). The second term manifests as a *poroelastic correction* to the stress on the fault

$$\Delta\sigma_h = \alpha\frac{1-2\nu}{1-\nu}\Delta p \tag{5.8}$$

that induces additional compression upon the fault/fracture and tends to stabilize it against pressure increase [*Segall and Fitzgerald*, 1996].

The evolution of effective stress may also be obtained by more elaborate analysis methods that take into account the fully coupled spatial deformations, both elastic and plastic, due to pore pressure. For example, *Jha and Juanes* [2014a, 2014b] achieve this using a finite element based formulation. Similarly, in the context of geothermal stimulation, *Dempsey et al.* [2013] considered the impact of both pressure- and thermal-induced changes using a coupled finite element model.

The role of faults in providing both seals and flow conduits has been studied extensively in the context of hydrocarbon production. For example, *Wiprut and Zoback* [2002] discuss evidence of fault reactivation in the North Sea due in part to elevated pore pressures. *Dholakia et al.* [1998] found that hydrocarbon transport within the Monterey Formation was controlled by shearing along preexisting discontinuities. *Zhang et al.* [2007] considered the influence of fault reactivation upon permeability evolution during oil production.

Faults are believed to act as seals in several actual or proposed CO_2 storage fields. For example, the CO_2 storage target at Snøvhit (the Tubåen Formation) is believed to be compartmentalized by a number of intersecting sealing faults. As a result, pressure buildup has been more rapid than would be expected in a homogeneous reservoir [*Pham et al.*, 2011; *Chiaramonte et al.*, 2014].

In the specific case of CO_2 sequestration, many sequestration targets are effectively closed on one or more sides by noncritically stressed (impermeable or sealing) faults. *Streit and Hillis* [2004] describe in detail how fault stability and sustainable fluid pressures can be estimated for a range of sequestration sites by evaluating the tendency for fault reactivation during injection. *Lucier and Zoback* [2008] present a geomechanics-based workflow that includes the fault reactivation concept as a key element for characterizing leakage potential.

Cappa and Rutqvist [2011] considered a range of hypothetical stress environments and fault orientations. They investigated the potential for injection-induced slip on the faults and the associated seismicity. Their study specifically focused upon unfavorable circumstances, favoring fault slip, and they concluded that site selection and injection pressure monitoring are key to limiting fault reactivation.

Several authors have analyzed proposed and actual sequestration sites for potential fault reactivation. For example, *Chiaramonte et al.* [2008] considered a potential sequestration project at Teapot Dome and determined the bounds of pressure perturbations that would be expected to avoid fault leakage. *Morris et al.* [2011] performed analyses of the influence of fault orientation and pore pressure on fault stability within the Krechba reservoir. *Morris et al.* [2011] determined that the data at the time were consistent with reactivation and pressurization of a preexisting fault that cut one of the injectors. Using more data collected at a later date, *White et al.* [2014] concluded that the observations were more likely consistent with reactivation and coalescence of preexisting fracture pathways within the reservoir and lower caprock. As such, *White et al.* [2014] performed a combined study of fracture/fault reactivation and hydraulic fracture extension motivated by surface deformation observations, stress data, and seismic data (see Fig. 5.2).

5.4. FRACTURE PERMEABILITY AND LEAKAGE

When the vertical fluid pressure gradient is larger than hydrostatic or the injected fluid is less dense than the resident fluids, injected fluid may flow upward through fractures in overlying caprocks. These fractures may be preexisting or induced discrete fractures or complex networks of interconnected fractures within the damage zone surrounding a reactivated fault. In either case, for CO_2 storage, where the injected fluid is a separate phase, the injection-induced pressure increases can extend far beyond the CO_2 plume, resulting in three distinct leakage scenarios: (i) the injected phase enters fractures in the overlying caprocks, displacing some of the resident brine, and migrates upward as a separate phase; (ii) reservoir fluid, which may contain dissolved CO_2, migrates upward through existing or induced fractures; or (iii) simultaneous migration of reservoir fluid and CO_2. In each case, fracture apertures play an important role in the potential for leakage and the resulting leakage rates.

Direct leakage of CO_2 through a fractured caprock will be limited to the region in which the plume of supercritical CO_2 is in direct contact with the caprock. The potential for CO_2 to migrate through fractures is controlled by the fracture aperture, b, the contact angle between the CO_2-brine interface and the fracture surface, θ, and the CO_2-brine interfacial tension, γ. In the case where CO_2 is the non-wetting fluid (i.e., $\theta < 90°$), CO_2 will only enter a caprock fracture when the pressure difference between the CO_2 and fracture-filling brine exceeds the threshold capillary pressure, P_c, which can be quantified using the Laplace-Young equation:

$$P_c = \frac{\gamma \cos\theta}{b} \tag{5.9}$$

Though it is common to assume that CO_2 will be a non-wetting fluid in most settings, experimental observations of CO_2 injection into sandstone cores suggest the potential for the development of mixed wettability conditions, particularly near the critical point for CO_2 [*Plug and*

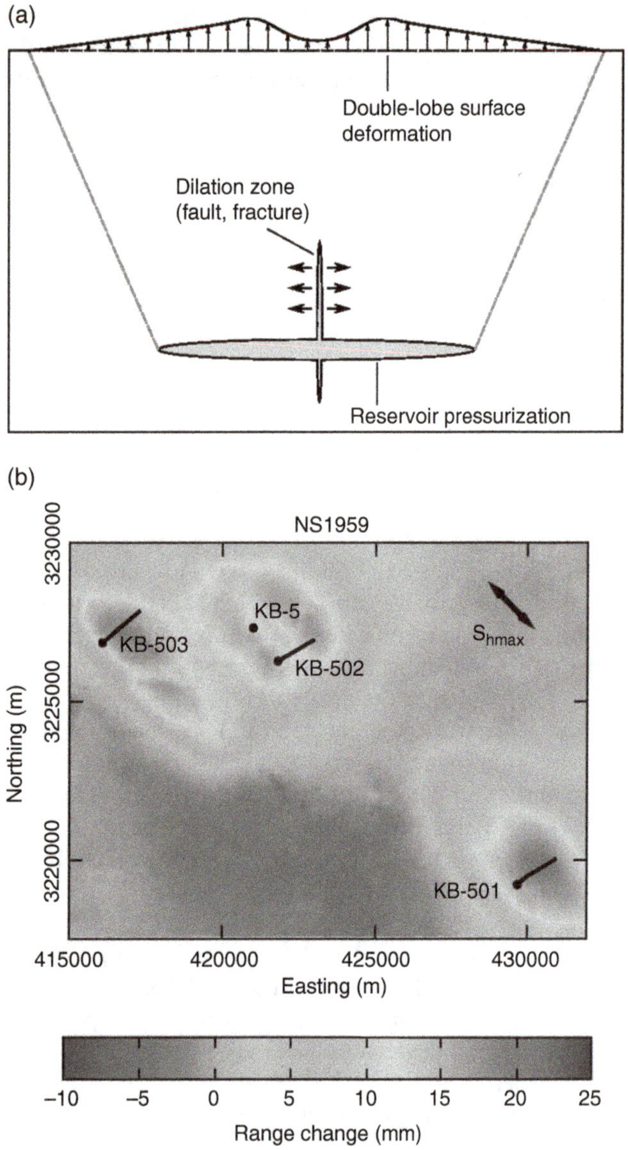

Figure 5.2 *White et al.* [2014] explored a range of fracture/fault reactivation and hydraulic fracture extension scenarios (a) motivated by seismic data, stress data, and surface deformation observations (b). They concluded that the double-lobe surface deformation above the KB-502 well (b) is consistent with a dilation of vertical hydrofractures within the lower portion of the caprock (a). Reprinted with permission from *White et al.* [2014].

Bruining, 2007]. When CO_2 is the non-wetting fluid, leakage requires that the CO_2 pressure continues to exceed the capillary threshold as it migrates upward through interconnected fractures. This process can result in leakage through localized interconnected large-aperture regions and, thus, defies representation by two-phase continuum models. Invasion percolation models have been shown to effectively predict the migration of wetting and non-wetting fluids through single variable-aperture fractures under conditions where viscous forces are negligible [*Glass et al.*, 1998]. This approach has been extended to simulate upward migration of CO_2 from the Sleipner field in Norway [*Cavanagh and Haszeldine*, 2014] by modeling P_c as a parameter that varies in different shale layers. The results from invasion percolation simulations at the reservoir scale were more consistent with field observations than those obtained using a continuum two-phase flow model. These results highlight the challenges associated with using reservoir simulators to predict leakage through caprock fractures.

Pressure increases caused by large injection volumes will extend far beyond the CO_2 plume, resulting in vertical pressure gradients favorable for upward flow of reservoir fluids in a region much larger than the plume. In this case, leakage is driven by fluid pressure gradients, not buoyancy and capillary forces. The magnitude of the upward flow, or leakage rate, per unit width of fracture (q) through a single fracture will depend upon the effective hydraulic aperture of the fracture, b_{eff}, using the cubic law:

$$\mathbf{q} = -\frac{b_{\text{eff}}^3}{12\mu}\nabla P \tag{5.10}$$

where the reduced pressure $P = p + \rho g z$, p is the fluid pressure, ρ is the fluid density, g is gravitational acceleration, z is elevation, and μ is the dynamic viscosity of the fluid [e.g., *Zimmerman and Bodvarsson*, 1996]. Note that in this context, fracture permeability is defined as $k_f = b_{\text{eff}}^2/12$. For rough-walled fractures, (5.10) can be used to describe flow through the fracture by replacing b with b_{eff}, which, due to tortuosity caused by surface roughness and contact regions between fracture surfaces, is always smaller than the mechanical aperture, b [*Zimmerman and Bodvarsson*, 1996]. Furthermore, assuming (5.10) applies locally and imposing continuity ($\nabla \cdot q = 0$), leads to the local cubic law [e.g., *Brown*, 1987]:

$$\nabla \cdot \left(\frac{b^3}{12\mu} \nabla P \right) = 0 \tag{5.11}$$

which has been shown to predict flow through variable-aperture fractures reasonably well when $Re < 1$, $\text{Re}\frac{\langle b \rangle}{\lambda_b} < 1$, and $\text{Re}\frac{s_b}{\langle b \rangle} < 1$, where $\text{Re} = \frac{\rho \langle \mathbf{q} \rangle}{\mu}$ is the Reynolds number and $\langle b \rangle$, s_b, and λ_b are the mean, standard deviation, and correlation length of the fracture aperture field, respectively [*Brush and Thomson*, 2003].

5.4.1. Mechanical Alteration of Permeability

While fluid-driven fractures may initially be fully open, as fluid pressure within the fracture decreases and effective stress increases, the fracture surfaces will close.

However, fluid-driven opening-mode fractures will maintain some non-negligible permeability as they close because the rough faces of the fractures will not be perfectly matched. As rough fracture surfaces come into contact, the initial contacting asperities deform, leading to increased contact area and fracture aperture that is very sensitive to the effective normal stress, σ_n. *Bandis et al.* [1983] summarized a wide range of experimental measurements and developed an empirical relationship between applied normal stress, σ_n, and fracture closure, which we modify to state in terms of the mean of the change in mechanical aperture, $\langle \Delta b \rangle$:

$$\sigma_n = -\frac{\langle \Delta b \rangle}{a + \beta \langle \Delta b \rangle} \tag{5.12}$$

where a and β are empirical parameters for which a/β is the maximum closure and $1/a$ is the initial fracture stiffness, or $-(d\sigma_n/d\langle \Delta b \rangle)$. They used this model of fracture closure to predict changes in b_{eff} using the following empirical relationship:

$$b_{eff} = \frac{\mathrm{JRC}^{2.5}}{\left(-\langle \Delta b \rangle / b_{eff}\right)^2} \tag{5.13}$$

where *JRC*, the *joint roughness coefficient*, is an empirical measure of the impact of surface roughness on shear deformation of fracture surfaces and, thus, provides an indirect measure of surface roughness.

Subsequent efforts focused on developing mechanistic models of flow through fractures with variable surface roughness and contact areas relied upon direct measurements of surface roughness. *Brown* [1987] used (5.11) to quantify the relative impact of surface roughness and the resulting discrepancies between b and b_{eff}. *Zimmerman and Bodvarsson* [1996] presented an expression relating b_{eff} to b and contact area between the fracture surfaces, c:

$$b_{eff} = \left[\langle b \rangle^3 \left(1 - 1.5 \frac{s_b^2}{\langle b \rangle^2} \right) (1 - 2c) \right]^{1/3} \tag{5.14}$$

This model includes the influence of contact area on fracture permeability but requires an independent estimate of contact area. *Detwiler et al.* [2005] modified this expression to quantify the impact of an entrapped nonaqueous phase (e.g., CO_2) on flow through a variable-aperture fracture by modifying the last term to account for the area of the fracture occupied by the entrapped non-wetting phase.

Mechanistic models have been developed to quantify the development of contact area with increasing stress and the subsequent influence on b_{eff} [*Unger and Mase*, 1993; *Pyrak-Nolte and Morris*, 2000]. These models represent individual asperities on each fracture surface as elastic posts confined between elastic half-spaces and explicitly represent the micro-mechanics that control the mechanical deformation of two contacting rough surfaces subjected to a normal stress. *Pyrak-Nolte and Morris* [2000] used this approach to demonstrate the importance of the correlation structure of fracture surfaces on deformation and the resulting evolution of b_{eff} under applied normal stresses. Because the developing contact area depends upon both the applied stress and the roughness of the contacting fracture surfaces, it is not possible to develop a simple relationship between σ_n and c, but these mechanistic models may be used to constrain this relationship for particular rocks.

Reactivated faults typically consist of a fault core with very low permeability surrounded by a zone of damaged rock with much larger permeability [*Faulkner et al.*, 2010]. Due to the high density of fractures and microfractures likely surrounding the damaged zone, it is impractical to model flow through discrete fractures within a reactivated fault zone. However, treating the zone as an effective continuum with a suitably constrained anisotropic permeability tensor has been shown to represent observations quite well [*Jeanne et al.*, 2013].

5.4.2. Geochemical Alteration of Permeability

The evolution of b_{eff} is further complicated when reservoir fluids or injected fluids are out of chemical equilibrium with the minerals in the caprock. Geochemical reactions between the fluid and minerals lead to dissolution or precipitation of minerals along fracture surfaces or in the adjacent porous matrix. These reaction processes alter the strength of contacting asperities, which subsequently affects the mechanical properties of the fracture. Alteration of the fracture aperture caused by these coupled geochemical and geomechanical mechanisms changes flow paths and influences subsequent transport of dissolved ions and can result in significant changes to b_{eff} over time.

For the idealized case of a fracture in a rock with negligible porosity and permeability, flow occurs only through the fracture. If the rock is composed of a single mineral and the flowing fluid is either undersaturated or oversaturated with respect to the mineral, mass transfer between the fluid and rock surfaces will alter the topography of the fracture surfaces. Absent significant stresses (and corresponding deformation), dissolution or precipitation will increase or decrease fracture aperture and alter flow paths within the fracture. Early efforts to model these reactive transport processes in the context of cave formation used one-dimensional flow and reactive transport models to simulate fracture permeability evolution [*Groves and Howard*, 1994; *Dreybrodt*, 1996]. A focus of these studies was quantifying the initiation timescale or the time required for a small conduit to grow sufficiently large that flow through the conduit became turbulent, causing a marked increase in the dissolution rate of the conduit walls. Results from these one-dimensional models suggest that breakthrough time depends strongly on the kinetics of the dissolution reaction.

Hanna and Rajaram [1998] noted that though flow through fractures is often represented as a Darcy-like one-dimensional process, aperture variability in fractures leads to the formation of preferential flow paths and potential for reaction-front instabilities similar to observations in porous media [*Ortoleva et al.*, 1987]. They used a two-dimensional reactive transport equation to approximate transport of dissolved ions and the resulting dissolution or precipitation reaction on the fracture surfaces:

$$\frac{\partial(bc)}{\partial t}+\nabla\cdot(\mathbf{q}c)-\nabla\cdot(b\mathbf{D}\nabla c)=R(c) \tag{5.15}$$

where v, the local depth-averaged fluid velocity, is calculated using the local cubic law, c is the depth-averaged concentration of the ion controlling the reaction, D is the local dispersion tensor, and $R(c)$ is the concentration-dependent rate of the dissolution or precipitation reaction which controls the local rate of change of fracture aperture:

$$\frac{\partial b}{\partial t}=\frac{R(c)}{\alpha\rho_r} \tag{5.16}$$

where α is a stoichiometric coefficient relating the relative mass of the ionic species tracked in the transport equation to the molar mass of the dissolving mineral and ρ_r is the density of the dissolving minerals. In most geological settings, the timescale for aperture change is much longer than the timescale for equilibration of the reactive transport process, so it is reasonable to solve the flow and reactive transport problems as quasi steady state.

Numerical simulations using (5.11) and (5.15) demonstrate that reaction-front instabilities are inherent to these reactive flow problems and these instabilities result in reductions in breakthrough times of up to an order of magnitude [*Hanna and Rajaram*, 1998]. These computational results are consistent with dissolution experiments in monomineralic fractures in which the formation of preferential flow paths caused by dissolution was directly observed [*Durham et al.*, 2001; *Detwiler et al.*, 2003]. Subsequent evaluation of the two-dimensional modeling approach through comparison to controlled experiments [*Detwiler and Rajaram*, 2007] demonstrates that despite the simplification to two dimensions, this model predicts the evolution of fracture aperture and transmissivity quite well. The characteristics of the reaction-front instabilities depend strongly upon the relative value of the timescales associated with advection (τ_a) and diffusion (τ_d) of dissolved ions and surface reactions (τ_r). These timescales can be quantified as $\tau_a = L\langle b\rangle/\mathbf{q}$, $\tau_d = \langle b\rangle^2/D_m$, and $\tau_r = b/k$. Two important dimensionless parameters relate these timescales: the Damköhler number ($Da = kL/\mathbf{q}$) and Peclet number ($Pe = \mathbf{q}/D_m$).

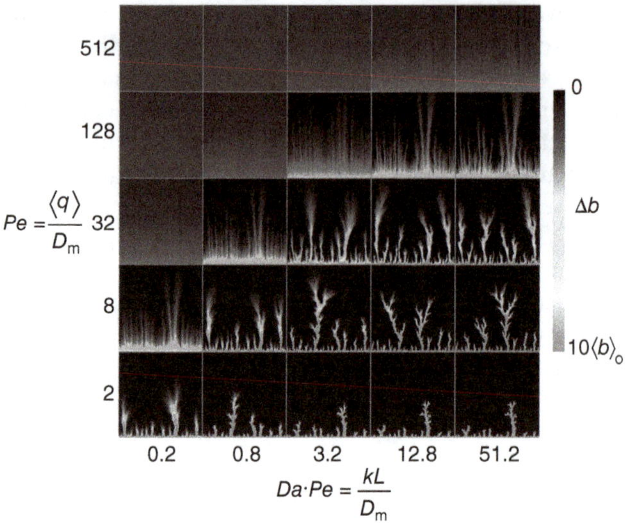

Figure 5.3 Fracture dissolution simulations carried out over a broad range of *Da* and *Pe*. Flow of an undersaturated fluid from bottom to top in each simulation led to change in fracture aperture, Δb, caused by dissolution of the fracture surfaces. Values of k and $\langle q\rangle$ were varied, but all other parameters including the initial aperture field were identical for each simulation. Each simulation ran until $\langle b\rangle = 2\langle b\rangle_o$. These results show the broad range of plausible dissolution-induced fracture alteration scenarios, which range from very uniform dissolution when reaction kinetics are slow and/or flow rates are high (i.e., large *Pe* and small $Da \cdot Pe$) to very compact dissolution channels when reaction kinetics are fast and flow rates are low such that local dissolution rates are limited by mass transport rather than local reaction kinetics (i.e., small *Pe* and large $Da \cdot Pe$). Uniform dissolution may result in fracture closure in fractures subjected to significant normal stress, whereas the formation of dissolution channels is more likely to generate a persistent leakage pathway.

Figure 5.3 highlights the importance of *Da* and *Pe* during dissolution of fractures caused by reactive fluid flow. Dissolution simulations based on (5.11) and (5.15) result in a range of different behaviors depending on the relative values of *Pe* and *Da*. Note that $Da \cdot Pe$ is a modified Damköhler number that relates local reaction rates to diffusive transport from the fracture surfaces and is independent of velocity. These simulation results highlight the importance of the development of reaction-front instabilities, particularly for larger values of $Da \cdot Pe$, where development of preferential flow paths eventually leads to runaway permeability increases. Linear stability

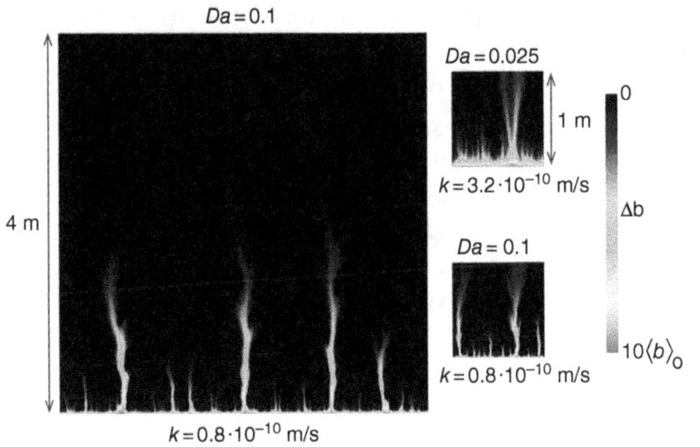

Figure 5.4 A demonstration of the influence of fracture length scale, L, on the formation of dissolution-induced preferential flow paths. Preferential flow paths in the 4 m × 4 m fracture with $k = 0.8 \cdot 10^{-10}$ m/s differ from those in a 1 m × 1 m portion of the same fracture at the same value of $\langle q \rangle$ and k, but are very similar to those in the 1 m × 1 m fracture with $k = 3.2 \cdot 10^{-10}$ m/s. Due to the direct dependence of Da on both k and L, concurrently decreasing L and increasing k leads to no change in Da and very similar dissolution pathways. The scaling behavior of the fracture dissolution process has significant implications when using laboratory-scale observations to infer field-scale fracture alteration.

analysis suggests that these reaction-front instabilities develop regardless of initial aperture variability and that the rate of propagation of instabilities is relatively insensitive to the amount of initial aperture variability but strongly dependent upon Da [*Szymczak and Ladd*, 2012]. Furthermore, because Da depends directly upon the fracture length, this analysis suggests a strong scale dependence of the formation of preferential flow paths. Figure 5.4 shows results of simulations in a 4 m × 4 m fracture compared to simulations in a 1 m × 1 m subset of the same fracture all at the same flow rate, but different values of k. The results show that the preferential flow paths created by dissolution depend not on the reaction rate coefficient, k, but on Da; thus, in a longer fracture, reaction-front instabilities and the formation of preferential leakage paths become more likely. It is particularly important to consider this scaling behavior when interpreting results from core-scale experiments.

Fewer studies have considered fluids that are oversaturated with respect to the rock minerals. In such cases, mineral precipitation reduces fracture apertures. Unlike dissolution, this results in a negative feedback in which precipitation occurs preferentially in large-aperture regions, which ultimately causes the fracture to seal along its entrance. However, under conditions in which the solubility of the fluid changes due to mixing of two fluids with different chemical compositions [*Tartakovsky et al.*, 2008], flow through a temperature gradient [*Chaudhuri et al.*, 2008], or spatial heterogeneity of the mineralogic composition of the rock [*Li et al.*, 2007; *Jones and Detwiler*, 2016], localized mineral precipitation may occur. In the case of mixing-induced precipitation, the process is self-limiting because precipitation along the mixing zone between the two fluids reduces diffusive exchange between the two fluids [*Zhang et al.*, 2010]. The other two scenarios are more likely to cause mineral precipitation during leakage through a fractured caprock. Because mineral solubility is temperature dependent, geothermal gradients induce a vertical solubility gradient. *Chaudhuri et al.* [2008] presented simulation results demonstrating the formation of horizontal bands of precipitated minerals and fracture permeability loss for minerals with prograde solubilities (e.g., silica). Alternatively, for minerals with retrograde solubility (e.g., calcite), combined mineral dissolution and precipitation led to the formation of vertically oriented preferential flow paths and runaway permeability growth.

Experimental observation and analysis of permeability alteration in idealized fractures provide useful insights, but more recent experimental studies in fractured rock cores have highlighted a number of complexities that arise in more realistic settings. Reactive flow experiments in rock cores under simulated reservoir conditions (applied confining stress and elevated temperature) favoring mineral dissolution resulted in b_{eff} increasing [*Elkhoury et al.*, 2013], decreasing [*Ellis et al.*, 2013], and changing nonmonotonically (i.e., decreasing then increasing) [*Polak et al.*, 2004]. These studies demonstrate the potential influences of mechanical stresses and chemical alteration of fracture surfaces, mineral heterogeneity within the host rock, and complexities resulting from secondary reactions that influence overall reaction kinetics. Due to the complexity of such experiments, most studies involve only one or two experiments, making it difficult to draw generalized conclusions about how these coupled processes may affect permeability over time. Furthermore, because both reactive alteration and geomechanical deformation are strongly scale dependent, care must be taken when extrapolating the results of these laboratory experiments to reservoir scales.

Mechanistic computational models provide a means for studying how coupled geochemical and mechanical processes affect fracture permeability at larger scales. Different models have been developed to explain the range of dissolution-induced changes to b_{eff} in fractures subjected to significant normal stress. *Yasuhara et al.* [2004] suggest that pressure solution at contacts between fracture surfaces leads to enhanced dissolution of contacting asperities. When the dissolution rate at contacts exceeds dissolution rates in the open regions of the fracture, the net result is decreased fracture aperture and permeability loss. Combining these processes in a lumped parameter model demonstrated that the range of experimental observations could be qualitatively reproduced. Alternatively, *Ameli et al.* [2014] presented a mechanistic model that combines solution of a two-dimensional representation of advection, diffusion, and reactions with a micro-mechanical model of fracture deformation. Results suggest that even in the absence of pressure solution, dissolution can lead to permeability decreases. This occurs under transport-limited conditions, where local dissolution rates become inversely proportional to fracture aperture [*Detwiler and Rajaram*, 2007; *Szymczak and Ladd*, 2012].

Heterogeneity of the mineralogic composition of rocks can also significantly influence permeability alteration in fractured rock. Reactive flow experiments in fractured cores with fixed displacement show a strong dependence of evolution of surface roughness, and fracture aperture, on local mineral heterogeneity [*Gouze et al.*, 2003; *Noiriel et al.*, 2013]. Under reaction-limited conditions, variations in local reaction rates lead to a roughening of the surfaces as more soluble regions dissolve more quickly. In some cases, less soluble minerals become completely detached from the surface due to dissolution of the surrounding minerals. Even in the absence of confining stress, mineral heterogeneity can result in dissolution-induced permeability decreases. *Ellis et al.* [2013] hypothesized that mobilization of less soluble minerals in a sample from a polymineralic carbonate caprock caused an order-of-magnitude reduction in permeability. *Elkhoury et al.* [2015] demonstrated the influence of adding a confining stress during reactive flow through fractured cores from a dolomitic anhydrite caprock. Preferential dissolution of the anhydrite left unconsolidated dolomite in the fracture, which deformed significantly under the imposed confining stress. The resulting permeability alteration was sensitive to flow rate: at high flow rate (reaction-limited conditions), the permeability decreased by two orders of magnitude, and at low flow rate (transport-limited conditions), the permeability remained relatively constant, despite dissolution and corresponding displacement.

Experimental observations of the influence of geochemical reactions on b_{eff} suggest that these effects defy representation by constitutive relationships. However, insights gained from laboratory-scale experiments can aid in the development of representative mechanistic models that may be useful in at least determining the sign of changes in b_{eff}, which is of significant importance in assessing the long-term potential for leakage from a storage site.

5.5. DISCUSSION

Predicting the development of permeability in caprocks intended to contain fluids remains a challenge to the safe implementation of subsurface injection for fluid disposal or isolation. Despite ongoing efforts to develop improved understanding of the fundamental processes and corresponding predictive capabilities, uncertainty in the potential for long-term leakage from deep formations is unavoidable. Thus, successful management of CO_2 storage sites requires extensive geomechanical and geochemical characterization prior to injection to establish baselines and throughout injection and post-closure to build confidence in storage site performance.

Prior to injection, the in situ stress state and elastic properties of the reservoir, underburden, and overburden must be well characterized. This characterization is critical for predicting not only the creation and propagation of injection-induced fractures but also the associated deformation signatures that they would induce. The in situ stress state, combined with detailed characterization of fractures and faults, will aid in identifying the role of fractures and faults as either fast flow conduits or flow barriers that will lead to compartmentalization.

Monitoring for deformation of the reservoir and underburden should be performed and compared with forward models to identify anomalous deformation that may be indicative of fracture creation and extension. Deformation tilt, InSAR, and GPS technologies have complementary advantages in this application. Tilt, for example, can be very precise, but is prone to drift over time and is typically too expensive to deploy over the entire injection field. GPS is not subject to drift, but is less accurate than tilt and makes only discrete, localized measurements of deformation. InSAR can provide surface deformation data over a large area at relatively low costs, but is only appropriate for land operations [*Bürgmann et al.*, 2000]. Given a suitably rich characterization, it is possible to use anomalous deformation observations as an indicator of fluid-driven fracturing events. In the absence of complete characterization ahead of time, monitoring during injection may reveal the presence and role of fractures and faults. For example, in the case of the Krechba reservoir, it was only after some period of injection that fast flow conduits were recognized through observations of anomalous deformation indicative of a pressurized, vertically oriented fracture or fractured zone over one of the injectors [*Vasco et al.*, 2010].

Devices for detection of microseismic activity, such as geophones, should be employed to provide measurements that can be used to help constrain the location of fracturing. New fiber-based deformation monitoring techniques show some promise for providing relatively cheap seismic monitoring over a larger extent than can otherwise be achieved, but currently suffer from low signal to noise compared with geophones [*Daley et al.*, 2013]. The microseismic monitoring technology should be in place some time before injection commences and placed near the injectors in order to monitor for fracturing events most effectively. It is not unlikely that public concern regarding potential larger felt events may require such monitoring to be performed as part of permitting in some jurisdictions. It is particularly important that a baseline of microseismic activity be established in order that events triggered by injection activity can be more readily distinguished from those naturally occurring. Note also that induced events may occur after injection has ceased and the pressure perturbation has diffused away from the injection area and encountered critically stressed fractures and faults at significant distance from the injection point [*Nicholson and Wesson*, 1992].

We have seen that both fluid pressure and temperature strongly influence the propensity for injection-induced fracturing. Consequently, during injection, the downhole properties (pressure and temperature at a minimum) of the injected fluid should be directly measured and monitored rather than only inferred from a model of flow in the wellbore. This will reduce the probability of inadvertently fracturing the lower caprock due to poorly constrained thermal stress perturbations. In addition, pressure response can reveal that fractures/faults are acting as flow barriers. For example, at Snøvhit, it was only after some period of injection that compartmentalization of the reservoir was fully recognized through observations of anomalous injection pressure [*Chiaramonte et al.*, 2014].

In addition to monitoring deformations associated with injection, directly monitoring the evolution of the CO_2 plume has value in assessing the potential for leakage. Electrical resistance tomography (ERT) has been demonstrated to have potential as a tool for monitoring two-phase fluid flow in the laboratory [*Breen et al.*, 2012] and, more recently, in the field to monitor CO_2 injection [*Carrigan et al.*, 2013; *Wagner et al.*, 2015]. These studies employed crosswell ERT, which involves challenges associated with deploying and maintaining electrodes at significant depths in contact with corrosive reservoir fluids. While there are challenges deploying this technology with wide areal coverage, this downhole technique has the advantage of significantly better resolution than surface-based observations.

Prior to injection, the geochemical composition of reservoir fluids and overlying freshwater aquifers should be characterized to provide a baseline for assessing injection-induced changes over time. The mineralogy of important geologic seals and sediments in overlying aquifers should also be assessed to provide input to related modeling of reactive transport and permeability alteration that may result in the event of leakage of reservoir fluids or injected CO_2 from target formations. During and after completion of injection, monitoring of overlying aquifers should be targeted to identify changes in fluid chemistry indicating either the presence of CO_2 (e.g., pH decrease) or changes in fluid chemistry indicating reactions with aquifer sediments (e.g., elevated metal concentrations). The establishment of a baseline water chemistry is critical not only for subsequent leakage detection but also to establish attribution. In the context of hydraulic fracturing, it has been observed that the lack of an adequate baseline has made it difficult to determine if subsequently identified water quality issues are related to injection or some other activity.

More work is required to identify appropriate mitigation steps in the event leakage is detected. Typical current approaches have been to cease injection. However, given that the pressure in the reservoir will remain elevated post shut-in, this may not be sufficient to minimize the leakage rate. Alternatively, silica gels have been proposed as suitable materials for controlled reduction of formation permeability [*Ezzedine et al.*, 2012]. Approaches have been developed to intervene in well leakage that could also be employed for fracture-based or wellbore-based leakage from CO_2 storage [*Genedy et al.*, 2014]. In the petroleum industry, fiber-based diversion technologies are used to slow flow through open fractures in order to divert fluid and build pressure in other locations [*Heaton et al.*, 2016].

5.6. CONCLUDING REMARKS

Safely storing fluids such as CO_2 in deep geologic formations requires careful evaluation and monitoring for the potential for leakage to shallow aquifers and the atmosphere. Leakage through hundreds of meters of low-permeability caprock is unlikely in the absence of interconnected fractures through the caprock formation. We have reviewed processes associated with fracture creation and reactivation and permeability evolution over time. Associated with these processes are a number of key parameters including in situ stress state of the formation and the presence of preexisting faults, fluid pressure gradients within overlying caprocks, and the chemical composition of resident and injected fluids and host rock. These parameters will play a critical role in the long-term viability of any proposed storage facility and, thus, must be constrained both prior to initiation of any storage activities and monitored throughout the lifespan of the facility.

Predicting the potential for fracture propagation requires constraining in situ stresses and reasonable estimates of the mechanical properties of the underburden, reservoir, and overburden. During the initial injection phase of a storage project, pore pressures will increase, leading to the potential development of leakage pathways associated with fracture propagation or fault reactivation. Injection will also lead to the development of vertical pressure gradients that may further favor upward fluid migration. Subsequently, after the completion of the injection phase, pore pressure will gradually decrease, leading to fracture closure and permeability decrease. The rate of permeability decrease will depend upon the rate of pressure dissipation in the reservoir and the hydromechanical properties of the fracture. The potential for concurrent geochemical alteration of fracture surfaces will depend upon fluid composition, mineralogy of the rock formations, and the rates of flow and transport through fractures. Numerous experimental studies have shown that the interaction between geochemical and geomechanical processes challenges our ability to predict the rates (and even the sign) of permeability changes. However, modeling efforts and scaling analyses strongly suggest that under field conditions favoring mineral dissolution, the formation of reaction-front instabilities, which can lead to development of preferential flow channels, is more likely. Developing new approaches for upscaling these coupled processes from laboratory-scale observations to field-scale applications that will lead to enhanced predictions of the relative rates of these processes remains a fundamental challenge to assessing the long-term viability of CO_2 storage systems.

The permeability of fractures (either preexisting or induced) depends upon the hydromechanical characteristics of the fractures, the stress distribution, and pore pressures within the formation.

ACKNOWLEDGMENTS

This work performed under the auspices of the US Department of Energy by Lawrence Livermore National Laboratory under contract **DE-AC52-07NA27344**. This work was funded by the Laboratory Directed Research and Development Program at **LLNL** under project tracking code **15-ERD**-010.

REFERENCES

Ahmadov, R., A. Aydin, M. Karimi-Fard, and L. Durlofsky (2007), Permeability upscaling of fault zones in the Aztec sandstone, valley of fire state park, Nevada, with a focus on slip surfaces and slip bands, *Hydrogeol. J.*, *15*, 1239–1250.

Ameli, P., J. Elkhoury, J. Morris, and R. Detwiler (2014), Fracture permeability alteration due to chemical and mechanical processes: A coupled high-resolution model, *Rock Mech. Rock. Eng.*, *47*, 1563–1573.

Bandis, S., A. Lumsden, and N. Barton (1983), Fundamentals of rock joint deformation, *Int. J. Rock Mech. Min. Sci.*, *20*(6), 249–268, doi:10.1016/0148-9062(83)90595-8.

Barton, C. A., M. D. Zoback, and D. Moos (1995), Fluid flow along potentially active faults in crystalline rock, *Geology*, *23*(8), 683–686.

Bissell, R. C., D. W. Vasco, M. Atbi, M. Hamdani, M. Okwelegbe, and M. H. Goldwater (2011), A full field simulation of the In Salah gas production and CO_2 storage project using a coupled geo-mechanical and thermal fluid flow simulator, *Energy Procedia*, *4*, 3290–3297.

Breen, S. J., C. R. Carrigan, D. J. LaBrecque, and R. L. Detwiler (2012), Bench-scale experiments to evaluate electrical resistivity tomography as a monitoring tool for geologic CO_2 sequestration, *Int. J. Greenhouse Gas Control*, *9*, 484–494, doi:10.1016/j.ijggc.2012.04.009.

Brown, S. R. (1987), Fluid flow through rock joints: The effect of surface roughness, *J. Geophys. Res.*, *92*(B2), 1337–1347.

Brush, D. J., and N. R. Thomson (2003), Fluid flow in synthetic rough-walled fractures: Navier-stokes, stokes, and local cubic law simulations, *Water Resour. Res.*, *39*(4), doi:10.1029/2002WR001346.

Bürgmann, R., P. A. Rosen, and E. J. Fielding (2000), Synthetic aperture radar interferometry to measure earth's surface topography and its deformation, *Annu. Rev. Earth Planet. Sci.*, *28*, 169–209.

Cappa, F., and J. Rutqvist (2011), Impact of CO_2 geological sequestration on the nucleation of earthquakes, *Geophys. Res. Lett.*, *38*(17), l17313, doi:10.1029/2011GL048487.

Carrigan, C. R., X. Yang, D. J. LaBrecque, D. Larsen, D. Freeman, A. L. Ramirez, W. Daily, R. Aines, R. Newmark, J. Friedmann, and S. Hovorka (2013), Electrical resistance tomographic monitoring of CO_2 movement in deep geologic reservoirs, *Int. J. Greenhouse Gas Control*, *18*, 401–408, doi:10.1016/j.ijggc.2013.04.016.

Carroll, S. A., E. Keating, K. Mansoor, Z. Dai, Y. Sun, W. Trainor-Guitton, C. Brown, and D. Bacon (2014), Key factors for determining groundwater impacts due to leakage from geologic carbon sequestration reservoirs, *Int. J. Greenhouse Gas Control*, *29*, 153–168, doi:10.1016/j.ijggc.2014.07.007.

Cavanagh, A. J., and R. S. Haszeldine (2014), The Sleipner storage site: Capillary flow modeling of a layered CO_2 plume requires fractured shale barriers within the utsira formation, *Int. J. Greenhouse Gas Control*, *21*, 101–112, doi:10.1016/j.ijggc.2013.11.017.

Chaudhuri, A., H. Rajaram, and H. Viswanathan (2008), Alteration of fractures by precipitation and dissolution in gradient reaction environments: Computational results and stochastic analysis, *Water Resour. Res.*, *44*(10), doi:10.1029/2008WR006982.

Chiaramonte, L., M. D. Zoback, S. J. Friedmann, and V. Stamp (2008), Seal integrity and feasibility of CO_2 sequestration in the teapot dome eor pilot: Geomechanical site characterization, *Environ. Geol.*, *54*(8), 1667–1675.

Chiaramonte, L., J. A. White, and W. Trainor-Guitton (2014), Probabilistic geomechanical analysis of compartmentalization at the Snøvhit CO_2 sequestration project, *J. Geophys. Res. Solid Earth*, *120*, 1195–1209.

Chuprakov, D., O. Melchaeva, and R. Prioul (2013), Hydraulic Fracture Propagation Across a Weak Discontinuity Controlled by Fluid Injection, in *Effective and Sustainable Hydraulic Fracturing*, edited by Rob Jeffrey (Ed.), InTech, doi:10.5772/55941. Available from: http://www.intechopen.com/books/effective-and-sustainable-hydraulic-fracturing/hydraulic-fracture-propagation-across-a-weak-discontinuity-controlled-by-fluid-injection.

Daley, T. M., B. M. Freifeld, J. Ajo-Franklin, S. Dou, R. Pevzner, V. Shulakova, S. Kashikar, D. E. Miller, J. Goetz, J. Henninges, and S. Lueth (2013), Field testing of fiber-optic distributed acoustic sensing (DAS) for subsurface seismic monitoring, *Lead. Edge*, *June*, 936–942.

Dempsey, D., S. Kelkar, K. Lewis, S. Hickman, N. Davatzes, D. Moos, and E. Zemach (2013), Modeling shear stimulation of the desert peak egs well 27–15 using a coupled thermal-hydro-logical-mechanical simulator, in *47th US Rock Mechanics/Geomechanics Symposium 2013*, pp. 1006–1019, San Francisco, CA, June 23–26, 2013.

Detwiler, R. L., and H. Rajaram (2007), Predicting dissolution patterns in variable aperture fractures: Evaluation of an enhanced depth-averaged computational model, *Water Resour. Res.*, *43*(4), W04403.

Detwiler, R. L., R. J. Glass, and W. L. Bourcier (2003), Experimental observations of fracture dissolution: The role of peclet number on evolving aperture variability, *Geophys. Res. Lett.*, *30*(12), doi:10.1029/2003gl017396.

Detwiler, R. L., H. Rajaram, and R. J. Glass (2005), Satiated relative permeability of variable-aperture fractures, *Phys. Rev. E*, *71*, 031114, doi:10.1103/PhysRevE.71.031114.

Dholakia, S. K., A. Aydin, D. D. Pollard, and M. D. Zoback (1998), Fault-controlled hydrocarbon pathways in the Monterey formation, California, *AAPG Bull.*, *82*(8), 155–1574.

Dreybrodt, W. (1996), Principles of early development of karst conduits under natural and man-made conditions revealed by mathematical analysis of numerical models, *Water Resour. Res.*, *32*(9), 2923–2935, doi:10.1029/96WR01332.

Dunphy, R., and R. D. Campagna (2011), Fractures, elastic moduli and stress: Geological controls on hydraulic fracture geometry in the Horn River Basin, in *Recovery – 2011 CSPG CSEG CWLS Convention*, Calgary, Alberta, May 9–11, 2011.

Durham, W. B., W. L. Bourcier, and E. Burton (2001), Direct observation of reactive flow in a single fracture, *Water Resour. Res.*, *37*, 1–12.

Eaton, B. A. (1969), Fracture gradient prediction and its application to oilfield operations, *J. Petrol. Tech.*, *October*, 1353–1360.

Elkhoury, J. E., P. Ameli, and R. L. Detwiler (2013), Dissolution and deformation in fractured carbonates caused by flow of CO_2-rich brine under reservoir conditions, *Int. J. Greenhouse Gas Control*, *16S1*, S203–S215, doi:10.1016/j.ijggc.2013.02.023.

Elkhoury, J. E., R. L. Detwiler, and P. Ameli (2015), Can a fractured caprock self-heal?, *Earth Planet Sci. Lett.*, *417*, 99–106, doi:10.1016/j.epsl.2015.02.010.

Ellis, B. R., J. P. Fitts, G. S. Bromhal, D. L. McIntyre, R. Tappero, and C. A. Peters (2013), Dissolution-driven permeability reduction of a fractured carbonate caprock, *Environ. Eng. Sci.*, *30*(4), 187–193, doi:10.1089/ees.2012.0337.

Ezzedine, S. M., J. Hunt, W. Bourcier, S. Roberts, and J. J. Roberts (2012), Impact of silica gel deployment on subsurface flow and heat extraction from enhanced geothermal systems, in *Proceedings of the Thirty-Seventh Workshop on Geothermal Reservoir Engineering*, Stanford University, Stanford, California, January 30–February 1.

Faulkner, D., C. Jackson, R. Lunn, R. Schlische, Z. Shipton, C. Wibberley, and M. Withjack (2010), A review of recent developments concerning the structure, mechanics and fluid flow properties of fault zones, *J. Struct. Geol.*, *32*(11), 1557–1575, doi:10.1016/j.jsg.2010.06.009, fault zones.

Fisher, M. K., and N. R. Warpinski (2012), Hydraulic-fracture-height growth: Real data, *SPE Prod. Oper.*, *27*(1), 275–278.

Fu, P., Y. Hao, S. D. C. Walsh, and C. R. Carrigan (2015), Thermal drawdown-induced flow channeling in fractured geothermal reservoirs, *Rock Mech. Rock. Eng.*, *49*(3), 1001–1024.

Garagash, D. I., E. Detoutnay, and J. I. Adachi (2011), Multiscale tip asymptotics in hydraulic fracture with leak-off, *J. Fluid Mech.*, *669*, 260–297.

Genedy, M., J. Stormont, E. Matteo, and M. R. Taha (2014), Examining epoxy-based nanocomposites in wellbore seal repair for effective CO_2 sequestration, in *12th International Conference on Greenhouse Gas Control Technologies, GHGT-12, Energy Procedia*, vol. 63, edited by Dixon, T and Herzog, H and Twinning, S, pp. 5798–5807, doi:10.1016/j.egypro.2014.11.612, *12th International Conference on Greenhouse Gas Control Technologies (GHGT)*, Austin, TX, October 5–9, 2014.

Glass, R. J., M. J. Nicholl, and L. Yarrington (1998), A modified invasion percolation model for low-capillary number immiscible displacements in horizontal rough-walled fractures: Influence of local in-plane curvature, *Water Resour. Res.*, *34*(12), 3215–3234, doi:10.1029/98WR02224.

Gouze, P., C. Noiriel, C. Bruderer, D. Loggia, and R. Leprovost (2003), X-ray tomography characterization of fracture surfaces during dissolution, *Geophys. Res. Lett.*, *30*(5), doi:10.1029/2002GL016755.

Groves, C. G., and A. D. Howard (1994), Minimum hydrochemical conditions allowing limestone cave development, *Water Resour. Res.*, *30*(3), 607–615, doi:10.1029/93WR02945.

Guglielmi, Y., F. Cappa, J.-P. Avouac, P. Henry, and E. Elsworth (2015), Seismicity triggered by fluid injection–induced aseismic slip, *Science*, *348*(6240), 1224–1226, doi:10.1126/science.aab0476.

Hanna, R. B., and H. Rajaram (1998), Influence of aperture variability on dissolutional growth of fissures in karst formations, *Water Resour. Res.*, *34*(11), 2843–2853, doi:10.1029/98WR01528.

Heaton, N., J. Rondon, T. Do, A. Valenca, J. Bruns, C. Ohlson, K. Eisele, J. Paneitz, B. Rollins, B. Betts, and S. Cleaver (2016), Stimulation optimization using engineered diversion workflow to increase wellbore contact in cemented completions, in *SPE Argentina Exploration and Production of Unconventional Resources Symposium*, Buenos Aires, Argentina, 1–3 June.

Jeanne, P., Y. Guglielmi, and F. Cappa (2013), Hydromechanical heterogeneities of a mature fault zone: Impacts on fluid flow, *Groundwater*, *51*(6), 880–892, doi:10.1111/gwat.12017.

Jeffrey, R. G., and A. Bunger (2009), A detailed comparison of experimental and numerical data on hydraulic fracture height growth through stress contrasts, *SPE J.*, *14*(3), 413–422, doi:10.2118/106030-PA.

Jha, B., and R. Juanes (2014a), Coupled modeling of multiphase flow and fault poromechanics during geologic CO_2 storage, *Energy Procedia*, *63*, 3313–3329.

Jha, B., and R. Juanes (2014b), Coupled multiphase flow and poromechanics: A computational model of pore pressure effects on fault slip and earthquake triggering, *Water Resour. Res.*, *50*(5), 3776–3808, doi:10.1002/2013WR015175.

Johnson, J. W., J. J. Nitao, and J. P. Morris (2005), Reactive transport modeling of cap-rock integrity during natural and engineered CO_2 storage, in *Carbon Dioxide Capture for Storage in Deep Geologic Formations*, edited by D. C. Thomas, pp. 787–813, Elsevier Science, Amsterdam, doi:10.1016/B978-008044570-0/50134-3.

Jones, T. A., and R. L. Detwiler (2016), Fracture sealing by mineral precipitation: The role of small-scale mineral heterogeneity, *Geophys. Res. Lett.*, doi:10.1002/2016gl069598, 2016GL069598.

Khan, S., H. Han, S. Ansari, M. Vishteh, and N. Khosravi (2011), Caprock integrity analysis in thermal operations: An integrated geomechanics approach, in *WHOC-609, Proceedings for the 2011 World Heavy Oil Congress*, Edmonton, Alberta, March 14–17, 2011.

Li, L., C. A. Peters, and M. A. Celia (2007), Effects of mineral spatial distribution on reaction rates in porous media, *Water Resour. Res.*, *43*(1), doi:10.1029/2005WR004848.

Lucier, A., and M. Zoback (2008), Assessing the economic feasibility of regional deep saline aquifer CO_2 injection and storage: A geomechanics-based workflow applied to the rose run sandstone in eastern Ohio, USA, *Int. J. Greenhouse Gas Control*, *2*, 230–247.

Morris, A. P., D. Ferrill, and D. Henderson (1996), Slip tendency and fault reactivation, *Geology*, *24*, 275–278.

Morris, J. P., Y. Hao, W. Foxall, and W. McNab (2011), A study of injection-induced mechanical deformation at the In Salah CO_2 storage project, *Int. J. Greenhouse Gas Control*, *5*(2), 270–280.

Morris, J. P., P. Fu, R. R. Settgast, C. S. Sherman, M. Friedrich, and P. Leonard (2016), The combined influence of stress barriers and natural fractures upon hydraulic fracture height growth, in *50th US Rock Mechanics / Geomechanics Symposium*, Houston, TX, June 26–29, 2016.

Nicholson, C., and R. L. Wesson (1992), Triggered earthquakes and deep well activities, *Pure Appl. Geophys.*, *139*(3/4), 561–578.

Noiriel, C., P. Gouze, and B. Made (2013), 3D analysis of geometry and flow changes in a limestone fracture during dissolution, *J. Hydrol.*, *486*, 211–223, doi:10.1016/j.jhydrol.2013.01.035.

Ortoleva, P., E. Merino, C. Moore, and J. Chadam (1987), Geochemical self-organization i: Reaction-transport feedbacks and modeling approach, *Am. J. Sci.*, *287*(10), 979–1007, doi:10.2475/ajs.287.10.979.

Pamukcu, Y., S. Hurter, L. Jammes, D. Vu-Hoang, and L. Pekot (2011), Characterizing and predicting short term performance for the in salah krechba field {CCS} joint industry project, *Energy Procedia*, *4*, 3371–3378, doi:10.1016/j.egypro.2011.02.259, 10th International Conference on Greenhouse Gas Control Technologies.

Pham, T., T. Maast, H. Hellevang, and P. Aagaard (2011), Numerical modeling including hysteresis properties for CO_2 storage in tubåen formation, Snøvhit field, barents sea, *Energy Procedia*, *4*, 3746–3753.

Plampin, M. R., R. N. Lassen, T. Sakaki, M. L. Porter, R. J. Pawar, K. H. Jensen, and T. H. Illangasekare (2014), Heterogeneity-enhanced gas phase formation in shallow aquifers during leakage of CO_2-saturated water from geologic sequestration sites, *Water Resour. Res.*, *50*(12), 9251–9266, doi:10.1002/2014wr015715.

Plug, W.-J., and J. Bruining (2007), Capillary pressure for the sand–CO_2–water system under various pressure conditions. Application to {CO_2} sequestration, *Adv. Water Resour.*, *30*(11), 2339–2353, doi:10.1016/j.advwatres.2007.05.010.

Plumb, R. A., S. Edwards, G. Pidcock, and D. Lee (2000), The mechanical earth model concept and its application to high-risk well construction projects, in *IADC/SPE Drilling Conference*, New Orleans, LA, February 23–25, SPE paper 59128.

Polak, A., D. Elsworth, J. Liu, and A. S. Grader (2004), Spontaneous switching of permeability changes in a limestone fracture with net dissolution, *Water Resour. Res.*, *40*(W03502).

Pyrak-Nolte, L., and J. Morris (2000), Single fractures under normal stress: The relation between fracture specific stiffness and fluid flow, *Int. J. Rock Mech. Min. Sci.*, *37*(1–2), 245–262, doi:10.1016/S1365-1609(99)00104-5.

Qiuguo, L., X. Zhang, K. Al-Ghammari, L. Mohsin, and A. Al Rawahi (2013), 3D geomechanical modeling, wellbore stability analysis improve field's performance, *World Oil, October*, 53–61.

Rutqvist, J., B. Freifeld, K.-B. Min, D. Elsworth, and Y. Tsang (2008), Analysis of thermally induced changes in fractured rock permeability during 8 years of heating and cooling at the Yucca Mountain drift scale test, *Int. J. Rock Mech. Min. Sci.*, *45*, 1373–1389.

Segall, P., and S. D. Fitzgerald (1996), A note on induced stress changes in hydrocarbon and geothermal reservoirs, *Techtonophysics*, *289*, 117–128.

Siirila, E. R., A. K. Navarre-Sitchler, R. M. Maxwell, and J. E. McCray (2012), A quantitative methodology to assess the risks to human health from {CO_2} leakage into groundwater, *Adv. Water Resour.*, *36*, 146–164, doi:10.1016/j.advwatres.2010.11.005, special issue on Uncertainty Quantification and Risk Assessment.

Simonson, E. R., A. S. Abou-Sayed, and R. J. Clifton (1978), Containment of massive hydraulic fractures, *SPE J.*, *18*(1), 27–32.

Streit, J. E., and R. R. Hillis (2004), Estimating fault stability and sustainable fluid pressures for underground storage of CO_2 in porous rock, *Energy*, *29*, 1445–1456.

Szymczak, P., and A. J. C. Ladd (2012), Reactive-infiltration instabilities in rocks. Fracture dissolution, *J. Fluid Mech.*, *702*, 239–264, doi:10.1017/jfm.2012.174.

Tartakovsky, A. M., G. Redden, P. C. Lichtner, T. D. Scheibe, and P. Meakin (2008), Mixing-induced precipitation: Experimental study and multiscale numerical analysis, *Water Resour. Res.*, *44*(6), doi:10.1029/2006WR005725.

Uehara, S., and T. Shimamoto (2004), Gas permeability evolution of cataclasite and fault gouge in triaxial compression and implications for changes in fault-zone permeability structure through the earthquake cycle, *Tectonophysics*, *378*, 183–195.

Unger, A. J. A., and C. W. Mase (1993), Numerical study of the hydromechanical behavior of two rough fracture surfaces in contact, *Water Resour. Res.*, *29*(7), 2101–2114, doi:10.1029/93WR00516.

Vasco, D. W., A. Rucci, A. Ferretti, F. Novali, R. C. Bissell, P. S. Ringrose, A. S. Mathieson, and I. W. Wright (2010), Satellite-based measurements of surface deformation reveal fluid flow associated with the geological storage of carbon dioxide, *Geophys. Res. Lett.*, *37*(3), doi:10.1029/2009GL041544, l03303.

Verdon, J. P., J.-M. Kendal, A. L. Stork, R. A. Chadwick, D. J. White, and R. C. Bissell (2013), Comparison of geomechanical deformation induced by megatonne-scale CO_2 storage at Sleipner, Weyburn, and In Salah, *PNAS*, pp. E2762–E2771.

Vilarrasa, V., and L. Laloui (2015), Potential fracture propagation into the caprock induced by cold CO_2 injection in normal faulting stress regimes, *Geomech. Energy Environ.*, *2*, 22–31.

Wagner, F. M., P. Bergmann, C. Rücker, B. Wiese, T. Labitzke, C. Schmidt-Hattenberger, and H. Maurer (2015), Impact and mitigation of borehole related effects in permanent crosshole resistivity imaging: An example from the ketzin {CO_2} storage site, *J. Appl. Geophys.*, *123*, 102–111, doi:10.1016/j.jappgeo.2015.10.005.

Warpinski, N. R., and L. W. Teufel (1987), Influence of geologic discontinuities on hydraulic fracture propagation, *J. Pet. Technol.*, *39*, 209–220.

White, J. A., L. Chiaramonte, S. Ezzedine, W. Foxall, Y. Hao, A. Ramirez, and W. McNab (2014), Geomechanical behavior of the reservoir and caprock system at the In Salah CO_2 storage project, *PNAS*, *111*(24), 8747–8752.

Wiprut, D. J., and M. D. Zoback (2002), Fault reactivation, leakage potential, and hydrocarbon column heights in the northern North Sea, in *Hydrocarbon Seal Quantification*, vol. *11*, pp. 203–219, NPF Special Publication.

Yasuhara, H., D. Elsworth, and A. Polak (2004), Evolution of permeability in a natural fracture: Significant role of pressure solution, *J. Geophys. Res. Solid Earth*, *109*(B3), doi:10.1029/2003JB002663.

Zhang, X., N. Koutsabeloulis, and K. Heffer (2007), Hydromechanical modeling of critically stressed and faulted reservoirs, *Am. Assoc. Pet. Geol. Bull.*, *91*(1), 31–50.

Zhang, C., K. Dehoff, N. Hess, M. Oostrom, T. W. Wietsma, A. J. Valocchi, B. W. Fouke, and C. J. Werth (2010), Pore-scale study of transverse mixing induced $CaCO_3$ precipitation and permeability reduction in a model subsurface sedimentary system, *Environ. Sci. Technol.*, *44*(20), 7833–7838, doi:10.1021/es1019788.

Zimmerman, R., and G. Bodvarsson (1996), Hydraulic conductivity of rock fractures, *Transp. Porous Media*, *23*(1), 1–30, doi:10.1007/BF00145263.

6

Effect of Fracture Density on Effective Permeability of Matrix-Fracture System in Shale Formations

Li Chen[1,*], Jeffrey De'Haven Hyman[1,2], Zhou Lei[1], Ting Min[3], Qinjun Kang[1], Esteban Rougier[1], and Hari Viswanathan[1]

ABSTRACT

Fractures play an important role in providing preferred flow pathways in low-permeability shale matrix and significantly enhance its permeability. They can improve gas production for shale gas development but can also increase the risks for leaking of CO_2 or brine from geological storage sites. In this chapter, a generalized lattice Boltzmann (LB) model for fluid flow through porous media is adopted to simulate fluid flow in matrix-fracture systems and to predict the effective permeability k_{eff}. Discrete fracture-matrix networks (DFN) are constructed using line fractures and elliptical fractures in 2D and 3D systems, respectively. Power law relationships are observed between k_{eff} and fracture density in the DFN. Further, the combined finite discrete element method (FDEM) is adopted to simulate fracture propagation process. Dynamic evolution of k_{eff} during the propagation is simulated using the LB model. The results show that power law relationship between k_{eff} and fracture density is also obeyed.

6.1. INTRODUCTION

Tight rocks or low-permeability formations such as mudstones and shales (clay mineral-rich rocks) are crucial to geological carbon sequestration (GCS) by acting as seals or caprocks. These sealing properties are achieved by the fact these rocks have pore sizes on the order of a few micrometers to a few nanometers which makes their permeability extremely low, on the order of a few nano-Darcy. For example, *Javadpour et al.* [2007] measured 152 shale samples from 9 shale reservoirs and reported a permeability peak around 5.4×10^{-5} mD with 90% samples having permeability less than 1.5×10^{-4} mD. *Wang and Reed* [2009] plotted the permeability of shales in North America with porosity of 0–0.1 and found that the permeability ranges from 1×10^{-12} mD to 1×10^{-2} mD. When used as seals or caprocks, the low permeability of these rocks will help prevent the upward migration of brine and gas and trap the stored CO_2 over geological timescales. However, these rocks may be damaged due to various underground hydrogeochemical and mechanical processes, resulting in fractures or fracture networks in the otherwise intact formations. Therefore, to ensure safe long-term GCS, it is crucial to understand the transport property of the matrix-fracture systems in these low-permeability media.

On the other hand, the successful exploitation of shale gas reservoirs in North America has led to a global boom of research and development of shale gas. Advanced techniques such as horizontal well drilling and hydraulic fracturing have significantly increased our access to shale gas in previously inaccessible reservoirs. Due to the low permeability of shale gas reservoirs, economic shale gas

[1] *Earth and Environmental Sciences Division, Los Alamos National Laboratory, Los Alamos, NM, USA*

[2] *Center for Nonlinear Studies, Theoretical Division, Los Alamos National Laboratory, Los Alamos, NM, USA*

[3] *State Key Laboratory for Mechanical Behavior of Materials, Xi'an Jiaotong University, Xi'an, PR China*

**Current address: School of Energy & Power Engineering, Xi'an Jiaotong University, Xi'an, PR China*

Geological Carbon Storage: Subsurface Seals and Caprock Integrity, Geophysical Monograph 238, First Edition.
Edited by Stéphanie Vialle, Jonathan Ajo-Franklin, and J. William Carey.

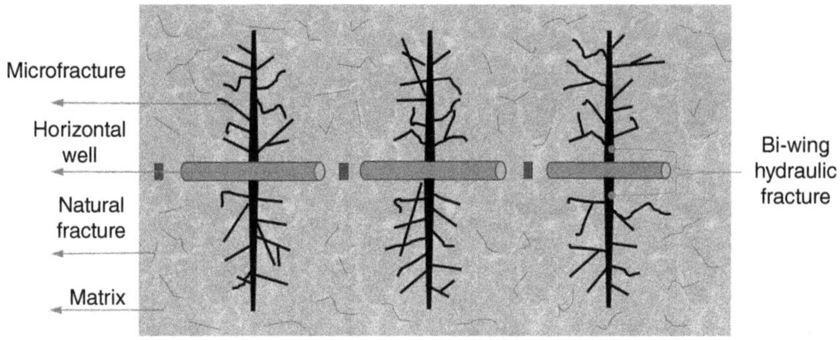

Figure 6.1 Schematic of multi-scale fracture system in the shale gas reservoir after hydraulic fracturing operations. Natural fracture, microfracture, and hydraulic fractures coexist.

production is only possible when a fracture network in the otherwise impermeable shale is present. To create such networks, multistage hydraulic fracturing along a horizontal well is the most widely adopted technique [*King*, 2010]. Low-viscosity fluid is often used to promote the fracture complexity or increase the contact area between fractures and matrix. Figure 6.1 shows a conceptual image of how fractures propagate away from the horizontal well. The declining fracturing fluid pressure away from the main bi-wing fractures results in microfractures with lower fracture density. The changing fracture density leads to varying effective permeability away from the hydraulic fractures. The microfractures and natural fractures are usually lumped together with matrix as matrix blocks in the dual/triple-continuum model, and a constant permeability is adopted in the matrix blocks. Under such formulation, the effects of the varying fracture density in the matrix blocks are not considered.

Currently, water is widely employed as the high-pressure fluid for fracturing in shale. In regions encountering water availability issues and with the increasing concerns of groundwater pollution caused by contaminants added into the fracturing water, using nonaqueous fracturing fluid such as supercritical CO_2 currently is under consideration. CO_2 shows a greater adsorptive capacity relative to CH_4 [*Kang et al.*, 2011; *Heller and Zoback*, 2014]. For example, at a pressure of 1000 psi, *Heller and Zoback* [2014] found in Barnett Shale that adsorbed amount of CO_2 is about two times greater than that of CH_4. Therefore, using CO_2 as fracturing fluid can facilitate desorption of methane and enhance the production and also has the benefit of providing the potential for CO_2 sequestration [*Middleton et al.*, 2015].

It is clear that fluid flow in matrix-fracture systems in low-permeability shale formations is a problem common to both GCS and shale gas production. A better understanding of this problem is critical for mitigating risks of GCS and improving long-term production of shale gas. The objective of this study is to perform a physics-based study of fluid flow in matrix-fracture systems of shale formations. Particularly, we focus on the effect of fracture density on the effective permeability of matrix-fracture systems. The predicted relationship between fracture density and permeability can be upscaled into a large-scale reservoir model. In this study, first synthetic two-dimensional (2D) and three-dimensional (3D) matrix-fracture systems are prepared using a stochastic discrete fracture-matrix networks (DFN) generator. Then a representative elementary volume (REV)-scale lattice Boltzmann (LB) model for fluid flow in tight porous media is adopted to predict the effective permeability for the matrix-fracture systems. A power law relationship between fracture density and permeability is found based on the simulation results. To test this relationship in a real-world situation, a finite discrete element method (FDEM) calculation is used where the dynamic evolution of fractures is simulated and calibrated on an experiment. LB model simulations of the effective permeability are performed as the fracture density evolves and the power law relationship between fracture density and permeability is also obtained.

6.2. NUMERICAL METHOD

Over the past three decades, the lattice Boltzmann method (LBM) has been developed as a powerful numerical method for fluid flow and transport processes [*Chen and Doolen*, 1998]. Due to its underlying kinetic nature, the LBM is particularly useful in fluid flow applications involving interfacial dynamics and complex boundaries, for example, multiphase or multicomponent flows in porous media [*Chen et al.*, 2014]. The LBM considers flow as a collective behavior of pseudo-particles residing on a mesoscopic level and solves the Boltzmann equation using a small number of velocities adapted to a regular grid in space [*Chen and Doolen*, 1998]. Fluid flow and transport in porous media are usually observed physically and treated theoretically at two different scales:

pore scale and REV scale [*Bear*, 1972]. A REV of a porous medium is the smallest volume on which large fluctuations of observed quantities (such as porosity and permeability) no longer occur and on which the continuum models are based. Several REV-scale LBM models have been proposed, which enhance the capacity of LBM for larger-scale applications [*Freed*, 1998; *Guo and Zhao*, 2002; *Kang et al.*, 2002]. These models recover the common continuum equations for fluid flow in porous media such as the Darcy equation or extended Darcy equations. In the present study, the generalized REV-scale LB model developed by *Guo and Zhao* [2002] for fluid flow through porous media is adopted to simulate the fluid flow in the reconstructed matrix-fracture system. In this model, the generalized Navier-Stokes equation proposed by *Nithiarasu et al.* [1997] for isothermal incompressible fluid flow in porous media is solved:

$$\nabla \cdot \mathbf{u} = 0 \tag{6.1a}$$

$$\frac{\partial \mathbf{u}}{\partial t} + (\mathbf{u}\cdot\nabla)\frac{\mathbf{u}}{\varepsilon} = -\frac{1}{\rho}\nabla(\varepsilon p) + \upsilon\nabla^2\mathbf{u} + \mathbf{F} \tag{6.1b}$$

where t is the time, ρ is the volume averaged fluid density, p is the volume averaged pressure, $\mathbf{u}$ is the superficial velocity, ε is the porosity, and υ is the fluid viscosity. The force term $\mathbf{F}$ represents the total body force due to the presence of the porous media and other external body forces:

$$\mathbf{F} = -\frac{\varepsilon\upsilon}{k}\mathbf{u} - \frac{\varepsilon F_\varepsilon}{\sqrt{k}}|\mathbf{u}|\mathbf{u} + \varepsilon\mathbf{G} \tag{6.2}$$

where $\mathbf{G}$ is the external force. The first term on RHS is the linear (Darcy) drag force, and the second term is the nonlinear (Forchheimer) drag force. The geometric function $F\varepsilon$ and the permeability k are related to the porosity of the porous medium. For example, for a porous medium composed of solid spherical particles, they can be calculated by Ergun's equation [*Guo and Zhao*, 2002]. The quadratic term (Forchheimer term) becomes important when the fluid flow is relatively strong and inertial effects become relevant. It can be neglected for fluid flow with Reynolds number much lower than unity, which is the case in the present study as extremely slow shale gas transport in the matrix-fracture system is simulated. The corresponding LB model is as follows [*Guo and Zhao*, 2002]:

$$f_i(\mathbf{x}+\mathbf{e}_i\Delta t, t+\Delta t) - f_i(\mathbf{x},t) = -\frac{1}{\tau}\left(f_i(\mathbf{x},t) - f_i^{eq}(\mathbf{x},t)\right) + \Delta t F_i \tag{6.3a}$$

$$f_i^{eq} = \omega_i\rho\left[1 + \frac{3}{c^2}(\mathbf{e}_i\cdot\mathbf{u}) + \frac{9}{2\varepsilon c^4}(\mathbf{e}_i\cdot\mathbf{u})^2 - \frac{3}{2\varepsilon c^2}\mathbf{u}^2\right] \tag{6.3b}$$

$$F_i = \omega_i\rho\left(1-\frac{1}{2\tau}\right)\left[\frac{3}{c^2}(\mathbf{e}_i\cdot\mathbf{F}) + \frac{9}{\varepsilon c^4}(\mathbf{e}_i\cdot\mathbf{u})(\mathbf{e}_i\cdot\mathbf{F}) - \frac{3}{c^2}(\mathbf{u}\cdot\mathbf{F})\right] \tag{6.3c}$$

$$\rho = \sum_i f_i, \quad \mathbf{u} = \frac{\sum_i f_i\mathbf{e}_i + (\Delta t/2)\rho\varepsilon\mathbf{G}}{\rho + (\Delta t/2)(\varepsilon\upsilon/k)\rho} \tag{6.3d}$$

$$\upsilon = \frac{1}{3}\left(\tau - \frac{1}{2}\right)\frac{\delta x^2}{\delta t} \tag{6.3e}$$

where $fi(\mathbf{x},t)$ is the ith density distribution function at the lattice site $\mathbf{x}$ and time t and Δt is the lattice time step. The dimensionless relaxation time τ is related to the viscosity as described in Eq. (6.3e). ω_i is the weight factor. For 2D simulations, the D2Q9 lattice model (nine velocity directions at a given point in 2D space) is used, while for 3D cases, D3Q19 lattice model is adopted. For more details of the LB model, one can refer to *Guo and Zhao* [2002].

Note that there are three permeability terms discussed in the present study: the matrix permeability k_m, the fracture permeability k_f, and the effective permeability k_{eff}. The former two are input parameters for each node in the matrix-fracture system. The latter one is the effective permeability of the entire system. According to Eqs. (6.1–6.3), it is clear that to predict k_{eff}, both porosity and permeability of the matrix and the fracture need to be known. In the original model of *Guo and Zhao* [2002], permeability of each node in the domain is calculated by porosity using the Ergun equation. The porosity, although serving as an input parameter in the REV-scale LB model, actually has no influence on the predicted k_{eff}. For the steady state low Reynolds number (*Re*) gas flow, the unsteady term, the convective term, and the viscous term in Eq. (6.1b) can be neglected; further, the quadratic term in Eq. (6.2) also can be neglected as discussed in Section 6.3. Based on the above discussion, and substituting Eq. (6.2) into Eq. (6.1), Eq. (6.1b) can be simplified as

$$-\nabla p = \frac{\rho\upsilon}{k}\mathbf{u} \tag{6.4}$$

which is the well-known Darcy equation. This indicates that for the REV-scale LB model adopted in the present study, only the values of k_m and k_f affect k_{eff}. This point is also proved by our preliminary simulations conducted with changing porosity but fixed k_m and k_f. In the present study, k_m and k_f are directly assigned to matrix and fracture nodes in the domain, respectively, without using any porosity-permeability relationship to calculate the permeability of each node. k_{eff} of the matrix system is then predicted using the REV-scale LB model.

Nanosize pores are pervasive in the shale matrix. In these nanosize pores, the Knudsen number (*Kn*, defined as the ratio of the gas mean free path to the pore size) is relatively high, and the gas molecules tend to slip on the solid surface, leading to fluid flow regimes beyond the continuum viscous flow, including slip flow regime ($0.01 < Kn < 0.1$), transition flow regime ($0.1 < Kn < 10$), and free molecular flow regime ($Kn > 10$) [*Beskok and Karniadakis*, 1999]. Gas slippage in porous media leads to higher values of effective permeability, which is called apparent permeability [*Klinkenberg*, 1941; *Javadpour et al.*, 2007]. Recent experimental studies of natural gas in tight porous rocks [*Ziarani and Aguilera*, 2012], as well as high-resolution nanoscale simulations [*Chen et al.*, 2015a, 2015c], found that the apparent permeability can be 100 times higher than the intrinsic permeability, emphasizing the importance of gas slippage for fluid flow in tight porous rocks. Recently, we developed a generalized REV-scale LB model for fluid flow through porous media with slippage [*Chen et al.*, 2015b, 2015d], in which the apparent permeability substitutes for the permeability term in Eq. (6.2). The model was applied to simulate fluid flow in shale matrix with organic matter and inorganic materials. The results demonstrate significant importance of slippage on shale gas transport, especially in smaller pores and under lower pressures. In the present study, to focus on the effects of fracture density, the slippage effect is not considered. This is reasonable because connected fractures and fracture networks serve as the preferred pathways and dominate the flow and slip effect is not significant in fractures due to their relatively large size, especially under high pressure.

6.3. RESULTS AND DISCUSSIONS

6.3.1. 2D Matrix-Fracture System

The 2D matrix-fracture system is represented by a square domain, Ω, of dimensions $L \times L$. Initially, the domain Ω contains only matrix Ω_m. Fractures are represented as straight lines and are randomly placed into the domain. Independent, identically distributed random numbers sampled from a uniform distribution are used to generate two points with coordinates (x_1, y_1) and (x_2, y_2). A line is thus determined by the two points and represents the fracture. Points on this line are thus belonged to the fracture. Other fractures are inserted into the matrix in the same way until a prescribed fracture density is obtained. The fracture density is defined as the ratio between the total fracture length and the entire area of the matrix-fracture system:

$$\chi = \frac{L_f}{L^2} \tag{6.5}$$

Length, orientation, and width of each fracture can be controlled and can be specified to obey certain distributions. For example, a power law length distribution can be adopted to determine the length of fractures. In the present study, the width of the fracture is unity, the length is not required to be less than the domain size, and the orientation angle covers the entire range of 0°–360°. Note that although fracture systems in shale form in ways that lead to preferred orientations, multiple tectonic events and changes in stress field with geologic time also tend to create complex multi-oriented fracture systems.

In the matrix-fracture system with a specified fracture density, for a node **x**, it either belongs to the matrix Ω_m or to the fractures Ω_f, and the corresponding properties are different:

$$\Gamma(\mathbf{x}) = \begin{cases} \Gamma_m & \mathbf{x} \in \Omega_m \\ \Gamma_f & \mathbf{x} \in \Omega_f \end{cases} \tag{6.6}$$

Γ can be structural parameters (porosity ε, specific surface area s, tortuosity τ, etc.) or transport properties (permeability k, diffusivity, thermal conductivity, etc.).

Figure 6.2 shows the generated 2D matrix-fracture system with different example fracture densities. The domain size is $L \times L = 200 \times 200$ lattices, with physical length of each lattice as 200 μm, leading to a physical length of the domain as 4 cm. Under lower fracture density, fractures are isolated from each other. As fracture density increases, the connectivity between fractures gets increasingly better.

Permeability is a tensor quantity and only its principal components are predicted in this study. Shale gas flow along one direction (x direction) is simulated using the REV-scale LB model introduced in Section 6.2. The fluid flow is driven by a pressure difference $\Delta p = 2 \times 10^{-6}$ (lattice units) applied at $x = 0$ and $x = L$. Δp is sufficiently low to guarantee low *Re* flow; thus Darcy law can be adopted to calculate k_{eff}. Periodic boundary conditions are adopted for the y direction. Since the matrix-fracture systems generated are isotropic, only the permeability in x direction is simulated.

Figure 6.3 shows the velocity field predicted for a sample with fracture density of 0.1, $k_f = 1 \times 10^{-14}$ m², and $k_m = 1 \times 10^{-20}$ m². As shown in Figure 6.3, velocity magnitude is relatively higher in the fractures compared with that in the matrix, indicating the fractures are the preferred flow path for shale gas as expected.

Figure 6.4 shows the effective permeability predicted under different fracture densities. Due to the stochastic nature of the generated matrix-fracture systems, 50 samples are generated for each fracture density. During the simulations, permeability of the matrix k_m is fixed as 1×10^{-20} m² (or 10 nano-Darcy) which is typical of shale permeability. Fracture permeability k_f is varied in the

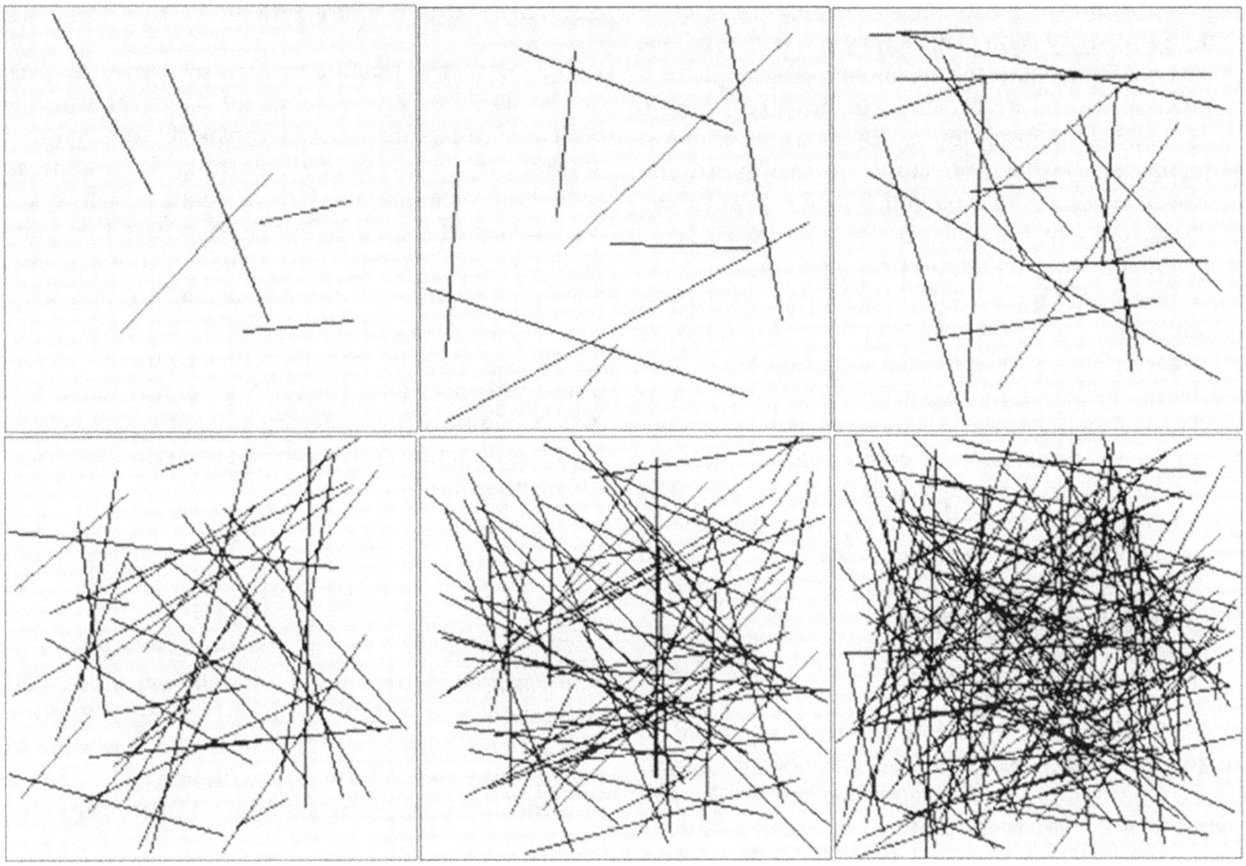

Figure 6.2 2D matrix-fracture system with increasing fracture density. The fracture is represented by lines embedded in the matrix. The fracture density is 0.001, 0.01, 0.1, 0.2, 0.4, and 0.6 from top left to bottom right.

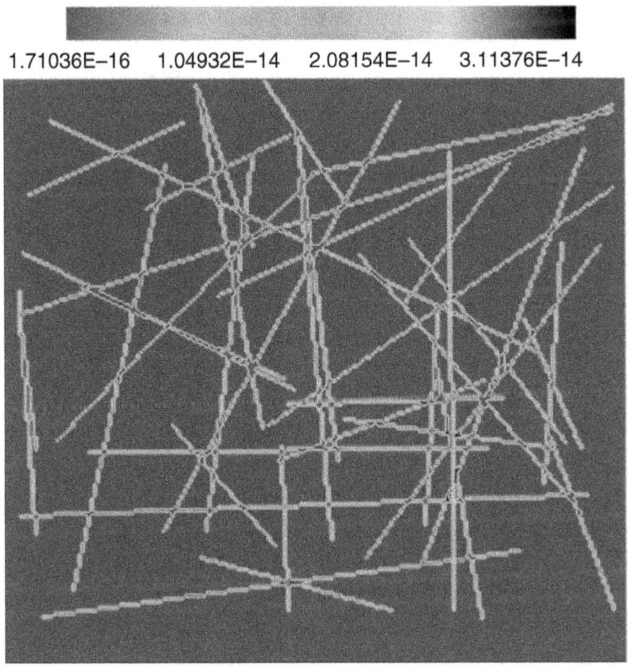

Figure 6.3 Velocity magnitude $|\mathbf{u}| = \sqrt{u_x^2 + u_y^2}$ in a matrix-fracture system with fracture density as 0.1 and $k_f/k_m = 10^6$.

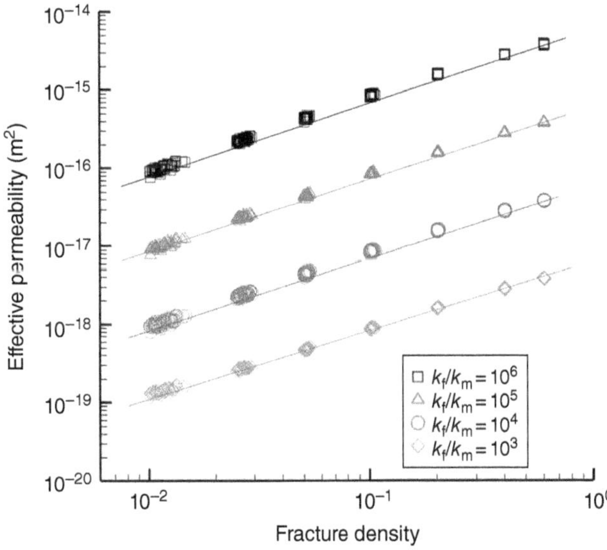

Figure 6.4 Relationship between fracture density and effective permeability under different k_f/k_m.

range of 1×10^{-14} to 1×10^{-17} m^2. Note that the fracture width in this study is one lattice (physical length of one lattice δ is 200 μm); thus the maximum permeability of a fracture is 3.33×10^{-9} m^2, according to the cubic law $\delta^2/12$. The k_f studied is relatively low, which is reasonable due to two factors. One is that the fractures are usually tortuous, and the other one is that natural fractures in shale are usually partially or fully mineralized [*Gale et al.*, 2007, 2014], both of which reduce the fracture permeability. As can be seen from Figure 6.4, for a lower fracture density, the values of k_{eff} are more scattered, due to less fractures in the matrix leading to higher heterogeneity. As the fracture density increases, such scatter reduces, and it can be found that when fracture density is higher than 0.1, values of k_{eff} for the 50 samples almost coincide. Besides, as fracture density or k_f increases, k_{eff} increases, as expected. The most important observation from Figure 6.4 is that the relationship between k_{eff} and fracture density is linear on a log-log scale. This indicates that this relationship can be described by a power law function:

$$k_{eff} = a\chi^n \tag{6.7}$$

Table 6.1 lists a and n fitted for different k_f/k_m based on Eq. (6.7). As k_f/k_m increases, the exponent coefficient increases. Such a power law relationship between k_{eff} and fracture density has been identified by previous studies [*Yazdi* et al., 2011; *Jafari and Babadagli*, 2012]. In most previous studies of matrix-fracture system, the matrix permeability is assumed to be low enough to be negligible [*Long and Witherspoon*, 1985; *Yazdi* et al., 2011], and thus, if there are no fractures percolating the domain, the effective permeability will be zero. For shale matrix with embedded natural fractures, transport in matrix sometimes may be as important as that in the natural fractures for two reasons. First, the organic matter in some shales has high porosity (as high as 0.5), and in some shales, interparticle pores may also be abundant, both of which can provide well-connected flow network in shale matrix [*Loucks et al.*, 2012]. Second, the natural fractures are usually mineralized, whose permeability is not always significantly higher than the surrounding shale matrix. Therefore, both the fluid flow in matrix and fractures are solved in the present study, which is an advantage of the REV-scale LB model.

Table 6.1 Fitted Values of a and n in Eq. (6.7).

$\log(k_f/k_m)$	a	n
3	5.535×10^{-18}	0.791
4	6.703×10^{-17}	0.922
5	6.925×10^{-16}	0.943
6	6.925×10^{-15}	0.946

6.3.2. 3D Matrix-Fracture System

The 3D matrix-fracture systems are generated based on the 3D DFN [*Hyman et al.*, 2014, 2015a], which has been used to explore flow and transport through fracture networks [*Hyman et al.*, 2015b]. In the DFN approach, geological data are used to generate a network of interconnected fractures. The geometry and properties of individual fractures are represented explicitly as planar polygons in three dimensions [*Hyman et al.*, 2015a]. Fractures in the network are assigned a shape, location, width, and orientation based on distributions of the geological data. The DFN generated here consists of 855 elliptical fractures drawn from two families of fractures. Fracture radii are drawn from truncated power law distributions:

$$r = r_0\left[1-\delta+\delta\left(\frac{r_0}{r_1}\right)^{\alpha}\right]^{1-\alpha} \tag{6.8}$$

with upper and lower cutoffs as $r_0 = 0.1$ μm and $r_1 = 1$ μm and exponent $\alpha = 2.28$, in accordance with geological data [*O'Malley et al.*, 2016]. δ in Eq. (6.8) is a random number from the uniform distribution in (0,1). The fracture orientations are sampled from a Fisher distribution:

$$f(x,\mu,\kappa) = \frac{\kappa \exp(\kappa\mu^T x)}{4\pi \sinh(\kappa)} \tag{6.9}$$

where μ is a vector that defines the mean orientation and κ is the concentration for each fracture family, where mean directions of the families are orthogonal to one another and have concentration parameters of 10 and 20. Fracture locations are distributed uniformly through the domain. Random subsets of the entire DFN are selected to explore the effects of varying fracture density. Each of these subsets is mapped into a uniform hexahedral mesh where a location is assigned a width-based permeability if the voxel intersects the subset of the DFN; otherwise it is assigned the effective matrix permeability. Figure 6.5 shows the 3D matrix-fracture system generated using the DFN.

Figure 6.6 shows the relationship between fracture density and permeability for 3D matrix-fracture systems with different fracture densities predicted by the LBM simulations. The fracture density is defined similarly to Eq. (6.5) but in 3D case is the ratio between the total fracture area and the entire volume of the matrix-fracture system. Again, the power law relationship between k_{eff} and χ and effective permeability is found.

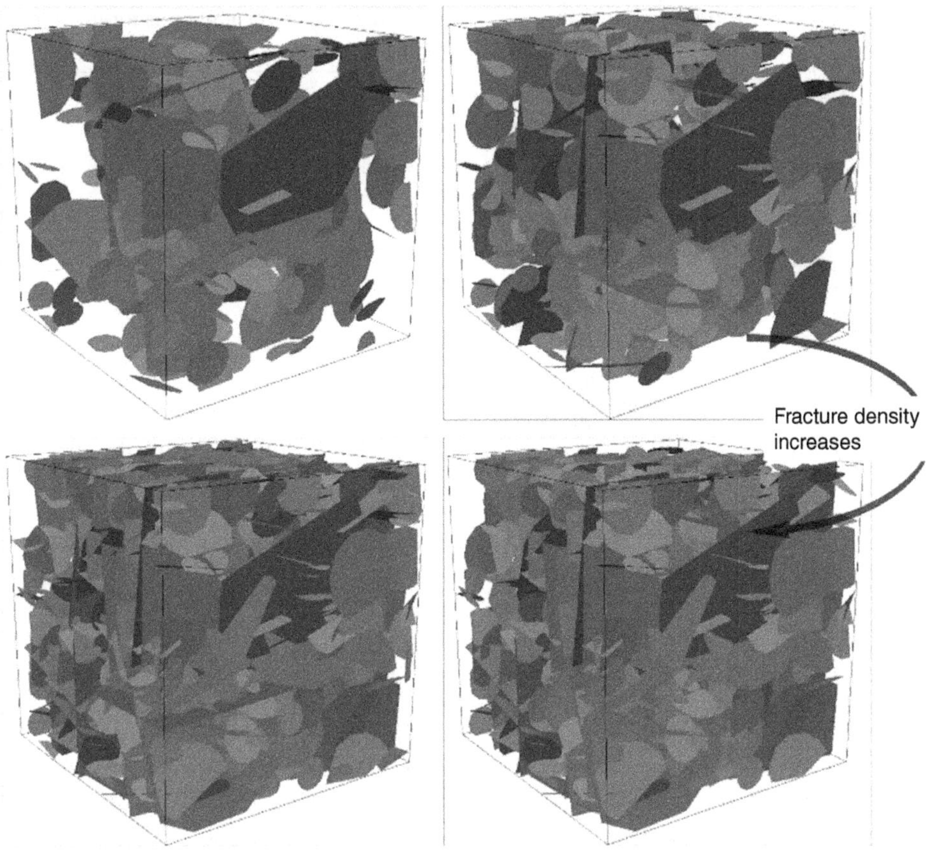

Figure 6.5 3D matrix-fracture systems generated using the DFN.

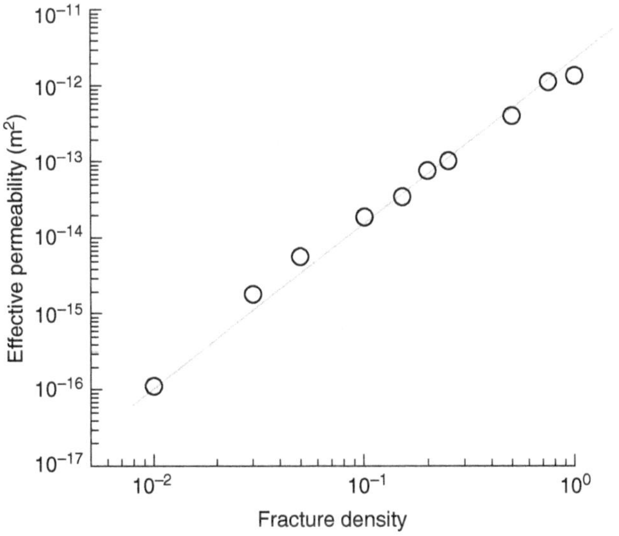

Figure 6.6 Relationship between fracture density and effective permeability with $k_f/k_m = 10^8$.

6.3.3. Dynamic Evolution of Fractures

Most studies have considered the permeability of artificial fractures (sawn or split samples) or artificially separated natural fractures using triaxial or shear box devices, while very few studies have been conducted under in situ conditions with simultaneous fracture and permeability measurements at reservoir conditions. There is an extensive amount of literature on brittle and ductile behavior in shale and the various factors that are being used to predict its possible mechanical behavior in response to stress. However, much less is known about the relationship between these mechanical properties and the permeability of damaged shale.

In order to gain more insight on this type of problem, LANL's experimental team has designed and conducted triaxial core flooding experiments [*Carey et al.*, 2015]. The results obtained from these experiments were used on a combined FDEM study aimed at simulating complex

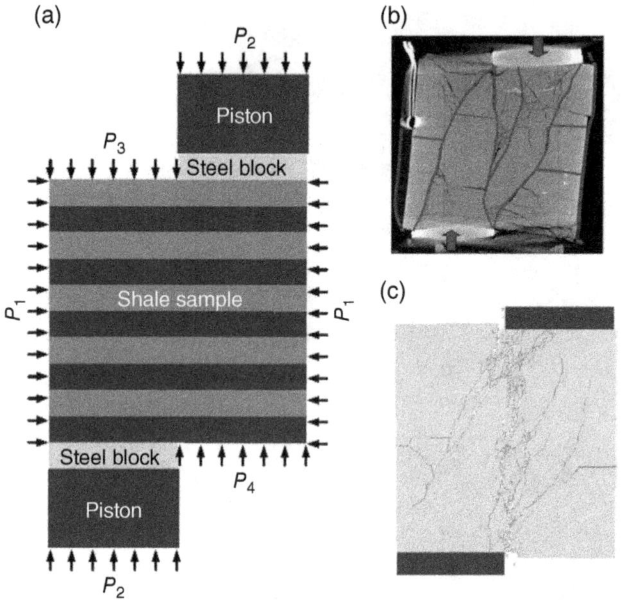

Figure 6.7 Fracture propagation process. (a) Model setup for the numerical simulations, (b) experimental results [*Carey et al.*, 2015], and (c) FDEM results for case 1. Reprinted with permission of Elsevier.

fracture and fragment processes in shale under a variety of different boundary conditions [*Carey et al.*, 2015].

The setup of the FDEM model used in the simulations, which replicates the setup of the accompanying experiment conducted, is shown in Figure 6.7a. The rock sample is compressed in an asymmetric way by the action of two semicircular rigid anvils (direct shear configuration). As shown in the figure, four different pressures are applied to the system: (i) pressure p_1 = 3.45 MPa is the confining pressure in the triaxial experiments, (ii) pressure p_2 = 6.89 MPa is the pressure applied by the action of the axial pistons under initial hydrostatic conditions, and (iii) pressure p_3 = 1.24 MPa and (iv) pressure p_4 = 1.38 MPa represent the pore pressures at either end of the sample. The shale rock sample was modeled by using 10 identical horizontal layers, shown with two different colors. Within each layer, the material properties are assumed to be homogeneous, but the interlayers have smaller strength than the layers (half the strength of the materials inside the inlayers). Rather than using the measured properties of the shale samples, the FDEM analysis was conducted by using material parameters corresponding to those of a generic shale rock, as shown in Table 6.2. The simulation contains two main stages. In the first stage, the system is preloaded with the confining pressures mentioned above. A dynamic relaxation approach is used to get a steady state prestressed situation for the shale sample. In the second stage, the two pistons are moved with a very small constant velocity, and the resultant fracture propagation process is modeled.

Table 6.2 Material Parameters for the Interlayers of the Shale Rock Sample.

Parameter	Values
Young's modulus (MPa)	12,000
Poisson's ratio	0.15
Density (kg/m^3)	2400.0
Tensile strength (MPa)	9.7
Shear strength (MPa)	33.7
Maximum normal displacement (m)	0.0001
Maximum tangential displacement (m)	0.00005

A comparison between the final fracture patterns obtained from the numerical simulation against the experimental observations indicates a quite reasonable agreement. As shown in Figure 6.7b and c, the main features of the fracture patterns observed in the experiment are reproduced by the FDEM simulation results, including the central fracture line, the two surrounding fracture arcs, and the densification of fracture networks near the anvils.

Since it is hard to get the fracture evolution from the experiments, the fracture propagation process obtained from the FDEM simulation is used here to study the influence of the fracture evolution on the fluid permeability. A sequence of the fracture propagation process obtained from the FDEM simulations is shown in Figure 6.8a–e, which shows the complex fracture network. The LB model in Section 6.2 is adopted to simulate fluid flow in such matrix-fracture structure. Pressure drop is applied at the y direction with a periodic boundary condition at the x direction. Figure 6.8f shows the dynamic evolutions of permeability during the fracture propagation process, in which $k_f/k_m = 10^6$ with $k_m = 10^{-14}$ m^2. It can be seen that the relationship between effective permeability and fracture density still obeys the power law.

6.4. CONCLUSION

A generalized LB model for fluid flow through porous media is adopted to simulate fluid flow in matrix-fracture systems and to predict the effective permeability k_{eff}. 2D and 3D DFN systems are constructed using line fractures and elliptical fractures, respectively. Power law relationships are found between effective fracture permeability and fracture density in the DFN. FDEM is employed to simulate fracture propagation process, and the simulation results are in good agreement with the experimental results. Dynamic evolution of effective fracture permeability during the fracture propagation process is predicted using the LB model. It is found that power law relationship between effective permeability and fracture density is still obeyed.

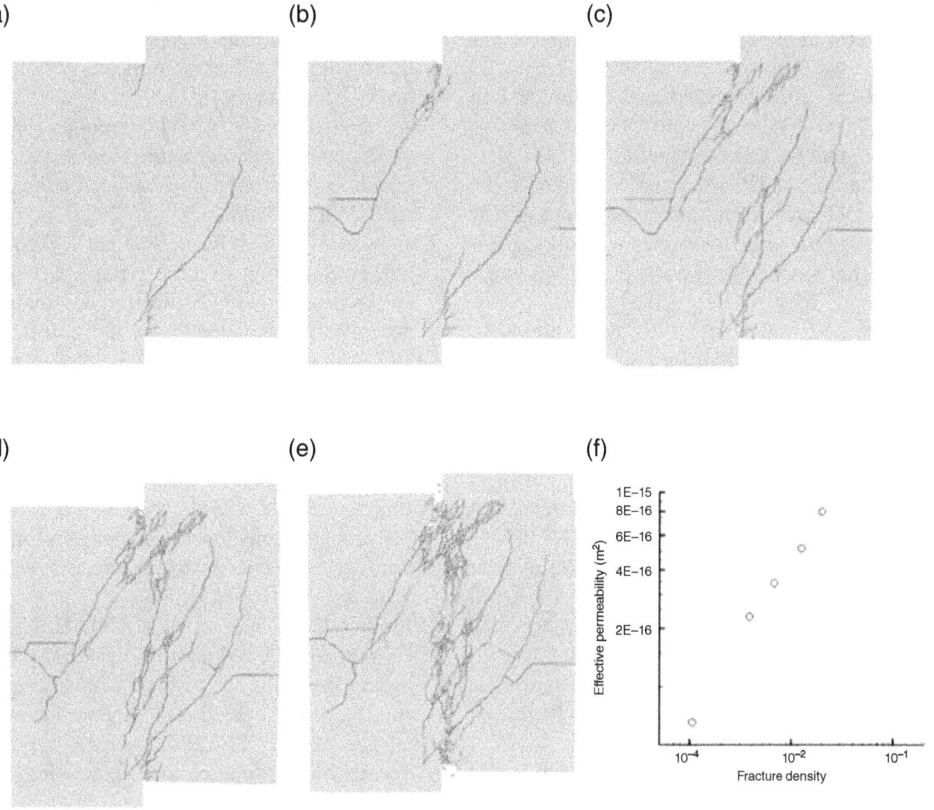

Figure 6.8 (a–e) Evolution of the fracture patterns obtained from the FDEM simulation and (f) relationship between fracture density and effective permeability with $k_f/k_m = 10^6$.

ACKNOWLEDGMENT

The authors acknowledge the support of LANL's LDRD Program, Institutional Computing Program, and a DOE NETL Unconventional Oil & Gas Project.

REFERENCES

Bear, J. (1972), *Dynamics of Fluids in Porous Media*, Elsevier, New York.

Beskok, A., and G. Karniadakis (1999), Report: a model for flows in channels, pipes, and ducts at micro and nano scales, *Microscale Thermophysical Engineering*, *3*(1), 43–77.

Carey, J. W., Z. Lei, E. Rougier, H. Mori, and H. Viswanathan (2015), Fracture-permeability behavior of shale, *Journal of Unconventional Oil and Gas Resources*, *11*, 27–43.

Chen, S. Y., and G. D. Doolen (1998), Lattice Boltzmann method for fluid flows, *Annual Review of Fluid Mechanics*, *30*, 329–364.

Chen, L., Q. Kang, Y. Mu, Y.-L. He, and W.-Q. Tao (2014), A critical review of the pseudopotential multiphase lattice Boltzmann model: methods and applications, *International Journal of Heat and Mass Transfer*, *76*, 210–236.

Chen, L., Q. Kang, R. Pawar, Y.-L. He, and W.-Q. Tao (2015a), Pore-scale prediction of transport properties in reconstructed nanostructures of organic matter in shales, *Fuel*, *158*, 650–658.

Chen, L., Q. Kang, Z. Dai, H. S. Viswanathan, and W. Tao (2015b), Permeability prediction of shale matrix reconstructed using the elementary building block model, *Fuel*, *160*, 346–356.

Chen, L., L. Zhang, Q. Kang, H. S. Viswanathan, J. Yao, and W. Tao (2015c), Nanoscale simulation of shale transport properties using the lattice Boltzmann method: permeability and diffusivity, *Scientific Reports*, *5*, 8089.

Chen, L., W. Fang, Q. Kang, J. De'Haven Hyman, H. S. Viswanathan, and W.-Q. Tao (2015d), Generalized lattice Boltzmann model for flow through tight porous media with Klinkenberg's effect, *Physical Review E*, *91*(3), 033004.

Freed, D. M. (1998), Lattice Boltzmann method for macroscopic porous media modeling, *International Journal of Modern Physics C*, *9*(8), 1491–1503.

Gale, J. F. W., R. M. Reed, and J. Holder (2007), Natural fractures in the Barnett shale and their importance for hydraulic fracture treatments, *AAPG Bulletin*, *91*(4), 603–622.

Gale, J. F. W., S. E. Laubach, J. E. Olson, P. Eichhubl, and A. Fall (2014), Natural fractures in shale: a review and new observations, *AAPG Bulletin*, *98*(11), 2165–2216.

Guo, Z., and T. S. Zhao (2002), Lattice Boltzmann model for incompressible flows through porous media, *Physical Review E*, *66*, 036304.

Heller, R., and M. Zoback (2014), Adsorption of methane and carbon dioxide on gas shale and pure mineral samples, *Journal of Unconventional Oil and Gas Resources*, *8*, 14–24.

Hyman, J. D., C. W. Gable, S. L. Painter, and N. Makedonska (2014), Conforming Delaunay triangulation of stochastically generated three dimensional discrete fracture networks: a feature rejection algorithm for meshing strategy, *SIAM Journal on Scientific Computing*, *36*(4), A1871–A1894.

Hyman, J. D., S. Karra, N. Makedonska, C. W. Gable, S. L. Painter, and H. S. Viswanathan (2015a), DFNWorks: a discrete fracture network framework for modeling subsurface flow and transport, *Computers & Geosciences*, *84*, 10–19.

Hyman, J. D., S. L. Painter, H. Viswanathan, N. Makedonska, and S. Karra (2015b), Influence of injection mode on transport properties in kilometer-scale three-dimensional discrete fracture networks, *Water Resources Research*, *51*(9), 7289–7308.

Jafari, A., and T. Babadagli (2012), Estimation of equivalent fracture network permeability using fractal and statistical network properties, *Journal of Petroleum Science and Engineering*, *92–93*, 110–123.

Javadpour, F., D. Fisher, and M. Unsworth (2007), Nanoscale gas flow in shale gas sediments, *Journal of Canadian Petroleum Technology*, *46*(10), 55–61.

Kang, Q., D. Zhang, and S. Chen (2002), Unified lattice Boltzmann method for flow in multiscale porous media, *Physical Review E*, *66*(5), 056307.

Kang, S. M., E. Fathi, R. J. Ambrose, I. Y. Akkutlu, and R. F. Sigal (2011), Carbon dioxide storage capacity of organic-rich shales, *SPE Journal* *16*(4), 842–855.

King, G. (2010), Thirty years of gas shale fracturing: what have we learned? SPE 133456, in *SPE annual technical conference and exhibition*, edited, Florence, Italy.

Klinkenberg, L. J. (1941), The permeability of porous media to liquids and gases, paper presented at Drilling and Production Practice, American Petroleum Institute, New York, 1 January 1941.

Long, J. C. S., and P. A. Witherspoon (1985), The relationship of the degree of interconnection to permeability in fracture networks, *Journal of Geophysical Research, Solid Earth*, *90*(B4), 3087–3098.

Loucks, R. G., R. M. Reed, Stephen C. Ruppel, and U. Hammes (2012), Spectrum of pore types and networks in mudrocks and a descriptive classification for matrix-related mudrock pores, *AAPG Bulletin*, *96*(6), 1071–1098.

Middleton, R. S., J. W. Carey, R. P. Currier, J. D. Hyman, Q. Kang, S. Karra, J. Jiménez-Martínez, M. L. Porter, and H. S. Viswanathan (2015), Shale gas and non-aqueous fracturing fluids: opportunities and challenges for supercritical CO_2, *Applied Energy*, *147*, 500–509.

Nithiarasu, P., K. N. Seetharamu, and T. Sundararajan (1997), Natural convective heat transfer in a fluid saturated variable porosity medium, *International Journal of Heat and Mass Transfer*, *40*(16), 3955–3967.

O'Malley, D., S. Karra, R. P. Currier, N. Makedonska, J. D. Hyman, and H. S. Viswanathan (2016), Where does water go during hydraulic fracturing?, *Groundwater*, *54*(4), 488–497.

Wang, F., and R. Reed (2009), Pore networks and fluid flow in gas shales. SPE124253, paper presented at Annual Technical Conference and Exhibition, New Orleans, LA, 4–7 October 2009.

Yazdi, A., H. Hamzehpour, and M. Sahimi (2011), Permeability, porosity, and percolation properties of two-dimensional disordered fracture networks, *Physical Review E*, *84*(4), 046317.

Ziarani, A. S., and R. Aguilera (2012), Knudsen's permeability correction for tight porous media, *Transport in Porous Media*, *91*, 239–260.

7

Gas-Water-Mineral Reactivity in Caprocks: Measurements, Estimates, and Observations

Julie K. Pearce and Grant K.W. Dawson

ABSTRACT

Geochemical reactions during CO_2 storage can dynamically change caprock properties, CO_2 migration, water quality, and CO_2 trapping. We reviewed insights from geochemical CO_2-fluid-rock reactions of natural analogues and experimental and modeling studies of caprock cores. Natural analogue observations can validate precipitated minerals in modeling studies and have also highlighted the importance of the reservoir-seal interface. Laboratory experiments and core characterization have provided data which can validate dissolution processes and provide reactive surface area estimates for upscaling. Less well understood processes include the effect of mobilized elements such as Ca from calcite dissolution on ion exchange in coexisting clays to mobilize trace metals to solution. Impure gas streams may precipitate oxide or sulfate minerals at the reservoir-caprock interface with the potential to plug porosity in reactive seals. Combined geochemical CO_2-fluid-rock and poro-perm studies (including impurity gases) are needed to interpret and validate models. Our understanding of caprock reactivity will be advanced by using multidisciplinary approaches, coupling different experimental approaches with modeling, and upscaling to field trials and natural analogues.

7.1. INTRODUCTION

The purpose of this chapter is to provide an overview of geochemical caprock reactivity during CO_2 storage. We present examples from a target CO_2 storage site in Queensland, Australia, and international sites including from Germany, the United States, Poland, Norway, Spain, and Algeria. We discuss studies at various spatial scales and timescales on mineral reactivity in caprock cores. Researchers have approached the challenge of understanding geochemical CO_2-water-rock reactivity from several angles which encompass different spatial scales and timescales. At one extreme lies characterizing geological formations with natural accumulations of CO_2, that is, natural analogues. At the other extreme are short lab based experimental studies ranging from determining mineral dissolution kinetics and thermodynamics to provide input data for models to reactivity of whole core material from target or pilot sites. Bridging the gap in timescale are geochemical models at the lab to site scale, with pilot and operational injection trials. Experimentally determined single mineral dissolution kinetics have been published widely and reviewed previously [*Palandri and Kharaka*, 2004; *Black et al.*, 2014]. Scaling up to rock core necessitates reactive surface areas as a model input. The reactive surface area is related to the accessibility of the mineral surface, texture, and habit, for example, pore rimming or cementing, and may dynamically change by dissolution and precipitation processes. The review by *Black et al.* [2014] reported wide ranges of (reactive) surface areas, activation energies, and rates of single mineral dissolution used for model input. Kaszuba and coworkers have also reviewed extensively mineral dissolution data and pure mineral CO_2-water-rock interactions [*Kaszuba et al.*, 2013]. Validation data

School of Earth and Environmental Sciences, University of Queensland, Queensland, Australia

Geological Carbon Storage: Subsurface Seals and Caprock Integrity, Geophysical Monograph 238, First Edition.
Edited by Stéphanie Vialle, Jonathan Ajo-Franklin, and J. William Carey.

for models including reactive surface areas and precipitating minerals have been recently provided by experiments and characterization of whole core from target, trial, or operational storage sites, along with natural analogues and field trials. Several previous reviews of caprock integrity pointed out the lack of consideration of geochemical reaction processes and their potential to affect caprock integrity, along with factors such as entry pressure or CO_2 contact angle [*Wollenweber et al.*, 2009; *Busch et al.*, 2010; *Fleury et al.*, 2010; *Shukla et al.*, 2010]. Historically, the reactivity of CO_2 storage reservoir rock has been the focus of field, natural analogue, and experimental and computational modeling studies, with less attention on caprock reactivity. Hence, this chapter will focus on reactivity of caprock cores from active or targeted storage sites.

Supercritical CO_2 injected into a reservoir dissolves into formation water or brine, generating carbonic acid (dissolution trapping) with initial lowering of the pH by at least 1 unit. The lowered pH has been shown to cause dissolution of carbonates and reactive silicates, potentially initially increasing rock porosity [*Lamy-Chappuis et al.*, 2014]. Subsequent mineral precipitation may decrease porosity or occlude pore throats as can the migration of fine material such as clays [*Pudlo et al.*, 2015]. Disequilibrium may also induce ion exchange with clays, and additionally, redox changes or desorption can mobilize or immobilize metals and nonmetals [*Viswanathan et al.*, 2012; *Black and Haese*, 2014]. These effects are generally site specific to lithology and local conditions.

Formation water can also dissolve into the supercritical CO_2, and this water-saturated or *wet* supercritical CO_2 (+SO_2) has also been shown to have reactivity to basalt cores, and to minerals including phyllosilicates, plagioclase, and carbonates, all of which are often present in caprock [*Schaef et al.*, 2011; *Shao et al.*, 2011; *Garcia et al.*, 2012a; *Glezakou et al.*, 2012; *Wang et al.*, 2013]. This wet supercritical phase and its reactivity to caprock however received relatively little attention so far. A related issue is dry-out as the CO_2 plume migrates upward to the base of the caprock and spreads laterally. If the near wellbore dries out due to formation water dissolving into the moving CO_2 plume, this will limit any reactions.

A good quality sandstone reservoir for CO_2 storage is generally quartzose with high porosity and permeability often in the range of 15–30% and 10–20,000 mD for good injectivity. For example, the Precipice Sandstone of the Surat Basin, Australia, has an average reported porosity of 16 ± 5% and horizontal permeability of 320 ± 905 mD (500–1000 mD in the Mimosa Syncline); the Kingfish Formation, Gippsland Basin, Australia, has reported porosity in the range of 15–32% and permeability of 10–20,000 mD; and the Frio Formation, Gulf Coast, USA, C injection zone has a mean porosity of 32% and permeability of 2000–3000 mD [*Kharaka et al.*, 2006; *Gibson-Poole et al.*, 2008; *Kellett et al.*, 2012]. Additionally, cored samples from the Ketzin pilot site, Upper Triassic Stuttgart Formation, North German Basin, Germany, have reported porosity in the range of 13–26%, and the Waarre Sandstone of the CO2CRC demonstration site in a depleted gas field in Otway Basin, Australia, has a porosity of ~20% [*Förster et al.*, 2010; *Jenkins et al.*, 2012]. Several trial CO_2 storage or enhanced recovery operations are capped by anhydrite or halite, sometimes over a carbonate reservoir [*Hangx et al.*, 2013; *Shevalier et al.*, 2013]. The enhanced oil recovery (EOR) site, the Weyburn field, Williston Basin, Canada, for example, has a reservoir porosity of 10–26% and permeability of 3–50 mD, overlain by an anhydrite caprock.

Caprock is traditionally thought of as an overlying formation of shale, mudstone, or siltstone which is clay rich with a low permeability to prevent vertical movement of CO_2. The average permeability of the Evergreen Formation, Surat Basin, at the Chinchilla 4 well, for example, was reported to be 87 ± 245 mD; however, permeability is extremely variable through the interbedded sands and shales [*Farquhar et al.*, 2015]. More recent measurements on Evergreen Formation core from the West Wandoan 1 well show that permeability is variable but generally much lower at 9–59 μD in some depths [*Coote*, 1984; *Golab et al.*, 2015a]. Along with clay-rich caprock, other low-permeability formations also act as caprocks, sometimes sealed by diagenetic alteration. Evaporite caprocks commonly contain mainly halite, anhydrite, or gypsum. Carbonate caprocks include limestone and marl, often mainly composed of calcite or other carbonates such as dolomite. Basalts also act as caprock, mainly containing the silicates plagioclase feldspar, pyroxene, and also olivine. The CarbFix project in Iceland, for example, has injected CO_2 and H_2S dissolved in wastewater from a geothermal plant into basaltic lavas and hyaloclastites at 400–800 m depth and observed mineral trapping within 2 years [*Sigfusson et al.*, 2015]. It is sealed by 200 m of low-permeability hyaloclastites (basaltic glass).

Griffiths provides an excellent overview of the characteristics of potential CO_2 storage seals in the United States, including the major mineral components [*Griffith et al.*, 2011]. Calcite, quartz, dolomite, illite, feldspars, glauconite, and kaolinite were the most commonly occurring; however, as pointed out by those authors, mixed layer clays are likely to occur more often than documented owing to misidentification.

Katzuba and coworkers demonstrated that standard shale samples had geochemical reactivity to CO_2-brine in experimental simulations of an aquifer-aquitard system [*Kaszuba et al.*, 2003, 2005]. While a reactive caprock may initially seem counterintuitive or undesirable, geochemical reactivity drives CO_2 ionic and mineral trapping. Mineral trapping requires a source of cations, for example, Fe^{2+}, Ca^{2+}, and Mg^{2+}, to combine with dissolved CO_2 to eventually form carbonate minerals such as siderite, calcite, dolomite, and ankerite. Silicate dissolution provides the cations; for example, chlorite may be converted to siderite (Eqs. 7.1 and 7.2):

$$Fe_4MgAl_2Si_3O_{10}(OH)_8 + 10H^+ \leftrightarrow 4Fe^{2+} + Mg^{2+} + Al_2Si_2O_5(OH)_4 + SiO_2(aq) + 7H_2O \quad (7.1)$$

$$4Fe^{2+} + Mg^{2+} + 5HCO_3 \leftrightarrow 5Fe_{0.8}Mg_{0.2}CO_3 + 5H^+ \quad (7.2)$$

The concept of an optimized storage reservoir has been proposed through studies of natural CO_2 accumulations [*Watson and Gibson-Poole*, 2005]. The conceptual CO_2 storage reservoir is overlain by a geochemically reactive formation topped by a low-permeability caprock. Practically, storage reservoirs are often a complexity of lithologies with sections of higher porosity and permeability sandstone interbedded by shales, mudstones, siltstones, and coals, sometimes predicted to result in horizontal migration of CO_2 and baffling which can increase residual and mineral trapping. This process has been predicted in the Surat and Gippsland basins, Australia, and North German Basin, Germany [*Gibson-Poole et al.*, 2008; *Farquhar, S*, 2016]. A modern term is the storage complex, which contains at least one but generally two or more low permeability units in addition to the low-permeability units stratigraphically above the storage complex, with several potential high-permeability injection horizons. The Shell Quest project, for example, includes the basal Cambrian Sand as the storage zone, which is overlain by the primary seal, the Middle Cambrian Shale (in the middle of the storage complex), and the Lower and Upper Lotsberg Salts, all excellent seals. The Prairie Evaporite is further up the stratigraphic column, with other low-permeability units through the stratigraphic column.

Adding to the complexity of lithology, storage reservoir temperatures span a wide range of ~35–100°C depending on storage depth and local geothermal gradients. Generally, CO_2 storage sites are selected at conditions above 31.1°C and 7.38 MPa, at ~1–3 km depths where CO_2 exists as a supercritical fluid. Shallower low-temperature storage sites however do exist including the CarbFix site mentioned above at 400–800 m depth with formation water temperatures of ~20–33°C. Deep saline aquifers or depleted gas reservoirs (plus enhanced recovery options) have generally been considered for storage sites. It should be noted that the Surat Basin, discussed below, is a freshwater target for CO_2 storage in Australia, with sites considered for storage in the Gippsland Basin, Australia, potentially having a freshwater intrusion [*Hodgkinson and Grigorescu*, 2012].

7.2. GEOCHEMICAL REACTIVITY

7.2.1. Natural Analogues

Sites of natural CO_2 accumulations, that is, natural analogues of CO_2 storage, give important insights into long-term reactions of reservoir or caprock with CO_2 which cannot be attained in the lab and which should be used to validate model outputs. The excellent review by Bickel et al. and references within describe many of these studies [*Bickle et al.*, 2013]. Broadly, dissolution of carbonates, plagioclase feldspar, phyllosilicates (e.g., biotite, muscovite, and chlorite), K-feldspar, and iron oxides has been observed with the precipitation of carbonates, kaolin, smectite, illite, and quartz/colloidal silica. The type of carbonates precipitated depends on reservoir conditions (e.g., pressure, temperature, pH) and also on the cation constituents available from alteration of silicates or oxides. Fe-rich or Fe-Mg-rich siderite has been formed in Fe-rich sediments such as red beds or from alteration of annite, as have also ankerite ($Ca[Fe,Mg,Mn](CO_3)_2$) and dolomite. Calcite (± Mg, Mn) is also generally observed in Fe-poor reservoirs. Dawsonite precipitation is a rather contentious issue, with some modeling studies predicting its formation in sandstone reservoirs [*Hellevang et al.*, 2005; *Xu et al.*, 2005]. It has been observed in some natural analogues at high CO_2 pressures replacing Na-feldspars, for example [*Moore et al.*, 2005; *Uysal et al.*, 2011]; however, it has not been observed in lab CO_2-water-rock studies likely owing to its instability at low partial pressures of CO_2. Dawsonite stability in CO_2-charged reservoirs is further discussed in *Hellevang et al.* [2011].

A more recent natural analogue study of the Kapuni field, New Zealand, suggests that chlorite there has been altered to kaolinite, silica, and Fe-rich carbonates, as was also previously observed in the Ladbroke Grove field, Australia [*Watson et al.*, 2004; *Higgs et al.*, 2013]. Higgs and coworkers also observed calcite, siderite, and dolomite formation and feldspar alteration. Less feldspar reaction but more mineral precipitation occurred in fine-grained units of decreasing porosity, and also interestingly, carbonate precipitation was observed at gas-water contacts. In the Pretty Hill Formation, Otway Basin, Australia, different parts of the formation have been exposed to low, moderate, and high CO_2 contents [*Higgs et al.*, 2015]. Fe-rich chlorite is the major clay mineral still present where CO_2 content is low, with significant chlorite alteration and minor plagioclase reaction in areas exposed to moderate CO_2 content. Calcite, siderite, and ankerite cements in the moderate CO_2 zones are associated with either heterogeneity in the rock or the gas-water contact, with siderite as a replacement phase of grain coating clays. In the high CO_2 content Garvoc 1 well, chlorite and feldspars were completely reacted to form mainly siderite and kaolinite. The carbonates observed in the high CO_2 zone have several stages of formation, with siderite often predating calcite. Overall, the most advanced feldspar reaction was observed at the contact between the reservoir rock and fine-grained baffles and seal units where CO_2-rich fluids had pooled.

Conversion of biotite/chlorite to siderite/ankerite and kaolin, and also plagioclase conversion to calcite was observed in core from the Chinchilla 4 well of the Surat Basin, Australia, a proposed previous site of CO_2 accumulation. Significant siderite cementation occurred in the Westgrove Ironstone Member of the Evergreen Formation [Farquhar, S. M., S. Golding, G. Dawson, J. K. Pearce, A. Golab, R. Heath, and J. S. Esterle (2018), Fluid-flow history, diagenesis and evidence of CO_2 accumulation as a natural analogue for CO_2 storage in the Surat Basin, In preparation.]. Porosity and permeability were at maxima in the lower Precipice Sandstone reservoir and at minima in the overlying Evergreen Formation, with a strong negative correlation between clay content and permeability noted. The Precipice Sandstone is also a low-salinity target for CO_2 storage, and Farquhar et al. predicted the dissolution of carbonates, and reactive silicates, and the precipitation of Fe (Mn-Mg-Ca) carbonates and smectite in baffle sections of the Evergreen Formation.

Mineralized faults in the Carmel Formation which caps the Navajo Sandstone below the Entrada Sandstone were characterized near Green River, Utah, in the Paradox Basin [*Chen et al.*, 2016]. The Carmel Formation consists of sandstone, siltstone, shale, mudstone, and bedded gypsum. CO_2-charged brines leak to surface in the area via faults, abandoned petroleum, and water wells. The CO_2 is subcritical at low temperatures of 15–18°C with the base of the Carmel Formation at 199 m. Secondary gypsum and carbonates fill fractures, with gypsum from two sulfate sources: dissolution of preexisting gypsum beds in the Carmel Formation and sulfate-rich brines from evaporites in the underlying Paradox Formation. The secondary gypsum and carbonates in fractures decreased permeability and appeared to trap and inhibit CO_2 and sulfur leakage. This is an important observation for potential co-sequestration of CO_2 with SO_2 from power plants.

The importance of faults, the region around the reservoir-seal interface, areas of heterogeneity, and the gas-water interface are clear. The alteration of Fe-rich clays to carbonate mineral trapping CO_2 has been observed at the timescales of natural analogues, but what other roles do they play and what may be the effect of co-injection of SO_2?

7.2.2. Experimental and Modeling Studies

Several experimental and combined modeling studies on clay-rich caprock reactivity are discussed below in terms of the observed dissolution and precipitation of minerals. Water chemistry has been generally measured previously to assess the potential for ionic and mineral trapping and to identify minor dissolution and precipitation or other processes such as ion exchange which may not be apparent from examining and characterizing rock surface changes. A few site-scale modeling studies have also incorporated mineral reactivity. Experimental studies on fractured carbonate or evaporite cores have either observed preferential flow path formation via mineral dissolution or reduction of permeability through mineral precipitation or mechanical closure.

7.2.2.1. Mineral Dissolution and Precipitation in Clay-Rich Caprocks of Sandstone Reservoirs

The Precipice Sandstone (reservoir), Surat Basin, Australia, is predominantly quartz with some kaolinite and dickite, feldspars, and trace amounts of minerals including micas, carbonates, sulfides, and oxides [*Horner et al.*, 2014; *Farquhar et al.*, 2015; *Pearce et al.*, 2015b]. A clay-rich low-porosity section exists around 1205–1218 m in core from the Chinchilla 4 well (Fig. 7.1) and around 1207–1212 m in core from the West Wandoan 1 well. The overlying Evergreen Formation (caprock) generally contains more potentially reactive minerals, for example, clays such as Fe-rich chlorite, feldspars, and sporadic carbonate cements including calcite, siderite, and ankerite [*Farquhar et al.*, 2013, 2015]. The stratigraphy of the Surat Basin was recently modified to reclassify the fine-grained sandstone at the base of the Evergreen Formation as part of the Precipice Sandstone [*Ziolkowski et al.*, 2014]. The Evergreen Formation 1138 m core discussed below would in the new classification be part of the upper Precipice Sandstone, with the top of the Precipice Sandstone hence vertically variably clay rich and potentially reactive transitioning into the Evergreen Formation. The traditional stratigraphy is however used here for consistency with publications discussed here.

An example of the potentially wide variability in grain size between reservoir and caprock and also within formations can be observed in Figure 7.2. The Precipice Sandstone core at 1192.9 m, for example, is mainly poorly sorted quartz grains with visible porosity. Sections of the Evergreen Formation core, for example, at 1138 and 897.9 m, and the Hutton Sandstone at 868 m contain reactive minerals including Fe-rich chlorite, siderite, and ankerite (bright grains; Fig. 7.2), plagioclase mainly present as albite at the Chinchilla 4 well, and K-feldspars along with organic matter. Some of these reactive minerals exist in abundances below the level of detection by

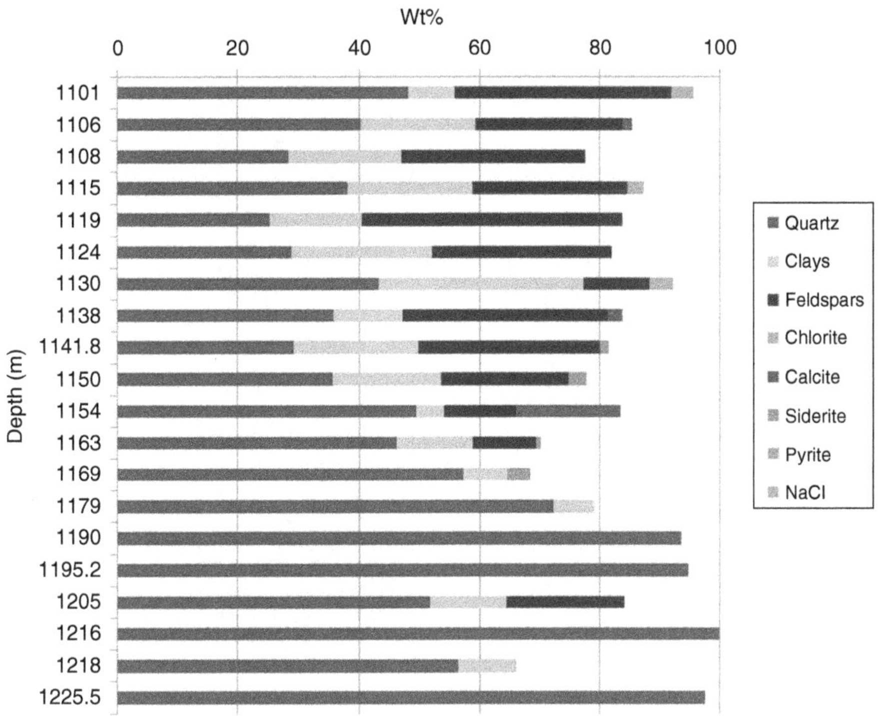

Figure 7.1 Mineral content (wt%) from bulk XRD of core between 1101 and 1225.5 m depth from the Chinchilla 4 well, Surat Basin. According to the original stratigraphy, cores from 1190 to 1225.5 m are classified as the Precipice Sandstone, and 1179 m and above are classified as the Evergreen Formation. Created based on data from the supplementary material of *Horner et al.* [2014].

commonly used characterization methods such as XRD or QEMSCAN; for example, ~1% calcite is present in Evergreen core from the Chinchilla 4 well at 1138 m. Carbonates are generally sporadic, for example, ~0–15%, at specific depths in the Chinchilla 4 well core. The Hutton Sandstone at 799.5 m, for example, is ~15% Mn-calcite cemented (bright areas; Figures 7.2 and 7.3).

Experimental scCO_2-water reactions have been performed at 120 bar and 60°C for 16 days after an initial baseline water-rock equilibration with cores from the Surat Basin Chinchilla 4 well [*Farquhar et al.*, 2015]. Figure 7.3 illustrates the core mineral contents, and those minerals which dissolved are labeled D. Compared to the Precipice Sandstone reservoir rock, the Evergreen Formation caprock and overlying Hutton Sandstone cores were more reactive, with dissolving minerals generally calcite, ankerite, siderite, and chlorite. Evergreen Formation core from the Chinchilla 4 well at 897.9 m contains sporadic fine-grained ankerite or siderite sometimes intermingled with chlorite (Fig. 7.4). On CO_2-water reaction of rock with trace carbonates, Fe-rich chlorite generally underwent incongruent dissolution or leaching of X-ray dense ions (Fig. 7.4c and d). The Evergreen core from 1138 m additionally contains ~1% calcite, with trace phosphates, chlorite, and Ti oxides (Fig. 7.5). On reaction with CO_2-water, the core disaggregated, some Fe was leached from the chlorite, and calcite cement was dissolved; subsequent clay stabilization with KCl however prevented disaggregation. Increasing the reactive surface areas of clays such as chlorite was necessary to match the observed water chemistry in geochemical modeling. In reaction of the calcite-cemented core Hutton 799 m however, calcite dissolved, and solution pH was buffered to ~6.5 with no corrosion of silicates directly observed on the experiment timescale. Pore-filling and rimming clays were uncovered by dissolution of calcite cement. Micro-computed tomography (CT) with a resolvable porosity of 2.2 μm for a 4 mm diameter sub-plug was performed before and after CO_2-water reaction [*Golab et al.*, 2014]. Resolvable porosity of the calcite-cemented core increased from 1.1 to 2.2% after reaction, with created porosity extending through the sub-plug. The calculated porosities of two Evergreen Formation core sub-plugs also changed slightly after reaction; however, as can be seen in Figure 7.6, these can be attributed at the instrument resolution to mainly disconnected porosity creation and leaching of X-ray dense ions such as Fe from chlorite. This implies that opened porosity in reacted calcite-cemented core may potentially allow CO_2 or CO_2-charged brine migration, whereas reaction of core containing minor or trace carbonates is less likely to lead to fluid migration in the short term. No directly observable

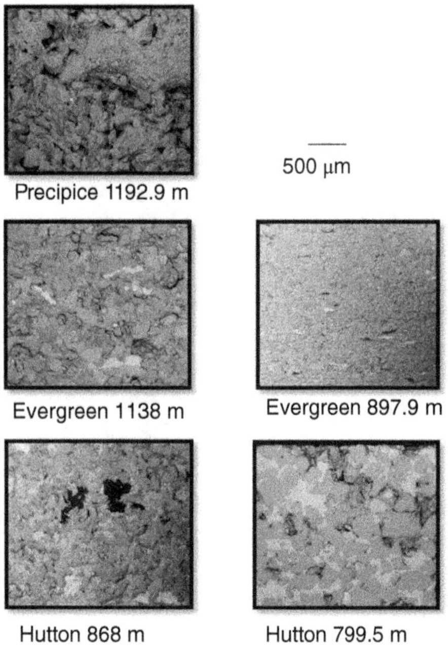

Figure 7.2 SEM images of core from the Chinchilla 4 well, Surat Basin. Note the variability in grain size and packing between sections of the Precipice Sandstone reservoir, the Evergreen Formation caprock, and the overlying Hutton Sandstone.

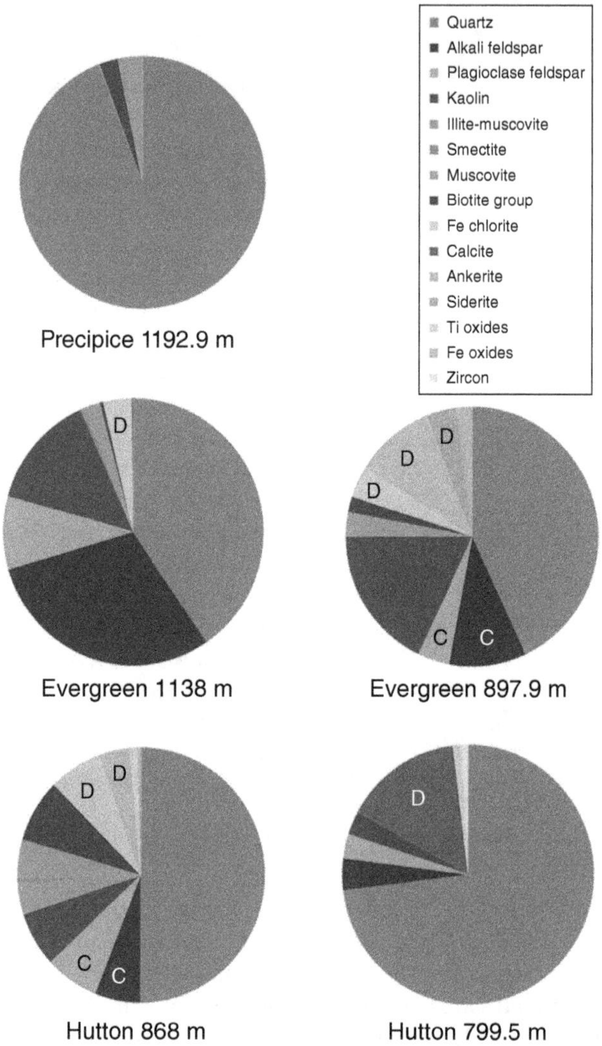

Figure 7.3 Mineralogy of core from the Chinchilla 4 well, Surat Basin, created from data in *Farquhar et al.* [2015]. Minerals labeled D dissolved during experimental $scCO_2$-water reaction. Dissolution of 1% calcite content in the 1138 m core was also observed (not shown), along with dissolution of trace amounts of fine-grained siderite and ankerite from all cores. Minerals labeled D dissolved and C additionally corroded during experimental $scCO_2$-SO_2-water reaction based on data from *Pearce et al.* [2015b].

minerals precipitated during the short experiments; kaolinite however was predicted to precipitate in kinetic geochemical modeling of the reaction of both Evergreen Formation and Hutton Sandstone cores using The Geochemist's Workbench (GWB) software. QEMSCAN and micro-CT were determined to be excellent methods to quantify and visualize mineralogy and pore networks in sandstones. However, for tight shales and mudstone caprocks, a component of unclassified minerals remained as the electron beam penetrated the surface, resulting in mixed mineral compositions. Additionally, a large component of the porosity in the tight shale, mudstone, or siltstone cores may be below the instrument resolution. Golab and coworkers recently improved micro-CT porosity characterization for core from the Surat Basin by characterizing both the resolved porosity, pores greater than a voxel, and sub-resolution porosity, where the voxel contains a mix of pore and clay [*Golab et al.*, 2015b]. Techniques such as small angle neutron scattering (SANS) have also been used to characterize accessible and non-accessible tight porosity in mudstones, coals, and shales and will likely play a role in future caprock characterization.

For the cores characterized by Farquhar and Golab above, except for the highly calcite cemented cores which underwent large changes on reaction, dissolution and corrosion of trace minerals often could not be detected and quantified by advanced and standard methods such as QEMSCAN and XRD alone as changes were below detection. This required additional direct observation of the mineral surfaces before and after reaction by SEM-EDS and characterization of water chemistry. Similar limitations with XRD have been noted in other studies [*Liu et al.*, 2012; *Wilke et al.*, 2012].

The accessibility of reactive minerals to CO_2 or CO_2-rich fluids in heterogeneous caprocks with variable grain size, porosity, and permeability is a key variable relating to reactive surface areas which should be considered as

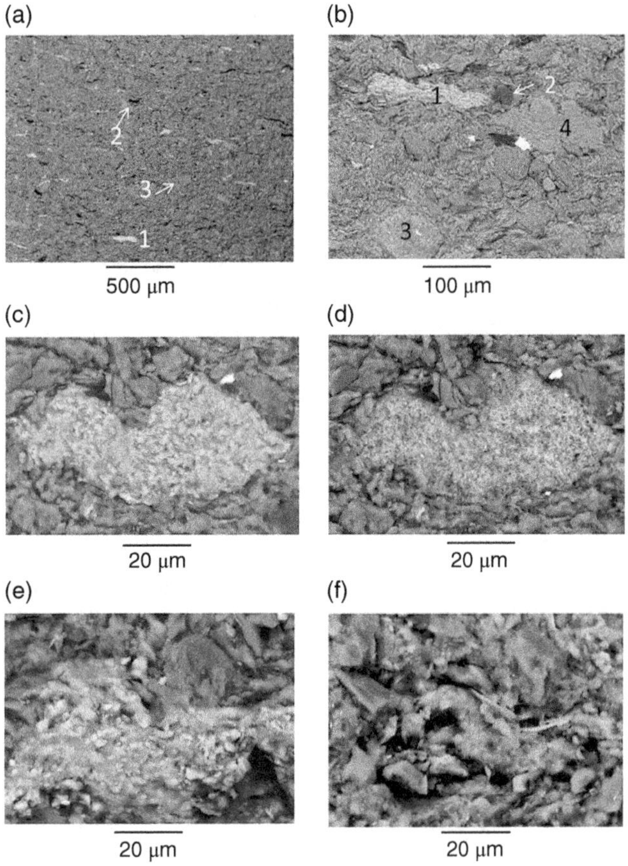

Figure 7.4 SEM images of Evergreen Formation 897.9 m (Chinchilla 4 well) caprock. (a and b) Horizontally extruded mixed ankerite and Fe-Mg chlorite (1), 2 = organic matter/coal, 3 = albite, and 4 = K-feldspar. (c) Ankerite and chlorite in the core before and (d) after CO_2-water reaction. (e) Before and (f) after CO_2-SO_2-water reaction.

input for predictive models. The accessibility can change over the course of CO_2 injection as carbonates dissolve, revealing pore space and clays increasing surface roughness, or as precipitates coat surfaces. The coupling of evolving 2D and 3D core imaging technologies before and after or in situ during experiments could be used to characterize dynamic accessibility.

The injected CO_2 stream may contain trace impurity gases from processes such as coal combustion or cement processing [*Bacon and Murphy*, 2011; *Garcia et al.*, 2012b]. These include SO_2, O_2, NOx, and so on, which are reactive when dissolved in water; however, the gas-water-rock reactivity of impure gas streams has had far less attention than pure CO_2 [*Talman*, 2015]. Reactions of the same core depth sections from the Surat Basin as described by Farquhar and coworkers have also been performed with scCO_2-SO_2-water [*Pearce et al.*, 2015b]. A concentration of 0.16% SO_2 in CO_2 caused the generation of sulfuric acid, lowering the solution pH to ~2 (Fig. 7.7) with additional corrosion of plagioclase (albite) and K-feldspar from the caprock cores as labeled C in Figure 7.3, and more extensive corrosion and incongruent dissolution of Fe-rich chlorite (Figs. 7.4e, f and 7.5d). Trace FeS precipitation occurred in reaction of the Evergreen 897.9 m and Hutton 868 m cores. The dissolution of calcite cement from the Hutton 799 m core however buffered pH to ~6, resulting in no observable silicate dissolution on the experiment timescale. Calcite cement dissolution revealed clays and open porosity (Fig. 7.5e and f). Precipitation of gypsum on quartz and K-feldspar grains occurred in the presence of the SO_2 gas (Fig. 7.5f inset) (Eqs. 7.3–7.5):

$$4SO_2 + 4H_2O \leftrightarrow 3H_2SO_4 + H_2S \tag{7.3}$$

$$CaCO_3 + H_2SO_4 \leftrightarrow Ca^{2+} + SO_4^{2-} + H_2CO_3 \tag{7.4}$$

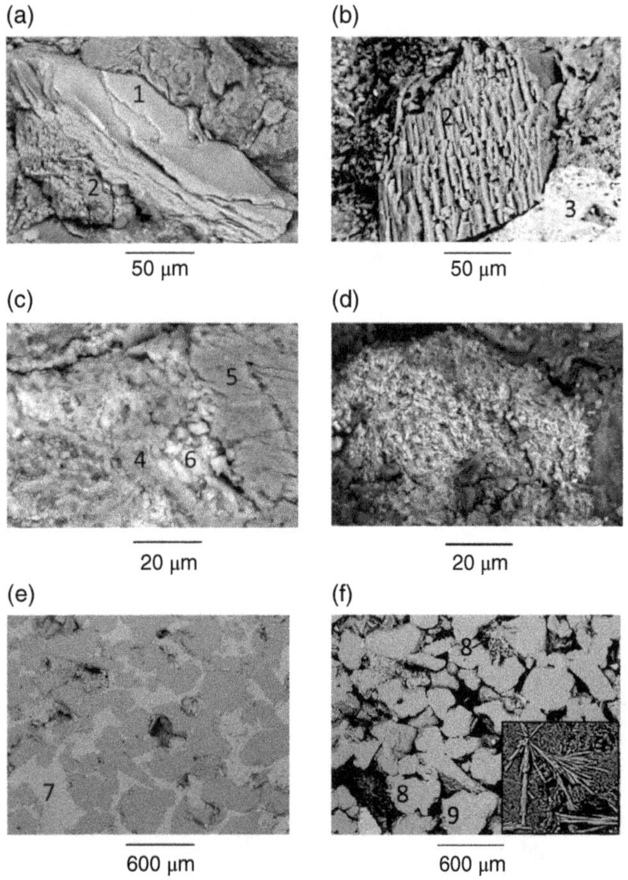

Figure 7.5 SEM images of (a–c) Evergreen Formation 1138 m (Chinchilla 4 well) core: Fe-Mg-Ti mica (1), K-feldspar (2), Ca-Fe-phosphate containing Ce and La (3), Fe-Mg chlorite (4), kaolin (5), and Ti oxide on chlorite (6). (d) After CO_2-SO_2-water reaction with core disaggregated and Fe leached from chlorite, bright Ti oxide is still present. (e) Hutton Sandstone 799 m pre and (f) post CO_2-SO_2-water reaction with calcite cement (7) dissolved revealing clays (8) and gypsum precipitated on K-feldspar surface (9); inset magnified view of gypsum with image width 150 μm.

$$Ca^{2+} + 2H_2O + SO_4^{2-} \leftrightarrow CaSO_4.2H_2O \quad (7.5)$$

Analysis of water chemistry during the reactions of Farquhar and Pearce of Evergreen 1138 m core showed higher concentrations of dissolved Ca and Mn from dissolution of the 1% calcite, with Ca and Mn correlated (Fig. 7.7). Dissolved Ca during reaction of Evergreen 897.9 m core from dissolving trace amounts of ankerite was also measured. A higher concentration of dissolved Al, Si, K, and Fe was observed from caprock cores reacted in SO_2-CO_2-water than pure CO_2-water (Fig. 7.7). This was owing to greater dissolution of silicates, mainly Fe chlorite, plagioclase, and K-feldspar, in the presence of co-injected SO_2. Trace amounts of siderite and ankerite also contributed to dissolved Fe in solution from Evergreen 897.9 m. With longer reaction time or the addition of co-injected O_2 in the gas mixture, several elements including Fe and Ca subsequently decreased through precipitation of Fe oxides, clays, or gypsum [*Pearce et al.*, 2013].

Geochemical modeling of the SO_2-CO_2-water reactions predicted precipitation of pyrite and elemental S for the Evergreen 897.9 m, and Hutton 868 m core, and also gypsum for the Hutton 799 m core. Separately, long-term geochemical modeling of an average mineralogical composition of the Evergreen Formation over 10 and 100 years with GWB predicted dolomite, smectite, quartz, and jarosite precipitation decreasing porosity on reaction of SO_2-CO_2 [*Frank and Kirste*, 2014]. The incorporation of reactive minerals including those present in trace amounts with correct compositions was important for accurate modeling, that is, the use of albite rather than labradorite and Fe-rich rather than Mg-rich chlorite. For example, Kirste showed in reactive transport modeling of the Precipice Sandstone with TOUGHREACT that the inclusion of observed trace amounts of carbonate was important, resulting in predicted buffering of solution pH within water quality guidelines even with co-injection of SO_2 [*Kirste et al.*, 2015].

Similar caprock and reservoir core sections from the Chinchilla 4 well have also been reacted with both O_2 and SO_2 impurities in CO_2, resulting in stronger water acidification and silicate dissolution; however, subsequent Fe oxide precipitation removed dissolved ions from solution [*Pearce et al.*, 2013]. The lithology of core was also shown to somewhat control pH and reactivity in O_2-SO_2-CO_2-brine reaction of West Wandoan 1 well cores from a different location in the Surat Basin [*Pearce et al.*, 2015a].

Enhanced dissolution of silicates such as chlorite and plagioclase provides higher concentrations of cations (e.g., Ca, Fe, Mg) available for ionic and potentially mineral trapping of CO_2 when pH is buffered. However, dissolution could also lead to changes in porosity and CO_2 migration in the short term. The precipitation of gypsum (or oxide minerals in the case of co-injection of O_2 or NOx) may be a competing sink for the dissolved cations but may positively lead to porosity plugging and self-sealing since gypsum, for example, has a higher molar volume than calcite or ankerite. Recent work on West Wandoan 1 well caprock and reservoir core from the Surat Basin has highlighted both porosity and permeability changes linked to mineral dissolution from reaction with low-salinity CO_2-brine over 1 month at reservoir conditions [*Dawson et al.*, 2013]. Significant differences in rock mechanical properties when reacted in the presence of even small amounts of NOx, SO_2, or O_2 have been demonstrated for reservoir rock elsewhere. Further coupling of geochemical gas-water-rock experiments with characterization of porosity, permeability, or rock property changes of caprock is needed.

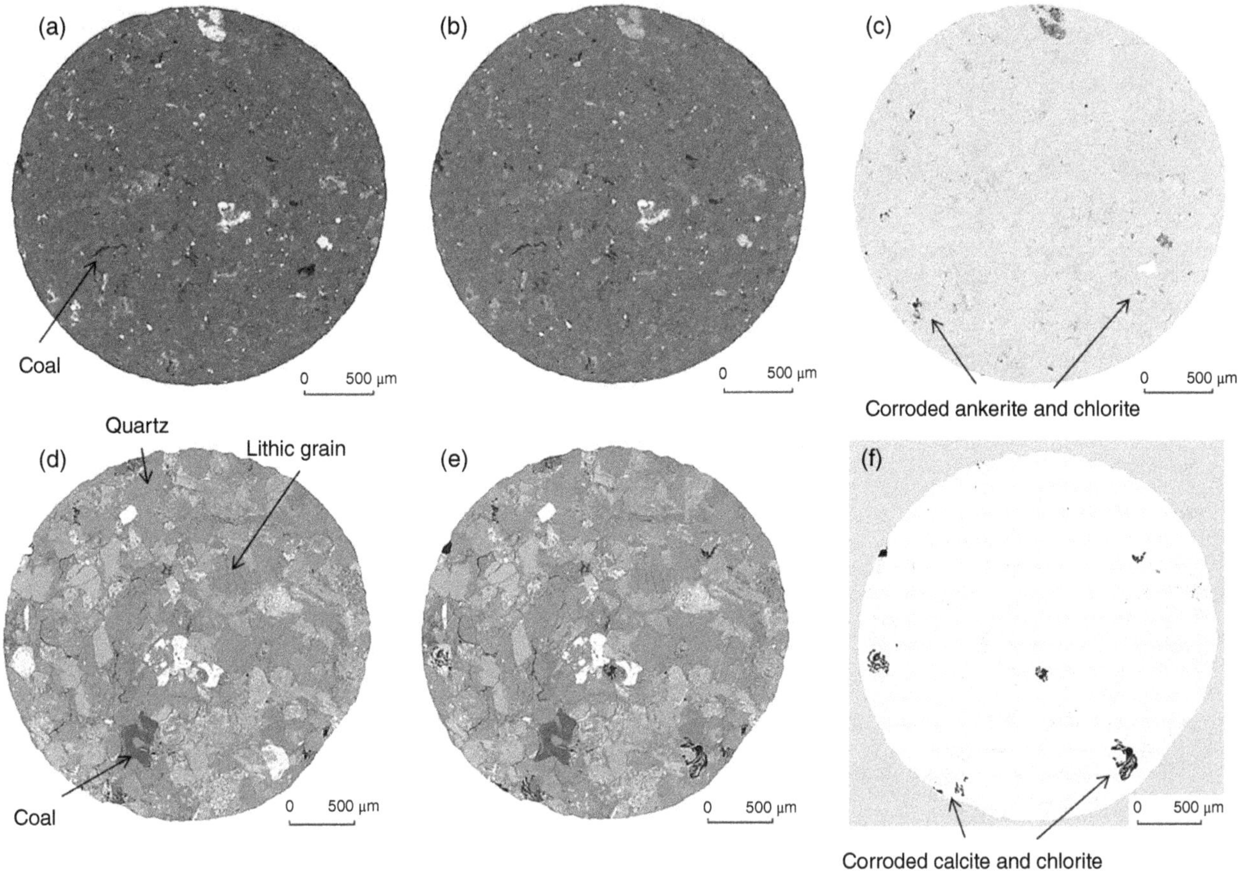

Figure 7.6 Micro-CT images of sub-plugs of caprock core (Chinchilla 4 well) at 2.2 μm resolution: Evergreen Formation 897.9 m (a) before reaction, (b) after CO_2-water reaction, (c) difference image where dark areas indicate loss of material or X-ray density. Evergreen Formation 1138 m (d) before reaction, (e) after CO_2-water reaction (stabilized with KCl), and (f) difference image. Modified with permission from *Golab et al.* [2014].

Recently, anions which are formed from co-injection of impurity gases, for example, sulfate, have been shown to increase plagioclase dissolution, bicarbonate has been reported to affect chlorite dissolution, and these anions also have the potential to desorb trace metals or non-metals as predicted for bicarbonate [*Black and Haese*, 2014; *Min et al.*, 2015]. The injected gas stream therefore can affect caprock reactivity through more complicated processes than simply acidification and mineral dissolution especially where trace oxide and sulfide minerals are present. These processes deserve further attention.

Batch experimental and geochemical modeling studies generally span a wide range of temperature and pressure conditions and timescales. Utilized temperatures either simulate in situ reservoir conditions, or some authors use elevated temperatures to *accelerate* reactions. Several authors of CO_2-brine reactions of caprock cores highlight the initial reactivity of carbonates but also that of clays, especially Fe-rich clays such as chlorite, biotite-muscovite, and illite, with water chemical analysis often needed to identify incongruent dissolution behavior, ion exchange, or dissolution of small amounts of carbonates.

Several core sections of the Stuttgart Formation from the Ktzi 202 observation well of the Ketzin pilot site in Germany were subject to batch reaction with pure CO_2-brine at 40°C and 5.5–7.5 MPa and rock-brine ratios of 2 : 1–4 : 1 [*Fischer et al.*, 2013]. Sandstones were reacted for 15–40 months immersed in brine, and also siltstones were reacted for 2–6 months with the bottom section immersed in brine and the top section in the CO_2 headspace, that is, wet $scCO_2$. An example of the mineralogical composition of sandstone from 628.4 to 628.6 m is shown in Fig. 7.8, along with that of the top and bottom sections of a siltstone sample from 625.8 to 626 m directly above the CO_2 storage reservoir representing the transition to the overlying Weser and Arnstadt Formation anhydritic mudstone caprocks. The siltstone samples were similar in mineralogy to the sandstone samples, but with smaller grain size, with Ba/Sr sulfate cements, and with no dolomite cements. After reaction, anhydrite, K-feldspar, and the anorthite component of the plagioclase

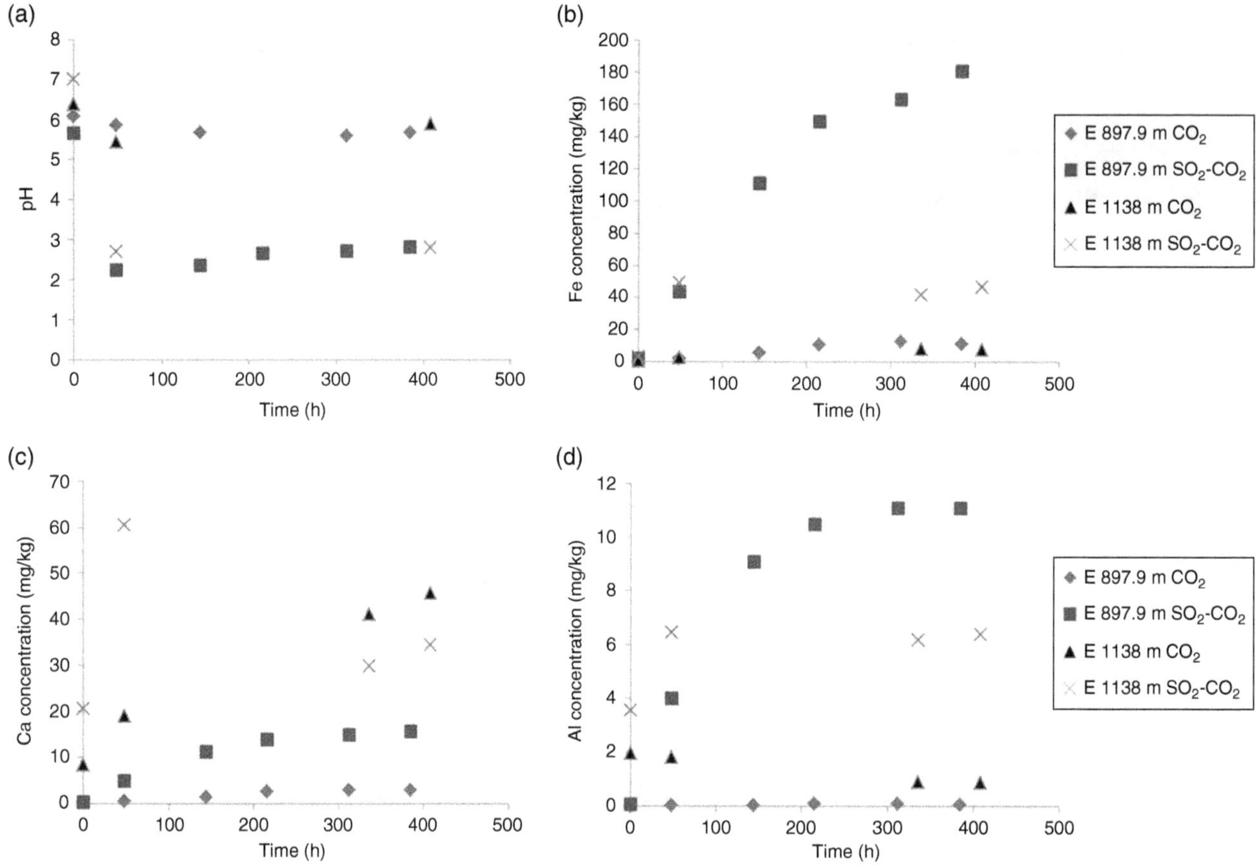

Figure 7.7 Water chemistry during CO_2-water or SO_2-CO_2-water reaction of Chinchilla 4 well Evergreen Formation cores (E = Evergreen) from two depths with concentrations normalized to 1 g of reacted core for comparison. (a) solution pH during reaction, (b) Fe concentration, (c) Ca concentration, and (d) Al concentration. Time zero represents a N_2-water-rock soak prior to CO_2 injection. Created from data in *Farquhar et al.* [2015] and *Pearce et al.* [2013, 2015a].

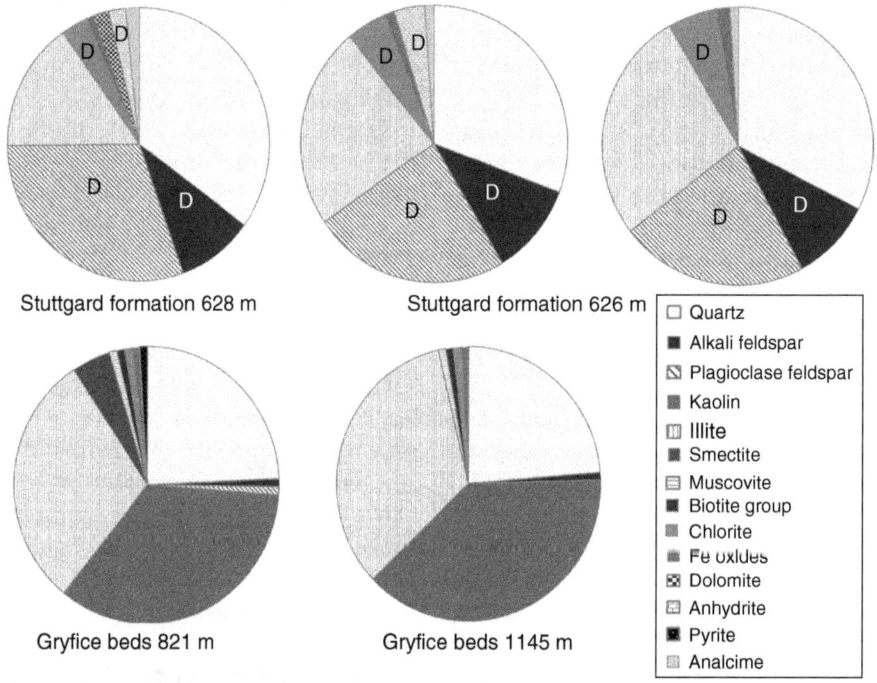

Figure 7.8 Mineralogy of sandstone (628 m) and siltstone (626 m top and bottom section) cores from the Stuttgart Formation, North German Basin, Germany. Where chlorite = both chlorite and biotite with an Fe (Mg) composition, plagioclase is albite and albite-anorthite, and illite is illite-muscovite. Minerals labeled D were directly observed to dissolve during experimental CO_2-brine reaction. Mineralogy of caprock cores from the Gryfice Beds, Chabowo anticline, Poland, are also shown. Created from data in *Fischer et al.* [2013] and *Wdowin et al.* [2014a].

were corroded along with incongruent chlorite dissolution and possible hematite dissolution. Pitting on quartz grains in the siltstone was also observed. Surprisingly, no dolomite corrosion from the sandstone could be observed. Geochemical modeling with PHREEQC predicted dissolution of the anorthite component of plagioclase and the cement phases (analcime, anhydrite, dolomite) and additionally precipitation of clay minerals (chlorite, illite, kaolinite). Fisher noted the difficulty in separating the natural heterogeneity of rock samples from changes on reaction, as has been pointed out by several researchers in this area. The need to use proxies for clays in geochemical modeling owing to a lack of existing model input data was also highlighted.

The Komorowo Beds Sandstone in the Chabowo anticline has been recently studied for potential CO_2 storage in Poland [*Wdowin et al.*, 2014a]. Mineralogies of two core depth sections of the overlying Gryfice beds claystone caprocks from bores Chabowo 1 and 3 are shown in Figure 7.8. Porosity in the Gryfice beds is variable in the range of 8–19%, and permeabilities are in the range of 38.5–435 mD. Kaolin and illite-muscovite dominate the mineralogies, with chloritized biotite and muscovite also present. Both reservoir sandstones and caprocks were reacted in CO_2-brine at 25°C and 6 MPa for 18 months in batch reactors. Disaggregation of the caprock core from 821 m was observed with halite precipitation from the brine. Corrosion of feldspars, pyrite, and mica and precipitation of kaolinite were observed in reaction of the reservoir sandstones. Dissolved concentrations of Ca, Mg, Na, and K were attributed to corrosion of clays, micas, and feldspars from the claystone caprocks. Target reservoir and caprocks from the Zaosie anticline in Poland were characterized texturally including with SEM, petrography, and XRD before and after batch reaction with CO_2 and complex brine at 6 MPa and 20–25°C for 20 months [*Wdowin et al.*, 2013]. Claystone core from 912.4 m and mudstone core from 1436 m were compared. The claystone contained mainly quartz and mica (biotite, muscovite) with matrix clays including kaolinite, illite, and montmorillonite. Micas were altered by illitization and chloritization, with also trace K-feldspar, iron oxides, zircon, and rutile present. After reaction, no corrosion or alteration was observed except halite precipitation from the brine. Water chemistry was however not analyzed in the experiments to observe minor reactions. The mudstone was mainly quartz, mica, kaolinite, illite, and organic matter, with plagioclase montmorillonite and chlorite also reported. Kaolinitization, illitization, and chloritization of micas were observed along with the presence of iron oxides and zircon. After reaction, chlorite was absent and the average pore size had increased.

Wdowin also texturally and mineralogically characterized reservoir sandstones and two caprock cores from target storage sites in the Polish lowlands and reacted them in brine for 3 months at 55°C and a CO_2 pressure of 10–12 MPa [*Wdowin et al.*, 2014b]. Upper Cretaceous limestone caprock from 865 m and Middle Jurassic dolostone from 1019 m were compared. The limestone was predominantly calcite with detrital quartz, mica, illite-smectite, pyrite, and iron oxides. The authors did not report any mineralogical changes after reaction, except for the presence of halite, calcite, and aragonite sediments in the reactor. Average pore diameter however decreased, indicating potential self-sealing. The dolostone contained ferruginous dolomite or ankerite stained by iron oxides with kaolinite, illite, quartz, pyrite, and muscovite. Only small changes were noted after reaction with halite precipitated from brine, a slight increase in pore diameter, and sediments containing calcite, aragonite, strontianite, and nacrite in the reactor.

Various experiments at 80 and 150°C and 1 and 15 MPa CO_2 for 30–365 days were performed on crushed caprock from the Chinle Formation, Utah, outcrop (Fig. 7.8) as an analogue for the Colorado Plateau. The Comblanchien reservoir-caprock at 1980 m from the Paris Basin and purified clay extracts were also assessed [*Credoz et al.*, 2009]. On reaction of the Chinle Formation, pure calcite dissolved, but Fe-Mg substituted calcite was more stable, with subsequent precipitation of a mixed Ca-Fe-Mg carbonate of undetermined composition. Modeling with CRUNCH additionally predicted destabilization or ion exchange of Ca-montmorillonite and illite and the precipitation of kaolinite and Na-montmorillonite.

Carroll and coworkers reacted the Eau Claire shale, the caprock formation of the Archer Daniels Midland (ADM) demonstration site for the Illinois Basin-Decatur Geologic Carbon Sequestration Project, with CO_2-brine at 51°C and 19.5 MPa for 30 days in static vessels [*Carroll et al.*, 2013]. Their core contained quartz, feldspar, and three clays such as kaolinite, a low-Fe illite-smectite, and an Fe-rich clay (modeled as annite). Dissolved concentrations of Si, Fe, and Al increased to about 1.5, 4, and 10 times higher than during reaction of the reservoir rock Mt. Simon Sandstone. They modeled a best fit to the experiments using PHREEQC, with incongruent dissolution of annite, illite, and K-feldspar (microcline) as pH decreased by two units and subsequent formation of montmorillonite, amorphous silica, and kaolinite. Dissolved Fe from the clay minerals had the potential to contribute significantly to mineral trapping through precipitation of Fe carbonates. Importantly, the Fe-rich clay needed significant adjustments to thermodynamic constraints, and incongruent reaction terms were needed to accurately match models to the experimental data, highlighting the need for model calibration. Eau Claire caprock (1542–1550 m) reported to contain quartz, orthoclase, illite, chlorite, and traces of anhydrite and pyrite was also reacted elsewhere at an initially very elevated temperature of 200°C at 30 MPa CO_2-brine for 23 days, and then cooled to 56°C (reservoir condition) for an additional 8 days [*Liu et al.*, 2012]. The corrosion of K-feldspar and precipitation of pore-filling and bridging illite-smectite and siderite near preexisting pyrite was reported. However, the initial temperature of 200°C was very high compared to reservoir conditions, presumably used to increase reaction kinetics.

Gothic shale core from the Paradox Basin, Utah, USA, which caps a carbonate reservoir site of CO_2 EOR and the pilot site Southwest Regional Partnership on Carbon Sequestration, was reacted with CO_2-brine at 25 MPa and an elevated temperature of 160°C to accelerate reactions to simulate 200 years [*Marcon and Kaszuba*, 2015]. The shale core was organic rich, with major quartz, calcite, dolomite, mica, and clay matrix with authigenic pyrite. Powders and fragments of the shale core were reacted 50:50 with calcite, dolomite, and 1% pyrite to simulate the reservoir-seal interface. GWB modeling of the experiments with the core described by 25% illite, 25% smectite, and 10% of each of calcite, dolomite, pyrite, quartz, and mica assisted interpretation. Dissolved Co, Cu, Ni, Pb, V, and Zn concentrations statistically increased during shale CO_2-brine reaction, and Pb, Co, and Cr subsequently decreased. Pb however was already elevated during brine-rock reaction before injecting CO_2. They observed precipitation of As- and Co-containing sulfides, smectite, and vermiculite, with sulfides also saturated in models. At end of the experiments, the authors reported that Fe, Zn, Pb, and Cr still exceeded EPA drinking water limits if assuming brine leakage to overlying freshwater aquifers was possible. The authors pointed out that although the shale is a large source of metals, the reservoir-shale interface also provides a sink through precipitation and adsorption.

Gothic shale from the Aneth unit at 1640 m reported to contain quartz, calcite, dolomite, montmorillonite, and trace pyrite was separately reacted in brine at 75°C and 10 MPa of CO_2 [*Jung et al.*, 2013]. Additional reactions with 1, 4, and 8 vol% O_2 in CO_2 were performed. In the presence of co-injected O_2, pyrite oxidation generated sulfuric acid and lowered (predicted) pH with ~30% higher dissolved concentrations of Ca, Mg, and Sr from increased dissolution of calcite and dolomite. The authors speculated that mobilized U(IV) and U(VI) concentrations were owing to desorption from core in the presence of bicarbonate from dissolved CO_2 (as had been observed elsewhere). Precipitation of gypsum, Fe oxides, and barite were observed. Lower mobilized U concentrations in the presence of O_2 were attributed through modeling with PHREEQC to coprecipitation with Fe oxides rather than adsorption.

Two studies observed more reactivity of caprock clays in the presence of dissolving wellbore cement or carbonate minerals. This may have been owed to Ca-induced ion exchange.

A shale end member from the Tournasin units, Krechba Field (In Salah), Algeria, and separately the shale in the presence of well cement were reacted in CO_2-brine at 95°C and 10 MPa for 31 days after a brine-rock baseline soak [*Carroll et al.*, 2011]. On shale-CO_2-brine reaction, an increase in dissolved Si, Fe, and Al was measured, but no mineral dissolution was directly observed by SEM (Fig. 7.9), only submicron precipitates. Geochemical models using PHREEQC best matched the experimental result with chlorite, illite, albite, quartz, and carbonates partly dissolving and with boehmite, smectite, $Fe(OH)_3$, and amorphous silica precipitation. Siderite and chlorite dissolution both contributed to dissolved Fe. In contrast, in the experiment in the presence of well cement, a high alkalinity and Ca concentration were measured in the brine from cement dissolution, with sheet silicates altered around the edges of sheets, and a Ca carbonate precipitated along with smectite. A wellbore model showed that the cement carbonation dominated reactions. The authors noted that compositional variations of carbonates, chlorite, and illite were not considered in the models owing to a lack of available input data.

Figure 7.9 also shows calcite-cemented caprock core from the Adventdalen Group of the Norway Central Tertiary Basin. On reaction with CO_2-brine at 250°C and 11 MPa for up to 35 days, plagioclase, chlorite, illite, and carbonates dissolved or were converted to smectite in the carbonate-rich shale [*Alemu et al.*, 2011]. New ankerite, calcite, and smectite were precipitated. Modeled fluids were saturated with carbonates and smectite. However, the clay-rich Jannusfillet Subgroup core reacted at 80, 150, and 200°C displayed no dissolution features, and only siderite dissolution was observed at 250°C. Modeled fluids were saturated with carbonates, chlorite, illite, kaolinite, and smectite. The very elevated temperature of 250°C was presumably used to accelerate reaction kinetics.

A limited number of site-scale geochemical models have focused on caprock reactivity with CO_2-brine. Simulations by Gaus suggest that caprock integrity was enhanced by carbonate, chalcedony, and kaolin precipitation over 100 years at Sleipner [*Gaus et al.*, 2005]. Gheradi and coworkers predicted redistribution of calcite within 0.1 m of the reservoir-caprock interface controlling porosity in a carbonate-rich shale of a depleted gas reservoir using TOUGHREACT [*Gherardi et al.*, 2007]. The simulated caprock initially contained 30% calcite and 3% dolomite, with 20% quartz, 20% muscovite, 15% Na-smectite, 6% chlorite, 4% kaolinite, and 3% illite. Calcite was the main reacting mineral, with also dolomite, illite, chlorite, and Na-smectite dissolving and precipitating over 1000 years at 45°C and 105 bar. Illite provided dissolved Fe for precipitation of ankerite and siderite, and dawsonite formation was

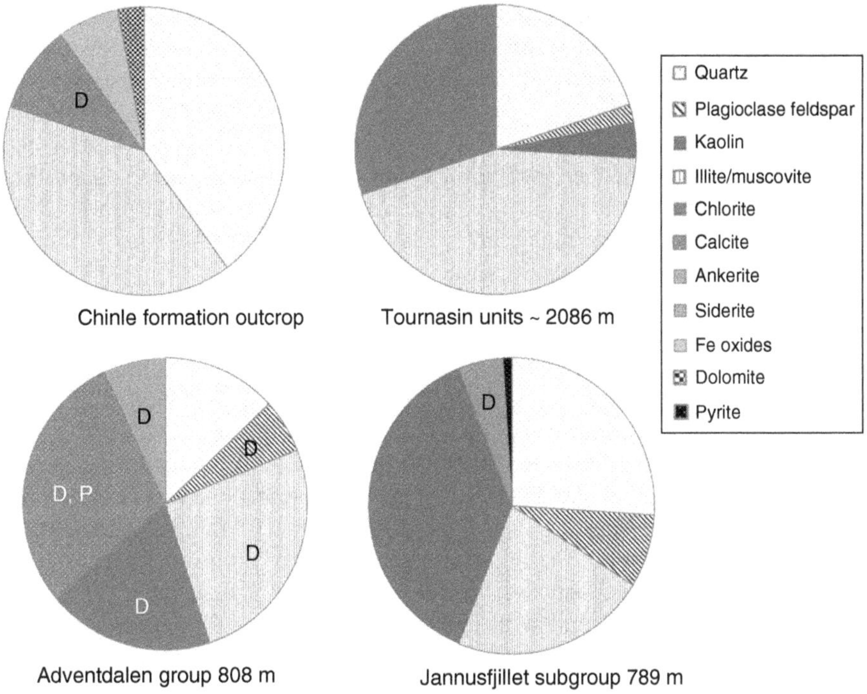

Figure 7.9 Mineralogy of caprock cores from the Chinle Formation outcrop, Utah, USA; the Tournasin units, Krechba Field site (In Salah), Algeria; the Adventdalen Group; and the Jannusfillet Subgroup, Central Tertiary Basin, Norway. Minerals labeled D were directly observed to dissolve during experimental CO_2-brine reaction. P indicates minerals which also precipitated. Note that smectite precipitation was additionally reported after reaction of the Adventdalen Group. Created from data in *Alemu et al.* [2011], *Armitage et al.* [2010], *Carroll et al.* [2011], and *Credoz et al.* [2009].

also predicted. The authors noted that predicted calcite dissolution was most significant where a CO_2-dominated phase migrated quickly into the caprock through fractures or high-porosity zones. Tian and coworkers compared the reactivity of a clay-rich shale with a mudstone caprock of a sandstone reservoir [*Tian et al.*, 2014]. The reservoir and clay-rich shale were based on Jianghan Basin, China, modeled with TOUGHREACT at 47°C and 101 bar. The porosity and permeability of both caprock types were 7.5% and 3×10^{-17} m^2 with physical parameters based on the clay-rich shale to compare only the differences in mineralogy. The mudstone contained mainly quartz and kaolinite with illite, muscovite, and siderite. The clay-rich shale contained mainly illite, with quartz, Ca-smectite, chlorite, gypsum (modeled as anhydrite), albite, K-feldspar, calcite, and pyrite. On CO_2-brine reaction, the pH decreased to 4.3 with the mudstone and was buffered to 4.8 with the clay-rich shale. Sealing occurred ~0.6 m into the caprock from the reservoir-caprock interface, with porosity eventually decreasing in both cases. However, self-sealing occurred at the beginning of the reaction of the mudstone, while for the clay-rich shale porosity increased and only decreased after 100 years. Precipitation of siderite and then subsequently ankerite and magnesite was predicted to decrease porosity for the mudstone reaction. Illite, kaolinite, and muscovite were also reactive minerals contributing to dissolved cations. For the clay-rich shale, anhydrite and illite provided Ca and Mg for precipitation of calcite, magnesite, and Ca-smectite (with kaolinite as an intermediate mineral). Precipitation at the interface of a reservoir and seal was also predicted elsewhere [*Balashov et al.*, 2015]. A shale caprock with mineralogy similar to the Marcellus Formation was simulated over a sandstone reservoir with the program MK76 at 75°C and 300 bar CO_2 with 0.24 ppb O_2. The shale was mainly quartz, illite, and chlorite, with calcite, pyrite, microcline, and 5% porosity. The authors reported the majority of alteration occurred within 1 m of the reservoir and caprock interface. Microcline dissolution and kaolinite replacement by illite occurred along with chlorite and calcite conversion to dolomite and ankerite. A clay-rich carbonate caprock was considered in 1D reactive transport simulations of diffusion and two-phase flow over 1000–10,000 years [*Fleury et al.*, 2010]. Dissolution in the first few decimeters from the caprock base increased porosity from 15 to 16.5%. The two-phase front however did not exceed 10 m infiltration in the most pessimistic simulation, with calcite precipitation reducing porosity.

Note that several natural analogue studies also highlighted the importance of the reservoir and caprock interface region and areas of heterogeneity with respect to carbonate precipitation as discussed earlier.

7.2.2.2. Mineral Dissolution and Precipitation in Carbonate or Evaporite Caprocks

Several studies have been performed on the potential for CO_2 reactivity to limestone, marl, dolomite, or evaporite (anhydrite, gypsum, or halite) caprock core. Mainly, these assessed fracture evolution or reopening, wormhole formation, and resulting changes in fracture permeability.

Fitts and Peters reviewed the potential for CO_2 leakage through caprocks at reactivated fractures and new flow paths formed by mineral reactivity [*Fitts and Peters*, 2013]. They point out that while observations at the Michigan site (and Sleipner site), for example, showed that the Amherstburg Formation (and Nordland Shale) were effective seals, caprock integrity risk assessments have been historically based on physical integrity and hydraulic flow assuming mineral reactivity does not affect integrity. They built a reactive transport simulation taking into account only calcite as a reactive phase as 5, 20, and 50 vol% of a caprock. They showed that permeability could potentially triple over 100 years if calcite dissolution was favorable through a 100 m flow path and highlighted that calcite was the most important mineral in terms of its fast reactivity. The authors also point out that several experimental studies have generated preferential flow paths and increases in surface roughness through mineral reactivity which may introduce complexities not accounted for in conventional fracture flow models.

The Amherstburg Formation is considered as the primary caprock for the Midwest Regional Carbon Sequestration Partnership's project in Otsego County, Michigan, USA. A dolomitic limestone core from the Amherstburg Formation sampled at 928 m depth was artificially fractured [*Ellis et al.*, 2011]. The core consisted of greater than 90% calcite and dolomite in approximately equal amounts. The remaining content was identified as quartz, K-feldspar, 2–5% clays, and pyrite by XRD and EDS mapping. CO_2-acidified brine was flowed through the sample at 10 ml per hour for 7 days at 27°C and 10 MPa. Real-time scans with a medical CT scanner and before and after micro-CT scans showed an increase in the fracture aperture with the greatest increase in the area closest to the outlet. Backscattered electron SEM and EDS analysis revealed that calcite was the main mineral dissolving along the fracture. Unaltered silicate-rich areas remained and this increased the surface roughness. The authors suggested that the increased surface roughness may reduce the hydraulic flow through the fracture and offset the effects of an increased aperture size. Subsequent similar experiments by the same group with reaction over 3 days at an initial flow rate of 3 ml/h revealed an 80% reduction in flow rate and decrease in permeability [*Ellis et al.*, 2013]. Although dissolved Ca in the effluent fluid decreased, no calcite precipitation was directly observed. Instead, calcite dissolution appeared to mobilize less soluble rock particles which were transported in the flow path and caused mechanical closure of the fracture.

A marl sample from an outcrop of the Camino Formation which acts as the caprock of the limestone reservoir of the CIUDEN pilot demonstration in Hontomin, Spain, was recently artificially fractured for flow-through reactions [*Dávila et al.*, 2016]. The sample contained 71% calcite, 10% quartz, 7% illite, 6.5% albite, 2% gypsum, 3% clinochlore (Mg chlorite), 0.5% anhydrite, and 0.2% pyrite, with 7% porosity and before fracturing a permeability of $<10^{-18}$ m^2. Sulfate-rich brine was used to replicate the reservoir groundwater along with two sulfate-free brines at three flow rates of 0.2, 1, and 60 ml/h. Cores were reacted in a flow-through setup with a total pressure of 15 MPa and CO_2 pressure of 6 MPa at 60°C. With the sulfate-free brine, mainly dissolution of calcite and gypsum and minor dissolution of clinochlore and albite occurred. However, with sulfate-rich brines, calcite, clinochlore, and albite dissolved, and gypsum precipitated (Fig. 7.10a). A high-porosity zone was formed along the fracture walls with the remaining non-dissolved silicates. Generally, wormhole formation was observed by micro-CT at higher flow rates, and more uniform dissolution was observed at low flow rates. The fracture permeability did not change significantly with sulfate-free brine. In reactions with sulfate-rich brine, permeability decreased at slow flow rates with gypsum precipitation sealing the fracture. However, permeability increased through predominant calcite dissolution at the highest flow rate. Davila also generally observed some fracture clogging by grain migration as did Ellis and coworkers.

A sub-core was drilled parallel to bedding from the Midale Marly sequence Three Fingers evaporite formation, the lower part of the caprock for the IEA-GHG Weyburn-Midale Monitoring and Storage Project at Saskatchewan, Canada. The evaporite caprock core (which was not fractured) was reacted by core flooding at 60°C and 24.8 MPa confining pressure with a complex brine (near equilibrium with anhydrite and saturated with calcite and dolomite), and subsequently the brine equilibrated with 3 MPa CO_2 for 13 days [*Smith et al.*, 2013]. The core contained variable amounts of mainly anhydrite, dolomite, and porosity in different zones along the sample, with more anhydrite at the inlet side. Minor quartz, K-feldspar, and trace amounts of aluminosilicate clays and pyrite were also identified with total silica content <6%. On addition of CO_2, brine pH decreased by 2 units to 5.2, although this was buffered by carbonate dissolution

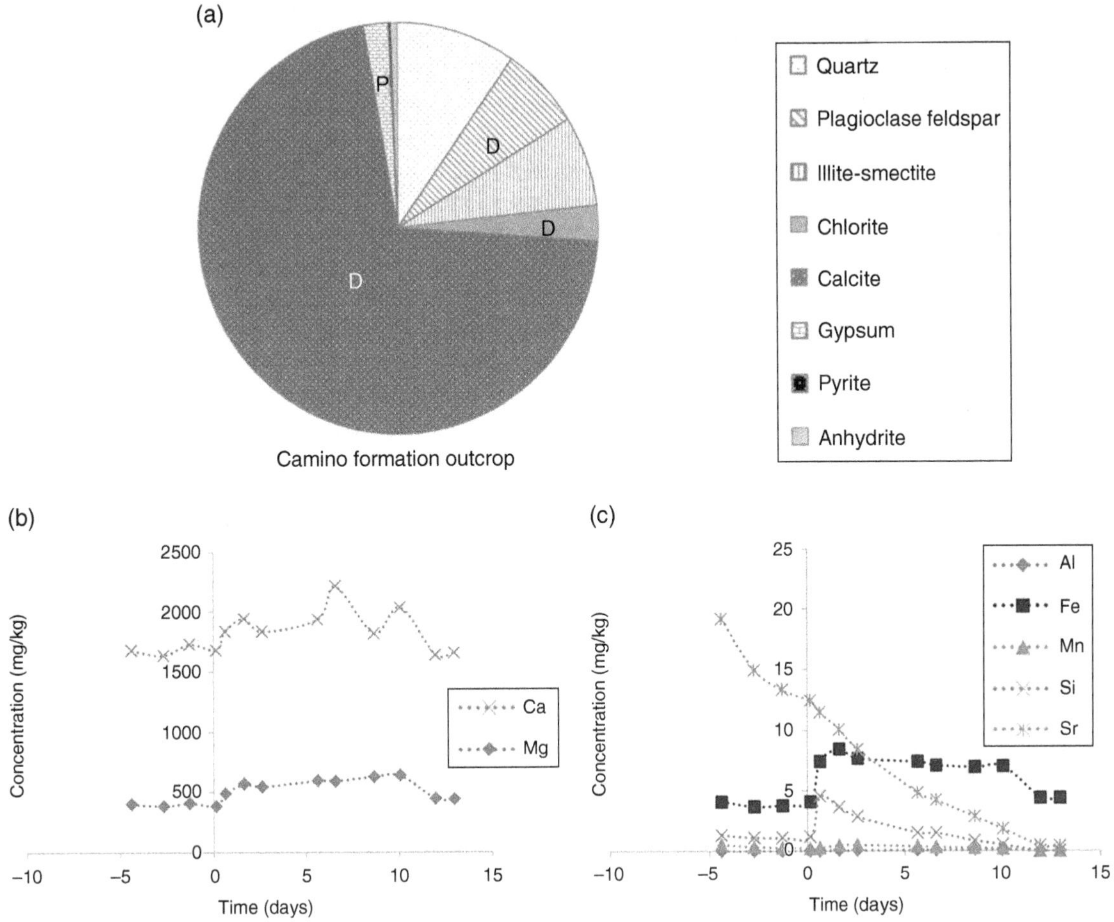

Figure 7.10 (a) Mineralogy of marl caprock from the Camino Formation outcrop, Hontomin, Spain, where chlorite is clinochlore and the plagioclase is albite. Minerals labeled D were observed to dissolve during experimental CO_2-sulfate brine reaction via SEM and water chemistry. P indicates gypsum also precipitated in reactions with sulfate-rich brine. Created from data in *Davila et al.* [*2016*]. (b and c) Water chemistry during brine and CO_2-brine reaction of Three Fingers evaporite core; 0 days denotes when CO_2 was added. Created from data in *Smith et al.* [2013].

when compared to the original CO_2-brine pH of 4.2. Solution chemistry showed that preferential dolomite dissolution occurred with increases in dissolved Ca, Fe, and Mg after CO_2 addition to the brine (Fig. 7.10b and c). However, the elevated sulfate in the initial brine kept anhydrite unreactive. Wormhole formation was observed by micro-CT, with a preferential channel branching out into preexisting microfractures, porosity, and preexisting areas of dolomite aligned with bedding. Flow rate increased gradually from 0.24 ml/h after 6 days' reaction and significantly after 10 days. Permeability increased by approximately three orders of magnitude before wormhole breakthrough. A residual skeleton of anhydrite remained around wormhole edges, although some anhydrite particles and silicate grains were flushed out or trapped in dead-end channels. The authors did not see any evidence of mineral precipitation.

These studies show that geochemical reactivity not only affects permeability through dissolution and precipitation but also affects physical (fines) migration and hence mechanical blocking of fractures. Brine composition and flow rate are important controls on mineral dissolution and precipitation.

7.3. SUMMARY AND CONCLUSIONS

Caprocks often contain minerals more reactive than reservoir rocks. Their low permeabilities however can prevent fluid infiltration and mineral accessibility. The reservoir-caprock interface and fractures are predicted and observed key areas of reactivity. Potentially reactive components include carbonates, clays, sulfates, plagioclase, and, especially in the case of co-injection of impurity gases, sulfide or oxide minerals. CO_2 storage fluid-rock interaction studies will move forward to understand not only dissolution-precipitation processes in whole rocks but also ion exchange and desorption/absorption processes. This will be coupled to predicting the effects on rock mechanics, porosity-permeability, and water quality from injecting CO_2 (plus impurity gases). Fe-rich clays and plagioclase provide cations for mineral trapping, with clays, plagioclase, and carbonates contributing to mobilized metals and nonmetals. The knowledge of clay or plagioclase compositions, for example, albite versus more reactive Ca-containing labradorite, has been shown to be important for model input. Reactive surface areas of clays have also been shown to be important for accurate geochemical modeling predictions. Reactive surface areas will also change over the course of CO_2 injection by, for example, dissolution of calcite cements revealing underlying minerals and increasing surface roughness. Dissolution of carbonates buffers pH, apparently preventing extensive dissolution of silicates. However, liberated ions such as Ca from calcite dissolution can also release metals from clays through inducing ion exchange. These processes and potentially also desorption of metals by bicarbonate, sulfate, or nitrate, along with redox processes, can initially mobilize metals. However, these processes remain poorly understood in the CO_2 storage context.

Stored gas streams from sources such as coal combustion or cement processing may contain variable concentrations of SO_2, NOx, and O_2 which are more reactive than CO_2. Studies have shown higher solution acidification, further dissolution of silicates, changes to physical properties of rocks, and different precipitation products including sulfate and oxide minerals. There remains however limited reactivity studies of caprock including either reactive or inert (e.g., N_2, Ar) impurity gases. Lower purification of gas streams has been proposed as an option to lower capture costs. This would increase the concentration of impurity gases and necessitate assessments of geochemical effects on caprock porosity and permeability and mobilized metals.

7.4. FUTURE DIRECTIONS AND RECOMMENDATIONS

1. Further natural analogue work focused on caprocks, fractures, the reservoir-caprock interface, and the wellbore-caprock interface should be performed and used to constrain mineral precipitation rates and products in geochemical models. Sites containing natural-sulfur-bearing volcanic gases should be included to give further insights into SO_2 co-injection. Characterizing natural deposits such as travertines, those from acid mine drainage, acid gas injection, and weathered outcrop environments would also inform as to the expected reactions under a range of redox conditions.

2. Clay composition specific model input data are clearly still needed. Additionally, inputting detailed caprock mineralogies, for example, trace amounts of carbonate, oxides, and sulfides, as well as correct mineral compositions, for example, Fe-rich or Mg-rich chlorite, albite, or anorthite, improves model predictions. This could be implemented by running a series of models encompassing several possible caprock compositions to provide the level of uncertainly or by upscaling from models validated by site-specific experimental CO_2-fluid-rock data. Detailed core characterization combining several techniques is preferable, for example, QEMSCAN constrained by XRD and SEM-EDS with rock digestion data and microprobe.

3. Estimates of reactive surface areas are often storage site specific and can differ from BET surface areas by orders of magnitude. They can be better estimated or modified by direct inspection of core mineral habits through techniques such as SEM. For example, calcite cement will have a lower reactive surface area than fine-grained calcite or minerals corroded by previous diagenesis. Reactive surface areas change over time with mineral precipitation covering surfaces or dissolution revealing new mineral surfaces and creating surface roughness. Mineral surface accessibility and dynamic changes may be further understood through imaging techniques such as micro-CT.

4. Geochemical (fluid analysis) data assist interpretation of dynamic changes to geomechanical and poro-perm properties of caprock. Reactions not only influence the creation of porosity by, for example, calcite dissolution but can also influence fines migration and mechanical pore clogging. Combined geochemical and geomechanical or poro-perm studies of the effect of CO_2-fluid injection on caprock are needed.

5. Understanding the sources (e.g., clays, oxides, or sulfides) and mode of mobilization of organics, metals, and metalloids such as through dissolution, ion exchange, or desorption from caprock along with their fate is becoming important for predictions of mineral trapping estimates and field site interpretations (especially in low-salinity aquifers or where drinking water aquifers overlie storage sites). Assessments should include where appropriate impurity gases which affect system acidity, redox, and so on, and which form anions (e.g., sulfate, nitrate) that in themselves affect mineral reactions.

6. There are still a limited number of caprock reactivity studies using impure CO_2 streams including inert gases such as N_2 and Ar along with reactive gases such as SO_2, O_2, and H_2S. Reactions including the impurities NO and NO_2 are particularly lacking. Generally, more work is also needed on the dissolution and speciation of impurity gases at CCS conditions for validation of model predictions.

7. Further work is still needed on caprock reactivity with wet $scCO_2$ and also with high rock- water ratio conditions

more representative of the subsurface. This may be especially required in situations where a buoyant gas cap (plus less soluble impurity gases) can contact caprock.

ACKNOWLEDGMENTS

The authors thank Sue Golding and Susan Farquhar and also acknowledge many fruitful years of collaboration with the groups of Victor Rudolf, Dirk Kirste, and Alexandra Golab. Two anonymous reviewers are thanked for their comments which substantially improved this chapter.

REFERENCES

Alemu, B. L., P. Aagaard, I. A. Munz, and E. Skurtveit (2011), Caprock interaction with CO_2: a laboratory study of reactivity of shale with supercritical CO_2 and brine, *Appl. Geochem.*, *26*(12), 1975–1989.

Armitage, P. J., R. H. Worden, D. R. Faulkner, A. C. Aplin, A. R. Butcher, and J. Iliffe (2010), Diagenetic and sedimentary controls on porosity in lower carboniferous fine-grained lithologies, krechba field, algeria: a petrological study of a caprock to a carbon capture site, *Mar. Pet. Geol.*, *27*, 1395–1410.

Bacon, D. H., and E. M. Murphy (2011), Managing chemistry underground: is co-sequestration an option in selected formations?, *Energy Procedia*, *4*, 4457–4464.

Balashov, V. N., G. D. Guthrie, C. L. Lopano, J. A. Hakala, and S. L. Brantley (2015), Reaction and diffusion at the reservoir/shale interface during CO_2 storage: impact of geochemical kinetics, *Appl. Geochem.*, *61*, 119–131.

Bickle, M. J., N. Kampman, and M. Wigley (2013), Natural analogues, in *Geochemistry of Geologic Carbon Sequestration*, edited by D. J. DePaolo, D. R. Cole, A. Navrotsky, and I. C. Bourg, vol. *77(1)*, pp. 15–71, GeoScienceWorld, Chapter in Reviews in Mineralogy and Geochemistry.

Black, J. R., and R. R. Haese (2014), Chlorite dissolution rates under CO_2 saturated conditions from 50 to 120°C and 120 to 200 bar CO_2, *Geochim. Cosmochim. Acta*, *125*, 225–240.

Black, J. R., S. A. Carroll, and R. R. Haese (2014), Rates of mineral dissolution under CO_2 storage conditions, *Chem. Geol.*, doi:10.1016/j.chemgeo.2014.09.020(0).

Busch, A., A. Amann-Hildenbrand, P. Bertier, M. Waschbuesch, and B. M. Krooss (2010), The significance of caprock sealing integrity for CO_2 storage, in *SPE International Conference on CO2 Capture, Storage, and Utilization*, SPE-139588-MS, Society of Petroleum Engineers, New Orleans, 10–12 November.

Carroll, S. A., W. W. McNab, and S. C. Torres (2011), Experimental study of cement – sandstone/shale – brine – CO_2 interactions, *Geochem. Trans.*, *12*(1), 1–19.

Carroll, S. A., W. W. McNab, Z. Dai, and S. C. Torres (2013), Reactivity of mount Simon sandstone and the Eau Claire shale under CO_2 storage conditions, *Environ. Sci. Technol.*, *47*(1), 252–261.

Chen, F., A. V. Turchyn, N. Kampman, D. Hodell, F. Gázquez, A. Maskell, and M. Bickle (2016), Isotopic analysis of sulfur cycling and gypsum vein formation in a natural CO_2 reservoir, *Chem. Geol.*, *436*, 72–83.

Coote, S. M. (1984), Stratigraphic drilling report – GSQ Charleville 1., *Queensland Gov. Min. J.*, *85*(997), 418–426.

Credoz, A., O. Bildstein, M. Jullien, J. Raynal, J.-C. Pétronin, M. Lillo, C. Pozo, and G. Geniaut (2009), Experimental and modeling study of geochemical reactivity between clayey caprocks and CO_2 in geological storage conditions, *Energy Procedia*, *1*(1), 3445–3452.

Dávila, G., L. Luquot, J. M. Soler, and J. Cama (2016), Interaction between a fractured marl caprock and CO_2-rich sulfate solution under supercritical CO_2 conditions, *Int. J. Greenh. Gas Control*, *48*, 105–119.

Dawson, G. K. W., C. Khan, S. M. Farquhar, D. Biddle, X. Jiang, J. K. Pearce, V. Rudolph, and S. D. Golding (2013), Initial batch geochemistry and first 4D geomechanical and permeability experiment results, ANLEC Project 3-1110-0101 Report.

Ellis, B., C. Peters, J. Fitts, G. Bromhal, D. McIntyre, R. Warzinski, and E. Rosenbaum (2011), Deterioration of a fractured carbonate caprock exposed to CO_2-acidified brine flow, *Greenhouse Gases Sci. Technol.*, *1*(3), 248–260.

Ellis, B., J. Fitts, G. Bromhal, D. McIntyre, R. Tappero, and C. Peters (2013), Dissolution-driven permeability reduction of a fractured carbonate caprock, *Environ. Eng. Sci.*, *30*(4), 187–193.

Farquhar, S. M., G. K. W. Dawson, J. S. Esterle, and S. D. Golding (2013), Mineralogical characterisation of a potential reservoir system for CO_2 sequestration in the Surat Basin, *Aust. J. Earth Sci.*, *60*(1), 91–110.

Farquhar, S. M., J. K. Pearce, G. K. W. Dawson, A. Golab, D. Kirste, D. Biddle, and S. D. Golding (2015), A fresh approach to investigating CO_2 storage: experimental CO_2-water-rock interactions in a freshwater reservoir system, *Chem. Geol.*, *399*(Measuring and predicting the geochemical impacts of CO_2 storage on reservoir rocks), 98–122.

Farquhar, S (2016),CO_2–Water–Rock Interactions in Low-Salinity Reservoir Systems . Ph.D. Thesis, University of Queensland, Brisbane, Australia.

Fischer, S., A. Liebscher, M. De Lucia, and L. Hecht (2013), Reactivity of sandstone and siltstone samples from the Ketzin pilot CO_2 storage site-laboratory experiments and reactive geochemical modeling, *Environ. Earth Sci.*, *70*(8), 3687–3708.

Fitts, J., and C. A. Peters (2013), Caprock fracture dissolution and CO_2 leakage, in *Rev. Mineral. Geochem.*, 459–479.

Fleury, M., et al. (2010), Evaluating sealing efficiency of Caprocks for CO_2 storage: an overview of the geocarbone-integrity program and results, *Oil Gas Sci. Technol. Rev. IFP*, *65*(3), 435–444.

Förster, A., R. Schöner, H. J. Förster, B. Norden, A. W. Blaschke, J. Luckert, G. Beutler, R. Gaupp, and D. Rhede (2010), Reservoir characterization of a CO_2 storage aquifer: the upper Triassic Stuttgart formation in the northeast German Basin, *Mar. Pet. Geol.*, *27*(10), 2156–2172.

Frank, A. B., and D. Kirste (2014), SO_2 co-injection during geologic CO_2 sequestration: modelling the geochemical impact, in *CO2CRC Research Symposium*, CO2CRC, Torquay, Victoria, Australia, 25–26 November 2014.

Garcia, D. J., H. Shao, Y. Hu, J. R. Ray, and Y.-S. Jun (2012a), Supercritical CO_2-brine induced dissolution, swelling, and secondary mineral formation on phlogopite surfaces at 75-95°C and 75 atm, *Energy Environ. Sci.*, *5*(2), 5758–5767.

Garcia, S., R. J. Rosenbauer, J. Palandri, and M. M. Maroto-Valer (2012b), Sequestration of non-pure carbon dioxide streams in iron oxyhydroxide-containing saline repositories, *Int. J. Greenh. Gas Control*, *7*, 89–97.

Gaus, I., M. Azaroual, and I. Czernichowski-Lauriol (2005), Reactive transport modelling of the impact of CO_2 injection on the clayey cap rock at Sleipner (North Sea), *Chem. Geol.*, *217*(3–4), 319–337.

Gherardi, F., T. F. Xu, and K. Pruess (2007), Numerical modeling of self-limiting and self-enhancing caprock alteration induced by CO_2 storage in a depleted gas reservoir, *Chem. Geol.*, *244*(1–2), 103–129.

Gibson-Poole, C. M., L. Svendsen, J. Underschultz, M. N. Watson, J. Ennis-King, P. J. van Ruth, E. J. Nelson, R. F. Daniel, and Y. Cinar (2008), Site characterisation of a basin-scale CO_2 geological storage system: Gippsland Basin, southeast Australia, *Environ. Geol.*, *54*(8), 1583–1606.

Glezakou, V.-A., B. Peter McGrail, and H. Todd Schaef (2012), Molecular interactions of SO_2 with carbonate minerals under co-sequestration conditions: a combined experimental and theoretical study, *Geochim. Cosmochim. Acta*, *92*, 265–274.

Golab, A., S. Sommacal, C. Goodwin, J. Khor, A. Carnerup, N. Dodd, G. Dawson, and S. Golding (2014), Technical report for Sub-project 5: Undertake time series (4D) imaging and conventional experimental studies to measure geochemical reactivity and dissolution trapping capacity of core material using supercritical CO_2, Lithicon Australia, Report for ANLEC R&D.

Golab, A., A. Arena, J. Khor, C. Goodwin, B. Young, A. Carnerup, and F. Hussain (2015a), Milestone 1.4 Final report of RCA and SCAL data on plugs from West Wandoan-1 Well, FEI Lithicon, Report for ANLEC R&D.

Golab, A., et al. (2015b), Milestone 2.9: final report of digital core analysis results for plug samples from west Wandoan-1 well, FEI-Lithicon, Report for ANLEC R&D.

Griffith, C., D. Dzombak, and G. Lowry (2011), Physical and chemical characteristics of potential seal strata in regions considered for demonstrating geological saline CO_2 sequestration, *Environ. Earth Sci.*, *64*(4), 925–948.

Hangx, S., A. van der Linden, F. Marcelis, and A. Bauer (2013), The effect of CO_2 on the mechanical properties of the captain sandstone: geological storage of CO_2 at the goldeneye field (UK), *Int. J. Greenh. Gas Control*, *19*, 609–619.

Hellevang, H., P. Aagaard, E. H. Oelkers, and B. Kvamme (2005), Can Dawsonite permanently trap CO_2?, *Environ. Sci. Technol.*, *39*, 8281–9287.

Hellevang, H., J. Declercq, and P. Aagaard (2011), Why is Dawsonite absent in CO_2 charged reservoirs?, *Oil Gas Sci. Technol. Rev. IFP*, *66*(1), 119–135.

Higgs, K. E., R. H. Funnell, and A. G. Reyes (2013), Changes in reservoir heterogeneity and quality as a response to high partial pressures of CO_2 in a gas reservoir, New Zealand, *Mar. Pet. Geol.*, *48*, 293–322.

Higgs, K. E., R. R. Haese, S. D. Golding, U. Schacht, and M. Watson (2015), The Pretty Hill formation as a natural analogue for CO_2 storage; an investigation of mineralogical and isotopic changes associated with sandstones exposed to low, intermediate and high CO_2 concentrations over geological time, *Chem. Geol.*, *399*(Measuring and predicting the geochemical impacts of CO_2 storage on reservoir rocks), 36–64.

Hodgkinson, J., and M. Grigorescu (2012), Background research for selection of potential geostorage targets – case studies from the Surat Basin, Queensland, *Aust. J. Earth Sci.*, *60*(1), 71–89.

Horner, K. N., U. Schacht, and R. R. Haese (2014), Characterizing long term CO_2-water-rock reaction pathways to identify tracers of CO_2 migration during geological storage in a low-salinity siliciclastic reservoir system, *Chem. Geol.*, doi:10.1016/j.chemgeo.2014.09.021.

Jenkins, C. R., et al. (2012), Safe storage and effective monitoring of CO_2 in depleted gas fields, *Proc. Natl. Acad. Sci.*, *109*(2), E35–E41.

Jung, H. B., W. Um, and K. J. Cantrell (2013), Effect of oxygen co-injected with carbon dioxide on gothic shale caprock–CO_2–brine interaction during geologic carbon sequestration, *Chem. Geol.*, *354*, 1–14.

Kaszuba, J. P., D. R. Janecky, and M. G. Snow (2003), Carbon dioxide reaction processes in a model brine aquifer at 200°C and 200 bars: implications for geologic sequestration of carbon, *Appl. Geochem.*, *18*(7), 1065–1080.

Kaszuba, J. P., D. R. Janecky, and M. G. Snow (2005), Experimental evaluation of mixed fluid reactions between supercritical carbon dioxide and NaCl brine: relevance to the integrity of a geologic carbon repository, *Chem. Geol.*, *217*(3–4), 277–293.

Kaszuba, J. P., B. Yardley, and M. Andreani (2013), Experimental perspectives of mineral dissolution and precipitation due to carbon dioxide-water-rock interactions, *Rev. Mineral. Geochem.*, *77*(1), 153–188.

Kellett, J. R., B. M. Radke, T. R. Ransley, J. G. Bell, and G. A. Stewart (2012), Hydrogeological framework, in *Water Resource Assessment for the Surat Region, A Report to the Australian Government from the CSIRO Great Artesian Basin Water Resource Assessment*, edited by B. D. Smerdon and T. R. Ransley, pp. 67–92, CSIRO Water for a Healthy Country Flagship, Australia.

Kharaka, Y. K., D. R. Cole, S. D. Hovorka, W. D. Gunter, K. G. Knauss, and B. M. Freifeld (2006), Gas-water-rock interactions in Frio formation following CO_2 injection: implications for the storage of greenhouse gases in sedimentary basins, *Geology*, *34*(7), 577–580.

Kirste, D. M., J. K. Pearce, S. Golding, and A. Frank (2015), Reactive transport modelling of CO_2 with SO_2 and O_2 as impurities for geological storage: Upscaling from the bench-top to the reservoir, paper presented at Goldschmidt, Prague, CZ, 16–21 August 2015.

Lamy-Chappuis, B., D. Angus, Q. Fisher, C. Grattoni, and B. W. D. Yardley (2014), Rapid porosity and permeability changes of calcareous sandstone due to CO_2-enriched brine injection, *Geophys. Res. Lett.*, *41*(2), 399–406.

Liu, F., P. Lu, C. Griffith, S. W. Hedges, Y. Soong, H. Hellevang, and C. Zhu (2012), CO_2–brine–caprock interaction: reactivity experiments on Eau Claire shale and a review of relevant literature, *Int. J. Greenh. Gas Control*, *7*, 153–167.

Marcon, V., and J. P. Kaszuba (2015), Carbon dioxide–brine–rock interactions in a carbonate reservoir capped by shale: experimental insights regarding the evolution of trace metals, *Geochim. Cosmochim. Acta*, *168*, 22–42.

Min, Y., J. D. Kubicki, and Y.-S. Jun (2015), Plagioclase dissolution during CO_2–SO_2 Cosequestration: effects of sulfate, *Environ. Sci. Technol.*, *49*(3), 1946–1954.

Moore, J., M. Adams, R. Allis, S. Lutz, and S. Rauzi (2005), Mineralogical and geochemical consequences of the long-term presence of CO_2 in natural reservoirs: an example from the Springerville–St. Johns field, Arizona, and New Mexico, U.S.A, *Chem. Geol.*, *217*(3–4), 365–385.

Palandri, J. L., and Y. K. Kharaka (2004), A compilation of rate parameters of water-mineral interaction kinetics for application to geochemical modeling, USGS Open File Report 2004-1068, 64 pp.

Pearce, J. K., G. K. W. Dawson, S. M. Farquhar, and S. Golding (2013), SO_2 and O_2 co-injection with potential carbon storage target sandstone from a fresh-water aquifer, *Mineral. Mag.*, *77*(5), 551–635.

Pearce, J. K., G. Dawson, A. Golab, L. Knuefing, and S. Golding (2015a), Geochemical CO_2-SO_2-O_2-fluid-rock interactions, paper presented at Goldschmidt, Prague, CZ, 16–21 August.

Pearce, J. K., D. M. Kirste, G. K. W. Dawson, S. M. Farquhar, D. Biddle, S. Golding, and V. Rudolph (2015b), SO_2 impurity impacts on experimental and simulated CO_2-water-reservoir rock reactions at carbon storage conditions, *Chem. Geol.*, *399*(Measuring and predicting the geochemical impacts of CO_2 storage on reservoir rocks), 65–86.

Pudlo, D., S. Henkel, V. Reitenbach, D. Albrecht, F. Enzmann, K. Heister, G. Pronk, L. Ganzer, and R. Gaupp (2015), The chemical dissolution and physical migration of minerals induced during CO_2 laboratory experiments: their relevance for reservoir quality, *Environ. Earth Sci.*, *73*(11), 7029–7042.

Schaef, H. T., B. P. McGrail, and A. T. Owen (2011), Basalt reactivity variability with reservoir depth in supercritical CO_2 and aqueous phases, *Energy Procedia*, *4*, 4977–4984.

Shao, H., J. R. Ray, and Y.-S. Jun (2011), Effects of salinity and the extent of water on supercritical CO_2-induced phlogopite dissolution and secondary mineral formation, *Environ. Sci. Technol.*, *45*(4), 1737–1743.

Shevalier, M., M. Nightingale, B. Mayer, I. Hutcheon, K. Durocher, and E. Perkins (2013), Brine geochemistry changes induced by CO_2 injection observed over a 10 year period in the Weyburn oil field, *Int. J. Greenhouse Gas Control*, *16*(Supplement 1), S160–S176.

Shukla, R., P. Ranjith, A. Haque, and X. Choi (2010), A review of studies on CO_2 sequestration and caprock integrity, *Fuel*, *89*(10), 2651–2664.

Sigfusson, B., et al. (2015), Solving the carbon-dioxide buoyancy challenge: the design and field testing of a dissolved CO_2 injection system, *Int. J. Greenh. Gas Control*, *37*, 213–219.

Smith, M. M., Y. Sholokhova, Y. Hao, and S. A. Carroll (2013), Evaporite caprock integrity: an experimental study of reactive mineralogy and pore-scale heterogeneity during brine-CO_2 exposure, *Environ. Sci. Technol.*, *47*(1), 262–268.

Talman, S. (2015), Subsurface geochemical fate and effects of impurities contained in a CO_2 stream injected into a deep saline aquifer: what is known, *Int. J. Greenh. Gas Control*, *40*, 267–291.

Tian, H., T. Xu, F. Wang, V. V. Patil, Y. Sun, and G. Yue (2014), A numerical study of mineral alteration and self-sealing efficiency of a caprock for CO_2 geological storage, *Acta Geotech.*, *9*(1), 87–100.

Uysal, I. T., S. D. Golding, R. Bolhar, J.-x. Zhao, Y.-x. Feng, K. A. Baublys, and A. Greig (2011), CO_2 degassing and trapping during hydrothermal cycles related to Gondwana rifting in eastern Australia, *Geochim. Cosmochim. Acta*, *75*(19), 5444–5466.

Viswanathan, H., Z. Dai, C. Lopano, E. Keating, J. A. Hakala, K. G. Scheckel, L. Zheng, G. D. Guthrie, and R. Pawar (2012), Developing a robust geochemical and reactive transport model to evaluate possible sources of arsenic at the CO_2 sequestration natural analog site in Chimayo, New Mexico. *Int. J. Greenh. Gas Control*, *10*, 199–214.

Wang, X., V. Alvarado, N. Swoboda-Colberg, and J. P. Kaszuba (2013), Reactivity of dolomite in water-saturated supercritical carbon dioxide: significance for carbon capture and storage and for enhanced oil and gas recovery, *Energy Convers. Manag.*, *65*, 564–573.

Watson, M., and C. M. Gibson-Poole (2005), Reservoir selection for optimised geological injection and storage of carbon dioxide: a combined geochemical and stratigraphic perspective, in *Fourth Annual Conference on Carbon Capture and Sequestration DOE/NETL*, Virginia, 2–5 May 2005.

Watson, M. N., N. Zwingmann, and N. M. Lemon (2004), The Ladbroke Grove-Katnook carbon dioxide natural laboratory: a recent CO_2 accumulation in a lithic sandstone reservoir, *Energy*, *29*(9–10), 1457–1466.

Wdowin, M., R. Tarkowski, and M. Manecki (2013), Petrographic-mineralogical and textural changes in reservoir and sealing rocks (Zaosie anticline) as a result of a long-term experiment in CO_2-brine-rock interactions, *Gospod Surowcami Min– Miner. Resour. Manag.*, *29*(4), 137.

Wdowin, M., R. Tarkowski, and W. Franus (2014a), Determination of changes in the reservoir and cap rocks of the Chabowo anticline caused by CO_2–brine–rock interactions, *Int. J. Coal Geol.*, *130*, 79–88.

Wdowin, M., R. Tarkowski, and W. Franus (2014b), Supplementary studies of textural and mineralogical changes in reservoir and caprocks from selected potential sites suitable for underground CO_2 storage, *Arab. J. Sci. Eng.*, *39*(1), 295–309.

Wilke, F. D. H., M. Vásquez, T. Wiersberg, R. Naumann, and J. Erzinger (2012), On the interaction of pure and impure supercritical CO_2 with rock forming minerals in saline aquifers: an experimental geochemical approach, *Appl. Geochem.*, *27*(8), 1615–1622.

Wollenweber, J., S. a. Alles, A. Kronimus, A. Busch, H. Stanjek, and B. M. Krooss (2009), Caprock and overburden processes in geological CO_2 storage: an experimental study on sealing efficiency and mineral alterations, *Energy Procedia*, *1*(1), 3469–3476.

Xu, T. F., J. A. Apps, and K. Pruess (2005), Mineral sequestration of carbon dioxide in a sandstone-shale system, *Chem. Geol.*, *217*(3–4), 295–318.

Ziolkowski, V., J. Hodgkinson, M. McKillop, M. Grigorescu, and J. L. McKellar (2014), Sequence stratigraphic analysis of the lower Jurassic succession in the Surat Basin, Queensland – preliminary findings, Queensland Minerals and Energy Review Series, Department of Natural Resources and Mines, Queensland.

8

Fluid-Rock Interactions in Clay-Rich Seals: Impact on Transport and Mechanical Properties

Elin Skurtveit[1,2], Rohaldin Miri[2], and Helge Hellevang[2,3]

ABSTRACT

Fluid-rock interaction in low-permeable clay-rich seal units is an important topic for the evaluation of the long-term seal integrity during geological storage of CO_2. In low-permeable sealing units, the diffusion of CO_2 into the matrix is a slow process, and studies of CO_2-initiated fluid-rock interaction in seals are challenging. In this paper, we present an overview of CO_2 transport mechanism and fluid-rock interaction processes that might alter mechanical and transport properties of seals. The review includes theoretical considerations and simulations, experimentally demonstrated processes, and field examples of flow and fluid-rock interaction in intact clay-rich seals as well as for fractures. For clay-rich seals dominated by minerals like quartz, illite, and smectite, the reactivity due to drop in pH is found to be low, and most reaction observed is found to involve calcite. Only minor porosity changes are observed, and implications for flow and CO_2 transport are uncertain due to limited data available. Swelling and shrinking property of smectites due to CO_2 sorption and CO_2 alterations within fractures in clay-rich seal is hardly addressed in the literature.

8.1. INTRODUCTION

The interdependency between fluid-rock interactions, fluid flow, and mechanical properties of seals is an important aspect for understanding the long-term effects and risk of CO_2 sequestration. The requirement of a good seal is to keep CO_2 from migrating out of the reservoir, but unforeseen events of damage to the seal integrity, changing pressure conditions, or the need for remediation of damaged seals requires thorough understanding of fluid-rock processes within sealing units and their impact on fluid transport and mechanical properties. Clay-rich seals defined as shales and mudrock, characterized by their high content of clay minerals and small pore throats, provide good seals for CO_2 storage due to the high entry pressure and corresponding high capillary sealing. Typical properties of sealing units summarized by *Nordgård Bolås et al.* [2005] show pore throat diameter ranging from 9 to 1000 nm and corresponding permeabilities of $<10^{-6}$ and up to 10^{-3} mD (10^{-21}–10^{-18} m^2). This covers the typical range of seal properties for existing CO_2 storage sites: for example, caprock permeability for the In Salah storage site is reported to be of the order of 10^{-22}–10^{-20} m^2 [*Armitage et al.*, 2011]; at Sleipner, the pore throat radius is reported to be in the range of 14–40 nm and brine permeability in the range of $3–10 \times 10^{-19}$ m^2 [*Harrington et al.*, 2009]; and caprock for the Snøhvit CO_2 storage site is reported to have an average permeability of 1–23 mD (10^{-16} m^2) with many low-permeable shale layers [*Chiaramonte et al.*, 2015].

CO_2 transport mechanisms and alteration processes discussed in this paper are summarized in Figure 8.1, showing increasing transport rates to the right. For an intact clay-rich sealing unit, the molecular-scale diffusion in pore water is typically a very slow process, whereas displacement-controlled volumetric Darcy flow in a connected pore

[1]Norwegian Geotechnical Institute (NGI), Oslo, Norway
[2]Department of Geosciences, University of Oslo, Oslo, Norway
[3]The University Centre in Svalbard (UNIS), Longyearbyen, Norway

Geological Carbon Storage: Subsurface Seals and Caprock Integrity, Geophysical Monograph 238, First Edition.
Edited by Stéphanie Vialle, Jonathan Ajo-Franklin, and J. William Carey.

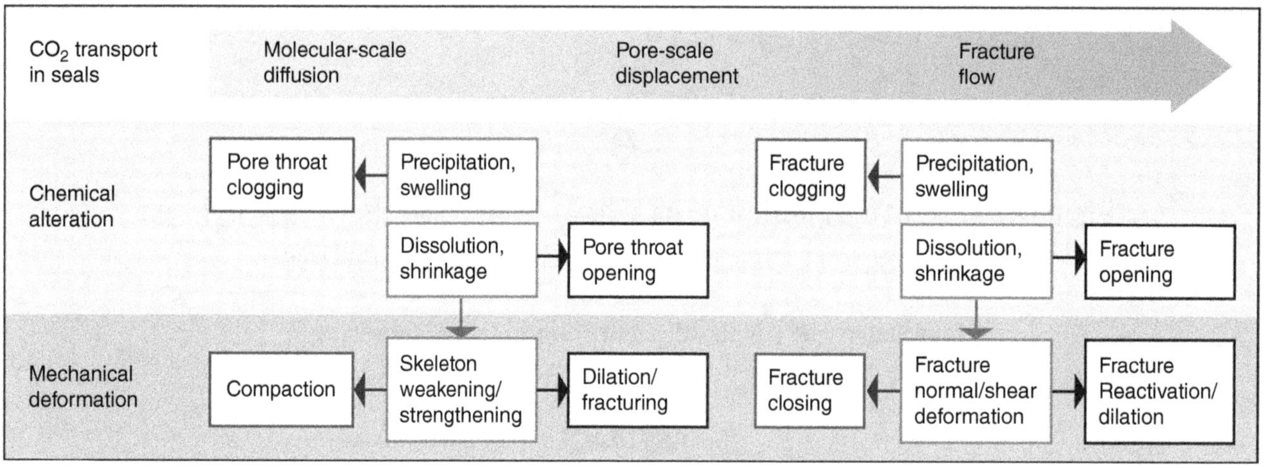

Figure 8.1 CO_2 transport mechanisms with increasing transport rate to the right and schematic illustration of fluid-rock interaction mechanism and their implications for transport of CO_2 through low-permeable clay-rich units.

Figure 8.2 Field example of fluid-rock interaction showing fracture parallel bleaching in permeable siltstone unit within the Entrada Formation, Utah. The bleached zone is found as a 1–10 cm thick zone along the fracture.

network is related to exceeding the capillary entry pressure [*Song and Zhang*, 2012] and may provide higher transport rates (Fig. 8.1). Further, outcrop-scale fractures are observed to provide conduits for migration of CO_2-saturated water (Fig. 8.2) in field outcrops like Green River, Utah, USA [e.g., *Kampman et al.*, 2012; *Shipton et al.*, 2004]. The role of microfractures in shales is less well understood [*Ougier-Simonin et al.*, 2016], providing a transition between pore-scale displacement and observed outcrop-scale fracture transport.

Fluid-rock interaction within a diffusive CO_2 front can be observed for CO_2 reservoir-seal boundaries [*Lu et al.*, 2009] and along faults and fractures within a leaking CO_2 system [e.g., *Shipton et al.*, 2004; *Ogata et al.*, 2014]. Dynamic interaction processes between fluid and rock involve processes of mineral dissolution, precipitation, and sorption. These processes may influence the transport properties for CO_2 in a sealing unit directly or cause alterations of the mechanical properties that induce deformation and changes in the transport properties (Fig. 8.1). In order to understand the complex interplay between flow, transport, chemical reactions, and mechanical changes, comparison of observations from various approaches is useful. Laboratory experiments give valuable input on the understanding of CO_2 transport mechanisms [e.g., *Wollenweber et al.*, 2010; *Skurtveit et al.*, 2012] and the fluid-rock reaction potential [e.g., *Alemu et al.*, 2011; *Liu et al.*, 2012; *Szabó et al.*, 2016] within laboratory limitations in time and conditions. Natural CO_2 field analogues can provide useful information about the long-term effects of materials being exposed to CO_2 or CO_2-enriched fluids and allow for back-calculation of reaction systems and rates [e.g., *Lu et al.*, 2009; *Kampman et al.*, 2014a, 2014b]. Geochemical kinetics of a CO_2-fluid-rock system can be simulated using reactive models [e.g., *Pham et al.*, 2011; *Hellevang and Aagaard*, 2013; *Hellevang et al.*, 2013; *Balashov et al.*, 2015; *Szabó et al.*, 2016] and provide information of long-term reaction potential in a system. Slow processes are challenging to quantify, and calibration of laboratory-determined transport and reaction rates with simulation and field observations is therefore critical to get a realistic model for the dynamic processes and implications for safe storage of CO_2 [*Kampman et al.*, 2014b]. Chemical interactions between injected CO_2, in situ fluids, and the sealing unit might be positive for the seal integrity if they act to limit the fluid transport [e.g., *Balashov et al.*, 2015] and reduce the potential for mechanical failure. On the other hand, fluid-rock interactions might also in some cases enhance the potential of fracturing of shales and increase the permeability [e.g., *Gherardi et al.*, 2007; *Armitage et al.*, 2013].

The occurrence of natural gas trapped under mudstone and shale structural or stratigraphic seals for millions of years is good evidence for the long-term integrity of this type of rock [*Van der Meer*, 2005]. Storage of CO_2 however differs from that of natural gas for a couple of reasons: First, natural gas is quite unreactive as a separate fluid phase or when dissolved in brines. CO_2, on the other hand, forms weak carbonic acid, lowering the pH typically to values between 3 and 5. The highest is observed when carbonate mineral dissolution buffers the pH change [*Pham et al.*, 2011]. This pH drop leads to dissolution of silicate minerals present in the reservoir and the precipitation of secondary silicates (typically kaolinite and silica polymorphs) and secondary carbonates, mostly FeMgCa carbonates, and in some special cases, models predict dawsonite [*Pham et al.*, 2011; *Hellevang and Aagaard*, 2013; *Hellevang et al.*, 2013], although debated as it is rare in natural analogues. Second, the aqueous solubility of natural gas and CO_2 are different, and this affects their potential to diffuse into the seal. The solubility of natural gas components in brine, at the same temperature and pressure, is orders of magnitude lower than for CO_2 [*Miri et al.*, 2014b]. In addition, gas solubility depends on the partial pressure rather than the total pressure, and because natural gas is generally a multicomponent mixture [*McCain*, 1990], their partial pressures may be quite low. CO_2, on the other hand, is injected as a nearly pure phase, and the partial pressure will be close to the total pressure. Third, CO_2 will typically be in a supercritical state, with high density and low viscosity.

The solubility of molecular CO_2 is controlled by the fluid pressure and temperature and the salinity of the aqueous solution and can be generally expressed as [*Hellevang*, 2006]

$$x_{CO_2} = \frac{P y_{CO_2} f_i}{K_H \gamma} \exp\left\{-\frac{v}{RT}\left(P - P_{sat}\right)\right\}, \tag{8.1}$$

where x and y denote molar fractions of CO_2 in the aqueous and gas/supercritical phases, respectively; P is the total pressure; fi and γ are the fugacity and activity coefficients for CO_2, respectively; K_H is Henry's law constant; v is the molar volume of CO_2; R is the universal gas constant; T is the absolute temperature; and subscript sat refers to the saturation pressure. It is not straightforward to see how the solubility is affected by temperature, pressure, and salinity, since both the fugacity and activity coefficients and Henry's law constant are temperature and pressure dependent. For typical CO_2 storage depths, the solubility increases with pressure and decreases with temperature, whereas the activity coefficient of CO_2 is unity for freshwater and increases with ionic strength, and dissolved salts therefore reduce the CO_2 solubility [e.g., *Spycher and Pruess*, 2005; *Miri et al.*, 2014a]. Equation (8.1) and Henry's law only dictate the solubility of molecular CO_2 (hereafter referred to as $CO_{2,aq}$), whereas the total solubility (TIC) also depends on pH and the speciation of carbon into carbonate (CO_3^{2-}) and bicarbonate (HCO_3^-) ions. However, if speciation is ignored and pH is considered low (3–5), typical for normal CO_2 storage settings [*Pham et al.*, 2011], the TIC can be well approximated by Eq. (8.1).

The CO_2 injected into a reservoir is expected to be supercritical CO_2 in the near vicinity of the injection well, surrounded by a mixing zone of supercritical CO_2 and formation water (brine) with a buffered pH. Further out from the injection, the CO_2 plume is expected to be fully saturated with water and acidified, whereas at the reservoir-caprock interface, the CO_2 plume is expected to be fully saturated with water or possible as a mixed zone. Dry supercritical CO_2 directly in contact with the caprock is considered unlikely, although possible if injection is close to the reservoir-caprock interface. As outlined above, the main mechanism for CO_2 transport into the caprock is expected to be molecular diffusion in the pore water from the zones where CO_2 is fully saturated with water or in a mixed phase, whereas advective flow in pore network or fractures will also be considered, although considered a less likely scenario for sites with proper pressure management.

The solubility of water in CO_2 is a function of temperature and pressure, and in the case of an upward migration through the seal, pressure and temperature conditions will change, and CO_2 will lose or take up water on the way [*Miri et al.*, 2014a]. The solubility of water in CO_2 is shown for a range of temperatures (0–75°C) and pressures (1–300 bars) in Figure 8.3. Upward migration will lead to complex changes where the solubility will first decrease, before it will increase at the lower temperatures and pressures (Fig. 8.3). If CO_2 leakage rate is high, within fractures or fault zones, where the flow velocity might be high compared to CO_2 flow in porous media, adiabatic expansion of the gas leading to a further cooling compared to the shallower depth may also be expected [*Mao et al.*, 2017]. However, this is not discussed in further detail in this paper.

Ensuring both short- and long-term integrity of seals is important for successful geological storage of CO_2. Key challenges related to the long-term integrity of clay-rich

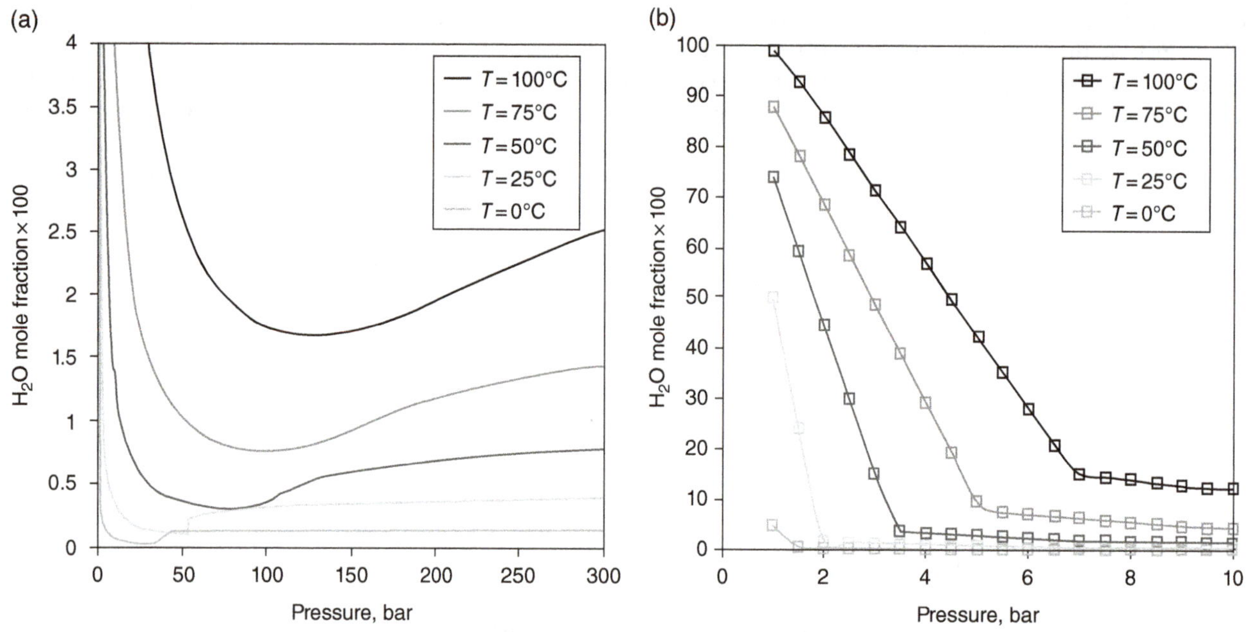

Figure 8.3 Solubility of H_2O in CO_2 as a function of temperature (0–100°C) and pressure. Reprinted with permission from *Miri et al.* [2014a]. Copyright (2014a) American Chemical Society. (a) Pressure range of 0–300 bar. (b) Details for pressure range of 0–10 bar, relevant during upward migration.

seals are to understand the mechanism for CO_2 transport into seals and the complex interplay between CO_2 fluid-rock interaction, alteration, and deformation that can alter the transport properties over time as outlined in Figure 8.1. Key challenges addressed in this paper include a review of theoretical models and laboratory experiments addressing interaction between CO_2 and clay-rich caprock material. Section 8.2 discusses CO_2 transport into clay-rich seals at the scale of molecular diffusion, pore-scale displacement, and fracture flow. Seal reaction such as precipitation, dissolution, and CO_2 adsorption will be reviewed with respect to CO_2 transport and deformation in Section 8.2.1, whereas Section 8.2.2 presents examples of fluid-rock interaction processes within selected sealing units of the North Sea. In the conclusion, the current knowledge and need for further research is summarized.

8.2. TRANSPORT OF CO_2 AND CO_2-SATURATED WATER

8.2.1. Diffusion-Dominated Matrix Transport

Dissolved CO_2 is transported into the seal by diffusion. The advancement rate of a diffusion front depends on the gradients of the dissolved inorganic carbon species and rock-specific factors such as the tortuosity and the potential of rock to consume CO_2 through chemical reactions such as CO_2 sorption and mineral dissolution and growth. The effect of the tortuosity and reactions on the diffusion rates of the carbon species can be seen in the reaction-diffusion equation:

$$\frac{\partial C_{\text{i}}}{\partial t} = D_{\text{e,i}} \nabla^2 C_{\text{i}} - R_{\text{c}}, \tag{8.2}$$

where C_{i} is the concentration of dissolved carbon species i, t is time, $D_{\text{e,i}} = \phi D_{\text{i}}^0 / \tau$ is the effective diffusion coefficient, D^0 is the bulk water diffusion coefficient, R_{c} is a reaction term, with positive values indicating uptake into the solid framework, ϕ is porosity, and τ is tortuosity. A more tortuous path offers more contact time between CO_2 and resident brine, resulting in better mixing and enhanced solubility in the system [*Rathnaweera et al.*, 2016]. Similarly, reactions reduce the mass of the free-phase CO_2, further retarding the rate of plume migration [*Xu et al.*, 2003, 2005].

The diffusion coefficient for CO_2 in a brine-saturated caprock sample may be measured experimentally. Experimental investigation of CO_2 diffusion in the Muderong Shale, Australia, provided effective diffusion coefficients of $3.08–4.81 \times 10^{-11}$ m²/s at reservoir conditions (T = 45–50°C and P < 20 MPa) [*Busch et al.*, 2008]. Repetitive CO_2 diffusion experiments on clay-rich marlstone from the Upper Cretaceous caprock sequence of the Münsterland Basin show an increase in the effective diffusion coefficient from 7.8×10^{-11} to 1.2×10^{-10} m², indicating a change in transport properties during the experiment [*Wollenweber et al.*, 2010]. Diffusion into intact mudstones/shales is a very slow process, and uncertainties are related to the upscaling of experimental data. Diffusion rates may also be estimated from studying natural analogues. One such analogue is CO_2 diffusion from the North Sea Brae Formation into the Kimmeridge Clay [*Lu et al.*, 2009]. Here carbon isotopes indicate that dissolved CO_2 has reached approximately 12 m into the caprock matrix over 70–80 million years. Assuming CO_2 a conservative tracer, this suggests an effective diffusion coefficient of 10^{-14} m^2/s (Fig. 8.4). Such a low effective diffusion coefficient indicates either very low values of the tortuosity or that the mudstones acted as a sink for the CO_2 species, thereby slowing down the transport. The Kimmeridge Clay is rich in organic matter [*Tribovillard et al.*, 1994], and CO_2 interaction with the mudstone is therefore one likely reason for the slow transport. The lower diffusion rate of the field analogue (Kimmeridge Clay) compared to the laboratory rates might be material dependent but could also be an indication that the diffusion under in situ condition and for long term is restricted by some factors that are less active in the laboratory experiments. Possible sample damage due to dehydration or unloading of laboratory samples might also contribute to the higher diffusivity observed at laboratory scale.

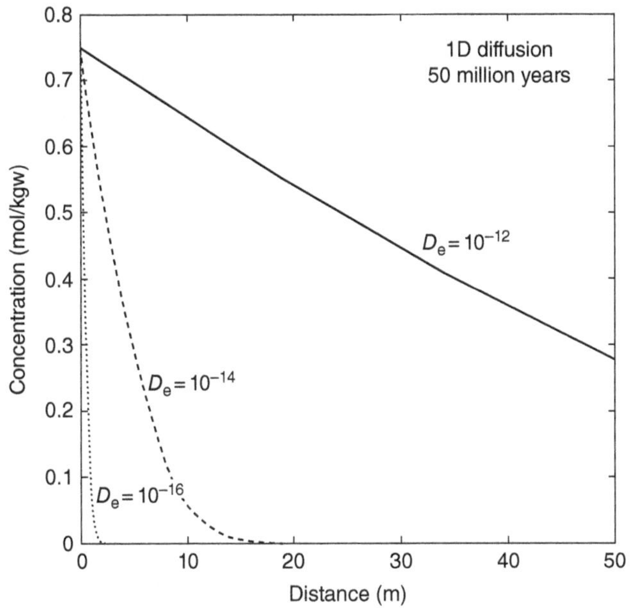

Figure 8.4 Simulated 1D diffusion of CO_2 over 50 million years varying the effective diffusion coefficient (see Eq. 8.2). The diffusion coefficient of CO_2 penetrating into the Kimmeridge Clay (12 m over 70–80 million years) can be approximated to be 10^{-14} m^2/s based on this model.

Diffusion fronts are sometimes observed to be sharp and sometimes diffuse, which provide valuable insights regarding timescale and length scale of the underlying mechanisms of CO_2 transport and the nature of the CO_2-brine-rock reactions. The type of diffusion front can be explained by the relative rates of diffusion and reactions (Eq. 8.2). If reactions are very fast compared to diffusion ($R >> D_{e,i} \nabla^2 C_i$), a sharp front will develop, whereas diffuse fronts develop when reactions are very slow compared to diffusion ($R << D_{e,i} \nabla^2 C_i$). Both diffuse and sharp diffusion fronts are easily observed in bleached siltstone in Utah [*Busch et al.*, 2014] (Fig. 8.5). Diffusion in mudstones is commonly very slow, and reactions can be fast compared to the diffusion, so sharp fronts are expected in shales and mudstones.

8.2.2. Capillary Entry Pressure and Displacement Flow

Gaseous or supercritical CO_2 is prevented from entering intact (non-fractured) mudstones or shales due to high capillary entry pressures. The capillary entry

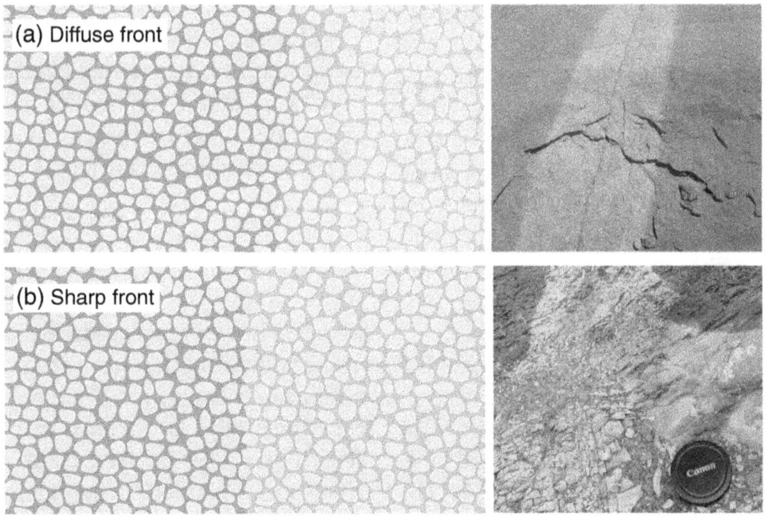

Figure 8.5 Difference between diffuse (a) and sharp (b) reaction fronts.

pressure, P_c, can be obtained from the Young-Laplace equation [*Washburn*, 1921]:

$$P_c = P_{CO_2} - P_w = \frac{2\gamma \cos\theta}{r}, \quad (8.3)$$

where r is the median pore throat radius, γ is the interfacial tension between the wetting and non-wetting fluids, and θ is the wettability angle between the wetting and non-wetting fluid and the host rock. In general, clay-rich seals with lower permeability (i.e., smaller median pore throat sizes) have higher capillary entry pressures, and therefore, they pose higher resistance against CO_2 invasion. The CO_2-water interfacial tension has been measured for relevant CO_2 storage conditions by *Chiquet et al.* [2007].

Capillary entry pressure for CO_2 can also be measured directly in the laboratory. Comparison of methods for determining capillary entry pressure is discussed by *Boulin et al.* [2013], showing differences in time and accuracy. For CO_2, experimentally measured capillary entry pressure (or displacement pressure) is in the range of 0.1–5 MPa for a selection of intact mudrock samples [*Hildenbrand et al.*, 2004] and 3.5–4.3 MPa for shale from the Draupne Formation in the Troll East area [*Skurtveit et al.*, 2012]. The measured CO_2 displacement pressure is found to be lower than for natural gas (CH_4) and nitrogen (N_2) [*Hildenbrand et al.*, 2004]. Effective Darcy permeability for CO_2 has been studied in the same experiments as for CO_2 entry pressure. High confining pressure was applied in order to avoid hydro-fracturing of the sample, and Darcy flow was imposed upon shale samples in the experiments by *Hildenbrand et al.* [2004], where a slight decrease in effective CO_2 permeability (range of 10^{-18}–10^{-24} m^2) was measured compared to the water permeability (range of 10^{-19}–10^{-21} m^2). In these experiments, a high pressure gradient (exceeding the expected breakthrough pressure) was imposed across the sample, and the resulting gas flux was monitored by means of pressure change. No mechanical changes (i.e., fracturing) were reported. *Skurtveit et al.* [2012] measured effective CO_2 permeability of the order of 10^{-21} m^2, within the same order of magnitude as for brine permeability; however, the effective CO_2 permeability was found to be dependent on volumetric dilation in the sample, and a microfracture-dominated flow was interpreted.

8.2.3. Fracture Transport

The CO_2 entry pressure for fractures can also be provided by Eq. (8.3), but with the pore throat radius replaced by b, the initial fracture aperture [*Wang and Peng*, 2014]. The most common model to describe single-phase flow through microfractures is given by the Navier-Stokes equations which express conservation of momentum and mass over the fracture with impermeable walls. The laminar flow of an incompressible Newtonian fluid with constant viscosity may be written as [*Bird*, 2002]

$$\rho(u.\nabla)u = \mu\nabla^2 u - \nabla e,\ \nabla.u = 0, \quad (8.4)$$

where ρ is the fluid density, μ is the fluid viscosity, u is the velocity vector, and $e(x,y,z)$ is the fluid potential (i.e., the mechanical energy per unit mass). Assuming that (i) the fracture is composed of two parallel plates separated by a small aperture, (ii) the variability in fracture aperture is minimal, and (iii) the inertial forces are much smaller compared with the viscous and pressure forces, Eq. (8.3) reduces to the so-called local cubic law (LCL) [*Zimmerman and Bodvarsson*, 1996; *Oron and Berkowitz*, 1998]:

$$\nabla.\left(\frac{b^3}{12\mu f}\nabla e\right) = \nabla.\left(T_f \nabla e\right) = 0, \quad (8.5)$$

where Tf is the transmissivity of fracture and f is the correction accounted for fracture roughness. Several studies have considered the validity of the LCL, and it has been shown that for the Darcian flow (Reynolds << 1), inertial term can be safely ignored [*Walsh*, 1981; *Renshaw*, 1995; *Zimmerman and Bodvarsson*, 1996; *Brush and Thomson*, 2003]. Furthermore, it has been shown that the assumption (ii) can be relaxed if aperture are measured as an average over a certain length. Equation (8.5) shows that fracture conductivity is proportional to the cube of mean aperture and that flow rate (Q) is directly proportional to the fluid potential and can be written as

$$Q = -\frac{b^3}{12\mu f}\nabla e, \quad (8.6)$$

As Eq. (8.6) shows, the fracture transmissivity decreases as the aperture size reduces, indicating that fracture flow of CO_2 is mainly controlled by the aperture size. The ability for a fracture to deform and change its aperture is given by the normal stiffness K_n given as

$$K_n = \frac{d\sigma}{d\delta}, \quad (8.7)$$

where $d\sigma$ is the change in stress and $d\delta$ is the corresponding fracture deformation [*Jaeger et al.*, 2009]. Characteristic nonlinear behavior for fracture stiffness is demonstrated experimentally as well as in conceptual models [e.g., *Bandis et al.*, 1983; *Myer*, 2000; *Pyrak-Nolte and Morris*, 2000] and explained by the increasing contact area in the fracture as the normal load increases. There is a large amount of work dealing with the correlation between applied stress conditions, fracture aperture, porosity, and the corresponding fracture permeability [e.g., *Barton et al.*, 1985; *Cho et al.*, 2013] as well as for permeability in fault zones [e.g., *Faulkner et al.*, 2010] that will not be further discussed in this review.

A complex interplay of chemical and mechanical factors controls the fracture aperture. Experimental evidence supports the movement of CO_2 in low-permeability shale matrices via fractures created from excess gas pressure. Direct laboratory measurements of CO_2 entry, breakthrough, and flow in an initially brine-saturated shale caprock have been described in *Angeli et al.* [2009] and *Skurtveit et al.* [2012], where the CO_2 breakthrough was recognized by a marked dilation of the test sample. Effective CO_2 permeability as a function of strain could be fitted to a model where the measured dilation (strain) represents a change in fracture aperture [*Olivella and Alonso*, 2008]. The effective CO_2 permeability observed after the CO_2 breakthrough was found to be more sensitive to changes in volumetric deformation than the absolute brine permeability, and it could be fitted to a power law dependency [*Skurtveit et al.*, 2012]. This supports experimental work by *Harrington and Horseman* [1999], suggesting that microfractures link up to form a distinct pathway through the low-permeable seal units in their gas breakthrough experiments on Boom clay and bentonite. *Edlmann et al.* [2013] suggested a critical fracture aperture: below this critical fracture aperture, there is little or no CO_2 flow along the fracture, nor in gaseous or supercritical state, close to or at the critical aperture, only gaseous CO_2 will flow, whereas above the critical fracture aperture, both $scCO_2$ and gaseous CO_2 will potentially flow. Yet quantitative description of the threshold aperture size is not provided. The variation of the fracture aperture size induced by chemical interaction between CO_2 and the host rock (e.g., mineralization, dissolution, desiccation, swelling, etc.) may enhance/decrease the sealing capacity. The advection of CO_2 after breakthrough supports water evaporation into $scCO_2$, increasing capillary suction, leading to additional sediment contraction and consequently formation of capillary-driven fractures [*Espinoza and Santamarina*, 2012; *Schaef et al.*, 2012]. In addition, the two-phase flow of CO_2 and brine in the fracture network modifies the capillary entry pressure of the fracture network and the sorptive chemistry within the shale matrix. In order to describe the two-phase flow of CO_2 and brine in the fracture network, it is needed to take into account the characteristics of multiphase flow such as relative permeabilities and capillary pressure between water and CO_2 that may affect the local deformation of the shale caprocks [*Gherardi et al.*, 2007; *Wang and Peng*, 2014]. This can be solved using a dual-porosity model in which the discontinuous nature of porosity and permeability is avoided by replacing them locally by their average values. Moreover, it may be assumed that there is only viscous flow in the fracture network (sink) and the matrix will act as a source term in the fracture flow equation [*Warren and Root*, 1963]:

$$\varnothing_{\mathrm{f}} \frac{\partial S_{\alpha \mathrm{f}}}{\partial t} + T_{\mathrm{mf}_\alpha} = -\nabla.u + R_\alpha, \tag{8.8}$$

where S is phase saturation, t is time, $\varnothing$is the fracture porosity, u is phase fluid Darcy velocity, R is fluid source/sink term, and T is the transfer function defining the interaction between matrix and fracture (see next section). The subscript α refers to either the wetting (water) phase or the non-wetting (CO_2) phase. The subscripts m and f represent the matrix and the fracture, respectively. Darcy's velocity of phase α is given by

$$u_\alpha = -\lambda_\alpha k_{\mathrm{f}} \nabla e_\alpha, \tag{8.9}$$

where k_{f} is the fracture absolute permeability given by the cubic law and $\lambda\alpha$ is the fluid mobility and is equal to the ratio of the relative permeability $k_{\mathrm{r}}\alpha$ and the dynamic viscosity.

Only limited experimental data are available on permeability of fractured shale and even less on permeability of CO_2 in fractures. *Carey et al.* [2015] measured the peak permeability of 900 mD for well-defined bedding parallel fractures, compared to 30 mD for the more complex fracture patterns formed across the bedding in a triaxial and direct shear device for the calcite-rich Utica shale core from Ohio and Pennsylvania. A naturally fractured and carbonate-cemented fracture from Kimmeridge shale, UK, was tested for flow properties in a direct shear device after artificially separated and demineralized (carbonate mineralization removed by acid) [*Gutierrez et al.*, 2000]. The test showed that increasing normal contact stress across the fracture lowered the water permeability in an exponential way. However, the fractures were not entirely healed, and permeability remained higher than the matrix permeability for the test period of hours. In general, fracture transmissibility is strongly controlled by the ductility index, defined as the effective mean stress normalized to the tensile strength of the intact rock [*Ishii*, 2015]. This means that the experimentally measured fracture permeability is highly dependent on the effective stress condition used in addition to the shale properties. For ductile shales and clays, the consolidation, creep, and swelling may close fractures during an experiment [*Zhang*, 2011] and the capillary entry pressure for fractures to be close to that of the matrix. This highlights the need for use of relevant pressure condition when considering CO_2 transport properties for fractures.

8.2.4. Fracture-Matrix Interactions

In the case of a two-phase displacement flow in a fracture network, fracture-matrix interactions may alter the dynamic of the flow process and even immobilize CO_2 that has migrated into the caprock along the fractures. Due to high fracture permeability, the fluids in the interconnected fracture network will quickly be displaced by CO_2, while the shale matrix remains water saturated [*Saidi*, 1983; *Wang and Peng*, 2014]. The fracture-matrix flow term, *Tmf* in Eq. (8.8), represents the volumetric flux of the fluid from the matrix blocks into the fractures (per unit time and per unit volume of the reservoir), and it is given as [*Barenblatt et al.*, 1960]

$$-T_{\mathrm{mf}_\alpha} = \phi_{\mathrm{m}} \frac{\partial S_{\alpha \mathrm{m}}}{\partial t}, \tag{8.10}$$

During a CO_2 invasion into the fracture network, different mechanisms may come into play, such as clay swelling, gravity drainage, CO_2 diffusion and sorption, and matrix drying [*Busch et al.*, 2010]. Generally, the pressure in the matrix, P_{m}, follows the overburden gradient, while the pressure in the fracture network, P_{f}, is controlled by the height of CO_2 column in contact with caprock. Considering the high compressibility of CO_2 in either gaseous or a supercritical state, the pressure differential between the fractures and the matrix creates a driving force leading to matrix depletion [*Saidi*, 1983; *Lim and Aziz*, 1995], given as

$$T_{\mathrm{mf}_\alpha}^{\mathrm{expansion}} = \sigma \lambda_\alpha \left[P_{\mathrm{f}} - P_{\mathrm{m}} \right], \tag{8.11}$$

where σ is the shape factor which represents the mean flow path between the matrix and its corresponding fracture [*Warren and Root*, 1963]. When CO_2 invades the fractures surrounding the water-saturated matrix, water may drain downward due to the density difference between CO_2 and water. This mechanism is referred to as gravity drainage. However, the capillarity (matrix) will act against this mechanism and retain water in the matrix. Since the shale matrix usually has high capillary threshold, it is very unlikely that CO_2-water gravity drainage occurs during CO_2 invasion into the caprock [*Festoy and Golf-Racht*, 1989; *Rossen and Shen*, 1989]. In addition to the gravity drainage flow, the pressure-driven flow (Eq. 8.11) will occur only when the CO_2 pressure exceeds the summation of reservoir pressure and capillary entry pressure. Nevertheless, in a capillary sealing case, other mechanisms will come into play, transferring CO_2 into the fracture-matrix system.

CO_2 may exist in several forms in the caprock: (i) as a free phase displacing water in the fracture network, (ii) as an absorbed/adsorbed phase onto the matrix, and (iii) dissolved in brine diffusing into the shale matrix. The role of free-phase CO_2 is already included in Eq. (8.8). The mass of CO_2 adsorbed in the fracture network can be calculated based on the Langmuir isotherm [*Wang and Peng*, 2014]. However, it may be reduced if the invaded CO_2 is sufficiently dry to partly evaporate the adsorbed water films. Nevertheless, the extent of dehydration will depend on the water content of the CO_2, and complete evaporation of adsorbed water films on the fracture surface is not very likely to occur in a CO_2 storage scenario due to strong adhesive solid–liquid interactions [*Espinoza and Santamarina*, 2012; *Giesting et al.*, 2012; *Schaef et al.*, 2012]. CO_2 diffusion and sorption into the matrix is an important mechanism which may reduce the risk of leakage given the high adsorption capacity of clay minerals [*Busch et al.*, 2008]. *De Jong et al.* [2014] have performed several unconfined volumetric strain measurements in smectite-bearing fault material and found that CO_2 penetrating into fracture/joint walls can be expected to cause swelling of a few percent (≈3%), reducing fracture apertures and thus reducing bulk permeability (≈11% calculated using Eq. (8.6)), thereby improving seal integrity. The mass exchange rate of gas sorption from the shale matrix to the fracture network depends on the difference between the matrix gas content and the equilibrium gas, and it is expressed by *Wang and Peng* [2014] as

$$T_{\mathrm{mf}_\alpha}^{\mathrm{diffusion}} = \frac{-\rho_{\mathrm{ga}} \rho_{\mathrm{c}}}{\sigma D} \left[m_{\mathrm{b}} - m_{\mathrm{e}} \left(P_{\mathrm{f}} \right) \right], \tag{8.12}$$

where ρ_{ga} is the gas density under the standard conditions and ρ_{c} is the bulk density of the shale; m_{b} and m_{e} are the current and equilibrium gas content at fracture pressure, P_{f}, respectively; and D is the effective diffusion coefficient of gas. *Busch et al.* [2008] have reported the sorption capacity of the Muderong shale and various clay minerals (kaolinite, illite, smectite) from Western Australia to about 1.0 mmol/g. *Wollenweber et al.* [2010] have performed similar sorption experiments on marlstone and reported sorption capacities of 0.27 mmol/g, which is slightly lower than the sorption capacity reported by *Busch et al.* [2008] and assigned to the higher organic content of the samples investigated by *Busch et al.* [2008]. Likewise, significantly higher sorptive uptake is reported for coals owing to high organic matter content [e.g., *Weniger et al.*, 2010; *Chareonsuppanimit et al.*, 2012]. In addition, the sorption capacity is a function of specific surface area (see Eq. 8.12) accessible for CO_2 to be in contact with matrix blocks. Smaller fracture spacing in shale implies smaller matrix block sizes and higher specific surface areas, hence intensified exchange rates for the sorbed gas.

8.3. SEAL REACTIONS AND IMPLICATIONS FOR FLOW AND DEFORMATION

8.3.1. CO_2-Shale Reactivity

As CO_2 diffuses into the rock matrix from the base of the seal or along a non-sealing fracture or heterogeneity, minerals will react and secondary phases may form [e.g., *Gaus et al.*, 2005]. The rate of the reactions will be controlled by both the diffusivity and the sink/source reactions as outlined in Eq. (8.2) (Section 8.2). In order to define the reaction, major minerals that will affect CO_2 diffusion over long timescales need to be defined, and the mineral kinetic parameters (kinetic constants, reactive surface areas) must be known for all mineral phases taking part in the reactions. There has been progress in understanding the parameterization and uncertainty of kinetic simulations [*Hellevang and Aagaard*, 2013; *Hellevang et al.*, 2013], but especially reactive surface areas are still very difficult to estimate.

Compared to kinetic modeling studies for reservoirs, only a limited number of kinetic models are addressing seals with a clay-dominated composition (see Table 8.1 for an overview). Reactive diffusive transport modeling calculations for clay-dominated seals predict a decrease in porosity due to kaolinite precipitation [*Gaus et al.*, 2005] and chlorite-to-ankerite transformation [*Balashov et al.*, 2015], whereas the reactivity of quartz and clay minerals (illite and smectite) is low. However, modeling by *Tambach et al.* [2015] predicts porosity increase at the reservoir contact mainly due to siderite dissolution. Simulations by *Gherardi et al.* [2007] on carbonate-rich shales suggest that during fully liquid-saturated conditions in a diffusion-controlled regime, pH will be buffered and calcite precipitation occurs, whereas calcite dissolution can occur in the caprock if a free, water-rich CO_2-dominated phase migrates into the caprock through discontinuities. Although differences in alteration are observed for a diffusion-dominated caprock matrix system and a fractured caprock with flow and possibly free CO_2 phase, the reactions observed are mainly calcite dominated and might not be of high relevance for the clay-rich seals.

The geochemical reactions to be expected from mineral reactions due to the reduced pH during dissolution of CO_2 into the pore fluid may be addressed in experimental work. Reaction experiments on shale and mudstone as summarized in Table 8.2 show only limited reactions in the shale, mainly nucleation and growth of carbonate and smectite [*Kaszuba et al.*, 2005; *Carroll et al.*, 2011], whereas observed illite-smectite transformation into illite might be due to heating [*de Lima et al.*, 2011]. In general, carbonate-rich shale is more reactive than clay-rich shale [*Alemu et al.*, 2011], showing dissolution of plagioclase and clay minerals (illite and chlorite) and precipitation of smectite.

Variations in matrix porosity can also be observed within geochemical alteration experiments on cuttings [*Mouzakis et al.*, 2016] and diffusion experiments [*Busch et al.*, 2008; *Wollenweber et al.*, 2010]. Experimental work on low-permeable caprock is time consuming and challenging due to the slow transport properties. Detection of chemical alteration and porosity changes with following changes in transport properties over time is not straightforward, and there is some spread in the reported experimental results (Table 8.2). Diffusion and sorption experiments by *Busch et al.* [2008] show dissolution of silicates and precipitation of carbonates to have measurable effects on the porosity, permeability, and diffusion properties of the Muderong Shale (Australia), with a tendency to enhance the transport properties. However, the effect of porosity increase on transport properties might also be related to the preferential dissolution of pores or pore throats. Caprock samples from the In Salah injection site show dissolution of siderite and chlorite from both pores and pore throats, and the increase in pore throat radii is used to explain the factor 8 permeability increase observed after the flooding with CO_2-saturated water [*Armitage et al.*, 2013].

Matrix properties, such as mechanical strength and stiffness, control matrix deformation response to stress. Conventional triaxial tests [e.g., *Berre*, 2011] are the most common method to address the mechanical strength of intact rock samples. There are a limited number of tests that have addressed potential modifications to the mechanical properties of rock samples due to CO_2 interaction by comparing the deformation moduli and mechanical strength of altered and unaltered samples, but most experimental work is on reservoir sandstone [e.g., *Le Guen et al.*, 2007; *Hangx et al.*, 2013] and is summarized in *Rohmer et al.* [2016]. The design of experimental programs capturing alteration-induced geomechanical changes is even more challenging for caprocks due to the low permeability and slow transport of CO_2 solutions into the rock. The long-term fate of CO_2 alteration of sealing units has been addressed using the Green River natural analogue in Utah [*Kampman et al.*, 2013; *Busch et al.*, 2014]. The study, comparing mechanical strength of unreacted and reacted (bleached) samples, concluded that the variation in depositional environment and porosity imposed a strong control on rock strength, whereas mechanical changes related to the observed bleaching cannot be documented [*Busch et al.*, 2014]. However, mechanical degradation related to long-term CO_2 exposure has been documented using indentation and scratch testing of the Entrada Sandstone and Summerville Siltstone, Utah [*Sun et al.*, 2016]. This micro-mechanical test method provides an evaluation of

Table 8.1 Overview of Simulations Addressing Reaction in Clay-Rich Caprocks and Effects on Porosity and Transport.

Formation/field	Method	Caprock minerals	Observed reactions	Effects on porosity and transport	Reference
Nordland Shale Sleipner, North Sea	Kinetic batch modeling PHREEQC and reactive transport modeling Conditions: 37°C and 101.3×10^5 Pa	Mass %: illite (24.7), kaolinite (18), quartz (21.5) plagioclase (12.4), smectite (8.8), calcite (1), chlorite (4.1), K-feldspar (2.1), pyrite (2.8), siderite (1.6), mixed layer clay (1.4)	Dissolution of calcite, feldspar alteration; apart from kaolinite precipitation, the reactivity of the clays (illite, smectite, etc.) is low	A decrease in porosity is <3% in 15 000 years, and improved sealing is modeled, whereas in some scenarios carbonate dissolution might induce a slight porosity increase at the bottom of the caprock, but without ability to migrate into the caprock	*Gaus et al.* [2005]
Chlorite- and illite-containing seal, close to the Marcellus Formation (USA) in composition	Reactive diffusion model MK76. Conditions 348.15 K and 30 MPa	Vol.%: quartz (38.41), illite (33.13) chlorite (13.76), calcite (4.93), microcline and pyrite	Transformation of chlorite to ankerite is the dominant reaction occluding the shale porosity. Quartz is almost inert. Reactions of feldspars and clays depend strongly on their reaction rate constants	A decrease in porosity is modeled from the initial 5% to 0% in 6–14,000 years	*Balashov et al.* [2015]
Shale from depleted gas reservoir offshore the Netherlands	1D kinetic diffusion modeling with PHREEQC for caprock	Vol.%: quartz (63.1), illite (10.1), dolomite (11.4), anhydrite (6.5), albite (2.9), K-feldspar (3.9), pyrite (0.3), siderite (1.7)	pH is partially buffered by dissolution of siderite, albite, and microcline, which are transformed into Fe-illite and quartz	Porosity increases from 5.0 to 5.7% at the reservoir contact during 10.000 years, and the CO_2 diffusion is 6.4–12.9 m into the caprock for the same time	*Tambach et al.* [2015]
Carbonate-rich shales from onshore gas reservoir in the north of Italy	Diffusion in the aqueous phase and gas and/or liquid advection. Batch simulations using TOUGHREACT / TOUGH2	Calcite (0.3), dolomite (0.03), quartz (0.2), muscovite (0.19), smectite (0.15)	Dissolution and precipitation reactions involving calcite dominates. Clay dissolution of illite, chlorite, and muscovite and precipitation reactions (Na-smectite) less significant	For a free CO_2-dominated phase in fractures, significant calcite dissolution and porosity enhancement are predicted. For diffusion, some calcite precipitation is predicted which leads to further sealing of the storage reservoir	*Gherardi et al.* [2007]

Table 8.2 Overview of Experimental Work Addressing Reaction in Clay-Rich Caprocks and Effects on Porosity and Transport.

Formation/field	Method	Caprock minerals	Observed reactions	Effects on porosity and transport	Reference
Geochemical batch reaction work					
Silurian Maplewood Shale, an argillaceous shale from Monroe County, New York, USA	Experimental work in reaction cell	Vol.%: clay minerals like phyllosilicates illite, mica (65), quartz (27), feldspar (5), chlorite (2), pyrite	Nucleation and growth of siderite on shale suggests the aquitard is a reactive component in the system	Not discussed	*Kaszuba et al.* [2005]
Shale caprock, Krechba Field, In Salah, Algeria	Cement-rock-brine-CO_2 experiments	Wt.%: illite (44), chlorite (30), quartz (20), albite, dolomite, kaolinite, siderite	Little indication of alteration of the shale by CO_2-rich brines; only small precipitates on the shale surface. Added cement gives more extensive clay dissolution and precipitation of smectite and calcium carbonate	Not discussed	*Carroll et al.* [2011]
Palermo shale caprock, Paraná Basin, Southern Brazil	Experimental work, pressurized cells, reservoir conditions	—	Transformation from illite-smectite to illite	Not discussed	*de Lima et al.* [2011]
Carbonate-rich shale, De Geerdalen Fm, and clay-rich shale, De Geerdalen Fm, Svalbard	Batch reaction experiments	Carbonate-rich shale Wt.%: quartz (13), calcite (29), illite (22), chlorite (38), ankerite (7), plagioclase/albite (6) Clay-rich shale Wt.%: quartz (26), illite (26), chlorite (19), plagioclase/albite (8), siderite (5), pyrite	Carbonate-rich shale: dissolution and re-precipitation of carbonates; dissolution of plagioclase, illite, and chlorite; and the formation of smectite Clay-rich shale: no significant mineralogical alterations except for dissolution of silicate minerals	Not discussed	*Alemu et al.* [2011]
Marine Tuscaloosa Shale, Mississippi	Fixed volume reactors	Quartz (60%), feldspar (14%), chlorite (9%), kaolinite (5%), illite (5%), calcite (5%)	Mineral dissolution under CO_2 sequestration conditions	Total porosity increased slightly but connected porosity decreased	*Mouzakis et al.* [2016]
Combined transport and geochemical work on intact samples					
Muderong Shale, Australia	Experimental work on diffusive transport and gas sorption	Illite-smectite (27%), smectite in I-S (20%), mica/illite (8%), kaolinite (26%), quartz (27%), chlorite (5%), siderite (2%), orthoclase (3%), pyrite (2%)	Dissolution of silicates and precipitation of carbonates	Tendency to enhance the transport properties	*Busch et al.* [2008]
Combined transport and geochemical work on fractured samples					
Fractured claystone, Upper Toarcian Formation, Tournemire, France	25°C in flow-through reactor with confinement of 0.12 MPa	Vol.%: calcite (25), siderite (2), quartz (25), clay minerals (45), and pyrite (3) Clay fraction: kaolinite (24), micas (muscovite) (10), interstratified illite-smectite (10), and chlorite (1)	Calcite and some quartz grain dissolution, altering of fracture surface	Sole seepage of CO_2-brine through a fracture does not alter permeability, while cycling flow of CO_2-gas and CO_2-brine increases fracture aperture and permeability	*Andreani et al.* [2008]
Mudstone, Krechba Field, In Salah, Algeria	Flow-through experiments (permeameter)	Quartz, detrital mica, detrital clay (likely Fe-rich 7 Å clay and illite-smectite) with minor feldspar and oxide phases	Dissolution of the minerals chlorite and siderite	Increasing pore throat radii, porosity increased from 7 to 10%, and increasing permeability by approximately a factor of 8	*Armitage et al.* [2013]

changes in mechanical properties, but the mechanical parameters measured for this type of test are not directly suitable as input to mechanical models.

8.3.2. CO_2-Smectite Interactions and Clay Swelling

Smectite is a group of minerals consisting of an octahedral sheet sandwiched between two tetrahedral sheets and found to varying degree in mudstones and shales. In the interlayers between the sandwiches, surfaces are negative, and positively charged ions (cations) are therefore attracted. In the presence of water, these cations tend to hydrate and increase in volume, and this leads to an expansion of the whole smectite structure. The swelling/shrinking property of smectites is commonly seen in soils. When a clay-rich soil dries out, it shrinks and cracks into polygonal patterns, whereas the soil expands and the cracks close when water is added. Similar fracturing may potentially happen at the base of a seal if dry or near-dry CO_2 is allowed to interact with the shale/mudstone and dry out the smectites [*De Jong et al.*, 2014; *Busch et al.*, 2016]. This will depend on the water content of the CO_2, but the exact shrinking/swelling property is complex and is not a linear function of water content. Although still debated, *Loring et al.* [2014] showed that the sorbed water content of Na-montmorillonite increases nearly linear with water content in the CO_2, whereas the smectite swells in a stepwise manner with three distinct plateau d_{001} (d-spacing perpendicular to the sheets) values. The reverse reactions were not attempted so any hysteresis effects are not known. The stepwise swelling is also known from traditional experiments using air moisture content and is referred to as crystalline swelling, where there are a number of discrete stable values of layer spacing [e.g., *Morodome and Kawamura*, 2009]. Also, CO_2 itself contributes to the volume increases of smectites. CO_2 will be incorporated into the layered structure of smectites in the same interlayers as the cations (a process referred to as intercalation), and the structure will swell [*Fripiat et al.*, 1974; *Michels et al.*, 2015]. The size of the interlaminar ions provides restrictions on the degree of CO_2 inclusion, and only the smectites saturated with the larger monovalent cations (including K- and Na-smectites) will allow complete or nearly complete CO_2 inclusion [*Fripiat et al.*, 1974]. Clay swelling as a self-limiting process within a caprock in a CO_2 system is not well understood, although some attempts have been made following the CO_2 sorption in clays discussed by *Busch et al.* [2008]. Clay swelling experiments [*Giesting et al.*, 2012; *Schaef et al.*, 2012; *De Jong et al.*, 2014] show that $scCO_2$ has the potential for mechanical swelling of smectite as function of the clays initial hydration stage. For smectite-rich caprocks, swelling may influence fracture closure and healing [*Bastiaens et al.*, 2007; *Zhang*, 2011], whereas the effect of swelling stress induced by CO_2 sorption is less well understood [*Busch et al.*, 2016].

8.3.3. Carbonate-Cemented Fractures

Both carbonate- and gypsum-cemented veins related to paleo-leakage and active CO_2 leakage are observed in the Green River area, Utah [*Dockrill and Shipton*, 2010; *Kampman et al.*, 2012]. Calcite precipitates rapidly from solutions supersaturated with $CaCO_3$ (if not inhibited by, e.g., Mg) and responds rapidly to changing thermodynamic conditions. Brines in the Green River area are found to be Mg rich, which might explain the formation of aragonite over calcite [*Kampman et al.*, 2012]. Calcite is getting more stable as temperature increases, and a heating of a saturated aqueous solution will therefore lead to precipitation; however, for the aragonite veins in Green River, precipitation is found to take place due to outgassing of the local system [*Kampman et al.*, 2012]. The calcite solubility is also a function of CO_2 pressure, and more calcite can be dissolved at higher CO_2 pressures, whereas the source for calcium in the system can be local from, for example, feldspar dissolving from rock matrix or transported into the system by the fluid. Carbonate veins can also be present in clay-rich caprocks, like the Kimmeridge shale [*Gutierrez et al.*, 2000]. These veins may provide pathways for CO_2 within the caprock if not fully sealed or if reactivated. If water-saturated CO_2 (or a CO_2-charged solution with high CO_2 partial pressure) is migrating into calcite-filled fractures, reopened by an overpressure, the existing calcite will to some extent dissolve and lead to permeability increase. However, because of the fast kinetics, solutions are soon saturated with calcite at higher aqueous Ca^{2+}, and the dissolution will cease.

Experimental work focusing on carbonate cement in fractures is mainly concerned with the dissolution of calcite and addressing the alteration in flow properties due to CO_2-saturated fluids in typical caprock materials [*Andreani et al.*, 2008; *Ellis et al.*, 2011; *Ellis et al.*, 2013]. Based on experimental results, it is observed that the effects of calcite dissolution, increasing porosity, and changes in transport properties are complex. The alteration experiments on fractured claystone by *Andreani et al.* [2008] show dissolution of quartz and calcite increasing the fracture porosity, whereas the permeability remains unchanged. This lack of effect on permeability is attributed to remaining clay framework and limited net effects on the fracture aperture. Cyclic CO_2-brine and CO_2-gas flow on the same material showed an increase in aperture explained by clay particle decohesion, and clay particles are wrenched from the fracture surface [*Andreani et al.*, 2008].

8.3.4. Dry CO_2: Fracture-Brine Interactions

Dry CO_2 has the potential to dry out fracture walls, and if the fracture water is brine, salt may precipitate. The risk of dry CO_2 in contact with the caprock is debated and not considered likely.

To what extent such dry-outs may occur in the caprock depends on the initial water content of the CO_2 and the volume of CO_2 that migrates through the fracture. The degree of dry-out also depends on the possibility of water to be supplied from the shale/mudstone matrix. Whether total dry-out or not may occur, salt may form rapidly as soon as the saturation limit of the water with respect to halite (NaCl) is reached. This has been shown numerically for the near-well area in the reservoir [*Miri et al.*, 2015] and also suggested from injectivity losses in CO_2 injection operations at Snøhvit [*Grude et al.*, 2014] and Ketzin [*Baumann et al.*, 2014]. Salt formation in fractured systems has been less studied. *Miri et al.* [2015] studied the process of salt precipitation at the pore scale by performing experiments in a microchip as shown close up in Figure 8.6. The microchip can be considered a dual-porosity medium, where the simulated fracture serves as a region dried by CO_2 and the connected porous medium reflects the water-saturated matrix. Salt growth was initiated at the interface, but directed into the CO_2-rich phase. Rapid nucleation leads to the formation of massive porous aggregates of micrometer-sized crystals. This growth then proceeded to fill the entire fracture width, and fracture flow was significantly reduced. In the case of complete clogging, the dry-out may provide a strong negative feedback to the flow, hindering the likelihood of CO_2 leakage. The experiments were, however, performed in a microfluidic system representing a fractured permeable sand rather than shale/mudstone, and similar experiments must also be done on a tight rock before any conclusions can be made.

8.3.5. Shear Fractures and Frictional Properties

Effects of CO_2 alteration on material properties like cohesion and friction of faults and fractures should be evaluated for longtime storage. Data from laboratory tests and field observations show that friction coefficients generally vary between 0.6 and 0.85 [*Byerlee*, 1978; *Morrow et al.*, 1992], although values as low as 0.2 have also been reported for clay material in the literature. Natural faults and shear fractures often contain gouge, so the friction is strongly dependent on the mineral composition of the gouge [*Moore and Lockner*, 1995; *Tembe et al.*, 2010; *Samuelson and Spiers*, 2012] and decreases with increasing the bulk clay content. The type of fluid in the sliding surface will also influence friction, with water-wet surfaces decreasing the friction coefficient. Influence of $scCO_2$- and CO_2-saturated brine on frictional properties and slip velocities of fault gauges has been investigated experimentally. Rate and state frictional experiments were performed for clastic caprock material dominated by illite, quartz, and quartz-rich reservoir material [*Samuelson and Spiers*, 2012], showing that $scCO_2$ had no clear influence on frictional strength of dry or brine-saturated gouges. For a full overview of CO_2-related work on fault stability and reactivation, see *Rohmer et al.* [2016] and discussions therein.

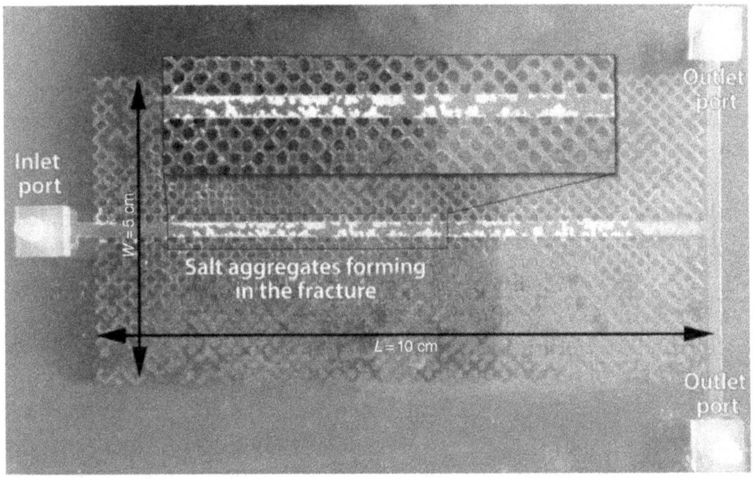

Figure 8.6 Pore-scale visualization of salt precipitation in the preferential pathway (fracture) of the heterogeneous microchip. After *Miri et al.* [2015]. A thick film of brine remains after CO_2 invasion at 5 s after injection. After 50 min, this film evaporates and the drying front shapes a meniscus at the interface between matrix and fracture. After 75 min, aggregation growth starts and occupies the fractures.

8.4. EXAMPLES FROM SEALING UNITS IN THE NORTH SEA

In the North Sea, the Sleipner CO_2 injection project has successfully injected more than 16 Mt. CO_2 since it was started in 1996 [*Furre et al.*, 2017]. At Sleipner, CO_2 is injected into the Utsira sand, and the regional extensive Pliocene to Pleistocene Nordland Shale is the caprock. The Utsira sand has been considered to have sufficient capacity to store CO_2 emissions from large part of the European industry (150 Gt/year) for several decades [*Ramírez et al.*, 2011]. Recent Norwegian feasibility studies have identified storage potential in the Sognefjord Formation, east of the Troll gas field in the North Sea. The primary seal for this unit is the Upper Jurassic Draupne Formation. In the following, the Nordland Shale and Draupne Formation will be discussed in more detail.

8.4.1. Nordland Shale

The Nordland Shale is only weakly consolidated because of the shallow burial (maximum ~800–1000 m) and consists mainly of clay and silt with some minor sands and with weak bedding [*Gregersen and Johannessen*, 2001; *Pillitteri et al.*, 2003]. The seal has been divided into three units: the lower seal draping the Utsira sand, the middle-down-lapping seal, and finally the upper seal with a truncated lower boundary [*Gregersen and Johannessen*, 2001]. *Bøe* and *Zweigel* [2001] estimated the pore throat radius and found that a CO_2 column of more than 800 m is required for capillary failure, and no free-phase CO_2 transport is therefore expected into the lower intact seal. A similar result was found by *Harrington et al.* [2009], concluding that only pressure-induced flow pathways (i.e., fractures) are relevant for seal breach because of the high gas breakthrough pressures for the intact caprock. However, the observed transport of CO_2 through thin intra-reservoir mudstone layers [*Zweigel et al.*, 2004] has raised discussions about the transport of CO_2 through the shale barriers. Diffusive transport through the seal will be slow, and significant amounts of plagioclase and chlorite in the caprock [*Bøe and Zweigel*, 2001] may react and have potential to bind CO_2 [*Pham et al.*, 2011; *Hellevang and Aagaard*, 2013], further slowing down diffusive transport of CO_2 through the seal.

The present-day CO_2 injection at Sleipner generates an overpressure that is <0.1 MPa [*Chadwick et al.*, 2012], making reservoir and seal pressurization a low risk. In addition, triaxial tests suggest that the lower seal is elastic-perfectly plastic, with Young's moduli measured to 0.19–0.29 GPa and Poisson's ratio of 0.18–0.25 [*Pillitteri et al.*, 2003], indicating that fracturing is unlikely. The vertical effective stress was found to be significantly larger than the horizontal, and it was also found that the vertical effective stress was larger than explained by the present-day overburden. This suggests over-consolidation of the sediments related to the Quaternary ice sheet loading in the North Sea. Capillary flow simulations including a fractured shale barrier were presented to replicate the monitored plume development [*Cavanagh and Haszeldine*, 2014], suggesting fracturing that predates the CO_2 injection explained as transient pore pressure hydro-fracturing during the deglaciation. Current plume monitoring program identifies a total of 9 CO_2 horizons separated by thin shale layers, typically 1–1.5 m thick, and one thicker layer of ca 6 m was identified [*Furre et al.*, 2017]. The data from Sleipner CO_2 project have been used for a wide variety of flow simulations [e.g., *Cavanagh*, 2013]. Geological assumptions about the nature and number of feeder channels for CO_2 from the injection point and up to the top layer where CO_2 accumulates are highly uncertain for current simulations [*Furre et al.*, 2017] and represent a key challenge for the future.

8.4.2. Draupne Formation

The Upper Jurassic Draupne Formation forms a several hundred meter thick source rock, but also a good seal for deeper reservoirs. Equivalent Upper Jurassic formation is the Kimmeridge Clay Formation, but also Spekk, Mandal, and Tau formations are approximate equates [*Faleide et al.*, 2015]. Recently, 9 m of well-preserved core material from the Draupne Formation within the Ling Depression was made available for CO_2-related research on seal integrity in the North Sea. The low permeability and high CO_2 capillary breakthrough pressure suggest this to be an excellent caprock [*Skurtveit et al.*, 2012; *Skurtveit et al.*, 2015]. Mineralogy of intact shale shows a bulk composition of quartz, microcline, and clay, with the clay phase dominated by kaolinite and smectite and total organic carbon around 7–8%. Along a natural shear fracture observed in the upper section of the core, increased calcite, pyrite, and siderite content is found compared to the intact shale mineralogy. Main uncertainty for the sealing capacity evaluation of this formation is related to the observed natural shear fracture, where the capillary entry pressure could be lower and the permeability could be higher than for the intact material. There are no indications that the observed shear fractures are common for the Draupne Formation, nor that they are continuous throughout the thickness of the seal; however, the samples provide excellent core material for further investigation of natural shear fractures in clay-rich seal. Potential reactions for CO_2 percolation into fractured Draupne Formation include calcite dissolution [e.g., *Alemu et al.*, 2011] that might alter the porosity

[*Andreani et al.*, 2008], whereas mobilization of less soluble particles or swelling of smectite may result in clogging and sealing of the fracture as demonstrated experimentally by *Bastiaens et al.* [2007] and *Ellis et al.* [2013].

8.5. CONCLUDING REMARKS AND FUTURE DIRECTIONS

Dedicated research on CO_2 storage and related fluid-rock interaction processes has provided insight and new data on transport and reactions to be expected for both reservoirs and sealing units. This review, focusing on flow and alteration processes in clay-rich seals (Fig. 8.1), showed that the most likely transport of CO_2 into clay-rich seal is by diffusion and only the lower 1–10 m of a typical clay seal will be affected. Flow of CO_2 into the seal is only likely if the seal is damaged due to preexisting fractures or faults within the seal. However, the long-term effects of alteration are an essential part of a storage system, and fluid-rock interaction processes may be important for leakage mitigation and as remediation strategies [*Kim and Santamarina*, 2013; *Tongwa et al.*, 2013; *Druhan et al.*, 2014; *Vialle et al.*, 2016].

Reaction experiments and kinetic modeling show reactivity between CO_2-saturated fluids (acidic) and clay-rich caprock material, involving kaolinite precipitation and chlorite-to-ankerite transformation. However, for dominating clay-rich caprock minerals like quartz, illite, and smectite, the reactivity due to drop in pH is low. Carbonate-rich shale is found to be more reactive than clay-rich shale; however, reactions may also be buffered depending on the fluid composition and water-rock ratio. For the experiments and models available, only minor porosity changes are observed, and implications for flow and CO_2 transport are uncertain due to limited data available, slow reaction rates, and low flow rates in clay-rich seals. Important alteration processes applicable for clay-rich caprocks include the swelling and shrinking property of smectites due to CO_2 sorption, where limited data are available. Clay swelling due to CO_2 raises an interesting discussion about the possibilities for self-limiting processes in clay-rich seal; however, a better understanding of the difference in amount of swelling between CO_2 and water and possible interaction effects is needed to fully understand the implications for CO_2 transport. Special cases of CO_2 transport in fractured caprock discussed in this paper include carbonate-cemented fractures, the possibility for salt precipitation resulting from entering of dry CO_2, and CO_2 effects on frictional properties of clay-rich gauge. The main finding is that there is very little work dedicated to effects of CO_2 in fractured clay-rich seals. It can be argued that damaged seals with extensive fracturing and faulting should be avoided; however, in order to perform proper risk evaluation of clay-rich caprocks, the process involved needs to be understood and discussed.

This review identifies several challenges related to the dynamic interaction and coupling of processes and research approaches. The available experimental data are difficult to upscale and extrapolate both in time and space. The time frame during laboratory experiments is different from reservoir conditions, and pressure and temperature conditions are adjusted to speed up the reactions. Alteration experiments might have unrealistic supply of reactive fluids, and the heterogeneities of a system might not be captured in a single experiment. Some important research gaps and processes that require a better understanding in order to secure the long-term safety of CO_2 sequestration are suggested for further investigation:

- Extrapolation between laboratory experiments and field observations to verify the rates and complexity used for kinetic modeling of reaction potential within sealing units and to document the net effect of alteration on transport properties
- More experimental data on clay-rich caprocks and models specifically addressing effect of clay swelling and dissolution of organic matter in a brine-CO_2 system
- Fracture transmissibility and the hydromechanical-chemical coupling of single fractures and fracture network for two-phase flow involving CO_2
- Long-term effects of CO_2 on frictional stability of clay-rich gauge and their mechanical and transport properties

ACKNOWLEDGMENTS

This work has been funded by Research Council of Norway (RCN) under the CO_2 seal bypass project (grant no. 244049) and by the FME SUCCESS Centre (grant 193825/S60). FME SUCCESS is a consortium with partners from industry and science, hosted by Christian Michelsen Research AS. The work by Elin Skurtveit is also supported by NGI internal funding. Bahman Bohloli is thanked for helpful comments on an early version of the manuscript. Two anonymous reviewers and editor, Stèphanie Vialle, are thanked for their detailed input significantly improving the manuscript.

REFERENCES

Alemu, B. L., P. Aagaard, I. A. Munz, and E. Skurtveit (2011), Caprock interaction with CO_2: A laboratory study of reactivity of shale with supercritical CO_2 and brine, *Applied Geochemistry*, *26*(12), 1975–1989, doi:10.1016/j.apgeochem.2011.06.028.

Andreani, M., P. Gouze, L. Luquot, and P. Jouanna (2008), Changes in seal capacity of fractured claystone caprocks induced by dissolved and gaseous CO_2 seepage, *Geophysical Research Letters*, *35*(14), L14404.

Angeli, M., M. Soldal, E. Skurtveit, and E. Aker (2009), Experimental percolation of supercritical CO_2 through a caprock, *Energy Procedia*, *1*(1), 3351–3358.

Armitage, P., D. Faulkner, R. Worden, A. Aplin, A. Butcher, and J. Iliffe (2011), Experimental measurement of, and controls on, permeability and permeability anisotropy of caprocks from the CO_2 storage project at the Krechba field, Algeria, *Journal of Geophysical Research - Solid Earth*, *116*(B12), B12208.

Armitage, P., D. Faulkner, and R. Worden (2013), Caprock corrosion, *Nature Geoscience*, *6*(2), 79–80.

Balashov, V. N., G. D. Guthrie, C. L. Lopano, J. A. Hakala, and S. L. Brantley (2015), Reaction and diffusion at the reservoir/shale interface during CO_2 storage: Impact of geochemical kinetics, *Applied Geochemistry*, *61*, 119–131.

Bandis, S., A. Lumsden, and N. Barton (1983), Fundamentals of rock joint deformation, paper presented at International Journal of Rock Mechanics and Mining Sciences & Geomechanics Abstracts, Elsevier.

Barenblatt, G., I. P. Zheltov, and I. Kochina (1960), Basic concepts in the theory of seepage of homogeneous liquids in fissured rocks [strata], *Journal of Applied Mathematics and Mechanics*, *24*(5), 1286–1303.

Barton, N., S. Bandis, and K. Bakhtar (1985), Strength, deformation and conductivity coupling of rock joints, paper presented at International Journal of Rock Mechanics and Mining Sciences & Geomechanics Abstracts, Elsevier.

Bastiaens, W., F. Bernier, and X. L. Li (2007), SELFRAC: Experiments and conclusions on fracturing, self-healing and self-sealing processes in clays, *Physics and Chemistry of the Earth, Parts A/B/C*, *32*(8), 600–615.

Baumann, G., J. Henninges, and M. De Lucia (2014), Monitoring of saturation changes and salt precipitation during CO_2 injection using pulsed neutron-gamma logging at the Ketzin pilot site, *International Journal of Greenhouse Gas Control*, *28*, 134–146.

Berre, T. (2011), Triaxial testing of soft rocks, *Geotechnical Testing Journal*, *34*(1), 61–75.

Bird, R. B. (2002), Transport phenomena, *Applied Mechanics Reviews*, *55*(1), R1–R4.

Bøe, R., and P. Zweigel (2001), Characterisation of the Nordland Shale in the Sleipner area by XRD analysis – a contribution to the Saline Aquifer CO_2 Storage (SACS) project. Confidential SINTEF Report 33.0764.00/01/01.

Boulin, P., P. Bretonnier, V Vassil, A. Samouillet, M. Fleury, and J. Lombard (2013), Sealing efficiency of caprocks: Experimental investigation of entry pressure measurement methods, *Marine and Petroleum Geology*, *48*, 20–30.

Brush, D. J., and N. R. Thomson (2003), Fluid flow in synthetic rough-walled fractures: Navier-Stokes, stokes, and local cubic law simulations, *Water Resources Research*, *39*(4), 1085.

Busch, A., S. Alles, Y. Gensterblum, D. Prinz, D. N. Dewhurst, M. D. Raven, H. Stanjek, and B. M. Krooss (2008), Carbon dioxide storage potential of shales, *International Journal of Greenhouse Gas Control*, *2*(3), 297–308.

Busch, A., A. Amann-Hildenbrand, P. Bertier, M. Waschbuesch, and B. M. Krooss (2010), The significance of caprock sealing integrity for CO_2 storage, paper presented at SPE International Conference on CO_2 Capture, Storage, and Utilization, Society of Petroleum Engineers.

Busch, A., N. Kampman, S. Hangx, J. Snippe, M. Bickle, P. Bertier, H. Chapman, C. Spiers, R. Pijnenburg, and J. Samuelson (2014), The Green River natural analogue as a field laboratory to study the long-term fate of CO_2 in the subsurface, *Energy Procedia*, *63*, 2821–2830.

Busch, A., P. Bertier, Y. Gensterblum, G. Rother, C. Spiers, M. Zhang, and H. Wentinck (2016), On sorption and swelling of CO_2 in clays, *Geomechanics and Geophysics for Geo-Energy and Geo-Resources*, *2*(2), 111–130.

Byerlee, J. (1978), Friction of rocks, *Pure and Applied Geophysics*, *116*(4–5), 615–626.

Carey, J. W., Z. Lei, E. Rougier, H. Mori, and H. Viswanathan (2015), Fracture-permeability behavior of shale, *Journal of Unconventional Oil and Gas Resources*, *11*, 27–43.

Carroll, S. A., W. W. McNab, and S. C. Torres (2011), Experimental study of cement-sandstone/shale-brine-CO_2 interactions, *Geochemical Transactions*, *12*(1), 9.

Cavanagh, A. (2013), Benchmark calibration and prediction of the Sleipner CO_2 plume from 2006 to 2012, *Energy Procedia*, *37*, 3529–3545.

Cavanagh, A., and R. S. Haszeldine (2014), The Sleipner storage site: Capillary flow modeling of a layered CO_2 plume requires fractured shale barriers within the Utsira formation, *International Journal of Greenhouse Gas Control*, *21*, 101–112.

Chadwick, R., G. Williams, J. Williams, and D. Noy (2012), Measuring pressure performance of a large saline aquifer during industrial-scale CO_2 injection: The Utsira Sand, Norwegian North Sea, *International Journal of Greenhouse Gas Control*, *10*, 374–388.

Chareonsuppanimit, P., S. A. Mohammad, R. L. Robinson, and K. A. Gasem (2012), High-pressure adsorption of gases on shales: Measurements and modeling, *International Journal of Coal Geology*, *95*, 34–46.

Chiaramonte, L., J. A. White, and W. Trainor-Guitton (2015), Probabilistic geomechanical analysis of compartmentalization at the Snøhvit CO_2 sequestration project, *Journal of Geophysical Research - Solid Earth*, *120*(2), 1195–1209.

Chiquet, P., D. Broseta, and S. Thibeau (2007), Wettability alteration of caprock minerals by carbon dioxide, *Geofluids*, *7*(2), 112–122.

Cho, Y., E. Ozkan, and O. G. Apaydin (2013), Pressure-dependent natural-fracture permeability in shale and its effect on shale-gas well production, *SPE Reservoir Evaluation & Engineering*, *16*(2), 216–228.

De Jong, S., C. Spiers, and A. Busch (2014), Development of swelling strain in smectite clays through exposure to carbon dioxide, *International Journal of Greenhouse Gas Control*, *24*, 149–161.

Dockrill, B., and Z. K. Shipton (2010), Structural controls on leakage from a natural CO_2 geologic storage site: Central Utah, USA, *Journal of Structural Geology*, *32*(11), 1768–1782.

Druhan, J. L., S. Vialle, K. Maher, and S. Benson (2014), A reactive transport model for geochemical mitigation of CO_2 leaking into a confined aquifer, *Energy Procedia*, *63*, 4620–4629.

Edlmann, K., S. Haszeldine, and C. McDermott (2013), Experimental investigation into the sealing capability of naturally fractured shale caprocks to supercritical carbon dioxide flow, *Environment and Earth Science*, *70*(7), 3393–3409.

Ellis, B. R., G. S. Bromhal, D. L. McIntyre, and C. A. Peters (2011), Changes in caprock integrity due to vertical migration of CO_2-enriched brine, *Energy Procedia*, *4*, 5327–5334.

Ellis, B. R., J. P. Fitts, G. S. Bromhal, D. L. McIntyre, R. Tappero, and C. A. Peters (2013), Dissolution-driven permeability reduction of a fractured carbonate caprock, *Environmental Engineering Science*, *30*(4), 187–193.

Espinoza, D. N., and J. C. Santamarina (2012), Clay interaction with liquid and supercritical CO_2: The relevance of electrical and capillary forces, *International Journal of Greenhouse Gas Control*, *10*, 351–362.

Faleide, J. I., K. Bjørlykke, and R. H. Gabrielsen (2015), Geology of the Norwegian continental shelf, in K. Bjørlykke (ed.), *Petroleum Geoscience: From Sedimentary Environments to Rock Physics*, pp. 603–637, Springer Verlag, Berlin Heidelberg.

Faulkner, D., C. Jackson, R. Lunn, R. Schlische, Z. K. Shipton, C. Wibberley, and M. Withjack (2010), A review of recent developments concerning the structure, mechanics and fluid flow properties of fault zones, *Journal of Structural Geology*, *32*(11), 1557–1575.

Festoy, S., and V. Golf-Racht (1989), Gas gravity drainage in fractured reservoirs through new dual-continuum approach (includes associated papers 20296 and 20390), *SPE Reservoir Engineering*, *4*(3), 271–278.

Fripiat, J., M. Cruz, B. Bohor, and J. Thomas Jr (1974), Interlamellar adsorption of carbon dioxide by smectites, *Clays and Clay Minerals*, *22*(1), 23–30.

Furre, A.-K., O. Eiken, H. Alnes, J. N. Vevatne, and A. F. Kiær (2017), 20 Years of monitoring CO_2-injection at Sleipner, *Energy Procedia*, *114*, 3916–3926, doi:10.1016/j.egypro.2017.03.1523.

Gaus, I., M. Azaroual, and I. Czernichowski-Lauriol (2005), Reactive transport modelling of the impact of CO_2 injection on the clayey cap rock at Sleipner (North Sea), *Chemical Geology*, *217*(3), 319–337.

Gherardi, F., T. Xu, and K. Pruess (2007), Numerical modeling of self-limiting and self-enhancing caprock alteration induced by CO_2 storage in a depleted gas reservoir, *Chemical Geology*, *244*(1), 103–129.

Giesting, P., S. Guggenheim, A. F. K. van Groos, and A. Busch (2012), Interaction of carbon dioxide with Na-exchanged montmorillonite at pressures to 640bars: Implications for CO_2 sequestration, *International Journal of Greenhouse Gas Control*, *8*, 73–81.

Gregersen, U., and P. Johannessen (2001), The Neogene Utsira sand and its seal in the Viking Graben area, North Sea saline aquifer CO_2 storage (SACS) project, phase 2 task 1.7 geology, *Geological Survey of Denmark and Greenland Report*, *100*(2001), 1–2.

Grude, S., M. Landrø, and J. Dvorkin (2014), Pressure effects caused by CO_2 injection in the Tubåen Fm., the Snøhvit field, *International Journal of Greenhouse Gas Control*, *27*, 178–187.

Gutierrez, M., L. Øino, and R. Nygård (2000), Stress-dependent permeability of a de-mineralised fracture in shale, *Marine and Petroleum Geology*, *17*(8), 895–907.

Hangx, S., A. van der Linden, F. Marcelis, and A. Bauer (2013), The effect of CO_2 on the mechanical properties of the captain sandstone: Geological storage of CO_2 at the Goldeneye field (UK), *International Journal of Greenhouse Gas Control*, *19*, 609–619.

Harrington, J., and S. Horseman (1999), Gas transport properties of clays and mudrocks, *Geological Society, London, Special Publications*, *158*(1), 107–124.

Harrington, J. F., D. J. Noy, S. T. Horseman, D. J. Birchall, and R. A. Chadwick (2009), Laboratory study of gas and water flow in the Nordland Shale, Sleipner, North Sea, in M. Grobe, J. C. Pashin, and R. L. Dodge (eds.), *Carbon Dioxide Sequestration in Geological Media – State of the Science*, AAPG Studies in Geology 59, p. 521–543.

Hellevang, H. (2006), Interactions between CO_2, saline water and minerals during geological storage of CO_2, PhD thesis, The University of Bergen.

Hellevang, H., and P. Aagaard (2013), Can the long-term potential for carbonatization and safe long-term CO_2 storage in sedimentary formations be predicted?, *Applied Geochemistry*, *39*, 108–118.

Hellevang, H., V. T. Pham, and P. Aagaard (2013), Kinetic modelling of CO_2–water–rock interactions, *International Journal of Greenhouse Gas Control*, *15*, 3–15.

Hildenbrand, A., S. Schlömer, B. Krooss, and R. Littke (2004), Gas breakthrough experiments on pelitic rocks: Comparative study with N_2, CO_2 and CH_4, *Geofluids*, *4*(1), 61–80.

Ishii, E. (2015), Predictions of the highest potential transmissivity of fractures in fault zones from rock rheology: Preliminary results, *Journal of Geophysical Research - Solid Earth*, *120*(4), 2220–2241.

Jaeger, J. C., N. G. Cook, and R. Zimmerman (2009), *Fundamentals of Rock Mechanics*, Hoboken, NJ, John Wiley & Sons.

Kampman, N., N. M. Burnside, Z. K. Shipton, H. J. Chapman, J. A. Nicholl, R. M. Ellam, and M. J. Bickle (2012), Pulses of carbon dioxide emissions from intracrustal faults following climatic warming, *Nature Geoscience*, *5*(5), 352–358.

Kampman, N., A. Maskell, M. Bickle, J. Evans, M. Schaller, G. Purser, Z. Zhou, J. Gattacceca, E. Peitre, and C. Rochelle (2013), Scientific drilling and downhole fluid sampling of a natural CO_2 reservoir, Green River, Utah, *Scientific Drilling*, *16*, 33–43.

Kampman, N., M. Bickle, A. Maskell, H. Chapman, J. Evans, G. Purser, Z. Zhou, M. Schaller, J. C. Gattacceca, and P. Bertier (2014a), Drilling and sampling a natural CO_2 reservoir: Implications for fluid flow and CO_2-fluid–rock reactions during CO_2 migration through the overburden, *Chemical Geology*, *369*, 51–82.

Kampman, N., M. Bickle, M. Wigley, and B. Dubacq (2014b), Fluid flow and CO_2–fluid–mineral interactions during CO_2-storage in sedimentary basins, *Chemical Geology*, *369*, 22–50.

Kaszuba, J. P., D. R. Janecky, and M. G. Snow (2005), Experimental evaluation of mixed fluid reactions between supercritical carbon dioxide and NaCl brine: Relevance to the integrity of a geologic carbon repository, *Chemical Geology*, *217*(3), 277–293.

Kim, S., and J. C. Santamarina (2013), CO_2 breakthrough and leak-sealing – experiments on shale and cement, *International Journal of Greenhouse Gas Control*, *19*, 471–477.

Le Guen, Y., F. Renard, R. Hellmann, E. Brosse, M. Collombet, D. Tisserand, and J. P. Gratier (2007), Enhanced deformation of limestone and sandstone in the presence of high fluids, *Journal of Geophysical Research - Solid Earth, 112*(B5), B05421.

Lim, K., and K. Aziz (1995), Matrix-fracture transfer shape factors for dual-porosity simulators, *Journal of Petroleum Science and Engineering, 13*(3), 169–178.

de Lima, V., S. Einloft, J. M. Ketzer, M. Jullien, O. Bildstein, and J.-C. Petronin (2011), CO_2 geological storage in saline aquifers: Paraná Basin caprock and reservoir chemical reactivity, *Energy Procedia, 4*, 5377–5384.

Liu, F., P. Lu, C. Griffith, S. W. Hedges, Y. Soong, H. Hellevang, and C. Zhu (2012), CO_2–brine–caprock interaction: Reactivity experiments on Eau Claire shale and a review of relevant literature, *International Journal of Greenhouse Gas Control, 7*, 153–167.

Loring, J. S., E. S. Ilton, J. Chen, C. J. Thompson, P. F. Martin, P. Bénézeth, K. M. Rosso, A. R. Felmy, and H. T. Schaef (2014), In Situ study of CO_2 and H_2O partitioning between Na–Montmorillonite and variably wet supercritical carbon dioxide, *Langmuir, 30*(21), 6120–6128.

Lu, J., M. Wilkinson, R. S. Haszeldine, and A. E. Fallick (2009), Long-term performance of a mudrock seal in natural CO_2 storage, *Geology, 37*(1), 35–38.

Mao, Y., M. Zeidouni, and R. Askari (2017), Effect of leakage pathway flow properties on thermal signal associated with the leakage from CO_2 storage zone, *Greenhouse Gases: Science and Technology, 7*(3), 512–529.

McCain, W. D. (1990), *The Properties of Petroleum Fluids*, Tulsa, OK, PennWell Books.

Michels, L., J. O. Fossum, Z. Rozynek, H. Hemmen, K. Rustenberg, P. A. Sobas, G. N. Kalantzopoulos, K. Knudsen, M. Janek, and T. Plivelic (2015), Intercalation and Retention of Carbon Dioxide in a Smectite Clay Promoted by Interlayer Cations. Scientific Reports, 5.

Miri, R., P. Aagaard, and H. Hellevang (2014a), Examination of CO_2–SO_2 solubility in water by SAFT1. Implications for CO_2 transport and storage, *The Journal of Physical Chemistry B, 118*(34), 10214–10223.

Miri, R., H. Hellevang, A. Braathen, and P. Aagaard (2014b), Phase relations in the Longyearbyen CO_2 lab reservoir-forecasts for CO_2 injection and migration, *Norwegian Journal of Geology, 94*(2–3), 217–232.

Miri, R., R. v. Noort, P. Aagaard, and H. Hellevang (2015), New insights on the physics of salt precipitation during injection of CO_2 into saline aquifers, *International Journal of Greenhouse Gas Control, 43*, 10–21.

Moore, D. E., and D. Lockner (1995), The role of microcracking in shear-fracture propagation in granite, *Journal of Structural Geology, 17*(1), 95–114.

Morodome, S., and K. Kawamura (2009), Swelling behavior of Na-and Ca-montmorillonite up to 150 C by in situ X-ray diffraction experiments, *Clays and Clay Minerals, 57*(2), 150–160.

Morrow, C., B. Radney, and J. Byerlee (1992), Frictional strength and the effective pressure law of Montmorillonite and illite clays, *International Geophysics, 51*, 69–88.

Mouzakis, K. M., A. K. Navarre-Sitchler, G. Rother, J. L. Bañuelos, X. Wang, J. P. Kaszuba, J. E. Heath, Q. R. Miller, V. Alvarado, and J. E. McCray (2016), Experimental study of porosity changes in shale caprocks exposed to CO_2-saturated brines I: Evolution of mineralogy, pore connectivity, pore size distribution, and surface area, *Environmental Engineering Science, 33*(10), 725–735.

Myer, L. R. (2000), Fractures as collections of cracks, *International Journal of Rock Mechanics and Mining Sciences, 37*(1), 231–243.

Nordgård Bolås, H. M., C. Hermanrud, and G. M. Teige (2005), Seal capacity estimation from subsurface pore pressures, *Basin Research, 17*(4), 583–599.

Ogata, K., K. Senger, A. Braathen, and J. Tveranger (2014), Fracture corridors as seal-bypass systems in siliciclastic reservoir-cap rock successions: Field-based insights from the Jurassic Entrada Formation (SE Utah, USA), *Journal of Structural Geology, 66*, 162–187, doi:10.1016/j.jsg.2014.05.005.

Olivella, S., and E. E. Alonso (2008), Gas flow through clay barriers. *Géotechnique 58*(3), 157–176.

Oron, A. P., and B. Berkowitz (1998), Flow in rock fractures: The local cubic law assumption reexamined, *Water Resources Research, 34*(11), 2811–2825.

Ougier-Simonin, A., F. Renard, C. Boehm, and S. Vidal-Gilbert (2016), Microfracturing and microporosity in shales, *Earth-Science Reviews, 162*, 198–226.

Pham, V., P. Lu, P. Aagaard, C. Zhu, and H. Hellevang (2011), On the potential of CO_2–water–rock interactions for CO_2 storage using a modified kinetic model, *International Journal of Greenhouse Gas Control, 5*(4), 1002–1015.

Pillitteri, A., P. Cerasi, J. Stavrum, P. Zweigel, and R. Bøe (2003), Rock mechanical tests of shale samples from the cap rock of the Utsira Sand in well 15/9-A11, A contribution to the saline aquifer CO_2 Storage (SACS) project, *Rep. 33.5324.00/06/03*, SINTEF Petroleum Research.

Pyrak-Nolte, L., and J. Morris (2000), Single fractures under normal stress: The relation between fracture specific stiffness and fluid flow, *International Journal of Rock Mechanics and Mining Sciences, 37*(1), 245–262.

Ramírez, A., R. Hoefnagels, M. van den Broek, N. Strachan, A. Fidje, K. Espegren, P. Seljom, M. Blesl, T. Kober, and P. E. Grohnheit (2011), A Comparison of national CCS strategies for Northwest Europe, with a focus on the potential of common CO_2 storage at the Utsira formation, *Energy Procedia, 4*, 2401–2408.

Rathnaweera, T., P. Ranjith, M. Perera, and A. Haque (2016), Influence of CO_2–brine Co-injection on CO_2 storage capacity enhancement in deep saline aquifers: An experimental study on Hawkesbury sandstone formation, *Energy & Fuels, 30*(5), 4229–4243.

Renshaw, C. E. (1995), On the relationship between mechanical and hydraulic apertures in rough-walled fractures, *Journal of Geophysical Research - Solid Earth, 100*(B12), 24629–24636.

Rohmer, J., A. Pluymakers, and F. Renard (2016), Mechano-chemical interactions in sedimentary rocks in the context of CO_2 storage: Weak acid, weak effects?, *Earth-Science Reviews, 157*, 86–110.

Rossen, R., and E. Shen (1989), Simulation of gas/oil drainage and water/oil imbibition in naturally fractured reservoirs, *SPE Reservoir Engineering, 4*(4), 464–470.

Saidi, A. (1983), Simulation of naturally fractured reservoirs, paper presented at SPE Reservoir Simulation Symposium, Society of Petroleum Engineers.

Samuelson, J., and C. J. Spiers (2012), Fault friction and slip stability not affected by CO_2 storage: Evidence from short-term laboratory experiments on North Sea reservoir sandstones and caprocks, *International Journal of Greenhouse Gas Control*, *11*, S78–S90.

Schaef, H. T., E. S. Ilton, O. Qafoku, P. F. Martin, A. R. Felmy, and K. M. Rosso (2012), In situ XRD study of Ca 2+ saturated montmorillonite (STX-1) exposed to anhydrous and wet supercritical carbon dioxide, *International Journal of Greenhouse Gas Control*, *6*, 220–229.

Shipton, Z. K., J. P. Evans, D. Kirschner, P. T. Kolesar, A. P. Williams, and J. Heath (2004), Analysis of CO_2 leakage through 'low-permeability' faults from natural reservoirs in the Colorado plateau, east-central Utah, *Geological Society, London, Special Publications*, *233*(1), 43–58.

Skurtveit, E., E. Aker, M. Soldal, M. Angeli, and Z. Wang (2012), Experimental investigation of CO_2 breakthrough and flow mechanisms in shale, *Petroleum Geoscience*, *18*(1), 3–15, doi:10.1144/1354-079311-016.

Skurtveit, E., L. Grande, O. Y. Ogebule, R. H. Gabrielsen, J. I. Faleide, N. H. Mondol, R. Maurer, and P. Horsrud (2015), Mechanical testing and sealing capacity of the Upper Jurassic Draupne Formation, North Sea, 49th U.S. Rock Mechanics/Geomechanics Symposium, 28 June–1 July 2015, San Francisco, CA.

Song, J., and D. Zhang (2012), Comprehensive review of caprock-sealing mechanisms for geologic carbon sequestration, *Environmental Science & Technology*, *47*(1), 9–22.

Spycher, N., and K. Pruess (2005), CO_2-H_2O mixtures in the geological sequestration of CO_2. II. Partitioning in chloride brines at 12–100 C and up to 600 bar, *Geochimica et Cosmochimica Acta*, *69*(13), 3309–3320.

Sun, Y., M. Aman, and D. N. Espinoza (2016), Assessment of mechanical rock alteration caused by CO_2–water mixtures using indentation and scratch experiments, *International Journal of Greenhouse Gas Control*, *45*, 9–17.

Szabó, Z., H. Hellevang, C. Király, E. Sendula, P. Kónya, G. Falus, S. Török, and C. Szabó (2016), Experimental-modelling geochemical study of potential CCS caprocks in brine and CO_2-saturated brine, *International Journal of Greenhouse Gas Control*, *44*, 262–275.

Tambach, T. J., M. Koenen, L. J. Wasch, and F. van Bergen (2015), Geochemical evaluation of CO_2 injection and containment in a depleted gas field, *International Journal of Greenhouse Gas Control*, *32*, 61–80.

Tembe, S., D. A. Lockner, and T. F. Wong (2010), Effect of clay content and mineralogy on frictional sliding behavior of simulated gouges: Binary and ternary mixtures of quartz, illite, and montmorillonite, *Journal of Geophysical Research - Solid Earth*, *115*(B3), B08206.

Tongwa, P., R. Nygaard, A. Blue, and B. Bai (2013), Evaluation of potential fracture-sealing materials for remediating CO_2 leakage pathways during CO_2 sequestration, *International Journal of Greenhouse Gas Control*, *18*, 128–138.

Tribovillard, N.-P., A. Desprairies, E. Lallier-Vergès, P. Bertrand, N. Moureau, A. Ramdani, and L. Ramanampisoa (1994), Geochemical study of organic-matter rich cycles from the Kimmeridge Clay Formation of Yorkshire (UK): Productivity versus anoxia, *Palaeogeography, Palaeoclimatology, Palaeoecology*, *108*(1–2), 165–181.

Van der Meer, B. (2005), Carbon dioxide storage in natural gas reservoir, *Oil & Gas Science and Technology*, *60*(3), 527–536.

Vialle, S., J. L. Druhan, and K. Maher (2016), Multi-phase flow simulation of CO_2 leakage through a fractured caprock in response to mitigation strategies, *International Journal of Greenhouse Gas Control*, *44*, 11–25.

Walsh, J. (1981), Effect of pore pressure and confining pressure on fracture permeability, paper presented at International Journal of Rock Mechanics and Mining Sciences & Geomechanics Abstracts, Elsevier.

Wang, J., and Y. Peng (2014), Numerical modeling for the combined effects of two-phase flow, deformation, gas diffusion and CO_2 sorption on caprock sealing efficiency, *Journal of Geochemical Exploration*, *144*, 154–167.

Warren, J., and P. J. Root (1963), The behavior of naturally fractured reservoirs, *Society of Petroleum Engineers Journal*, *3*(3), 245–255.

Washburn, E. W. (1921), The dynamics of capillary flow, *Physical Review*, *17*(3), 273.

Weniger, P., W. Kalkreuth, A. Busch, and B. M. Krooss (2010), High-pressure methane and carbon dioxide sorption on coal and shale samples from the Paraná Basin, Brazil, *International Journal of Coal Geology*, *84*(3), 190–205.

Wollenweber, J., S. Alles, A. Busch, B. Krooss, H. Stanjek, and R. Littke (2010), Experimental investigation of the CO_2 sealing efficiency of caprocks, *International Journal of Greenhouse Gas Control*, *4*(2), 231–241.

Xu, T., J. A. Apps, and K. Pruess (2003), Reactive geochemical transport simulation to study mineral trapping for CO_2 disposal in deep arenaceous formations, *Journal of Geophysical Research - Solid Earth*, *108*(B2), 2071.

Xu, T., J. A. Apps, and K. Pruess (2005), Mineral sequestration of carbon dioxide in a sandstone–shale system, *Chemical Geology*, *217*(3), 295–318.

Zhang, C.-L. (2011), Experimental evidence for self-sealing of fractures in claystone, *Physics and Chemistry of the Earth, Parts A/B/C*, *36*(17), 1972–1980.

Zimmerman, R. W., and G. S. Bodvarsson (1996), Hydraulic conductivity of rock fractures, *Transport in Porous Media*, *23*(1), 1–30.

Zweigel, P., R. Arts, A. E. Lothe, and E. B. Lindeberg (2004), Reservoir geology of the Utsira Formation at the first industrial-scale underground CO_2 storage site (Sleipner area, North Sea), *Geological Society, London, Special Publications*, *233*(1), 165–180.

9

Coupled Processes in a Fractured Reactive System: A Dolomite Dissolution Study with Relevance to GCS Caprock Integrity

Jonathan Ajo-Franklin, Marco Voltolini, Sergi Molins, and Li Yang

ABSTRACT

Predicting the temporal evolution of fractures in impermeable sealing units above geological carbon storage reservoirs is crucial to understanding leakage risk as well as optimal selection of storage sites. Process models are required which couple geochemical, hydrological, and mechanical effects to predict whether transmissive fractures increase or decrease in permeability during exposure to CO_2 or CO_2-saturated brines. We present results from a unique flow-through experiment conducted to image fracture evolution in a dolomite sample with dynamic synchrotron X-ray microtomography (SXR-micro-CT) at appropriate stress states. The results of the experiment show localized aperture enlargement and development of a microporous weathered zone due to rapid dissolution of calcite crystals, consistent with prior observations. Surface roughening and fines mobilization were also observed during dissolution. Confining stress cycles conducted after initial flow confirmed that the asperities remained strong even after exposure to aqueous CO_2 for prolonged periods. Geochemical analysis of effluent samples shows an increasing rate of dolomite dissolution with time, likely due to early buffering by calcite dissolution and increasing reactive surface area. The resulting measurements are consistent with a simplified reactive transport model and will be used to calibrate high-resolution models of near-fracture processes in the context of geologic carbon storage.

9.1. INTRODUCTION

The geological storage of carbon dioxide relies on a variety of mechanisms to ensure that injected CO_2 remains deep in the subsurface beneath potable aquifers and interaction with the surface environment. The integrity of sealing units, typically low-permeability rocks, for example, shales, tight carbonates, or anhydrites, is crucial for CO_2 containment, particularly at early times when the buoyancy of $scCO_2$ might drive upward migration, or in cases where the reservoir unit is significantly overpressured during injection.

In such situations, the transport behavior of flaws in the sealing unit, particularly either natural or induced fractures, becomes crucial to understanding the security of injected CO_2. Fractures can provide high-permeability conduits for $scCO_2$ or carbonated brine movement; however, the long-term evolution of such conduits is highly uncertain and dependent on an intricate coupling between geomechanical, hydrological, and geochemical processes. Scenarios can be imagined where the fracture aperture expands and increases in permeability due to dissolution of reactive phases in the sealing unit (presuming carbonated brine movement). Alternatively, chemical dissolution could weaken fracture contact points, referred to as asperities, maintaining the open fracture aperture, resulting in fracture collapse under horizontal stresses and self-sealing behavior. Both dissolution and precipitation could also coexist, with sealing induced by precipitation downstream of the entry point of carbonated brine. Both enhancing and sealing scenarios have been demonstrated in prior laboratory experiments

Lawrence Berkeley National Laboratory, Berkeley, CA, USA

Geological Carbon Storage: Subsurface Seals and Caprock Integrity, Geophysical Monograph 238, First Edition.
Edited by Stéphanie Vialle, Jonathan Ajo-Franklin, and J. William Carey.

[e.g., *Ellis et al.*, 2011; *Elkhoury et al.*, 2015]. The integrated path of permeability during $scCO_2$ or carbonated brine movement determines the net loss of CO_2 from the reservoir.

In natural systems, examples can be found of both self-sealing and self-enhancing behaviors in fracture systems conducting fluids. In near-surface environments, fractures in carbonate formations often show aperture expansion due to dissolution induced by rainwater infiltration and atmospheric CO_2, a process responsible for karst development [e.g., *Groves and Howard*, 1994]. In contrast, fractures and fault material exhumed from deep low-permeability formations tend to show fractures sealed due to precipitation of gypsum, calcite, or quartz [e.g., *Cobbold et al.*, 2013], moderated by several processes including pressure solution [e.g., *Renard et al.*, 2000]. One of several significant differences between these environments is stress state and fluid pressure, which varies from between close to zero in the near-surface settings to 40–50 MPa lithostatic for GCS repositories at moderate depths. This suggests that operating close to in situ stress and pressure conditions may be useful in replicating sealing processes.

The Crystal Geyser site (Paradox Basin, Utah), one of the few natural analogues for a leaky seal in a CO_2-bearing system, demonstrates a combination of behaviors [*Shipton et al.*, 2004, 2005]. At the site, CO_2-charged brine migrates upward through a section of the Little Grand Wash Fault from a permeable sandstone reservoir and penetrates several clay-rich sealing formations. Upon reaching the near surface, CO_2 degassing and the consequent increase in pH lead to calcite precipitation and occlusion of permeable fractures. However, instead of halting CO_2 migration, this process is hypothesized to cause lateral translation of the surface CO_2 vent in space. The long-term persistence of CO_2 transport from depth, as evidenced by paleo-CO_2 vents at the surface, suggests that in this specific case, fault permeability is changing slowly if at all and is certainly not self-sealing over the century to millennium scale.

Even in the absence of stress, the evolution of fractures in reactive materials during dissolution can be quite complicated. A variety of studies have effectively demonstrated that in the absence of mechanical compaction, the impact of dissolution on fracture aperture enlargement is predominantly controlled by competition between mass transfer (advective as well as diffusive) and reaction rate at the rock surface with behaviors in this space characterized by a *phase diagram* of enhancement behavior. *Szymczak and Ladd* [2009] provide a comprehensive modeling study of this behavior and decompose behavior phase space in terms of Damhköhler (*Da*) number and Péclet (*Pe*) number where these two dimensionless numbers are the ratio of reaction rate to mean fluid velocity and the ratio between mass flux and diffusion coefficient, respectively. *Szymczak and Ladd* [2009], as well as many others [e.g., *Detwiler et al.*, 2003], show that boundaries exist in Da/Pe space separating uniform dissolution, face retreat, and wormhole evolution in fractured systems, where *wormholes* refer to local zones of dissolution which feed back to high influent fluxes and dissolution rates, a phenomenon studied in porous media by *Fred and Fogler* [1998].

9.2. PRIOR EXPERIMENTS

As mentioned, a variety of previous experiments have probed the evolution of fractures during dissolution with attention to permeability modification in the context of GCS. Typically, these experiments have examined changes in fracture aperture using a combination of X-ray micro-computed tomography (micro-CT) and surface profilometry with secondary measurements of permeability and aqueous chemistry. Of these measurements, micro-CT and profilometry are conducted off-line or in a simple before/after mode, preventing detailed temporal monitoring of fracture evolution. The exception to this is the remarkable transparent reactive fracture studies reported by *Detwiler et al.* [2003] which use a Hele-Shaw cell and a fracture model composed of potassium dihydrogen phosphate (KDP) crystals to directly resolve aperture enhancement with optical imaging; unfortunately, this approach is not transferrable to geological materials. We should note that the experiments we review investigate single-phase reactive fluids, largely carbonated brine, rather than two-phase ($scCO_2$/brine) systems. While multiphase flow would likely be present in the context of GCS and such flow processes have been studied in fractures [e.g., *Karpyn et al.*, 2007], few experiments to date have probed reactive $scCO_2$/brine mixtures in the laboratory at the aperture scale.

Noiriel et al. [2007] present an experiment where a small fractured limestone sample was subjected to injection of CO_2-saturated brine (partial pressure of 0.1 MPa) and monitored using differential pressure and repeated off-line microtomography. In this case, fracture aperture and permeability was observed to increase during dissolution, and the absence of applied normal load prevented any compaction. *Noiriel et al.* [2013] conducted a similar experiment with more extensive 3D analysis of fracture alteration and observe a similar increase in aperture as well as an increase in surface roughness due to variable matrix dissolution rates.

Ellis et al. [2011] conducted reactive core flood experiments and subjected artificially fractured samples from the Amherstburg limestone (50/50 calcite/dolomite) to prolonged carbonated brine flow at close to in situ pressures (10 MPa pore, 14 MPa confining). In the experiment, fracture aperture is observed continuously using medical CT with postmortem analysis using micro-CT and electron microscopy. During the course of the experiment, fracture cross-sectional area increased by 2.7 times over a week of flow. They also observed development of a significant altered zone and roughening due to variable dissolution rates in zones of differing mineralogy.

Ellis et al. [2013] conducted a similar experiment but directly measured an order-of-magnitude permeability reduction over the course of the experiment despite very similar experimental parameters. In this case, the reduction occurred during dissolution and was attributed to mobilization of less reactive fines as well as some erosion of asperities which maintained fracture aperture.

Elkhoury et al. [2013] present results from a core flood experiment on an artificially fractured sample from the lower Midale formation (Weyburn field). Their experiments, conducted at close to reservoir conditions (14 MPa pore, 28 MPa confining, ~61°C), also show development of enhanced dissolution zones, wormholes constrained in the fracture plane, and permeability enhancement during flow of carbonated brine. They utilize an optical profilometer and alignment jig to allow for time-lapse monitoring of fracture aperture ex situ.

Elkhoury et al. [2015] present a similar set of long-term (up to 6 months) dissolution studies examining fracture alteration in a dolomitic anhydrite subjected to carbonated brine injection. They find that at low flow rates, permeability enhancement and local dissolution occur; however, at higher flow rates, creation of a weak weathered zone led to eventual fracture compaction and sealing. This remarkable observation is true despite nearly equal mass loss between the two samples.

This list is not exhaustive; other studies involving dissolution observe widely contrasting behavior including aperture expansion [*Gouze et al.*, 2003], closure [*Polak et al.*, 2003], and various combinations during multicomponent experiments [*Polak* et al. 2004; *Liu et al.*, 2005; *Luquot et al.*, 2013; *Davila et al.*, 2016]. To summarize the existing results, a rich set of recent studies have demonstrated both self-enhancing and self-sealing behaviors in GCS-relevant seal materials subjected to carbonated brine flow. As observed by both *Ellis et al.* [2011, 2013] and *Elkhoury et al.* [2013, 2015], a key factor that may lead to self-sealing behavior during carbonated brine flow is the mechanical weakening of the near-fracture region due to dissolution, a process which involves both asperity erosion and creation of weaker weathered materials flanking the fracture aperture. In the case of *Elkhoury et al.* [2015], this was attributed to the presence of bimodal mineral dissolution rates, allowing generation of a microporous zone with distinct mechanical properties. The temporal evolution of both the near-fracture weathered zone and stress concentrations due to changes in aperture is at present poorly constrained due to the low temporal sampling in prior experiments.

9.3. MODELS FOR COUPLED PROCESSES IN FRACTURES

Notably, the coupled hydrological/geochemical models currently used in field-scale simulations are not yet capable of capturing the dynamics observed at the μm to mm range in the laboratory studies discussed previously. Commonly used field-scale simulators (e.g., TOUGHREACT, CrunchFlow) rely on coupling classical hydraulic aperture laws based on plane parallel fractures with mass loss due to dissolution [*Steefel and Lasaga*, 1994; *Xu et al.*, 2012]. One significant missing component is the absence of Damköhler-Péclet number-dependent fracture permeability relationships. These models are also not tightly coupled to mechanical models and predict the retreat of fracture faces proportional to mineral mass loss. They do not represent colloid mobilization and fines transport and only predict fracture closure in scenarios where downstream precipitation occludes flow. At present, they are not capable of representing self-sealing processes caused by the mechanisms observed at the laboratory scale and are likely inappropriate for field-scale prediction of GCS leak evolution.

The most mechanistic models developed to date for coupled processes within fractures parameterize the fracture hydrology, chemistry, and mechanics at the mm to micron scale, appropriate for predicting system response at the core scale [e.g., *Ameli et al.*, 2014; *Kim and Santamarina*, 2015; *Lang et al.*, 2015]. The leading edge of these efforts is currently the work of *Ameli et al.* [2014] who present a coupled computational model integrating a vertically averaged micro-continuum reactive transport solver [*Detwiler and Rajaram*, 2007], local aperture modification, and mechanical compression under normal loading conditions [*Pyrak-Nolte and Morris*, 2000]. Their model assumes elastic deformation of the two fracture halves and effectively captures asperity erosion. Qualitatively, their approach is able to capture fracture sealing behavior at high flow rates as observed by *Elkhoury et al.* [2015]. However, because near-fracture weathered zone evolution and mechanical compaction (nonelastic) is not captured, the approach would likely need to be altered to quantitatively match experimental data sets currently being generated. Contact models with plasticity and compaction are now being developed [e.g., *Li et al.*, 2015] but have yet to be fully coupled to flow and reactive chemistry.

9.4. CONTEXT FOR OUR STUDY

As can be seen from our review of recent studies relevant to fracture evolution under GCS conditions, a crucial missing component is laboratory data sets capable of capturing the temporal evolution of both the microporous weathered zone and the distribution of supporting asperities in systems at close to in situ stress conditions. In this study, we present the first fracture dissolution experiment where aperture and weathered zone extent are dynamically monitored at intermediate time points without depressurization, allowing characterization of system evolution and providing a necessary validation data set for the next generation of coupled fracture simulators.

These measurements are enabled by a novel high-pressure triaxial core holder developed specifically for SXR-micro-CT; the resulting measurement set provides a spatially resolved map at multiple time points. We study an artificially fractured dolomite sample with a sizable calcite fraction, a system with a bimodal reaction rate distribution driven by mineralogy, similar to the dolomite sample discussed in *Elkhoury et al.* [2015]. In our experiment, we observe both flow-controlled aperture enlargement and development of a microporous zone, in concordance with prior work [*Ellis et al.*, 2011, 2013; *Elkhoury et al.*, 2015]. Despite the extensive dissolution observed, the contact regions of the fracture retained sufficient strength to prevent closure, as demonstrated during a confining stress cycle.

9.5. MATERIALS AND EXPERIMENTAL PROCEDURE

The sample used for the experiment was a core (~3/8″ in diameter and ~1″ long) cut from a piece of dolomite from the Devonian Duperow Formation, a regionally extensive reservoir and seal in the Williston Basin with significant fracturing [e.g., *Babcock and Lageson*, 2013]. The Duperow is also both the reservoir and primary seal for proposed GCS projects on Kevin Dome in north central Montana. The sample was collected at the Gibson Dam site, where a fresh outcrop allowed the collection of minimally altered specimens. A picture of the sample with the mini cores and a map of the site where the sample was collected are shown in Figure 9.1. Fragments from the same piece were used for preliminary characterization via X-ray powder diffraction (XRPD) and Scanning electron microscopy with energy dispersive spectroscopy (SEM-EDS) to obtain quantitative information about the phases present and to evaluate their morphology and the rock texture.

The sample for the micro-CT experiment was split into two halves by placing it lengthwise between two blades and compressing it. To keep the fracture open and to

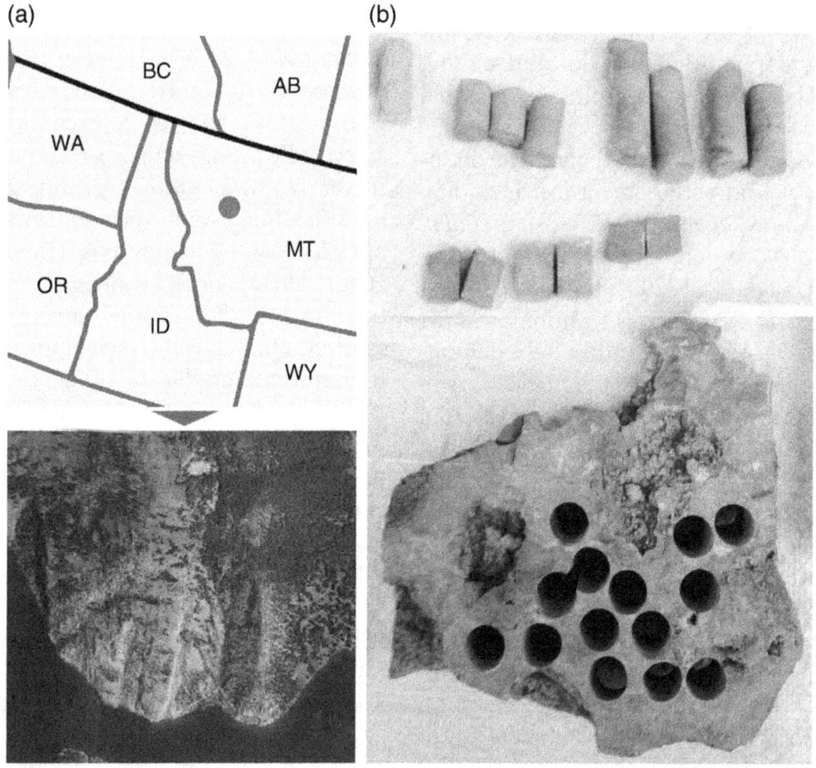

Figure 9.1 (a) Map showing the location of the Gibson Dam site in Montana (top) and a satellite view of the outcrop (bottom). In (b) the actual sample of Duperow dolomite and a selection of micro-cores both whole and split in halves are displayed.

allow a flow of the fluid during the experiment, the two halves were shifted vertically by ~1 mm; finally, the ends were remachined flat in order to be placed properly in the HP flow cell. This small offset created a variable aperture along the sample with well-defined contact areas and could be considered similar to a shear-induced fracture propped by asperity translation.

The experimental setup at the 8.3.2. beamline at the Advanced Light Source (ALS) included a high-pressure triaxial core holder developed at the Energy Geosciences Division of the Earth and Environmental Sciences Area, Lawrence Berkeley National Laboratory [*Voltolini et al.*, 2017]. The core holder, constructed of Al 7075-T651, is capable of axial and confining stress to 24 MPa (3500 psi), sufficient to simulate shallow GCS sites. The sample was jacketed using PTFE heat-shrink tubing and inserted in the triaxial cell. The influent solution was prepared by equilibrating a mixture of deionized water (55 ml) and CO_2 (48 ml, at 3.5 MPa/500 psi) at room temperature and at 10.34 MPa (1500 psi) for 12 h.

Confining pressure was provided by a high-pressure syringe pump (Isco 65DM) used in constant pressure mode at 11 MPa (~1600 psi). Axial load was slightly higher than confining pressure (~14 MPa) to simulate lithostatic stress during leakage at a shallow GCS site. A second syringe pump was used in constant flow mode (5 μl/min) to inject the influent. The pore pressure 9.6 MPa (~1400 psi) was maintained by a back pressure regulator in the outlet line. Samples of the effluent were collected approximately every 8 h during flow (flow was interrupted during the tomographic scans) to monitor the variation of the chemistry of the solution with the reaction. A schematic of the system is shown in Figure 9.2. One should note that the differential stress was relatively low during the flow portion of the experiment; this choice was made to preserve integrity of the thin confining jacket over the course of the experiment. The sample was slowly saturated with the influent solution before acquisition of the baseline image volume.

The experiment provided a series of volumes with the sample at different reaction times (time of effective reaction): 0 (baseline), 20, 56, and 113 h. After the reaction portion of the experiment was completed, confining pressure was cycled to higher levels to probe asperity strength and fracture closure. Confining stress was increased from 11 MPa (~1600 psi) by steps to 13.1 MPa (1900 psi), to 14.5 MPa (2100 psi), and finally to 15.85 MPa (2300 psi), keeping pore pressure constant. Axial load was kept at a constant ratio of 1.27 to confining stress. Scans were conducted after each pressure step.

After the completion of the micro-CT experiment, the sample was recovered for postmortem studies, including high-resolution SXR-micro-CT, SEM-EDS analysis, and inductively coupled plasma mass spectrometry (ICP-MS) of the effluent. The details of the single measurements and the results are documented below.

9.6. ANALYTICAL TECHNIQUES

To obtain a comprehensive knowledge of the complex system investigated, different experimental techniques were employed to provide constraints on mineralogy, microstructure, and fracture evolution.

9.6.1. X-Ray Powder Diffraction (XRPD)

A section from a Duperow dolomite core cut from the same outcrop sample was crushed and powdered to perform XRPD analysis. The instrument used was a Rigaku SmartLab®, and data were collected from 4° to 70° of 2θ using Cu Kα radiation. After manual identification of

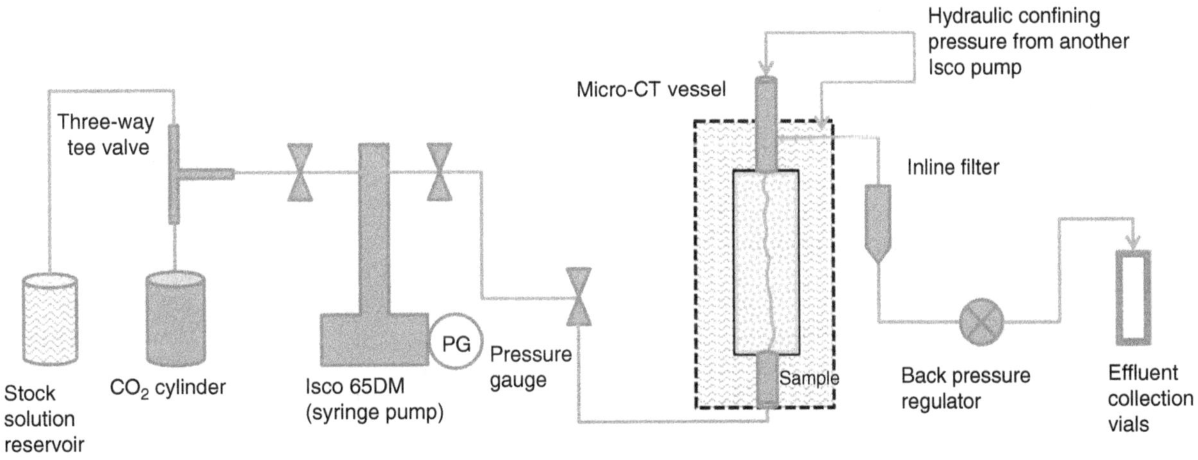

Figure 9.2 Schematics of the HP plumbing system used during the in situ SXR-micro-CT experiment.

Table 9.1 Weight Fractions of the Phases Identified via XRPD.

Mineral phase	Weight (%)
Quartz	<1
Smectite clay	1.8
Calcite	9.5
Dolomite	86.9
Plagioclase	1.8

the phases present, a Rietveld analysis of the collected XRPD profile [e.g., *Young*, 1993] was performed to obtain their weight fractions. The results are listed in Table 9.1.

As it is possible to appreciate from the results, this sample is composed mostly of dolomite, with a significant amount of calcite, plus minor amounts of non-carbonate phases, such as quartz, clay (~14 Å smectite), and feldspar (plagioclase). This analysis provided the starting point for planning the experiment, highlighting an interesting system with phases with different reactivity in a context of acidified brine flow. While for a bulk analysis of the sample XRPD is an excellent technique, to check the morphology of the phases and the microstructure of the sample, imaging techniques are needed.

9.6.2. Scanning Electron Microscopy with Energy Dispersive Spectroscopy (SEM-EDS)

A technique of choice to investigate the nanostructure of a material is scanning electron microscopy (SEM) with electron dispersive spectroscopy (EDS) used to provide spatial constraints on mineral phases. This technique was used to obtain information about the evolution of the morphology of the different phases as well as to image the formation of new textures induced by the dissolution process on the surface of the fracture. The analytical SEM used in this study is a variable pressure system based on a tungsten filament source (Zeiss EVO LS10®), and imaging was conducted using both secondary electron (SE) and backscattered electron (BSE) detectors. The instrument is also equipped with an EDS detector (Oxford Instruments X-Max® 20 mm²), allowing us to perform semiquantitative chemical analyses on the most interesting areas of the sample.

9.6.3. Synchrotron X-Ray Micro Tomography (SXR-Micro-CT)

The main experimental technique used for this work was SXR-micro-CT. With the physical setup described in the experimental section, we were able to perform an experiment which monitored dissolution induced by CO_2-saturated water in the vertically fractured dolomite core described previously. All imaging was conducted with the sample maintained at realistic stress conditions and without removal of the core. The scope of this experiment was to obtain 3D images of the sample at different dissolution stages as well as during a confining stress cycle at the end of the experiment.

The experiment was performed at X-ray micro-imaging beamline 8.3.2 [*MacDowell et al.*, 2012] at the ALS, Lawrence Berkeley National Laboratory. During the experiment, data were acquired in white light mode (polychromatic beam) directly utilizing the flux from the associated super bend magnet. The beam was pre-hardened using a 6 mm Al filter (6061) and a 1 mm Cu filter. This procedure filters (by means of absorption) the low-energy part of the spectrum from the source, optimizing the transmission through the sample as well as minimizing soft X-ray heating of the core holder. The detection chain utilized a 1 mm Ce-doped YAG crystal as a scintillator. The resulting optical projection was magnified using a long working distance lens and recorded using a high-resolution sCMOS detector (PCO.Edge, 2560 × 2160 pixels).

Each scan included 2049 projections over a 180° continuous rotation, with an exposure time of 240 ms for each projection. The pixel size of the projections was 6.71 μm, with a resulting horizontal field of view (FOV) of ~17 mm. Given the limited height of the X-ray beam, seven vertical tiles were collected in order to scan the entire rock core. The reconstruction of the volumes from the projections was carried out with a conventional filtered back-projection procedure [*Kak and Slaney*, 2001] using the software Octopus® [*Dierick et al.*, 2004].

The high-resolution scans done on selected portions of the sample, after the in situ experiment, were performed using a different configuration utilizing monochromatic light (33 keV), a 0.5 mm LuAG scintillator, 5× optics, and the same detector. This configuration allowed the collection of data with a 1.34 μm pixel size and lateral and vertical FOVs of ~3.4 and 2.9 mm, respectively, thus providing greater spatial resolution in sample sections with evolving microporous components.

9.6.4. Inductively Coupled Plasma Mass Spectrometry (ICP-MS)

During the in situ SXR-micro-CT experiment, samples of the effluent were collected in PET vials approximately every 8 h of reaction to monitor the chemistry of the reacted fluid. The collected fluids were analyzed after the experiment via ICP-MS (Perkin Elmer DRCII), after acidification with 2% ultrapure HNO_3, to obtain the molarity of different cations. The results of this analysis are plotted in Figure 9.3 and show evident changes in time in the amounts of Mg^{2+} and Ca^{2+} in solution with small amounts of secondary cations.

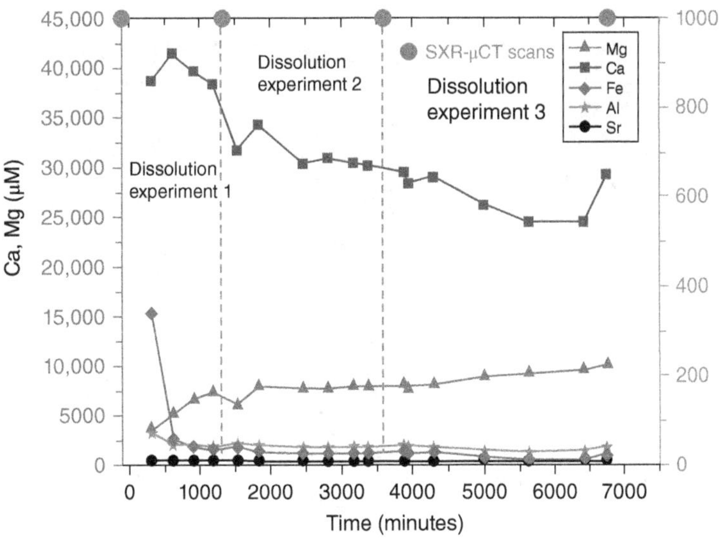

Figure 9.3 Curves from ICP-MS plotting the concentrations of the main cation species in the effluent.

9.7. RESULTS

9.7.1. In Situ Dissolution Monitored via SXR-Micro-CT

The in situ SXR-micro-CT experiment provided four primary data sets showing the evolution of the dolomite sample in time. The main results from qualitative observation of the data showed several interesting features: the development of a weathered layer on the surface of the fracture, the development of a transport-enhanced zone of dissolution (i.e., wormhole), and the absence of fracture closure from confining pressure due to mechanical stability of the supporting asperities. These features are visible in Figure 9.4. In Figure 9.4, vertical cutaways of the sample volume, perpendicular with respect to the fracture plane, are shown in panel (a). It is possible to appreciate the development of a weathered layer on the surface of the fracture, as well as some modifications in the fracture aperture, increasing with the dissolution time and developing faster close to the inlet (bottom of the sample: the flow direction was bottom up). Two mineral phases are also visible in the sample with different absorption levels, translated as gray values in figure; dolomite is dark gray, while calcite is light gray/white. A calcite veinlet is visible close to the inlet, a large biogenic calcite fragment is visible in the upper half of the sample, and scattered small calcite grains are visible throughout the whole sample.

In Figure 9.4 panel (b), a horizontal cut close to the inlet, highlighted in yellow (see e-book version) in the baseline volume, is shown to visualize a part of the sample where the reaction features are more extensive. In this panel, the main features of system development are visible, including a well-developed fracture-constrained wormhole, labeled in green. The modifications of the fracture aperture are not immediate; before the enlargement of the fracture, a 100–250 μm darker gray layer on the fracture surface develops. This darker gray color is due to partial volume effects in the data [e.g., *Ketcham and Carlson*, 2001], meaning that the system is developing at a scale smaller than the actual resolution of the measurement. In our case, this translates into having for each voxel (6.71 μm in size, isotropic) a summed contribution of dolomite + calcite + porosity, considering only X-ray attenuation. A decrease in calcite and/or dolomite, with the subsequent increase of the porosity (lowest attenuation), will make the single voxel darker. In addition to this partial volume effect, the weathered layer also displays newly created pores with sizes larger than the measurement resolution. These dissolution behaviors open new questions about how weathered layers develop. In this work, we have attempted to address sub-resolution features with a combination of SEM-EDS analysis and high-resolution SXR-micro-CT scans, as discussed below.

Concerning the aperture changes of the fracture, we show in Figure 9.5 renderings of the medial axis of the fracture. These planes have been labeled with the local aperture values, calculated using a local thickness transform [*Dougherty and Kunzelmann*, 2007] of the pore space in the fracture, aligned on the fracture's medial axis. In this plot, the contact points of the fracture are transparent. From this figure, it is possible to appreciate that the geometry of the fracture (the shape of its medial axis) does not change, except for a small section close to the inlet, while the aperture changes are quite evident (color map on the medial axis plane), with the start and growth

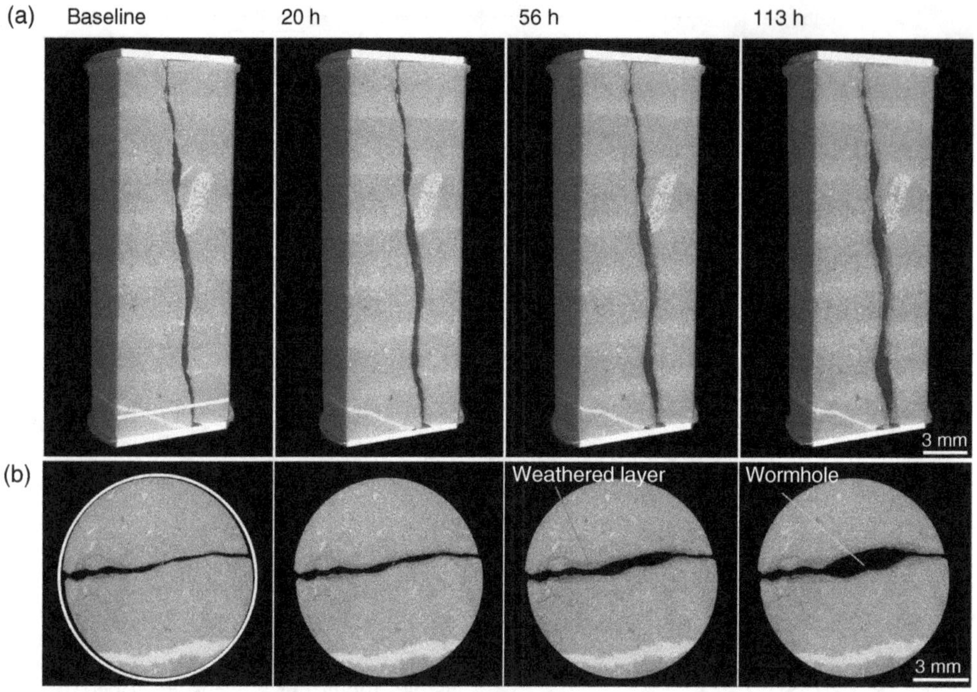

Figure 9.4 Panel (a) shows the volume cut images of the dissolution sequence (reaction times at the top of the figures). The sample is vertically cut approximately perpendicular with respect to the fracture plane. Dolomite is the dark gray phase, while calcite is light gray/white. The flow direction is from bottom to the top. Panel (b) shows a horizontal cut of the sample (marked with the light line in the baseline volume). The cut is close to the inlet, where the dissolution processes are more evident. Labeling shows the wormhole and the weathered layer developed on the surface of the fracture.

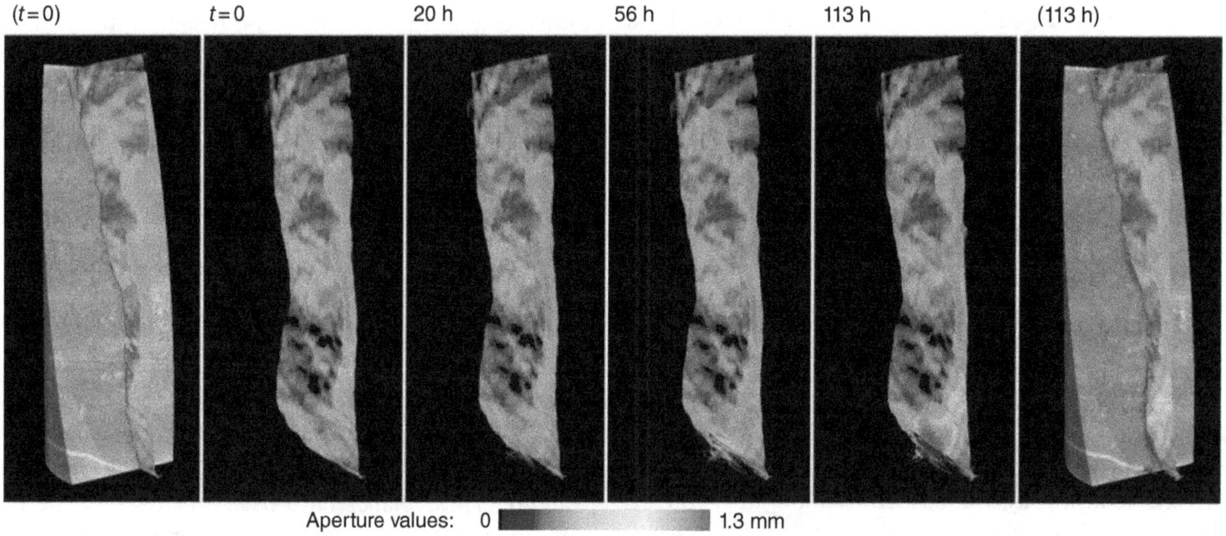

Figure 9.5 Analysis of the medial axis of the fracture of the sample at different stages. The medial axis is labeled with a color proportional to the local aperture.

of a wormhole, generating at the inlet and then bending on the right side of the sample where a preferential flow path is present.

To better visualize the dissolution process occurring at the fracture, Figure 9.6 shows the changes in aperture at each stage in fracture development. These images were generated by differencing sequential aperture projections after image registration to avoid artifacts due to the sample motion. From these images, it is possible to see that during the first stage, rapid dissolution occurs near

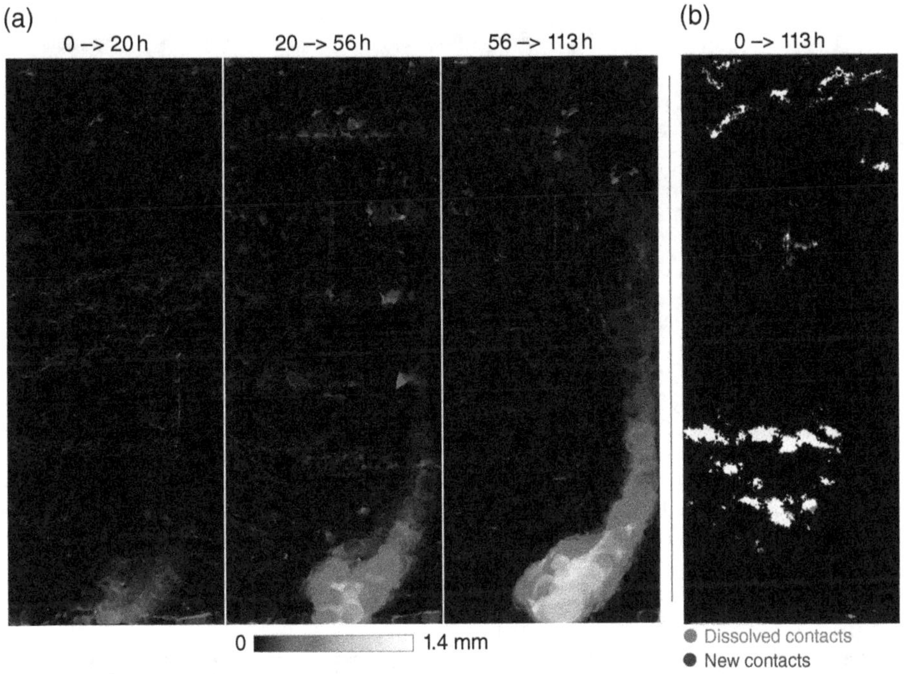

Figure 9.6 Panel (a) depicts changes in aperture due to dissolution after each step of the experiment. Panel (b) shows variation in the contact points between the baseline and the last measurements. Red: contacts which dissolved or collapsed. Blue: new contacts. White: contacts present at both the beginning and at the end of the experiment. (*See electronic version for color representation of the figure.*)

the inlet; however, more diffuse dissolution throughout the whole fracture is also visible. The same is true for the second step, but here the development of the wormhole becomes more and more evident, connecting the inlet with the portion of the fracture with a larger aperture and more favorable flow path; additionally, some parts of the fracture where the calcite is present (right side of the image) show more pronounced dissolution as well. During the last stage, the spatially diffuse dissolution becomes virtually absent, but strong dissolution along the wormhole, where the bulk of the dissolution occurs, continues.

Using the projection of the aperture values, it is also possible to check the evolution of macroscopic asperities that support the fracture. As mentioned previously, the development of a weak weathered zone near the fracture surface as well as high applied stress could lead to fracture closure. This is not what was observed in our experiment: after 113h of dissolution, the contact points appear to be largely unmodified in area or structure. In Figure 9.6, panel (b), we have plotted the variation of the geometry of the contact points from the baseline to the end of the experiment. Marked in red are the contact areas that disappeared (due to dissolution), in blue are the new contact points (local shrinkage and/or deposit of insoluble material), and in white are the contact points in common before and after the flow (see e-book version for color information). As can be seen, the variation of the contact zones is small. A limited amount of contact dissolution appears close to the wormhole, where it curves, and even less dissolution, close in extent to the experiment resolution, in the front of some contact areas. New contact points develop (again, to a very small extent), and they seem mostly related to the migration/accumulation of insoluble phases, since they are close to constriction points in the fracture plane. The new contacts tend to face the inlet and to be in the part of the sample closer to the outlet, further supporting the concept of the deposit of fines transported by the flow.

Finally, time-lapse SXR-micro-CT volumes were acquired during the confining stress cycle at the conclusion of the experiment to probe the strength of weathered asperities in the system. In Figure 9.7, a cross section of the fracture with contact points is shown at baseline confining stress state of 11 MPa (~1600 psi) and maximum confining stress state of 15.85 MPa (2300 psi) corresponding to effective stress conditions of 1.38 MPa (200 psi) and 6.2 MPa (900 psi). After a rigid-body-type registration of the two data sets [*Thevenaz et al.*, 1998], we have calculated the difference image of the two images; the lower panel in Figure 9.7 shows the difference between reconstructed fracture images for the two stress states. As can be seen, the difference is largely reconstruction noise, indicating that minimal deformation of contacts occurred under loading.

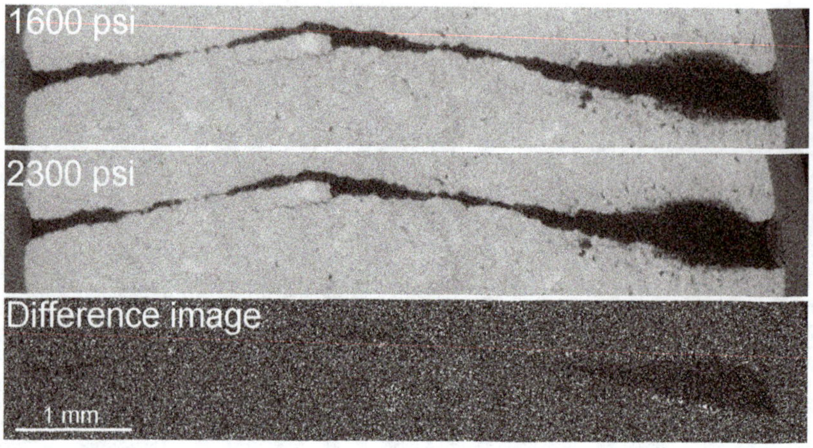

Figure 9.7 Detail of the whole fracture at about 1 cm from the inlet, horizontal slice. A sequence showing the fracture at 11 MPa (1600 psi) and 15.85 MPa (2300 psi) of confining pressure is shown. No reduction in fracture aperture is observed, as is shown by the difference image in the bottom panel.

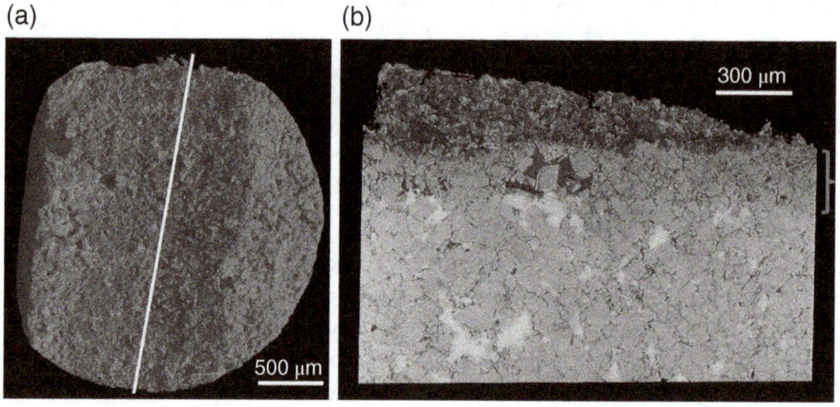

Figure 9.8 High-resolution tomography of a sample fragment with the wormhole. Panel (a) is a surface rendering, showing the face of the fracture with a portion of the wormhole. Panel (b) shows a section (marked line on the volume on the left) of the sample along the wormhole to show the weathered layer.

9.7.2. High-Resolution SXR-Micro-CT After the In Situ Dissolution Experiment

After taking the in situ SXR-micro-CT experimental setup apart, half of the dolomite sample was trimmed using a thin diamond blade to make samples aimed at observing in 3D, at higher resolution, the most interesting parts of the sample, specifically the weathered zone close to the zone of enhanced dissolution. In Figure 9.8, we show two views of the sample cut across the dissolution zone. A surface rendering of the sample, viewed from the fracture, is shown in Figure 9.8, panel (a), where the wormhole surface is clearly visible, due to its peculiar shape, color (darker, since it is made of the less X-ray-attenuating phases), and texture (a mixture of flakes and granules). The light line indicates the virtually sectioned plane, shown in Figure 9.8 panel (b). From the cut surface, it is clear that an altered zone with a thickness of ~300 μm is present, highlighted by the bracket.

Several observations can be made from this data set. First, the surface of the fracture in the enhanced dissolution zone is covered with loose material of low density, the residual of the dissolution of the carbonate rocks. This material is composed mainly of clays, quartz, feldspar, and residual dolomite fragments, either in flakes or small granules. Just below this surface, a zone where calcite has been completely dissolved is present. In this sample, calcite often appears both in small vugs and filling space between dolomite rhombohedra acting as a cementing phase, as is visible in Figure 9.8 panel (b), likely formed after the dolomitization process of the rock. The dissolution of calcite creates larger voids where the patches were present, but also dramatically increases the exposed surface of the dolomite phase where this cementing calcite is dissolved. In this weathered zone, some dolomite dissolution is visible: the extent of dolomite dissolution is smaller, because of the lower solubility of dolomite at the experimental conditions, but still noticeable. The dissolution in dolomite proceeds faster at the grain contacts and crystal defects, generating a specific texture in the weathered crystals with a likely increase in the potentially reactive surface area available for dolomite. In this figure, it is possible to see a large vug close to the surface where the calcite has been partially dissolved, leaving euhedral dolomite crystals exposed. A larger vug, unweathered and still completely filled with calcite (lighter gray in color), is present farther from the surface and is shown for comparison.

To summarize, the high-resolution SXR-micro-CT shows a specific pattern, variable with the depth, in the dissolution of this material. The surface layer is a thin (~50 μm) deposit of clays, quartz, feldspars, and mobilized small dolomite fragments very porous and made of fine-grained material. Below this surface deposit we find an altered layer where the calcite is removed; this zone has a structure consisting of partially dissolved dolomite crystals. The new porosity created in this layer is due mainly to three different mechanisms:

1. Fast dissolution of calcite in *large* (50–500 μm) pockets, leaving behind the larger pores and exposing dolomite to the reactant.
2. Fast dissolution of calcite at the dolomite grain contacts, exposing a potentially large reactive surface area for dolomite.
3. Slow dolomite dissolution: The dissolution mechanism of dolomite is complicated and creates a significant population of micropores due to dissolution at grain interfaces and crystal defects.

9.7.3. SEM-EDS Analysis of the Sample Recovered After the Dissolution

After the in situ SXR-micro-CT experiment, the sample was recovered and SEM-EDS analysis was performed on one half of the fracture face. The first step was to collect a high-resolution image of the complete fracture surface. In Figure 9.9, a tiled BSE image of the fracture is shown. To assist visualization of the fracture and to help with the correlation of BSE gray-scale values and dissolution features, we have overlaid the final aperture map in color, obtained from the SXR-micro-CT experiment. The different features are now much more evident, and it becomes easy to locate the contact points in gray (labeled with a *c*), the inlet (*i*), and the wormhole (*w*) highlighted by the higher aperture values (warmer colors). If we consider the gray-scale values of the BSE signal, which are related to the *Z* numbers/density of the materials imaged, we can see that the wormhole looks darker than the surrounding areas. This is due to dissolution of the carbonates, leaving a residue of less soluble phases such as quartz, plagioclase, and clays. This residual material, other than being made of less dense mineral phases, is also loosely packed, further increasing the difference in density and therefore backscattered signal strength.

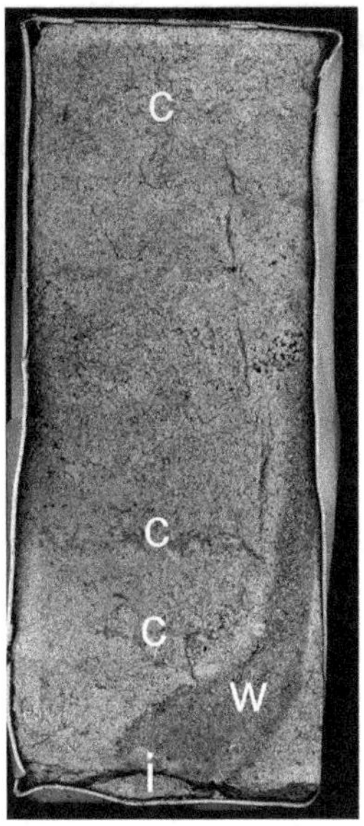

Figure 9.9 SEM/BSE image of the whole surface of the fracture. The aperture map of the fracture calculated from the SXR-micro-CT data has been superimposed.

Figure 9.10 provides higher-resolution sections detailing components of the alterated fracture. In the center panel, the full BSE image of the sample is shown. The letters in the flanking image panels represent the respective SE images on the sides, taken at higher magnification. The different phases have also been verified with EDS analysis to assure a reliable identification. Figure 9.10 panel (a) depicts a residual clay deposit found on the wormhole surface; in this case, it appears to be a loosely flocculated smectite aggregate as would be expected for recently mobilized clays. Figure 9.10 panel (b) shows the face of the calcite veinlet close to the inlet. The dissolution of calcite shows this typical pattern with large steps retreating, forming semi-regular etch pits. Some residual fines from the surrounding material are also deposited on the surface. On the wormhole surface, different insoluble phases can be identified; Figure 9.10 panel (c) shows an example of dissolution-resistant quartz crystals. Outside the wormhole, dissolution and weathering are present to a smaller extent. In Figure 9.10 panel (d), we can see a residual particle of detrital mica partially separated from the dolomite matrix. Around the mica, we can see that the dolomite features many small and irregular dissolution pits, highlighting a different dissolution mechanism than calcite.

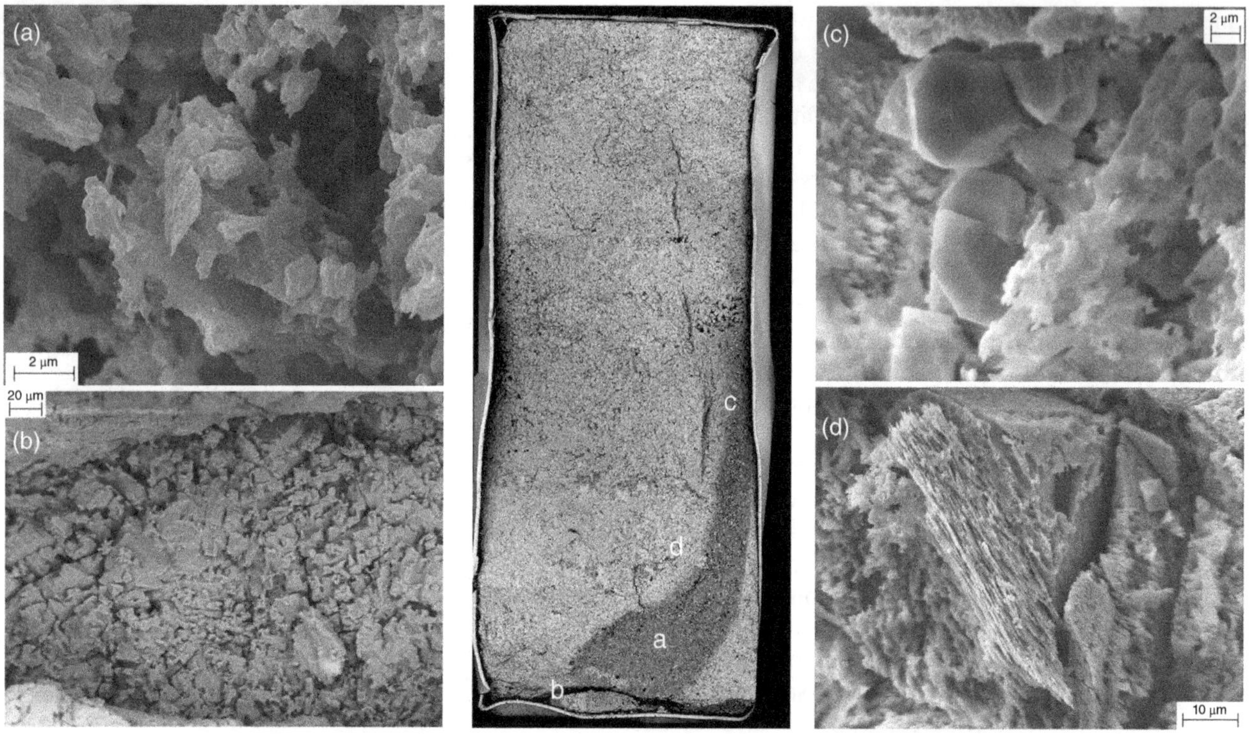

Figure 9.10 SEM images of different features on the wormhole surface. In the BSE/SEM image of the whole fracture surface in the center, different areas have been labeled. Corresponding SE/SEM pictures of the specific areas showing (a) clays deposited on the wormhole, (b) calcite veinlet with dissolution features, (c) quartz crystals on the wormhole surface, (d) residual mica exposed, and partially dissolved dolomite outside the wormhole.

Figure 9.11 Combined SE/BSE image of a surface showing the dissolution features of the dolomite phase.

The dissolution mechanism of dolomite in this sample is crucial, since the phase is the dominant component of the sample. In Figure 9.11, we show an SEM image of a surface where the dissolution of calcite left a vug with exposed dolomite crystals. This image was selected to illustrate the impact of reactant availability on dissolution mechanism; the surface crystals were exposed to rapid advection, while inside the vug, solute transport was likely diffusion dominated. The transition between advective and diffusive transport with depth would explain the residual calcite, even in an open vug and facing the fracture (as in the section of Fig. 9.8b) far from the crack surface, as well as the difference in the dissolution features on the dolomite crystals, as visible in Figure 9.11. On the surface, where the flow rates are high and the supply of reactant for the dissolution process is likewise high, the dissolution proceeds by creating a large number of small and irregular etch pits. For crystal surfaces deeper inside the vug, we can see that the dissolution features gradually change from the clusters of small etch pits to large dissolution steps. This seems to suggest that where the dissolution is controlled predominantly by diffusion, the dissolution mechanism also changes, requiring more high-energy surface sites to trigger the dissolution process. Similar mechanisms for calcite dissolution, where a larger amount of reactant molecules available triggers the nucleation and growth of new etch pits, starting with a *pure* step retreat dissolution mechanism, have been previously demonstrated; one example is the in situ atomic force microscopy results presented by *Offeddu et al.* [2014].

9.7.4. ICP-MS Analysis of the Effluent

Observations concerning calcite and dolomite dissolution are further confirmed by ICP-MS analysis of cations in the effluent. In Figure 9.3, it is possible to observe an increase in effluent Mg^{2+} molarity in the fluid as the dissolution process continues, while an inverse behavior is noticed for Ca^{2+}. If we link the molarity of Mg^{2+} in the effluent to the extent of dolomite dissolution, safely assuming that dolomite is the main phase providing Mg^{2+} and Ca^{2+} to calcite (Mg/Ca in dolomite is assumed to be constant in the sample), we can confirm that the dissolution of calcite slows down with the time, while the dolomite dissolution rate increases.

9.7.5. Reactivity Evolution Model

The development of a weathered zone is one of the critical observations made in this work. This weathered zone is characterized by calcite depletion due to the relatively fast dissolution of calcite in contrast with the slower dissolution of dolomite. The development of this weathered zone depends on the carbonate content as well as on the heterogeneity of the minerals and their spatial patterning (this work; *Ellis et al.*, 2011; *Elkhoury et al.*, 2015) and has an impact on both the evolution of fracture geometry and observed effluent concentrations of reaction products (Fig. 9.3). In particular, the progressive increase in Mg^{2+} concentrations in the effluent with time seems to indicate that depletion of calcite in the weathered zone increases the apparent dolomite reactivity with time. Furthermore, the decrease in calcium concentrations indicates a decrease

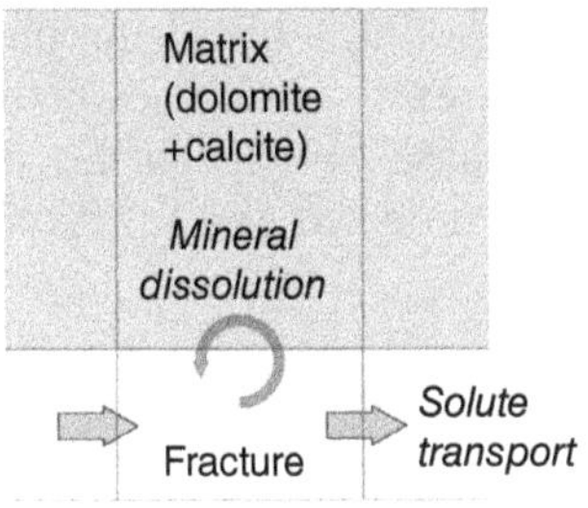

Figure 9.12 Schematic that outlines the conceptual model used in the 1D reactive transport simulation.

in calcite reactivity. While comprehensive modeling of the experiment is beyond the scope of this work, to further support this observation, we perform exploratory modeling of the experiment presented here. The objective here is simply to capture the evolution of the reactivity of the minerals by adjusting their reactive surface area and using observed effluent concentrations as constraints.

A simplified 1D representation of the fracture along the flow direction is proposed, where the fracture is assumed to have a uniform aperture equal to the average obtained from the SXR-micro-CT characterization (Fig. 9.12). The volumetric flow rate is set to the experimental 5 μl/min. The geochemical system is composed of five components (H^+, $CO_{2(aq)}$, Ca^{2+}, Mg^{2+}) and eight aqueous complexation reactions. The total carbon concentration of the inlet solution is that used in the experiments (1.2 mol/kg), which using the model of *Duan and Sun* [2003] implies a partial pressure of 50.2 bars and a pH of 3.1. The rate expression and constants of *Chou et al.* [1989] are used to calculate calcite dissolution at the fracture wall, while the rate expression and constants of *Pokrovsky et al.* [2005] are used for dolomite rates. Initially, the entire surface of the fracture is assumed to be reactive with dolomite and calcite, assigned a reactive surface area proportional to their volume fraction in the sample (Table 9.1). With time, the surface area evolves as a result of the dissolution of each mineral and is updated according to a power relationship:

$$A_{s,i}(t) = A_{o,i}\left(\frac{V_i(t)}{V_{o,i}}\right)^n \tag{9.1}$$

where $A_{s,}\ i$ [m²/m³ bulk] and Vi [m³/m³ bulk] are time-dependent surface areas and volume fractions of mineral i, respectively, with a zero subscript denoting initial state. In Eq. (9.1), each mineral phase is independent of the other components in the mineral assemblage in the sense that modification of phase 1 has no impact on the surface area (and hence dissolution rate) of phase 2. Here we consider a heuristic formulation for including this interaction. Namely, we consider the scenario where relatively faster dissolution of calcite (phase 1) results in increased surface area of a more slowly dissolving matrix phase (dolomite, phase 2). In lieu of Eq. (9.1), the surface area of phase 2 is modeled as a function of the phase 1 volume fraction:

$$A_2 = A_{o,2} + \psi_2(V_1)A_{o,2} \tag{9.2}$$

where the function ψ_2 depends on the microstructural distribution of the two phases with respect to each other and could range from *non-blocking* behavior (close to 1) or entirely coating (close to 0). A simple coating model might be

$$\psi_2(V_1) = 1 - \frac{V_1}{V_{o,1}} \tag{9.3}$$

which represents a case where the coating phase occludes the surface area of the matrix phase relative to the coating's initial volume fraction. As the coating phase dissolves, ψ approaches unity and the surface area of phase 2 approaches the non-coated case. Physically, this could represent a complete coating which dissolves as a front with respect to the matrix, uncovering the surface area of phase 2. The simulator CrunchFlow [*Steefel et al.*, 2014] is modified to include this formulation.

Simulation results using Eq. (9.1) with $n = 2/3$ for calcite (value of n often used in porous media applications; e.g., *Lichtner*, 1996) and Eqs. (9.2 and 9.3) for dolomite show a relatively good agreement with measurements (Fig. 9.13). However, this decrease is not as rapid initially as is apparent from effluent solution data. Simulation results with $n = 2$ capture better the early decrease in effluent calcium concentrations as well as the gradual slowdown of the rate at late times (Fig. 9.13). The need for a value greater than 1 for n reveals that the well-mixed assumption implicit in the porous media treatment ($n = 2/3$) does not apply and that transport limitations between fracture and matrix through the weathered zone may be affecting the rate evolution. Simulation results for $n = 2$ also show magnesium effluent concentrations closer to measured values. The increase, however, is not related to the mechanism described by Eqs. (9.2 and 9.3), which should predict a decrease in dolomite reactivity as less calcite is being dissolved in this scenario. Instead, relatively slower calcite dissolution results in lower calcium concentrations, which allows for undersaturated conditions with respect to dolomite to extend further into the fracture length. This results in faster dolomite dissolution. In both scenarios ($n = 2/3$, and $n = 2$), the effluent solution is oversaturated with respect to dolomite, but in the latter one, it is much closer to equilibrium (Fig. 9.13). While Ca^{2+} release from calcite affects dolomite dissolution, the release of Mg^{2+} and possibly of other tracer cations could have played an inhibitory role on calcite dissolution [e.g., *Harstad and Stipp*, 2007], accounting in part for the slowdown of the calcite rate at late times. This effect was not modeled explicitly but could have in part contributed to the improved fit with $n = 2$.

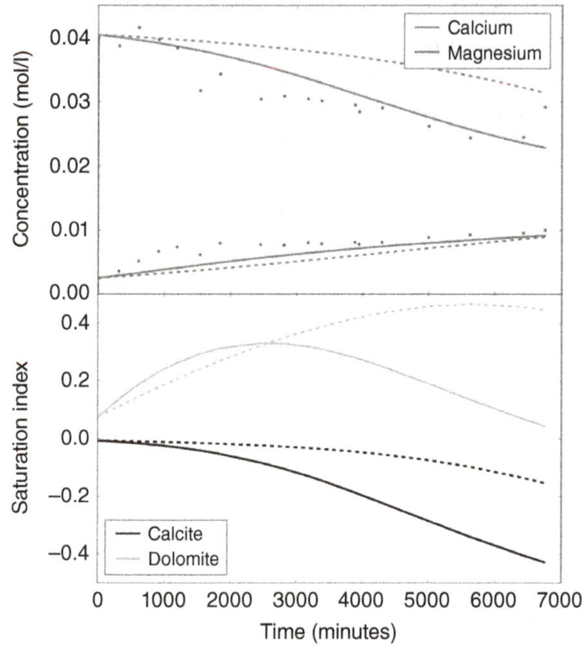

Figure 9.13 Simulation results obtained under two scenarios: dashed lines indicate a sublinear decrease of reactivity with respect to calcite volume as usually employed in reactive transport model ($n = 2/3$), and solid lines indicate a superlinear decrease in reactivity ($n = 2$), likely indicative of mass transfer limitation between matrix and fracture. Panel (a) shows calcium and magnesium concentrations, and panel (b) shows saturation indices with respect to calcite and dolomite in the effluent solution.

9.8. DISCUSSION

The observations from the SXR-micro-CT fracture dissolution sequence, especially when considering the fracture aperture evolution in Figure 9.6, have led us to conclude that for the extent of dissolution considered in this experiment, the fractures are self-enhancing. The wormholing effect, with the enlargement of the fracture aperture, is leading to increased flow along the fracture and therefore an increased permeability. While permeability was not directly measured, fracture areal cross section is increasing or stable at all locations. An obvious limitation in this experiment was the relatively short duration of injection conducted (~113 h) which was due to synchrotron access constraints; despite the fact that no closure was observed, initial weathering beneath contact zones suggests that some compaction might have happened given a

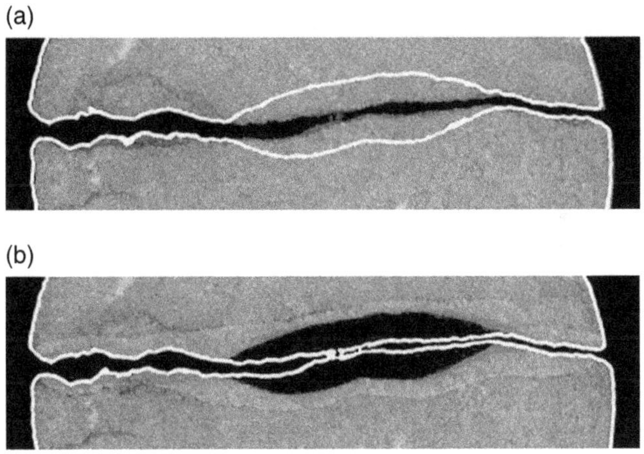

Figure 9.14 Comparing the evolution of the fracture morphology: in the two horizontal slices in gray scale the sample before (a) and after (b) the dissolution process. The heavy line indicates the fracture outline from the corresponding sections, while the shaded overlay indicates the weathered zone.

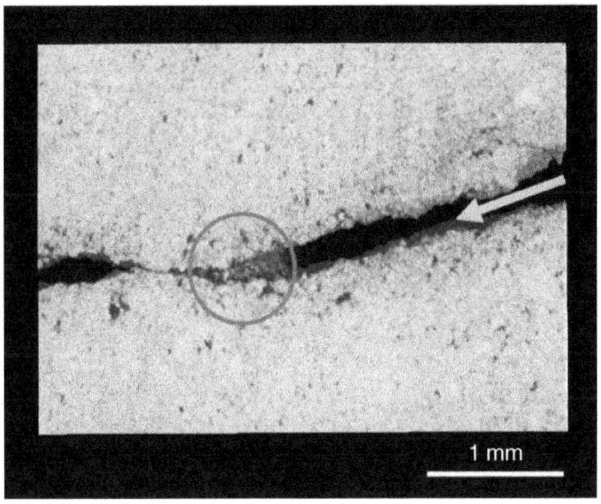

Figure 9.15 High-resolution micro-CT slice perpendicular with respect to the crack plane showing a choke point with accumulation of insoluble fines (circled). The arrow indicates the flow direction.

longer period of observation. To generalize this process for larger dissolution extents, modeling needs to be employed. The present experiment can provide an excellent starting point for modelers, as well as a validation data set for coupled fracture-scale simulators. A first effort in modeling the short-term chemical behavior of this sample is presented in this work and provides a promising first step in developing predictive simulators for fracture behavior in GCS. The short time evolution of effluent chemistry was effectively captured by a simple fracture representation. The next step in model development will require representation of the fracture plane as well as weathered zone mechanics to allow accurate prediction of closure and obstruction during fracture evolution.

The enlargement of the aperture of the fracture is presented in finer detail in Figure 9.14 where a horizontal slice (about 4 mm from the inlet) from the sample during the baseline scan and at the end of the experiment, after 113 h of reaction, is shown. In each image, the contour calculated from the other image has been overlaid to show the dissolved volume causing the wormhole. In Figure 9.14b, the weathered layer has been highlighted in red. One observation is that the thickness of the weathered layer is somewhat thinner toward the center of the region of dissolution enhancement in comparison to the regions which experience lower flow velocities. This is likely due to the faster flow providing a steady supply of reactants to the surface, thereby enhancing complete dissolution of both calcite and dolomite. On the sides of the wormhole, the size of the weathered layer is the thickest, having still a good supply of reactant, but a process more controlled by diffusion rather than advection, favoring more the dissolution of faster-dissolving species, for example, calcite.

Prior studies have suggested that mobilization of fine particles, particularly clays, can play a significant role in the evolution of fracture permeability during dissolution. In this study, residual clay present in the matrix stayed largely stationary on the fracture surface during dissolution. However, hints of some mobilization were visible at one constriction point as can be seen in Figure 9.15. In this case, a small low contrast aggregate was visible in one of the time-lapse volumes but did not seem to be sufficient to occlude the primary fracture zone. Higher flow velocities present near a fracture inlet may also contribute to greater mobilization. Likewise, in formations with larger clay fraction, the fines mobilization mechanism may be a more significant contributor to fracture occlusion.

An alternative view of fracture evolution is provided by the statistical evaluation of aperture distribution at the level of individual patches. Prior studies, including *Ellis et al.* [2011] and *Detwiler and Rajaram* [2007], show a broadening in aperture distribution (larger outliers) and a shift toward higher and larger mean aperture. Figure 9.16 shows histograms of local fracture aperture values (panel a) for the sequence of tomograms as well as simple statistics on the resulting distributions (panel b). The segmented fracture aperture maps discussed previously were used for this calculation, and only the lower half of the sample, which exhibits a larger dissolution signature, was considered. As can be seen in panel (a), later times (F1–F3) exhibit a longer statistical tail corresponding to local apertures within the enhanced dissolution pathways. The bulk of the fracture exhibits only small changes as can be seen by the stability of the central histogram peak at ~250 μm. Panel (b) depicts a variety of statistical metrics

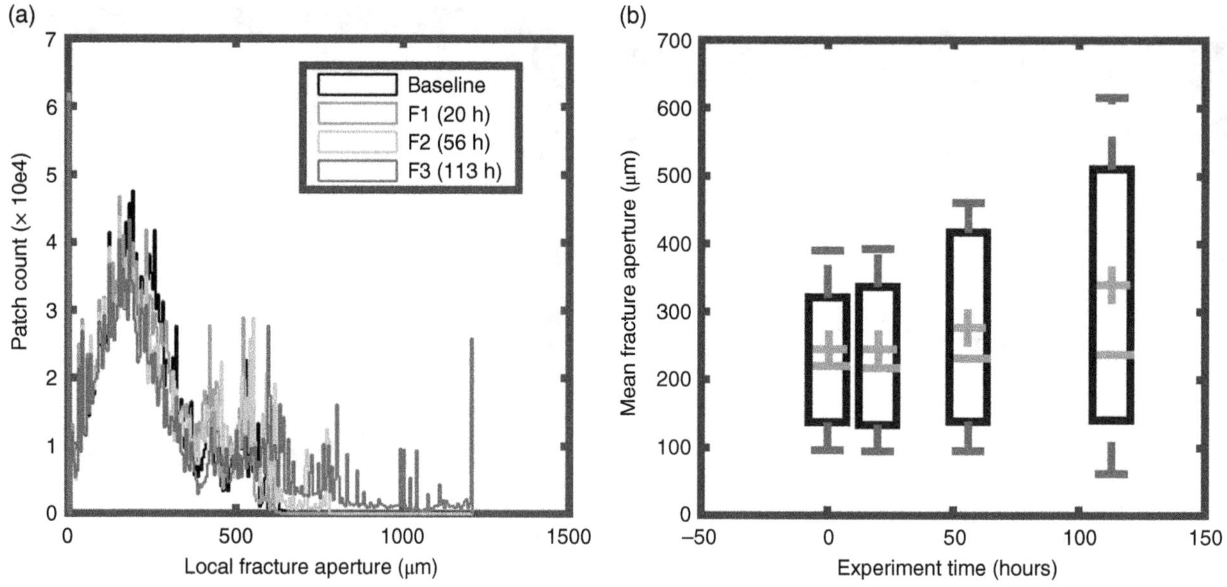

Figure 9.16 Aperture statistics for the sequence of imaged fractures. Panel (a) depicts histograms of local aperture for the bottom half of the fracture. Panel (b) depicts statistical measures of the distribution including median (center bar), mean (cross), 25/75% quartile (box), and 10/90% outlier spread (outer marks).

for the different time points including mean aperture (cross), median aperture (bar), 25/75% quartile (box), and 10/90% outliers (outer lines). The mean fracture aperture increases as a function of dissolution time from ~240 to 320 μm. The mean is however predominantly impacted by the outlier values from the enhanced dissolution zone; the median fracture aperture is relatively stable. The quartile and outlier metrics also show the impact of the enhanced dissolution region. We should note that this statistical analysis demonstrates that, as expected, enhanced dissolution or wormhole generation has distinct statistical properties from uniform dissolution where a more uniform histogram shift would be expected.

From a geochemical perspective, the development of the weathered zone is the result of a combination of the rock microstructure (calcite cementing the dolomite crystals, filling interstices, and small vugs), flow properties, and especially the difference in dissolution rates between the two primary mineral phases. Concerning dolomite, from the SEM images, we observed fast dissolution of the dolomite crystals on the surface of the wormhole with a mechanism promoting the formation and growth of a large number of pinhole-like etch pits, highlighting an environment very favorable for dolomite dissolution. This density of etch pits per surface area unit also significantly increased the potential reactive surface available on the wormhole pathway on the fracture. In contrast, the dolomite crystals observed inside the weathered layer, where the availability of the carbonic acid is more limited, exhibit a different dissolution mechanism. SEM images show a clear step retreat dissolution process, likely implying that the energy involved in the dissolution process of the crystal surface is not large enough to induce the formation of a large number of etch pits; in this case, a mechanism relying on preexistent higher-energy sites on the surface, such as steps, becomes dominant. When a weathered layer is established, the role of calcite becomes extremely important: the calcite dissolves faster on the surface of the fracture, leaving a surface enriched in dolomite (and insoluble phases) which now faces less competition for the reactant. In the weathered layer, some calcite is still present, but the supply of reactant is slower; therefore, the calcite still leads the dissolution process, but the overall rates are decreased. The calcite also acts as a strong buffer in this diffusion-dominated zone, making the dissolution of carbonates even slower. This behavior is confirmed by the ICP-MS analysis of the effluent (Fig. 9.3); for the first hours of reaction, during which the weathered zone starts to develop and calcite is still available for the reaction on the fracture surface, the amount of Ca^{2+} (linked to the amount of calcite dissolved) is the largest. With the evolution of the system, less and less calcite is present close to the advection-dominated volume of the sample; hence, we observe a faster dissolution of the dolomite left on the surface. In this phase, the amount of Mg^{2+} increases with time, while the amount of calcite dissolving, limited by the diffusion-dominated weathered zone, steadily decreases, decreasing the total concentration of Ca^{2+} in the effluent.

The development of a weathered layer on the surface of the fracture also has important implications for the geomechanics of the fracture. The dissolution of the calcite cementing the dolomite crystals, and the partial dissolution of dolomite, generates a layer likely having lower compressive strength and different elastic properties. Therefore, if this leached layer can develop on the contact areas of the fracture, it could have a fundamental role in fracture aperture evolution, since a steady weakening of the contact areas would promote compaction and self-sealing behavior. On the other hand, a pronounced wormholing effect with minimal weakening of the contact areas would make the system leaning more toward a self-enhancing one. From the observation in Figure 9.6b, it is evident that the morphology of the contact areas is not changing significantly. A change in morphology does not necessarily imply that a weakening is not present, but just that the present differential pressure is still too small to enable the collapse of the contact points to close the fracture. As shown previously, Figure 9.7 demonstrates that even with stress concentration at asperities, the weathering process is insufficient at short times to facilitate collapse of the near-fracture zone at moderate effective stress levels. This appears to be due to both the low weathering rate within the contact regions and the relatively strong microstructure in the weathered zone, consisting of bonded dolomite crystals. Further studies will be required to better constrain the relationship between porosity increase and reductions in compressive strength which drive this aspect of the coupling.

From the modeling perspective, the simulations presented here are exploratory in nature and do not represent the rich level of geometric complexity observed in the tomographic data sets. Nevertheless, the simulations made it possible to identify the main controls on effluent chemistry at early times, namely, buffering by calcite dissolution and reactivity changes, considered here via surface area evolution but likely associated with mass transfer limitations as suggested by the SEM-EDS observations described earlier. The transport control on rates highlights the importance to develop models that characterize the evolution of transport properties in altered layers. To explore these mechanisms in more detail is beyond the scope of this manuscript, but the multi-technique, high-resolution data sets presented in this work provide a unique opportunity to bring to bear advanced simulation approaches such as those based on direct numerical simulation on image data at the pore scale [*Steefel et al.*, 2013; *Molins*, 2015]. In combination with continuum-scale models, there is the opportunity to improve our conceptual understanding of evolving reactivity in fractured systems.

9.9. CONCLUSION

We have presented results from a unique experiment monitoring the state of a fracture during CO_2-induced dissolution at close to in situ GCS conditions. The experiment provides geometric constraints on the temporal evolution of fracture aperture, near-fracture weathering, and resulting effluent chemistry. We observe multiple domains within the fracture with morphologically distinct dolomite dissolution patterns, likely driven by differences in flow regime. For this particular case, the near-fracture region remains mechanically strong, and flow-enhanced dissolution is observed over the duration of the experiment. A relatively simple reactive transport model, which includes a novel coating model for representing dissolution feedbacks to reactive surface area, is capable of capturing effluent chemistry over the course of the experiment. The resulting data set will also provide an excellent validation data set for the next generation of fully coupled fracture-scale simulators currently being developed.

Our experiment also points to several gaps in knowledge which should be addressed to better represent coupled processes relevant to seal evolution in CO_2-rich environments. The first is experimental constraints for longer timescales when diffusion-mediated weathering in zones of contact has the potential to weaken these regions. The second is experimental constraints for longer length scales over which fracture aperture scaling, fines mobilization, and downstream re-precipitation of carbonates should play a more significant role. The third is a valid representation for the small-scale coupling between dissolution and compressive strength in the near-fracture region in multimineralic systems; this relationship will undoubtedly be complex and rock specific but at present is both unrepresented in models and unconstrained experimentally. With these challenges in mind, predictive computational models describing fracture evolution under stress with all of the required couplings appear to be within reach.

ACKNOWLEDGMENTS

The authors would like to acknowledge Lori Babcock (Montana State University) for obtaining the Duperow outcrop sample used in this experiment. Primary support for this research was provided by the Center for Nanoscale Controls on Geologic CO_2, an Energy Frontier Research Center, funded by the US DOE, Office of Basic Energy Sciences (contract DE-AC02-05CH11231). X-ray microtomography experiments were performed with the assistance of Dula Parkinson and Alastair MacDowell at the Advanced Light Source, Beamline 8.3.2, which is supported by the Office of Science, Office of Basic Energy Sciences, of the US DOE (contract DE-AC02-05CH11231).

REFERENCES

Ameli, P., Elkhoury, J.E., Morris, J.P., and R.L. Detwiler, (2014), "Fracture Permeability Alteration due to Chemical and Mechanical Processes: A Coupled High-Resolution Model,"

Rock Mechanics and Rock Engineering, Vol. *47*, No. 5, pp. 1563–1573.

Babcock, L., and D.R. Lageson, (2013), "Fracture and joint analysis of upper Devonian dolomite across southwest and north-Central Montana: 3D modeling of structurally controlled diagenesis," *Geological Society of America, Abstracts with Programs*, Vol. *45*, No. 7, p. 172.

Chou, L., R.M. Garrels, and R. Wollast, (1989), "Comparative study of the kinetics and mechanisms of dissolution of carbonate minerals," *Chemical Geology*, Vol. *78*, No. 3–4, pp. 269–282, doi:10.1016/0009-2541(89)90063-6.

Cobbold, P.R., Zanella, A., Rodrigues, N., and H. Loseth, (2013), "Bedding-parallel fibrous veins (beef and cone-in-cone): worldwide occurrence and possible significance in terms of fluid overpressure, hydrocarbon generation, and mineralization," *Marine and Petroleum Geology*, Vol. *43*, pp. 1–20.

Davila, G., Luquot, L., Soler, J.M., and J. Cama, (2016), "Interaction between a fractured marl caprock and CO_2-rich sulfate solution under supercritical CO_2 conditions," *International Journal of Greenhouse Gas Control*, Vol. *48*, No. 1, pp. 105–119.

Detwiler, R.L., and H. Rajaram, (2007), "Predicting dissolution patterns in variable aperture fractures: evaluation of an enhanced depth-averaged computational model, *Water Resources Research*, Vol. *43*, W04403.

Detwiler, R.L., Glass, R.J., and W.L. Bourcier, (2003), "Experimental observations of fracture dissolution: the role of Peclet number on evolving aperture variability," *Geophysical Research Letters*, Vol. *30*, No. 12, p. 1648.

Dierick, M., Masschaele, B., and L. Van Hoorebeke, (2004). "Octopus, a fast and user-friendly tomographic reconstruction package developed in LabView®," *Measurement Science and Technology*, Vol. *15*, No. 7, p. 1366.

Dougherty, R., and K.-H. Kunzelmann, (2007), "Computing local thickness of 3D structures with ImageJ," *Microscopy and Microanalysis*, Vol. *13*, No. S02, pp. 1678–1679.

Duan, Z., and R. Sun, (2003), An improved model calculating CO_2 solubility in pure water and aqueous NaCl solutions from 273 to 533 K and from 0 to 2000 bar, *Chemical Geology*, Vol. *193*, No. 3–4, pp. 257–271, doi:10.1016/S0009-2541(02)00263-2.

Elkhoury, J.E., Ameli, P., and R.L. Detwiler, (2013), "Dissolution and deformation in fractured carbonates caused by flow of CO_2-rich brine under reservoir conditions," *International Journal of Greenhouse Gas Control*, Vol. *16S*, pp. S203–S205.

Elkhoury, J.E., Detwiler, R.L., and P. Ameli, (2015), "Can a fractured caprock self-heal?," *Earth and Planetary Science Letters*, Vol. *417*, pp. 99–106.

Ellis, B.R., and C.A. Peters, (2015), "3D mapping of calcite and a demonstration of its relevance to permeability evolution in reactive fractures," *Advances in Water Resources*, in press, doi:10.1016/j.advwatres.2015.07.023.

Ellis, B., Peters, C., Fitts, J., Bromhal, G., McIntyre, D., Warzinski, R., and E. Rosenbaum, (2011), "Deterioration of a fractured carbonate caprock exposed to CO_2-acidified brine flow," *Greenhouse Gases: Science and Technology*, Vol. *1*, pp. 248–260.

Ellis, B.R., Fitts, J.P., Bromhal, G.S., McIntyre, D.L., Tappero, R., and C.A. Peters, (2013), "Dissolution-driven permeability reduction of a fractured carbonate Caprock," *Environmental Engineering Science*, Vol. *30*, No. 4, pp. 187–193.

Fred, C.N., and H.S. Fogler, (1998), "Influence of transport and reaction on wormhole formation in porous media, fluid mechanics and transport phenomena," *AIChE Journal*, Vol. *44*, No. 9, pp. 1933–1949.

Gouze, P., Noiriel, C., Bruderer, C., Loggia, D., and R. Leprovost, (2003), "X-ray tomography characterization of fracture surfaces during dissolution," *Geophysical Research Letters*, Vol. *30*, No. 5, pp. 71:1–71:4.

Groves, C.G., and A.D. Howard, (1994), "Early development of karst systems. 1. Preferential flow path enlargement under laminar flow," *Water Resources Research*, Vol. *30*, pp. 2837–2846.

Harstad, A.O., and S.L.S. Stipp, (2007), "Calcite dissolution: effects of trace cations naturally present in Iceland spar calcites," *Geochimica et Cosmochimica Acta*, Vol. *71*, No. 1, pp. 56–70, doi:10.1016/j.gca.2006.07.037.

Kak, A.C., and M. Slaney, (2001), *Principles of Computerized Tomographic Imaging*. Society for Industrial and Applied Mathematics, Philadelphia, PA.

Karpyn, Z.T., Grader, A.S., and P.M. Halleck, (2007), "Visualization of fluid occupancy in a rough fracture using micro-tomography," *Journal of Colloid and Interface Science*, Vol. *307*, pp. 181–187.

Ketcham, R.A., and W.D. Carlson, (2001), "Acquisition, optimization and interpretation of X-ray computed tomographic imagery: applications to the geosciences," *Computers & Geosciences*, Vol. *27*, No. 4, pp. 381–400.

Kim, S., and J.C. Santamarina, (2015), "Geometry-coupled reactive fluid transport at the fracture scale: application to CO_2 geologic storage," *Geofluids*, Vol. *16*, No. 2, pp. 329–341.

Lang, P.S., Paluszny, A., and R.W. Zimmerman, (2015), "Hydraulic sealing due to pressure solution contact zone growth in siliciclastic rock fractures," *Journal of Geophysical Research - Solid Earth*, Vol. *120*, pp. 4080–4101.

Li, B., Zhao, Z., Jiang, Y., and L. Jing, (2015), "Contact mechanism of a rock fracture subjected to normal loading and its impact on fast closure behavior during initial stage of fluid flow experiment," *International Journal for Numerical and Analytical Methods in Geomechanics*, Vol. *39*, pp. 1431–1449.

Lichtner, P.C., (1996), "Continuum formulation of multicomponent-multiphase reactive transport," *Reviews in Mineralogy and Geochemistry*, Vol. *34*, No. 1, pp. 1–81.

Liu, J., Polak, A., Elsworth, D., and A. Grader, (2005), "Dissolution-induced preferential flow in a limestone fracture," *Journal of Contaminant Hydrology*, Vol. *78*, pp. 53–70.

Luquot, L., Abdoulghafour, H., and P. Gouze, (2013), "Hydrodynamically controlled alteration of fracture Portland cements flowed by CO_2-rich brine," *International Journal of Greenhouse Gas Control*, Vol. *16*, pp. 167–179.

MacDowell, A.A., Parkinson, D.Y., Haboub, A., Schaible, E., Nasiatka, J.R., Yee, C.A., Jameson, J.R., Ajo-Franklin, J.B., Brodersen, C.R., and A.J. McElrone, (2012), "X-ray microtomography at the advanced light source." In *SPIE Optical Engineering+ Applications*, p. 850618. International Society for Optics and Photonics.

Molins, S., (2015), "Reactive interfaces in direct numerical simulation of pore-scale processes," *Reviews in Mineralogy and Geochemistry*, Vol. *80*, No. 1, pp. 461–481, doi:10.2138/rmg.2015.80.14.

Noiriel, C., Gouze, P., and B. Made, (2007), "Time-resolved 3D characterization of flow and dissolution patterns in a single rough-walled fracture," in *IAH Selected Papers: Groundwater in fractured rocks*, pp. 629–642 (eds. K. Krasny and J.M. Sharp). Taylor & Francis, London.

Noiriel, C., Gouze, P., and B. Made, (2013), "3D analysis of geometry and flow changes in a limestone fracture during dissolution," *Journal of Hydrology*, Vol. *486*, pp. 211–223.

Offeddu, F.G., Cama, J., Soler, J.M., and C.V. Putnis, (2014), "Direct nanoscale observations of the coupled dissolution of calcite and dolomite and the precipitation of gypsum," *Beilstein Journal of Nanotechnology*, Vol. *5*, No. 1, pp. 1245–1253.

Pokrovsky, O.S., S.V. Golubev, and J. Schott, (2005), "Dissolution kinetics of calcite, dolomite and magnesite at 25°C and 0 to 50 atm pCO_2," *Chemical Geology*, Vol. *217*, No. 3–4, pp. 239–255, doi:10.1016/j.chemgeo.2004.12.012.

Polak, A., Elsworth, D., Yasuhara, H., Grader, A.S., and P.M. Halleck, (2003), "Permeability reduction of a natural fracture under net dissolution by hydrothermal fluids," *Geophysical Research Letters*, Vol. *30*, No. 20, p. 2020.

Polak, A., Elsworth, D., Liu, J., and A.S. Grader, (2004), "Spontaneous switching of permeability changes in a limestone fracture with net dissolution," *Water Resources Research*, Vol. *40*, W03502.

Pyrak-Nolte, L., and J. Morris, (2000), "Single fractures under normal stress: the relation between fracture specific stiffness and fluid flow," *International Journal of Rock Mechanics and Mineral Science*, Vol. *37*, pp. 245–262.

Renard, F., Gratier, J.P., and B. Jamtveit, (2000), "Kinetics of crack-sealing, intergranular pressure solution, and compaction around active faults," *Journal of Structural Geology*, Vol. *22*, No. 10, pp.1395–2407.

Shipton, Z., Evans, J.P., Kirschner, D., Kolesar, P.T., Williams, A.P., and J. Heath, (2004), "Analysis of CO_2 leakage through 'low-permeability' faults from natural reservoirs in the Colorado Plateau, east-central Utah," in *Geological Storage of Carbon Dioxide*, pp. 43–58 (eds. S.J. Baines and R.H. Worden), Geological Society, London.

Shipton, Z., Evans, J.P., Dockrill, B., Heath, J., Williams, A., Kirchner, D., and P.T. Kolesar, (2005), "Natural leaking CO_2-charged systems as analogs for failed geologic storage reservoirs," in *Carbon Dioxide Capture for Storage in Deep Geologic Formations, Volume 2*, pp. 699–712 (eds. D.C. Thomas and S.M. Benson), Elsevier, San Diego, CA.

Steefel, C.I., and Lasaga, A.C., (1994), "A coupled model for transport of multiple chemical species and kinetic precipitation/dissolution reactions with applications to reactive flow in single phase hydrothermal systems," *American Journal of Science*, Vol. *294*, pp. 529–592.

Steefel, C.I., and P.C. Lichtner, (1998), "Multicomponent reactive transport in discrete fractures: I. Controls on reaction front geometry," *Journal of Hydrology*, Vol. *209*, No. 1–4, pp. 186–199, doi:10.1016/S0022-1694(98)00146-2.

Steefel, C.I., Molins, S., and D. Trebotich, (2013), "Pore scale processes associated with subsurface CO_2 injection and sequestration," *Reviews in Mineralogy and Geochemistry*, Vol. *77*, No. 1, pp. 259–303, doi:10.2138/rmg.2013.77.8.

Steefel, C.I. et al., (2014), "Reactive transport codes for subsurface environmental simulation," *Computers & Geosciences*, Vol. *19*, No. 3, pp. 445–478, doi:10.1007/s10596-014-9443-x.

Szymczak, P., and A.J.C. Ladd, (2009), "Wormhole formation in dissolving fractures," *Journal of Geophysical Research*, Vol. *114*, B06203.

Thevenaz, P., Ruttimann, U.E., and M. Unser, (1998), "A pyramid approach to subpixel registration based on intensity," *IEEE Transactions on Image Processing*, Vol. *7*, No. 1, pp. 27–41.

Voltolini, M., Kwon, T.-H., and J.D. Ajo-Franklin, (2017), "Visualization and prediction of supercritical CO_2 distribution in sandstones during drainage: an in situ synchrotron X-ray micro-computed tomography study," *International Journal of Greenhouse Gas Control*, Vol. *66*, pp. 230–245.

Xu, T., Spycher, N., Sonnenthal, E., Zheng, L., and K. Pruess, (2012), "TOUGHREACT user's guide: a simulation program for non-isothermal multiphase reactive transport in variably saturated geologic media, version 2.0," *LBNL Report*.

Young, R.A., (1993), "Introduction to the Rietveld method," *The Rietveld Method*, Vol. *5*, pp. 1–38.

10

Leakage Processes in Damaged Shale: In Situ Measurements of Permeability, CO_2 Sorption Behavior, and Acoustic Properties

J. William Carey[1], Ronny Pini[2], Manika Prasad[3], Luke P. Frash[1], and Sanyog Kumar[3]

ABSTRACT

Caprock integrity is one of the chief concerns in the successful development of a CO_2 storage site. In this chapter, we provide an overview of the permeability of fractured shale, the potential for mitigation of CO_2 leakage by sorption to shale, and the detection by acoustic methods of CO_2 infiltration into shale. Although significant concerns have been raised about the potential for induced seismicity to damage caprock, relatively little is known about the permeability of the damaged shale. We present a summary of recent experimental work that shows profound differences in permeability of up to three orders of magnitude between brittle and ductile fracture permeability. In the ductile regime, it is possible that shale caprock could accommodate deformation without a significant loss of CO_2 from the storage reservoir. In cases where CO_2 does migrate through damaged shale caprock, CO_2 sorption onto shale mineralogy may have a mitigating impact. Measured total storage capacities range from 1 to 45 kg-CO_2/tonne-shale. Once CO_2 is in the caprock, changes in the acoustic properties of CO_2-saturated shale that are predicted by Gassmann fluid substitution calculations show a significant reduction of the bulk modulus of CO_2-saturated shale.

10.1. INTRODUCTION

During geologic sequestration of CO_2, most storage scenarios will require an impermeable geologic layer (commonly known as caprock) to prevent the escape of the buoyant supercritical CO_2 plume from the injection reservoir [*Metz et al.*, 2005]. The most important types of caprock are shale and evaporite although low-permeability carbonates and sandstones can also serve as seals [*Grunau*, 1987]. Methods of assessing caprock integrity have been developed as an important part of oil and gas exploration, where a good seal is as necessary to the occurrence of hydrocarbon deposits as the presence of good source rocks [*Downey*, 1984; *Grunau*, 1987]. The suitability and integrity of a caprock depends on a combination of low permeability, a capillary barrier to flow of gas and oil (i.e., the *non-wetting* fluid phases), and the overall continuity and thickness of the formation [*Downey*, 1984; *Grunau*, 1987]. Ideally, the caprock should not have been subjected either to tectonic events or a production history that would have created faults and fractures [e.g., *Hawkes et al.*, 2005].

CO_2 injection into reservoirs will change the state of stress in both the reservoir and caprock formations due to poromechanical, thermal, and chemical effects. As a result of induced stress, the caprock can be mechanically damaged, preexisting sealing faults and fractures can be reactivated, or new fracture systems can be created. Such changes are of particular concern when they impact the caprock because they could initiate fluid migration out of the storage reservoir.

[1]*Earth and Environmental Sciences, Los Alamos National Laboratory, Los Alamos, NM, USA*

[2]*Department of Chemical Engineering, Imperial College, London, UK*

[3]*Petroleum Engineering Department, Colorado School of Mines, Golden, CO, USA*

Geological Carbon Storage: Subsurface Seals and Caprock Integrity, Geophysical Monograph 238, First Edition.
Edited by Stéphanie Vialle, Jonathan Ajo-Franklin, and J. William Carey.

For these reasons, recent work has called into the question the viability of CO_2 storage in the light of potential injection-induced damage. *Zoback and Gorelick* [2012] observe that in many places the Earth's crust is *critically stressed* such that injection of CO_2 would result in small-magnitude earthquakes. The point at issue is that despite their relatively low magnitude, these events could result in fault activation across the caprock and compromise storage security. Others have disagreed with this conclusion on the basis of earthquake locations, the historical record of oil and gas reservoirs, and considerations related to coupled mechanics and fluid flow [*Juanes et al.*, 2012; *Vilarrasa and Carrera*, 2015]. Nevertheless, the threat of injection-induced earthquakes remains a significant concern for the risk assessment of a CO_2 sequestration operation.

This debate, however, has not addressed the potential *consequences* of the presence of fractures through caprock. In order for damage to the caprock to have an impact, a significant and persistent flow of CO_2 must occur. In this work, we focus on the question of permeability and transport of CO_2 through fractured shale. We consider a scenario in which injection processes, natural tectonic activity, or perhaps even an existing damage zone has created a potential pathway for CO_2 migration from the reservoir through a shale caprock. We consider how the coupled processes involved in geomechanical damage might affect leakage risk, namely, enhanced permeability, (immiscible) CO_2 flow, fracture closure, and CO_2 trapping by sorption.

Likewise, there are few studies that have developed methods of detecting damaged caprock or the presence of CO_2-saturated caprock. CO_2 flow into and through caprock will necessarily be accompanied by fluid exchange. Such processes may be associated with a change in seismic and electrical properties. We consider the magnitude of changes in these geophysical properties that might be used to monitor caprock damage. Potential changes of geophysical properties can be attributed to at least two effects: the effect of a change of saturating fluid from connate water or hydrocarbon to CO_2 and the effect of geomechanical damage.

This chapter consists of a literature review of what is known of the factors governing potential leakage of CO_2 through shale caprock, focusing on the properties of fractured or damaged shale. We discuss three subtopics in more detail based on recent experimental work by the authors: (i) fracture permeability behavior of shale, (ii) sorption of CO_2 in shale, and (iii) acoustic properties of CO_2-reacted shale. Other recent reviews of caprock integrity and CO_2 sequestration include those of *Fitts and Peters* [2013], *Shukla et al.* [2010], and *Song and Zhang* [2013] and a recent review of the geomechanical stability of caprock by *IEAGHG* [2015].

10.2. CAPROCK INTEGRITY FAILURE

A good caprock isolates water-immiscible, buoyant fluids (supercritical CO_2, oil, and gas) in the subsurface. There are clearly many examples of good caprock as illustrated by the many accumulations of oil, gas, and even CO_2 (e.g., CO_2 reservoirs of the Southwestern United States; *Allis et al.*, 2001). However, caprock might also have poor integrity and could have allowed fluids to escape as evidenced by subsea oil leaks [*Hornafius et al.*, 1999], leaking oil reservoirs [*Macgregor*, 1996], and travertine surficial deposits formed above CO_2 reservoirs [*Moore et al.*, 2005; *Keating et al.*, 2014]. Identifying good caprock involves establishing the basic geometric features of a caprock (vertical and lateral containment of the reservoir; *Biddle and Wielchowsky*, 1994), determining that the capillary properties of the caprock will block entry of the immiscible phase of interest, and determining that stresses arising from tectonic and/or hydrocarbon production operations have not previously damaged the caprock [*Macgregor*, 1996; *Hawkes et al.*, 2005; *Cartwright et al.*, 2007].

In studies of caprock integrity for CO_2 sequestration, much work has focused on measuring caprock permeability and on assessing the quality of the capillary barriers of undamaged shale [e.g., *Schlömer and Krooss*, 1997; *Hildenbrand et al.*, 2002; *Armitage et al.*, 2011; *Heath et al.*, 2012; *Wollenweber et al.*, 2010; *Song and Zhang*, 2013]. Typical absolute permeability values range from 0.001 to 1 μD [e.g., *Hildenbrand et al.*, 2002]. However, as noted by *Hildenbrand et al.* [2002] and others, injection pressures must exceed the capillary breakthrough pressure before CO_2 flow can occur (from 0.1 to 6.7 MPa for the samples considered in *Hildenbrand et al.* [2002]). Values of the capillary breakthrough pressure can be used to compute the maximum thickness of a CO_2 plume beneath a seal such that breakthrough of the gas phase is not possible [e.g., *Naylor et al.*, 2011]. Thus, limiting the thickness of the CO_2 plume limits the potential of transport processes of CO_2 through the caprock to only diffusion.

Much less work has been done on the permeability of *fractured* shale and, in particular, on the consequences of caprock fractures (i.e., by measuring the permeability of the resulting fracture system) either at the field scale [*Cartwright et al.*, 2007] or laboratory scale [*Gutierrez et al.*, 2000; *Carey et al.*, 2015]. Some studies indicate that the predominant risk for caprock integrity is fracture flow [e.g., *Edlmann et al.*, 2013; *Pawar et al.*, 2015]. Interestingly, *Skurtveit et al.* [2012] find that even in studies of capillary breakthrough of CO_2 through nominally *intact* shale samples, deformation occurs as a result of dilation and microfracture formation.

10.2.1. Origin of Faults in Reservoir-Caprock Systems

A critical operational parameter for CO_2 storage is the maximum sustainable reservoir overpressure that can be achieved without fault activation [*Sibson*, 2003]. Injection into a reservoir can reactivate faults due to an increase of pore pressure within a fault that reduces the effective stress normal to the fault plane. Pore pressure changes within a reservoir will also modify the total stress state as shown by the significant changes to the horizontal minimum stress reported for measurements in the field [*Addis*, 1997]. These changes would tend to increase fault slip potential in some normal fault regimes and most thrust fault environments [*Hawkes et al.*, 2005]. Faults in the caprock can also be induced by the dilation or compaction of the underlying reservoir in response to injection or production, which is likely to be particularly pronounced for elastically soft reservoirs [*Hawkes et al.*, 2005]. While the reservoir experiences a change in pore pressure that results in changes to the horizontal stress, the caprock does not; this can lead to the development of shear stresses at the interface between the reservoir and the caprock, which could also damage wells [*Hawkes et al.*, 2005]. Hydraulic fractures are an additional mode of potential failure; these could originate due to pore pressure at the top of the reservoir exceeding the least principal stress in the caprock [e.g., *Finkbeiner et al.*, 2001]. In general, the formation of hydraulic fractures in the caprock would only be possible in the absence of pre-existing, low-cohesion fractures of suitable orientation [*Morris et al.*, 1996; *Sibson*, 2003]. Such fractures can be more easily stimulated at lower injection pressures than hydraulic fractures.

The link between fractures and fluid flow through caprock is well established [e.g., *Hickman et al.*, 1995]. Several studies on hydrocarbon migration have found that fluid pressure in such reservoirs is bounded by the minimum horizontal stress; this suggests that an equilibrium is maintained in such hydrocarbon systems through episodic faulting and fluid migration [*Watts*, 1987; *Hunt*, 1990; *Dewhurst et al.*, 1999; *Ingram and Urai*, 1999; *Finkbeiner et al.*, 2001; *Hermanrud and Bols*, 2002]. A useful concept put forward by *Finkbeiner et al.* [2001] is that of a *dynamic capacity model* representing the critical pore pressure at which a caprock fails either by activation of an existing fault or by a newly formed hydraulic fracture into the caprock. As discussed above, this critical pore pressure would generally be lower in the case of fault activation. Given the arguments that the Earth's crust is critically stressed with abundant fractures [*Zoback*, 2007], this further suggests that the limits of injection pressure would be lower than those set based on criteria for hydraulic fracture failure of the caprock. Such behavior is reported in *Townsend and Zoback* [2000] who provide evidence that much of the crust maintains a hydrostatic pore pressure due in part to fault activation that allows redistribution of fluids [also see *Sibson*, 2003].

The picture for fracture development in caprock is complex. On the one hand, it is clear that shale caprock provides long-term sealing behavior for oil and gas reservoirs and that therefore it could provide similar storage for CO_2. On the other hand, it is also clear that reservoirs leak and that leakage can be initiated by fluid-pressure-induced fault activation. The two logical conclusions from this are that (i) injection pressures that can trigger faults have to be avoided a priori and that (ii) the potential leakage of CO_2 along faults (either because of existing faults or stimulated faults) has to be considered as part of the risk assessment of a storage operation. *The missing component in this analysis is a consideration of the consequences of leakage on fractures or faults*: that is, what is the permeability of faults, and what do we know of the time dependence of permeability and the effect of the particular properties of supercritical CO_2 on fault transmissivity? Such information is key to the development of contingency plans that have to be put in place in case such an event occurs.

10.2.2. Permeability of Fractured and Faulted Shale

Faults can be transmissive, sealing, vertically transmissive but laterally sealed, and episodic in behavior [*Aydin*, 2000]. The occurrence of fault-compartmentalized oil reservoirs clearly shows that the mere existence of faults does not preclude long-term accumulation of buoyant hydrocarbon or supercritical CO_2 for that matter [e.g., *Fisher and Knipe*, 2001]. In fault-bounded reservoirs, the most important sealing mechanism is the juxtaposition of an impermeable unit against the reservoir although the faults themselves may act as seals [*Downey*, 1984; *Fisher and Knipe*, 2001]. Several methods of predicting fault transmissivity are based on clay mineralogy and include clay smear potential, shale smear factor, and shale gouge ratio, as summarized in *Edlmann et al.* [2013] and *Fisher and Knipe* [2001], in which increasing clay content promotes sealing behavior.

Faults have complex structures that are often multistranded and may consist of a fault core surrounded by a damage zone. *Aydin's* [2000] review of fault structures indicates that fault cores in porous sandstone, consisting of comminuted material, are typically less permeable than the original rock by up to four orders of magnitude, whereas the damage zone might have enhanced permeability up to two orders of magnitude. For example, *Agosta et al.* [2007] show that fault zones in carbonate host rocks can be separated into two main units with distinct hydraulic characteristics: a sealing fault core that prevents fluids from crossing the fault and a cracked zone surrounding the core that allows fluid flow parallel to the fault. The latter can also occur due to the formation of opening-mode fractures within the cemented fault rocks. In *Vialle et al.'s* [2016] discussion of fault structures, they describe work in low-porosity rocks that shows contrasting behavior with an elevated permeability fault core and reduced permeability damage zone. *Aydin* [2014] reviews fault structures in shale including the likely low permeability of shear bands. The key message here is that the hydraulic properties of faults are likely to be highly anisotropic, with faults in some cases acting as barriers and in other cases conduits to flow.

The permeability resulting from shale deformation is likely to be strongly influenced by whether strain is accommodated by brittle or ductile behavior [*Dewhurst et al.*, 1999; *Ingram and Urai*, 1999; *Nygård et al.*, 2006]. Brittle shear deformation is characterized by dilation and the possibility of formation of fracture apertures, thus increasing permeability. On the contrary, ductile behavior is characterized by compaction (net porosity reduction) and would likely result in more limited changes or even reduction in permeability. This transition from dilation to contraction can be expressed by the concept of critical state behavior of deforming sediments [*Schofield and Wroth*, 1968; *Jones and Addis*, 1986]. Two approaches to predict critical state behavior are based on the concept of the over-consolidation ratio (OCR) (defined by the ratio of the maximum vertical stress experienced by the shale to the present-day vertical stress) and the brittleness index (defined in terms of unconfined strength; *Ingram and Urai*, 1999; *Nygård et al.*, 2006). (We note that there are many different definitions of brittleness index in the literature.) *Nygård et al.* [2006] find the transition to ductile behavior at $OCR > 2.5$, suggesting that brittle behavior is observed in shale that is at substantially shallower depths in relation to the maximum depth of burial.

The mineralogy of shale must also play a role in ductility and fracture transmissivity. *Bourg* [2015] argues that there is a critical transition in brittle to ductile behavior in shale as clay content increases above 33%. In support of this concept, *Bourg* [2015] found that shale gas reservoirs had lower clay content than shale formations proposed for radioactive waste repositories and CO_2 sequestration. This is also consistent with experimental observations including those of *Sone and Zoback* [2013] who found increasing ductility and viscoelastic creep with clay content of shale gas samples. Creep is likely to play an essential role in understanding the long-term permeability of fractured shale with such self-sealing processes, a focus of work in the European nuclear waste repository community [e.g., *Bock et al.*, 2010].

Measurements of permeability of fracture-damaged shale are relatively scarce perhaps because of challenges posed by anisotropy, heterogeneity, and sensitivity of shale to moisture. Some work has focused on investigations of manufactured fractures or samples with natural fractures [e.g., *Bernier et al.*, 2007; *Davy et al.*, 2007; *Gutierrez et al.*, 2000; *Cho et al.*, 2013; *Edlmann et al.*, 2013; *Zhang*, 2013; *Zhang et al.*, 2013]. For example, *Edlmann et al.* [2013] showed that a shale sample with natural fractures would not transmit supercritical CO_2, but was permeable to gaseous CO_2, which they attributed to increased fracture apertures at lower confining pressure.

An even smaller set of laboratory studies have made measurements of shale fractured at in situ conditions in order to gain insight into fracture permeability behavior at reservoir conditions [*Nygård et al.*, 2006; *Bernier et al.*, 2007; *Zhang and Rothfuchs*, 2008; *Monfared et al.*, 2012; *Carey et al.*, 2015; *Menaceur et al.*, 2015]. *Bernier et al.* [2007] made measurements of hydraulic fracture conductivity using a hollow cylinder method and observed an increase of permeability of four to five orders of magnitude before self-sealing processes reduced these values. In triaxial compression experiments, *Bernier et al.* [2007] observed a *decrease* in permeability of Boom Clay but a significant increase (unspecified) in the permeability of Opalinus Clay. In similar tests, *Nygård et al.* [2006] and *Zhang and Rothfuchs* [2008] also observed increases in permeability, which the former did not quantify and the latter observed a four order of magnitude increase to 1 μD. *Menaceur et al.* [2015] and *Monfared et al.* [2012] used a hollow cylinder method on Callovo-Oxfordian and Boom shale, respectively, and found no increase in permeability after fracture formation. Results by *Carey et al.* [2015] using direct shear methods will be discussed below.

Field measurements of fault permeability in shale were conducted by *Nussbaum and Bossart* [2004] for the Main Fault in the Mont Terri underground rock laboratory. They found that the fault zone had permeability similar to the impermeable Opalinus clay matrix. *Guglielmi et al.* [2015] reactivated the strike-slip Tournemire fault by injection of fluid using a straddle packer system. They calculate a two order of magnitude increase in permeability following fault reactivation.

10.2.3. Geochemical Interaction of CO_2 with Fractured Shale

The thickness and impermeability of shale indicate that CO_2-induced alteration of the bulk material is unlikely to lead to the formation of significant leakage pathways [e.g., *Fitts and Peters*, 2013]. This is not to say that CO_2 has no effect on shale: *Armitage et al.* [2013] found that flow of CO_2-saturated water through a chlorite-siderite-cemented siltstone resulted in an eightfold increase in permeability from about 10^{-20} to 10^{-19} m^2. However, the most important geochemical effects on fracture permeability are likely to be the result of dissolution and precipitation processes [e.g., *Detwiler*, 2008]. Under geological confinement, dissolution can lead to either fracture closure as asperities collapse or localized aperture opening (channel formation) depending on relative rates of chemical reaction and fluid movement [*Detwiler*, 2008]. There has been relatively little work on the coupled geochemical-hydrologic effects of flow of CO_2 and brine on fractured caprock. Most of such experimental work has focused on fractures in carbonate [e.g., *Noiriel et al.*, 2009; *Ellis et al.*, 2011; *Elkhoury et al.*, 2013].

In addition to mineral reactions, the geochemistry of CO_2 sorption may be an important component of CO_2 migration through shale. According to *Bourg et al.* [2015], the role of CO_2 sorption is poorly understood. Studies of CO_2 sorption on shale include those of *Heller and Zoback* [2014] who measured CO_2 sorption up to a pressure of 4 MPa and found that it was two to three times greater than methane (12–75 SCF CH_4/ton, i.e., 0.25–1.6 kg CH_4/tonne) on four different gas shale samples. This is much lower than the approximately 770 SCF CO_2/ton (44 kg/tonne) measured by *Busch et al.* [2008] at a pressure of 12 MPa on Muderong Shale. We provide additional discussion of sorption processes and their role in potentially mitigating CO_2 leakage through caprock below.

Finally, CO_2 has the potential of altering fracture development and propagation through chemical reactions or stress corrosion (*Anderson and Grew*, 1977; also see a recent review of the effect of CO_2 on fault friction properties by *IEAGHG* [2015]). *Rinehart* [2014] found that chlorite-cemented sandstone was weakened on exposure to supercritical CO_2. Experimental work of *Samuelson and Spiers* [2012] examined slip in fault gouges in shale and sandstone and found that the coefficient of friction was not impacted by the presence of CO_2, at least in short-term experiments. *Masoudi et al.* [2012] injected shale core with liquid CO_2 and found a small decrease in the friction coefficient and small increase in the cohesion, while Young's modulus increased.

10.3. EXPERIMENTAL STUDY OF FRACTURE PERMEABILITY IN SHALE

In this section, we describe experiments that were designed to measure the permeability of fractured shale in order to provide insight into the potential leakage of CO_2 from damaged shale caprock. The experiments were conducted with a triaxial core flood device that simultaneously measures permeability and creates fractures under in situ reservoir conditions. The goal was to develop insight into factors (pressure, temperature, stress conditions, shale mineralogy, shale textures) that control the magnitude of fluid leakage through damaged shale.

The experiments were conducted in a custom-designed triaxial core flood system combined with in situ tomography capabilities [*Carey et al.*, 2015; *Frash et al.*, 2016a]. The system is capable of delivering maximum pore pressure and horizontal stress of 34.5 MPa (5000 psi) and maximum axial stress of 490 MPa (71,000 psi) and temperature of 100°C. Fluid injection of water, supercritical CO_2, or oil (or a controlled combination of any two phases) is possible. Cylindrical core samples of Utica Shale, courtesy of Chesapeake Energy, were used in the experiments with length and diameter of about 2.5 cm (1″) and mineralogy that was roughly 28% carbonate, 22% quartz + feldspar, and 50% clay [c.f., *Carey et al.*, 2015]. Experiments were conducted in compression, direct shear, and hydraulic fracture configurations (Fig. 10.1).

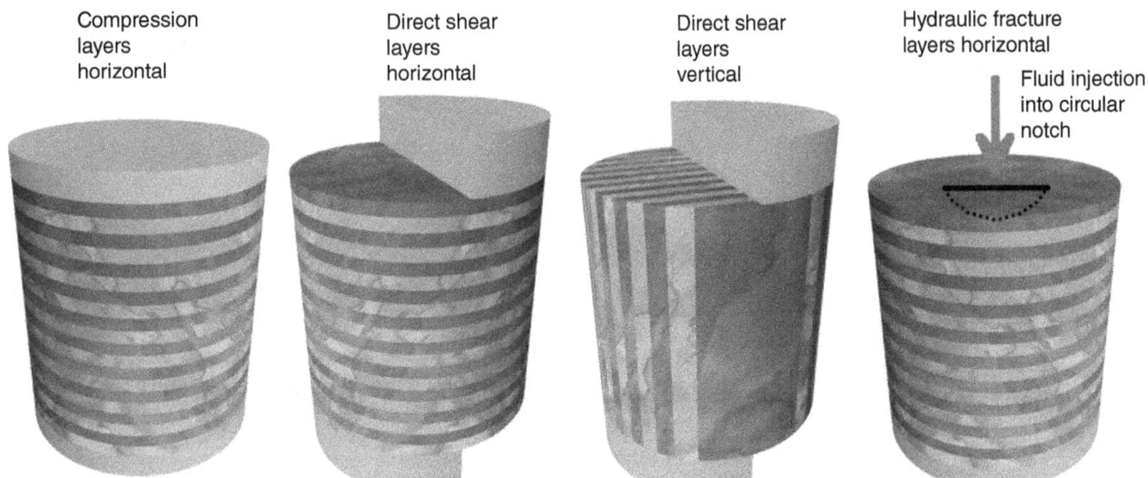

Figure 10.1 Schematic diagram of compression (left image), direct shear (two middle images with bedding planes perpendicular on the left and parallel to the load on right), and hydraulic fracture (right image) deformation modes in the triaxial core flood experiments. The shale core appears as a striped cylinder with bedding planes, either horizontal or vertical. The platens are shown on top and bottom as a solid shade.

Compression experiments were based on conventional triaxial methods of squeezing the 1″ diameter specimen between 1″ diameter platens. Direct shear experiments were conducted with two opposing semicircular platens that generated a shear plane through the length of the shale specimen. The bedding planes of the shale specimen were oriented either perpendicular or parallel to the axis of the applied load bearing down on the semicircular platens. Hydraulic fracture experiments were conducted by cutting a circular notch into the face of the shale specimen and injecting at fluid pressures greater than the confining pressure.

The experiments were conducted at 20–50°C with confining pressures ranging from 3.4 to 22 MPa (500–3200 psi). An injection pressure (e.g., 1.4 MPa or 200 psi) using water was maintained at the inlet face of the core sample. Permeability was continuously monitored by a combination of the measured pressure drop and flow rate of water. Deformation was induced by constant flow of the axial pump (resulting in about 10 microstrains/s) and measured using a linear variable displacement transducer (LVDT) that recorded displacement of the axial piston. X-ray tomography data were collected as described in *Carey et al.* [2015] with a resolution of about 25 μm. Measurements of permeability included changing the injection pressure or flow rate as well as the confining pressure and were conducted over periods ranging from 2 to 5 h.

10.3.1. Fracture Behavior of Shale

Compression, direct shear, and hydraulic fracture experiments produce distinctly different fracture systems that are strong functions of the orientation of the bedding planes (Figs. 10.2 and 10.3). Compression experiments produced shear fracture apertures that were generally narrower than those observed in direct shear. Although short sample lengths were used, fractures in compression experiments were not clearly connected between the upper and lower surfaces of the specimens, resulting in uncertainty in the permeability measurements. More details on the compression experiments are given in *Carey et al.* [2014].

Direct shear experiments, in contrast, produced fractures that were well connected between the upper and lower specimen faces and were very suitable for permeability measurements. Fractures in both horizontal and vertical bedding specimens had relatively large apertures, resulting in permeability values that were significantly higher than the compression experiments (see below). Hydraulic fracture experiments produced simple fractures that closed upon release of the injection pressure with residual apertures that were less than 25 μm. Additional details on hydraulic fracture experiments are in *Frash et al.* [2016b].

In direct shear experiments, fractures that propagated parallel to the bedding planes were simpler (less bifurcations,

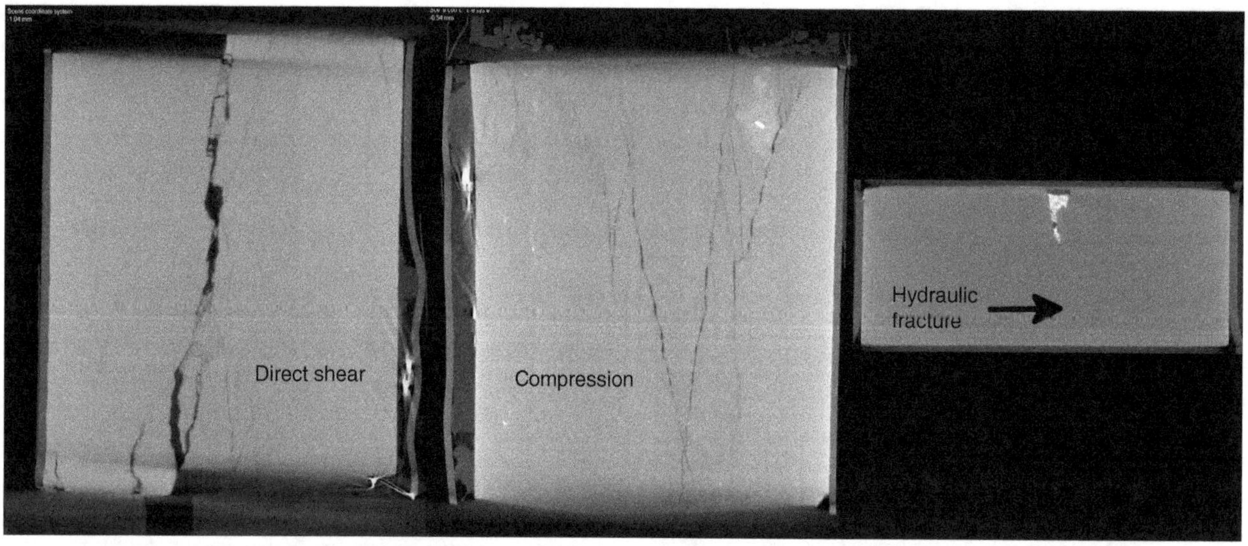

Figure 10.2 X ray tomography fracture patterns generated in Utica Shale (1″ diameter) using three different triaxial deformation modes. Left: direct shear experiments produce a single dominant, wide-aperture vertical fracture. Middle: compression experiments generate complex, narrow-aperture fractures with top-to-bottom asymmetry. Right: hydraulic fracture experiments produce minute (< 25 μm) single-plane fractures. Bedding planes are vertical in the direct shear and compression experiments and horizontal in the hydraulic fracture experiment. The hydraulic fracture experiment involved proppants (bright grains at top). Modified from *Carey et al.* [2015] and *Frash et al.* [2016b].

fewer strands) than fractures that crossed bedding planes (cf., Figs. 10.2 and 10.3). *Carey et al.* [2015] interpreted stress-strain behavior to indicate that significant strain in the horizontal bedding specimen (Fig. 10.3, left) was accommodated by slip along the bedding planes and that this limited the development of through going fractures and permeability.

Permeability of fractured shale (referenced to the 1″ diameter core) was continuously measured using water during the fracture experiments as a function of the initial confining pressure (Fig. 10.4). Compression experiments yielded maximum permeability values ranging from 0.1 to 22 mD with fractures parallel to bedding having generally higher permeability. The permeability of direct shear experiments ranged from a low of 30 mD for the horizontal bedding samples to a high of 900 mD for vertical bedding samples [*Carey et al.*, 2015]. The hydraulic fracturing experiment yielded a low permeability value of 0.1 mD [*Frash et al.*, 2016b]. *Carey et al.*

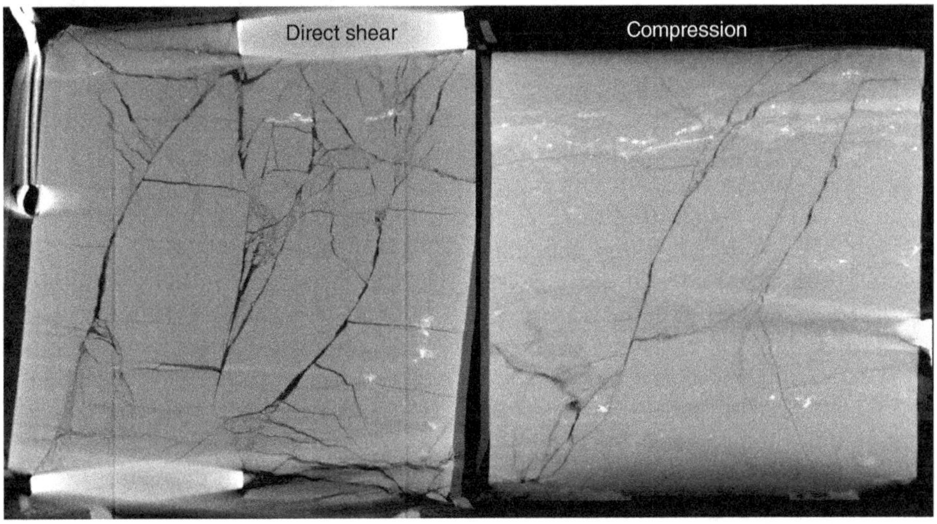

Figure 10.3 X-ray tomography fracture patterns in Utica Shale (1″ diameter) for samples with bedding perpendicular to the load (horizontal with respect to the image) studied in direct shear (left) and compression experiments (right). Bedding plane displacements are evident in both tests, while the direct shear fractures provide good connectivity between the upper and lower faces of the shale specimen. The direct shear platens are visible on the left. Modified from *Carey et al.* [2014, 2015].

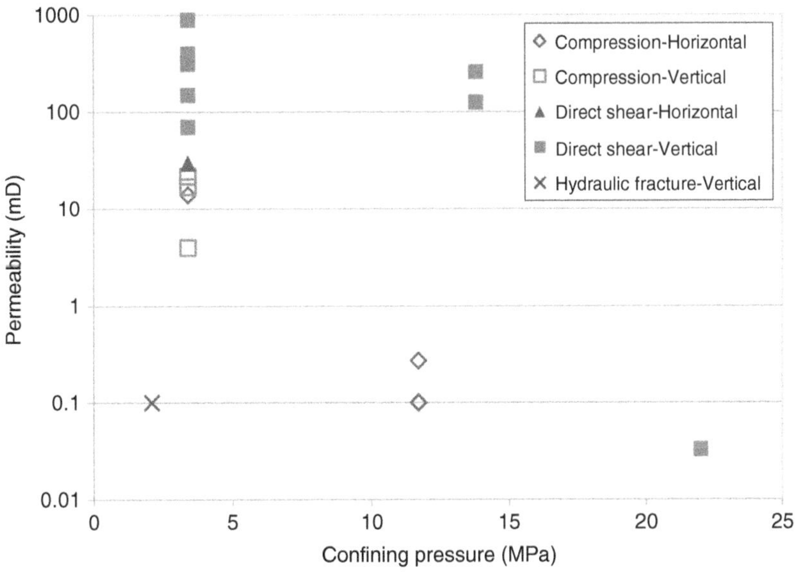

Figure 10.4 Summary of permeability measurements made on Utica Shale using compression, direct shear, and hydraulic fracture methods. Permeability decreases with confining pressure and varies with experiment type.

[2015] also found that permeability and the peak stress achieved before failure (strength) were a strong function of the angle between the vertical bedding plane and the direct shear plane (Fig. 10.5). Peak strength was a linear function of this angle ranging from 25 to 95 MPa, and the permeability was characterized by a distinct maximum around an angle of 45° between bedding and fracture plane.

After formation of fractures, *Carey et al.* [2015] explored fracture compliance by changing the confining pressure and measuring the resulting changes in permeability. They found an exponential decay in permeability in which $k = P_0 e^{-}aP$, where P_0 was the permeability of the unloaded fracture, P was the confining pressure in MPa, and a varied from 0.1 to 0.2. The exponential character of the data suggested that while fracture apertures decreased with applied stress, a return to matrix permeabilities would require quite high stresses.

In recent work, we have investigated the changing character of fractures observed under in situ experimental conditions. The experimental result shown in Figure 10.4 at 22 MPa confining pressure was particularly interesting. The fractured specimen permeability was only slightly greater than the pre-fracture permeability. In situ X-ray tomography showed that despite extensive deformation, fractures were difficult to resolve at the 25 μm resolution of the tomographic images. These results were consistent with a brittle-to-ductile transition that resulted in dilative open fracture formation occurring at low pressures (3.4 MPa) and contractive deformation at high pressure (22 MPa). The tomography shows that significant shortening of the sample was accommodated without fracture opening. The permeability measurements support the concept that deformation at these conditions *can* occur without significant development of transmissivity. Upon decompression to atmospheric conditions, a much more significant fracture system developed and had permeabilities that were an order of magnitude greater. The results highlight the challenges in the interpretation of the permeability of fractured materials recovered from high pressure conditions. Additional details are in *Frash et al.* [2016a].

10.4. CO_2 SORPTION BEHAVIOR IN SHALE

Mudrock or shale caprock contains very narrow pores with nanometer dimensions that are the primary cause for their low (matrix) permeability. Most importantly, these nanopores create intimate fluid-rock interactions that could lead to the physical adsorption of gases, that is, to the trapping of gas molecules at near-liquid-like densities attached to surfaces of pores in the material. Thus, CO_2 sorption in shale has the potential to mitigate CO_2 leakage processes. The mechanism of adsorption is widely exploited for the purification of industrial gas streams, where materials such as activated carbon or zeolites are routinely used that possess large micropore and mesopore volumes and, accordingly, high surface areas [500–4000 m²/g; *Ruthven*, 1984]. In a shale rock, micropores are typically associated with minerals, such as clays, and/or organic material, such as kerogen [*Schettler and Parmely*, 1991]. Micropores (< 2 nm) and mesopores (2–50 nm) in various shales have been correlated to the dominance of the illite-smectite type of clays in the rock [*Kuila and Prasad*, 2013]; so-called intraparticle organic pores with diameters as small as 4 nm have been observed

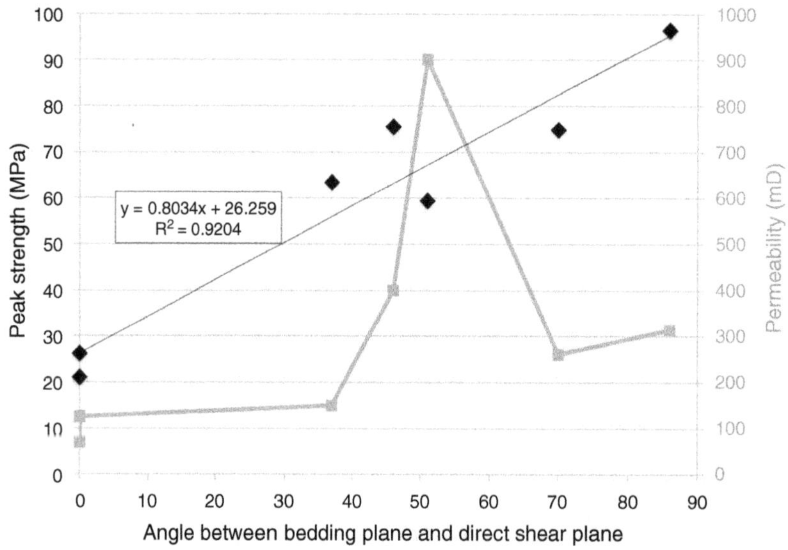

Figure 10.5 Permeability and strength of Utica Shale fractured in direct shear experiments as a function of the angle between the bedding plane and the shear plane (cf., Fig. 10.1). Modified from *Carey et al.* [2015].

in samples from the Barnett Shale and contribute to grain porosities as high as 20–30% [*Loucks et al.*, 2009]. Remarkably, some results suggest that 50–80% of the total amount of natural gas found in shale plays is trapped as an adsorbed phase in the micropores of the rock [*Curtis*, 2002]; however, other studies suggest that the contribution of adsorbed gas to total production may be small when the latter is driven by a reduction in reservoir pressure only [*Heller and Zoback*, 2014].

To better appreciate the extent of gas sorption on shale, we compare CO_2 adsorption isotherms measured at 50°C and pressures up to 20 MPa on two commercially available sorbent materials (activated carbon and zeolites), a bituminous coal, and a shale (Fig. 10.6). The adsorption capacity of *engineered* materials is significantly larger (~30–37 wt%) than the values obtained for natural materials (2 and 8 wt% for shale and coal, respectively). However, as shown by the filled symbols in the same figure, this margin is greatly reduced when the total volume of the system (e.g., the fixed-bed adsorption column or the geologic unit) is considered, thus highlighting the significance of adsorption in (dense) microporous rocks. Accordingly, we argue that to fully evaluate the sealing effectiveness of (damaged) shale caprock, a better understanding is needed of the mechanisms of adsorption.

Measuring adsorption at conditions representative of shale caprock requires achieving elevated pressures (> 10 MPa) and temperatures (> 40°C). While under these conditions the behavior of the free gas phase can be described by an equation of state, the thermodynamic treatment of the adsorbed phase is more complex. For a gas below the critical point (for CO_2 this is 31.1°C, 7.39 MPa), the adsorption mechanism is controlled by the vapor-liquid equilibrium, and, accordingly, the density of adsorbed fluid takes a value near that of the saturated liquid. On the other hand, supercritical fluids show no discontinuities in the transition to a dense phase, and the density in its adsorbed state is not well defined. The debate regarding the mechanism of adsorption and its appropriate representation under such conditions is still matter of controversy [*Zhou and Zhou*, 2009]. A recent publication that reviews various adsorbate/adsorbent systems shows that values for the adsorbed density can range between the liquid and the solid state of the adsorbate even for similar pressure and temperature conditions and large discrepancies were found between different measurement techniques [*Pini*, 2014a]. In practice, because the molecular volume of the adsorbed phase cannot be determined accurately in microporous material, *incremental* rather than *absolute* quantities are used to report adsorption at elevated pressures [*Gumma and Talu*, 2010]. Such measures require the definition of a reference state; when the skeletal (inaccessible to adsorption) volume of the solid material is known, the so-called excess adsorbed amount is obtained from an adsorption measurement that is defined as the difference between the actual amount adsorbed (n^A) and the amount of homogeneous bulk fluid with density ρ_M that would be present in the (unknown) volume occupied by the adsorbed phase [V^A; i.e., as $n^{EX} = n^A - \rho_M V^A$; *Sircar*, 2001]. We refer here to the skeletal volume as the impenetrable volume of the adsorbent, that is, the value obtained from a pycnometer measurement with helium. Yet concerns are being raised with regard to finding an appropriate methodology to measure the inaccessible volume of a microporous material [*Do and Do*, 2007; *Do et al.*, 2010; *Gumma and Talu*, 2010; *Pini*, 2014a], as this may significantly affect the behavior of the adsorption isotherm at elevated pressures (or densities), a condition that is indeed readily achieved in the subsurface.

As an example of general validity, CO_2 adsorption data on a dry Eagle Ford Shale sample are presented in Figure 10.7 that have been measured at 50°C and over a wide range of pressure (0–20 MPa) in a Rubotherm magnetic suspension balance. The gray-filled circles are raw excess adsorption data obtained by using a solid volume, V^S, of the shale measured by a (traditional) helium experiment (ρ^S = 2.87 g/cm³); the empty symbols refer to the same data, but where V^S has been obtained by X-ray diffraction (XRD) (circles, ρ^S = 2.64 g/cm³, by assuming values for the density of the main mineral constituents [calcite-quartz-clays] obtained by quantitative

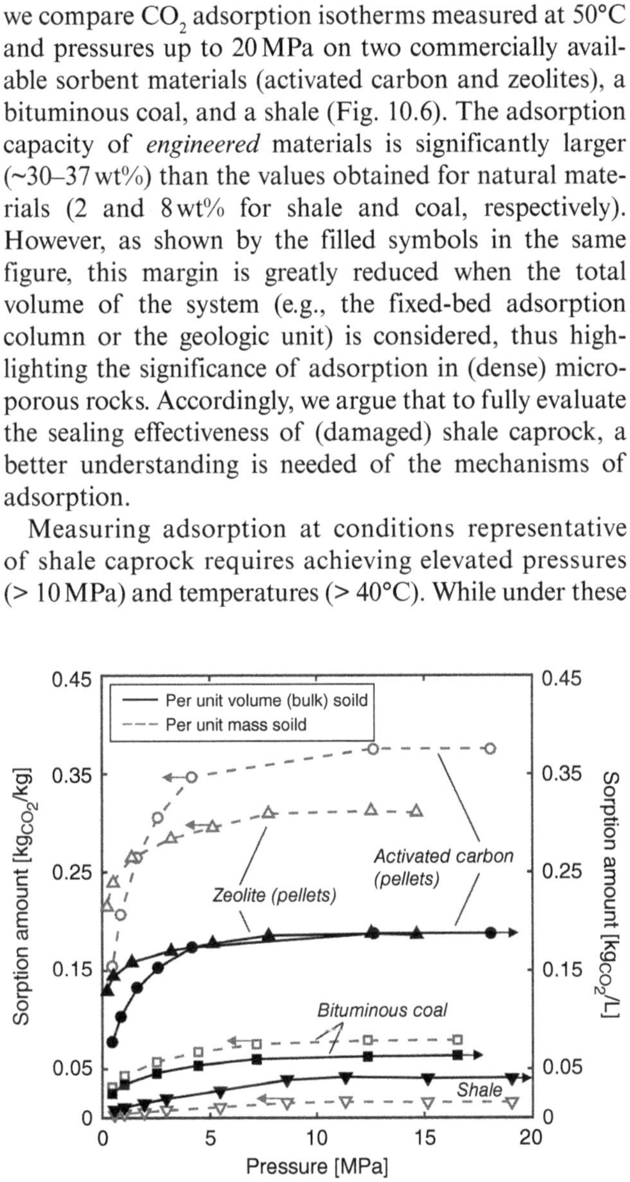

Figure 10.6 CO_2 sorption on various sorbent materials at 50°C as a function of gas pressure: activated carbon [*Pini et al.* 2006; Filtrasorb 400, Chemviron Carbon, ρ_B = 0.5 g/cc; V^A = 0.344 cc/g], 13X zeolite [*Pini*, 2014a; Z10-03, Zeochem, ρ_B = 0.6 g/cc; V^A = 0.27 cc/g)], a bituminous coal [*Pini et al.*, 2010; Australia – A1, ρ_B = 0.8 g/cc; V^A = 0.055 cc/g], and a shale (this study, ρ_B = 2.5 g/cc).

XRD and a kerogen density of 1.3 g/cm³) or by choosing its value such that the isotherm is positive (squares, *nonzero excess*) over the entire pressure range ($\rho^S = 2.59$ g/cm³). As expected from the definition of excess adsorption, the isotherms are characterized by a maximum and then decrease linearly when sufficiently large fluid densities are reached, that is, when adsorption saturation is likely to be attained. It can be seen that the choice of the parameter V^S affects significantly the shape of the isotherm, a result that has been reported also for activated carbon [*Malbrunot et al.*, 1997] and zeolite [*Pini*, 2014a]. In the case of a weakly adsorbing material, such as shale, this effect is exacerbated and leads to a significant portion of the adsorption isotherm having negative values. While the existence of negative adsorption values has been questioned in the literature [*Malbrunot et al.*, 1997; *Do et al.*, 2010], the definition of excess adsorption as an incremental quantity does not necessarily preclude it [*Gumma and Talu*, 2010]. In fact, one could argue that at large densities, steric effects associated with the confined space of a micropore lead to a less effective packing of the molecules as compared to the (free) bulk phase. Nevertheless, the discrepancy between the three sets of data raises concerns with respect to the interpretation of adsorption experiments at elevated pressures with materials that possess relatively low adsorption capacity. Absolute sorption values, n^A, can be calculated by estimating the volume of adsorbed phase, V^A, from the slope of the linear region of the isotherm, that is, when adsorption saturation has been reached, and by assuming that this value is independent of pressure. For the three excess isotherms reported, V^A takes a value of 0.06, 0.03, and 0.02 cm³CO_2/g rock, respectively. Note that upon forcing the adsorbed phase to be constant in volume, we assume that it entirely fills the (micro)pore space, irrespective of the gas pressure. As shown by the black-filled symbols in the figure, this graphical approach is not affected by the choice of the parameter V^S, thus increasing the reliability of these estimates. For the Eagle Ford sample considered here, CO_2 sorption increases up to a saturation value of about 0.4 mol-CO_2/kg (~300 SCF/ton); in comparison, CO_2 adsorption capacities in the range of 0.7–0.8 and 0.7–2 mol-CO_2/kg have been observed at similar conditions for samples of Devonian and Permian shales, respectively [*Weniger et al.*, 2010].

Mathematically, the fluid storage capacity of a caprock can be expressed as

$$S_{CO_2} = \rho_{CO_2}\left[\frac{\phi}{\rho_B} - V^A\right] + n^A \qquad (10.1)$$

where S_{CO_2} (mol-CO_2/kg-rock) is the gas-in-place per unit weight of material, ρ_{CO_2} (mol-CO_2/L-CO_2) is the gas-phase molar density, ρ_B (kg rock/L bulk rock) is bulk rock density, V^A (mol-CO_2/L-rock) is the adsorbed phase volume, ϕ (−) is the total (effective) porosity, and n^A (mol-CO_2/kg-rock) is the adsorbed amount. The first term on the right-hand side of Eq. (10.1) represents the amount of free gas in the pore space of the rock (described by an equation of state, such as a real gas law) and accounts for the reduction of the pore space due to gas adsorption. The second term is the amount of adsorbed gas, which can be obtained through the measurement of an adsorption isotherm, such as the one presented in Figure 10.7. Application of Eq. (10.1) to the Eagle Ford sample considered here leads to the situation depicted in Figure 10.8, where the total CO_2 uptake (converted to SCF/ton) is plotted as a function of the fluid pressure. In the figure, the dashed and dash-dot curves represent the total storage capacity of the rock, as given by either one of the two mechanisms alone. As expected, adsorption is particularly effective below 1200 psia (pounds per square inch absolute; as the adsorbed phase is still much denser than the gas phase), while the contribution of free gas increases with pressure and eventually overcomes adsorption. Unfortunately, common practices to estimate volume capacities often improperly account for gas adsorption by neglecting the volume occupied by the adsorbed fluid (i.e., by assuming $V^A = 0$ or $\rho^A \to \infty$; *Ambrose et al.*, 2012).

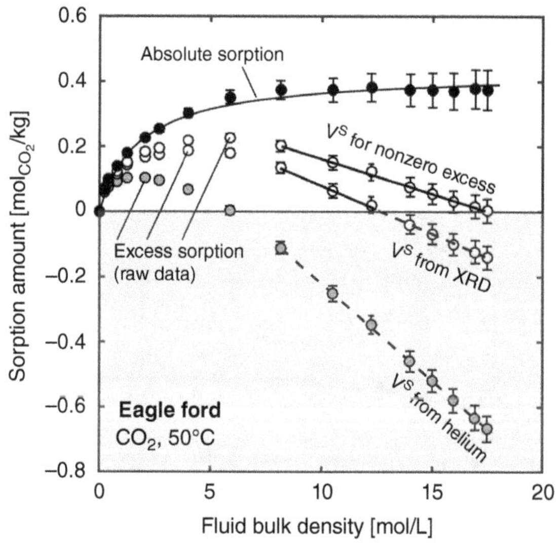

Figure 10.7 Experimental data set from a high-pressure CO_2 adsorption experiment carried out at 50°C on an Eagle Ford sample (dried at 105°C, size fraction: 180–425 µm). The sample has a TOC content of about 7 wt% and a total pore volume of 0.04 cm³/g (as measured by N_2 adsorption at 77 K). Three sets of excess adsorption isotherms are shown that differ in the value assumed for the skeleton volume of the sample; absolute sorption data (black-filled symbols) are estimated by following a graphical approach that uses the slope of the linearly descending part of the isotherm.

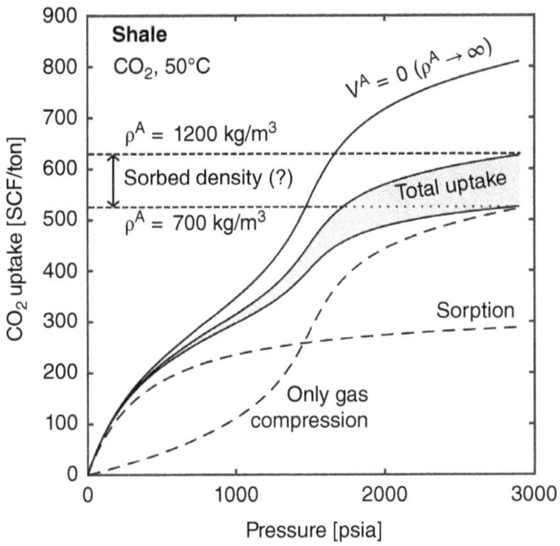

Figure 10.8 Theoretical CO_2 storage capacity of a shale caprock based on experimental data from a sample of Eagle Ford Shale. The dashed and dash-dot lines represent contribution from gas compression in the entire pore space ($\phi \approx 9\%$) and from gas adsorption in the micropores (described by a curve fitted to the black-filled symbols in Figure 10.7), respectively. The solid curves have been estimated from Eq. (10.1) by assuming distinct values of the adsorbed phase density (or, accordingly, volume), namely, $\rho^A = 560, 700, 1200, \infty$ kg/m^3.

As shown in the figure, such an approach, which is obtained as the simple sum of gas compression and gas sorption (dashed and dash-dot lines in Fig. 10.8, respectively), leads to a significant overestimation of the storage capacity. Yet, when V^A is accounted for, assumptions are made regarding its value (or equivalently the adsorbed density). As discussed by *Pini* [2014b], values between 600 and 1400 kg/m^3 have been reported in the literature for the adsorbed gas density observed in clay and shale samples. We have shown in Figure 10.7 that the method of estimating the adsorbed phase volume from the measured high-pressure adsorption isotherms creates an additional source of uncertainty ($\rho^A = 560$–700 kg/m^3). The solid curves in Figure 10.8 represent total storage capacity (*total uptake*) estimates based on four distinct values of ρ^A (or V^A), namely, $\rho^A = 560$, 700, 1200 kg/m^3 and $\rho^A \to \infty$, with the area shaded in purple referring to the region where most of the estimates are found. It can be seen that knowledge of the density of the adsorbed phase is quite important as its effect on the total uptake is quite large; also, in agreement with results reported above, the contribution of adsorption is dominant below about 1500 psi, and the appearance of a negative excess adsorbed amount (the case with $\rho^A = 560$ kg/m^3) leads to a total uptake capacity at elevated pressures that is less than the amount that would be stored by simple gas compression without adsorption.

10.5. ACOUSTIC PROPERTIES OF THE CO_2-SHALE SYSTEM

One approach to monitoring and characterizing CO_2 infiltration into caprock utilizes changes in the acoustic properties of CO_2-saturated rock. Numerous studies have been conducted to study the effect of replacing water or hydrocarbon with CO_2 in potential reservoir rocks [e.g., *Batzle et al.*, 2005; *Adam et al.*, 2006; *Vanorio et al.*, 2010; *Yam and Schmitt*, 2011; *Njiekak et al.*, 2013; *Sharma et al.*, 2013]. Changes in seismic *P*- and *S*-wave velocities due to rock-fluid interactions have been used to develop poroelastic models of rocks [*Dvorkin and Nur*, 1993; *Berryman and Wang*, 1995; *Diallo et al.*, 2003; *Khazanehdari and Sothcott* 2003]. These studies have been able to detect changes in seismic signatures due to displaced reservoir fluids in mostly clastic reservoirs. Application of these rock physics methods to caprocks has been more limited, largely due to the limited experimental data on mudrocks. As compared with conventional reservoir rocks, most caprocks are characterized by a relatively higher degree of heterogeneity of the pore topology as well as mineral composition, containing a mixture of clay, quartz, and carbonate minerals and organic matter. This may result in a nonuniform rock-fluid interaction and heterogeneous saturation profile in the pores due to the preferential coverage of different fluids for specific rock components [*Sharma et al.*, 2013]. For example, *Kumar et al.* [2015] have shown that organic-rich shales are characterized by selective adsorption of water in the inorganic clay pores and not the organic pores. Thus, in a multi-mineralic rock, water is likely to selectively saturate clay-bound pores, while the organic pores will be saturated by the nonaqueous phase.

Here we consider the isotropic formulation of the theoretical Gassmann fluid substitution to quantify the effect of preferential fluid saturation in organic-rich caprock. *Kumar et al.* [2015] have quantified mesopore and micropore volumes of various storage and caprocks for different sorbents (Table 10.1). We hypothesize a case in which CO_2 is injected in a water-saturated reservoir overlain by a water-saturated caprock. With CO_2 invasion, the new saturation profile in the caprock would be as follows. At the injection pressure, CO_2 can be expected to saturate the pores of the caprock that are comparable in size with the pores in the reservoir. However, pore sizes are much smaller in caprocks [*Kuila and Prasad*, 2013] and are dominated by mesopores and micropores. Thus, the displacement of the aqueous phase from the smaller pores with the invading CO_2 will take place only if CO_2 pressure is larger than the capillary pressure of the CO_2-aqueous phase in the pores. The capillary pressure, in turn, is dependent upon the degree of affinity of the pores toward the aqueous phase. This preference was

Table 10.1 Effect of CO_2 Invasion on the Bulk Modulus (K) of Caprock Modeled Using Gassmann Fluid Substitution.

	ϕ_{Hi+Ho}			ϕ_{Hi}			ϕ_{Ho}									
Sample	Mes	Mic	Tot	Mes	Mic	Tot	Mes	Mic	Tot	K_o	K^*	K_{init}	K_f^{ads}	K_f^{gas}	K_{final}^{ads}	K_{final}^{gas}
UBS1	8.6	1.8	10.4	5.5	0.8	6.3	3.0	1.1	4.1	32.5	27.7	35.7	0.46	0.42	29.3	29.1
UBS5	6.1	1.3	7.4	3.3	0.4	3.7	2.8	0.9	3.8	34.4	30.8	43.4	0.38	0.35	32.6	32.4
UBS13	7.9	1.5	9.4	1.4	0.2	1.7	6.4	1.3	7.8	37.3	32.2	42.3	0.24	0.23	33.2	33.1
LBSG	9.0	2.2	11.2	3.6	0.3	3.9	5.4	1.9	7.4	37.1	31.1	39.2	0.30	0.28	32.6	32.0
7219	13.5	2.5	16.0	8.9	0.5	9.4	4.5	2.0	6.6	30.7	23.7	28.6	0.46	0.41	24.7	24.6
11005	8.0	3.6	11.6	2.4	0.4	2.8	5.5	3.2	8.8	34.3	28.6	36.0	0.28	0.24	29.5	29.4

Estimates of hydrophilic (ϕ_{Hi}) and hydrophobic (ϕ_{Ho}) mesopores and micropores were obtained from hexane and water vapor adsorption. The initial bulk modulus for water-saturated rock (K_{init}) and the final bulk modulus after invasion with CO_2 (K_{final}) were calculated by using a frame bulk modulus (K^*) that was a weighted average of the mineral bu k moduli (K_o); a Reuss average was used for the fluid bulk modulus (K_f) [*Smith et al.*, 2003] after CO_2 invasion. The effect of supercritical CO_2 sorption in microɔores was modeled assuming that the CO_2 fluid modulus in hydrophobic micropores was the same as in the condensed liquid phase at near-critical conditions (0.29 MPa). In the remaining pores, it was assumed to be the same as that of its gas phase (0.19 MPa).

determined by quantifying separate water and hexane vapor adsorption experiments; the results obtained are listed in Table 10.1. The porosity obtained from water vapor is smaller than that with hexane vapor, which reflects the saturation preference of organic matter for the nonaqueous fluid. Similar to the preferential hexane saturation, we postulate that the initially water-saturated mesopores and macropores should be displaced by the injected CO_2. Thus, in our hypothesis, the patchy saturation profile after CO_2 injection can be summarized as follows:

1. Macro-sized pores occupied by gaseous CO_2 phase.
2. Clay-bound mesopores and micropores occupied by water. CO_2 can only exist in these pores if it is dissolved in the water.
3. The organic matter pores (difference between the hexane-derived and water-derived pores) are occupied with CO_2. This CO_2 can be supercritical if the conditions exceed the critical point of CO_2. As a result, the organic mesopores will be filled with supercritical CO_2 gas, whereas the organic micropores are filled with an adsorbed CO_2 phase with a density of 1.03 g/cm^3 corresponding to that of the van der Waals volume [*Humayun and Tomasko*, 2000], which is 23.45 mmol/cm^3.

In our fluid substitution model for predicting the seismic properties of caprock, we followed the abovementioned saturation profile. We ignored the effect of CO_2 dissolved in the aqueous phase because of the low solubility of CO_2 in water, giving negligible changes in the acoustic properties of water. Figure 10.9 shows the significant reduction in bulk modulus of the caprock after CO_2 invasion.

In the above analysis, CO_2 is considered as an inert gas that does not interact with the mineral matrix. However, core damage after CO_2 injection has been quantified in numerous rocks [e.g., *Browne*, 1978; *Adam et al.*, 2006, 2013; *Andreani et al.*, 2009; *Sharma*, 2015]. In carbonate rocks, core damage due to dissolution is followed by carbonate re-precipitation [*Sharma*, 2015]. Formation damage due to CO_2 interaction with the caprock has largely been ignored. Given the multi-mineralic nature of most caprocks, preferential surface coverage depending on mineral-fluid interactions could lead to selective mineral alterations depending on CO_2 access and reactivity with the host minerals. Thus, caprock might experience selective damage depending on fluid access and coverage that could result in additional changes to shale caprock acoustic properties.

10.6. CONCLUSIONS

Caprock integrity is one of the chief concerns in the successful development of a CO_2 storage site. In this chapter, we discuss fracture permeability relationships in shale, CO_2 sorption by shale, and the acoustic properties of CO_2-saturated shale. These sections provide an overview of the permeability of fractured shale, the potential for mitigation of CO_2 leakage by sorption to shale, and the detection by acoustic methods of CO_2 infiltration into shale. A review of the literature reveals that while significant concerns have been raised about the potential for induced seismicity to damage the caprock and cause leakage, relatively little is known about the permeability

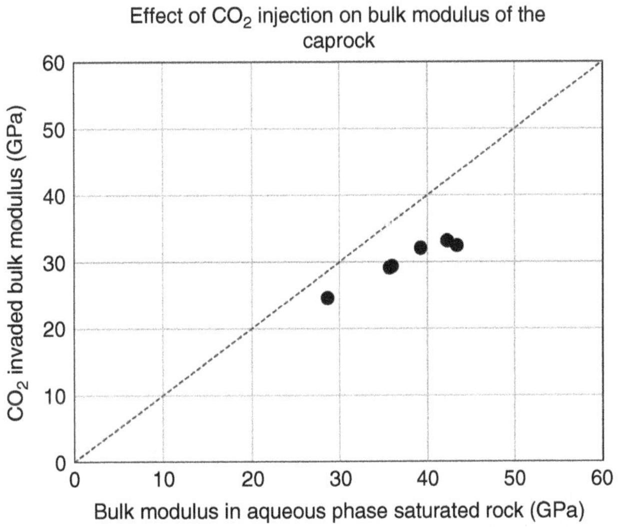

Figure 10.9 Comparison of the bulk moduli of initial and final bulk modulus of the caprock. The bulk modulus can be reduced significantly with CO_2 invasion in the caprock.

of the damaged shale. Previous studies, motivated in part by the role of shale as caprock to hydrocarbon reservoirs, have identified a critical transition between brittle and ductile deformation of shale that corresponds to a change from dilation to contraction during shear deformation. This transition is clearly a function of temperature, pressure, previous burial/diagenetic history, and the shale mineralogy and textural features.

Shale that deforms in the ductile regime would be expected to have significantly lower permeability than shale experiencing brittle fracturing. Knowledge of the permeability of damaged shale in both regimes is necessary if we are to evaluate the potential consequences of induced seismicity. Relatively little is known of this permeability either in the field or laboratory. We present a summary of recent experimental work conducted by the authors that show profound differences in permeability of up to three orders of magnitude between brittle and ductile fracture permeability. In the ductile regime, it is possible that shale caprock could accommodate deformation without a significant loss of CO_2 from the storage reservoir.

Our work also examined the effect of changes in confining pressure on the permeability of fractured shale. The results show a modest degree of permeability reduction with increased pressure and provide a measure of fracture compliance. Perhaps of even greater interests is the question of creep and self-healing of fractured shale. Studies conducted as part of the European nuclear waste disposal program clearly show strong evidence of self-healing in favorable shale types. In the absence of long-term experiments on creep, the experiments on the effect of confining pressure may serve as a proxy for the potential of creep to reduce permeability. Presumably, fractures with greater compliance would also be likely to experience greater creep over long periods of time.

In cases where CO_2 does migrate through damaged shale caprock, CO_2 sorption onto shale mineralogy may have a mitigating impact. Measured adsorption values on shale reported here range from 0.4 to 2 mol-CO_2/kg-shale and are significantly smaller than CO_2 sorption onto commercial sorbents. However, the tremendous mass of shale available in the subsurface implies that significant CO_2 may be captured by shale. The total storage capacity of caprock is the sum of the free gas phase present in the pores and the adsorbed gas on the pore walls. Adsorption is dominant at lower pressures (below about 10 MPa), and the free phase becomes more important as pressure and the density of the free phase increase. Measured total storage capacities range from 1 to 45 kg-CO_2/tonne-shale.

Once CO_2 is in the caprock, changes in the acoustic properties of CO_2-saturated shale might provide a means of detecting or monitoring CO_2 leakage. Although this approach has been applied extensively to reservoir rocks (porous sandstones and carbonates), application to caprock with elastic anisotropy is much more limited. Theoretical work described here suggests that during CO_2 invasion of caprock, macro-sized pores and organic-lined mesopores and micropores would be filled with CO_2, while mesopores and micropores lined by clay would retain water. The resulting Gassmann fluid substitution calculations show a significant reduction of the bulk modulus of CO_2-saturated shale. However, the impact of reaction-induced changes to the acoustic properties of shale has yet to be determined.

The recent development of shale gas resources has created an opportunity to focus greater attention on the fracture and fluid flow properties of shale and how shale performs both as a hydrocarbon resource and as caprock. In our summary of the literature and presentation of recent experimental results, it is clear that much remains to be learned of the permeability of fractured shale, the conditions under which CO_2 sorption in shale mitigates CO_2 leakage, and the detection of CO_2 leakage through shale. There are a wide variety of shale types, and developing a predictive capability in all three areas is a grand challenge. One of the key questions is fracture compliance and the timescale for creep and self-healing to occur and shut down the permeability. To the extent that this timescale is short, CO_2 leakage from damaged caprock could be temporally limited. In addition, other lines of evidence indicate that fluid flow through caprock may shut off below a critical fluid pressure necessary to maintain fracture apertures and permeability.

ACKNOWLEDGMENTS

This work is supported by the US Department of Energy (DOE) National Energy Technology Laboratory (NETL) under Grant Number DEFE0023223, which is managed and administered by the Colorado School of Mines and funded by DOE/NETL and cost-sharing partners, and by Grant Number FE-715-16-FY17, which is managed and administered by Los Alamos National Laboratory. We sincerely thank the US Department of Energy for funding this research.

REFERENCES

Adam, L., M. Batzle, and I. Brevik, (2006), Gassmann's fluid substitution and shear modulus variability in carbonates at laboratory seismic and ultrasonic frequencies, *Geophysics*, *71*, F173 F183.

Adam, L., K. van Wijk, T. Otheim, and M. Batzle, (2013), Changes in elastic wave velocity and rock microstructure due to basalt-CO_2-water reactions, *J. Geophys. Res. Solid Earth*, *118*, 2169–9356.

Addis, M. A., (1997), the stress-depletion response of reservoirs, paper presented at *SPE Annual Technical Conference*, San Antonio, TX, 5–8 October 1997, p. 11.

Agosta, F., M. Prasad, and A. Aydin, (2007), Physical properties of carbonate fault rocks, Fucino basin (Central Italy): Implications for fault seal in platform carbonates, *Geofluids*, *7*, 19–32.

Allis, R., T. Chidsey, W. Gwynn, C. Morgan, S. White, M. Adams, and J. Moore, (2001), Natural CO_2 Reservoirs on the Colorado Plateau and Southern Rocky Mountains: Candidates for CO_2 Sequestration, paper presented at *Proceedings of the First Annual Carbon Capture & Sequestration Conference*, Alexandria, VA, May 2001.

Ambrose, R. J., R. C. Hartman, M. Diaz-Campos, I. Yucel Akkutlu, and C. H. Sondergeld, (2012), Shale gas-in-place calculations part I: New pore-scale considerations, *SPE J.*, *17*, 219–229.

Anderson, O. L., and P. C. Grew, (1977), Stress corrosion theory of crack propagation with applications to geophysics, *Rev. Geophys.*, *15*, 77–104.

Andreani, M., L. Luquot, P. Gouze, M. Godard, E. Hoisé, and B. Gibert, (2009), Experimental study of carbon sequestration reactions controlled by the percolation of CO_2-rich brine through peridotites, *Environ. Sci. Technol.*, *43*, 1226–1231

Armitage, P. J., D. R. Faulkner, R. H. Worden, A. C. Aplin, A. R. Butcher, and J. Iliffe, (2011), Experimental measurement of, and controls on, permeability and permeability anisotropy of caprocks from the CO_2 storage project at the Krechba field, Algeria, *J. Geophys. Res. Solid Earth*, *116*, B12208.

Armitage, P. J., D. R. Faulkner, and R. H. Worden, (2013), Caprock corrosion, *Nat. Geosci.*, *6*, 79–80.

Aydin, A., (2000), Fractures, faults, and hydrocarbon entrapment, migration and flow, *Mar. Pet. Geol.*, *17*, 797–814.

Aydin, A., (2014), Failure modes of shales and their implications for natural and man-made fracture assemblages, *AAPG Bull.*, *98*, 2391–2409.

Batzle, M., R. Hofmann, M. Prasad, G. Kumar, L. Duranti, and D.-H. Han, (2005), Seismic attenuation; observations and mechanisms, paper presented at *Society of Exploration Geophysicists,* 75th Annual Meeting, Vol.*24*, Houston, TX, 6–11 November 2005, pp. 1565–1568.

Bernier, F., A. L. Li, W. Bastiaens, L. Ortiz, M. Van Geet, L. Wouters, B. Freig, P. Blümling, J. Desrues, G. Viaggiani, C. Coll, S. Chanchole, V. De Greef, R. Hamza, L. Malinsky, A. Vervoort, Y. Vanbrabant, B. Debecker, J. Verstraelen, A. Govaerts, M. Wevers, V. Labiouse, S. Escoffier, J.-F. Mathier, L. Gastaldo, and C. Bühler, (2007), Fractures and self-healing within the excavation disturbed zone in clays (SELFRAC). *Technical Report EUR 22585*, European Commission Nuclear Science and Technology.

Berryman, J. G., and H. F. Wang, (1995), The elastic coefficients of double-porosity models for fluid transport in jointed rock, *J. Geophys. Res. Solid Earth*, *100*, 24611–24627.

Biddle, K. T., and C. C. Wielchowsky, (1994), Hydrocarbon traps, in *The Petroleum System-from Source to Trap*, edited by L. B. Magoon and W. G. Dow, vol. *60* of *AAPG Memoir*, pp. 219–235, American Association of Petroleum Geologists.

Bock, H., B. Dehandschutter, C. D. Martin, M. Mazurek, A. de Haller, F. Skoczylas, and C. Davy, (2010), Self-sealing of fractures in argillaceous formations in the context of geological disposal of radioactive waste, *Technical Report NEA No. 6184*, Nuclear Energy Agency. Organisation for Economic Co-Operation and Development.

Bourg, I. C., (2015), Sealing shales versus brittle shales: A sharp threshold in the material properties and energy technology uses of fine-grained sedimentary rocks, *Environ. Sci. Technol. Lett.*, *2*, 255–259.

Bourg, I. C., L. E. Beckingham, and D. J. DePaolo, (2015), The nanoscale basis of CO_2 trapping for geologic storage, *Env. Sci. Tech.*, *49*, 10265–10284.

Browne, P. R. L., (1978), Hydrothermal alteration in active geothermal fields, *Annu. Rev. Earth Planet. Sci.*, *6*, 229–250.

Busch, A., S. Alles, Y. Gensterblum, D. Prinz, D. N. Dewhurst, M. D. Raven, H. Stanjek, and B. M. Krooss, (2008), Carbon dioxide storage potential of shales, *Int. J. Greenh. Gas. Con.*, *2*, 297–308.

Carey, J. W., H. Mori, D. Brown, and R. Pawar, (2014), Geomechanical behavior of caprock and cement: Plasticity in hydrodynamic seals, *Ener. Proc.*, *63*, 5671–5679.

Carey, J. W., Z. Lei, E. Rougier, H. Mori, and H. S. Viswanathan, (2015), Fracture-permeability behavior of shale, *J. Unconven. Oil Gas Resour.*, *11*, 27–43.

Cartwright, J., M. Huuse, and A. Aplin, (2007), Seal bypass systems, *AAPG Bull.*, *91*, 1141–1166.

Cho, Y., O. Apaydin, and E. Ozkan, (2013), Pressure-dependent natural-fracture permeability in shale and its effect on shale-gas well production, *SPE Reserv. Eval. Eng.*, *16*, 216–228.

Curtis, J. B., (2002), Fractured shale-gas systems, *AAPG Bull.*, *86*, 1921–1938.

Davy, C. A., F. Skoczylas, J.-D. Barnichon, and P. Lebon, (2007), Permeability of macro-cracked argillite under confinement: Gas and water testing, *Phys. Chem. Earth Part A/B/C*, *32*, 667–680.

Detwiler, R. L., (2008), Experimental observations of deformation caused by mineral dissolution in variable-aperture fractures, *J. Geophys. Res. Solid Earth*, *113*, B08202.

Dewhurst, D. N., Y. Yang, and A. C. Aplin, (1999), Permeability and fluid flow in natural mudstones, in *Muds and Mudstones: Physical and Fluid Flow Properties*, edited by A. C. Aplin, A. J. Fleet, and J. H. S. Macquaker, vol. *158*, pp. 23–43, The Geological Society, London.

Diallo, M. S., M. Prasad, and E. Appel, (2003), Comparison between experimental results and theoretical predictions for P-wave velocity and attenuation at ultrasonic frequency, *Wave Motion*, *37*, 1–16.

Do, D. D., and H. D. Do, (2007), Appropriate volumes for adsorption isotherm studies: The absolute void volume, accessible pore volume and enclosing particle volume, *J. Colloid Interface Sci.*, *316*, 317–330.

Do D. D., H. D. Do, C. Fan, and D. Nicholson, (2010), On the existence of negative excess isotherms for argon adsorption on graphite surfaces and in graphitic pores under supercritical conditions at pressures up to 10,000 atm, *Langmuir*, *26*, 4796–4806.

Downey, M. W., (1984), Evaluating seals for hydrocarbon accumulations, *AAPG Bull.*, *68*, 1752–1763.

Dvorkin, J., and A. Nur, (1993), Dynamic poroelasticity: A unified model with the squirt and the biot mechanisms, *Geophysics*, *58*, 524–533.

Edlmann, K., S. Haszeldine, and C. McDermott, (2013), Experimental investigation into the sealing capability of naturally fractured shale caprocks to supercritical carbon dioxide flow, *Environ. Earth Sci.*, *70*, 3393–3409.

Elkhoury, J. E., P. Ameli, and R. L. Detwiler, (2013), Dissolution and deformation in fractured carbonates caused by flow of CO_2-rich brine under reservoir conditions, *Int. J. Greenh. Gas. Con.*, *16*(Supplement 1), S203–S215.

Ellis, B. R., G. S. Bromhal, D. L. McIntyre, and C. A. Peters, (2011), Changes in caprock integrity due to vertical migration of CO_2-enriched brine, *Enrgy. Proced.*, *4*, 5327–5334.

Finkbeiner, T., M. Zoback, P. Flemings, and B. Stump, (2001), Stress, pore pressure and dynamically constrained hydrocarbon columns in the South Eugene Island 330 field, northern Gulf of Mexico, *AAPG Bull.*, *85*, 1007–1031.

Fisher, Q. J., and R. J. Knipe, (2001), The permeability of faults within siliclastic petroleum reservoirs of the North Sea and Norwegian Continental Shelf, *Mar. Pet. Geol.*, *18*, 1063–1081.

Fitts, J. P., and C. A. Peters, (2013), Caprock fracture dissolution and CO_2 leakage, in *Geochemistry of Geologic CO_2 Sequestration*, edited by D. J. DePaolo, D. Cole, A. Navrotsky, and I. Bourg, vol. *77*, pp. 459–479, Mineralogical Society of America, Washington, DC.

Frash, L. P., J. W. Carey, T. Ickes, and H. S. Viswanathan, (2016a), High-stress triaxial direct-shear fracturing of Utica shale and in situ x-ray microtomography with permeability measurement, *J. Geophys. Res.*, *121*, 5493–5508.

Frash, L. P., J. W. Carey, and H. S. Viswanathan, (2016b), Notched specimen hydraulic fracturing method for conducting mechanical and hydrological experiments at triaxial reservoir conditions, paper presented at *50th US Rock Mechanics/Geomechanics Symposium*, Houston, TX, 26–29 June 2016.

Grunau, H. R., (1987), A worldwide look at the cap-rock problem, *J. Pet. Geol.*, *10*, 245–265.

Guglielmi, Y., F. Cappa, J.-P. Avouac, P. Henry, and D. Elsworth, (2015), Seismicity triggered by fluid injection–induced aseismic slip, *Science*, *348*, 1224–1226.

Gumma, S., and O. Talu, (2010), Net adsorption: A thermodynamic framework for supercritical gas adsorption and storage in porous solids. *Langmuir*, *26*, 17013–17023.

Gutierrez, M., L. E. Øino, and R. Nygård, (2000), Stress-dependent permeability of a de-mineralized fracture in shale, *Mar. Pet. Geol.*, *17*, 895–907.

Hawkes, C., P. McClellan, and S. Bachu, (2005), Geomechanical factors affecting geological storage of CO_2 in depleted oil and gas reservoirs, *J. Can. Pet. Technol.*, *44*.

Heath, J. E., T. A. Dewers, B. J. McPherson, M. B. Nemer, and P. G. Kotula, (2012), Pore-lining phases and capillary breakthrough pressure of mudstone caprocks: Sealing efficiency of geologic CO_2 storage sites, *Int. J. Greenh. Gas. Con.*, *11*, 204–220.

Heller, R., and M. Zoback, (2014), Adsorption of methane and carbon dioxide on gas shale and pure mineral samples, *J. Unconvent. Oil Gas Resour.*, *8*, 14–24.

Hermanrud, C., and H. M. N. Bols, (2002), Leakage from overpressured hydrocarbon reservoirs at Haltenbanken and in the northern North Sea, in *Hydrocarbon Seal Quantification Norwegian Petroleum Society Conference*, edited by A. G. Koestler and R. Hunsdale, vol. *11*, pp. 221–231, Norwegian Petroleum Society Special Publications, Elsevier.

Hickman, S., R. Sibson, and R. Bruhn, (1995), Introduction to special section: Mechanical involvement of fluids in faulting, *J. Geophys. Res. Solid Earth*, *100*, 12,12831–12840.

Hildenbrand, A., S. Schlömer, and B. M. Krooss, (2002), Gas breakthrough experiments on fine-grained sedimentary rocks, *Geofluids*, *2*, 3–23.

Hornafius, J. S., D. Quigley, and B. P. Luyendyk, (1999), The world's most spectacular marine hydrocarbon seeps (coal oil point, Santa Barbara Channel, California): Quantification of emissions, *J. Geophys. Res. Oceans*, *104*, 20,20703–20711.

Humayun, R., and D. L. Tomasko, (2000), High-resolution adsorption isotherms of supercritical carbon dioxide on activated carbon, *AICHE J.*, *46*, 2065–2075.

Hunt, J. M., (1990), Generation and migration of petroleum from abnormally pressured fluid compartments, *AAPG Bull.*, *74*, 1–12.

IEAGHG, (2015), Criteria of fault geomechanical stability during a pressure build-up, *Technical Report 2015/04*, IEA Greenhouse Gas.

Ingram, G. M., and J. L. Urai, (1999), Top-seal leakage through faults and fractures: The role of mudrock properties, in *Muds and Mudstones: Physical and Fluid Flow Properties*, edited by A. C. Aplin, A. J. Fleet, and J. H. S. Macquaker, vol. *158*, pp. 125–135, Geological Society, London, Special Publications.

Jones, M. E., and M. A. Addis, (1986), The application of stress path and critical state analysis to sediment deformation, *J. Struct. Geol.*, *8*, 575–580.

Juanes, R., B. H. Hager, and H. J. Herzog, (2012), No geologic evidence that seismicity causes fault leakage that would render large-scale carbon capture and storage unsuccessful, *Proc. Natl. Acad. Sci.*, *109*, E3623.

Keating, E., D. Newell, D. Dempsey, and R. Pawar, (2014), Insights into interconnections between the shallow and deep systems from a natural CO_2 reservoir near Springerville, Arizona, *Int. J. Greenh. Gas. Con.*, *25*, 162–172.

Khazanehdari, J., and J. Sothcott, (2003), Variation in dynamic elastic shear modulus of sandstone upon fluid saturation and substitution, *Geophysics*, *68*, 472–481.

Kuila, U., and M. Prasad, (2013), Specific surface area and pore-size distribution in clays and shales, *Geophys. Prospect.*, *61*, 341–362.

Kumar, S., M. Prasad, and R. Pini, (2015), Selective adsorptives to study pore structure and wetting behavior of self-resourcing shales, Society of Petrophysicists and Well-Log Analyst 56th Annual Logging Symposium, 18–22 July, Long Beach, CA.

Loucks, R., R. M. Reed, S. C. Ruppel, and D. M. Jarvie, (2009), Morphology, genesis, and distribution of nanometer-scale pores in siliceous mudstones of the Mississippian Barnett shale, *J. Sediment. Res.*, *79*, 848–861.

Macgregor, D. S., (1996), Factors controlling the destruction or preservation of giant light oilfields, *Pet. Geosci.*, *2*, 197–217.

Malbrunot, P., D. Vidal, J. Vermesse, R. Chahine, and T. K. Bose, (1997), Adsorbent helium density measurement and its effect on adsorption isotherms at high pressure, *Langmuir*, *13*, 539–544.

Masoudi, R., M. A. A. Jalil, D. J. Press, K.-H. Lee, C. Phuat Tan, L. Anis, N. B. Darman, and M. Othman, (2012), An integrated reservoir simulation-geomechanical study on feasibility of CO_2 storage in m4 carbonate reservoir, Malaysia, paper presented at *International Petroleum Technology Conference*, Bangkok, Thailand, 7–9 February 2012.

Menaceur, H., P. Delage, A. Tang, and N. Conil, (2015), On the thermo-hydro-mechanical behaviour of a sheared Callovo-Oxfordian claystone sample with respect to the EDZ behaviour, *Rock Mech. Rock. Eng.*, *49*, 1875–1888.

Metz, B., O. Davidson, H. de Coninck, M. Loos, and L. Meyer, (2005), *IPCC Special Report on Carbon Dioxide Capture and Storage*, Intergovernmental Panel on Climate Change. https://www.ipcc.ch/pdf/special-reports/srccs/srccs_wholereport.pdf.

Monfared, M., J. Sulem, P. Delage, and M. Mohajerani, (2012), On the THM behaviour of a sheared boom clay sample: Application to the behaviour and sealing properties of the EDZ, *Eng. Geol.*, *124*, 47–58.

Moore, J., M. Adams, R. Allis, S. Lutz, and S. Rauzi, (2005), Mineralogical and geochemical consequences of the long-term presence of CO_2 in natural reservoirs: An example from the Springerville-St. Johns field, Arizona, and New Mexico, USA, *Chem. Geol.*, *217*, 365–385.

Morris, A., D. A. Ferrill, and D. B. Henderson, (1996), Slip-tendency analysis and fault activation, *Geology*, *24*, 275–278.

Naylor, M., M. Wilkinson, and R. Haszeldine, (2011), Calculation of CO_2 column heights in depleted gas field from known pre-production gas column heights, *Mar. Pet. Geol.*, *28*, 1083–1093.

Njiekak, G., D. R. Schmitt, H. Yam, and R. Kofman, (2013), CO_2 rock physics as part of the Weyburn-Midale geological storage project, *Int. J. Greenh. Gas. Con.*, *16*, S118–S133.

Noiriel, C., L. Luquot, B. Made, L. Raimbault, P. Gouze, and J. Van Der Lee, (2009), Changes in reactive surface area during limestone dissolution: An experimental and modelling study, *Chem. Geol.*, *265*, 160–170.

Nussbaum, C., and P. Bossart, (2004), Compilation of K-values from packer tests in the Mont Terri rock laboratory, Mont Terri Technical Note, TN 2005-10, p. 29. Federal Office of Topogra-phy (swisstopo), Wabern, Switzerland.

Nygård, R., M. Gutierrez, R. K. Bratli, and K. Høeg, (2006), Brittle–ductile transition, shear failure and leakage in shales and mudrocks, *Mar. Pet. Geol.*, *23*, 201–212.

Pawar, R. J., G. Bromhal, J. W. Carey, B. Foxall, A. Korre, P. Ringrose, O. Tucker, M. Watson, and J. White, (2015), Recent advances in risk assessment and risk management of geologic CO_2 storage, *Int. J. Greenh. Gas. Con.*, *40*, 292–311.

Pini, R., (2014a), Interpretation of net and excess adsorption isotherms in microporous materials, *Microporous Mesoporous Mater.*, *187*, 40–52.

Pini, R., (2014b), Assessing the adsorption properties of mudrocks for CO_2 sequestration, *Energy Procedia*, *63*, 5556–5561.

Pini, R., S. Ottiger, A. Rajendran, G. Storti, and M. Mazzotti, (2006), Reliable measurement of near-critical adsorption by gravimetric method, *Adsorption*, *12*, 393–403.

Pini, R., S. Ottiger, L. Burlini, G. Storti, and M. Mazzotti, (2010), Sorption of carbon dioxide, methane and nitrogen in dry coals at high pressure and moderate temperature, *Int. J. Greenh. Gas. Con.*, *4*, 90–101.

Rinehart, A., (2014), *Impact of Microstructurally Heterogeneous Strength and Compliance on Macroscopic Rock Failure and Elastic Degradation*, Ph.D. thesis, University of New Mexico.

Ruthven, D. M., (1984), *Principles of Adsorption and Adsorption Processes*, Wiley, New York.

Samuelson, J., and C. J. Spiers, (2012), Fault friction and slip stability not affected by CO_2 storage: Evidence from short-term laboratory experiments on North Sea reservoir sandstones and caprocks, *Int. J. Greenhouse Gas Control*, *11*, 78–90.

Schettler, P. D., and C. R. Parmely, (1991), Contributions to total storage capacity in Devonian shales, paper presented at *SPE Eastern Regional Meeting*, Lexington, KY, 22–25 October.

Schlömer, S., and B. M. Krooss, (1997), Experimental characterisation of the hydrocarbon sealing efficiency of cap rocks, *Mar. Pet. Geol.*, *14*, 565–580.

Schofield, A., and P. Wroth, (1968), *Critical State Soil Mechanics*, McGraw-Hill Book Company, London.

Sharma, R., (2015), *Impact of Heterogeneity and Saturation on Elastic and Viscoelastic Properties of Carbonates*, Ph.D. thesis, Colorado School of Mines.

Sharma, R., M. Prasad, M. Batzle, and S. Vega, (2013), Sensitivity of flow and elastic properties to fabric heterogeneity in carbonates, *Geophys. Prospect.*, *61*, 270–286.

Shukla, R., P. Ranjith, A. Haque, and X. Choi, (2010), A review of studies on CO_2 sequestration and caprock integrity, *Fuel*, *89*, 2651–2664.

Sibson, R. H., (2003), Brittle-failure controls on maximum sustainable overpressure in different tectonic regimes, *AAPG Bull.*, *87*, 901–908.

Sircar, S., (2001), Measurement of Gibbsian surface excess, *AIChE J*, *47*, 1169–1176.

Skurtveit, E., E. Aker, M. Soldal, M. Angeli, and Z. Wang, (2012), Experimental investigation of CO_2 breakthrough and flow mechanisms in shale, *Pet. Geosci.*, *18*, 3–15.

Smith, T. M., C. H. Sondergeld, C. S. Rai, (2003), Gassmann fluid substitutions: A tutorial, *Geophysics*, *68*, 430–440.

Sone, H., and M. D. Zoback, (2013), Mechanical properties of shale-gas reservoir rocks–part 2: Ductile creep, brittle strength, and their relation to the elastic modulus, *Geophysics*, *78*, D393–D402.

Song, J., and D. Zhang, (2013), Comprehensive review of caprock-sealing mechanisms for geologic carbon sequestration, *Environ. Sci. Technol.*, *47*, 9–22.

Townsend, J., and M. D. Zoback, (2000), How faulting keeps the crust strong, *Geology*, *28*, 399–402.

Vanorio, T., G. Mavko, S. Vialle, and K. Spratt, (2010), The rock physics basis for 4D seismic monitoring of CO_2 fate: Are we there yet?, *Lead. Edge*, *29*, 156–162.

Vialle, S., J. L. Druhan, and K. Maher, (2016), Multi-phase flow simulation of CO_2 leakage through a fractured caprock in response to mitigation strategies, *Int. J. Greenh. Gas. Con.*, *44*, 11–25.

Vilarrasa, V., and J. Carrera, (2015), Geologic carbon storage is unlikely to trigger large earthquakes and reactivate faults through which CO_2 could leak, *Proc. Natl. Acad. Sci.*, *112*, 5938–5943.

Watts, N., (1987), Theoretical aspects of cap-rock and fault seals for single- and two-phase hydrocarbon columns, *Mar. Pet. Geol.*, *4*, 274–307.

Weniger, P., W. Kalkreuth, A. Busch, and B. M. Krooss, (2010), High-pressure methane and carbon dioxide sorption on coal and shale samples from the Paraná Basin, Brazil, *Int. J. Coal Geol.*, *84*, 190–205.

Wollenweber, J., S. Alles, A. Busch, B. M. Krooss, H. Stanjek, and R. Littke, (2010), Experimental investigation of the CO_2 sealing efficiency of caprocks, *Int. J. Greenh. Gas. Con.*, *4*, 231–241.

Yam, H., and D. R. Schmitt, (2011), CO_2 rock physics: A laboratory study, *InSite Mag.*, *30*, 13–16.

Zhang, C. L., (2013), Sealing of fractures in claystone, *J. Rock Mech. Geotech. Eng.*, *5*, 214–220.

Zhang, C.-L., and T. Rothfuchs, (2008), Damage and sealing of clay rocks detected by measurements of gas permeability, *Phys. Chem. Earth A/B/C*, *33*(Supplement 1), S363–S373.

Zhang, J., A. Kamenov, D. Zhu, and A. Hill, (2013), Laboratory measurement of hydraulic fracture conductivities in the Barnett shale, paper presented at *2013 SPE Hydraulic Fracturing Technology Conference,* The Woodlands, TX, 4–6 February, 2013.

Zhou, Y., and L. Zhou, (2009), Fundamentals of high pressure adsorption, *Langmuir*, *25*, 13461–13466.

Zoback, M. D., (2007), *Reservoir Geomechanics*, Cambridge University Press, Cambridge.

Zoback, M. D., and S. M. Gorelick, (2012), Earthquake triggering and large-scale geologic storage of carbon dioxide, *Proc. Natl. Acad. Sci.*, *109*, 10164–10168.

Part III
Monitoring Caprock Failure

11

In-Zone and Above-Zone Pressure Monitoring Methods for CO_2 Geologic Storage

Seyyed A. Hosseini, Mahmood Shakiba, Alexander Sun, and Susan Hovorka

ABSTRACT

This paper reviews the most recent developments in the area of pressure monitoring relevant to CO_2 geological storage in saline and depleted oil and gas reservoirs. We discuss some of the pressure monitoring advantages such as large area of coverage, well-developed pressure gauge technology, cost effectiveness, high sensitivity to small pressure changes, and so on and mention some of the possible challenges. We categorize the pressure monitoring into in-zone and above-zone monitoring and based on this categories review some technical aspects of recent technical and field developments.

11.1. INTRODUCTION

The work of *Darcy* [1856] established that pressure is a basic parameter needed to assess fluid flow in a porous medium. The application of pressure data, especially in hydrocarbon reservoir management, has a history dating back to the 1920s [*Girelli*, 2005]. Pressure data routinely have been collected in oil and gas reservoirs to characterize and track reservoir performance, including monitoring reservoir pressure during the preparation stage for miscible CO_2 injection (see the example of Weyburn as part of preparation for enhanced oil recovery [EOR] [*Elsayed et al.*, 1993]). However, in the first overview of the developing field of geologic storage, the *Intergovernmental Panel on Climate Change* [2005] limited its review of the use of pressure measurements of injection wells to obtain the maximum injection rate without exceeding the maximum geomechanically stable bottom-hole pressure. The limited appraisal of the application of pressure monitoring technology initially set low expectations for use of pressure in carbon capture and storage (CCS) monitoring that should now be revised upward based on both historic and recent practices.

Bureau of Economic Geology, University of Texas at Austin, Austin, TX, USA

In freshwater aquifers, pressure in the aquifer can be observed as a function of the height of the water column in a well completed in the zone of interest. However, in the deep subsurface, the presence of fluids of different densities, such as brines of different compositions and various hydrocarbons, adds to the complexity of measurements because (i) the top of the fluid in the well may be far below the land surface and (ii) the density of the fluid column in the well is needed to calculate pressure at reservoir depth [*Verma et al.*, 2013]. CO_2 has strong density variation based on temperature and, because of the geothermal gradient, will have highly variable density with depth in the well tubulars. Changes in density can be great over time as well fluids equilibrate with the surrounding rock and aquifer system. Because of these complexities, it is typically preferred in wells with complex fluids to measure pressure at or near the perforated interval, known as the *bottom-hole pressure*. Wellhead pressure also can be used for monitoring [*Ringrose et al.*, 2013] after first converting it to bottom-hole pressure by using the hydrostatic head of fluids in the well. This conversion can be an easy task if the wellbore is filled with brine, but tricky if the wellbore is filled with gas

Geological Carbon Storage: Subsurface Seals and Caprock Integrity, Geophysical Monograph 238, First Edition.
Edited by Stéphanie Vialle, Jonathan Ajo-Franklin, and J. William Carey.

mixtures. Even if a well is filled with brine, wellhead pressure data could start to variate if CO_2 breakthrough occurs, depending on how fast fluids replace the brine in tubing. The replacement process can take days to months depending on how quickly gases are charging into the brine-filled well [*Hosseini et al.*, 2013].

Pressure can be used for monitoring a CO_2 injection project in many ways. An important example of the use of pressure in monitoring is provided by the US EPA in its Class VI rule for CO_2 injection. This rule emphasizes the need for assessing predicted plume extent and elevated pressure in areas distant from the injection well (area of review) because of the risk of lifting saline water to damage freshwater resources through uncontrolled well completions. The rule, which is consistent with other rules controlling injection, implies that dedicated pressure monitoring points away from injection wells may be needed to calibrate models [*US EPA*, 2010].

Use of pressure surveillance for leakage detection has a specific application for CO_2 storage that is not widely used in other applications because of the importance of documenting retention in the intended zone by monitoring possible leakage pathways. If the pressure increase in the injection zone (IZ) causes leakage of fluids into overlying permeable zones through a natural open conduit or a flawed well completion, the hydraulic connection will become evident by pressure in the overlying zone. This type of monitoring has been developed and demonstrated for gas storage reservoirs [*Katz and Tek*, 1981]. Our work refers to a permeable zone overlying the IZ as an *above-zone monitoring interval* (AZMI).

Figure 11.1 depicts a schematic of IZ and AZMI monitoring in a CO_2 storage project. The response of an AZMI can be modeled by inputting the spatial and hydraulic properties of the IZ, the connection path, and the AZMI [*Sun and Nicot*, 2012; *Zeidouni and Pooladi-Darvish*, 2013; *Strandli et al.*, 2014; *Zeidouni*, 2014]. This formulation allows monitoring of pressure in the AZMI to demonstrate the degree of connection or isolation from the IZ. Modeling the response of the AZMI to leakage is important and valuable in documenting the anticipated negative finding: that fluids are not leaking out of the intended IZ above a specified rate and retention can therefore be considered effective. Note that direct measurement of pressure as a means of finding leakage would be meaningful only if the pressure signal is strong enough for interpretation. A combination of formation properties and leakage rate controls the magnitude of the increase in pressure in the monitoring zone [*Keating et al.*, 2014]. Prior modeling and characterization would be required to ensure that the most suitable formation is selected for pressure monitoring.

In the EPA's Class VI rules [*US EPA*, 2010], AZMI monitoring is specified as occurring in the first permeable zone above the seal. However, it is important to select and characterize the hydrologic properties of a large interval overlying the IZ to determine the appropriate zone for designing an effective monitoring deployment. A confining system may have more than one permeable zone within it; AZMI selection will have a strong impact on the conclusions made at the end of the project [*Zahid et al.*, 2012]. An AZMI very close to the IZ will respond quickly and to the smallest out-of-zone flow while also responding to geomechanical and geothermal changes [*Kim and Hosseini*, 2014, 2015]. In contrast, an AZMI at the top of the confining system but below freshwater may provide evidence of the

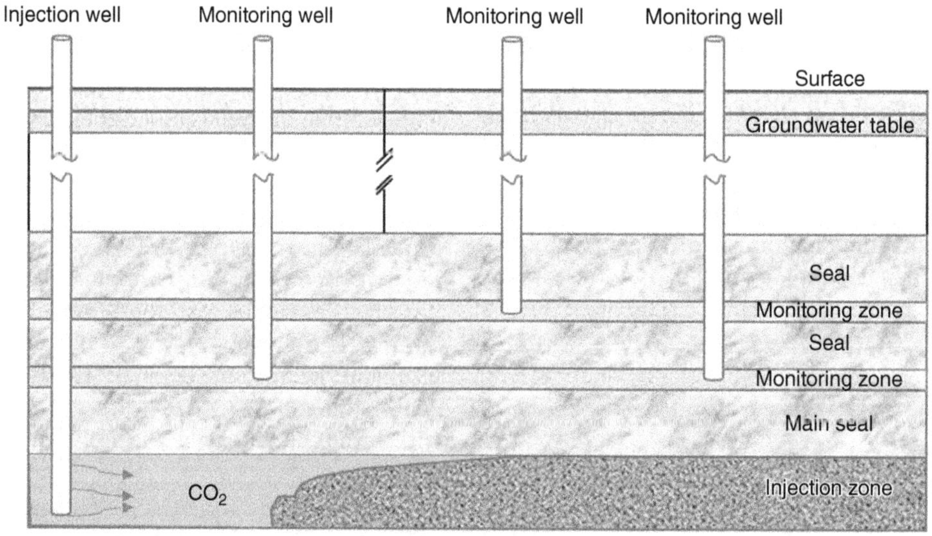

Figure 11.1 IZ and AZMIs. Pressure monitoring wells are drilled in permeable formations above IZ and separated by one or more layers of seal.

Table 11.1 Role of Pressure Monitoring in Selected R&D Projects Evaluating Technologies for Monitoring Geologic Storage.

Location	Project name	Injection well monitoring type E (IZ)	Observation well(s) monitoring type (IZ)	Observation well(s) monitoring type (AZMI)
South Liberty field, Texas, USA [*Hovorka et al.*, 2006]	Frio test I and II	BHP	BHP	Chemistry
Saskatchewan, Canada [*Whittaker*, 2010]	Weyburn field	Not reported	RFT Well testing Seismic response	No
Iwanohara, Japan [*Sato et al.*, 2011]	Nagaoka	BHP	BHP	No
Cranfield, MS, USA [*Hosseini et al.*, 2013]	SECARB Early Test	BHP	BHP	BHP
Decatur, IL, USA [*Strandli et al.*, 2014]	IBDP	Not reported	Westbay System Multiport	Westbay System Multiport
Brandenburg, Germany [*Liebscher et al.*, 2013]	Ketzin pilot site	BHP	BHP	BHP
Michigan, USA [*Kelley et al.*, 2014]	Pinnacle reef	BHP falloff tests	BHP	No
Norway [*Grude et al.*, 2014]	Tubåen Fm. CO_2 storage	BHP falloff tests	No	No
Algeria [*Ringrose et al.*, 2013]	In Salah	WHP	WHP	No
Alberta, Canada [*Bourne et al.*, 2014]	Quest Carbon Capture and Storage Project	BHP falloff tests	No	WHP
Australia [*Flett et al.*, 2008]	Gorgon Carbon Dioxide Injection Project	BHP	No	BHP
Montana, USA [*Steadman*, 2013]	PCOR test at Bell Creek field	WHP casing pressure	BHP	BHP

BHP, bottom-hole pressure; BHT, bottom-hole temperature; RFT, repeat-formation tester tool; WHP, wellhead pressure.

overall effectiveness of confinement, but any leakage response may be delayed for decades after the start of flow. Thickness, continuity, and boundary conditions of the AZMI must also be matched to monitoring goals. A thin zone will have a more rapid response to a leakage signal than will a thicker zone but also may have a higher chance of being hydrologically discontinuous. An AZMI with hydrologically closed boundaries will also respond to a signal with higher sensitivity; however, discontinuities that create closed boundaries may be associated with poor lateral continuity and reduce the effective areal coverage of the monitoring. Other necessary characterizations of AZMI include the nature of the fluids (brine or gas) and any use of the zone for production of disposal.

Most geologic storage projects reported in the literature, as well as upcoming projects that have made their monitoring plans public, have used IZ bottom-hole pressure either in the injection well(s) or in observation wells distant from injection. One exception was the first CO_2 injection project with greenhouse gas (GHG) reduction wells at Statoil's Sleipner field, which began injection in 1996, when the equipment was considered too unreliable for offshore deployment and the absence of both pressure and temperature measurements was a source of uncertainty [*Alnes et al.*, 2011] (temperature still remains a source of uncertainty in this field). Since that time, bottom-hole pressure has become an essential tool even in offshore projects, playing a significant role in reservoir management to assure safe storage at Statoil's next project at Snøhvit field [*Grude et al.*, 2014]. The long-term pressure increase at Snøhvit field was faster than expected, ultimately leading to the decision to move the injection [*Hansen et al.*, 2013] to the overlying Stø Formation. Research projects that have included a substantive component of pressure monitoring are partially inventoried in Table 11.1.

11.2. PRESSURE MONITORING ADVANTAGES AND CHALLENGES

Before reviewing some of the recent developments and applications of pressure data in the context of monitoring CO_2 storage in saline and CO_2-EOR operations, it is useful to discuss some of the advantages and disadvantages of pressure-based monitoring in the CCS context. Table 11.2 summarizes some of the pros and cons of pressure monitoring based on observations in several field projects.

While current pressure gauges measure down to 0.01 psi pressure change, every small pressure perturbation is not necessarily due to a leak. Several natural and artificial phenomena could contribute to observed pressure

Table 11.2 Advantages and Challenges of Pressure Monitoring.

Advantages	Challenges
Large area of coverage compared to other monitoring methods	Presence of pressure noises
Deployed at wide range of pressure and temperature	Higher data resolution and frequency needed (compared to typical operational needs)
Well-developed pressure-gauge technology	Drifting problem (both in time and pressure recordings)
Cheap compared to other methods	Field-operations interference (access to data)
Very sensitive to small pressure changes (high-resolution pressure gauges)	Power shortage (due to natural events, lightning strikes, floods, etc.)
Distinguishes CO_2 leak from brine leak (change in formation total compressibility)	Storage-memory shortage
Detection of small leaks (over long period of time, small leaks can accumulate to create detectable changes in formation total compressibility)	Unclear definition of abnormal pressure response

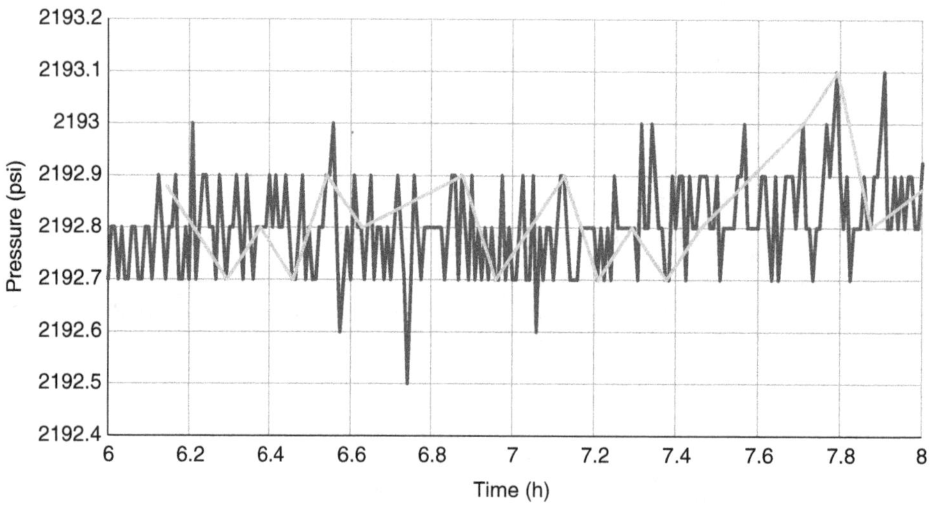

Figure 11.2 Measured pressure data for a well in a monitoring interval. Amplitude of noise is about 0.6 psi.

fluctuations (noise) in the subsurface, including changes in atmospheric pressure, Earth tides, ocean tides, and surface precipitation [*Chabora and Benson*, 2009], and can be filtered out. Pressure change associated with atmospheric pressure and surface precipitation could be of the order of 1 psi, while pressure change due to Earth and ocean tides is in the range of 0.03–0.3 psi [*Chabora*, 2009]. These natural fluctuations can be easily observed in pressure data if long-term sensitive measurements are available. *Chabora and Benson* [2009] also suggested a cutoff range to categorize detection potential as unlikely, marginal, or likely based on specifications of downhole gauges.

In addition to natural fluctuations, pressure variations associated with measurement instruments and operating equipment include drifting of recorded data. The amount and frequency of such pressure noises vary depending on the noise source and could be hard to quantify in some situations. Before pressure data can be used for leak monitoring and interpretation purposes, it is critical to filter the artificial and natural pressure fluctuations from the data. Figure 11.2 represents the original measured pressure for a well in a monitoring interval of Miocene sands in the Texas Gulf Coast. Pressure noise is clearly seen in both high-frequency (~30 s interval) and low-frequency (~5 min interval) measurements. The amplitude of the pressure noise (peak to valley) is 0.5 psi, and the average pressure is 2192.8 psi. The gauge resolution is approximately 0.1 psi. Although the pressure data of this AZMI may have tidal effects, they cannot be identified with a 0.1 psi resolution gauge; however, a 0.01 psi pressure gauge might be able to reveal the tidal behavior at this formation. Pressure data are very useful in detecting abnormal behavior in the system, but for each individual project, their limits in detecting leakage must be studied and characterized considering local and regional parameters. A same pressure monitoring network (such as same type of gauge or spacing) may have a different detection limit in different formations.

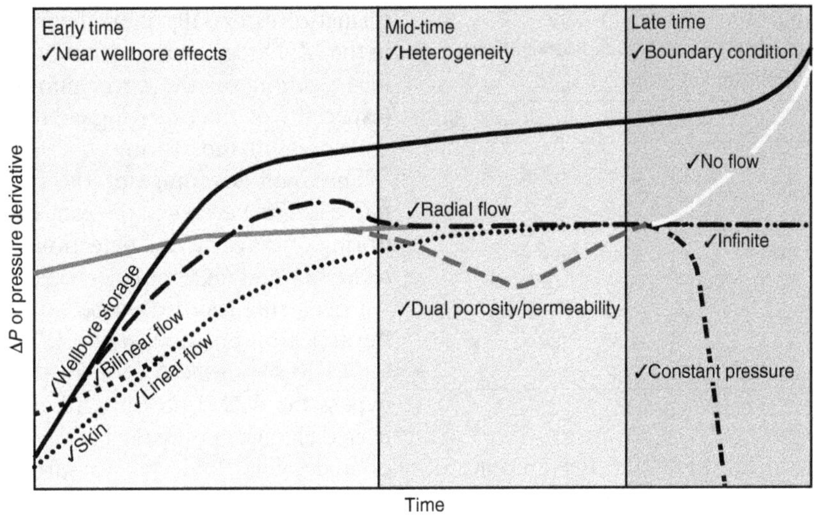

Figure 11.3 Curves of pressure derivatives (DER) used in formation characterization (source: www.fekete.com). BDF is closed boundary formations.

Another challenge in pressure monitoring is that even when pressure is measured with high accuracy, there is no clear definition (by current rules) of what kind of signal has to be defined as a threshold that would require an action. For example, if an increase in pressure of the monitoring zone is observed, at what point (a 5, 10, or 50 psi increase in pressure) should operations stop? When does the next monitoring plan have to be activated to clarify the observed abnormal pressure? This problem is common in many other monitoring technologies as well, and typically, the project only stops in the case of a high-impact incident. New rules are needed to make some of the investments in monitoring technologies meaningful beyond the research goals. Current practices heavily depend on the experience of the field operator. Note that some of the other disadvantages of pressure monitoring, for example, noise in collected data, are applicable to most other monitoring methods as well.

One major advantage of pressure monitoring is the large area of coverage because of the diffusive nature of the pressure waves (in a large storage formation, you may still need a few pressure gauges to cover the whole area of the review). Pressure waves travel deep into the formation and are affected by formation parameters averaged over a large area. Derived pressure data are rich in information and can be used for further reservoir characterization, including flow regime (including radial, multilayer, dual-porosity, and composite models). Late-time pressure data can be used to characterize the boundary condition of the system (such as no flow, infinite, and constant pressure), and early-time pressure data can be used to find wellbore issues (such as skin, storage, fractures, and turbulence) (Fig. 11.3). In addition, pressure measurement technology with a wide range of pressure is well developed, and many commercial companies carry pressure gauges with high resolution and accuracy at reasonable cost. Nowadays, most pressure gauges can be remotely streamed in real time to monitoring facilities miles away from the operation site, and a large amount of continuous pressure data can be measured and recorded for formation surveillance.

Successful interpretation of measured pressure data in the IZ and AZMI primarily depends on the accuracy and consistency of pressure gauges used at the bottom of monitoring wells. To detect a reasonably small amount of CO_2 leakage through pressure monitoring, sensitive pressure gauges are required. Early gauges were not able to sustain prolonged exposure to reservoir temperature and pressure and were deployed on slickline with various systems of recording data in memory cards for later readout. However, as pressure-gauge manufacturing technology has improved over time, high-resolution gauges compatible with high-pressure and high-temperature environments are available at reasonable cost. The reduced cost and improved reliability of digital gauges allow them to be deployed on wireline and installed permanently with surface readout (SRO) [*Nestlerode*, 1963; *Eck et al.*, 2000]. Installed gauges have a higher equipment cost but reduce operational costs since repeat deployments are eliminated; they also allow high-frequency monitoring and low-cost repeated testing. Although more expensive, gauges with higher sensitivity maximize the probability of detection while minimizing the size of the detected leak. Considering the diverse conditions of monitoring formations and the expected pressure fluctuations, the recommended pressure resolution is about 0.01 psi [*Chabora and Benson*, 2009].

It is worth mentioning that compared to other monitoring technologies, some discussed here, pressure

Table 11.3 General Comparison Between In-Zone and Above-Zone Pressure Monitoring. Check Marks Show the Method that Better Meets the Criteria.

Criteria	In zone	Above zone
Clean baseline		✓
Chance of detecting a leak	✓	
Ease of interpretation		✓
Brine vs. CO_2 leak detection		✓
Lower noise level		✓
Lower cost	✓	
Additional monitoring data		✓
Public assurance		✓

monitoring offers the most value for the amount of money spent. Geophysical methods such as crosswell seismic are typically limited in area of coverage and can be operated in intermittent intervals over the life of the project. The area of coverage for 4D seismic is large, but costs associated with it are many times greater than those of pressure monitoring. Moreover, seismic technology may not work in all formations, depending on the quality of the rocks and fluids.

Pressure monitoring methods can be categorized in different ways, including in zone versus above zone and active versus passive. The ultimate goal for all is to identify the leakage and its features, including the location and amount. In the next section, we will discuss in-zone versus above-zone methods. Table 11.3 shows the effectiveness of in-zone and above-zone pressure monitoring in a few categories such as providing assurance for the public, ease of their deployment in the field, and their associated costs. This table is based on a general understanding of pressure-based monitoring; tabulated results may not be valid for certain specific cases.

Many of the approaches mentioned in this chapter are only studied theoretically for potential use in CCS projects. Currently, the common practice of industry is to record pressure data and look for abnormal behavior (passive methods). Many of the proposed methods would require some action (similar to well-test analysis) or special type of data analysis during the project operation.

11.3. IN-ZONE PRESSURE MONITORING

This section reviews a number of recent developments and applications of pressure data in the context of monitoring CO_2 storage in saline and EOR operations. Pressure monitoring network optimization and leakage detection under uncertainty are discussed at the end of the section. Most of these approaches are well rooted in our fundamental understanding of pressure diffusion in porous media and have been optimized for use in CO_2 storage monitoring. Injection of CO_2 into the target formation essentially alters the trend of pressure observed in the IZ. Direct measurement of the pressure change and investigation of the corresponding reservoir response (especially of the overlying caprock) are the basis for in-zone pressure monitoring.

The main advantage of the in-zone monitoring technique is direct access to the zone of interest. In-zone monitoring offers early leak detection, although interpretation techniques might appear to be challenging in some cases. An occurring leak would be expected to initially impact the trend of pressure in the IZ (a response might take longer to see in the AZMI, or the leak could potentially bypass the AZMI pressure monitoring zone). Any unexpected change in pressure behavior might be an indication of a possible fluid leak to surrounding formations. To derive the baseline for IZ pressure, mass-balance calculations are usually performed. Normal pressure behavior can be predicted by keeping track of total fluid injections and extractions (inventory verification) and considering the size and storage capacity of the IZ. If a significant discrepancy is observed between the measured reservoir pressure and the predicted model, then a fluid leak could be occurring in the formation (assuming modeling work is done correctly, collected data are correct, etc.).

Another advantage of in-zone monitoring is the utilization of preexisting drilled and completed injection and production wells. Specifically, when the target formation is an oil and gas reservoir, numerous injection and production wells are usually available. These wells are mostly drilled and completed with respect to IZ depth and structure. Thus, they can be easily converted to monitoring wells without the need for further well-completion operations, which significantly lowers the cost of the monitoring project.

Some challenges of in-zone monitoring techniques include the specification of the pressure baseline. When CO_2 is injected into a formation, pressure variation is a function of several parameters such as injection rate, formation rock and fluid properties, reservoir boundaries, and well-completion type. Often, the exact measurement of the value and the distribution of these parameters is not available, and their impact on pressure is not clear. Thus, the calculation of the baseline pressure often bears some degree of uncertainty. In addition, the effect of other ongoing reservoir operations (such as far-field injection and productions in EOR projects) on measured pressure is not well determined. A departure of measured formation pressure from an established long-time baseline might be associated with an operating-condition change in another section of the reservoir. Furthermore, because of the injection process and the active nature of the IZ, the level of pressure noise is expected to be higher compared to that of above-zone pressure monitoring techniques.

Another challenge of in-zone monitoring techniques is mapping of leak migration if gas escapes the IZ. Since in-zone monitoring only provides information for the target injection formation, it fails to investigate the extent of leak migration to overlying formations. Often, multiple layers of permeable rock on top of the target zone might be separated by one or more layers of impermeable rock. If a leak occurs and gas migrates upward past the caprock (because of the lower density of gas compared to brine), vertical migration should be taken into account when assessing the risk of contamination of shallower drinking water resources. Having the capability to monitor and track the leak from the IZ is necessary to ensure the safety of drinking water resources and to design the best remediation strategy. In addition to in-zone techniques, extra monitoring steps in above-zone layers are required to increase public assurance of safety of CO_2 injection projects.

Another focus of in-zone pressure monitoring projects is confirmation of a safe pressure range during the operation. Measured pore pressure during the injection operation is used primarily to ensure that the caprock is geomechanically stable. A limit on pressure avoids hydraulically fracturing the confining layer and slipping geologic faults. Characterization and verification of such pressure bounds has been the focus of several research projects [*Rutqvist and Tsang*, 2002; *Hawkes et al.*, 2005; *Rutqvist et al.*, 2007; *Angeli et al.*, 2009]. As the mechanical strength of the seal is identified, the CO_2 injection rate is adjusted such that the formation pressure remains well below the derived pressure limit.

In addition to monitoring IZ absolute pressure to honor the caprock limit, other techniques have been developed to further analyze the measured pressure data in order to extract additional information. *Hu et al.* [2015] used time-lapse pressure tomography to monitor CO_2 plume evolution in the IZ (a deep saline aquifer). The pressure-tomography experiment is similar to the crossed ray paths of a seismic-tomography experiment. In pressure tomography, the travel time of a pressure signal between two wells (active well and observation well) is a function of the average diffusivity of the medium. Thus, by running several experiments with various settings, the distribution of diffusivity can be obtained for the interwell space. *Hu et al.* [2015] designed a synthetic numerical model for the pressure-tomography experiment, with multiple points of isolated pressure source and pressure receiver along the vertical direction of the injection well and observation well (Fig. 11.4). The pressure response observed from each test was then matched with the global pressure equation for the mixed phase (CO_2 and brine) by adjusting diffusivity coefficient. Combining all tests, *Hu et al.* [2015] derived a diffusivity tomogram, which was translated to CO_2 saturation using two-phase flow properties. Figure 11.5 shows a comparison of the synthetic CO_2 saturation profile versus predicted CO_2 saturation from the diffusivity tomogram.

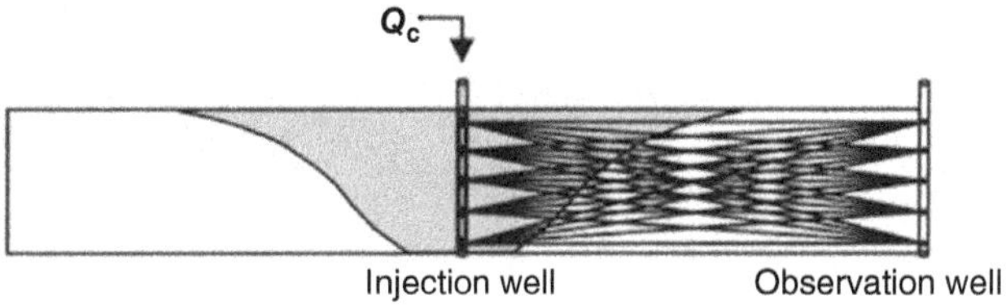

Figure 11.4 Configuration of pressure-tomography test. Multiple points of pressure source and pressure receiver lie along injection and observation wells, respectively [*Hu et al.*, 2015]. Reprinted with the permission of Elsevier.

Downhole pressure data can be used to extract a vast amount of information in CO_2 injection projects [*Mishra et al.*, 2013], including boundary condition and formation static properties. In this systematic approach, overall injectivity of the wells over time can be estimated, and boundary effects in formation can be identified. This type of analysis can be applied even post breakthrough time in monitoring wells.

These types of analysis are based mostly on the well-known fundamentals that have been developed in the oil and gas industry. For example, late-time pressure falloff response at monitoring wells can be interpreted to estimate intrinsic permeability. Effective compressibility values estimated using injection-falloff sequences at different times are used in the AEP Mountaineer CO_2 Capture and Geological Storage Project for tracking the CO_2-brine front, thus serving as an indirect plume monitoring technique [*Mishra et al.*, 2014]. For example, total compressibility measured during the CO_2 injection project was estimated multiple times over the life of project. Total compressibility values increase by one order of magnitude. This increase, measured from pressure data collected in monitoring wells, can be related to overall displacement of the formation brine with more compressible fluid (CO_2) and eventually translated to location of the CO_2 plume front (Fig. 11.6), potentially a very strong and cost-effective tool for monitoring the plume location in the subsurface.

11.4. ABOVE-ZONE PRESSURE MONITORING

Although in-zone pressure monitoring techniques provide a tool for directly assessing IZ integrity, they cannot rigorously confirm the confinement of the injected CO_2 plume in the target zone. Because of the high-pressure noise environment, obscured pressure baseline, and uncertainty associated with formation size and properties, interpreting measured pressure data and detecting CO_2 leakage early in the process are more challenging.

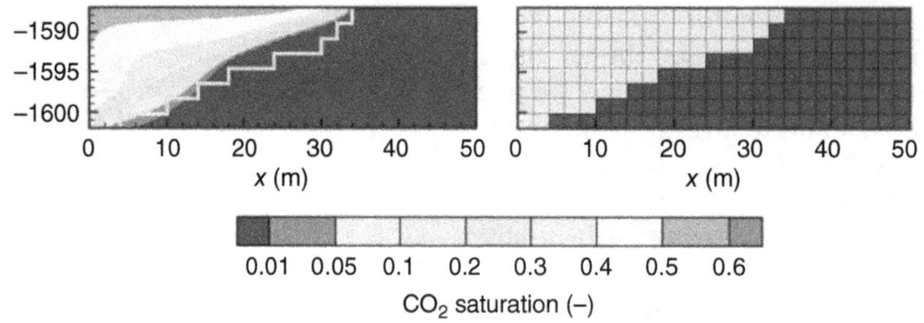

Figure 11.5 In diffusivity tomogram, CO_2 front is captured very well, but saturation values are homogenized.

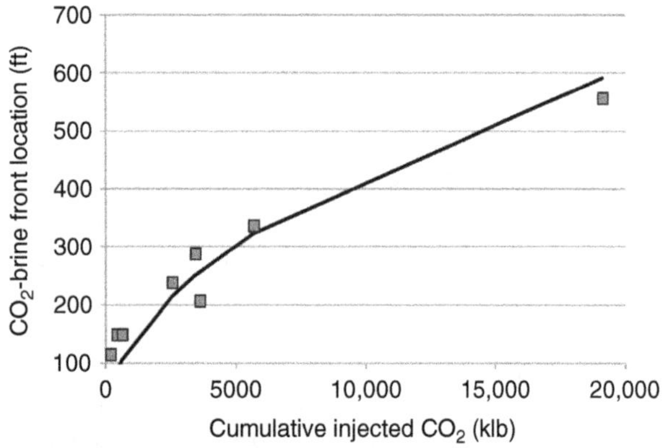

Figure 11.6 Estimated CO_2-brine front radius based on changes in total compressibility values estimated from pressure data.

To address this issue and raise the safety levels of CO_2 storage, AZMIs have received significant attention [*Park et al.*, 2012]. The body of literature on above-zone pressure monitoring and the corresponding analytical and numerical solutions are more extensive than those on in-zone pressure monitoring. Ideally, pressure in a formation above the IZ separated by a confining layer remains constant if no fluids escape from the underlying layer(s). Thus, the pressure baseline is supposed to be a flat line, and any further deviation could be considered the result of a leak. However, note that in some circumstances, pressure in the monitoring layer starts to change only because of geomechanical effects and not the leak itself. Since the monitoring zone is assumed to be a *silent* environment, the noise level in the above-zone layers created by field operations is much lower compared to that of the IZ. Because of the distance from the IZ, high-resolution pressure gauges (of the order of 0.01 psi) in the AZMI can provide high-quality data for detailed study of the formation behavior. This section discusses several research studies on above-zone pressure monitoring and summarizes their results.

Several analytical solutions in the literature describe pressure change in the monitoring well as a function of leakage rate and rock-fluid parameters in the injection and monitoring formations. *Zeidouni and Pooladi-Darvish* [2013] derived an analytical expression to calculate the pressure response in the observation well and the leakage rate. Their analytical model comprises two infinite-acting, homogeneous, isotropic aquifers separated by an impermeable layer. The injection was made in the active well at the bottom aquifer; the corresponding response was measured in the monitoring well at the top aquifer. To simulate the leak, a defective well, penetrating both aquifers, was considered at a location between the injector and monitoring wells. Figure 11.7 displays a schematic of the synthetic model.

Authors extended the solution to multilayer systems as well. Figure 11.8 shows the pressure rise and the leakage rate as a function of aquifer radius for a synthetic example [*Zeidouni*, 2014]. It is clear that as the radius of the AZMI increases, the pressure rise decreases and a more sensitive pressure gauge is required.

Although a leak can be detected by monitoring absolute formation pressure, additional steps of characterization are required to distinguish between gaseous-phase and liquid-phase leakage. *Hosseini and Alfi* [2015] applied type-curve analysis of reservoir pressure to calculate the total compressibility of the investigated zone in a numerical model. Because of the significant difference between the isothermal compressibility of the leaked gaseous phase (e.g., CO_2 or CH_4) and the original pore fluid liquid phase (brine), the measured total formation compressibility would change significantly if a gaseous phase leaked into the zone of interest and in the vicinity of the monitoring wells. Based on this concept, they calculated the total dimensionless compressibility (c_{tD}) (using a numerically generated pulse test to produce the synthetic well-test data) as a function of various rock and fluid properties. (Pulse tests are discussed in more detail in the following paragraphs.) They then performed several numerical simulation studies for various values of porosity (∅), initial total compressibility (c_{t0}), ratio of water and gas initial compressibility (c_w/c_g), reservoir thickness (h), formation volume factor (B_g), well spacing (d), and the amount of leaked gaseous phase (Q). They proposed a generalized model to relate the change in formation compressibility to the cumulative amount of gaseous leak

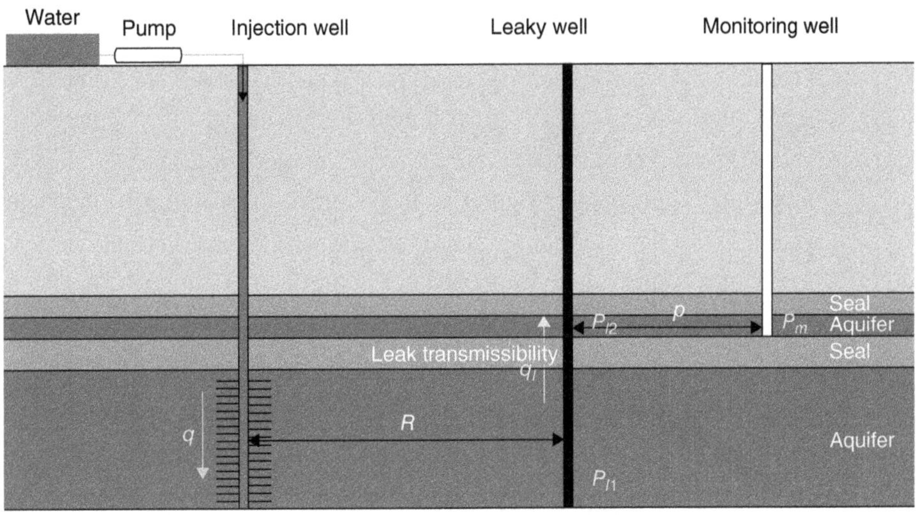

Figure 11.7 Synthetic model for leak between two aquifers separated by an impermeable layer [*Zeidouni and Pooladi-Darvish*, 2013]. Reprinted with permission from Elsevier.

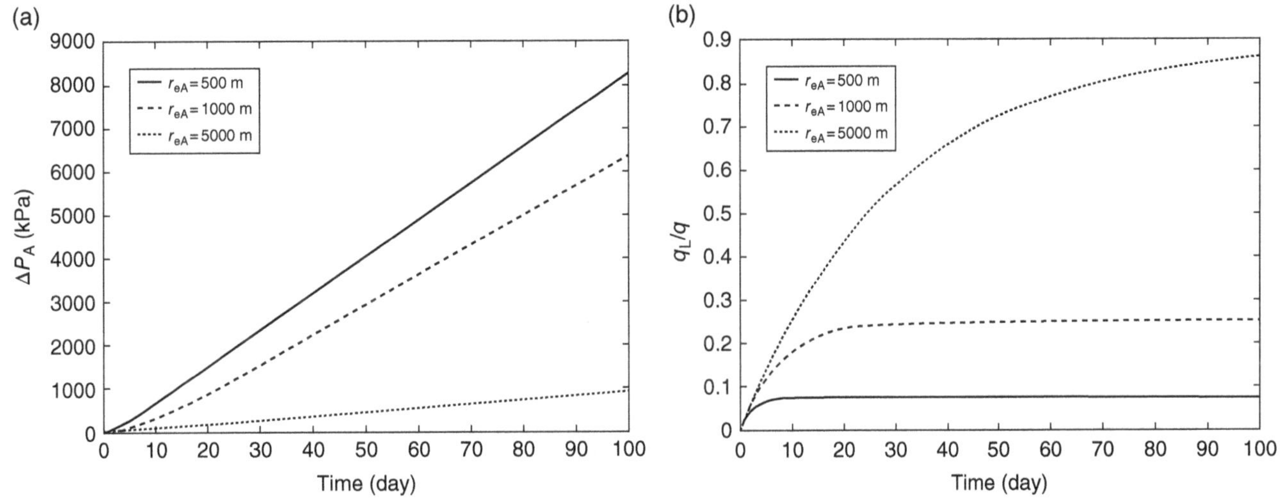

Figure 11.8 (a) Pressure change and (b) leakage rate variation as function of AZMI radius for a specific numerical example [*Zeidouni*, 2014]. 1000 kPa = 145 psi. Reprinted with permission from Elsevier.

(Eq. 11.1). In this equation, L-INDEX is a function of rock and fluid properties as well as of the amount of leaked compressible phase. Equation (11.2) defines the L-INDEX formulation, where $\gamma = 2.61 \times 10^{-6}$ is a unit conversion factor ($m^{2.6}$. Pa^{-1}. $tonnes^{-1}$). They suggest using this equation for L-INDEX values greater than 10.

$$c_{tD} = 0.0102 \times \mathrm{L-INDEX} \tag{11.1}$$

$$\mathrm{L-INDEX} = \gamma \frac{Q \times B_g}{d^{1.75} c_{t0}} \left(\frac{c_g}{c_w \O h} \right)^{0.85} \tag{11.2}$$

Figure 11.9 illustrates the plot of dimensionless compressibility as a function of L-INDEX. Data points are derived through numerical simulation examples. A linear behavior is observed in the plot. More parameters possibly could have been included in this approach, such as parameters related to the relative permeability curve. This approach is also similar to that of *Mishra et al.* [2014] in that it uses the change in total compressibility to track the amount of CO_2 in the subsurface; however, this new method is generalized for application in other projects by developing through application of newly proposed dimensionless parameters.

Pressure pulse testing is another approach used in the monitoring zone to detect leaks. The US EPA requires such a well-testing program for closure of deep-well injection under their Class I program for hazardous waste injection [*US EPA*, 2001]. We expand on the details of pulse testing (as one of the more common well-testing

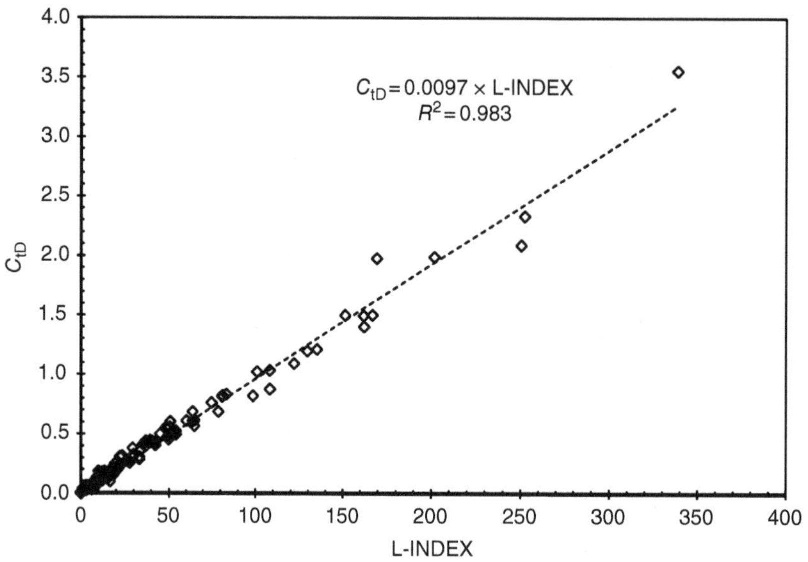

Figure 11.9 Measured dimensionless compressibility versus L-INDEX. Each data point is derived using numerical simulation examples. L-INDEX is a function of rock and fluid properties as well as of the cumulative leaked gaseous phase [*Hosseini and Alfi*, 2015]. Reprinted with permission from Wiley.

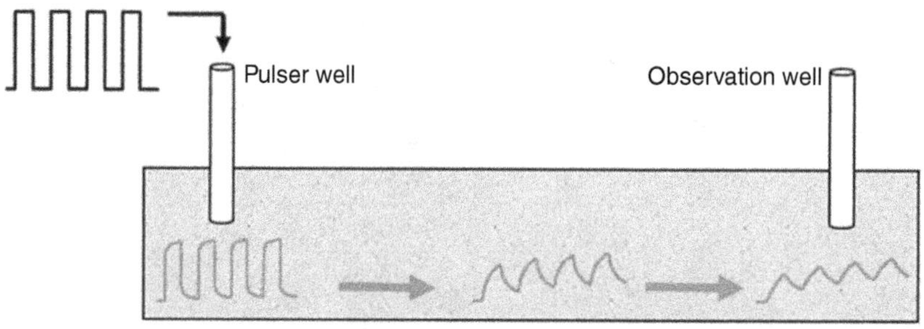

Figure 11.10 Schematic illustration of pulse test. The initially square-shaped signal of the injection pulse gives rise to a pressure signal at the bottom of the well. As pressure signal travels inside formation, shape of signal changes.

methods) because most of the monitoring methods discussed in this chapter use well testing to create pressure pulses. In pressure pulse testing, a pressure signal is first generated by periods of injection and shut-in from a pulsing well. The signal is then propagated inside the formation, and the corresponding response is recorded at the observation well(s). The recorded pressure is analyzed to measure rock and fluid properties (such as hydraulic diffusivity coefficient $D = \frac{k}{\mu\phi c_t}$) for the interwell zone. Figure 11.10 displays a schematic illustration of a pulse-test setup. In this figure, the injection of a square rate signal into the injection well creates a periodic pressure disturbance at the bottom hole of the pulsing well. As the pressure pulse diffuses inside the formation, the shape and the characteristics of the signal change. Quantification of these changes with respect to time as well as to the location of monitoring wells is applied in deriving the formation properties in the zone between the two wells.

Since pressure response in the pulse test is affected by all injection and shut-in periods, the superposition principle can be applied to model this behavior. According to this principle, the oscillatory behavior of the pressure signal at the pulsing and observation wells can be reproduced based on the summation of the solutions to the diffusivity equation for each injection and shut-in period. Equation (11.3) describes the pressure response at the observation well (Theis solution). In this equation, P is the pressure in the monitoring well, r is the distance between wells, t is time, q_j is the rate at each step, B is the formation volume factor (used to convert reservoir condition volume to surface condition volume), T is the transmissibility, D is the diffusivity, and N is the total number of steps (active and shut-in periods). C_0 and C_1

are unit conversion factors. *Ei* is the exponential integral function:

$$P(r,t) = C_0 \frac{B}{T} \sum_{j=1}^{j=N} (q_j - q_{j-1}) \left[E_i \left(C_1 \frac{Dr^2}{t} \right) \right] \quad (11.3)$$

If all other parameters are known, hydraulic diffusivity can be measured by matching Eq. (11.3) to the observed pressure response in the monitoring well. The temporal changes of the diffusivity coefficient could be analyzed further to detect and quantify a possible CO_2 leak to the overlying formations. Because of the significant difference in isothermal compressibility and viscosity of gaseous phase and brine, even a small leak can be detected using this method (diffusivity is inversely proportional to viscosity and compressibility $D = \frac{k}{\mu \phi c_t}$).

If a network of pulsing and observation wells is available in the AZMI, the location and proximity of the leak can be determined with pulse testing [*Shakiba and Hosseini*, 2016]. The derived diffusivity coefficient with the above technique would be minimum in the direction of the leak. Figure 11.11 illustrates a normalized diffusivity tomogram derived from a series of pulse tests between the pulsing well (middle) and the observation wells (at corners). Having such diffusivity tomograms available can accurately guide remediation efforts and significantly save the associated cost if a leak occurs in the AZMI. In Figure 11.11, darker areas (northeast section of figure) represent areas where diffusivity has reduced and a leak has possibly occurred. The current method is developed to simply detect the leakage, not to estimate its amount. Potentially, a relationship should exist between the amount of reduction in diffusivity and the amount of the leaked fluid that needs more investigation.

Another means of pressure measurement is the natural pressure signals, available because of the lunar-solar tide effect. If a very sensitive pressure gauge (with resolution of 0.01 psi) is used at the bottom of the well, the cyclic variation of pressure due to tidal effects can be measured. Similar to the pressure-pulse-test analysis, as the properties of the formation fluid change, so do the shape and characteristics of the tidal signals. *Sato* [2006] used pressure signals of Earth tides to measure the temporal change of the formation poroelastic parameter (χ) as a function of fluid distribution and properties. He used field-test data of an offshore CO_2 sequestration project to measure χ as CO_2 was injected into the formation. Figure 11.12 shows the pressure fluctuation due to tidal effects extracted from the original pressure trend. Accordingly, he measured χ as a function of time, which clearly shows the arrival of CO_2 at the observation well after 340 days (Fig. 11.13). Although this method eliminates the need for the periodic injection required by the pressure pulse test (and thus costs less), it cannot detect leaked gas until it is very close to the monitoring well. Obviously, if CO_2 breaks through the well, it can be readily detected without analysis of tidal effects (such as change in wellhead pressure or wellhead gas sample).

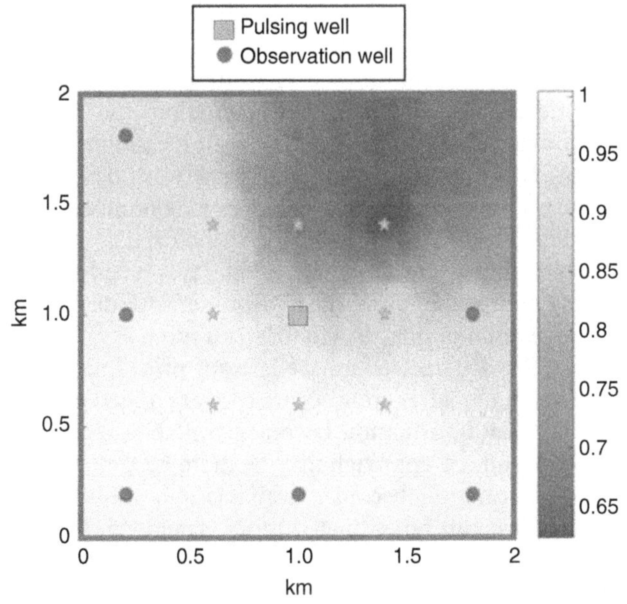

Figure 11.11 Normalized diffusivity tomogram for a pulse test between one pulsing well (square) and four observation wells (circles). Darker color shows region containing gas leak (lower diffusivity). Stars highlight points we have assigned the calculated value of diffusivity (midpoints between the wells). *(See electronic version for color representation of the figure.)*

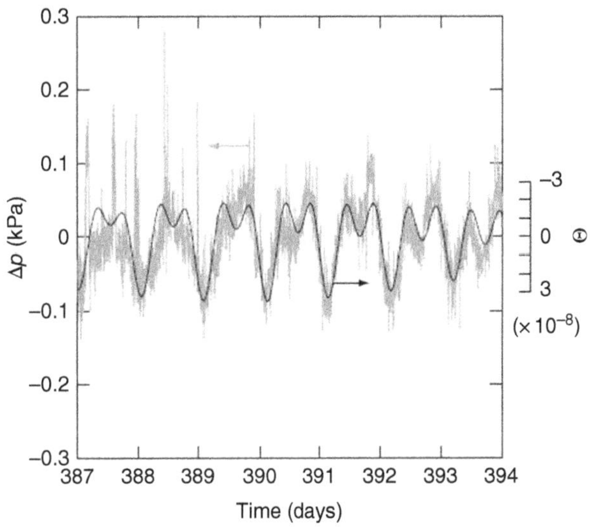

Figure 11.12 Pressure fluctuation at bottom hole due to tidal effects (left axis) [*Sato*, 2006]. Reprinted with the permission from Elsevier.

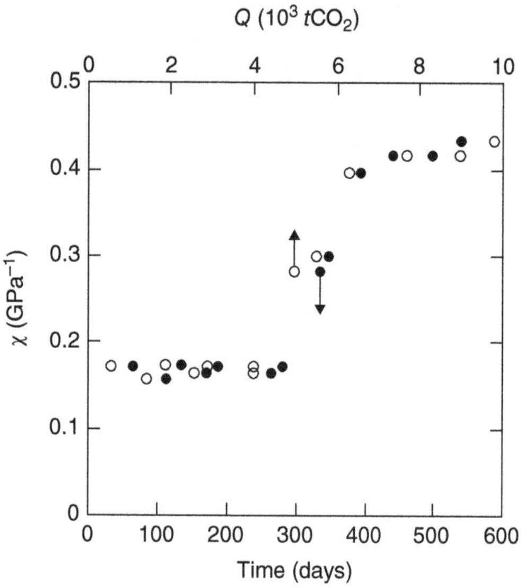

Figure 11.13 Formation poroelastic parameter (χ) versus time. Sharp increase in χ shows arrival of CO_2 [*Sato*, 2006]. Reprinted with the permission from Elsevier.

This method may have value in detection of very large CO_2 leakage scenarios far from the monitoring well, where CO_2 volume is large enough to impact the tidal effects in the well. More studies are needed to quantify this theory.

Pressure perturbation in the AZMI may not always indicate leakage. Pressure change could be related to geomechanical deformation due to the injection in underlying layers. *Kim and Hosseini* [2014] presented and interpreted the results of pressure and temperature monitoring of an AZMI to evaluate the migration of CO_2 from the underlying IZ in Cranfield, Mississippi. They applied analytical and numerical geomechanical techniques in an attempt to explain the observed pressure increase in the monitoring interval, considering no fluid communication between the IZ and the monitoring interval. Figure 11.14 shows the comparison between measured bottom-hole pressure and the numerical simulation results for an observation well. In the same field, another well with simultaneous IZ and AZMI permanent gauges is used for CCS monitoring [*Meckel et al.*, 2013; *Tao et al.*, 2013]

11.5. NETWORK OPTIMIZATION AND DETECTION UNDER UNCERTAINTY

Geologic carbon storage (GCS) monitoring is data intensive as a result of the proliferation of digital instrumentation and smart sensing technologies. Monitoring networks are prevalent in all stages of GCS operations. During site characterization, monitoring networks are used to establish baseline conditions. During CO_2 injection, sensors and geophysical surveys are used to track injected CO_2 as the plume moves within the storage reservoir. During the postinjection stage, routine monitoring is performed to ensure the long-term secure containment of injected CO_2.

GCS monitoring is also resource intensive, often requiring multidisciplinary teams to perform different monitoring, verification, and accounting (MVA) tasks throughout the life cycle of a storage project. As a consequence, operators are motivated to seek the optimization of monitoring networks to achieve the goals and/or regulations of GCS monitoring with limited resources.

Optimization of monitoring networks is a special type of optimal experiment design (OED) problem and has been widely studied in fields such as operations research, industrial engineering, and environmental monitoring [*Meyer et al.*, 1994; *Datta and Dhiman*, 1996; *Mahar and Datta*, 1997; *Kennedy and O'Hagan*, 2001; *Reed and Minsker*, 2004; *Khader and McKee*, 2014; *Prakash and Datta*, 2015]. OED is typically defined as the solution of a multiobjective optimization problem [*Sun and Sun*, 2015]:

$$\text{Minimize}\{f_1(D), f_2(D), \ldots, f_k(D)\}\text{subject to } g_l(D) \le 0, \quad (l = 1, \ldots, L), \tag{11.4}$$

where D represents a set of design/decision variables, $fk(D)$, $k = 1, \ldots, K$ are K objectives, and $gl(D)$, $l = 1, \ldots, L$ are L constraints. Common design variables (D) relevant to environmental monitoring include (i) the number and locations of sensors, (ii) the reliability of sensor networks (e.g., the maximum volume of pollutant leaked before first detection), and (iii) instrumentation and maintenance costs [*Sun et al.*, 2013]. If sensor location is one of the primary concerns of monitoring network design, Eq. (11.4) becomes a sensor placement or maximum coverage problem, which are commonly combinatorial in nature.

Depending on the design paradigm, two types of monitoring approaches exist: the a priori network design and the sequential design. In the former approach, all monitoring sites are selected on the basis of prior knowledge, whereas in the latter, monitoring sites are added one at a time as new information becomes available. Intuitively, the wait-and-see approach may be more cost effective as more information becomes available and the value of information can be evaluated more accurately. In many environmental monitoring problems, however, the perceived environmental damage is so high that the wait-and-see approach is precluded. The operators are often mandated to maximize the use of their present resources to minimize potential future risks incurred by an operation (e.g., radionuclide migration).

Sun et al. [2013] adopted a classic integer programming method to design above-zone, pressure-based GCS

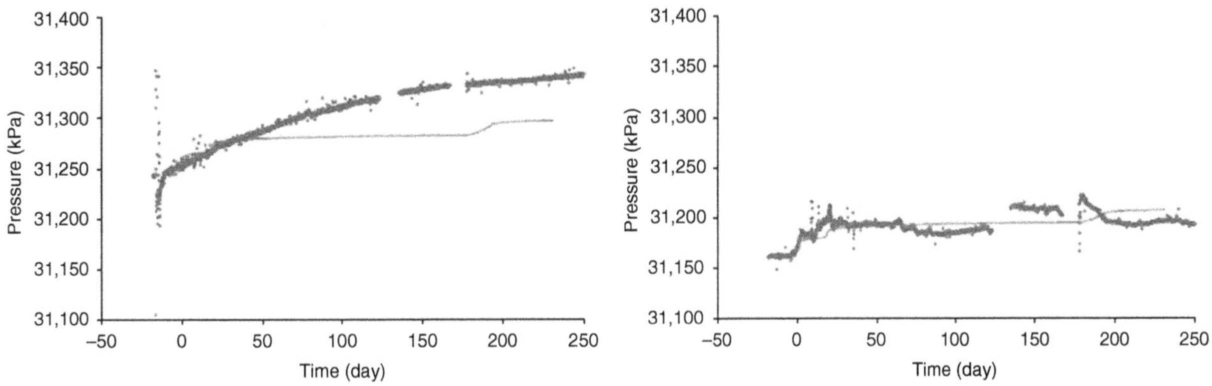

Figure 11.14 Comparison between measured bottom-hole pressures (dots) of observation well and corresponding numerical simulation results considering only geomechanical effects (continuous lines) [*Kim and Hosseini*, 2014]. The small increase in AZMI pressure can be attributed to geomechanical effects, not leakage. Reprinted with the permission from Wiley.

monitoring networks under model and parameter uncertainty. An equation was formulated to minimize both the total volume of leakage and the number of uncovered potentially leaky locations. The approach was demonstrated for selecting optimal monitoring locations in both homogeneous and heterogeneous formations. In both cases, the algorithm was able to identify appropriate sensor-placement locations to ensure that any predefined leakage from the formation could be detected. Their study indicates that (i) the forward model needs to be able to resolve pressure perturbations accurately, especially along the pathways between the injector, leakage, and observation points, and (ii) even designing the network using an analytical model can be computationally demanding under parametric uncertainty.

11.6. SUMMARY

GCS monitoring shares many attributes with conventional environmental monitoring. However, GCS monitoring is complicated by (i) lack of high-fidelity forward models during the network design stage, (ii) large-scale contrast between the area of review and limited radius of coverage of individual sensors, and (iii) high costs associated with deep subsurface monitoring. In the context of pressure-based monitoring, additional complicating factors may include (i) correct classification of anomalies, (ii) signal-to-noise ratio of sensors, and (iii) pressure-gauge malfunctions.

Both in-zone and above-zone pressure monitoring have great potential for adoption by industry as a primary part of the monitoring and verification plans for a GCS project because of the long history of using pressure data for reservoir management. Operators must be convinced, however, that pressure-data collection in zone and above zone can be designed and analyzed further for monitoring of CO_2 storage permanence, not only management of the flood (although the two sometimes happen to overlap). Operators tend to be more open to in-zone monitoring because it fits better into their overall field development plan, but the importance of above-zone monitoring to allow higher levels of confidence in CO_2 retention should be clarified for them.

Going forward, we believe that various monitoring network optimization methods need to be demonstrated for more commercial-scale GCS projects, thanks to the new advances made in large-scale optimization solvers and modeling approaches. The key will lie in the proper formulation of the design problem and the parameterization of geologic uncertainty. Enumeration of scenarios with the assistance of experts and prior knowledge will be valuable for designing under uncertainty.

Many new techniques based on the use of pressure data specifically designed for monitoring of CO_2 storage performance have been developed recently, some highlighted in this study. Most of these techniques, however, are based on synthetic scenarios and numerical simulations; a necessary step toward proving their advantages and limits is to apply them to pilot and field projects to evaluate their performance.

REFERENCES

Alnes, H., O. Eiken, S. Nooner, G. Sasagawa, T. Stenvold, and M. Zumberger (2011), Results from Sleipner gravity monitoring: updated density and temperature distribution of the CO_2 plume, *Energy Procedia*, *4*, 5504–5511.

Angeli, M., M. Soldal, E. Skurtveit, and E. Aker (2009), Experimental percolation of supercritical CO_2 through a caprock, *Energy Procedia*, *1*(1), 3351–3358.

Bourne S., C. Syrie, and M. Smith (2014), A risk-based framework for measurement, monitoring and verification of the

Quest CCS Project, Alberta, Canada, *International Journal of Greenhouse Gas Control*, *26*, 109–126.

Chabora, E. R. (2009), The utility of above-zone pressure measurements in monitoring geologically stored carbon dioxide, Ph.D. Dissertation, Stanford University, pp. 21–26.

Chabora E. R, and S. Benson (2009), Brine displacement of leakage detection using pressure measurements in aquifer overlying CO_2 storage reservoirs, *Energy Procedia*, *1*(1), 2405–2412.

Darcy, H. (1856), *Les fontaines publiques de la ville de Dijon*: Dalmont, Paris, 647 p.

Datta, B., and S. D. Dhiman (1996), Chance-constrained optimal monitoring network design for pollutants in ground water, *Journal of Water Resources Planning and Management*, *122*, 180–188.

Eck, J., U. Ewherido, J. Mohammed, R. Ogunlowo, J. Ford, L. Fry, S. Hiron, L. Osugo, S. Simonian, T. Oyewole, and T. Veneruso (2000), Downhole monitoring: the story so far, *Oilfield Review*, *11*(3), 20–33. http://www.slb.com/~/media/Files/resources/oilfield_review/ors99/win99/pages20_33.pdf

Elsayed, S. A., R. Baker, P. L. Churcher, and A. C. Edmunds (1993), Multidisciplinary reservoir characterization and simulation study of the Weyburn unit, *Journal of Petroleum Technology*, *45*(10), 930–937.

Flett M., G. Beacher, J. Brantjes, A. Burt, C. Dauth, F. Koelmeyer, R. Lawrence, S. Leigh, J. McKenna, R. Gurton, W. F. Robinson IV, and T. Tankersley (2008) Gorgon project: subsurface evaluation of carbon dioxide disposal under Barrow Island, *SPE Asia Pacific Oil and Gas Conference and Exhibition*, Perth, Australia, October 20–22, SPE116372-MS.

Girelli, A. (2005), History of the upstream oil industry, Encyclopedia of Hydrocarbons, v. V, Instituto della Enciclopedia Italiana, INI, 525–563. http://www.treccani.it/portale/opencms/handle404?exporturi=/export/sites/default/Portale/sito/altre_aree/Tecnologia_e_Scienze_applicate/enciclopedia/inglese/inglese_vol_5/appendix_ing.pdf&

Grude, S., M. Landrø, and J. Dvorkin (2014), Pressure effects caused by CO_2 injection in the Tubåen Fm., the Snøhvit field, *International Journal of Greenhouse Gas control*, *27*, 178–187.

Hansen, O., D. Gilding, B. Nazarian, B. Osdel, P. Ringrose, J.-B. Kristofferson, O. Eiken, and H. Hansen (2013), Snøhvit: the history of injecting and storing 1 Mt CO2 in the fluvial Tubåen Fm., *Energy Procedia*, *37*, 3565–3573.

Hawkes, C. D., P. J. McLellan, and S. Bachu (2005), Geomechanical factors affecting geological storage of CO_2 in depleted oil and gas reservoirs, *Journal of Canadian Petroleum Technology*, *44*(10), 52–61.

Hosseini, S. A., and M. Alfi (2015), Time-lapse application of pressure transient analysis for monitoring compressible fluid leakage, *Greenhouse Gases: Science and Technology*, *6*(3), 352–369.

Hosseini, S. A., H. Lashgari, J. W. Choi, J.-P. Nicot, J. Lu, and S. D. Hovorka (2013), Static and dynamic reservoir modeling for geological CO_2 sequestration at Cranfield, Mississippi, USA, *International Journal of Greenhouse Gas Control*, *18*, 449–462.

Hovorka, S. D., S. M. Benson, C. Doughty, B. M. Freifeld, S. Sakurai, T. M. Daley, Y. K. Kharaka, M. H. Holtz, R. C. Trautz, H. Nance, L. R. Myer, and K. G. Knauss (2006), Measuring permanence of CO_2 storage in saline formations: the Frio experiment, *Environmental Geosciences*, *13*, 105–121.

Hu, L., P. Bayer, P. Alt-Epping, A. Tatomir, M. Sauter, and R. Brauchler (2015), Time-lapse pressure tomography for characterizing CO_2 plume evolution in a deep saline aquifer, *International Journal of Greenhouse Gas Control*, *39*, 91–106.

Intergovernmental Panel on Climate Change (2005), *Carbon Dioxide Capture and Storage: Special Report of the Intergovernmental Panel on Climate Change*: New York, Cambridge University Press, *431* p.

Katz, D. L., and M. R. Tek (1981), Overview on underground storage of natural gas, *Journal of Petroleum Technology*, *33*, 943–951.

Keating E., Z. Dai, D. Demsey, and R. Pawar (2014), Effective detection of CO_2 leakage: a comparison of groundwater sampling and pressure monitoring, *Energy Procedia*, *63*, 4163–4171.

Kelley, M., M. Abbaszadeh, S. Mishra, S. Mawalkar, M. Place, N. Gupta, and R. Pardini (2014), Reservoir characterization from pressure monitoring during CO_2 injection into a depleted pinnacle reef – MRCSP commercial-scale CCS demonstration project, *Energy Procedia*, *63*, 4937–4964.

Kennedy, M. C., and A. O'Hagan (2001), Bayesian calibration of computer models, *Journal of the Royal Statistical Society. Series B Statistical Methodology*, *63*(3), 425–464.

Khader, A. I., and M. McKee (2014), Use of a relevance vector machine for groundwater quality monitoring network design under uncertainty, *Environmental Modelling & Software*, *57*, 115–126.

Kim, S., and S. A. Hosseini (2014), Above-zone pressure monitoring and geomechanical analyses for a field-scale CO_2 injection project in Cranfield, MS, *Greenhouse Gases: Science and Technology*, *4*(1), 81–98.

Kim, S., and S. A. Hosseini (2015), Hydro-thermo-mechanical analysis during injection of cold fluid into a geologic formation, *International Journal of Rock Mechanics and Mining Sciences*, *77*, 220–236.

Liebscher, A., F. Möller, A. Bannach, S. Köhler, J. Wiebach, C. Schmidt-Hattenberger, M. Weiner, C. Pretschner, K. Eber, and J. Zemke (2013), Injection operation and operational pressure-temperature monitoring at CO_2 pilot site Ketzin, Germany – design, results, recommendations, *International Journal of Greenhouse Gas Control*, *15*, 163–173.

Mahar, P. S., and B. Datta (1997), Optimal monitoring network and ground-water-pollution source identification, *Journal of Water Resources Planning and Management*, *123*, 199–207.

Meckel T., M. Zeidouni, S. D. Hovorka, and S. A. Hosseini (2013), Assessing sensitivity to well leakage from three years of continuous reservoir pressure monitoring during CO_2 injection at Cranfield, MS, USA, *International Journal of Greenhouse Gas Control*, *18*, 439–448.

Meyer, P. D., A. J. Valocchi, and J. W. Eheart (1994), Monitoring network design to provide initial detection of groundwater contamination, *Water Resources Research*, *30*, 2647–2659.

Mishra, S., M. Kelley, E. Zeller, N. Slee, N. Gupta, I. Bhattacharya, and M. Hammond (2013), Maximizing the value of pressure monitoring data from CO_2 sequestration projects, *Energy Procedia*, *37*, 4155–4165.

Mishra, S., Y.-D. Oruganti, N. Gupta, P.-R. Ganesh, C. McNeil, I. Bhattacharya, and G. Spitznogle (2014), Modeling CO_2 plume migration based on calibration of injection and post-injection pressure response at the AEP Mountaineer Project, *Greenhouse Gases: Science and Technology*, *4*, 331–356.

Nestlerode, W. A. (1963), The use of pressure data from permanently installed bottom hole pressure gauges, SPE Rocky Mountain Joint Regional Meeting, Denver, Colorado, May 27–28, SPE-590.

Park, Y.-C., D.-G. Huh, and C.-H. Park (2012), A pressure-monitoring method to warn CO_2 leakage in geological storage sites, *Environmental Earth Sciences*, *67*, 425–433.

Prakash, O., and B. Datta (2015), Optimal characterization of pollutant sources in contaminated aquifers by integrating sequential-monitoring-network design and source identification: methodology and an application in Australia, *Hydrogeology Journal*, *23*, 1089–1107.

Reed, P. M., and B. S. Minsker (2004), Striking the balance: long-term groundwater monitoring design for conflicting objectives, *Journal of Water Resources Planning and Management*, *130*, 140–149.

Ringrose, P. S., A. S. Mathieson, I. W. Wright, F. Selama, O. Hansen, R. Bissell, N. Saula, and J. Midgley (2013). The In Salah CO_2 storage project: lessons learned and knowledge transfer, *Energy Procedia*, *37*, 6226–6236.

Rutqvist, J., and C. F. Tsang (2002), A study of caprock hydromechanical changes associated with CO_2-injection into a brine formation, *Environmental Geology*, *42*(2–3), 296–305.

Rutqvist, J., J. Birkholzer, F. Cappa, and C. F. Tsang (2007), Estimating maximum sustainable injection pressure during geological sequestration of CO_2 using coupled fluid flow and geomechanical fault-slip analysis, *Energy Conversion and Management*, *48*(6), 1798–1807.

Sato, K. (2006), Monitoring the underground migration of sequestered carbon dioxide using Earth tides, *Energy Conversion and Management*, *47*(15), 2414–2423.

Sato, K., M. Saeko, H. Tadashi, O. Hiroshi, S. Hideki, W. Jiro, and Y. Tsukasa (2011), Monitoring and simulation studies for assessing macro- and meso-scale migration of CO_2 sequestered in an onshore aquifer: experiences from the Nagaoka pilot site, Japan, *International Journal of Greenhouse Gas Control*, *5*, 125–137.

Shakiba, M., and S. A. Hosseini (2016), Detection and characterization of CO_2 leakage by multi-well pulse testing and diffusivity tomography maps, *International Journal of Greenhouse Gas Control*, *54*, 15–28.

Steadman, E. N. (2013), Bell Creek MVA overview, CSLF CO_2 monitoring interactive workshop, Rome, Italy. https://www.cslforum.org/cslf/sites/default/files/documents/rome2013/Steadman-Workshop-Rome0413.pdf

Strandli, C. W., E. Mehnert, and S. M. Benson (2014), CO_2 plume tracking and history matching using multilevel pressure monitoring at the Illinois Basin–Decatur Project, *Energy Procedia*, *63*, 4473–4484.

Sun, A. Y., and J.-P. Nicot (2012), Inversion of pressure anomaly data for detecting leakage at geologic carbon sequestration sites, *Advances in Water Resources*, *44*, 20–29.

Sun, N. Z., and A. Sun (2015), *Model Calibration and Parameter Estimation: For Environmental and Water Resource Systems*: New York, Springer, 621 p.

Sun, A. Y., J.-P. Nicot, and X. Zhang (2013), Optimal design of pressure-based, leakage detection monitoring networks for geologic carbon sequestration repositories, *International Journal of Greenhouse Gas Control*, *19*, 251–261.

Tao, Q., S. L. Bryant, and T. A. Meckel (2013), Modeling above-zone measurements of pressure and temperature for monitoring CCS sites, *International Journal of Greenhouse Gas Control*, *18*, 523–530.

US EPA (2001), Class I Underground Injection Control Program: Study of the Risks Associated with Class I Underground Injection Wells; EPA 816-R-01-007.

US EPA (2010), Code of Federal Regulations, 40 CFR Parts 124, 144, 145, 146, and 147. Federal Requirements under the Underground Injection Control (UIC) Program for Carbon Dioxide (CO_2) Geologic Sequestration (GS) Wells. Final Rule; EPA Docket: EPA-HQ-OW-2008-0390 FRL-9232-7.

Verma, S., C. S. Oakes, N. Chugunov, T. S. Ramakrishnan, S. A. Hosseini, and S. Hovorka (2013), Reservoir fluid monitoring in carbon dioxide sequestration at Cranfield, *Energy Procedia*, *37*, 4344–4355.

Whittaker, S. (2010), IEA GHG Weyburn-Midale CO_2 Storage and Monitoring project, Regional Carbon Sequestration Partnerships Annual Review, 29 p. https://www.netl.doe.gov/publications/proceedings/10/rcsp/presentations/Tues%20am/Karen%20Cohen/Whittaker.%20WMP_Regional%20Partnership.pdf.

Zahid, K. M., S. A. Hosseini, V. Nunez-Lopez, and S. D. Hovorka (2012), Characterizing CO_2 storage reservoirs and shallow overburden for above-zone monitoring in Texas Gulf Coast EOR fields, *Greenhouse Gases: Science and Technology*, *2*(6), 460–473.

Zeidouni, M. (2014), Analytical model of well leakage pressure perturbations in a closed aquifer system, *Advances in Water Resources*, *69*, 13–22.

Zeidouni, M., and M. Pooladi-Darvish (2013), Leakage characterization through above-zone pressure monitoring: 1 – inversion approach. *Journal of Petroleum Science and Engineering*, *98*, 95–106.

12

Monitoring and Modeling Caprock Integrity at the In Salah Carbon Dioxide Storage Site, Algeria

Donald W. Vasco[1], Robert C. Bissell[2], Bahman Bohloli[3], Thomas M. Daley[1], Alessandro Ferretti[4], William Foxall[1], Bettina P. Goertz-Allmann[5], Valeri Korneev[1], Joseph P. Morris[6], Volker Oye[5], Abe Ramirez[6], Antonio Pio Rinaldi[7], Alessio Rucci[4], Jonny Rutqvist[1], Josh White[6], and Rui Zhang[8]

ABSTRACT

The In Salah gas development project, initiated in 2004, involves the long-term storage of waste carbon dioxide associated with natural gas production at several central Algerian fields. Long-term plans were to store up to 17 million tons of carbon dioxide during the life of the program. The project at In Salah was a pioneering effort in several respects. One feature that distinguishes In Salah from several other industrial-scale CO2 sequestration projects, such as Sleipner, Weyburn, Decatur, and Aquastore, is the generally smaller storage volume in terms of overall reservoir thickness and lower average porosities of around 15%. The unique features of the In Salah storage site point to the importance of wellbore, geophysical, and geochemical monitoring to ensure that the integrity of the storage complex is maintained both during and after injection. In 2005 a Joint Industry Project was set up, in an effort to use various geophysical, geochemical, and production techniques to monitor the fate of the injected carbon dioxide. Here we describe the efforts made to monitoring the fate of the injected carbon dioxide, the state of the caprock, the estimated outcome of the storage effort, and some of the uncertainties that remain.

12.1. INTRODUCTION

The In Salah gas development project, initiated in 2004, involved the long-term storage of waste carbon dioxide associated with natural gas production at several central Algerian fields. Rather than vent the extracted CO_2 into the atmosphere, the operators BP, Sonatrach, and Statoil chose to have the gas compressed, dehydrated, transported, and injected into a deep saline aquifer leg of the Krechba gas reservoir [*Mathieson et al.*, 2011]. As part of this effort, three state-of-the-art horizontal injection wells were drilled to a depth of 1500–1800 m. Geosteering technology was employed in order to follow the roughly 20 m thick reservoir. The wells were oriented perpendicular to the direction of maximum stress, the dominant fracture direction, as a way to maximize the injectivity. The long-term plan was to store up to 17 million tonnes of carbon dioxide during the life of the program [*Mathieson et al.*, 2010]. The In Salah project was a pioneering effort in several respects and involved the investment of over 100 million dollars by the member companies. One feature that distinguishes In Salah from several other industrial-scale CO_2 sequestration projects, such as Sleipner [*Chadwick et al.*, 2012], Weyburn [*White et al.*, 2011], Decatur [*Kaven et al.*, 2014], and Aquistore

[1] *Energy Geosciences Division, Lawrence Berkeley National Laboratory, University of California, Berkeley, CA, USA*
[2] *Carbon Fluids Limited, Bucks, UK*
[3] *Norwegian Geotechnical Institute, Oslo, Norway*
[4] *TRE ALTAMIRA, Milan, Italy*
[5] *NORSAR, Kjeller, Norway*
[6] *Atmospheric, Earth, and Energy Division, Lawrence Livermore National Laboratory, Livermore, CA, USA*
[7] *Swiss Seismological Service, ETH, Zurich, Switzerland*
[8] *School of Geosciences, University of Louisiana at Lafayette, Lafayette, LA, USA*

Geological Carbon Storage: Subsurface Seals and Caprock Integrity, Geophysical Monograph 238, First Edition.
Edited by Stéphanie Vialle, Jonathan Ajo-Franklin, and J. William Carey.

[*Worth et al.*, 2014], is the generally smaller storage volume in terms of overall reservoir thickness and lower average porosities of around 15%. With the exception of Weyburn, the other sites are characterized by generally thick sequences of high-porosity sandstones. Therefore, one might expect that the pressure buildup near the wells might be larger at In Salah, with consequences for reservoir deformation and caprock strain. The role of fractures in providing storage and increasing reservoir permeability is one factor that would likely impact the pressure distribution around the well. However, the detailed effect of existing fractures on flow was not established prior to the injection at In Salah. The unique features of the In Salah storage site point to the importance of wellbore, geophysical, and geochemical monitoring to ensure that the integrity of the storage complex is maintained both during and after injection. In 2005, a Joint Industry Project (JIP) was set up, in an effort to use various geophysical, geochemical, and production techniques to monitor the fate of the injected carbon dioxide [*Mathieson et al.*, 2011]. Here we describe the efforts made to monitor the fate of the injected carbon dioxide, the state of the caprock, the estimated outcome of the storage effort, and some of the uncertainties that remain.

12.2. STRUCTURAL SETTING AND BASELINE DATA

The reservoir is a thin Carboniferous sandstone, with porosities ranging from 10 to 18% and permeabilities of around 10 milli-Darcies. It is contained in a gentle northwest-trending anticline that lies below the approximately 950 m of mudstones that forms a seal for the storage interval (Fig. 12.1). This mudstone sequence is overlain by an additional 900 m sequence of interbedded sandstones, minor mudstones, and limestones, the Continental Intercalaire, and by the Continental Intercalaire sandstone that forms the regional pan-Saharan potable aquifer, a major source of freshwater. The anticline was formed by mid to late Carboniferous basin inversion during the Hercynian-Appalachian orogeny [*Coward and Ries*, 2003; *Guiraud et al.*, 2005]. The anticline lies above reactivated basement faults and was influenced by strike-slip faults propagating up from the underlying Devonian formations [*Ringrose et al.*, 2009]. The apex of the anticline hosted a significant volume of gas, indicating that the mudstone seal was effective in preventing flow out of the reservoir.

The characteristics of the reservoir and the overburden were determined by formation micro-imager (FMI) logs, suites of borehole logs from several wells, cores, mud loss data, and a regional seismic survey from 1997. A composite stratigraphy for the region is plotted in Figure 12.2. There are sets of well logs in 22 of the 24 wells in the area; however, only 4 wells had sonic logs that extended from the reservoir to the surface. Of these logs, only one, associated with well KB-10, contained shear wave information through most of the overburden. Another well, KB-502, had a density log extending from the surface to a depth of a few hundred meters above the reservoir (1650 m). The overburden pressure (P_o), or vertical stress, was determined by integrating this density log. The estimated vertical stress is in agreement with a value obtained using a constant density of 2.283 g/cc. This density is close to the earlier estimate of 2.3 g/cc obtained by an earlier internal JIP study from 1998. The uniaxial compressive strength (UCS) of the reservoir and overburden were estimated using laboratory data from core samples taken from the C10.3 formation and a correlation between the rock strength and the compressional wave slowness log data from well KB-2 [*Armitage et al.*, 2011]. In general, the tensile strength is assumed to be 1/10 of the UCS. Laboratory data from the University of Liverpool confirmed this for core samples from the C10.3 interval, a formation that lies just above the reservoir. Note that the properties of the C10.3 formation may not be representative of the entire stratigraphic section at In Salah, as the mechanical properties vary significantly with depth [*Morris et al.*, 2011; *Gemmer et al.*, 2012].

The stress regime is of a strike-slip nature, with the largest principal stress in the horizontal northwest-southeast direction (315/135°). The intermediate principal stress is oriented vertically, and the least principal stress is in the northeast-southwest direction. The orientation of the principal stresses, with the largest deviatoric stress in the horizontal plane, has led to numerous drilling problems [*T. Darling*, personal communication, BP internal report 2006]. The minimum horizontal principal stress (σ_{hmin}) was found by fitting a geomechanical model to observational data (Fig. 12.3) [*Bissell et al.*, 2011]. The estimates were derived from formation integrity tests (FIT), corresponding to the highest fluid pressure that could be applied without leak-off, and from leak-off tests (LOT), in which pressure is increased in a well until fluid is forced into the surrounding formation.

Because the estimates of the minimum horizontal principal stress in Figure 12.3 are inferred from LOT and FIT, and not from actual stress measurements, they are subject to large uncertainties. The maximum horizontal stress (σ_{hmax}), assumed to have the same depth trend as the vertical stress, was calibrated against wellbore failure data from image and caliper logs. For depths above 1620 m (1170 mTVD), it was found that $\sigma_{hmax}/\sigma_z = 1.09$, while for depth below 1620 m it was observed that $\sigma_{hmax}/\sigma_z = 1.12$. Both of these values are close to unity. *Bissell et al.* [2011] estimated a value of 29 MPa for the fracture pressure at well KB-502 at a depth of 1314mTVD using the formula (see Fig. 12.4)

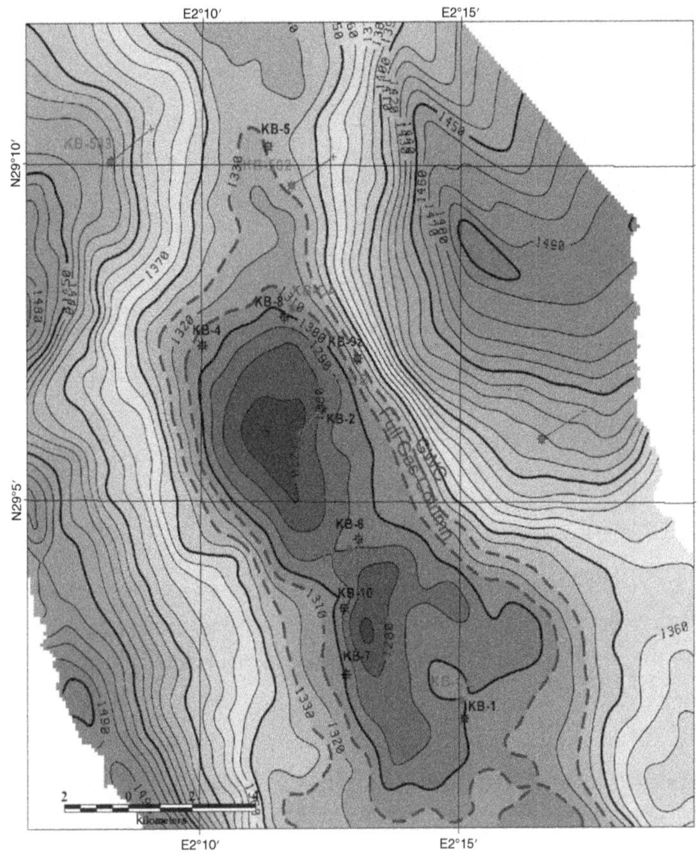

Figure 12.1 Topography of the top of the reservoir (C10.2) at In Salah storage site. The horizontal gas producers are signified by the green lines and green dots (wellheads). The three carbon dioxide injection wells (KB-501, KB-502, and KB-503) are designated by the blue lines. Several exploration wells are also plotted as black stars. The contour units are meters below a sea level datum. Figure after I. Wright CSLF International Projects Workshop, Berlin, 29 September 2005, courtesy of the In Salah Joint Industry Project. (*See electronic version for color representation of the figure.*)

$$P_{\text{fracture}} = 3\sigma_{\text{hmin}} - \sigma_{\text{hmax}} - P_{\text{o}} + 0.1\sigma_{\text{UCS}},$$

where the subscript UCS denotes the uniaxial compressive strength, defined earlier. The cooling of the formation due to the injection of the carbon dioxide may have reduced the fracturing pressure by 1.5–5.0 MPa [*Bissell et al.*, 2011]. The formula given above is only valid for a vertical well, and the estimates may differ from the fracture pressure for the horizontal section of a well. *Bissell et al.* [2011] also estimated the fracture pressure at KB-502 by plotting the well injectivity index versus flowing bottom-hole pressure, arriving at a value of 28.6 MPa for a depth of 1330 m TVD. Others such as *Shi et al.* [2012] and *White et al.* [2014] have also examined the fracture pressure, and the latter study obtained a value near 30 MPa for well KB-502.

Four wells had FMI logs run from the surface to the level of the reservoir. These data suggest that the injection horizon (C10.2) and the immediate overburden are naturally fractured with a northwesterly fracture orientation [*Iding and Ringrose*, 2010]. Significant uplift during the Hercynian and the Tertiary produced stress release joints. The FMI logs and mud loss data indicate that the reservoir contains two to three northwest-trending fractures per meter, with an aperture range of 0.1–1.0 mm and lengths ranging from 6 to 25 m [*Iding and Ringrose*, 2010]. In the lower caprock, just above the reservoir, the fractures trend in the same direction with a spatial density of 1–3 fractures per meter and aperture and length ranges of 0.1–1.0 mm and 6–25 m, respectively [*Iding and Ringrose*, 2010]. Core obtained at well KB-2, from the reservoir interval (C10.2) itself, displayed evidence of fracturing.

Reprocessing of the original 1997 seismic line confirmed that there are minor faults at the reservoir level and in the immediately overlying caprock. Furthermore, mud losses observed while drilling indicate the presence of conductive fractures or faults. In particular, large mud losses occurred

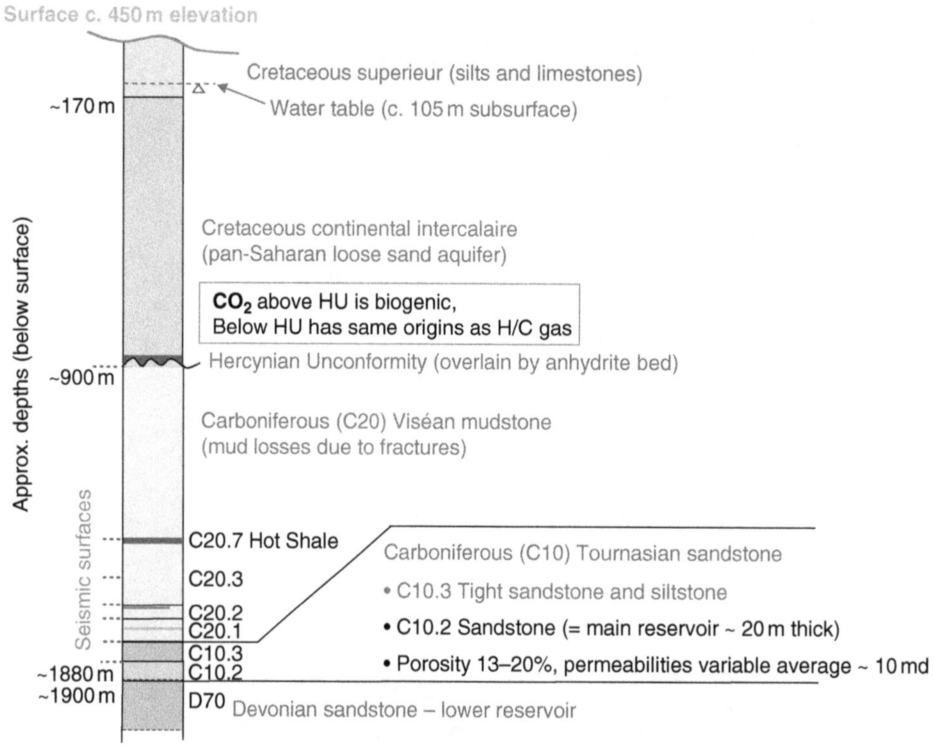

Figure 12.2 Simplified stratigraphic section for the In Salah region. The reservoir interval is the C10.2 sandstone. After Ringrose (personal communication, courtesy of the In Salah Joint Industry Project).

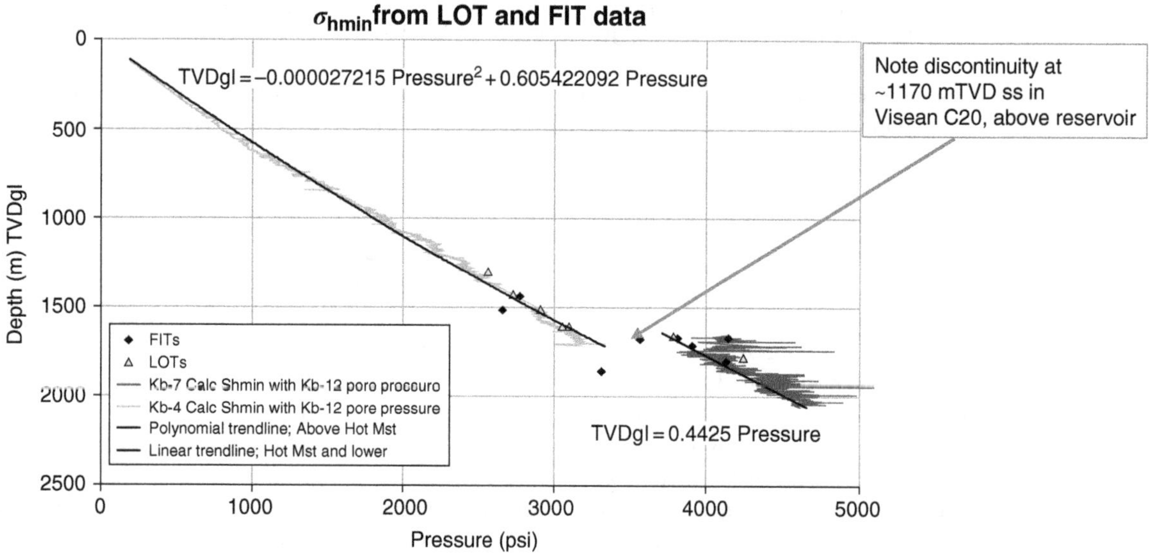

Figure 12.3 Estimated minimum horizontal principal stress, obtained by fitting a low-degree polynomial to well data. Figure from Geoscience Ltd. report to BP.

in wells KB-14, KB-501, and KB-502 in both the Intercalaire and the C20 mudstone units (see Fig. 12.2). Well KB-503 had mud loss in the C20 mudstone and the reservoir interval (C10.2). In the reservoir and lower overburden (mudstone), the faults are thought to be due to deeper structures over which the reservoir is flexed (Fig. 12.5). These faults induced northwest to southeast oriented fractures [*Ringrose et al.*, 2009; *Mathieson et al.*, 2011]. The deeper structures were verified by a later (2009) seismic survey. While drilling the injection well KB-502

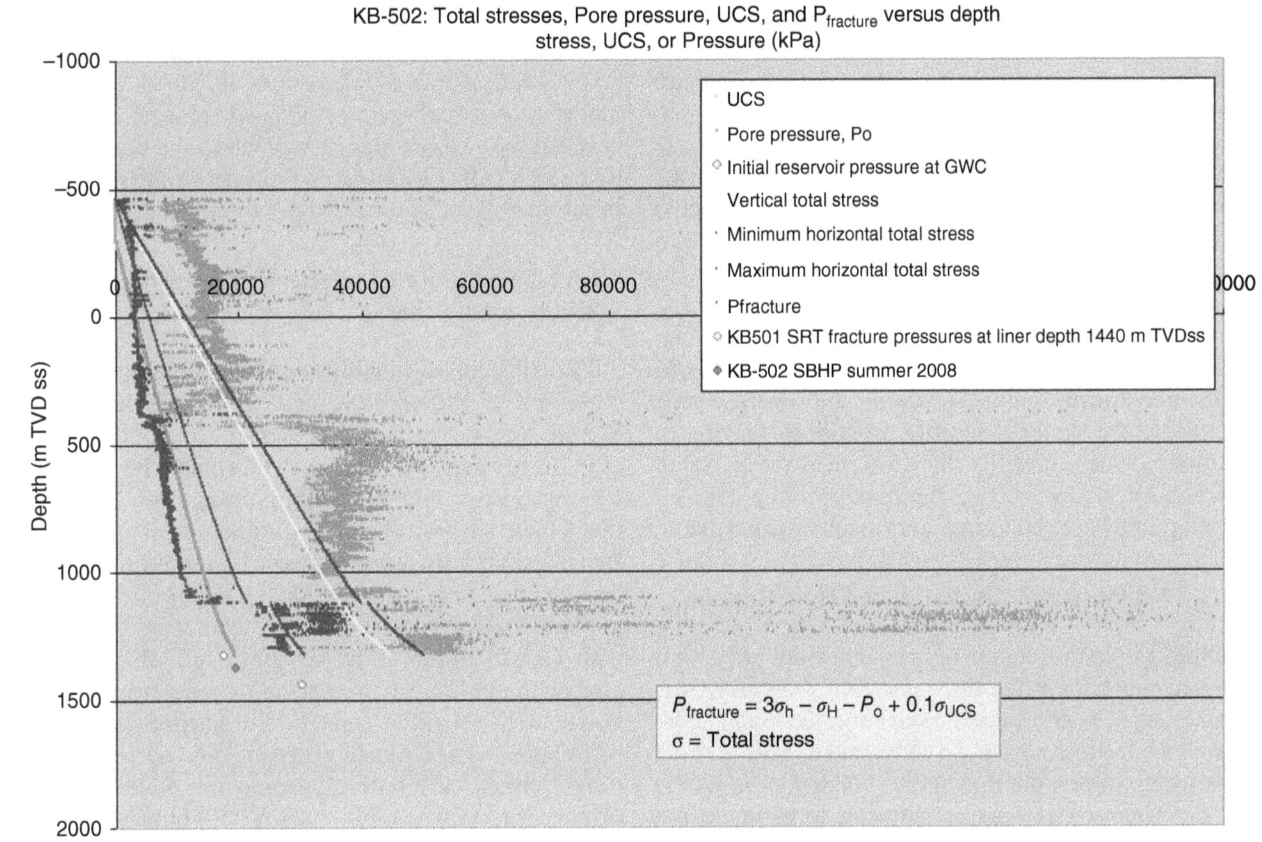

Figure 12.4 Estimated fracture pressure (blue dots) at well KB-502 along with the stresses that contribute to the estimates: the minimum horizontal principal stress σ_{hmin} or σ_h, the maximum horizontal principal stress σ_{hmax} or σ_H, the overburden or vertical stress P_o, and the uniaxial compressive strength σ_{UCS}. Figure courtesy of In Salah Joint Industry Project. TVD is the total vertical depth below sea level in meters. (*See electronic version for color representation of the figure.*)

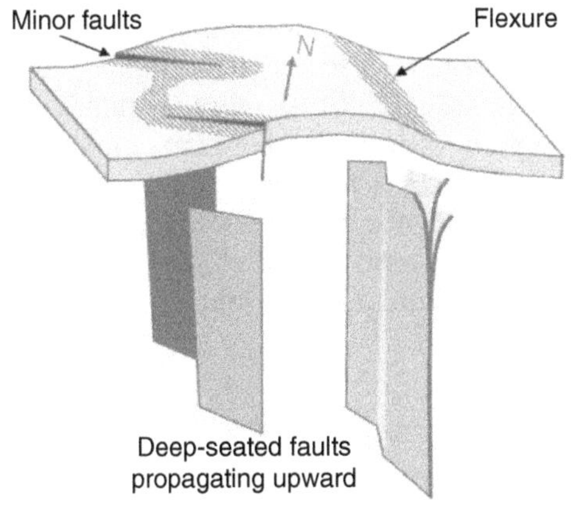

Figure 12.5 Structural setting of the In Salah area inferred from seismic data at the Carboniferous reservoir level. From *Ringrose et al.* [2009], reprinted with permission.

horizontally through the C10.2 section, the drill bit suddenly reentered the C10.3 formation, suggesting that it had passed through a steeply dipping fault with some vertical offset. The well trajectory was adjusted to return to the C10.2 formation after a few hundred meters of drilling.

12.3. PLANNED MONITORING

The monitoring program was designed around the perceived risks and the available monitoring technologies. The key risks were injection well problems, early breakthrough of carbon dioxide at the gas production wells, vertical leakage in the formation, and well leakage [*Mathieson et al.*, 2011]. Early data collection included wellhead pressure monitoring, fluid sampling, soil gas surveys conducted above each injection well, and fluid sampling from shallow aquifer water wells. Distinct perfluorocarbon tracers were added to the CO_2 injected in each well in order to fingerprint the fluids, should they be detected in other wells or at the surface. Due to initial modeling work, it was thought that because of the depth of the reservoir and its low porosity, surface-based geophysical monitoring would have difficulty detecting injection-related changes. For example, geomechanical calculations predicted surface displacements of the order of a few millimeters to a centimeter. However, it was recognized that monitoring any changes within the overlying mudstone seal and shallower Cretaceous sediments comprising the aquifer was important for assessing caprock integrity. To determine the potential cost-benefit of various monitoring tools, an initial set of 29 monitoring techniques were evaluated using a Boston Square methodology [*Mathieson et al.*, 2011]. Of the initial set, 10 types of observations were chosen to monitor the fate of the injected carbon dioxide and the integrity of the caprock. In the remainder of this chapter, we will discuss the acquisition and analysis of the data that was found to be the most important for understanding the movement of the injected carbon dioxide and changes in the caprock overlying the reservoir: well pressure and fluid sampling, interferometric synthetic aperture radar (InSAR), and repeat three-dimensional (3D) seismic and microseismic monitoring. We shall discuss the techniques roughly in the chronological order that they impacted the In Salah project. We will close with a discussion of the coupled modeling that was used to interpret much of the data.

12.3.1. Well Pressure, Flow Rates, and Fluid Sampling

Wellhead pressures, temperatures, and flow rates were measured routinely at the three injection wells (KB-501, KB-502, and KB-503) since the start of injection in 2004. The wellhead values have to be converted to bottom-hole values using a borehole flow model. *Bissell et al.* [2011] used the commercial Prosper software to estimate the bottom-hole pressures, calibrated by shut-in bottom-hole pressure measurements that were available for each of the three wells. They found that the estimated bottom-hole pressures in all three wells exceeded the formation fracture pressure (Fig. 12.6). This is not surprising, given that the original field development plan relied on the fractures in the reservoir to increase the injectivity of the wells in the low-porosity reservoir. The long reach of the three horizontal wells and the fracture-aided injectivity were assumed to be sufficient to accept the carbon dioxide at the specified flow rates.

Systematic wellhead and annulus fluid and tracer sampling programs were initiated in 2005 and 2006, respectively. Samples were generally collected every 2 months and monthly in areas of special interest. Five shallow aquifer wells (one beside each of the three injectors, one in a remote location as a control point, and one between wells KB-5 and KB-502) were drilled to augment the existing array of observation wells. In June of 2007, carbon dioxide was detected in appraisal well KB-5, located 1.3 km to the northwest of KB-502 (Fig. 12.1). KB-5 was drilled into the aquifer leg of the reservoir in 1980 and left open to the formation (not cemented). The wellhead had been vandalized and was found to be venting a small quantity, a few cubic feet per day, of carbon dioxide to the atmosphere. The perfluorocarbon tracer added to the carbon dioxide injected into KB-502 in June 2007 was detected at KB-5 in March 2008. Analysis of the pressure variations at wells KB-5 and KB-502 indicated that the wells are in pressure communication with the reservoir and are stable [*Ringrose et al.*, 2009]. A reservoir model with a high-permeability corridor of 1–4 Darcy between the wells produced a match to the breakthrough time. Well KB-5 was successfully decommissioned, and injection at KB-502 restarted in November 2009.

12.3.2. Satellite-Based Interferometric Synthetic Aperture Monitoring

The JIP involved collaboration between the commercial operators of the venture, such as BP, Sonatrach, and Statoil, and several research institutions, including the US Department of Energy's Lawrence Berkeley and Lawrence Livermore National Laboratories, NORSAR, the University of Liverpool, and others. Lawrence Berkeley Laboratory was tasked with monitoring the surface deformation over the In Salah injection sites. In mid-2007, Lawrence Berkeley Laboratory contracted with TRE of Milan, Italy, to acquire InSAR data associated with the In Salah carbon dioxide storage effort. Fortunately, satellites from the European Space Agency (ESA) had been routinely gathering data over the region, roughly every few months, for some time prior to the start of the earliest injection in 2004. TRE obtained synthetic aperture radar (SAR) scenes from the ESA archive, starting from 12 July 2003. The data were processed using a permanent scatterer algorithm [*Ferretti et al.*, 2000].

InSAR is a technique for measuring precise changes in backscattered radar waveforms over time [*Ferretti*, 2014]. The change in phase of two radar returns is of particular interest for monitoring deformation at the Earth's surface. In the permanent scatterer technique, the statistical analysis of a sequence of many radar images is used to identify objects with stable phase and amplitude characteristics. A time series for the complex return associated with each permanent scatterer is derived from the suite of radar images. Changes in the phase ($\Delta\phi$) for successive images are computed and attributed to several different contributions:

$$\Delta\phi = \Delta\varphi + \frac{4\pi}{\lambda}\Delta r + \Delta\alpha + \Delta n$$

where the first term $\Delta\varphi$ accounts for changes in the reflection characteristics of the image pixel containing the scatterer, λ is the wavelength, and Δr is the change in the sensor to target distance, the range change. The final two terms represent changes in the atmospheric and random noise contributions, respectively. Through processing techniques, it is possible to estimate the various contributions to the phase change [*Ferretti*, 2014]. The ultimate goal is a reliable estimate of the range

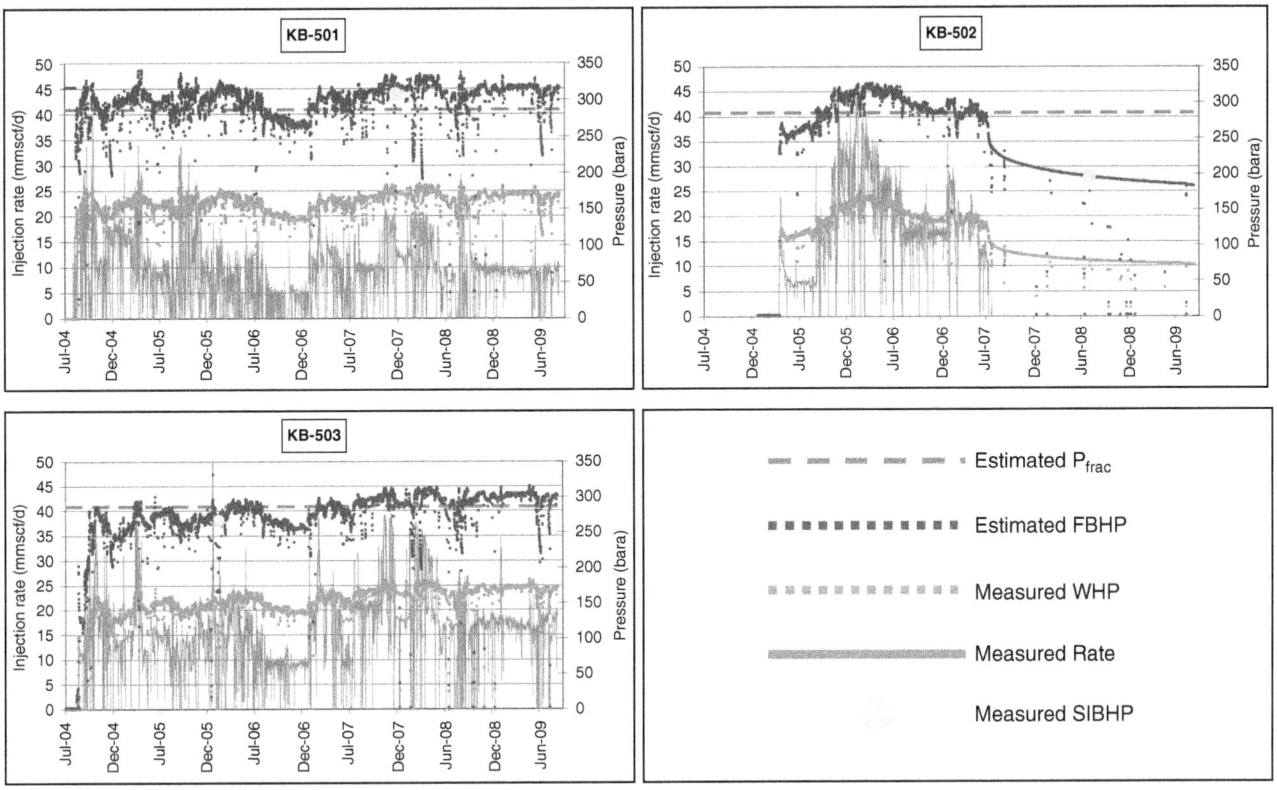

Figure 12.6 Estimated fracture pressure (P_{frac}), estimated flowing bottom-hole pressure (FBHP), measured wellhead pressure (WHP), measured flow rate, and measured shut-in bottom-hole pressure (SIBHP) for the three carbon dioxide injection wells at the In Salah site. From *Bissell et al.* [2011]. Reprinted with the permission of Elsevier.

change, as this provides the projection of the displacement of the scatterer along the line of sight to the satellite.

The permanent scatterer technique worked quite well at In Salah, with its boulder-strewn desert environment and limited amount of movable sand, and it was possible to obtain range changes for roughly 300,000 scatterers, accurate to a few millimeters in most areas (Fig. 12.7). The resulting estimates of range change displayed elongated anomalies over each injection well (KB-501, KB-502, and KB-503), indicating surface displacement of the order of 0.5 cm per year for the first few years of injection.

In their initial analysis, *Vasco et al.* [2008] focused on the range changes above KB-501 because this injector was sufficiently isolated from the others. The decrease in range associated with uplift over the injector caused the surface to move in the direction of the satellite. This movement was thought to be due to volumetric changes within the reservoir induced by the introduction of carbon dioxide. The time series of range changes from July 2003 to March 2007, some 41 images in all, were inverted for the volume change within the reservoir and used to map the propagation of pressure at depth. A strong preferential flow to the northwest was noted, and an inversion for hydraulic diffusivity within the reservoir revealed a narrow, high-permeability corridor that appeared to correlate with a fault inferred from seismic data [*Vasco et al.*, 2008].

It was subsequently noted by collaborators from Pinnacle Technologies that the double-lobed pattern of range change over KB-502 was reminiscent of the uplift over a vertical hydro-fracture [*Davis*, 1983; *Wright*, 1998]. Motivated by this suggestion, the modeling and inversion of the range change data associated with KB-502 was generalized to allow for both reservoir volume change and variable-aperture change within a vertical damage zone [*Vasco et al.*, 2010]. Aperture change is the opening of a fracture. For a fracture or damage zone, the aperture change is distributed over the many fractures comprising the zone. The large-scale properties of the damage zone, such as the azimuthal orientation and the position of the zone along the borehole, were determined by a grid search. The left panel of Figure 12.8 displays the fractional volume change in the reservoir layer; the vertical damage zone is indicated by the line of open rectangles. The distribution of aperture change over the damage zone, inferred from the range changes, is plotted in the right panel of Figure 12.8.

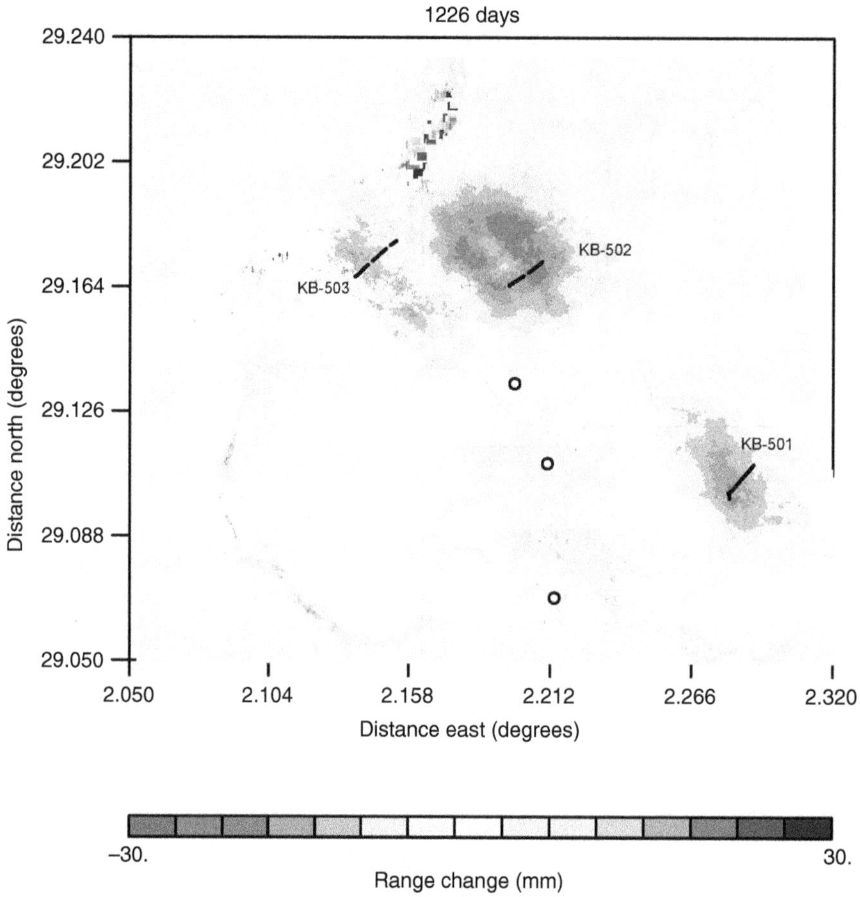

Figure 12.7 Range change after 1226 days of injection. The three injection wells are labeled, and the well traces in the reservoir are indicated by solid lines. The open circles denote three gas-producing wells.

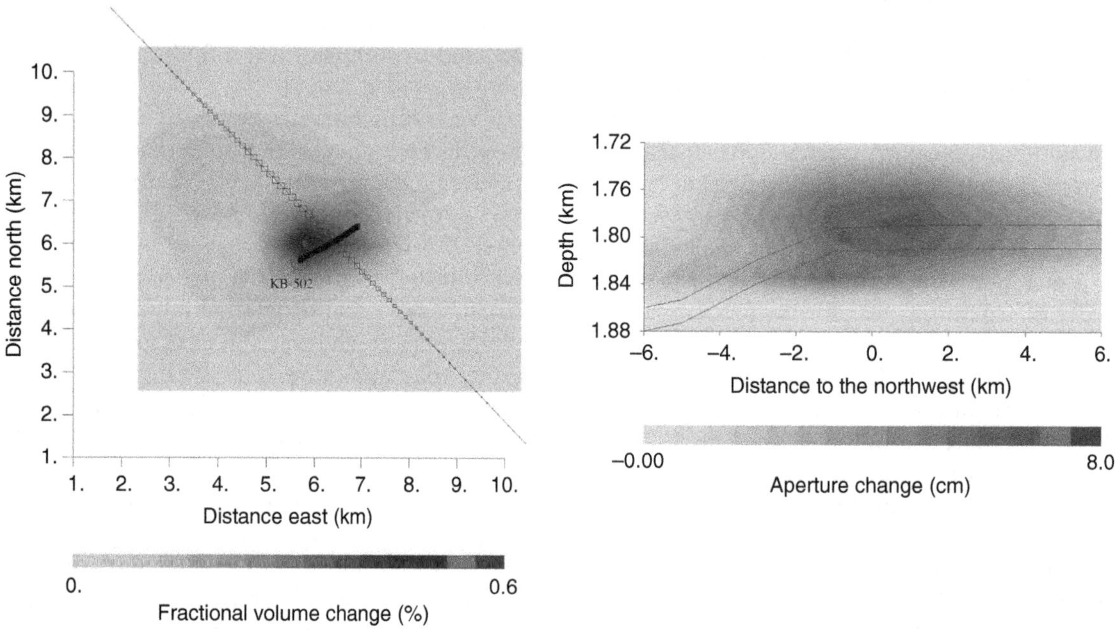

Figure 12.8 (Left panel) Fractional volume change within the reservoir interval due to the injection of carbon dioxide. The color scale indicates the volume change. The vertical damage zone is indicated by the line of open rectangles. (Right panel) Aperture change (fracture opening) distribution on the vertical damage zone that is compatible with the observed range changes. From *Vasco et al.* [2010]. Reprinted with permission from Wiley. (*See electronic version for color representation of the figure.*)

The aperture changes in the damage zone lie within 80 m of the depth of the reservoir.

Range change is the projection of the displacement vector along the line of sight to the satellite. It is possible to combine radar data from different acquisition geometries to approximate additional components of displacement [*Rucci et al.*, 2013]. Because all current SAR satellites follow near-polar orbits, every point on Earth can be imaged by two distinct acquisition geometries: one with the instrument orbiting from north to south (descending mode) and looking westward (for right-looking sensors) and the other with it moving from south to north (ascending mode) while looking eastward. To illustrate how the decomposition is performed, imagine a Cartesian reference system, where the three axes correspond to east-west (X), north-south (Y), and vertical (Z) directions. Consider the case in which two estimates of the target range change are available, obtained from both ascending and descending radar acquisitions, namely, r_a and r_d.

Given our knowledge of the satellite orbits, the line of sight of the radar antennas is known, as are the corresponding direction cosines of the velocity vectors r_a and r_d. If we wish to invert for the full 3D velocity vector, the problem is poorly posed, because there are three unknowns and only two equations. However, because the satellite orbit is almost circumpolar, the sensitivity to possible motion in the north-south direction is negligible. This allows us to rewrite the system as two equations that may be solved for the displacements in the vertical and east-west directions.

At the In Salah site, *Rucci et al.* [2013] applied the decomposition to data acquired by the ESA's Envisat satellite. Only a single time interval proved to be useful for the decomposition, from July 2004 to May 2008, so it was only possible to derive the displacements shown in Figure 12.9. The additional component of displacement provided useful information and enabled *Rucci et al.* [2013] to discriminate between source models. Specifically, it was not possible to match both components of displacement at any of the wells using models that only contained reservoir volume changes. It was possible to fit all observations using subvertical damage zones with spatially variable volume and aperture changes. In this solution, significant aperture change extends up to 100–400 m above the reservoir. For example, aperture and volume changes were required on a steeply dipping fault/fracture (damage) zone intersecting well KB-501, as shown in Figure 12.10.

Additional support for the existence of a subvertical damage zone at well KB-501 is provided by a reprocessed InSAR data set, obtained by combining the Envisat C-band data with X-band data from a newer configuration of SAR satellites. Two X-band satellite constellations, TerraSAR-X and COSMO-SkyMed, became available later in the project, and the InSAR monitoring was transitioned to them as the Envisat satellite was decommissioned. The X-band InSAR data had better temporal sampling, between 8 and 11 days, than the Envisat satellite. This allowed for better estimates of corrections to the phase estimates and routine decomposition into quasi-vertical and quasi-east-west components. The Envisat C-band data and the X-band data were merged into a common time series by TRE and provided to Lawrence Berkeley Laboratory. The range changes over well KB-501 suggest a subtle double-lobed pattern, as shown in Figure 12.11.

As noted above, it was possible to satisfy the InSAR range changes at all of the wells with models containing

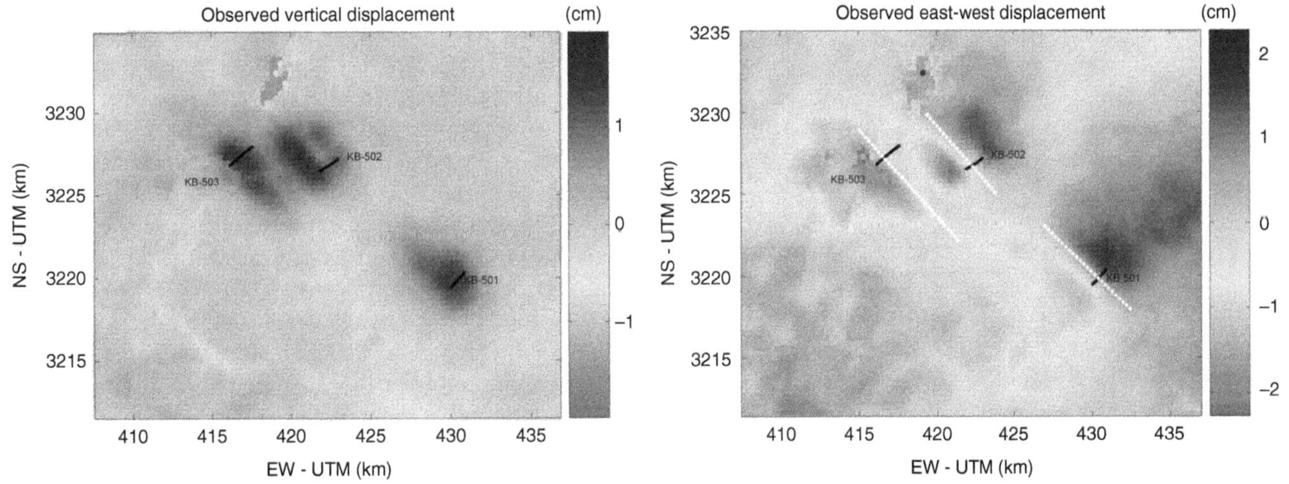

Figure 12.9 Decomposition of two orbital data sets into approximate vertical (left panel) and east-west (right panel) components. The surface projections of the damage zone traces associated with each injector are plotted as dashed white lines. From *Rucci et al.* [2013] and reprinted with permission.

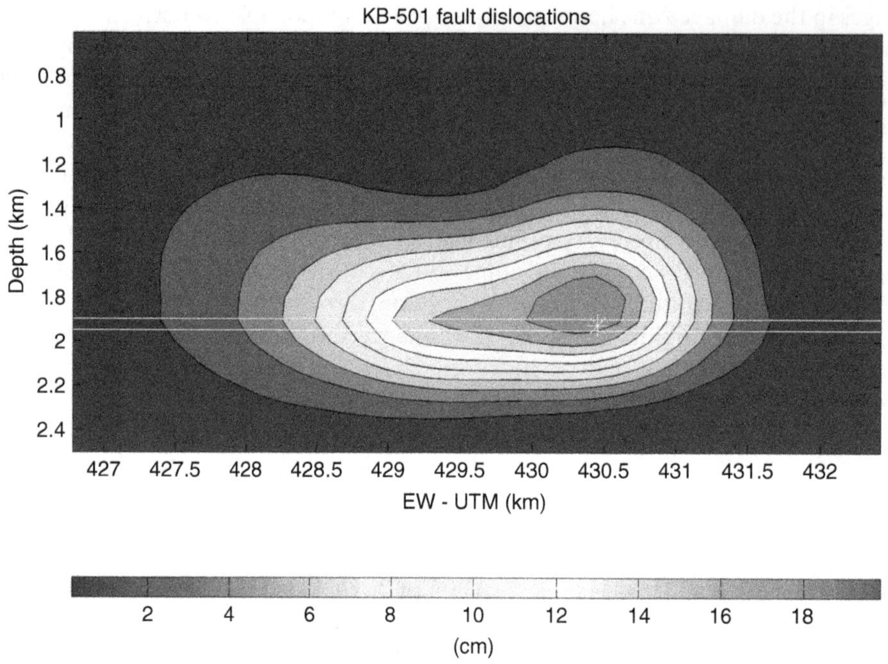

Figure 12.10 Opening (aperture change) on a steeply dipping damage zone intersecting well KB-501. The well intersection with the fracture plane is denoted by a white star. The reservoir interval is situated between the two parallel white lines. From *Rucci et al.* [2013], reprinted with permission.

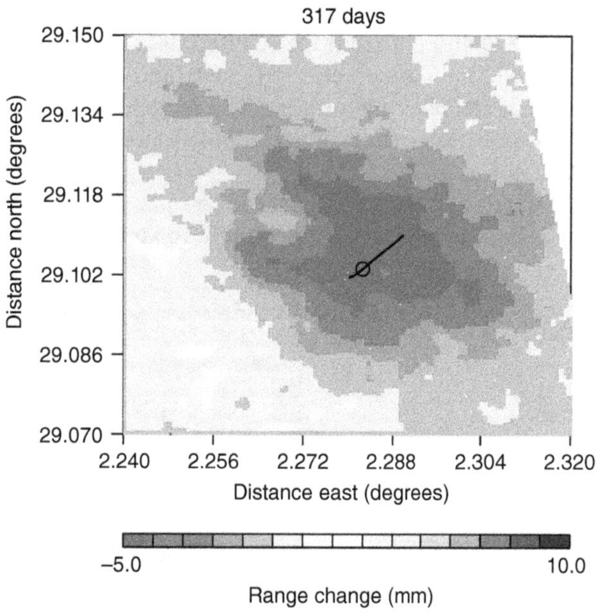

Figure 12.11 Range change above well KB-501, from a combined C-band and X-band data set.

volume and aperture changes near the reservoir interval. This conclusion proved true in all of the inversions that we ran using a layered elastic model appropriate for the In Salah region. Furthermore, as we shall see below, coupled modeling by *Morris et al.* [2011] and *Rinaldi and Rutqvist* [2013] also indicates that the range changes at KB-502 can be satisfied by reservoir volume change and changes on a vertical fault/fracture zone that only extends a few hundred meters above the reservoir (200 m in *Morris et al.*, 2011 and 350 m in *Rinaldi and Rutqvist*, 2013). However, geodetic data, such as InSAR range changes, generally have poor depth resolution, and one typically faces uniqueness issues when inverting for distributions of sources in depth. Thus, other solutions may be possible, depending upon the model parameterization and the constraints or penalty terms used in the inversion. An example of this is provided by the solution discussed in *Davis* [2011] that matches the range change over KB-502 between November 2003 and August 2009. The model consisted of four dislocations, with each dislocation specified by 10 parameters for a total of 40 degrees of freedom. The individual dislocation models are specified by their height, width, and length, azimuth, dip, centroid, and slip or dislocation vector. *Davis* [2011] found that 80% of the volume change lies near the injection depth. The remainder is in a vertical dislocation some distance above the reservoir in the C20 mudstone. Two assumptions mentioned in the paper are that the effects of layering can be neglected and that there is only tensile opening and/or shear slip and no volumetric expansion. The study does not consider a model of distributed sources as used in *Rucci et al.* [2013]. An earlier study by *Davis and Marsic* [2010] does find that layering is important for obtaining a reasonable depth estimate. They fit a dual dislocation model to range change data from KB-501. The model consisted of a vertical fracture zone extending upward from a horizontal tensile dislocation. In that case, a homogeneous half-space overburden model introduces an upward shift in the depth of the source of about 500 m (from 1800 m for the layered models to 1300 m for a homogeneous model).

A study by *Ramirez and Foxall* [2014] sought to quantify the uncertainty associated with estimates of the depth of penetration of a subvertical fracture zone into the caprock. Using a stochastic Markov chain Monte Carlo inversion method, *Ramirez and Foxall* [2014] derive a suite of models, representing the posterior distribution. The Markov chain Monte Carlo solution is a probability distribution function defined over the model space. The distribution may be used to estimate the probability that a certain model attribute is present in the set of solutions. For example, Figure 12.12 is a plot from *Ramirez and Foxall* [2014] that indicates the probability that a pressure-induced disturbance extends to different depths above the reservoir. They find that there is a high probability that injection-related changes propagated around 150 m above the reservoir and there is less than 50% probability that such changes reached the Hot Shale at a depth of 1500 m. As in the work by *Davis* [2011], this study used a homogeneous half-space to calculate the range changes.

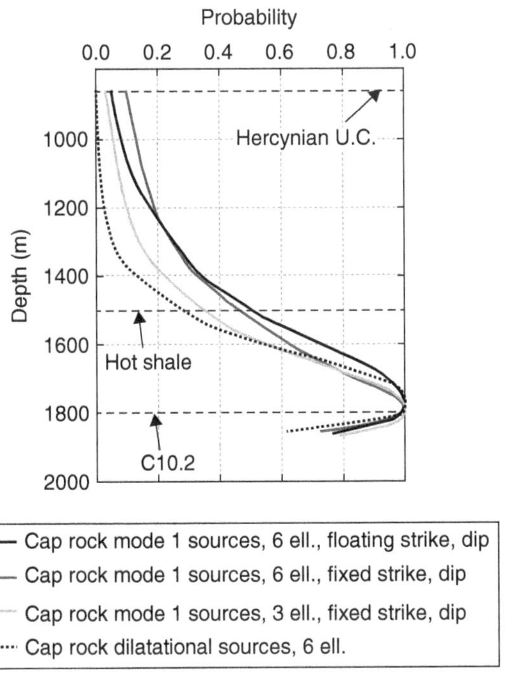

Figure 12.12 Probability that the pressure disturbance associated with the injection of carbon dioxide in KB-502 extends to different distances from the reservoir layer. From *Ramirez and Foxall* [2014]. Reprinted with permission of Elsevier.

12.3.3. Active Seismic Surveys

Due to the expense, logistical difficulties, and the expectation that the signal would be poor, time-lapse surface seismic was not a key component of the initial monitoring program. For example, there was no baseline survey specifically designed for time-lapse monitoring. However, following the detection of carbon dioxide in well KB-5 and the observed deformation associated with all three injectors evident in the InSAR observations, it was decided to conduct a full 3D surface seismic survey in May 2009. For both logistical reasons, and in order to maximize the possibility of extracting time-lapse changes, the general layout of the field program and the orientation of the seismic lines followed that of an earlier regional survey from 1997. To limit the cost of the survey, it only covered the northern and central region of the field and did not extend down to well KB-501 (Fig. 12.13). The first description of the seismic data and the earliest interpretation were presented by *Gibson-Poole and Raikes* [2010]. One issue that limited the utility of the stacked and processed data was the assumption of isotropic velocities in the overlying shale. At the very least, the velocities in the shale are transversely isotropic. Assuming isotropy introduces errors that are particularly severe for data at large offsets.

Gibson-Poole and Raikes [2010] were the first to note the spatial correlation between a sharp linear feature, evident in several seismic time horizons, and the range changes derived from the InSAR data. In particular, the trough defined by the seismic time delays aligns with the trough between the two lobes of range change anomaly at KB-502. The seismic velocity push-down is thought to be due to saturation and pressure changes associated with the injection of carbon dioxide. The sharp, parallel, linear boundaries and the width of the zone suggest a fault, fracture, or damage zone that focused the flow of the carbon dioxide. A similar linear feature was also observed to intersect well KB-503 and to extend several kilometers to the southeast [*White et al.*, 2014]. One unusual aspect of that feature was that the greatest amount of push-down was offset several kilometers up-structure from the injector, suggesting that this feature may be related to the migration of carbon dioxide up-dip, rather than due to fracturing and pressure change around the injector. The velocity push-down was not large enough to be interpreted in the earlier 1997 seismic survey.

One question that has bearing on the integrity of the caprock is the vertical extent of the velocity push-down and other features, such as amplitude changes, that might be due to the migration of carbon dioxide. In order to examine this in more detail, *Zhang et al.* [2015] calculated a number of post-stack attributes from the 2009 seismic data. They first examined the variation in the push-down

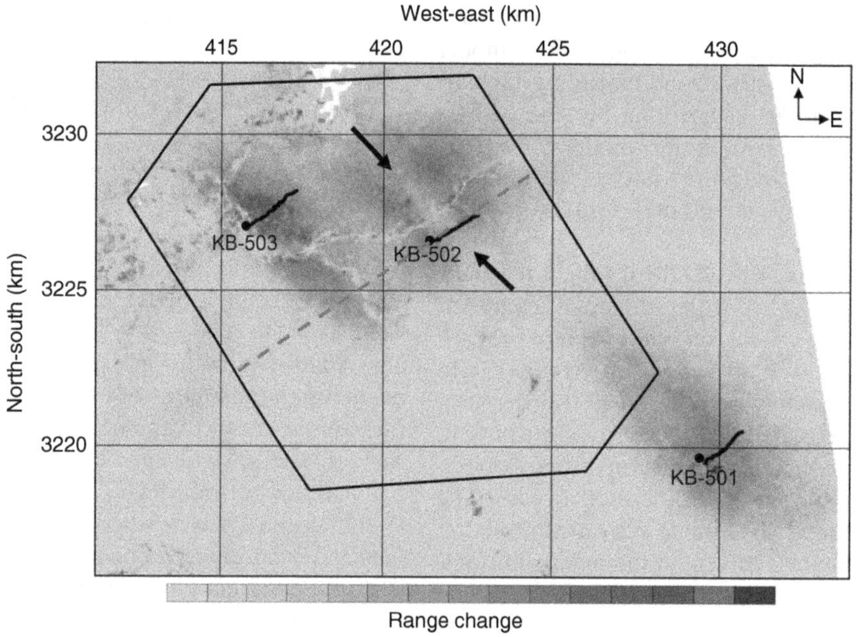

Figure 12.13 The extent of the 2009 seismic survey (solid polygon,) plotted with respect to the three CO_2 injectors and the observed range changes. The arrows denote ends of a linear push-down anomaly observed in the seismic time horizons associated with the C10.2, C20.1, and C20.4 formations, first noted by *Gibson-Poole and Raikes* [2010]. From *Zhang et al.* [2015], reprinted with permission. The colored background indicates the InSAR range change, varying from yellow (0.0 mm) to blue (30.0 mm). (*See electronic version for color representation of the figure.*)

with depth, considering the two-way time delays for the reservoir layer (C10.2) and for two seismic surfaces near the bottom of the shale seal, C20.1 and C20.4. The amplitude of the push-down appears to diminish at the top of layer C20.4 (Fig. 12.14).

The attributes examine by *Zhang et al. [*2015] were associated with the curvature of the seismic reflection amplitudes along particular time horizons. Graphical definitions of positive and negative curvature are given in Figure 12.15. The main point is that the trough of a linear depression is characterized by negative curvature, while at the edges of the trough the curvature is positive as indicated in panel (d) in Figure 12.15.

Curvature attributes are computed for three of the more prominent interfaces in the lower caprock (C20.1 and C20.4) and reservoir (C10.2) [*Zhang et al.*, 2015]. The set of points with the most negative curvature define a long linear trough (Fig. 12.16). The positive curvature attributes delineate two parallel lines bounding the edges of the trough. Furthermore, these features are relatively robust in the reservoir layer (C10.2) and for layer C20.1 but weaken considerably in layer C20.4. Based upon these results, *Zhang et al.* [2015] estimate that the damage zone is approximately 3500 m long, 80 m wide, and 350 m high. The trough demarcated by the lines of positive curvature correlates with the trough in the double-lobed pattern of range change (Fig. 12.17), similar to the findings of *Gibson-Poole and Raikes* [2010]. Note that there is a hint of a trough below the range change associated with injection at KB-503 (Fig. 12.17).

In spite of the limitations of the seismic survey such as the lack of a high-quality baseline survey for time-lapse analysis and the absence of corrections or accounting for anisotropy in the shale overburden, it provided valuable information on the structure at depth that may be controlling the movement of the carbon dioxide.

12.3.4. Microseismicity

Microseismic monitoring was an essential element of the planned program at In Salah. Unfortunately, the borehole array of seismometers near well KB-502 was not installed before the start of injection in 2004. In addition, problems with the installation further delayed the effort [*Oye et al.*, 2012; *Goertz-Allmann et al.*, 2014]. In spite of the setbacks in the initiation of the monitoring effort, more than 5000 microseismic events were detected between August 2009 and June 2011 (Fig. 12.18). Identification of a seismic event relied upon a master event waveform cross-correlation method [*Shearer*, 1997; *Waldhauser and Ellsworth*, 2000]. As is evident in Figure 12.18, most of the events can be correlated with the injection of carbon dioxide.

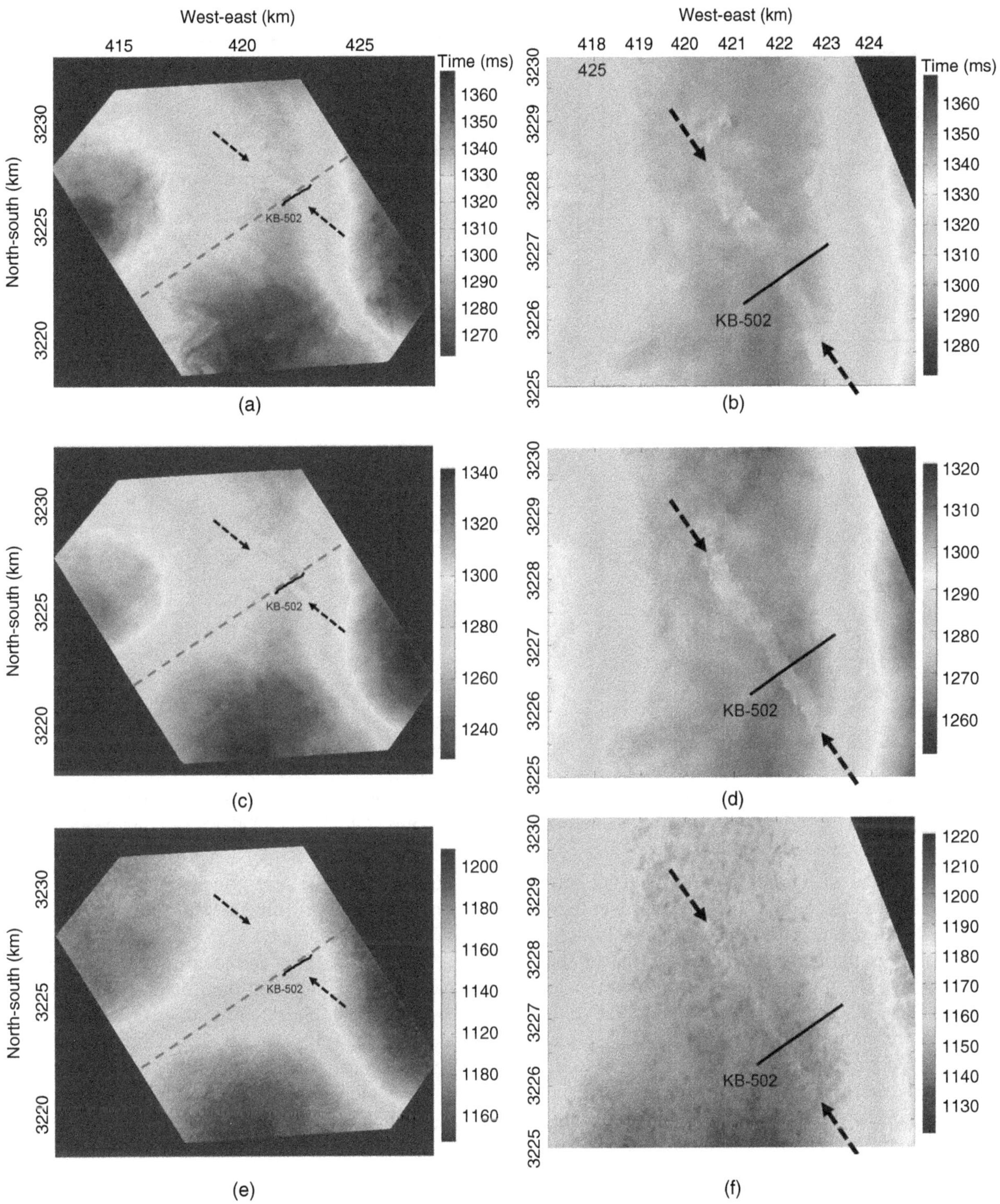

Figure 12.14 (Left column) Two-way travel time surfaces for the tops of formations C10.2 (a), C20.1 (c), and C20.4 (e). (Right column) Detailed view of the two-way travel time variations for these respective layers. From *Zhang et al.* [2015], reprinted with permission. The arrows identify the location of the linear push-down feature. The figure was modified from the original.

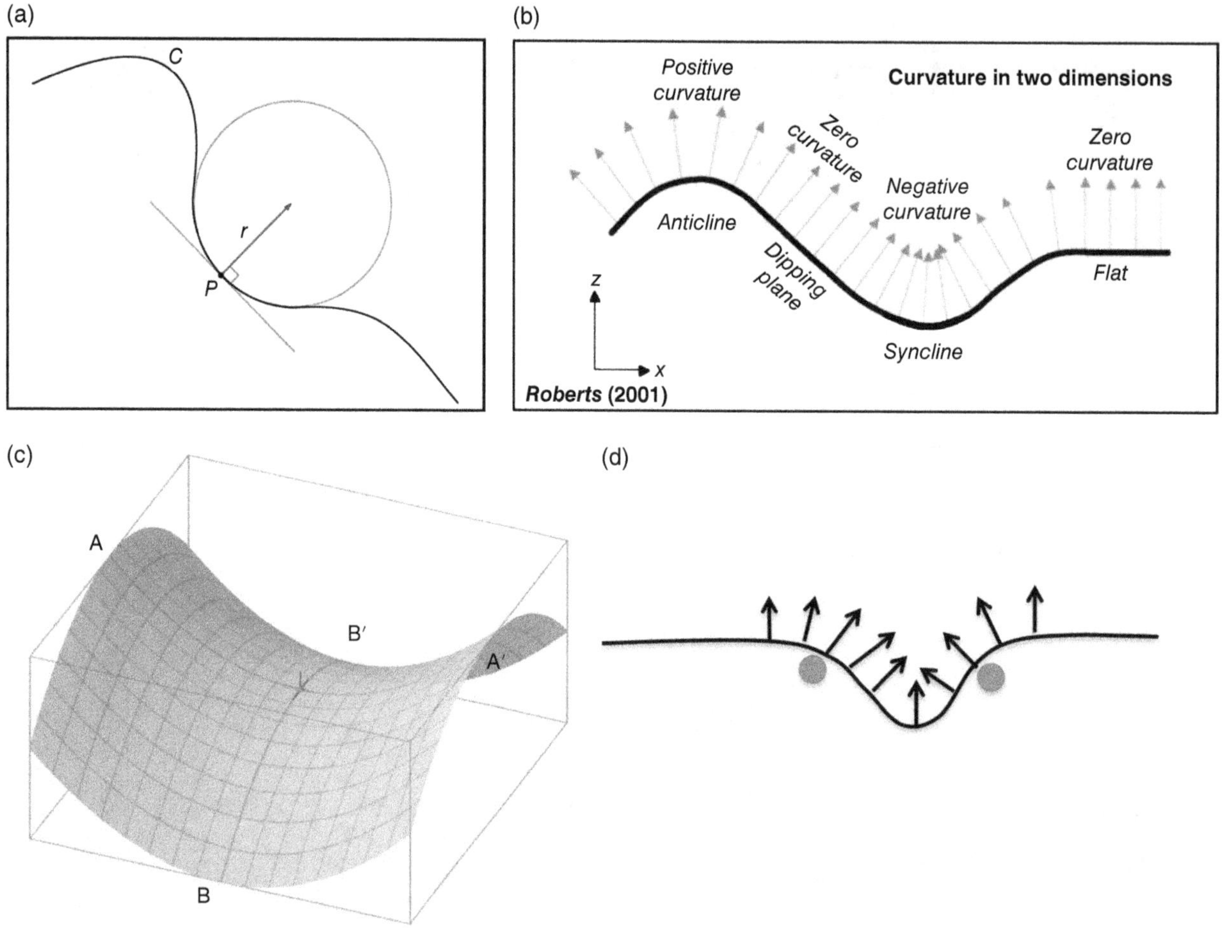

Figure 12.15 A graphical definition of curvature attributes. The curvature is $1/r$, where r is the radius of a sphere that provides the best local approximation to the surface. From *Zhang et al.* [2015], reprinted with permission.

In particular, there is a notable correspondence between seismic activity and variations in the pressure at the well. As the well pressure approaches the estimated fracturing pressure, there is an observable increase in the number of seismic events. The fracturing pressure was estimated in the same fashion as *Bissell et al.* [2011], based upon changes in well injectivity. The estimate can be seen in Figure 12.19, a plot of injection pressure against injection rate. Prior to fracturing, the pressure changes rather rapidly with changes in injection rate. After fracturing, the permeability and storage around the well have increased, and the rate of change in pressure with injection rate decreases. The break in slope in Figure 12.19 indicates the initiation of fracturing. The corresponding pressure is the fracture pressure, estimated to be 155 bars.

The injection generated sufficient seismic activity to allow for both characterization and monitoring. Due to instrument damage, likely from a constriction in the wellbore [*T. Daley*, personal communication], leading to problems including failed sensors and strong electrical noise, only a few sensors were active, and just the uppermost geophones provided reliable three-component data that could be used for analysis. This instrument was determined to be at a depth of approximately 80 m in a shallow borehole (KB-601) roughly 500 m to the northeast of injection well KB-502 (Fig. 12.20). Further details about the data processing and monitoring setup may be found in *Goertz-Allmann et al.* [2014].

Waveform data from a single receiver can only provide rough event locations with uncertainties given by *Oye et al.* [2012]. Event clusters are defined by combining distance estimates from the S-P arrival times with azimuth and inclination estimates from *P*-wave polarization analysis. Waveforms within a cluster are highly correlated (see some examples in Fig. 12.21) and probably have a similar source location and mechanism. An analysis of event clusters allows one to address questions of fluid migration, principal stress orientation, and seal integrity [*Goertz-Allmann et al.*, 2014]. Figures 12.20 and 12.21 show the locations of two of these clusters, A and B, together with the ray paths for the different phases. Shorter S-P arrival times of around 0.7 s indicate that

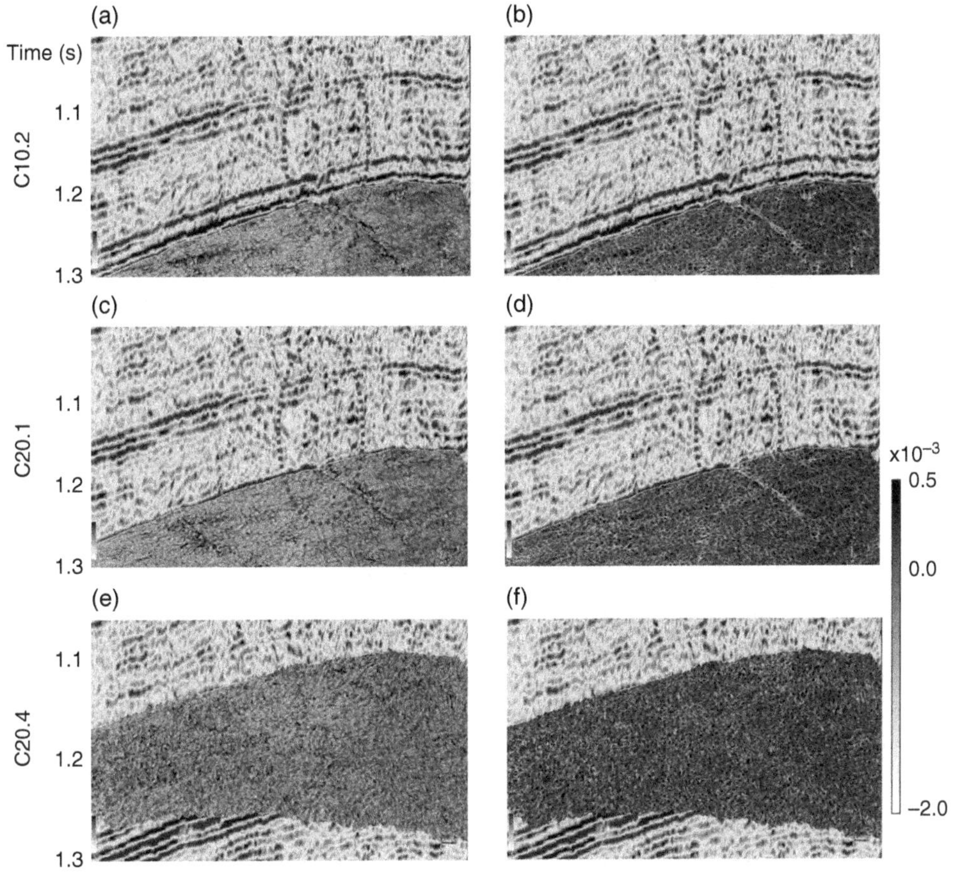

Figure 12.16 (Left panels) The foreground (gray scale) displays the most positive curvature for C10.2, C20.1, and C20.4, while the background (color scale) is a plot of the seismic reflection data. Note that the lower panels are for formations at shallower depths. (Right panels) The most negative curvature for C10.2, C20.1, and C20.4. The shading scale for the curvature measures in the foreground is shown at the right side of the figure: lighter shades indicate more negative curvature, and darker shades denote more positive curvature. See Figure 12.15 for an illustration of the curvature attributes. The red ellipses surround the areas of push-down, providing an indication of the vertical extent of the zone. From *Zhang et al.* [2015] and reprinted with permission. (*See electronic version for color representation of the figure.*)

cluster B is located closer to the receiver, and hence closer to injection point then is cluster A, with an S-P time of about 1 s. Both clusters are located to the northwest of the receiver, with azimuths between 285° and 300°.

Event depths are crucial in determining if fracturing is occurring in the caprock above the reservoir. Even with a sufficient receiver aperture for accurate event locations, the focal depths typically suffer from large uncertainties. In particular, it is not possible to determine event focal depths accurately with only one receiver. The incorporation of multiple arrivals, in addition to direct P and S observations, coupled with 3D ray tracing, can improve estimates of the depth of an event. An additional seismic phase is observed on the vertical component of waveforms from cluster A, between the direct P and S arrivals (see Fig. 12.21; some approximately 500 ms after the P-wave first onset). This phase appears to be a conversion from an S to a P wave at the Hercynian unconformity (see Fig. 12.2). The velocity model obtained from the full surface seismic reflection survey [*Mathieson et al.*, 2010; *Gemmer et al.*, 2012] is used for ray tracing, based upon a wavefront construction method [*Vinje et al.*, 1993]. If the observed phase is a conversion from S to P at the Hercynian unconformity, the result suggests event depths for cluster A of about 1.7 km, which is above the top of the reservoir, within the lower caprock. Note that the events cannot originate from shallower depths, assuming the model and the phase identifications are correct. Also, no correlation is observed between the occurrence of events in cluster A and the wellhead data at the injection well. Microseismic activity in clusters A' and B are highly correlated with the injection rate in well KB-502. In particular, during two time periods where the fracture pressure seems to have been exceeded, there is a burst of microseismic activity (Fig. 12.18).

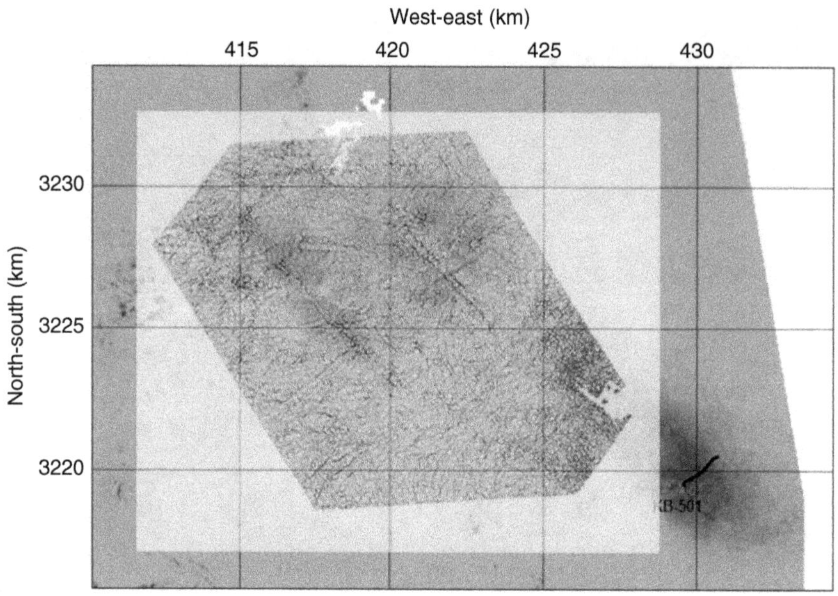

Figure 12.17 The textured surface indicates the most positive curvature attribute for amplitudes at the top of the C20.1 formation, a layer just above the reservoir. The plot simulates a surface illuminated from a distance light source. A color plot, where the colors vary according to the cumulative range change associated with 5 years of injection, underlies the textured attribute surface. The superposition of these figures indicates the correspondence of the seismic linear push-down zone and the InSAR double-lobed pattern, first pointed out by *Gibson-Poole and Raikes* (2010). From *Zhang et al.* [2015], reprinted with permission. (*See electronic version for color representation of the figure.*)

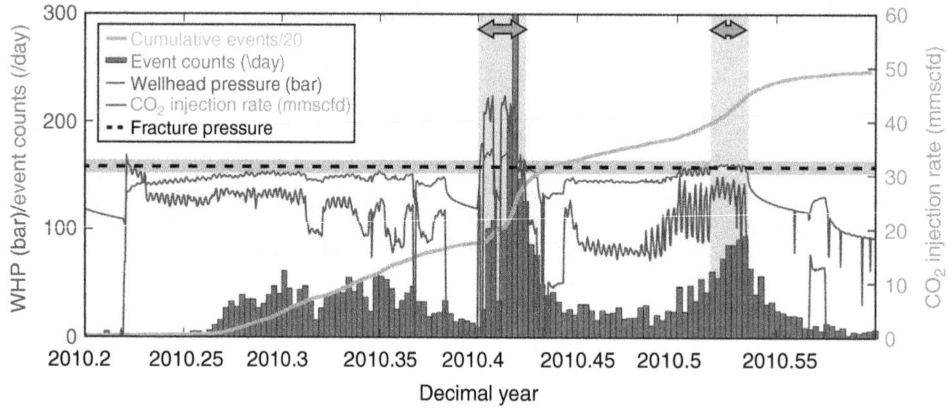

Figure 12.18 Histogram of detected microseismic events (red) compared to wellhead pressure (blue) and injection rate (green) during a selected time period. The orange line shows the cumulative number of events. The black dashed line and gray band shows the estimated fracture pressure (corrected to the wellhead) of 155 bar ± 5 bar. During two time periods, the fracture pressure was exceeded (yellow shaded areas/arrows). From *Goertz-Allmann et al.* [2014] and reprinted with permission. (*See electronic version for color representation of the figure.*)

A clear correlation between seismic activity and injection and well pressure data at In Salah shows that microseismic monitoring does provide a cost effective approach for monitoring seal integrity. The event depth is particularly critical for assessing the likelihood of fracturing within the caprock. Despite the limitations in the receiver setup, with only one reliable three-component station, one can roughly constrain event cluster locations by combining polarization angles, S-P travel time differences, and identification of converted phases (Fig. 12.22). An event cluster (A) is most likely associated with preexisting tectonic faults within the lower caprock above the reservoir. Events in this cluster may be due to the transfer of stress and not the direct result of the movement of injected carbon dioxide or the associated fluid pressure changes. An interpretation of the surface seismic data identified

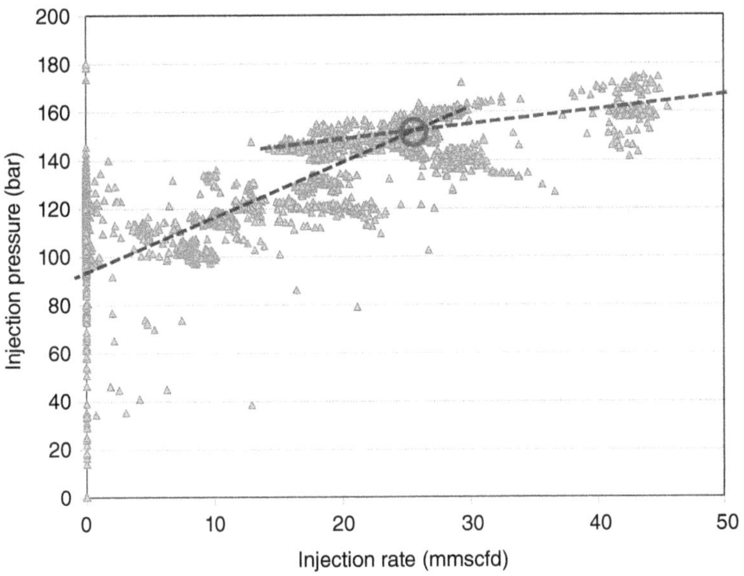

Figure 12.19 Injection pressure versus injection rate at the wellhead of KB-502. The dashed lines indicate fits to the data sets. The intersection of the lines indicates the fracture pressure. The red circle signifies the estimated fracture pressure at the wellhead, about 155 bars, equivalent to a downhole pressure of about 277 bars at a depth of 1800 m. From *Goertz-Allmann et al.* [2014] and reprinted with permission. (*See electronic version for color representation of the figure.*)

the end of a fault/fracture (damage) zone close to this cluster [*Goertz-Allmann et al.*, 2014]. These *stress-induced* events may either occur naturally or are triggered by the injected-induced deformation of the region [*Vasco et al.*, 2010], as the events occur at the edge of the uplifted region, where the displacement gradient is greatest. The two other event clusters most likely indicate the extent of either the carbon dioxide plume or its pressure front and may be associated with the opening of partially tensile fractures, or related shear fractures, due to high pore pressure. This interpretation is further corroborated by a geomechanical analysis of the wellhead pressure and fluid flow data, suggesting that the formation fracture pressure was exceeded during certain times of the injection (as in Figs. 12.6 and 12.18), associated with an increase in seismicity [*Goertz-Allmann et al.*, 2014]. Note that recent work by *Stork et al.* [2014, 2015] presents an additional analysis of the microseismic data from In Salah. Of the 9506 events that they detected, the vast majority of them are located at or below the injection interval. Only 11 less well constrained events are located above the injection interval, and they suggest that these are due to stress transfer rather than the migration of injected carbon dioxide.

12.3.5. Coupled Modeling

Coupled numerical simulations provided insight and furthered our understanding of the geomechanical changes within the caprock due to the injection of carbon dioxide at In Salah. Early coupled full field simulations were conducted using the commercial STARS simulator [*Bissell et al.*, 2011]. Initially, the deformation was attributed to reservoir volume change, but this was later modified to include a set of faults/fractures using strongly anisotropic and time-varying permeabilities. The model was history matched to the injectors flowing bottom-hole pressure, the shut-in bottom-hole pressure, and the observed InSAR range changes. *Rutqvist et al.* [2010, 2011] also carried out some of the first coupled simulations using TOUGH-FLAC [*Rutqvist et al.*, 2002; *Rutqvist*, 2011]. The package couples the TOUGH2 code for modeling multiphase, multicomponent fluid flow in a porous medium [*Pruess et al.*, 2011] with the FLAC3D simulator for geomechanical modeling. We shall discuss some of the more recent results of this simulator below, in relation to the question of the degree of penetration of a possible fracture zone into the mudstone caprock seal. Coupled modeling by *Morris et al.* [2011] highlighted the role that existing faults play in the migration of the injected carbon dioxide and in the geomechanical changes associated with the storage effort. Using both forward modeling and sensitivity studies, they note the contributions of both fault/fracture and reservoir flow to the observed double-lobed pattern of range change above injector KB-502. Work by *Gemmer et al.* [2012] highlights the importance of the elastic model of the overburden in generating the correct estimates of range change. Recent modeling and history matching by *Shi et al.* [2012] explore the role of carbon dioxide migration

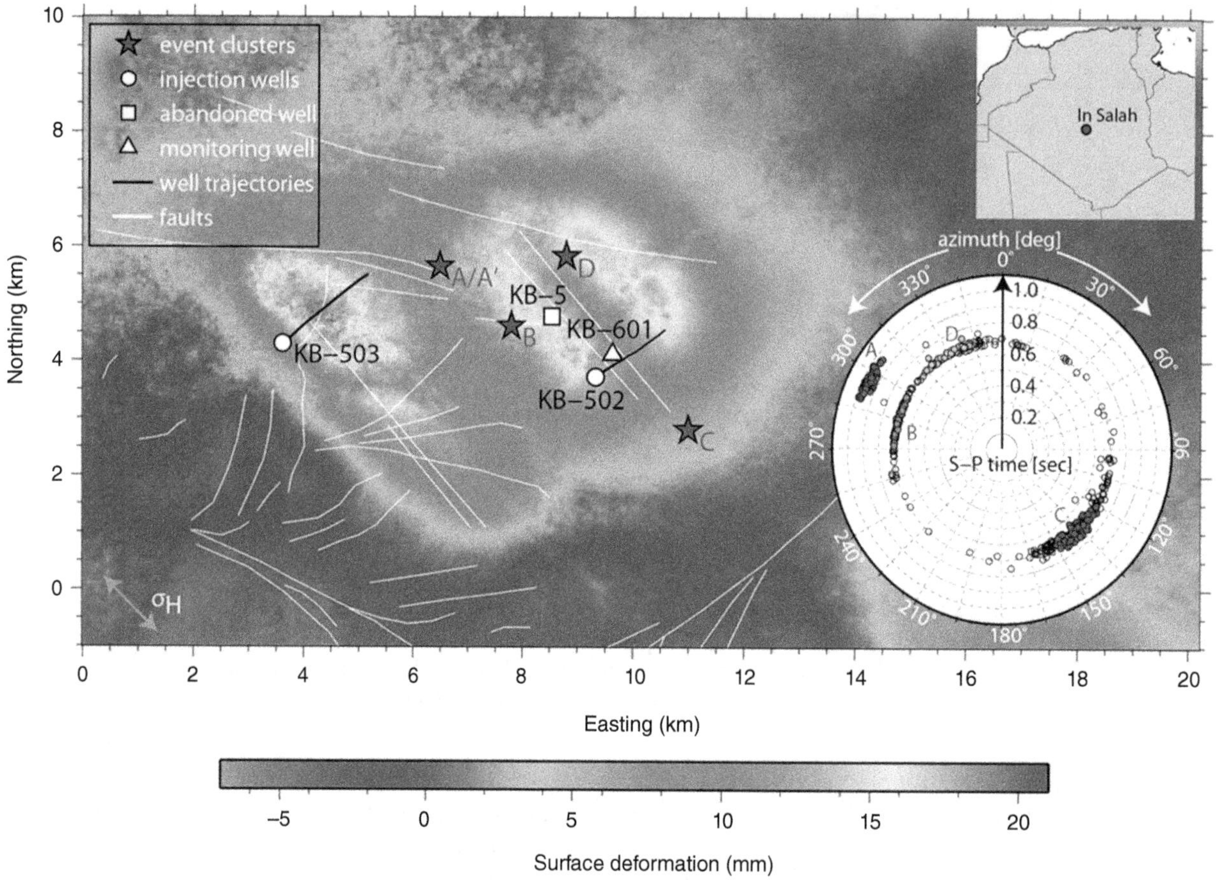

Figure 12.20 Krechba field, In Salah, Algeria, showing the location of two horizontal CO_2 injection wells (KB-502 and KB-503), one abandoned test well in the area (KB-5), and the vertical microseismic monitoring well (KB-601). The approximate locations of the major event clusters (A–D) are denoted by the red stars. Known faults are shown down to a depth of 2000 m (white lines). The direction of the maximum horizontal stress σ_H is indicated. The background color indicates the observed surface deformation. Polar coordinate plot displays the difference in azimuth between event clusters for all events (circles). The radius indicates the S-P travel time. The separation of event clusters A, B, C, and D is indicated by the colored circles defined by varying azimuth and S-P travel time. From *Goertz-Allmann et al.* [2014], reprinted with permission. (*See electronic version for color representation of the figure.*)

and pressure propagation in a fracture or fault damage zone in the lower caprock. These authors believe that the growth of the hypothesized fracture at KB-502 is confined to the main reservoir interval (C10.2) during the early injection period (before March, 2006) and then extends into the lower caprock after that. As with earlier studies by *Bissell et al.* [2011] and *Rutqvist et al.* [2011], the opening of the fracture zone is implemented using a time-varying transmissibility. Note that the entire 4 km fault/fracture zone in their model has a uniform transmissibility and the dynamics of lateral fracture propagation and growth is not captured.

All of the coupled modeling to date supports the notion of focused flow within a fault/fracture damage zone intersecting well KB-502. Furthermore, the models have restricted the zone to the reservoir and lower portion of the caprock. Recent work by *Rinaldi and Rutqvist* [2013] explored the issue of fault/fracture height and its implications for both surface displacement and the saturation distribution at depth.

In this work, *Rinaldi and Rutqvist* [2013] modified the model of *Rutqvist et al.* [2011] to include a more detailed fracture/fault zone (Fig. 12.23). The model consists of four main layers, listed in Table 12.1. The mechanical properties are close to estimates from a well log analysis by *Gemmer et al.* [2012]. With the exception of the reservoir permeability and bulk modulus, all of the hydraulic and mechanical parameters are held constant. The model contains a fracture zone that intersects the reservoir and extends for 3500 m along the northwest (x) direction, penetrating 350 m upward into the caprock (Fig. 12.23b). The fracture zone is based upon the analysis of the surface seismic data, noted above, that revealed a linear feature appearing to extend more than a hundred meters

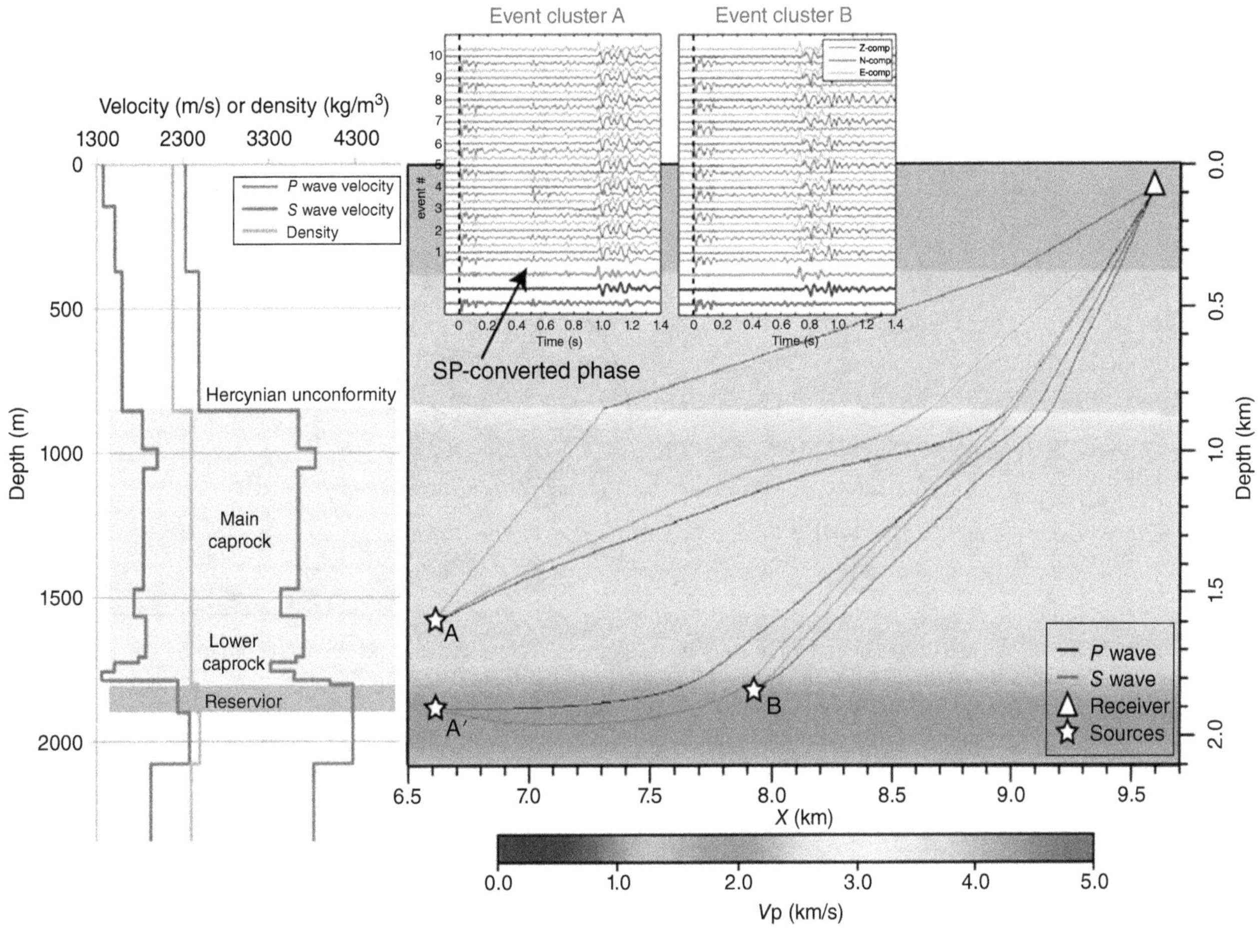

Figure 12.21 Potential event cluster locations A, A′, and B (stars) together with direct P-wave (black) and S-wave (red) ray path. The ray path of an S-to-P converted phase at the Hercynian unconformity is also shown for cluster A. The background color denotes the P-wave velocity. Insets show some example microseismic event waveforms for clusters A and B. A layered velocity and density model, extracted from the isotropic 3D stacking velocity model at the location of the injection borehole, is shown on the left together with the main layers (reservoir = red shading, lower caprock = green shading, and main caprock = blue shading). From *Goertz-Allmann et al.* [2014], reprinted with permission. (*See electronic version for color representation of the figure.*)

above the reservoir [*Gibson-Poole and Raikes*, 2010]. In *Rinaldi and Rutqvist* [2013], the fracture zone was assumed to reactivate after some months of injection. In the simulations shown here, the reservoir is subjected to a Mohr-Coulomb failure criterion, and its permeability and bulk modulus depend upon the stress condition. A similar criterion applies in the fracture zone, although we assign a larger fixed permeability to it upon reactivation and there is no stress-dependent permeability within the fracture zone. All others, layers behave in a poroelastic fashion.

As noted by *Iding and Ringrose* [2010], the reservoir is highly fractured. As a consequence, if the maximum and minimum principal stresses in the reservoir satisfy the Mohr-Coulomb failure criterion for a given friction angle (φ_{res}), then the reservoir permeability and bulk modulus are subject to change as a function of the mean effective stress. A relationship between fracture aperture and normal effective stress was derived by *Liu and Rutqvist* [2013]. Accounting for a cubic law, and referring *Liu and Rutqvist's* [2013] relationship to the initial state of stress and the mean effective stress, a stress-dependent permeability can be derived [*Rinaldi et al.*, 2014]:

$$\frac{\kappa_{hm}}{\kappa_i} = \left(\frac{b}{b_i}\right)^3 = \frac{\gamma_e + \gamma_t e^{\frac{\sigma'_m}{K_{t,f}}}}{\gamma_e + \gamma_t e^{\frac{\sigma'_{m,i}}{K_{t,f}}}}$$

where b and b_i are the current and initial apertures and κ_{hm} and κ_i are the permeabilities at the current and initial state of stress, respectively. $K_{t,f}$ refers to the bulk modulus of the reservoir fractures, and σ'_m is the effective mean

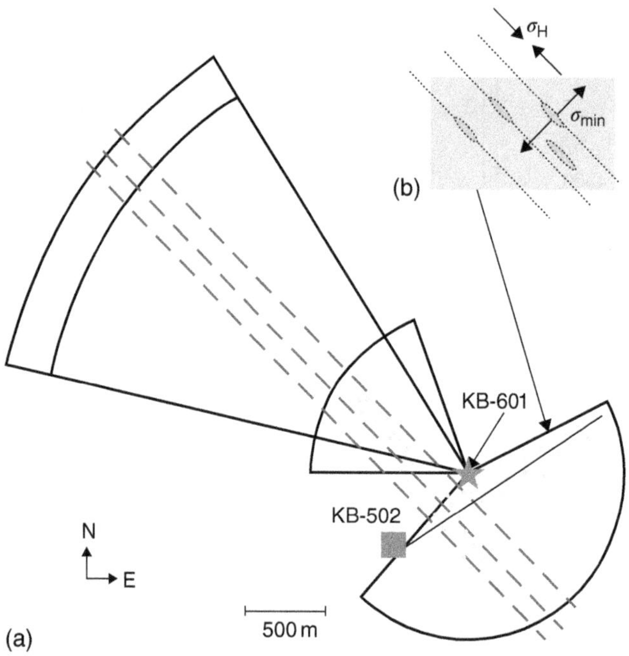

Figure 12.22 (a) Illustration of the potential source areas (green sectors) for seismic events recorded by the single receiver in well KB-601 (blue star). The dashed lines indicate the location of the damage zone observed in the seismic velocity push-down. (b) Blown-up image of events aligned with the minimum horizontal stress. Taken from *Stork et al.* [2015]. Reprinted with permission of Elsevier. Their work also highlights the variability in the locations, particularly the variation in the azimuth of the events from the functioning receiver (Fig. 12.22). (*See electronic version for color representation of the figure.*)

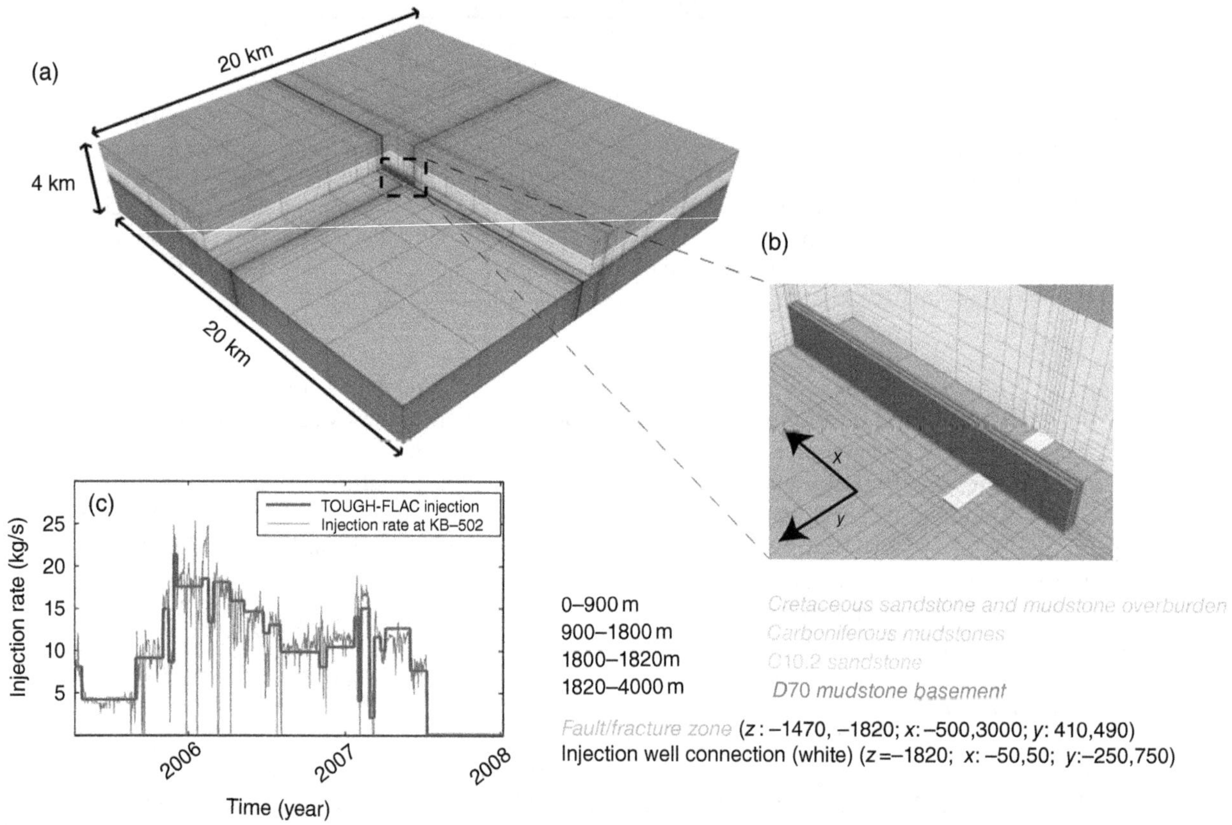

Figure 12.23 (a) Computational domain. (b) Enlargement over the injection well and the fracture zone. (c) Injection rate at KB-502 injection well (red line) and injection rate for TOUGH-FLAC model. From *Rinaldi and Rutqvist* [2013]. Reprinted with permission of Elsevier. (*See electronic version for color representation of the figure.*)

Table 12.1 Hydrogeological Properties.

	Depth (m)	ϕ_0 (–)	κ_0 (m²)	Friction
Upper aquifer	0–900	0.1	10^{-12}	—
Caprock	900–1800	0.01	10^{-21}	—
Reservoir	1800–1820	0.17	$\mathbf{0.8\times10^{-14}}$	27.87
Basement	>1820	0.01	10^{-19}	—

Rinaldi and Rutqvist [2013]. Reprinted with permission of Elsevier.
Stress-dependent parameters in bold.

Table 12.2 Deep Fracture Zone Properties.

Bulk modulus	E_x	0.14 GPa
	E_y	0.51 GPa
	E_z	1.16 GPa
Poisson's ratio	ν_{yz}	0.25
	ν_{xy}, ν_{xz}	0.18
Porosity	ϕ	1%
Permeability	κ	10^{-13} m
Friction	φ_{frac}	30.55

stress. γ_e and γ_t represent the unstressed volume fraction for the hard (matrix and pores) and soft (fracture) parts of a body rock, respectively. The stress-dependent permeability produces an excellent match between observed and simulated bottom-hole pressure (Fig. 12.24a). The permeability variations calculated by the model agree with those proposed by *Rinaldi and Rutqvist* [2013], although the new permeability changes are focused in a region few hundred meters around the injection well and are not distributed over the entire reservoir as in the previous model. The permeability changes may be related to the opening of small fractures caused by the injection and any associated effective stress changes within the reservoir [*Rinaldi and Rutqvist*, 2013]. Major differences are found after shut-in, probably related to phase or temperature changes within the borehole that the model cannot reproduce. In order to compare the results with the observed range change, consider the calculated line-of-sight displacements in Figure 12.24b and c. The comparison between simulated and observed line-of-sight surface displacements is along two profiles located at 500 and 1700 m, respectively, northwest and parallel to the injection well. The match is relatively good, although it overestimates uplift in regions far from the double lobe. The differences are likely due to the coarse nature of the mesh far from the injection well. Figure 12.24d shows the resulting transient evolution of the line-of-sight displacement at a single point, located above the injection well at about 600 m from the fracture zone along the well direction. The simulated result is in excellent agreement with the observed variation.

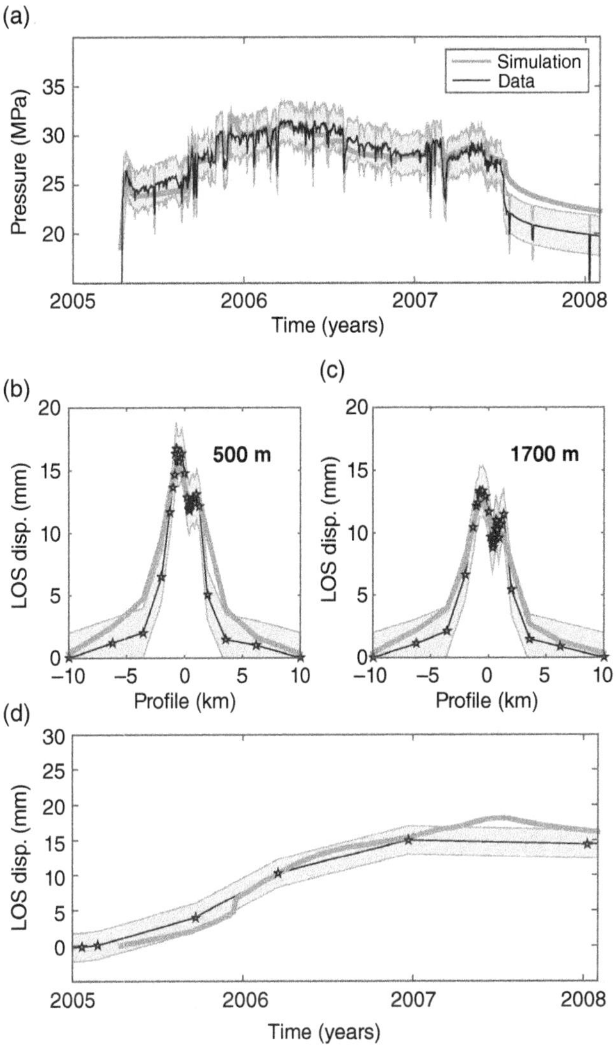

Figure 12.24 Comparison between simulation and observed data at KB-502. (a) Evolution of bottom-hole pressure. (b) Profile of ground uplift at 500 m after 618 days. (c) Profile of ground uplift at 1700 m after 618 days. (d) Temporal evolution of line-of-sight (LOS) displacement. From *Rinaldi and Rutqvist* [2013]. Reprinted with permission of Elsevier.

The results presented here are a generalization of the sensitivity analysis conducted by *Rinaldi and Rutqvist* [2013]. Specifically, their simulations are modified to account for the stress-dependent permeability in the reservoir. At issue is the height of the fracture zone above the reservoir and how it influences the surface deformation. Results of the sensitivity analysis for the line-of-sight displacements along the two profiles are shown in Figure 12.25a and b. As observed by *Rinaldi and Rutqvist* [2013], the surface displacement does not vary simply in proportion to the fracture zone height. If the fracture is confined within the caprock and does not penetrate into the shallow aquifer, then the surface deformation does seem to increase in correspondence to the height.

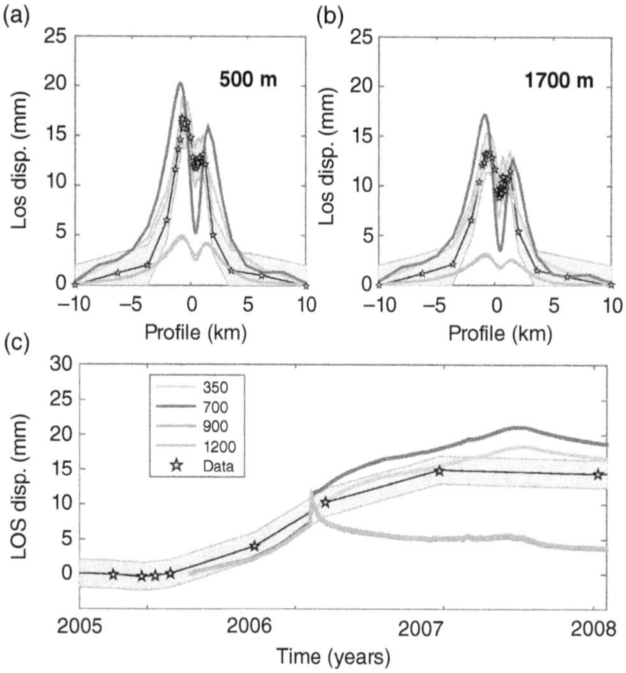

Figure 12.25 Sensitivity analysis of the height of fracture zone. (a and b) Effects of the fracture zone height (from the reservoir depth) along profile after 618 days at 500 and 1700 m from injection well, respectively. (c) Effect of fracture zone height on temporal evolution of ground uplift. From *Rinaldi and Rutqvist* [2013]. Reprinted with permission of Elsevier.

However, when the fracture zone penetrates through the caprock and extends into the aquifer (900 and 1200 m cases, green and cyan line, respectively, in Fig. 12.25), the displacement along the two profiles falls precipitously to a value less than 10 and less than 5 mm, respectively. These differences can also be seen in the temporal evolution of displacement for a point located above the injection well (Fig. 12.25c). For fracture zones extending 350 and 700 m above the reservoir, the displacement grows over time and then levels off and finally decays over time. As before, the higher fracture zone (700 m) has a larger displacement than the 300 m high fracture zone. However, for fracture zones that extend across the entire overburden seal, heights of 900 and 1200 m, there is a gradual initial increase from 2005 to 2006, followed by a rapid decrease as the caprock is penetrated and the aquifer is reached (Fig. 12.25c).

As explained by *Rinaldi and Rutqvist* [2013], this trend is not observed in the field data, suggesting that the fracture zone is unlikely to extend all the way through the caprock.

Rinaldi and Rutqvist [2013] also analyzed the distribution of carbon dioxide within the fracture zone. If the linear feature observed by the 3D seismic survey was the result of push-down caused by carbon dioxide replacing water, then a depth-limited fracture zone explains this behavior. Indeed, they show that only a fracture zone that is limited in depth within the caprock can produce a plume of carbon dioxide that spreads laterally for thousands of meters and is concentrated near the reservoir interval (Fig. 12.26).

12.4. CONCLUSIONS

The monitoring methods described in this chapter provided insight, and some unexpected results, regarding the migration of injected carbon dioxide at the In Salah gas development site. Significant resources were devoted to monitoring over the course of the project. The planned monitoring effort was extensive in scope, and the evaluation of techniques by the JIP was relatively systematic. One major limitation was a delay in the initiation of most monitoring efforts until after the start of injection and the lack of baseline data prior to injection. Luckily, InSAR data were routinely gathered, both prior and during injection, as part of a global monitoring effort by the ESA. However, the delay did compromise the microseismic monitoring efforts and meant that seismic events associated with the first few years of injection, and critical fracture zone growth, were simply not recorded. Also, problems in the installation of the borehole microseismic monitoring network severely limited its usefulness. The

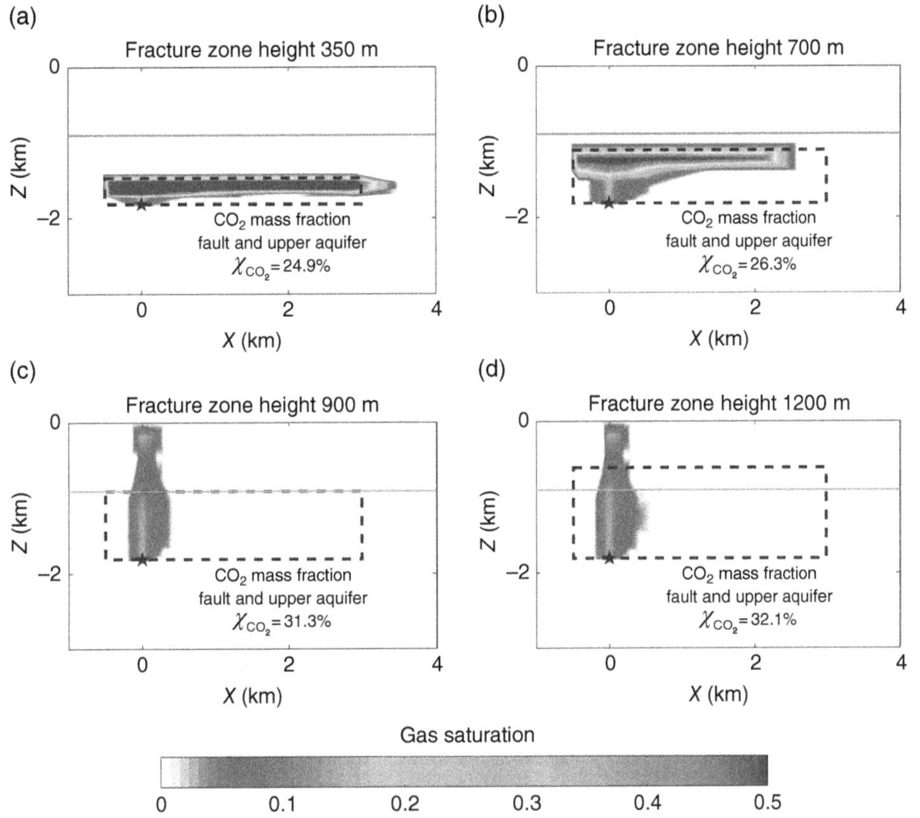

Figure 12.26 Carbon dioxide plume for fracture zones with differing heights. The fracture zone size is indicated by the dashed rectangle. The gas saturation is indicated by the color scale. The red line indicates the depth extent of the upper aquifer. From *Rinaldi and Rutqvist* [2013]. Reprinted with permission of Elsevier. (*See electronic version for color representation of the figure.*)

lack of baseline data impacted all subsequent developments, such as efforts to extract time-lapse seismic changes. Furthermore, the lack of baseline data may have contributed to the early termination of injection at In Salah, in an effort to act conservatively in the face of the resulting uncertainty.

As the monitoring progressed, and observations were gathered and interpreted, our understanding of the factors controlling the movement of the injected carbon dioxide has changed significantly. It was hypothesized that rather than flowing uniformly through the reservoir, the flow was strongly influenced by fault/fracture zones, later referred to as damage zones. Such features were not prominent in the existing, pre-injection seismic survey from 1997. Rather, the damage zones were inferred from interpretations of the InSAR data [*Vasco et al.*, 2008, 2010; *Ringrose et al.*, 2009; *Morris et al.*, 2011; *Rucci et al.*, 2013]. Utilizing the temporal resolution provided by the InSAR data, it was possible to identify long narrow fast flow paths [*Vasco et al.*, 2008]. In addition, the double-lobed pattern over well KB-502 indicated that a narrow tensile fault or fracture zone governed the pressure propagation away from the well. A subsequent seismic survey from 2009, several years after the start of the injection in 2004, displayed long, linear push-down features. These features are likely due to the accumulation of carbon dioxide, injected in wells KB-502 and KB-503, in the damage zones. The seismic push-down delineated a narrow zone with well-defined parallel sides (Fig. 12.14) that extended for over four kilometers from KB-502 [*Gibson-Poole and Raikes*, 2010; *Zhang et al.*, 2015]. No significant offset in layering was observed across these linear damage zones, making them difficult to detect in the seismic reflection data. Note that there is precedence for swarms of fractures, known as fracture corridors, that are difficult to detect seismically and yet control fluid flow [*Al-Mulhim et al.*, 2010]. Fluid flow data from a monitoring well (KB-5) provided confirmation that there was a fast flow path from well KB-502 in a north to northwest direction, with an early breakthrough of carbon dioxide at well KB-5. Some upward flow out of the reservoir was required to fit the InSAR data [*Vasco et al.*, 2010; *Rucci et al.*, 2013]. However, coupled modeling [*Rutqvist et al.*, 2011; *Shi et al.*, 2012; *Rinaldi and Rutqvist*, 2013] and a seismic attribute analysis [*Zhang et al.*, 2015] indicated the limited vertical extent of the

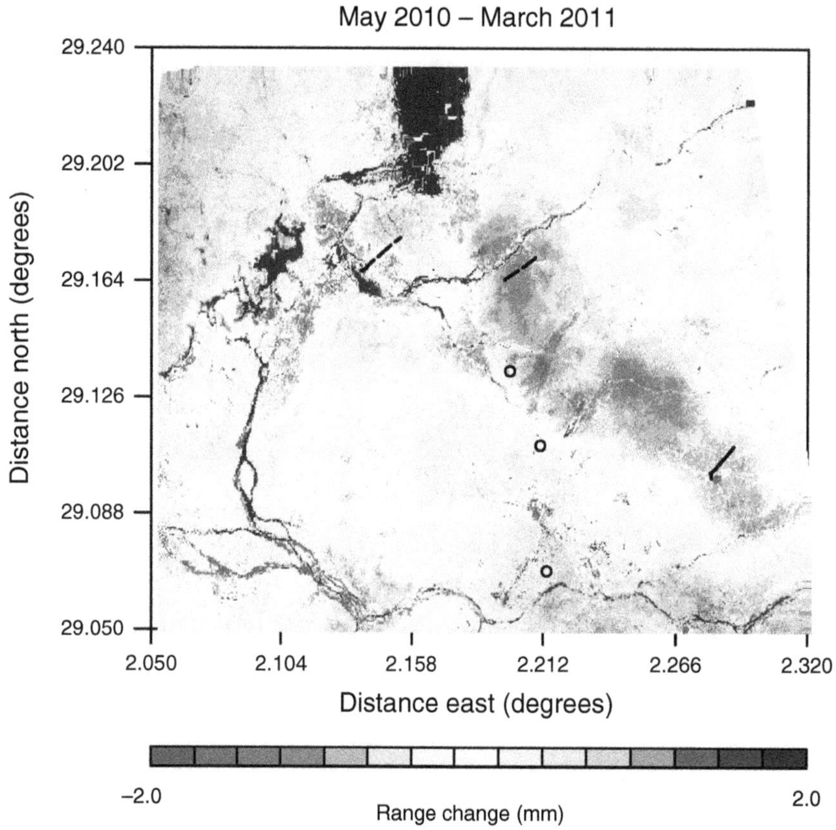

Figure 12.27 Range change between May 2010 and March 2011. The three injection wells are indicated by the black lines. A decrease in range change is most likely caused by uplift due to the injection of carbon dioxide. The effects of changes in the dried stream beds (wadiis) can be seen in this figure, as can the changes due to a lake bed at the northernmost edge of the figure.

fracture zone and that a breach of the caprock is unlikely. A stochastic inversion of the InSAR also suggests that the fracture zone is most likely restricted to the lower caprock [*Ramirez and Foxall*, 2014]. As detailed in this chapter, most models satisfy the observations with fluid intrusion into the caprock that is limited to a few hundred meters.

In spite of the progress in understanding the factors controlling flow and deformation at the In Salah storage site, key questions remain. The exact location of the injected carbon dioxide has not been imaged, and the upper boundary of the plume has not been constrained precisely. The techniques that could possibly have provided such information, microseismic monitoring and time-lapse seismic reflection surveys, were sufficiently compromised by the lack of baseline data that they could not help in this regard. There are also questions regarding the nature of the damage zones intersecting the wells. Are they preexisting fracture/fault zones that were subsequently reopened during the injection? Or are they simply fractures created during the injection that propagated for several kilometers from the wells? Detailed numerical modeling of the injection that includes the dynamic generation of fractures might be able to help answer this question. Also, longer-term monitoring can indicate if the pressure changes continue to follow a preferential flow path far from the well. For example, in Figure 12.27, we plot the range change associated with late-stage injection at In Salah. A long linear northwesterly trend of range decrease, associated with uplift, connects well KB-501 and KB-502, suggesting that a large-scale feature might be controlling the flow. Long-term monitoring can also help answer questions regarding the ultimate fate of the injected carbon dioxide and the ultimate stability of the storage complex. Some issues also remain concerning other monitoring technologies that were not considered at In Salah. For example, time-lapse electromagnetic or gravity field measurements might be of use in detecting the upward movement of carbon dioxide and monitoring the integrity of the seal but were not available at In Salah.

The In Salah gas field was an early test of industrial-scale CO_2 capture and geologic storage. The monitoring program was an integral part of the project and also an early example of large-scale deployment with cost-benefit analysis influencing choices. Overall, the monitoring program and subsequent modeling efforts were a success in understanding the flow and transport of injected CO_2 at the In Salah gas field. The importance of fracture flow became clear during the course of the project. With hindsight, we see the monitoring program would have benefited from more baseline data and from prioritizing fracture flow monitoring to increase effort, for example, in microseismic monitoring. Importantly, the In Salah project has demonstrated the need for adaptability in a monitoring program so that data acquired during injection is used to update the monitoring priorities as well as the geologic and reservoir models. Overall, the In Salah project stands as an important example in the use of monitoring and modeling to understand caprock integrity in geologic storage of CO_2.

ACKNOWLEDGMENTS

Portions of this work were performed under the auspices of the GEOSEQ project for the Assistant Secretary for Fossil Energy, Office of Coal and Power Systems, through the National Energy Technology Laboratory of the US Department of Energy by Lawrence Berkeley National Laboratory under contract number DE-AC02-05CH11231 and by Lawrence Livermore National Laboratory under Contract DE-AC52-07NA27344. All authors would like to thank the members of the Joint Industry Project and the Joint Venture, in particular BP, Sonatrach, and Statoil, for their cooperation and support at every stage of this work. Special thanks go out to Iain Wright, Allen Mathieson, Sue Raikes, John Midgley, Catherine Gibson-Poole, David Roberts, Stephen Cawley, Phil Ringrose, Lykke Gemmer, and the other members of the Joint Industry Project.

REFERENCES

Al-Mulhim, W. A., Al-Ajmi, F., Al-Shehab, M., and Pham, T. R. (2010). Best practices in Khurais Complex Development. *SPE/DGS Annual Technical symposium and Exhibition*, Al-Khobar, Saudi Arabia, 4–7 April 2010.

Armitage, P. J., Faulkner, D. R., Worden, R. H., Aplin, A. C., Butcher, A. R., and Iliffe, J. (2011). Experimental measurement of, and controls on, permeability and permeability anisotropy of caprocks from the CO_2 storage project at the Krechba field, Algeria, *Journal of Geophysical Research*, *116*, B12, doi:10.1029/2011JB008385.

Bissell, R. C., Vasco, D. W., Atbi, M., Hamdani, M., Okwelegbe, M., and Goldwater, M. H. (2011). A full field simulation of In Salah gas production and CO_2 storage project using coupled geo-mechanical and thermal fluid flow simulator, *Energy Procedia*, *4*, 3290–3297.

Chadwick, R. A., Williams, G. A., Williams, J. D. O., and Noy, D. J. (2012). Measuring pressure performance of a large saline aquifer during industrial-scale CO_2 injection: the Utsira sand, Norwegian North Sea, *International Journal of Greenhouse Gas Control*, *10*, 374–388.

Coward, M. P., and Ries, A. C. (2003). Tectonic development of north African basins, in *Petroleum Geology of Africa: New Themes and Developing Technologies*, Arthur, T. J., MacGregor, D. S., and Cameron, N. R. (eds.), Geological Society, London, Special Publications, *207*, 61–83

Davis, P. M. (1983). Surface deformation associated with a dipping hydrofracture, *Journal of Geophysical Research*, *88*, 5826–5834.

Davis, E. J. (2011). Interpretation of CO_2-sequestration-induced surface deformation over KB-502 at Krechba, Algeria, paper presented at the *SPE Annual Technical Conference and Exhibition*, Denver, CO, 30 October–2 November 2 2011.

Davis, E. J., and Marsic, S. D. (2010). Interpretation of CO_2-sequestration-induced surface deformation at Krechba, Algeria, paper presented at the *SPE International Conference of CO_2 Capture, Storage, and Utilization*, New Orleans, LA, 10–12 November 2010.

Ferretti, A. (2014). *Satellite InSAR Data – Reservoir Monitoring from Space*. EAGE Publications, The Netherlands.

Ferretti, A., Prati, C., and Rocca, F. (2000). Nonlinear subsidence rate estimation using permanent scatterers in differential SAR interferometry, *IEEE Transactions on Geoscience and Remote Sensing*, *38*(5), 2202–2212.

Gemmer, L., Hansen, O., Iding, M., Leary, S., and Ringrose, P. (2012). Geomechanical response to CO_2 injection at Krechba, In Salah, Algeria, *First Break*, *30*(6), 79–84.

Gibson-Poole, C. M., and Raikes, S. (2010). Enhanced understanding of CO_2 storage at Krechba from 3D seismic. *Proceedings of the 9th Annual Conference on Carbon Capture and Sequestration*, Pittsburgh, PA, 10–13 May 2010.

Goertz-Allmann, B. P., Kühn, D., Oye, V., Bohloli, B., and Aker, E. (2014). Combining high microseismic activity and geomechanics to interpret cap rock integrity at the In Salah, CO_2 storage site, *Geophysical Journal International*,*198*, 447–461, doi:10.1093/gji/ggu010.

Guiraud, R., Bosworth, W., Thierry, J., and Delplanque, A. (2005). Phanerozoic geological evolution of northern and Central Africa: an overview, *Journal of African Earth Sciences*, *43*, 83–143.

Iding, M., and Ringrose, P. (2010). Evaluating the impact of fractures on the performance of the In Salah CO_2 storage site,

International Journal of Greenhouse Gas Control, *4*, 242–248, doi: 10.1016/j.ijggc.2009.10.016.

Kaven, J. O., Hickman, S. H., McGarr, A. F., Walter, S., and Ellsworth, W. L. (2014). Seismic monitoring at the Decatur, IL, CO_2 sequestration demonstration site, *Energy Procedia*, *63*, 4264–4272.

Liu, H. H., and Rutqvist, J. (2013). Coupled hydro-mechanical processes with multiphase flow in a dual-continuum system: Formulation and an application, *Rock Mechanics and Rock Engineering*, *46*, 1103–1112.

Mathieson, A., Midgley, J., Dodds, K., Wright, I., Ringrose, P., and Saoul, P. (2010). CO_2 sequestration monitoring and verification technologies applied at Krechba, Algeria, *The Leading Edge*, *29*(2), 216–222.

Mathieson, A., Midgley, J., Wright, I., Saoula, N., and Ringrose, P. (2011). In Salah CO_2 storage JIP: CO_2 sequestration monitoring and verification technologies applied at Krechba, Algeria, *Energy Procedia*, *4*, 3596–3603.

Morris, J. P., Hao, Y., Foxall, W., and McNab, W. (2011). A study of injection-induced mechanical deformation at the In Salah CO_2 storage project, *International Journal of Greenhouse Gas Control*, *5*, 270–280.

Oye, V., Aker, E., Daley, T. M., Kuhn, D., Bohloli, B., and Korneev, V. (2012). Microseismic monitoring and interpretation of injection data from the In Salah CO_2 storage site (Krechba), Algeria, *Energy Procedia*, *37*, 4191–4198.

Pruess, K., Oldenburg, C., and Moridis, G. (1999). Tough2 Users Guide, Version 2.0, LBNL Report 43134, Berkeley, CA, USA.

Ramirez, A., and Foxall, W. (2014). Stochastic inversion of InSAR data to assess the probability of pressure penetration into the lower caprock at In Salah, *International Journal of Greenhouse Gas Control*, *27*, 42–58.

Rinaldi, A. P., and Rutqvist, J. (2013). Modeling of deep fracture zone opening and transient ground surface uplift at KB-502 CO2 injection well, In Salah, Algeria. *International Journal of Greenhouse Gas Control*, *12*, 155–167.

Rinaldi, A. P., Rutqvist, J., Finsterle, S., and Liu, H. H. (2014). Forward and inverse modeling of ground surface uplift at In Salah, Algeria. *Proceedings of the 48th US Rock Mechanics and Geomechanics Symposium*, Minneapolis, MN, 1–4 June 2014.

Ringrose, P., Atbi, M., Mason, D., Espinassous, M., Myhrer, O., Iding, M., Mathieson, A., and Wright, I. (2009). Plume development around well KB-502 at the In Salah CO_2 storage site, *First Break*, *27*, 85–89.

Rucci, A., Vasco, D. W., and Novali, F. (2013). Monitoring the geologic storage of carbon dioxide using multicomponent SAR interferometry, *Geophysical Journal International*, *193*(1), 197–208.

Rutqvist, J. (2011). Status of TOUGH-FLAC simulator and recent applications related to coupled fluid flow and crustal deformations, *Computational Geoscience*, *37*, 739–750

Rutqvist, J., Wu, Y. S., Tsang, C.-F., and Bodvarsson, G. (2002). A modeling approach for analysis of coupled multiphase fluid flow, heat transfer, and deformation in fractured porous rock, *International Journal of Rock Mechanics and Mining Sciences*, *39*, 429–442.

Rutqvist, J., Vasco, D. W., and Myer, L. (2010). Coupled reservoir-geomechanical analysis of CO_2 injection and ground deformations at In Salah, Algeria, *International Journal of Greenhouse Gas Control*, *4*, 225–230.

Rutqvist, J., Liu, H. H., Vasco, D. W., Pan, L, Kappler, K., and Majer, E. (2011). Coupled non-isothermal, multiphase fluid flow, and geomechanical modeling of ground surface deformations and potential for induced micro-seismicity at the In Salah CO_2 storage operation. *Energy Procedia*, *4*, 3542–3549.

Shearer, P. M. (1997). Improving local earthquake locations using the L1 norm and waveform cross correlation: application to the Whittier narrows, California aftershock sequence, *Journal of Geophysical Research*, *102*, 8269–8283.

Shi, J.-Q., Sinayuc, C., Durucan, S., and Korre, A. (2012). Assessment of carbon dioxide plume behavior within the storage reservoir and the lower caprock around the KB-502 injection well at In Salah, *International Journal of Greenhouse Gas Control*, *7*, 115–126.

Stork, A. L., Verdon, J. P., and Kendall, J.-M. (2014). Assessing the effect of velocity model accuracy on microseismic interpretation at the In Salah carbon capture and storage site, *Energy Procedia*, *63*, 4385–4393.

Stork, A. L., Verdon, J. P., and Kendall, J.-M. (2015). The microseismic response at the In Salah carbon capture and storage (CCS) site, *International Journal of Greenhouse Gas Control*, *32*, 159–171.

Vasco, D. W., Ferretti, A., and Novali, F. (2008). Estimating permeability from quasi-static deformation: temporal variations and arrival time inversion, *Geophysics*, *73*, O37–O52, doi:10.1190/1.2978164.

Vasco, D. W., Rucci, A., Ferretti, A., Novali, F., Bissell, R. C., Ringrose, P. S., Mathieson, A. S., and Wright, I. W. (2010). Satellite-based measurements of surface deformation reveal fluid flow associated with the geological storage of carbon dioxide, *Geophysical Research Letters*, *37*, L03303, 1–5, doi:10.1029/2009GL041544.

Vinje, V., Iversen, E., and Gjøystdal, H. (1993). Traveltime and amplitude estimation using wavefront construction, *Geophysics*, *58*, 1157–1166.

Waldhauser, F., and Ellsworth, W. L. (2000). A double-difference earthquake location algorithm: method and application to the northern Hayward fault, California, *Bulletin of the Seismological Society of America*, *90*, 1353–1368, doi:10.1785/0120000006.

White, D. J., Meadows, M., Cole, S., Ramirez, A., Hao, Y., Carle, S., Duxbury, A., Samson, C., Kendall, J.-M., Verdon, J. P., Dietiker, B., Johnson, J., and Morozov, I. (2011). Geophysical monitoring of the Weyburn CO_2 flood: results during 10 years of injection, *Energy Procedia*, *4*, 3628–3635.

White, J. A., Chiaramonte, L., Ezzedine, S., Foxall, W., Hao, Y., and Ramirez, A. (2014). Geomechanical behavior of the reservoir and caprock system at the In Salah CO_2 storage project.

Proceedings of the National Academy of Science, doi:10.1073/pnas.1316465111.

Worth, K., White, D., Chalaturnyk, R., Sorensen, J., Hawkes, C., Rostron, B., Johnson, J., and Young, A. (2014). Aquistore project measurement, monitoring and verification: from concept to CO_2 injection, *Energy Procedia*, *63*, 3202–3208, doi:10.1016/j.egypro.2014.11.345.

Wright, C. A. (1998). Tiltmeter fracture mapping: from the surface and now downhole, *Petroleum Engineer International*, *71*, 50–63.

Zhang, R., Vasco, D., Daley, T. M., and Harbert, W. (2015). Characterization of a fracture zone using attributes at the In Salah CO_2 storage project, Interpretation, May, SM37-SM46, doi:10.1190/INT-2014-0141.1.

13

Evaluation of Perfluorocarbons (PFCs) as Tracers for CO_2 Containment and Migration Monitoring

Matthew Myers and Cameron White

ABSTRACT

Perfluorocarbon tracers provide a unique opportunity for monitoring caprock integrity for CCS and complement other geochemical and geophysical techniques. In this chapter, a discussion on the various PFC sampling strategies is provided. The chapter focuses on the assumptions necessary to implement these tracers in the field, strategies for deploying tracers in an injection scenario effectively and methods for estimating a CO_2 leak rate. The limitations of PFC tracers are discussed along with the opportunities that they can provide to monitoring and verification (M&V) programs for both demonstration and potentially commercial-scale CCS projects.

13.1. INTRODUCTION

To achieve many of the aims of a carbon capture and storage (CCS) monitoring and verification (M&V) program, accurate assessments of the CO_2 leakage rate from the reservoir are essential. This leakage can come from a variety of factors such as through the caprock, old wellbores, fault lines, seal breaches, or capillary breakthrough [*Chiquet et al.*, 2007; *Wells et al.*, 2007]. Several studies have shown that with widespread deployment of CCS technologies, a total leakage rate below 0.01–0.1% per year is necessary to reduce atmospheric CO_2 levels and diminish the effects of long-term climate change [*Ha-Duong and Keith*, 2004; *Pacala*, 2003]. More recently, long-term modeling (i.e., 10^5 years) suggests that a total leakage rate of 0.001% per year is necessary to achieve an outcome similar to a low emission projection with no sequestration [*Shaffer*, 2010]. Nevertheless, the US Department of Energy (USDOE) has established criteria of a global leakage rate of less than 0.01% per year [*USDOE*, 2003]. As such, M&V methods and technologies need to be developed and implemented to detect leakage at or below these rates. Furthermore, the USDOE guidelines from 2003 specify that the total cost of a M&V program should be no more than 10% of the total sequestration cost.

Spectroscopic methods using atmospheric concentration or eddy-covariance-type measurements and flux-chamber-type measurements are often conducted to directly quantify reservoir leakage rates [*Krevor et al.*, 2010; *Lewicki et al.*, 2009]. However, as the background concentration of CO_2 in the atmosphere is high and dynamic (on both diurnal and seasonal time frames), careful baseline studies prior to injection are necessary for comparison with post-injection data. There are many possible external sinks and sources (e.g., biological processes, industrial discharges, and automotive emissions) for CO_2 that might interfere with these measurements. As a result, leak detection measurements are often plagued by large experimental uncertainties which further increase at lower leakage rates where other sources of variation (e.g., weather patterns, land use) tend to dominate any measurements.

There are many indirect approaches for detecting CO_2 leaking from a reservoir which can be broadly categorized

CSIRO Energy, Kensington, New South Wales, Australia

Geological Carbon Storage: Subsurface Seals and Caprock Integrity, Geophysical Monograph 238, First Edition.
Edited by Stéphanie Vialle, Jonathan Ajo-Franklin, and J. William Carey.

into qualitative (e.g., detecting whether a significant leak is occurring or not) and quantitative (e.g., how much is leaking) techniques. Chemical- or biological-based methods include monitoring for changes in the groundwater chemistry (e.g., bicarbonate concentration) of overlying bodies of water and changes in the ^{13}C isotopic signature of nearby soil gas and water [*Apps et al.*, 2011]. Hyperspectral imaging can be used to monitor for changes in the local flora [*Bellante et al.*, 2013]. Pressure monitoring in the injection and overlying zones provides an opportunity to monitor for leaks [*Hosseini et al.*, this issue; *Hovorka et al.*, 2011; *Wang and Small*, 2014]. Other geophysical monitoring techniques such as 4D seismic or InSAR have been used for storage reservoir monitoring pre-injection and post-injection [*Arts et al.*, 2004; *Eiken et al.*, 2011]. However, with the possible exception of larger leaks (i.e., through caprock or faults), these chemical, biological, and geophysical methods generally suffer from insufficient sensitivity and resolution to be useful for quantification of lower leak rates (i.e., <0.01% per year). However, by combining information from these different approaches using various statistical methods including Bayesian models, smaller leaks can potentially be detected [*Jenkins*, 2013; *Yang et al.*, 2012]. With these statistical models, a qualitative estimate of the magnitude of any detected leak might be possible. Quantifying the leak rate requires calibrating the measurement method which is possible through controlled release programs where the leak rate is known [*Lewicki et al.*, 2009; *Strazisar et al.*, 2009].

Artificial chemical tracers co-injected with CO_2 (e.g., perfluorocarbons (PFCs)) into a storage reservoir provide a unique and cost-effective opportunity to overcome these issues with sensitivity or resolution, potentially enabling a more accurate assessment of the leakage rate [*Myers et al.*, 2013]. The primary assumption with chemical tracers is that the leakage of the tracer would mimic the CO_2 leakage out of the primary container. Importantly, unlike other methods, the sensitivity of this method can be enhanced by increasing the amount of chemical tracer used [*Watson and Sullivan*, 2012]. By monitoring for the presence of the chemical tracer at the surface or an overlying body of surface water (i.e., assurance monitoring), the tracer flux over a chosen sampling area can be used to estimate the leakage rate of CO_2 over that sampling area. Information from many sampling areas can then be combined and extrapolated to determine an overall reservoir leakage rate. Performance monitoring achieved through sample acquisition from nearby wells can also be conducted to assess whether the injected CO_2 has migrated laterally or to another geological layer (intended or otherwise).

For this particular type of implementation of chemical tracers, PFCs (see Table 13.1) are the most sensitive, while isotopically labeled carbon and inert gases are also used [*Wells et al.*, 2007, 2013]. PFCs are a unique class of fluorinated hydrocarbon compounds characterized by exceptional chemical and thermal stability, very low water solubility, minimal toxicity, and an extremely low atmospheric mixing ratio. As an artificial chemical tracer with a very low background concentration, the presence of elevated levels of PFCs would have a strong correlation with leaking CO_2 and possibly lead to a low false alarm rate [*Simmonds et al.*, 2002; *Watson et al.*, 2007]. Furthermore, these compounds can be uniquely identified, and the concentration can be determined at extremely low levels (i.e., fL per L levels) using adsorbent tubes (for soil-gas or atmospheric monitoring) combined with gas chromatography (GC) techniques [*Nazzari et al.*, 2013]. Due to the large number of potential PFCs that can be used as tracers, many different transits between injector-producer well pairs can be resolved. Other tracers that have been used within CCS or the oil industry such as inert gases (e.g., krypton and xenon), sulfur hexafluoride, and fluorescent compounds do not have the unique combination of properties highlighted above for PFCs.

We have previously reviewed the use of chemical tracers for various applications related to CCS including the use of PFCs in M&V programs [*Myers et al.*, 2013]. The feasibility of using PFCs to estimate CO_2 leakage rates has been demonstrated at the West Pearl Queen site, the San Juan Basin in New Mexico, USA [*Wells et al.*, 2007, 2013], and the ZERT shallow release site in Montana, USA [*Strazisar et al.*, 2009]. This chapter will focus on the different methodologies and assumptions necessary to assess seal integrity toward CO_2 leakage in a reservoir and quantify CO_2 leak rates based on field tracer data.

13.2. INJECTION STRATEGIES

Within reasonable constraints (e.g., limit on the amount of chemical tracer injected and the number of tracer sampling stations), the strategy for injecting PFC tracers should aim to maximize the likelihood of detecting tracer (and by inference CO_2) leakage and provide an accurate assessment of the actual volume of CO_2 leakage. Aside from cost and other practical considerations, PFCs have a large global warming potential (approximately 7000 times the heat trapping capacity of CO_2) [*Shine et al.*, 2005]. Due to their extreme stability and persistence in the environment, any use will cause background concentrations to increase, limiting their utility as chemical tracers [*Watson and Sullivan*, 2012]. As a result, careful consideration needs to be given to its handling during injection and recovery to maximize its utility, prevent any accidental spillage, and maximize the accuracy of information garnered.

Table 13.1 Chemical Structure, Name, Molecular Weight, and Boiling Point for Selected PFC Chemical Tracers That Have Been Used for CCS M&V Applications.

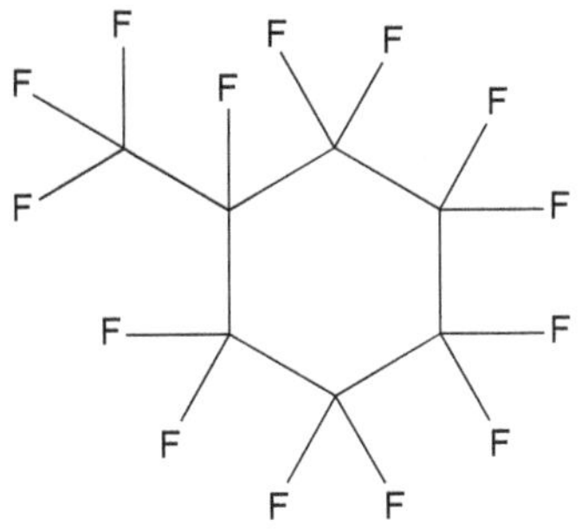

PMCH, perfluoromethylcyclohexane (C_7F_{14}), molecular weight: 350.05, boiling point: 76°C

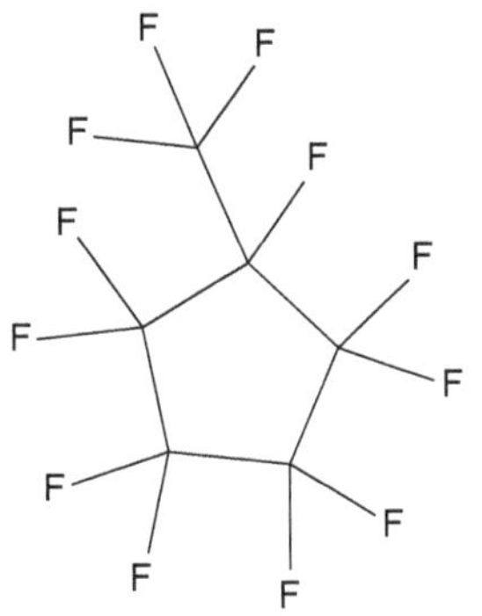

PMCH, perfluoromethylcyclopentane (C_6F_{12}), molecular weight: 300.05, boiling point: 48°C

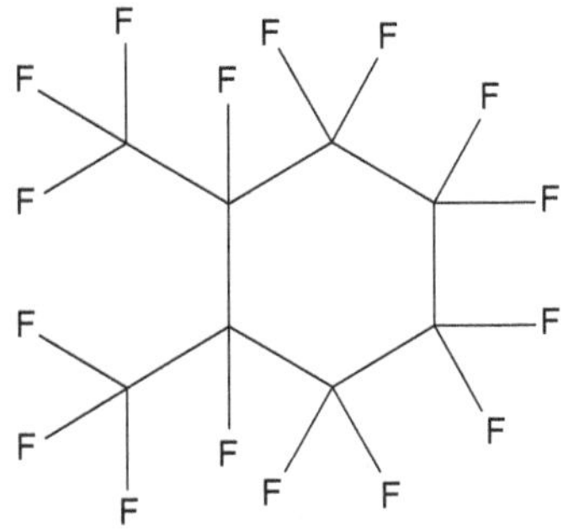

PDCH, perfluorodimethylcyclohexane (C_8F_{16}), molecular weight: 400.06, boiling point: 101°C

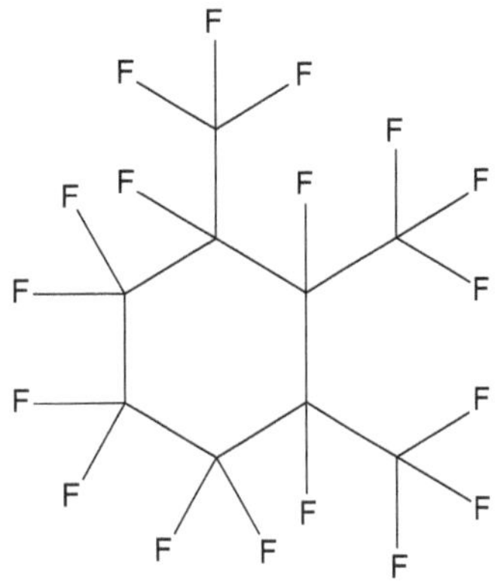

PTCH, perfluorotrimethylcyclohexane (C_9F_{18}), molecular weight: 450.07, boiling point: 127°C

Using a range of gas diffusion coefficients typical of sandy soils, subsurface diffusion, or dispersion models, it can be shown that to achieve a millionfold dilution of the tracer at the surface would require greater than 100 years even at a modest depth of 600 m [*Watson and Sullivan*, 2012]. Based on this, any detected leak would result from buoyant convective flow through fractures in the seal or poorly cemented wellbores and not through subsurface diffusion. While PFC tracers are typically co-injected (or *tagged*) with the CO_2 into the reservoir, important decisions regarding the amount of chemical tracer added and the timing of the injection need to be made. Under ideal circumstances, only the CO_2 that will be leaking through the seal should be *tagged*; however, this is impossible from a practical perspective given the near-radial flow of CO_2 from the wellbore into the reservoir and that any fractures are very unlikely to be radially symmetric. As a result, it is preferable that chemical tracers be continuously injected, *tagging* all of the injected CO_2; however, due to practical concerns (e.g., the expense of using very large amounts of tracer and higher likelihood of site contamination from continuous injection), this may not be feasible.

At the desired detection threshold (e.g., less than 0.01% per year), the amount of chemical tracer leaking needs to be sufficiently above the background amount to allow an accurate estimate of its contribution to the overall amount of chemical tracer sampled. By increasing the amount of chemical tracer injected, this contribution will become proportionally larger, increasing the confidence in the accuracy of the leak estimate. As an example, for the pilot-scale injection of CO_2 at West Pearl Queen, it was deemed impractical to inject PFC tracers with the entire 2090 tons of CO_2 that were injected [*Wells et al.*, 2007]. Instead, 20 tons of CO_2 was injected with 500 ml of tracer in three 12 h periods over several weeks. For the West Pearl Queen project, soil-gas samplers (containing sorbent material for PFCs) were placed in the ground for 2 months, and only samplers where the tracer amount exceeded the background amount (i.e., prior to injection) by twofold were considered for the leak rate calculation. The sampler tracer levels generally ranged from background levels to about eight times background levels. It is noteworthy that the estimated leak rate derived from these tracer results (averaged over a number of samplers and different PFC tracers) was 0.0085% per year which is just below the 0.01% per year target.

Extrapolating the concentration used at West Pearl Queen (i.e., 500 ml of tracer for 20 tons of CO_2) to a single 1000 MW power plant operating for 30 years that produces approximately 2.5×10^8 tonnes of CO_2 [*USEIA*, 2015] would require 12,400 tonnes of PFC tracer. This amount could be reduced to 120 tonnes of tracer if the amount of CO_2 *tagged* is limited to the same ratio used at West Pearl Queen (i.e., only 20 tons out of the 2090 tons injected was tagged). In either case, this amount of tracer would be prohibitively expensive at current prices, and any significant leak could significantly increase the amount of tracers currently in the atmosphere (approximately 2000 tonnes) [*Watson et al.*, 2007]. With the very high GWP associated with PFCs, the benefits of using PFC tracers on a large scale might be outweighed by the practical aspects of deployment, costs, and risks to the environment.

Using a surface dispersion model, Watson and Sullivan estimated that at 0.01% leakage per year with atmospheric sampling to achieve a level 1.5 times the background level, 19 tons would be required to *tag* the 30 year CO_2 emissions from a 1000 MW power plant [*Watson and Sullivan*, 2012]. This amount represents a lower (by at least one order of magnitude) concentration than that used at West Pearl Queen. In any case, significant amounts of tracers would be needed for deployment on a large scale. Extending this injection strategy to many sites could lead to substantial increases in atmospheric levels of PFCs, ultimately reducing their utility as tracers.

Due to the very sensitive nature of these measurements, special precautions should be exercised to prevent possible sampling contamination including using two physically separated teams for tracer injection and sampling, gathering samplers prior to injection of any tracers, and placing samplers the day after tracer injection concluded [*Wells et al.*, 2007; 2013]. Aside from issues related to tracer quantity, these precautions make continuous tracer injection impractical if monitoring is to occur simultaneously with CO_2 injection. Under these circumstances, it may be more productive to conduct tracer studies on a smaller portion of CO_2 at a particular site prior to expanding to commercial-scale operations [*Watson and Sullivan*, 2012]. This could be followed up by periodic injection of pulses of chemical tracers and subsequent monitoring during continued CO_2 injection. This could provide additional information if monitored appropriately to better understand rates of migration as zones become saturated with CO_2 [*Underschultz et al.*, 2011].

13.3. ASSUMPTIONS ABOUT THE BEHAVIOR OF PFC TRACERS

Injecting chemical tracers and then subsequently monitoring for their presence only provides information on the behavior and transport of the chemical tracer itself. Estimating the CO_2 leakage rate using PFC tracers requires certain assumptions, namely, that the dilution factor for the tracer in the reservoir is known and that the subsurface behavior of the tracer relative to the CO_2 can be accurately determined. At West Pearl Queen, for the leak estimate calculations, it was assumed that the tracer dilution factor equaled exactly the ratio of the injected amount of tracer and CO_2 and that the tracer and CO_2 leaked at exactly the same relative rates [*Myers et al.*, 2013; *Wells et al.*, 2007].

When PFCs are injected with CO_2 into a reservoir, the tracer is initially diluted in the wellbore followed by chromatographic dispersion and mixing with other fluids within the reservoir which will further affect the extent of dilution. This dilution ratio is used to determine the relative tracer-to-CO_2 gas composition and is one of the terms in the equation used to estimate the leak rate. In many instances, the assumption is that there is no further mixing and that the dilution factor is determined only by the relative amounts of injected CO_2 and tracer [*Wells et al.*, 2007, 2013]. This might be appropriate for an initial estimate of leakage rate for pilot-scale injection programs but may not be as suitable for large industrial-scale injections where a higher degree of scrutiny will likely exist through regulatory reporting. In these instances, it might be appropriate to conduct further studies to determine whether these assumptions are true and, if they are not, what changes should be made in order to achieve an accurate leak estimate.

At the K12-B gas field in the North Sea where CO_2 is separated from CH_4 and reinjected into an injection well, PFCs were injected as a pulse, and breakthrough was observed at the production well sometime later [*Vandeweijer et al.*, 2011]. Interestingly, a significant decline in PFC concentrations coincided with compressor failures that led to a temporary stop to CO_2 injection. The reasons for this are not exactly clear; however, the most likely explanation is probably related to a decline in reservoir pressure and a change in fluid flow paths. If the relative amounts of CO_2 and PFC tracers are affected by this, it would have a dramatic impact on the leak rate estimate, and it was not initially considered in their estimates. To the best of our knowledge, this effect has not been explained through laboratory experiments and remains a significant uncertainty. This is an example of where prior assumptions on tracer behavior may not be valid after implementation in an actual field setting.

In the likely case that PFC tracers are injected as a pulse with a portion of the overall injected CO_2, computational simulations (e.g., TOUGH2) could be used to predict the degree of mixing between *tagged* and *non-tagged* CO_2 through convection and diffusion. Even a relatively small rate of diffusion and convection within the reservoir could significantly impact the dilution factor and become increasingly inaccurate as the mixing progresses over time. The change in dilution factor within the reservoir may also be impacted by partitioning processes where the CO_2 and the chemical tracer behave differently in the presence of minerals, brine, or hydrocarbons.

13.4. LIMITATIONS AND CONCERNS OF USING PFC TRACERS

The sediment that comprises the overlying layers may interact with the PFC tracers, potentially affecting their transport relative to leaking CO_2. If PFC tracers adsorb either reversibly or irreversibly onto sediment, it could have a dramatic impact on the quantity, equilibration, and breakthrough time of tracers arriving at the surface. In the worst case, this could lead to a scenario where CO_2 is leaking to the surface and the tracer is adsorbed by the sediment, resulting in a *false negative* for the tracer test. In laboratory experiments using a sand-packed column at ambient pressure, temperatures ranging from 23 to 60°C, and with helium as a carrier gas, dry sand has been shown to strongly adsorb PFCs and retard their transport. However, with damp or moist sand, the PFCs traveled with the gas front [*Maxfield et al.*, 2005]. Subsequent research at ambient pressure conditions using CO_2 as a carrier gas gave similar results [*Zhong et al.*, 2014]. The results of the study by Zhong *et al.*, are summarized in Table 13.2 and Figure 13.1. Over several sediment types, a strong inverse correlation between moisture content and retention was observed. The larger PFC molecules generally exhibit stronger retention on sediment surfaces relative to smaller PFC molecules presumably due to differences in the intermolecular forces between the PFC molecule and the sediment. In some cases (Illinois dry sample), the PFC tracers were retained to such a degree that they did not elute during the time frame of the experiment, which was attributed to the presence of 8% w/w smectite with a higher surface area (~600–800 m²/g) relative to the other sediments. Ottawa sand containing no clay has a comparatively low surface area, and the PFCs eluted quickly even under dry conditions where a significant amount of retardation was exhibited. This large difference in behavior demonstrates the importance of sediment composition and moisture content on the ability to accurately detect and analyze a CO_2 leak. This influence will need to be accounted for in models analyzing leakage scenarios and subsurface behavior.

Another concern with PFC tracers is potential adsorption onto coal sediment above the storage reservoir. Activated carbon materials are often used for sampling PFC tracers over long periods of time [*Nazzari et al.*, 2013; *Straume et al.*, 1998]. The sorption is strongly irreversible at room temperature and requires a high temperature for desorption. Given similarities between coal and activated carbon materials, an assessment of both the moisture levels and the presence of coal in the sediment layers overlying the storage reservoir may be necessary as PFC sorption onto coal or dehydrated high surface area sediment may lead to a *false negative* for CO_2 leakage. This may be a significant issue in the near wellbore region which can become depleted of water due to the comparatively large amount of CO_2 passing through during injection. In this case, the timing of the tracer injection would need to be considered.

Deep saline aquifers are relatively common throughout the industrialized world and represent a significant opportunity in terms of potential CO_2 storage capacity [*Bachu*, 2003; *Bachu and Adams*, 2003]. However, the dramatic difference in the behavior of CO_2 (which is quite soluble) and PFC tracers (which are extremely insoluble) in water leads to questions concerning the validity of the

Table 13.2 Retardation Factors for a Variety of PFC Tracers Compared With SF_6 for a Selection of Wet and Dry Sediments.

Sediment	Water (wt%)	SF_6	PMCP	PMCH	PDCH	PTCH	$^{13}CO_2$
Hanford sediment, wet	11.9	1	0.94	0.96	0.96	0.99	NA
Illinois sediment, dry	0.21	1	NA	NA	NA	NA	NA
Illinois sediment, damp	6	1	1.07	1.12	1.17	1.31	NA
Illinois sediment, wet	23.9	1	0.95	0.97	0.96	0.97	NA
Ottawa sand, dry	0.02	1	1.05	1.16	1.11	1.27	1.57
Ottawa sand, damp	2.3	1	1.02	1.05	1.04	1.06	NA
Illinois sediment, wet	23.3	1	0.98	1	1	1.01	NA

NA denotes no tracer breakthrough detected.
The Hanford sediment is from the US Department of Energy site in Washington, USA, and the Illinois sediment is from the FutureGen-II site in Illinois, USA.
Zhong et al. [2014]. Copyright 2014. Reprinted with permission from Elsevier.

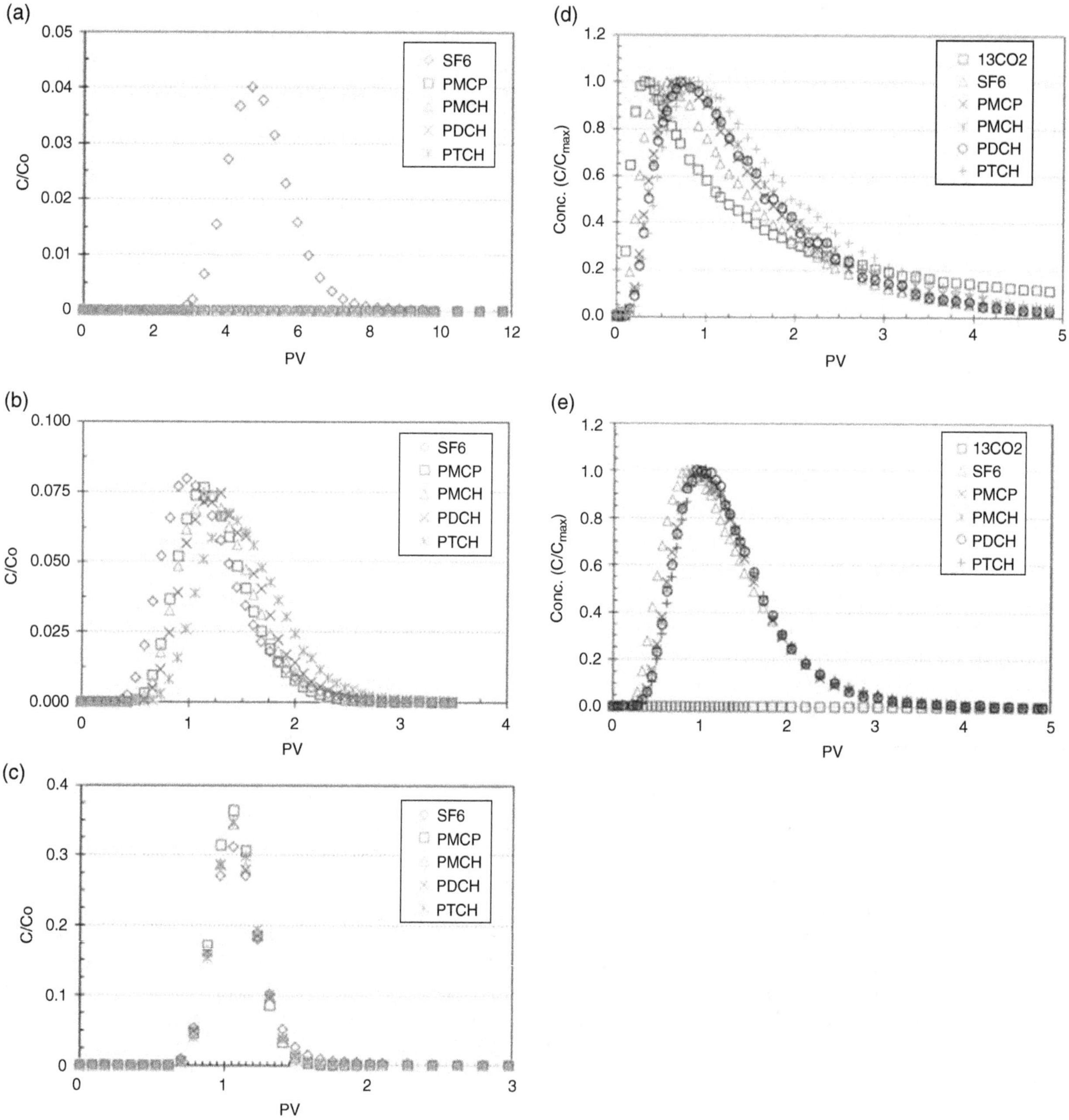

Figure 13.1 The influence of sediment type and moisture content on the breakthrough curves of a selection of PFC tracers compared to SF_6. (a) Illinois sediment, 0.21 wt% moisture content. (b) Illinois sediment, 6.0 wt% moisture content. (c) Illinois sediment, 23.9 wt% moisture content. (d) Dry Ottawa sand. (e) Damp Ottawa Sand. *Zhong et al.* [2014]. Copyright 2014. Reprinted with permission from Elsevier.

relative leak rate assumption. In a saline aquifer, the dissolution of CO_2 within the brine phase increases the fluid density, leading to gravity-induced convection which dominates over diffusion, causing the CO_2 to descend to deeper parts of the aquifer [*Emami-Meybodi et al.*, 2015]. Despite having liquid densities typically greater than 1.6 g/ml, the transport of the PFC tracers will be very different given their extremely low solubility in water and very low partition coefficient into water [*3M*, 2002; *Rauh et al.*, 2014]. This means that the PFC tracers may not be able to *tag* the CO_2 that is dissolved in the brine. Because the CO_2 sinks when dissolved in brine, the CO_2 most likely to leak from the storage reservoir will be in the pore space either structurally trapped or stored in the pore space through capillarity (i.e., residual saturation). It is unlikely that any CO_2 which is stripped of tracer through the dissolution and convection processes in brine would leak. However, this fractionation process at the gas-brine interface might substantially affect the dilution ratio (by increasing the relative amount of PFC tracers compared to CO_2) and any subsequent leak rate calculations.

Depleted gas reservoirs and enhanced oil recovery sites also represent a significant potential opportunity for CO_2 storage [*Emberley et al.*, 2005; *Jenkins et al.*, 2012]. Due to the very low intermolecular forces typically exhibited by PFCs, they are generally considered as conservative tracers (i.e., they do not partition into other fluid phases or undergo chemical reactions). However, PFCs readily partition into crude oil, implying that oil or condensates within the reservoir could potentially retard their transport within the reservoir, resulting in a depletion in the amount of PFCs associated with stored CO_2 [*Dugstad et al.*, 1992a]. High-pressure slim-tube experiments have been used to quantify how the transport of PFC tracers is impacted by oil under reservoir conditions [*Dugstad et al.*, 1992b, 1993]. Generally, larger PFC compounds exhibit a greater tendency to partition into the oil phase, and hence, their transport through the reservoir is more retarded. By using two or more PFC tracers of different sizes, it might be possible to ascertain the extent to which this partitioning occurs within the reservoir, allowing for more accurate calculations regarding the transport of the PFC tracers relative to CO_2. Use of multiple tracers in a single test tends to increase the data and information that can be used to better understand overall fluid behavior when a range of fluids are present in a system. This may include the use of water-soluble tracers as significant quantities of CO_2 can dissolve in water. Multiple tracers could also be used to tag different streams of injected CO_2 if more than one injector well or source of CO_2 is used.

Lastly, within the storage interval, the behavior of the PFC tracers relative to stored CO_2 and the effects of reservoir processes on the tracer dilution need to be considered. There are four main trapping mechanisms for CO_2 within a storage reservoir: structural, solubility, capillary, and mineral [*Orr*, 2009]. Within the reservoir, of the various trapping mechanisms, structurally trapped CO_2 is most likely to leak and is also likely to be the only CO_2 that arrives in a reasonable time frame to the surface because it has more potential to buoyantly migrate. For solubility trapping, the potential enrichment of PFC tracers in the CO_2 phase due to selective dissolution of CO_2 over PFCs in brine has been considered above. For capillary trapping where supercritical CO_2 is trapped within the pore space as a supercritical fluid, the PFC concentration within the trapped CO_2 and free CO_2 is not expected to change due to the very small concentration of the PFCs within both CO_2 phases. Mineral trapping, where CO_2 is incorporated into the minerals of the reservoir, could increase the relative amount of PFC to CO_2; however, the time frame for mineral trapping is typically very long (i.e., greater than 1000 years), so any effect on the dilution ratio is not expected over a reasonable time frame.

13.5. QUANTIFICATION OF PFC SORPTION WITHIN SEDIMENTS

For tracers within the reservoir, the adsorption of PFCs onto the sediment surface is a primary concern. To determine whether retention on different sediment types is significant, slim-tube experiments were conducted at CSIRO as part of a program to evaluate deployment at a potential commercial-scale CCS project. Slim-tube experiments were conducted where a length of tubing was packed with sediment and fluids were allowed to elute under reservoir conditions (i.e., high pressure and temperature). After elution from the slim tube, the water was trapped in a bubbler and the gas was passed through an ultralow flow rate back pressure regulator and flowed through a sparging tube containing Carbopack powder. Experimental results show that elution of different types of PFCs was impacted by different sediment mixtures (see Table 13.3). Adsorptive effects are a likely component of the observed changes where clays, coals, and other impurities were added to a sand base. But these differences might also be affected by trapping mechanisms within the pore space of the packed sediment as well, where some of the tracer is trapped within the residually trapped CO_2 between the sediment pores. By pushing CO_2 free of any tracer through the sediment after the first pulse of tracer is introduced, there is a gradual bleeding off or reduction in tracer (see Fig. 13.2) until the amount of tracer obtained asymptotes to zero.

Upon depressurization, any PFC in the CO_2 trapped within the pore space through capillarity elutes from slim tube, and up to 40% of the recovered PFC was recovered in this step. However, this loss of PFC to the capillarity-trapped CO_2 does not affect our leak calculation because if any of the trapped CO_2 mobilizes, the PFC will propagate with it. Sediment sorption, however, has direct implications concerning the relative concentrations of PFC that will reach the surface in the event of a leak in a CO_2 reservoir. In the slim-tube experiment, this sorption would manifest through a decreased ultimate recovery (combining the PFC recovered before and after sediment sorption) or significant retardation and tailing in the concentration breakthrough curve (see Fig. 13.2). For the sediments in this study, no significant retardation in the breakthrough was seen. Due to the difficulties in sampling extremely small amounts of PFC from high-pressure CO_2-water mixtures, it was not possible to quantitatively determine the recovery.

Table 13.3 Summary of the Recovery of Results From Slim-Tube Experiments at High Temperature and Pressure (65°C and 150 bar).

Tracer	Sediment	Pore volumes CO_2 injected	% of PFC recovered before depressurization	% of PFC recovered after depressurization
PMCH	Pure sand	3.48	76.6	23.4
1,3-PDCH	Pure sand	3.48	74.9	25.1
1,2-PDCH	Pure sand	3.48	77.4	22.6
PMCH	20:80 carbonate/sand	2.69	97.5	2.5
1,3-PDCH	20:80 carbonate/sand	2.69	95.5	4.5
1,2-PDCH	20:80 carbonate/sand	2.69	95.0	5
PMCH	50:50 sandstone/sand	2.97	80.5	19.4
1,3-PDCH	50:50 sandstone/sand	2.97	70.7	29.2
1,2-PDCH	50:50 sandstone/sand	2.97	80.4	19.5

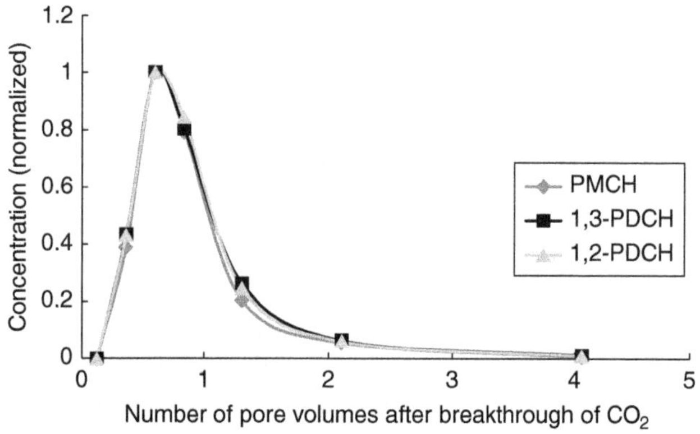

Figure 13.2 Normalized breakthrough curve for 2.5% kaolinite in sand at 80°C and 15 MPa of three tracers showing the initial breakthrough of the tracer followed by gradual tailing off after the initial breakthrough pulse.

It is good practice, however, to test whether this is the case for the relevant reservoir as this study only examined a limited number of sediment types. If there is significant sediment adsorption (either reversible or irreversible) of PFC within the reservoir, a scenario could occur where a leak could be taking place for a considerable amount of time before enough PFCs reach the upper layers of the reservoir where it would only then start migrating to the surface. Even if the PFC tracers reach the surface, the equilibration time could increase dramatically due to the large sediment surface area and trace levels of PFCs. In the worst case, the amount detected at the surface may no longer be representative of the amount of CO_2 being released, rendering accurate modeling difficult.

13.6. SAMPLING METHODOLOGIES AND CALCULATION TO DETERMINE CO_2 LEAKAGE RATE

As the rate of PFC leakage from the surface will be extremely small for even relatively large leaks, it is necessary to use absorbent material to collect gas samples over a period of time (i.e., up to several months possibly) for laboratory analysis [*Wells et al.*, 2007, 2013]. Atmospheric sampling has been proposed; however, varying wind conditions can dramatically affect whether the CO_2 plume is being sampled or background air is being sampled, adding a further complexity to the leakage estimate calculation [*Watson and Sullivan*, 2012]. Soil-gas sampling was used at West Pearl Queen and San Juan Basin; however, passive sampling required extended sampling time (up to 2 months) to adsorb adequate amounts for quantitative GC analysis [*Wells et al.*, 2007, 2013]. The use of active pumping to increase the sampling rate may decrease this time, but it might lead to anomalous behavior due to the strong advection and the possible depletion of gas originating from the subsurface near the sampler. Furthermore, flux chamber measurements are difficult due to diffusion gradients and pressure differential gradients that magnify over long sampling times. With these limitations, the use of both atmospheric and soil-gas sampling in combination may be beneficial. Collecting samples over a variety of spatial and temporal densities might lead to an understanding of how the flux rate varies from position to position and over time. This would lead to an improved understanding of the random and systematic sampling errors that may exist. In instances where multiple PFC tracers are used, it is important to consider the purity of the tracers which can be rather low due to the harsh conditions used in their synthesis. Also, the ability to separate different PFC tracers using GC has been compromised somewhat due to the lack of commercial availability for Carbograph PLOT columns, though Al_2O_3 PLOT columns deactivated with Na_2SO_4 have been shown to be a suitable alternative in many instances.

The simplest way to estimate the leak rate is first to use GC methods to determine the amount of PFC obtained at a sampler and then to extrapolate this amount to the entire surface area of the reservoir [*Wells et al.*, 2007, 2013]. The sampled surface area is estimated by determining the projected surface area from an assumed volume of sampled gas and a spherically shaped volume of sediment. This in itself can lead to significant errors as the flux will likely not be uniform over the surface area of the reservoir. Various statistical and averaging techniques and the use of multiple sampling sites could be used to reduce these errors and obtain a reasonable estimate of the overall tracer leak rate from the reservoir. The CO_2 leak rate is estimated by multiplying the measured tracer flux leak rate (from above), the PFC-in-CO_2 dilution factor, and the relative leak ratio. The previous sections examined assumptions regarding the tracer dilution factor and the leak ratio. For West Pearl Queen, the leak rate ratio was assumed to be unity (i.e., the transport behavior of the PFC and CO_2 was identical), and the dilution factor was simply the ratio of the amount of PFC to the amount of CO_2 injected during tracer injection. However, as discussed in the previous sections, these assumptions may not be valid, and a combination of laboratory experiments and computational modeling may lead to a better estimate of these parameters.

13.7. SUMMARY

In order to provide a long-term reduction in atmospheric CO_2 levels, a leakage rate from underground geological CO_2 sequestration sites of less than 1% per 100 years is considered necessary. The accuracy of the leak rate calculation from tracer tests relies on many assumptions, namely, that the relative leakage rates of the CO_2 and tracer can be quantified, the leak rate reaches an equilibrium value in a reasonable amount of time, the spatial and temporal variations in the tracer flux rates are accurately known, and there is no loss of tracer due to adsorption or selective partitioning. In many instances, the accuracy of these assumptions can be mitigated through both laboratory- and pilot-scale experiments and an understanding of the properties of the storage reservoir. This approach will aid in reducing uncertainty related to whether these assumptions are valid.

However, the degree to which PFCs can actually provide a quantitative and accurate estimate of the leak rate still needs to be carefully considered. As highlighted above, many assumptions are required, and a tracer injection and sampling program needs to be in place before leak rate estimates can be made. These estimates may be impacted by various circumstances specific to a storage reservoir or injection scenario that have not been previously considered (e.g., the anomalous loss of tracer concentration at the K12-B gas field in the North Sea when injection stopped temporarily). Though generally considered inert and unreactive, the adsorption characteristics of PFCs onto sediments remain an area that needs to be comprehensively investigated. Until further research into these effects is established, PFCs currently only provide a qualitative detection of a leakage scenario involving tagged CO_2. Even with significant amounts of research, the results may only be semiquantitative as it is likely impossible to account for all of the variation in the subsurface tracer behavior.

Despite these potential impediments, PFC tracers do have considerable utility in assessing any potential CO_2 leakage from a CCS storage site. It is still prudent however to use other monitoring techniques for any M&V program. In combination with other monitoring techniques, PFC chemical tracers have indeed the potential to provide greater certainty with a M&V program. Furthermore, with the exception of possible contamination of a site with PFCs, it is expected that by using tracers, the possibility for a *false positive* detection of leak CO_2 would be extremely low (e.g., the presence of elevated amounts of PFCs would be a very strong indication of leaking CO_2). On the other hand, the case against a *false negative* test is much weaker and requires addressing the issues outlined in this chapter. The utility of PFC tracers within the context of CCS is dependent on its extreme stability and very low atmospheric background concentrations. Due to the stability of PFCs, they have a very long lifetime in the atmosphere, and any PFCs that leak will tend to accumulate over time, ultimately limiting their utility as tracers. With this in mind, it may be most appropriate to limit the use of PFC tracers to pilot-scale CCS projects, research projects and possibly for commercial-scale projects on a limited basis. For cases where saline aquifers are used as a primary storage medium where CO_2 dissolves into the aqueous phase to a significant extent, water-soluble tracers could prove to be a very useful complement to water-insoluble PFCs in assessing containment. The ultimate aim of a M&V program is to develop assurance whether or not CO_2 is leaking from a reservoir and provide unbiased assessment of potential leakage scenarios. Through appropriate implementation of PFC tracers and other M&V technologies and the use of best practices for operating a CCS site, there is an opportunity to reduce industry, government, and community anxiety associated with CCS.

REFERENCES

3M, 2002. Perfluoro compound C5-C18, HPV Robust Summaries and Test Plans. EPA.

Apps, J.A. et al., (2011). Transient changes in shallow groundwater chemistry during the MSU ZERT CO_2 injection experiment. 10th International Conference on Greenhouse Gas Control Technologies, 4, Amsterdam, the Netherlands, 19–23 September 2010, pp. 3231–3238.

Arts, R. et al., (2004). Monitoring of CO_2 injected at Sleipner using time-lapse seismic data. *Energy*, *29*(9–10): 1383–1392.

Bachu, S., (2003). Screening and ranking of sedimentary basins for sequestration of CO_2 in geological media in response to climate change. *Environmental Geology*, *44*(3): 277–289.

Bachu, S., Adams, J.J., (2003). Sequestration of CO_2 in geological media in response to climate change: capacity of deep saline aquifers to sequester CO_2 in solution. *Energy Conversion and Management*, *44*(20): 3151–3175.

Bellante, G.J., Powell, S.L., Lawrence, R.L., Repasky, K.S., Dougher, T.A.O., (2013). Aerial detection of a simulated CO_2 leak from a geologic sequestration site using hyperspectral imagery. *International Journal of Greenhouse Gas Control*, *13*: 124–137.

Chiquet, P., Daridon, J.-L., Broseta, D., Thibeau, S., (2007). CO_2/water interfacial tensions under pressure and temperature conditions of CO_2 geological storage. *Energy Conversion and Management*, *48*(3): 736–744.

Dugstad, O., Bjornstad, T., Hundere, I., (1992a). Measurements and application of partition-coefficients of compounds suitable for tracing gas injected into oil-reservoirs. *Revue De L Institut Francais Du Petrole*, *47*(2): 205–215.

Dugstad, O., Bjornstad, T., Hundere, I.A., (1992b). Construction of a slim-tube equipment for gas tracer evaluation at simulated reservoir conditions. *Applied Radiation and Isotopes*, *43*(4): 527–535.

Dugstad, O., Bjornstad, T., Hundere, I.A., (1993). Measurements of gas tracer retention under simulated reservoir conditions. *Journal of Petroleum Science and Engineering*, *10*(1): 17–25.

Ha-Duong, M., Keith, D.W., (2004). Carbon storage: the economic efficiency of storing CO_2 in leaky reservoirs. In Sikdar S.K., Glavič P., Jain R. (eds) *Technological Choices for Sustainability*. Springer, Berlin, Heidelberg, pp. 165–182.

Eiken, O. et al., (2011). Lessons learned from 14 years of CCS operations: Sleipner, In Salah and Snohvit. 10th International Conference on Greenhouse Gas Control Technologies, *4*, Amsterdam, the Netherlands, 19–23 September 2010, pp. 5541–5548.

Emami-Meybodi, H., Hassanzadeh, H., Green, C.P., Ennis-King, J., (2015). Convective dissolution of CO_2 in saline aquifers: progress in modeling and experiments. *International Journal of Greenhouse Gas Control*, *40*: 238–266.

Emberley, S. et al., (2005). Monitoring of fluid-rock interaction and CO_2 storage through produced fluid sampling at the Weyburn CO_2-injection enhanced oil recovery site, Saskatchewan, Canada. *Applied Geochemistry*, *20*(6): 1131–1157.

Hosseini, S.A.S., Mahmood; Sun, Alexander; Hovorka, Susan, this issue. In zone and above-zone pressure monitoring for CO_2 geological storage.

Hovorka, S.D. et al., (2011). Monitoring a large volume CO_2 injection: year two results from SECARB project at Denbury's Cranfield, Mississippi, USA. *Energy Procedia*, *4*: 3478–3485.

Jenkins, C., (2013). Statistical aspects of monitoring and verification. *International Journal of Greenhouse Gas Control*, *13*: 215–229.

Jenkins, C.R. et al., (2012). Safe storage and effective monitoring of CO_2 in depleted gas fields. *Proceedings of the National Academy of Sciences of the United States of America*, *109*(2): E35–E41.

Krevor, S., Perrin, J.-C., Esposito, A., Rella, C., Benson, S., (2010). Rapid detection and characterization of surface CO_2 leakage through the real-time measurement of C signatures in CO_2 flux from the ground. *International Journal of Greenhouse Gas Control*, *4*(5): 811–815.

Lewicki, J.L. et al., (2009). Detection of CO_2 leakage by eddy covariance during the ZERT project's CO_2 release experiments. *Energy Procedia*, *1*(1): 2301–2306.

Maxfield, B.T., Ginosar, D.M., McMurtrey, R.D., Rollins, H.W., Shook, G.M., (2005). The effect of moisture content on retention of fluorocarbon tracers on sand. *Geothermics*, *34*(1): 47–60.

Myers, M., Stalker, L., Pejcic, B., Ross, A., (2013). Tracers – past, present and future applications in CO_2 geosequestration. *Applied Geochemistry*, *30*: 125–135.

Nazzari, M., Sciarra, A., Quattrocchi, F., (2013). A simple and sensitive gas chromatography-electron capture detection method for analyzing perfluorocarbon tracers in soil gas samples for storage of carbon dioxide. *International Journal of Greenhouse Gas Control*, *14*: 60–64.

Orr, F.M., (2009). Onshore geologic storage of CO_2. *Science*, *325*(5948): 1656–1658.

Pacala, S.W., (2003). Global constraints on reservoir leakage. Greenhouse Gas Control Technologies, Proceedings of the 6th International Conference, Kyoto, Japan, 1–11 October 2002, pp. 267–272.

Rauh, F. et al., (2014). A mid-infrared sensor for the determination of perfluorocarbon-based compounds in aquatic systems for geosequestration purposes. *Talanta*, *130*: 527–535.

Shaffer, G., (2010). Long-term effectiveness and consequences of carbon dioxide sequestration. *Nature Geosci*, *3*(7): 464–467.

Shine, K.P. et al., (2005). Perfluorodecalin: global warming potential and first detection in the atmosphere. *Atmospheric Environment*, *39*(9): 1759–1763.

Simmonds, P.G. et al., (2002). The background atmospheric concentrations of cyclic perfluorocarbon tracers determined by negative ion-chemical ionization mass spectrometry. *Atmospheric Environment*, *36*(13): 2147–2156.

Straume, A.G., Dietz, R.N., Koffi, E.N.d., Nodop, K., (1998). Perfluorocarbon background concentrations in Europe. *Atmospheric Environment*, *32*(24): 4109–4122.

Strazisar, B.R., Wells, A.W., Diehl, J.R., Hammack, R.W., Veloski, G.A., (2009). Near-surface monitoring for the ZERT shallow CO_2 injection project. *International Journal of Greenhouse Gas Control*, *3*(6): 736–744.

Underschultz, J. et al., (2011). CO_2 storage in a depleted gas field: an overview of the CO2CRC Otway project and initial results. *International Journal of Greenhouse Gas Control*, *5*(4): 922–932.

USDOE, (2003). Carbon Sequestration Technology Roadmap and Program Plan. US Department of Energy, Office of Fossil Energy, National Energy Technology Laboratory.

USEIA, (2015). US Energy Information Administration, http://www.eia.gov/tools/faqs/faq.cfm?id=74&t=11.

Vandeweijer, V. et al., (2011). Monitoring the CO_2 injection site: K12-B. *Energy Procedia*, *4*: 5471–5478.

Wang, Z., Small, M.J., (2014). A Bayesian approach to CO_2 leakage detection at saline sequestration sites using pressure measurements. *International Journal of Greenhouse Gas Control*, *30*: 188–196.

Watson, T.B., Sullivan, T., (2012). Feasibility of a perfluorocarbon tracer based network to support monitoring, verification, and accounting of sequestered CO_2. *Environmental Science & Technology*, *46*(3): 1692–1699.

Watson, T.B., Wilke, R., Dietz, R.N., Heiser, J., Kalb, P., (2007). The atmospheric background of perfluorocarbon compounds used as tracers. *Environmental Science & Technology*, *41*(20): 6909–6913.

Wells, A.W. et al., (2007). The use of tracers to assess leakage from the sequestration of CO_2 in a depleted oil reservoir, New Mexico, USA. *Applied Geochemistry*, *22*(5): 996–1016.

Wells, A.W., Diehl, J.R., Strazisar, B.R., Wilson, T.H., Stanko, D.C., (2013). Atmospheric and soil-gas monitoring for surface leakage at the San Juan Basin CO_2 pilot test site at pump canyon New Mexico, using perfluorocarbon tracers, CO_2 soil-gas flux and soil-gas hydrocarbons. *International Journal of Greenhouse Gas Control*, *14*: 227–238.

Yang, Y.M. et al., (2012). A Bayesian belief network (BBN) for combining evidence from multiple CO_2 leak detection technologies. *Greenhouse Gases-Science and Technology*, *2*(3): 185–199.

Zhong, L., Amonette, J.E., Mitroshkov, A.V., Olsen, K.B., (2014). Transport of perfluorocarbon tracers and carbon dioxide in sediment columns – evaluating the application of PFC tracers for CO_2 leakage detection. *Applied Geochemistry*, *45*: 25–32.

Part IV
Environmental Impacts and Remediation Techniques

14

Migration and Leakage of CO_2 From Deep Geological Storage Sites

Andreas Busch[1] and Niko Kampman[2]

ABSTRACT

Assessing the probability and risk of migration of CO_2 from a storage reservoir and its leakage at the surface forms a critical part of the evaluation of any geological storage site. Insights from small- and large-scale demonstration projects, experimentation, numerical simulation, and natural analogues provide insights into the mechanism and processes controlling the flows of CO_2; however, there are gaps and uncertainties in our understanding. Uncertainty surrounds the probability and significance of leakage by caprock failure, and significant gaps exist in our understanding and ability to model the processes of flow and dissolution that will retard leakage as CO_2 migrates through faults and fracture networks.

This study provides an overview of the physiochemical processes involved in fluid leakage from deep storage sites with a major focus on carbon storage. Our aim is to highlight the physical flow, geochemical, geomechanical, and hydrodynamic processes that govern the rates of CO_2 flow and to address the microscale, macroscale, and basin-scale geological features that influence these flows. We review published CO_2 fluxes from natural analogues relevant to deep storage sites and provide a summary of the learnings from a well-studied analogue for carbon storage in the area of Green River, Utah.

14.1. INTRODUCTION

Carbon capture and storage (CCS) is becoming a more mature technology with experience now gained from small-scale demonstration to large-scale commercial projects around the globe. Although the pace of implementation of new industrial-scale projects is slow, there is an increased confidence that CCS is feasible and safe and that the risk of large-scale CO_2 migration from a storage reservoir is low [*Michael et al.*, 2010]. Although governmental guidelines and regulations [*Commission of the European Communities*, 2009; *Det Norske Veritas*, 2010] are strict for CO_2 storage, the possibility that some volume of CO_2 will migrate from the storage target cannot be ruled out. The risk of migration from the storage target will depend on a number of factors including trap and seal geometry, reservoir faulting, aquifer dynamics, reservoir heterogeneity, migration pathways, and geochemical reactivity, among others. Since subsurface storage reservoirs are not engineered leak-tight volumes, migration risk cannot be ruled out in every case. Many natural CO_2 reservoirs demonstrate that secure storage with no detectable surface leakage over geological timescales is possible (Bravo Dome, USA [*Allis et al.*, 2001], Magnus Field, UK [*Worden and Smith*, 2004], and the Otway Basin, AUS [*Higgs et al.*, 2015], among others), while some natural CO_2 and gas reservoirs can demonstrate intermittent or continuous gas migration to surface along natural pathways over long timescales (10^3–10^4 years) (see below). The data available to characterize storage complexes usually originate from well and seismic data which have limited resolution and which have limited resolution, subject to interpretation.

[1]*Heriot-Watt University, Lyell Centre, Edinburgh, UK*
[2]*Shell Technology Centre Amsterdam, The Netherlands*

Geological Carbon Storage: Subsurface Seals and Caprock Integrity, Geophysical Monograph 238, First Edition.
Edited by Stephanie Vialle, Jonathan Ajo-Franklin, and J. William Carey.

Within the scientific literature, there are many studies discussing leakage from CO_2 reservoirs, examining the problem from different geological, physical, chemical, or engineering perspectives. The migration of CO_2 along abandoned wells is extensively studied, given the experience gained from decades of drilling activities in the oil and gas industry [*Bachu and Bennion*, 2009; *Gasda et al.*, 2004; *Gor et al.*, 2013; *Loizzo et al.*, 2011; *Tao et al.*, 2014; *Zhang and Bachu*, 2011]. Other mechanisms of migration which include fault/fracture leakage [*Collettini et al.*, 2008; *Dockrill and Shipton*, 2010; *Jeanne et al.*, 2013, 2014; *Jung et al.*, 2014; *Keating et al.*, 2013; *Rinaldi et al.*, 2014; *Shipton et al.*, 2004; *Tueckmantel et al.*, 2012], capillary leakage [*Amann-Hildenbrand et al.*, 2013; *Amann et al.*, 2011; *Busch and Amann-Hildenbrand*, 2013; *Hildenbrand et al.*, 2002, 2004; *Wollenweber et al.*, 2010] or diffusive leakage [*Berne et al.*, 2010; *Busch et al.*, 2008, 2010] are less well understood. This is because these features occur over a wide range of spatial and temporal scales and the quantification of rates and processes is difficult due to the imperfect characterization of geological structures. There are many areas around the globe where natural oil and gas seeps, pockmarks, and gas chimneys have been observed. Some of these have been studied in detail with the aim at quantifying rates of leakage or gas pool accumulation over timescales of up to 10^6 years [*Brown*, 2000; *Cathles et al.*, 2010; *Etiope and Milkov*, 2004; *Gal et al.*, 2011; *Hornafius et al.*, 1999; *Hovland*, 2002; *Logan et al.*, 2010; *McGinnis et al.*, 2011; *Whiticar and Werner*, 1981]. Such information aids understanding of the flux of CO_2 that can be expected at the surface, given certain gas accumulation volumes, geological terrains, and local structures.

It is clear that, given all the uncertainties related to potential leakage mechanisms and fluxes, quantification of leakage rates is challenging and site specific, and a more fundamental understanding of leakage mechanisms and their controls and associated risks, is required.

In this paper fracture and matrix leakage, and advective and diffusive flows are discussed separately. For leakage across the reservoir caprock, different processes will retard fluid leakage. These include dispersion, capillary trapping, or dissolution in shallow aquifers before getting to surface. We aim to give an overview on these processes. We then discuss natural CO_2 and CH_4 seeps, as well as natural CO_2 reservoirs worldwide to discuss their suitability as analogues for carbon storage. Finally, we use the Green River natural CO_2 reservoir in Utah, USA [*Kampman et al.*, 2014b], as a detailed case study, given the wealth of knowledge available there on the geological architecture, reservoir volumes, and surface fluxes.

14.2. MECHANISMS FOR FLUID ESCAPE FROM RESERVOIRS

There are numerous mechanisms for leakage from a natural or engineered/operated reservoir, and all of them are discussed in the literature in different levels of detail. The low density of CO_2, hydrocarbon, and other volatile gases (e.g., He, H_2S) relative to water drives the buoyant migration of supercritical liquids or gas in subsurface reservoirs and aquifers. CO_2 stored in geological reservoirs in its supercritical state will be approximately 80–20% less dense that the formation brine, depending on conditions of pressure, temperature, and salinity. On transition to a gas phase, this density difference increases to approximately 90%.

In natural settings, fluids can migrate from the storage reservoir to the surface or to other reservoirs due to transmissive discontinuities in the seal. For reservoirs selected as engineered storage sites, careful screening is required to evaluate potential geological leakage risks, anticipated volumes of CO_2 migration in the event of primary seal bypass, and the extent of CO_2 scavenged by trapping mechanisms operating in the overburden [*Bachu*, 2000].

14.2.1. Fracture/Fault Flow

It is well established that faults can act as conduits for fluids both along and across the plane of a fault or large fracture [*Faulkner et al.*, 2010]. Faults can act as hydraulic conduits between reservoirs for hydrocarbon migration or between the deep crust and sedimentary basins to allow along-fault flow of geothermal fluids. Natural CO_2 reservoirs are often charged by such fluids migrating along faults and originating from deep crustal or mantle sources [*Jeandel et al.*, 2010]. Natural gas reservoirs, on the other hand, are charged from more shallow sources with primary migration to the reservoir above or below the source rock, depending on pressure drive, secondary migration toward a trap, or tertiary migration through a seal along faults/fractures or the capillary network [*Tissot and Welte*, 1984]. This demonstrates that in principle both very deep and shallower crustal faults can be conductive, while fluxes need to be understood in order to distinguish between geological timescales (10^5–10^6 years) and timescales relevant to carbon storage (10^0–10^4 years). The permeability of fractures strongly depends on confining stresses and the relative orientation of fractures to the principal stress directions. Fracture permeability tends to increase toward the surface as a result of a reduction in confining stress.

Faults may also act as barriers to fluid flow, preventing fluids from migrating laterally, leading to compartmentalization of reservoirs. The concept of shale gouge ratio (SGR) has been used in the literature to predict the effectiveness of a fault as barrier for lateral flow; an overview is given elsewhere [*Manzocchi et al.*, 1998; *Yielding et al.*, 1997, 2010] and will not be discussed herein. Flow along faults (in a vertical direction) and fractures is however a more significant mechanism for transmitting fluids and represents a greater risk for leakage to surface. Quantifying such a process is challenging due to the complexities of characterizing fracture network connectivity across seals, fracture density, aperture sizes, relative permeability, or capillary entry pressures. Little data are available on these properties in the literature.

14.2.1.1. Diffusive Leakage Along Fractures

When gaseous CO_2 or CH_4 gets into contact with a seal containing faults and fractures and the capillary entry pressure of these conduits is not exceeded, the mode of transport will be diffusion (Fig. 14.1a). Assuming fracture apertures of the order of 10^{-3}–10^{-5} m [*Gale et al.*, 2014], effective diffusion coefficients for any gas in brine can be considered close to aqueous diffusion coefficients. These are of the order of 10^{-9} m^2 s^{-1} [*Maharajh and Walkley*, 1973], hence one to three orders of magnitude higher than those in narrow pore systems with nm-sized and tortuous pore systems. From Figure 14.2a, it becomes clear that, as a worst case scenario, diffusive leakage across a seal will only occur after 10^5 years with a shale thickness of 10 m. It seems however unlikely that seals with thicknesses of approximately 10 m would qualify as a sufficient barrier for CO_2 storage, and therefore much longer timescales can be expected.

14.2.1.2. Advective Leakage Along Fractures

For advective flow of gas to occur (Fig. 14.1b), the capillary pressure needs to exceed the capillary entry pressure of the fracture. For fracture apertures of the order of 10^{-3}–10^{-5} m, gas excess pressures exceed fracture entry pressures at values of 10^2–10^5 Pa, corresponding to gas column heights for CO_2 or CH_4 of less than a few meters under reservoir conditions. Capillarity is therefore an insufficient control to prevent open fractures from leakage. Considering a depleted hydrocarbon reservoir where pressures are below original hydrostatic pressures for at least a certain period of time but where the caprock is still at original hydrostatic pressure, the risk of advective leakage is considered small. This is because for advection to occur an excess gas pressure is required to displace

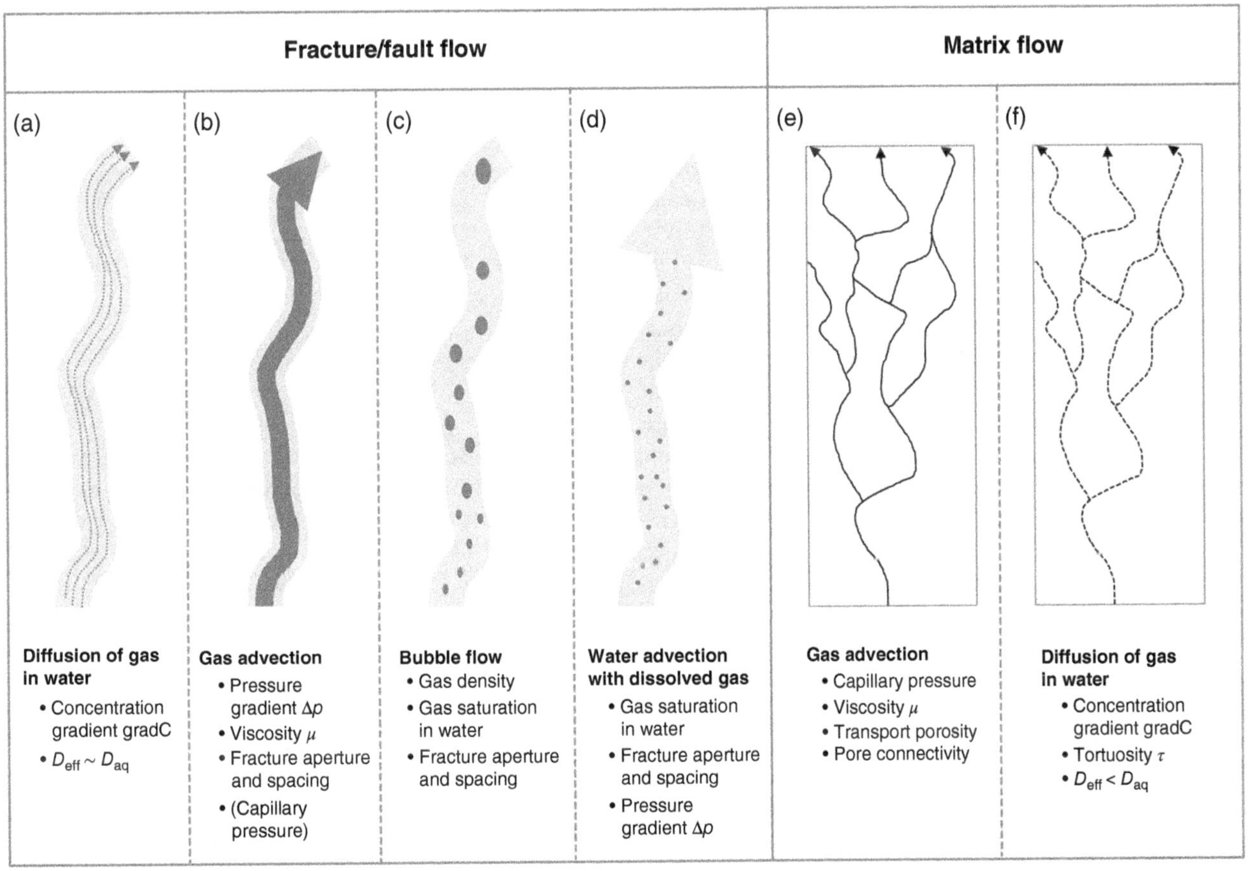

Figure 14.1 Leakage mechanisms along fault/fractures and through the caprock matrix.

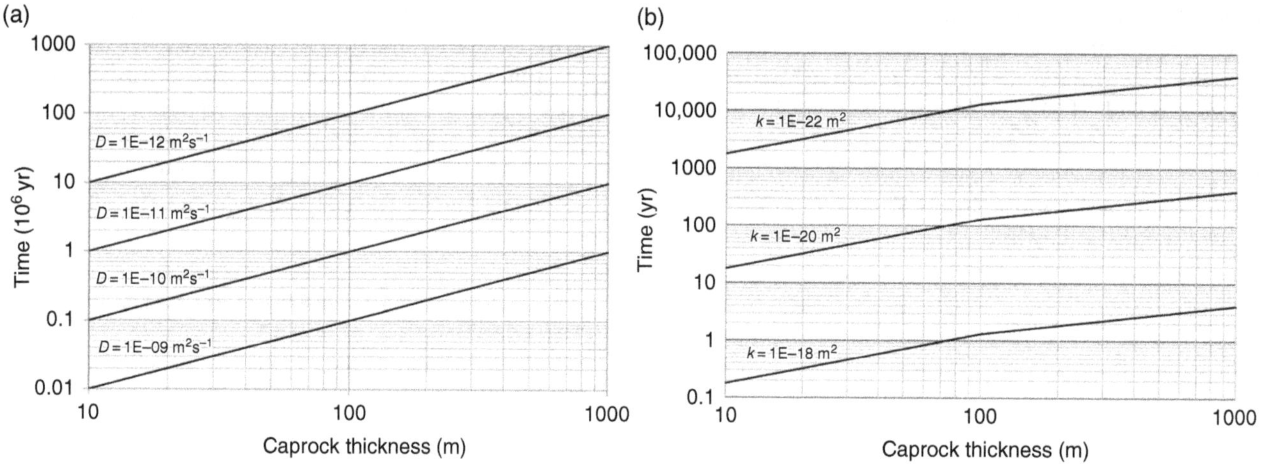

Figure 14.2 Diffusive (a) and advective (b) travel times through a caprock of varying thickness and permeability. Assumptions and calculation steps given in *Andreas Busch et al.* [2010]. The depth of the reservoir-caprock interface is at 2000 m; note that in (b) the time axis is in millions of years.

water in the caprock, which is not the case when the pressure gradient is inverted. In a saline aquifer where reservoir pressure after CO_2 injection is above hydrostatic, advective flow is likely for open or connected fault and fracture networks with an effective gas permeability that is considered lower than the single-phase brine permeability.

Although only rarely discussed in the context of CO_2 storage, a rather special case for gas advection is bubble flow along fractures and conduits. Near-surface examples include CO_2-driven cold-water geysers. Two examples are Crystal Geyser, Utah, and the Andernach Geyser in the Eifel area, Germany, where the former is driven by an open exploration well and the latter by a fracture network. For a thorough discussion of CO_2-driven cold-water geysers, see *Han et al.* [2013]. Cold water geysers are driven by the upward flow of gas-rich fluids; as the fluids flow to shallow depth a decrease in the hydrostatic pressure lowers solubility of the gas, gas bubbles nucleate, coalesce and rise buoyantly. As the density of the fluid column decreases the rate of degassing increases until a runaway degassing event occurs, driving an eruption. Gas leaves the solution via the nucleation of microbubbles (Fig. 14.1c) whose formation will be governed by gas solubility, surface energy effects, surface roughness and diffusion of gas in the solution. The increase in CO_2 solubility with decreasing temperature is small relative to the decrease in solubility with decreasing pressure, for temperature and pressure gradients in sedimentary basins, such that CO_2 solubility will strongly decrease at depths below 700 m. This is illustrated in Figure 14.3c for CO_2 [*Duan and Sun*, 2003] and Figure 14.3d for CH_4 [*Duan et al.*, 1992], documenting a decrease in gas solubility with decreasing depth. Gas bubbles will preferentially nucleate on fracture surfaces, and once they reach a critical size they will detach and migrate upwards, driven by the buoyancy force P_b. The velocity v of which depends on gas bubble radius r and density difference between the gas and surrounding fluid (brine): $\Delta\rho = (\rho_w - \rho_g)$, ignoring the effect of drag on the bubble velocity. At low Reynolds numbers, it can be calculated according to *Epstein and Plesset* [1950]:

$$v = \frac{2}{9} g \frac{(\rho_w - \rho_g)}{\mu_w} \left(\frac{3\mu_w + 3\mu_g}{2\mu_w + 3\mu_g} \right), \qquad (14.1)$$

For CH_4, having low viscosity and density compared to water, this equation can be simplified to

$$v = \frac{1}{3} g \frac{\rho_w}{\mu_w} \qquad (14.2)$$

where μ is viscosity, g is acceleration due to gravity, and the subscripts w and g refer to water and gas, respectively. Figure 14.4a shows the velocity of a gas bubble with depth assuming a formation radius of 0.1 mm at 1000 m depth [*Hey et al.*, 1994]. The fundamental controls on nucleation and detachment bubble radii are not well characterized. Nanobubbles will rise more slowly due to lower buoyancy on the one hand and because they can be affected by Brownian motion. Bubble velocity increases exponentially towards the surface as bubble density and viscosity decrease and the bubble volume increases. (Fig. 14.3a–d). As a result, rising gas bubbles will migrate more radially than the CO2 saturated fluid front. Subsequently, the rising bubbles pass through CO_2-undersaturated water and dissolve,

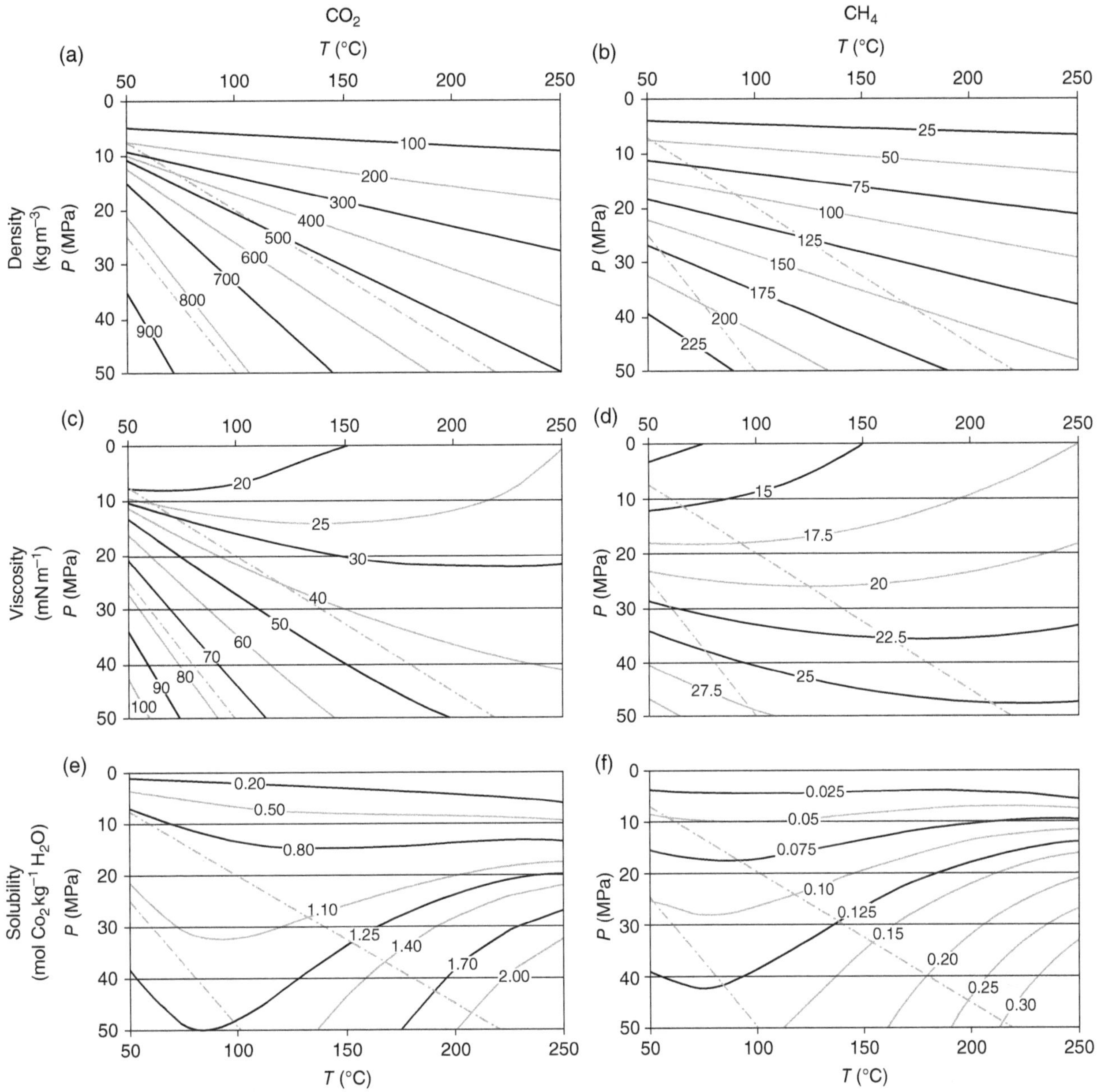

Figure 14.3 Pressure-temperature diagrams for density, viscosity, and solubility of CO_2 (left column) and CH_4 (right column). Solid curves show isolines within the p-T field. Also shown are geothermal gradients for 20 and 40°C km^{-1} depth, assuming hydrostatic pressures. Under normal hydrostatic conditions, the pressure gradient is 10 MPa km^{-1}, so 50 MPa refers to a depth of 5 km. Data sources: *Span and Wagner* [1996]: 2A and 2C; *Setzmann and Wagner* [1991]: 2B and 2D; *Duan and Sun* [2003]: 2E; and *Zhenhao Duan et al.* [1992]: 2F.

as illustrated by rising CO_2 bubbles in a seawater column [*Brewer et al.*, 2002] and which will be discussed further below. Due to its lower solubility, methane can be expected to rise higher in the water column than CO_2, such that migration of a bubble front drives fraction of the gas composition. In such a system, the advancement of a CO_2 exsolution bubble front will be pinned to the rate of advancement of the much slower CO_2-saturated brine, greatly limiting the potential for rapid gas release and leakage until CO_2-saturated brine has reached the surface. At this point, CO_2 bubbles will be able to migrate freely through the CO_2-saturated fluid, and the surface flux will increase. This form of reservoir-to-surface leakage will only be possible in a connected system with migration pathways or fracture apertures that are sufficiently large, to allow bubble or

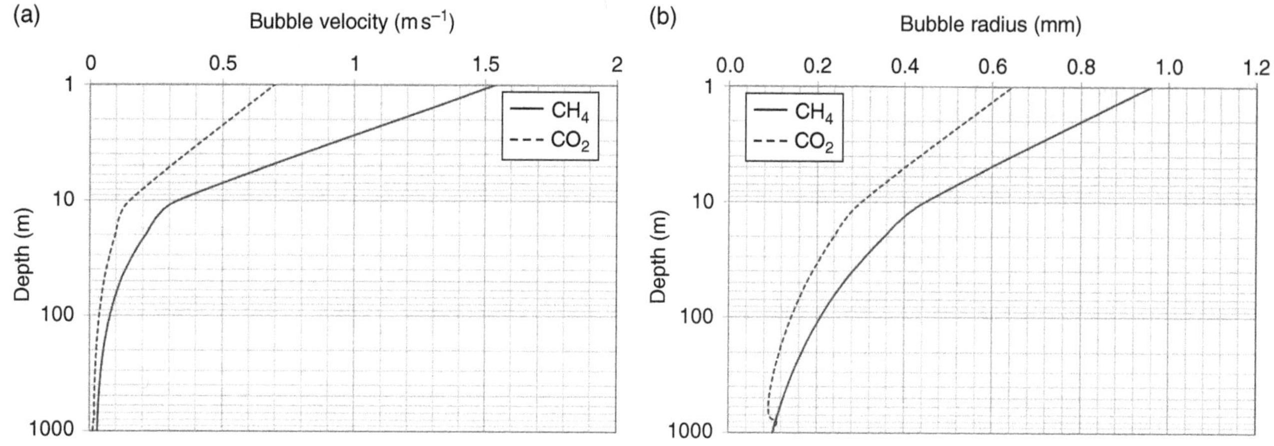

Figure 14.4 Semilogarithmic plot showing (a) bubble velocity and (b) bubble radius versus depth for CO_2 and CH_4.

slug migration. Figure 14.4b shows the development of bubble sizes for CH_4 and CO_2, formed at 1000 m depth at a radius $r = 0.1$ mm, when migrating to the surface. Changes in bubble radius were calculated according to the real gas law:

$$V_1P_1z_1 = V_2P_2z_2 \tag{14.3}$$

where V, P, and z are volume, hydrostatic pressure, and gas compressibility, respectively, and subscripts 1 and 2 denote different depths during upward migration, assuming spherical bubbles. Bubble radius can be calculated from its volume assuming spherical shape. Figure 14.4b illustrates how the bubble volume increases towards the surface, resulting in the density decrease that drives increasing bubble velocities. When bubbles encounter narrow fracture apertures, they will remain trapped by capillary forces until a sufficient coalesced gas column hight h is formed, with a sufficient buoyancy force to overcome the capillary forces. The minimum gas column h_{min} height required to access certain pore or aperture sizes is given by, for example [*Busch et al.*, 2010]:

$$h_{min} = \frac{2\gamma \cos(\theta)}{r \Delta\sigma g}, \tag{14.4}$$

where γ is the gas-brine interfacial tension (Pa.s), θ (°) is the contact angle, and r (m) is the pore size or fracture aperture radius.

The intrinsic permeability of single fractures is often estimated using a cubic law

$$k_{frac} = \frac{b^2}{12} \tag{14.5}$$

where b is the hydraulic fracture aperture.

14.2.2. Matrix Flow

14.2.2.1. Gas Advection

Gas leakage through the caprock matrix (Fig. 14.1e–f) has been studied in some detail in the past, by either focusing on concentration-driven molecular diffusion [*Busch et al.*, 2008; *Krooss et al.*, 1988; *Leythaeuser et al.*, 1980; *Schlömer and Krooss*, 1997] or by pressure-driven gas advection in the narrow pore space of caprocks [*Busch and Amann-Hildenbrand*, 2013; *Hildenbrand et al.*, 2002, 2004]. The relative permeability of gas and brine is difficult to quantify due to the generally low permeabilities of such rocks. Depending on gas saturations and rock type, effective gas permeability was reported up to two orders of magnitude lower compared to the single-phase permeability [*Busch and Amann-Hildenbrand*, 2013; *Hildenbrand et al.*, 2004]. Single-phase permeabilities in mudrocks or shales are typically of the order of 10^{-19}–10^{-21} m^2 (0.01–1 nD [*Busch and Amann-Hildenbrand*, 2013]). Considering such permeabilities, the time required for gas to bypass a seal of various thicknesses can be estimated (Fig. 14.2b) [*Busch et al.*, 2010]. It has been shown that leakage timescales are 10^0–10^3 years, depending on caprock thickness and effective permeability.

Capillary leakage phenomena commonly observed in sedimentary basins include gas chimneys that sometimes form pockmarks on the sea floor [*Cathles et al.*, 2010]. Such chimneys are usually fairly localized features, can include deformed and distured sediment and occur in areas where gas pressures are high and/or where the shaly overburden, including any sealing lithologies, is of potentially higher permeability/lower capillary entry pressure. Only when the capillary displacement pressure of a rock is low and a small gas column height create an excess pressure necessary for gas to migrate in the vertical direction. As the gas saturation increases high effective gas permeabilities across a seal can develop, however, the time take for matrix flow from km deep reservoirs to the surface through low permeability overburden, can be considered to be on the order of 10^5–10^6 years (Fig. 14.2b). [*Busch and Amann-Hildenbrand*, 2013].

14.2.2.2. Gas Diffusion

Molecular diffusion through the water saturated pore space is slow relative to rates of advection. Since diffusive transport rates are slow the mass transfer rates of CO_2 are sensitive to fluid-rock reactions. Such fluid rock reactions might increase (by mineral dissolution following increased porosity) or decrease (by CO_2 consumption through mineral reactions) transport velocities. For instance, the former was reported for a Carboniferous caprock in lab studies, implying extremely fast reactions [*Armitage et al.*, 2013], and the latter was reported from natural field observations on a clay-rich seal [*Kampman et al.*, 2016]. Assuming Fick's law and the assumptions made in *Krooss et al.* [1992], the time required for dissolved CO_2 to diffuse through a caprock can be calculated and is summarized graphically in Figure 14.2a. It is obvious that even through a thin caprock with 10 m thickness and just by pure aqueous diffusion (~$2\cdot10^{-9}$ m^2 s^{-1}; e.g., *Kampman et al.* [2016]), the time required is still >10^5 years. For effective diffusion coefficients of the order of 10^{-10}–10^{-12} m^2 s^{-1}, migration times increase to 10^5 or even 10^7 years, making diffusion through intact caprock an unlikely leakage mechanism from deep reservoirs.

14.2.3. Dispersion, Convective Mixing, Dissolution, and Trapping of Leaking CO_2

When gas migrates vertically along faults, fracture networks, or imperfections in wellbore cements, flow and mixing of fluids will take place in shallower formations, which will impede the transport of CO_2 to the surface. The geological overburden above typical storage targets in sedimentary basins often comprises sequences of aquifers and aquitards. During migration, flow and mixing processes operating across a range of scales will enhance rates of CO_2 dissolution (Fig. 14.5). Such processes include the following:

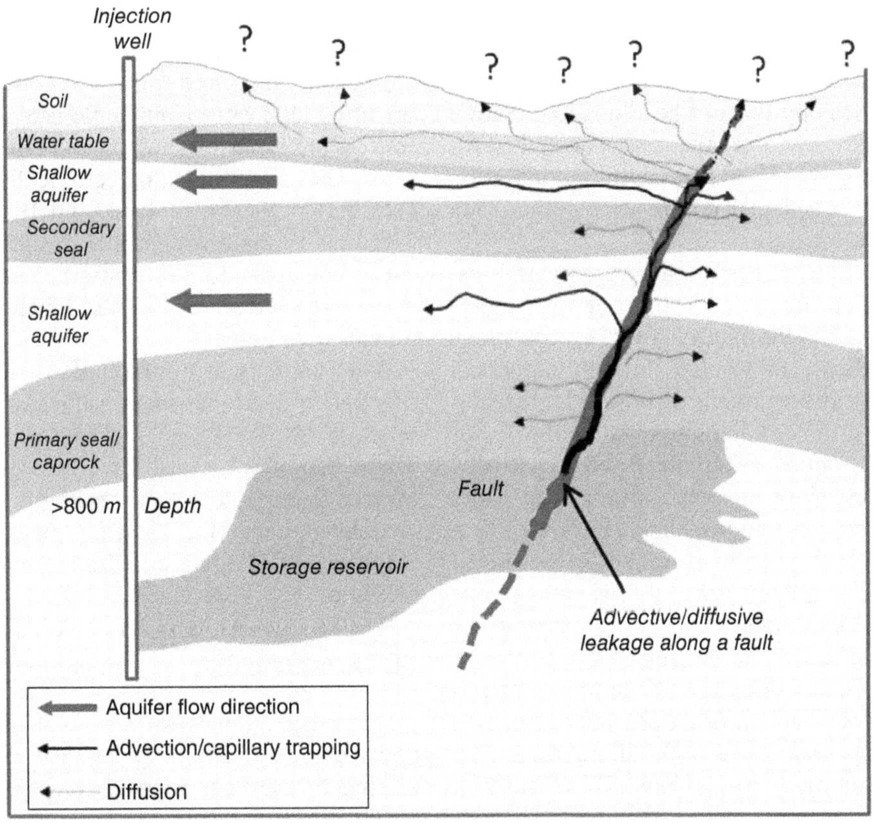

Figure 14.5 Illustration of a conceptual model for potential fault leakage to the surface. Transport mode in this illustration is by diffusion or advection along faults. Significant quantities of migrating fluids will be trapped in shallow permeable layers by diffusion or capillary trapping. Close to the surface, this will lead to massive dissipation and dilution.

• Macroscopic dispersive mixing of fluids in complex and tortuous fracture networks at tens to hundreds meter length scales

• Flow, dispersion, and mixing of brine and CO_2 with freshwater in aquifers penetrated by the fracture networks, arising from the intrinsic dispersivity of the porous matrix

• Gravity-driven convective mixing of dense CO_2-saturated water with CO_2-free low-density meteoric fluids in thief zones

These physical processes of flow will bring CO_2 and CO_2-charged brines into contact with increasing volumes of CO_2-free fluid, enhance mixing, and ultimately lead to more effective rates of CO_2 dissolution. The dense CO_2-saturated brines that form will tend to sink, rather than rise, relative to the background CO_2-free groundwaters, and flow laterally rather than vertically, governed by the topology of aquifer horizons and regional groundwater pressure gradients.

Figure 14.5 schematically shows flow, dispersion, and dissolution along a fault zone into lateral thief zones represented by high-permeability layers. These are represented by low vertical/horizontal permeability anisotropy. The surface CO_2 flux will therefore be reduced. In case leaking fluids get close to the surface in areas dominated by groundwater reservoirs, vadose zone soils (onshore) or marine sediment dispersion and dispersion rates can be expected to increase, leading to further dilution over large areas. In this case, leakage will not be a localized feature but rather occur over large areas that are technically difficult to monitor. This should be taken into consideration when planning for a suitable contingency monitoring strategy.

In general, advection, dispersion, convective, and dissolution processes are interrelated and not well understood on a basin scale, as the multiplicity of scale of these phenomena is difficult to resolve through numerical simulation. Numerical simulations of flow at large scales will tend to underestimate dissolution rates of CO_2 or other gases, as they cannot capture the heterogeneities of flow and the physics of mixing, which operates at scales below typical gridding resolutions. Similarly, flow of CO_2 and brine along a fracture network cannot be described by a single intrinsic or relative permeability, and variable flow velocities between fractures will tend to increase the contact area between brine and flowing CO_2. Conversely, accumulation of CO_2 gas in shallow formations may lead to formation of a gas column height sufficient to generate localized overpressurization and rock failure. As shown in Figure 14.3(c and d), a free gas phase from gas-rich brines forms by exsolution due to pressure reduction. Eventually, the fracture gradient of the rock may be exceeded, leading to the generation of a fracture network which enhances rates of CO_2 migration toward the surface. The competing effects of dissolution and accumulation on velocity of CO_2 flows at the basin scale are poorly understood and worth of much further investigation.

14.3. NATURAL ANALOGUES FOR FLUID LEAKAGE

14.3.1. Volcanic and Geothermal Gas Seeps Worldwide

Etiope and coworkers intensively studied and reviewed the global volcanic and nonvolcanic surface degassing of CO_2 and CH_4 [*Etiope*, 2009; *Etiope and Martinelli*, 2002; *Etiope and Milkov*, 2004; *Etiope et al.*, 2007, 2009; *Mörner and Etiope*, 2002]. For active subaerial volcanoes, they showed that CO_2 emissions can be of the order of 10^{-2}–10^{2} Mt yr^{-1}. Nonvolcanic gas vents in comparison are quantified with an output of 10^{-3}–10^{-1} Mt yr^{-1} (one exception is Lake Nyos in Cameroon, reaching 2.3 Mt yr^{-1} as a single event in 1986). In comparison, for soil degassing at or in the vicinity of subaerial volcanic activities, these output rates are lower and of the order of 10^{-2}–10^{0} Mt yr^{-1}, with the exception of two cases from Nicaragua where rates can be up to 10^{1} Mt yr^{-1}. Area normalized values for soil degassing range from 10^{-2} to 10^{5} kg m^{-2} yr^{-1}. Although CO_2 emissions from volcanic sources are often included in studies summarizing natural leakage rates [*Lewicki et al.*, 2007; *Pearce et al.*, 2004], these rates and fluxes are not directly analogue to potential leak rates from geological reservoirs. Volcanic sources are typically shallow and related to active volcanism. They are situated in tectonically active areas, characterized by active faulting and fracturing, high fluid pressures at shallow depth, and high geothermal gradients. Such areas do not typify the structural and permeability architecture of stable sedimentary basins targeted for carbon storage.

When looking at nonvolcanic fluxes from either geothermal or tectonic areas, values are generally of the order of 10^{-1}–10^{3} kg m^{-2} yr^{-1} [*Mörner and Etiope*, 2002]. These examples are also not representative for gas loss from engineered geological reservoirs since such gas vents are related to tectonically active zones, such as those at or in the vicinity of subduction zones or areas of continental collision.

14.3.2. Natural CO_2 Accumulations

Natural analogues for CO_2 leakage from engineered reservoirs are locations where naturally occurring CO_2 is leaking to the near surface. Such leaks are sometimes exploited commercially for carbonated waters (e.g., Eifel, German) or as a source of CO_2 for industrial applications. Recently, the EU-funded QICS project [*Kirk*, 2011] published a review of leakage rates for onshore and offshore sites. A comprehensive review of natural analogues is not the scope of this paper, and the reader is referred to *Kirk* [2011] and *Roberts et al.* [2015].

Natural accumulations of CO_2 in sedimentary basins, which currently discharge CO_2 to the surface, represent the most suitable analogues for understanding migration and surface leakage of CO_2 from geological reservoirs. There are a number of deep sedimentary CO_2 reservoirs, corresponding to storage depth for engineered storage, between 800 and about 2000 m, that have been discovered during petroleum exploration. Detailed studies are available for the Werkendam reservoir [*Hangx et al.*, 2015; *Koenen et al.*, 2013] and the Miller [*Lu et al.*, 2009, 2011] and Fizzy [*Heinemann et al.*, 2013; *Yielding et al.*, 2011] fields, all situated in the North Sea. Further examples are reservoirs on the US Colorado Plateau [*Allis et al.*, 2001] or in the Otway Basin, Australia [*Higgs et al.*, 2015; *Watson et al.*, 2004]. Leaking reservoirs found within sedimentary basins are typically at shallow depth; however, they are scarce in the public literature (Table 14.1). Many of the most suitable analogue sites can be found on the Colorado Plateau [*Allis et al.*, 2001; *Bickle et al.*, 2013; *Moore et al.*, 2005]. Of these sites, the Little Grand Wash (LGW) fault, near the town of Green River, Utah, has been the subject of the most numerous and detailed study [*Burnside*, 2010; *Busch et al.*, 2014; *Dockrill and Shipton*, 2010; *Jung et al.*, 2014; *Kampman et al.*, 2012]. Here CO_2 and CO_2-enriched waters have been leaking along the LGW and Salt Wash (SW) faults to the surface over the past approximately 10^5 years. This analogue will be discussed in more depth below.

Surface releases from natural CO_2 seeps are observed in different ways, such as bubbling waters, diffusive degassing, vents, springs, wells, or fumaroles [*Burnside et al.*, 2013; *Roberts et al.*, 2015], with the area over which leakage occurs can vary from the cm to the km scale or even larger in the case of paleo-leakages at Green River or the accumulation at Springerville-St. Johns, Arizona. At these locations, contemporary and ancient CO_2 leaks are preserved as extensive deposits of travertine. Table 14.1 shows natural CO_2 reservoirs. Only those at shallow depth seem to show surface releases.

14.3.3. Case Study: Little Grand Wash Fault

The natural CO_2 field at Green River, Utah, allows a number of observations to be made about the physiochemical processes that occur during CO_2 migration through faults, from deep geological reservoirs. At Green River, CO_2 and CO_2-charged brines escape to the surface through a series of natural springs and seeps along the LGW and the Salt Wash Graben (SWG) faults and through abandoned oil and water drill holes (Fig. 14.6). The geological setting, structure, and the evolution of this CO_2 field have been described by a number of studies [*Burnside et al.*, 2013; *Dockrill and Shipton*, 2010; *Heath*, 2004; *Heath et al.*, 2009b; *Kampman et al.*, 2009, 2012; *Shipton et al.*, 2004]. The site includes a major cold-water CO_2 geyser, Crystal Geyser, an uncapped petroleum exploration well drilled adjacent to the LGW fault in 1935, which has been the focus of a number of studies including measurements of the effused volumes of CO_2 [*Gouveia et al.*, 2005], temporal measurements of the eruption cycle [*Gouveia and Friedmann*, 2006; *Han et al.*, 2013; *Watson et al.*, 2014], and the link between changes in the effluent fluid chemistry eruption cycle and source reservoir [*Kampman et al.*, 2014b].

Green River is located in the north part of the Paradox Basin. The shallow dipping Paleozoic to Cretaceous stratigraphy comprises continental fluvial and eolian sandstones, shales, limestones, and evaporites, which are gently folded about a NNW trending anticline and cut by two steeply dipping, 25–30 km long, east-west striking fault sets, the LGW and SWG faults. Small reservoirs of free-phase CO_2 are thought to exist in the Permian Cutler Group and the underlying Paradox Formation which are penetrated by the fault at depths of around 1.5–2 km [*Kampman et al.*, 2014b]. The Paradox Formation is overlain by a number of regional aquifers, and local artesian groundwater conditions drive vertical migration of fluids, including deeply derived brines and CO_2, through the faults [*Kampman et al.*, 2009]. The CO_2 that emanates at the surface is sourced from degassing within the fault as fluids flow through a series of shallow Jurassic sandstone aquifers that have been the target of a recent scientific drilling program [*Kampman et al.*, 2013]. The $CO_2/^3He$ and $^3He/^4He$ ratio of the effused gases points to a mixed mantle and crustal origin for the CO_2 [*Wilkinson et al.*, 2009].

The site at Green River is an excellent analogue for the slow leakage of a deep CO_2 accumulation through a simple normal fault system. It illustrates the effectiveness of fluid flow and mixing on dissolving CO_2 migrating through the overburden and provides important insight into processes occurring during multiphase fluid in faults.

Movement on the LGW fault has been dated using $^{40}Ar/^{39}Ar$ dating of clay-rich fault gouge to 40 Ma ± 10 [*Pevear et al.*, 1997], corresponding with early Tertiary Laramide tectonism, a period attributed to the formation of most relatively large displacement faults in the region [*Doelling*, 1988; *McKnight*, 1940]. The structure of the faults is described in detail in *Dockrill and Shipton* [2010]. The maximum throws of the LGW and SWG faults are 260 and 366 m, respectively. Both faults comprise an argillaceous cataclastic fault core, surrounded by a damage zone in the sandstone host rock, comprising fracture sets that are broadly parallel to the strike of the main fault and most intense in the footwall of the fault. They are considered to have a low vertical transmissivity [*Jung et al.*, 2015; *Shipton et al.*, 2004]. The open fractures in

Table 14.1 Global Natural CO_2 Analogues Representative of Settings Suitable for Carbon Storage.

Location	On/ offshore	Leakage mechanism	Lithology	Leak rate (g/m²/h)	CO² reservoir				
					Depth (m)	P (MPa)	T (°C)	ϕ (%)	k (mD)
Matraderecske HUN	On	Faults/fractures	Tertiary karst systems that leak into weathered andesite and andesitic tuffs	~2000–4000 (close to fractures up to 17,000)	1000	10	40	n.d.	0.028–0.007
St. Johns-Springerville USA	On	Travertine mounts on the surface assumes periodic leakage along faults and fractures	Stacked Permian sandstone sealed by shale/anhydrite/silts	Unquantified	200–700	6	49	10	10
Florina Basin GRE	On	Diffusive along seal discontinuities	Miocene sands with alternating coal seams, overlain by claystones	Unquantified	180–260	n.d.	n.d.	n.d.	n.d.
Paradox Basin USA	On	Travertine mounts on the surface assumes periodic leakage along faults and fractures	Different Jurassic sandstone and carbonate reservoirs sealed by marine shale/silt layers	0.00215 ± 0.001 for Little Grand Wash Fault	Several 100 m	Differs per res. unit	Differs per res. unit	Differs per res. unit	Differs per res. unit
Montmiral FRA	On	n.d.	Mid-Jurassic carbonates sealed bu claystones and marls	No leak detected	2400–2480	24	103	6–12	0.3–120
McElmo Dome USA	On	n.d.	Carboniferous dolomite sealed by salt	No leak detected	1800–2600	24	71	3.5–25	(0.1)–23–(500)
Mihalyi-Repcelak HUN	On	n.d.	Tertiary (Pannonian) Conglomerates and sandstones sealed by pelites and claystones	No leak detected	1100–1300	12	42	n.d.	n.d.
Vert le Grand FRA	On	n.d.	Triassic continental arkosic sandstones sealed by Lower Jurassic shales	No leak detected	1800	18	75	n.d.	20–200
Bravo Dome USA	On	n.d.	Permian arkosic sandstones sealed by anhydrite and arkosic muds	No leak detected	580–900	8	41	20	42
Farnham Dome USA	On	n.d.	Jurassic Navajo sandstone sealed by Jurassic Carmel shale	No leak detected	900			12	>100

Ladbroke Grove AUS	On	n.d.	Early Cretaceous fluvial and lacustrine sediments sealed by shale and siltstone	No leak detected	3000	~30.8	~142	12.5	44.7
Miller Field UK	Off	n.d.	Jurassic submarine fan sandstone overlain by Kimmeridge mudrocks	No leak detected	3970–4090	40	110	15	n.d.
Magnus Field UK	Off	n.d.	Jurassic submarine fan sandstone overlain by Kimmeridge mudrocks	No leak detected	3200	32	110	20	~100
Blue Whale VIE	Off	n.d.	Miocene platform carbonates sealed by pelagic mudstones	No leak detected	1600	16	65	25	n.d.

All settings are in sedimentary basins, with reservoirs being sandstones or carbonates and seals being shales or evaporites. Mainly shallow reservoirs show leaks at some point in the geological history, while leaks for deep reservoirs are not documented (n.d.). Information summarized from various sources [*Allis et al.*, 2001; *Baines and Worden*, 2004; *Burnside et al.*, 2013; *Gaus et al.*, 2004; *Higgs et al.*, 2015; *Kampman et al.*, 2012, 2013, 2014a, 2014b; *Kirk*, 2011; *Lu et al.*, 2009, 2011; *Pearce et al.*, 2004; *Shipton et al.*, 2004, 2005; *Watson et al.*, 2004; *Worden and Barclay*, 2000]. Other CO_2 analogues that are related to active volcanism or tectonically active zones were disregarded since they are not considered to be representative for CO_2 storage and therefore also not representative as an analogue for CO_2 leakage.

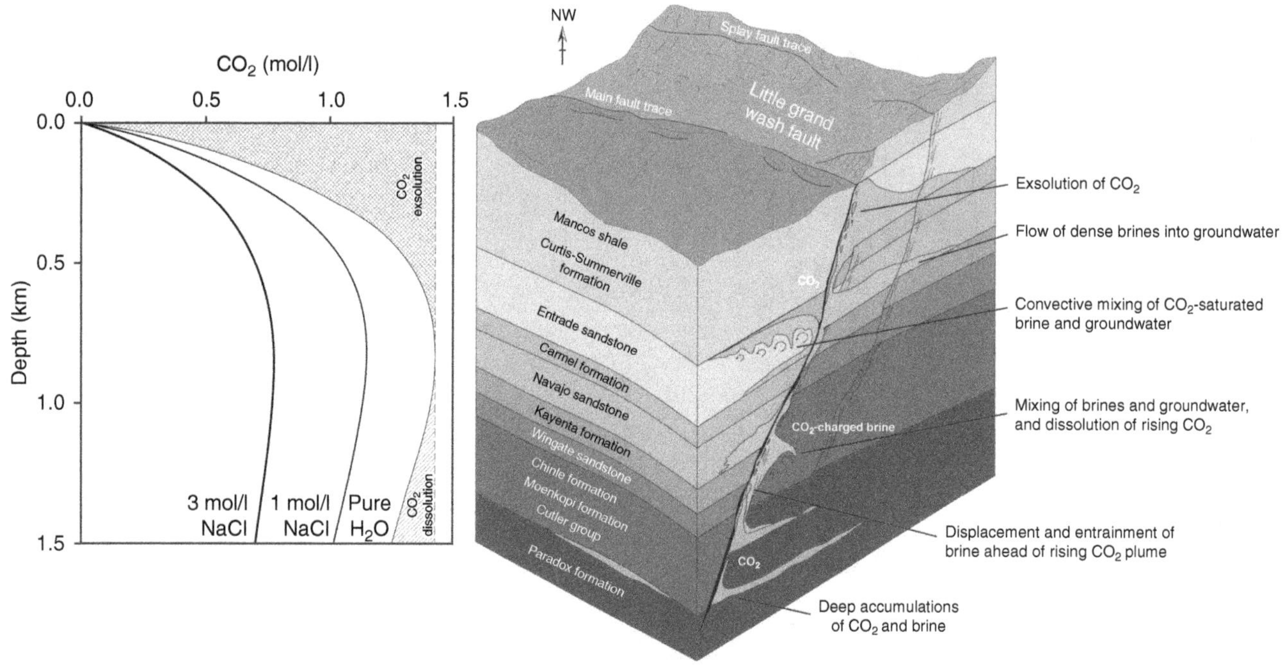

Figure 14.6 Conceptual model of the leaking CO_2 field at Green River, Utah, showing the sequence of processes that occurs during leakage of free-phase CO_2 from reservoirs in the Pennsylvanian Paradox Formation and its migration through the overburden. The behavior of the CO_2 during migration is a strong function of changes in CO_2 solubility with depth, as illustrated in the adjoining CO_2 solubility versus depth plot. From *Kampman et al.* [2014b]. Reprinted with permission from Elsevier.

the footwall provide the primary conduits through which CO_2 and CO_2-charged brines escape to the surface.

The surface CO_2 degassing and natural spring discharge is localized where these faults transect the apex of the anticline. Degassing of CO_2-charged brines flowing through the fault has led to the accumulation of extensive travertine deposits which drape the faults and networks of thick carbonate veins in the fault plane and fracture networks. U-Th dating of carbonate veins which crosscut the travertines has shown that the faults have been leaking CO_2 for up to 400 kyrs [*Burnside et al.*, 2013]. *Burnside et al.* [2013] estimate that approximately 20 Mt of CO_2 has escaped from the faults over the past 100 k.y., a rate of approximately 200 t yr^{-1}. For comparison, it is estimated that CO_2 now leaks from Crystal Geyser at 11000 t yr^{-1} [*Gouveia and Friedmann*, 2006; *Gouveia et al.*, 2005]. The discharge of CO_2 and brine from the fault has been episodic and cyclical, with a CO_2 flux in the geological past that was at times significantly higher than the modern flux of CO_2 [*Kampman et al.*, 2012]. Temporal changes in U-Th dated carbonate vein chemistry and travertine deposition rates have been used to demonstrate a correlation between increased discharge of CO_2 and brine from the faults and transitions in local glacial to interglacial climatic conditions. This is attributed to a reduction in effective normal stresses and the opening of tensile fractures in the fault, following regional crustal unloading of glaciers and pluvial lakes, and the increase in groundwater recharge rates and pore fluid pressures during the onset of wet interglacial conditions.

A complex distribution of travertine mounds has formed along the fault length, localized at intersecting fault and fracture planes, complexities which form high-permeability conduits. The recorded spatial-temporal pattern of leakage through the fault is complex [*Burnside et al.*, 2013], with some travertine mounds active for 15–20 k.y., while others are active for only a few tens or hundreds of years. The abandonment of some leakage sites suggests blocking of the flow path by carbonate mineralization, but leakage continued concurrently along other portions of the faults, suggesting that local fracture healing may not significantly inhibit overall brine and CO_2 discharge rates from the faults.

Modern ground surface CO_2 fluxes have been measured by *Allis et al.* [2005] and *Jung et al.* [2014]. The natural background soil-gas CO_2 flux in this arid and poorly vegetated area is around 0.005 $kg\,m^{-2}\,day^{-1}$. In the vicinity of the faults, the CO_2 fluxes range from 0.1 $kg\,m^{-2}\,day^{-1}$ to 5.9 $kg\,m^{-2}\,day^{-1}$. The CO_2 fluxes are elevated in the vicinity of CO_2-degassing springs, with a maximum recorded flux of 36 $kg\,m^{-2}\,day^{-1}$ in the soil around Crystal Geyser. The fluxes tend to increase in regions of complexity of the faults where multiple fault planes intersect, around zones of intense fracturing, and in the footwall of the faults. No anomalous CO_2 flux was detected more than 20 m from a fault trace. Flux measurements can vary by four orders of magnitude between sample points separated by only tens of centimeters. This highlights that the surface CO_2 flux is dominated by localized leakage from fractures, either exposed or buried beneath regolith, and that there is either no or negligible diffuse CO_2 flux.

The Jurassic sandstone aquifers from which the surface CO_2 flux is sourced have been the target of a recent scientific drilling program [*Kampman et al.*, 2013, 2014b]. The drilling targeted these aquifers within the footwall of the LGW fault, at the apex of the anticline, in the vicinity of Crystal Geyser. The drill penetrated an approximately 300 m thick sequence comprising the lower Jurassic Navajo Sandstone, a major regional aquifer, the overlying Carmel Formation, a regional seal, and the Upper Jurassic Entrada Sandstone which outcrops at surface. These shallow reservoirs are fed by active flows of deeply derived CO_2-charged brines through the faults. During drilling, CO_2-charged brines were encountered throughout the Navajo Sandstone, within a fracture zone penetrating the Carmel Formation, and sporadically through the Entrada Sandstone, at depths as shallow as approximately 50 m below surface. Pressurized fluids sampled from the Navajo Sandstone are CO_2 saturated at the base of the formation, suggesting both CO_2-saturated brines and free-phase CO_2 migrate through the fault zone. The geochemistry of fluids sampled at various depths from the drill hole reveals active mixing of dense CO_2-charged brines as they flow along the base of the formations, with meteoric waters flowing through the aquifers, and dispersion of these fluids in the regions around the faults. These CO_2-charged brine-meteoric water mixtures then flow laterally from LGW to the south, along regional potentiometric pressure gradients, where they then encounter and are further discharged from the SWG fault, as a series of springs and localized dry gas seeps. *Kampman et al.* [2009] modeled the along-flow geochemical gradients in the erupted fluids to show that sharp drops in pH reflected the addition of CO_2 at the two major fault systems and gradients in fluid chemistry downstream allowed calculation of the CO_2-promoted fluid-mineral reactions and their rates occurring in the Navajo Sandstone.

The element and isotope geochemistry of the shallow CO_2-charged groundwaters has been used to identify the source of the deeply derived brines as being Carboniferous limestone and evaporite formations at depths of greater than 2 km in this part of the basin [*Heath et al.*, 2009a; *Kampman et al.*, 2009, 2014b; *Wilkinson et al.*, 2009]. During the Quaternary, the Paradox fold and thrust belt, on which the SW and LGW faults lie on the periphery, experienced an intense period of brine expulsion from these deep formations, when extensive dissolution of salt from the Carboniferous Paradox Formation led to localized collapse of a series of salt-cored anticlines [*Baars and Stevenson*, 1981]. Free-phase CO_2 has been encountered in these formations and in the overlying Permian White Rim Sandstone during oil and gas explorations well drilling along the axis of the Green River anticline [*J. Beach*, Delta Petroleum, pers. comm.]. However, no free-phase CO_2 gas cap was encountered during the drilling of the shallow Jurassic reservoirs, suggesting that the majority of the CO_2 gas migrating from depth is effectively dissolved in groundwater during its passage to the surface and that the surface flux of CO_2 is dominantly derived from degassing of the fluids as they migrate vertically through the shallow Jurassic aquifers.

The migration and dissolution of a free CO_2 phase is supported by information recorded in fluids sampled during drilling of the anticline [*Kampman et al.*, 2014b]. Brines sampled from the anticline throughout the Permian to Jurassic stratigraphy show sequential dilution of their solute load as the saline Carboniferous brines mix with dilute meteoric groundwater on their passage to the surface. The elemental ratios of conservative ions (e.g., Br/Cl) are preserved during dilution of the brines, while the absolute concentrations are diluted approximately 50–200-fold in fluids sampled by the drill hole, relative to their concentrations in the original Carboniferous brines. However, the ratio of dissolved CO_2 to the conservative ions in the fluids sampled from the shallow aquifers, for example, CO_2/Cl, is significantly elevated relative to the concentrations expected for conservative dilution of the CO_2/Cl ratio in the original Carboniferous brines. This indicates that a significant fraction of the dissolved CO_2 in fluids flowing through the Navajo Sandstone, approximately 75%, was derived during fluid mixing and migration through the overburden. This suggests that free-phase CO_2 must be leaking from the deep Carboniferous reservoir along with the brine, but this free-phase CO_2 does significantly accumulate in the shallow reservoir in the footwall of the fault, suggesting that it is effectively dissolved during its passage to the surface. The Carboniferous source reservoirs are separated from the shallow Jurassic sediments by a thick sequence of shale aquitards with two intervening sandstone aquifers, the Permian White Rim Sandstone and Triassic Wingate Sandstone. Fluid flow and mixing where these aquifers intersect the faults, like that observed in the Navajo Sandstone, is likely to be an effective mechanism for bringing CO_2-undersaturated fluids into contact with migrating supercritical CO_2 and gas, resulting in effective dissolution of CO_2 in the groundwaters.

This process is illustrated by 2D numerical modeling of CO_2 and brine flow within the fault presented in *Jung et al.* [2015]. The authors model buoyant flow of supercritical and gaseous CO_2 through the LGW fault over a 1000-year period, from a supercritical reservoir of CO_2 in the Carboniferous strata, with a numerical model that includes CO_2 phase behavior and solubility in brine. By tuning the permeability of the faults in their model, the authors were able to quantitatively reproduce the observed surface CO_2 soil flux around portions of the LGW fault. The modeling showed that most supercritical CO_2 flowing through the LGW fault dissolves into ambient brine during upward migration along the fault plane. As a result, CO_2 travels predominantly dissolved in an aqueous phase. The dissolution of CO_2 leads to shrinking of the CO_2 plume along the flow paths; and therefore, free CO_2 migration is retarded. The authors suggest that the dissolution of CO_2 may occur even more intensely in the field situation, because their models assume a homogeneous salinity for all the aquifers, to which CO_2 solubility is highly sensitive, and do not include the active transport of low-salinity meteoric groundwaters into the vicinity of the fault, as is the case in nature, which would increase the efficiency of CO_2 dissolution.

In their model, the authors observed the flow of dense CO_2-saturated brines along the base of the aquifers in the northern footwall of the LGW, as was observed by *Kampman et al.* [2014b] for fluids from the drill hole. In the models, gaseous CO_2 pools in the hanging wall of the fault and flows to the south. This leads to overriding of buoyant CO_2 in the shallow aquifer horizons and the formation of a dense CO_2-saturated layer above less dense CO_2-free formation water. Such an arrangement would give rise to a density instability and convective mixing of the fluids that would increase the CO_2 dissolution rate; however, this could not be captured in their models due to the scale of the gridding. Based upon simplified analytical modeling using linear stability analysis, the authors postulate that this convective overturning may develop in very short timescales (months to years) and that this process would effectively dissolve all of the southward migrating CO_2 gas plumes observed in each shallow aquifer layer (Entrada, Navajo, Wingate) in their models in approximately 100–400 years. The inferences from modeling and field observations highlight the effectiveness with which groundwater flow and dispersive and convective mixing is attenuating the migration of a supercritical CO_2 plume by dissolving the CO_2 in groundwater.

A similar inference has been made by *Bickle and Kampman* [2013] who used a simple idealized flow model and field data to estimate the amounts of CO_2 transported away from the faults by regional groundwater flow and the amount sequestered in carbonate mineralization in the reservoir, as compared to that discharged from the faults and CO_2 springs. Based on the dimensions, CO_2 saturation, and groundwater flow rates in the Navajo Sandstone, they estimate that approximately 4000 t yr^{-1} of dissolved CO_2 is transported away from the faults by groundwater or approximately 20× the estimated rate of surface degassing of 200 t yr^{-1} from the LGW fault made by *Burnside et al.* [2013]. There are at least two deeper aquifers potentially comparable to the Navajo Sandstone (the Wingate and White Rim sandstones). If the fraction of dissolved CO_2 and the groundwater flow rate in each are comparable to that in the Navajo Sandstone such that 4000 t yr^{-1} of dissolved CO_2 is transported away from the fault in each aquifer, then the fraction of CO_2 reaching the surface would be 0.053 or 0.01% of the total estimated CO_2 migrating through the fault system. The authors also made an estimate of the net rate of CO_2 sequestration into carbonate minerals in the reservoir during flow between LGW and SWG from the calculations of *Kampman et al.* [2009]. These imply that approximately 9% of the CO_2 dissolved (400 t yr^{-1}) is precipitated in carbonate along this portion of the flow path. This is only a small part of the total amount dissolved, although more than this is naturally escaping at the surface. This is, however, the integrated mass over a limited portion of the ultimate flow path.

The Green River CO_2 field illustrates the importance of fluid flow, mixing, and dissolution during CO_2 migration through the overburden on impeding the passage of CO_2 to the surface. It represents one of the best studied analogues for a slowly leaking deep free phase accumulation of CO_2. Future studies of the site, including deeper drilling and fluid sampling and high-resolution numerical modeling, may shed further light on the length scales over which CO_2 dissolution and exsolution occur within the fault zone. Critically, a fundamental understanding of the length scales over which steady state flow and dissolution of CO_2 occurs is required to better predict the risk of surface escape during CO_2 migration through fault zones.

14.4. SUMMARY AND CONCLUSIONS

We examine a number of mechanisms for the migration of CO_2 from geological reservoirs and surface leakage of the stored CO_2. The flux of CO_2 if migration occurs from a storage complex will be dominated by flow through faults, fractures, abandoned wellbores, and imperfections in wellbore completions. There are some unresolved questions and uncertainties surrounding behavior and flow through seals; however, the absolute fluxes are anticipated to be small. There are numerous physicochemical and hydrodynamic processes that can hinder the flow of CO_2

or CO_2-charged waters to the surface. These include dispersive mixing of fluids in fractures and pores, gravity-driven convective mixing of dense fluids, regional mixing of flowing groundwaters and leaking fluids, capillary trapping of CO_2 in aquifers above the storage target, or accumulation in shallow traps and thief zones and mineral reactions. These processes occur across a multitude of scales and are difficult to capture through basin or reservoir-scale numerical simulations of flow. As such, their contribution to retarding leakage of CO_2 from storage reservoirs has most likely been significantly underestimated. It is important to note here that significant advances in computation or novel simulation methods are required to address these problems.

Physical constraints on leakage mechanisms and rates are typically derived from natural analogues. We find however that most analogues discussed in the literature for CO_2 storage are shallow reservoirs of CO_2-charged fluids or gas in tectonic or volcanic active areas which are not directly analogous to leakage of CO_2 from a storage operation in stable sedimentary basins. Few deep CO_2 geological reservoirs are reported to leak, either because there simply is no leak to report or because no quantifiable CO_2 flux reaches the surface. This suggests that the risk and probability of significant migration from a storage reservoir back to the atmosphere is small.

The flux of CO_2 in the event of a leak will be governed by the permeability, size, and density of fracture networks, independent of the size of the stored CO_2 volume. A deeper understanding of flow processes in fractures is required to fully address the risks of leakage. Detailed characterization of these systems for individual storage sites is desirable but likely unfeasible in every case, given the small scale of these geological structures and the inherent limits to the resolution of geophysical imaging techniques.

Monitoring for such leaks will be challenging. If leaks are diluted in the vertical column or over large areas in the near surface, monitoring strategies will need to adapt both in density and sensitivity. It remains to be tested whether migration out of a primary storage container will be detected by subsurface and well monitoring technologies before a leak would reach the seabed or the surface. If flow through fractures can be monitored, geophysical monitoring tools need to be tested and verified given the small overall saturation changes and the small percentage of fractures that would actually contribute to flow, for example [*Evans et al.*, 2005]. Ultimately, engineered or controlled gas release experiments are required to test the monitoring and remediation technologies and to aid in an understanding of the underlying mechanisms that will enhance or inhibit leakage.

ACKNOWLEDGMENTS

We thank M. Dean, R. Wentinck, and T. Tambach for valuable comments on an earlier version of this manuscript, two anonymous reviewers for very useful comments, and Shell Global Solutions for permission to publish this article. This study has been partly subsidized through the ERANET Cofund ACT (Project no. 271497). Support is greatly acknowledged.

REFERENCES

Allis, R., T. Chidsay, W. Gwynn, C. Morgan, S. White, M. Adams, and J. Moore (2001), Natural CO_2 reservoirs on the Colorado Plateau and Southern Rocky Mountains: candidates for CO_2 sequestration, Proceedings of the First National Conference on Carbon Sequestration, DOE-NETL CD DOE/NETL-2001/1144, 14–17 May 2001, Washington, DC.

Allis, R., D. Bergfeld, J. Moore, K. McClure, C. Morgan, T. Chidsey, J. Heath, and B. McPherson (2005), Implications of results from CO_2 flux surveys over known CO_2 systems for long-term monitoring, United States Geological Survey.

Amann, A., M. Waschbüsch, P. Bertier, A. Busch, B. M. Krooss, and R. Littke (2011), Sealing rock characteristics under the influence of CO_2, *Energy Procedia*, *4*, 5170–5177.

Amann-Hildenbrand, A., P. Bertier, A. Busch, and B. M. Krooss (2013), Experimental investigation of the sealing capacity of generic clay-rich caprocks, *International Journal of Greenhouse Gas Control*, *19*, 620–641, doi:10.1016/j.ijggc.2013.01.040.

Armitage, P. J., D. R. Faulkner, and R. H. Worden (2013), Caprock corrosion, *Nature Geoscience*, *6*(2), 79–80.

Baars, D., and G. Stevenson (1981), Tectonic evolution of the Paradox basin, Utah and Colorado.

Bachu, S. (2000), Sequestration of CO_2 in geological media: criteria and approach for site selection in response to climate change, *Energy Conversion and Management*, *41*(9), 953–970.

Bachu, S., and D. B. Bennion (2009), Experimental assessment of brine and/or CO_2 leakage through well cements at reservoir conditions, *International Journal of Greenhouse Gas Control*, *3*(4), 494–501.

Baines, S. J., and R. H. Worden (2004), The long-term fate of CO_2 in the subsurface: natural analogues for CO_2 storage, *Geological Society, London, Special Publications*, *233*(1), 59–85.

Berne, P., P. Bachaud, and M. Fleury (2010), Diffusion properties of carbonated caprocks from the Paris basin, *Oil and Gas Science and Technology, Revue de l'IFP*, *65*(3), 473–484.

Bickle, M., and N. Kampman (2013), Lessons in carbon storage from geological analogues, *Geology*, *41*(4), 525–526, doi:10.1130/focus0420132.1.

Bickle, M., N. Kampman, and M. Wigley (2013), Natural analogues, *Reviews in Mineralogy and Geochemistry*, *77*(1), 15–71, doi:10.2138/rmg.2013.77.2.

Brewer, P. G., E. T. Peltzer, G. Friederich, and G. Rehder (2002), Experimental determination of the fate of rising CO_2 droplets in seawater, *Environmental Science & Technology*, *36*(24), 5441–5446.

Brown, A. (2000), Evaluation of possible gas microseepage mechanisms, *AAPG Bulletin*, *84*(11), 1775–1789, doi:10.1306/8626c389-173b-11d7-8645000102c1865d.

Burnside, N. M. (2010), U-Th dating of travertines on the Colorado Plateau: implications for the leakage of geologically stored CO_2, Phd thesis, 306 pp, Glasgow University.

Burnside, N. M., Z. K. Shipton, B. Dockrill, and R. M. Ellam (2013), Man-made versus natural CO_2 leakage: a 400 k.y. History of an analogue for engineered geological storage of CO_2, *Geology*, *41*(4), 471–474.

Busch, A., and A. Amann-Hildenbrand (2013), Predicting capillarity of mudrocks, *Marine and Petroleum Geology*, *45*, 208–223, doi:10.1016/j.marpetgeo.2013.05.005.

Busch, A., S. Alles, Y. Gensterblum, D. Prinz, D. N. Dewhurst, M. D. Raven, H. Stanjek, and B. M. Krooss (2008), Carbon dioxide storage potential of shales, *International Journal of Greenhouse Gas Control*, *2*(3), 297–308.

Busch, A., A. Amann-Hildenbrand, P. Bertier, M. Waschbuesch, and B. M. Krooss (2010), The significance of caprock sealing integrity for CO_2 storage, in SPE International Conference on CO_2 Capture, Storage, and Utilization, Society of Petroleum Engineers, 10–12 November 2010, New Orleans, LA.

Busch, A., et al. (2014), The Green River natural analogue as a field laboratory to study the long-term fate of CO_2 in the subsurface, *Energy Procedia*, *63*, 2821–2830, doi:10.1016/j.egypro.2014.11.304.

Cathles, L. M., Z. Su, and D. Chen (2010), The physics of gas chimney and pockmark formation, with implications for assessment of seafloor hazards and gas sequestration, *Marine and Petroleum Geology*, *27*(1), 82–91, doi:10.1016/j.marpetgeo.2009.09.010.

Collettini, C., C. Cardellini, G. Chiodini, N. De Paola, R. E. Holdsworth, and S. A. F. Smith (2008), Fault weakening due to CO_2 degassing in the northern Apennines: short- and long-term processes, *Geological Society, London, Special Publications*, *299*(1), 175–194, doi:10.1144/sp299.11.

Commission of the European Communities(2009), Directive of the European Parliament and of the Council on the Geological Storage of Carbon Dioxide and Amending. Council Directives 85/337/EEC, 96/61/EC, Directives 2000/60/EC, 2001/80/EC, 2004/35/EC, 2006/12/EC and Regulation (EC) No 1013/2006.

Det Norske Veritas (2010), CO2QUALSTORE – Guideline for Selection and Qualification of Sites and Projects for Geological Storage of CO_2, 77 pp, Det Norske Veritas.

Dockrill, B., and Z. K. Shipton (2010), Structural controls on leakage from a natural CO_2 geologic storage site: Central Utah, U.S.A, *Journal of Structural Geology*, *32*(11), 1768–1782, doi:10.1016/j.jsg.2010.01.007.

Doelling, H. (1988), Geology of Salt Valley Anticline and Arches National Park, Grand County, Utah, in *Salt deformation in the Paradox region, Utah*, Utah Geological and Mineral Survey Bulletin (122), pp. 1–60.

Duan, Z., and R. Sun (2003), An improved model calculating CO_2 solubility in pure water and aqueous NaCl solutions from 273 to 533 K and from 0 to 2000 bar, *Chemical Geology*, *193*, 257–271.

Duan, Z., N. Møller, J. Greenberg, and J. H. Weare (1992), The prediction of methane solubility in natural waters to high ionic strength from 0 to 250°C and from 0 to 1600 bar, *Geochimica et Cosmochimica Acta*, *56*(4), 1451–1460, doi:10.1016/0016-7037(92)90215-5.

Epstein, P. S., and M. S. Plesset (1950), On the stability of gas bubbles in liquid-gas solutions, *The Journal of Chemical Physics*, *18*(11), 1505–1509, doi:10.1063/1.1747520.

Etiope, G. (2009), Natural emissions of methane from geological seepage in Europe, *Atmospheric Environment*, *43*(7), 1430–1443, doi:10.1016/j.atmosenv.2008.03.014.

Etiope, G., and G. Martinelli (2002), Migration of carrier and trace gases in the geosphere: an overview, *Physics of the Earth and Planetary Interiors*, *129*(3–4), 185–204, doi:10.1016/S0031-9201(01)00292-8.

Etiope, G., and A. V. Milkov (2004), A new estimate of global methane flux from onshore and shallow submarine mud volcanoes to the atmosphere, *Environmental Geology*, *46*(8), 997–1002.

Etiope, G., T. Fridriksson, F. Italiano, W. Winiwarter, and J. Theloke (2007), Natural emissions of methane from geothermal and volcanic sources in Europe, *Journal of Volcanology and Geothermal Research*, *165*(1–2), 76–86, doi:10.1016/j.jvolgeores.2007.04.014.

Etiope, G., A. Feyzullayev, and C. L. Baciu (2009), Terrestrial methane seeps and mud volcanoes: a global perspective of gas origin, *Marine and Petroleum Geology*, *26*(3), 333–344.

Evans, K. F., A. Genter, and J. Sausse (2005), Permeability creation and damage due to massive fluid injections into granite at 3.5 km at Soultz: 1. Borehole observations, *Journal of Geophysical Research. Solid Earth*, *110*(B4), doi:10.1029/2004JB003168.

Faulkner, D. R., C. A. L. Jackson, R. J. Lunn, R. W. Schlische, Z. K. Shipton, C. A. J. Wibberley, and M. O. Withjack (2010), A review of recent developments concerning the structure, mechanics and fluid flow properties of fault zones, *Journal of Structural Geology*, *32*(11), 1557–1575, doi:10.1016/j.jsg.2010.06.009.

Gal, F., B. Michel, B. Gilles, J. Frédéric, and M. Karine (2011), CO_2 escapes in the Laacher see region, east Eifel, Germany: application of natural analogue onshore and offshore geochemical monitoring, *International Journal of Greenhouse Gas Control*, *5*(4), 1099–1118.

Gale, J. F. W., S. E. Laubach, J. E. Olson, P. Eichhubl, and A. Fall (2014), Natural fractures in shale: a review and new observations, *AAPG Bulletin*, *98*(11), 2165–2216, doi:10.1306/08121413151.

Gasda, S. E., S. Bachu, and M. A. Celia (2004), Spatial characterization of the location of potentially leaky wells penetrating a deep saline aquifer in a mature sedimentary basin, *Environmental Geology*, *46*(6), 707–720.

Gaus, I., C. Le Guern, J. Pearce, H. Pauwels, T. Shepherd, G. Hatziyannis, and A. Metaxas (2004), Comparison of long term geochemical interactions at two natural CO_2 analogues: Montmiral (Southeast basin, France) and Messokampos (Florina basin, Greece) case studies, paper presented at 7th International Conference on Greenhouse Gas Control Technologies (GHGT-7), 5–9 September 2014, Vancouver, Canada.

Gor, G. Y., T. R. Elliot, and J. H. Prevost (2013), Effects of thermal stresses on caprock integrity during CO_2 storage, *International Journal of Greenhouse Gas Control*, *12*, 300–309, doi:10.1016/j.ijggc.2012.11.020.

Gouveia, F., and S. Friedmann (2006), Timing and prediction of CO_2 eruptions from Crystal Geyser, UT, Department of Energy.

Gouveia, F., M. Johnson, R. Leif, and S. Friedmann (2005), Aerometric measurement and modeling of the mass of CO_2 emissions from Crystal Geyser, Utah, UCRL-TR-211870, Lawrence Livermore National Laboratory, Livermore, CA.

Han, W. S., M. Lu, B. J. McPherson, E. H. Keating, J. Moore, E. Park, Z. T. Watson, and N. H. Jung (2013), Characteristics of CO_2-driven cold-water geyser, crystal geyser in Utah: experimental observation and mechanism analyses, *Geofluids*, *13*(3), 283–297.

Hangx, S., E. Bakker, P. Bertier, G. Nover, and A. Busch (2015), Chemical–mechanical coupling observed for depleted oil reservoirs subjected to long-term CO_2-exposure – a case study of the Werkendam natural CO_2 analogue field, *Earth and Planetary Science Letters*, *428*, 230–242, doi:10.1016/j.epsl.2015.07.044.

Heath, J. (2004), *Hydrogeochemical Characterization of Leaking Carbon Dioxide-Charged Fault Zones in East-Central Utah*, Utah State University, Logan, UT.

Heath, J. E., T. E. Lachmar, J. P. Evans, P. T. Kolesar, and A. P. William (2009a), Hydrogeochemical characterization of leaking carbon dioxide-charged fault zones in east-central Utah, with implications for geologic carbon storage, in *Carbon Sequestration and Its Role in the Global Carbon Cycle*, vol. *183*, Geophysical Monograph Series, American Geophysical Union, Washington, DC.

Heath, J. E., T. E. Lachmar, J. P. Evans, P. T. Kolesar, and A. P. Williams (2009b), Hydrogeochemical characterization of leaking, carbon dioxide-charged fault zones in east-central Utah, with implications for geologic carbon storage, *Carbon Sequestration and Its Role in the Global Carbon Cycle*, *183*, 147–158.

Heinemann, N., M. Wilkinson, R. S. Haszeldine, A. E. Fallick, and G. E. Pickup (2013), CO_2 sequestration in a UK North Sea analogue for geological carbon storage, *Geology*, *41*(4), 411–414, doi:10.1130/g33835.1.

Hey, M. J., A. M. Hilton, and R. D. Bee (1994), The formation and growth of carbon dioxide gas bubbles from supersaturated aqueous solutions, *Food Chemistry*, *51*(4), 349–357, doi:10.1016/0308-8146(94)90185-6.

Higgs, K. E., R. R. Haese, S. D. Golding, U. Schacht, and M. N. Watson (2015), The Pretty Hill Formation as a natural analogue for CO_2 storage: an investigation of mineralogical and isotopic changes associated with sandstones exposed to low, intermediate and high CO_2 concentrations over geological time, *Chemical Geology*, *399*, 36–64, doi:10.1016/j.chemgeo.2014.10.019.

Hildenbrand, A., S. Schlömer, and B. M. Krooss (2002), Gas breakthrough experiments on fine-grained sedimentary rocks, *Geofluids*, *2*, 3–23.

Hildenbrand, A., S. Schlömer, B. M. Krooss, and R. Littke (2004), Gas breakthrough experiments on pelitic rocks: comparative study with N2, CO_2 and CH4, *Geofluids*, *4*, 61–80.

Hornafius, J. S., D. Quigley, and B. P. Luyendyk (1999), The world's most spectacular marine hydrocarbon seeps (coal oil point, Santa Barbara Channel, California): quantification of emissions, *Journal of Geophysical Research, Oceans*, *104*(C9), 20703–20711.

Hovland, M. (2002), On the self-sealing nature of marine seeps, *Continental Shelf Research*, *22*(16), 2387–2394.

Jeandel, E., A. Battani, and P. Sarda (2010), Lessons learned from natural and industrial analogues for storage of carbon dioxide, *International Journal of Greenhouse Gas Control*, *4*(6), 890–909, doi:10.1016/j.ijggc.2010.06.005.

Jeanne, P., Y. Guglielmi, and F. Cappa (2013), Dissimilar properties within a carbonate-reservoir's small fault zone, and their impact on the pressurization and leakage associated with CO_2 injection, *Journal of Structural Geology*, *47*, 25–35, doi:10.1016/j.jsg.2012.10.010.

Jeanne, P., Y. Guglielmi, F. Cappa, A. P. Rinaldi, and J. Rutqvist (2014), The effects of lateral property variations on fault-zone reactivation by fluid pressurization: application to CO_2 pressurization effects within major and undetected fault zones, *Journal of Structural Geology*, *62*, 97–108, doi:10.1016/j.jsg.2014.01.017.

Jung, N. H., W. S. Han, Z. T. Watson, J. P. Graham, and K. Y. Kim (2014), Fault-controlled CO_2 leakage from natural reservoirs in the Colorado plateau, east-Central Utah, *Earth and Planetary Science Letters*, *403*, 358–367.

Jung, N.-H., W. S. Han, K. Han, and E. Park (2015), Regional-scale advective, diffusive, and eruptive dynamics of CO_2 and brine leakage through faults and wellbores, *Journal of Geophysical Research. Solid Earth*, *120*(5), 3003–3025, doi:10.1002/2014JB011722.

Kampman, N., M. Bickle, J. Becker, N. Assayag, and H. Chapman (2009), Feldspar dissolution kinetics and Gibbs free energy dependence in a CO_2-enriched groundwater system, Green River, Utah, *Earth and Planetary Science Letters*, *284*(3–4), 473–488.

Kampman, N., N. M. Burnside, Z. K. Shipton, H. J. Chapman, J. A. Nicholl, R. M. Ellam, and M. J. Bickle (2012), Pulses of carbon dioxide emissions from intracrustal faults following climatic warming, *Nature Geoscience*, *5*(5), 352–358.

Kampman, N., A. Maskell, M. Bickle, J. Evans, M. Schaller, G. Purser, Z. Zhou, J. Gattacceca, E. Peitre, and C. Rochelle (2013), Scientific drilling and downhole fluid sampling of a natural CO_2 reservoir, Green River, Utah, *Scientific Drilling*, *16*, 33–43.

Kampman, N., M. Bickle, M. Wigley, and B. Dubacq (2014a), Fluid flow and CO_2-fluid-mineral interactions during CO_2-storage in sedimentary basins, *Chemical Geology*, *369*, 22–50.

Kampman, N., et al. (2014b), Drilling and sampling a natural CO_2 reservoir: implications for fluid flow and CO_2-fluid-rock reactions during CO_2 migration through the overburden, *Chemical Geology*, *369*, 51–82, doi:10.1016/j.chemgeo.2013.11.015.

Kampman, N., et al. (2016), Observational evidence confirms modelling of the long-term integrity of CO_2-reservoir caprocks, *Nature Communications*, *7*, 12268.

Keating, E. H., D. L. Newell, H. Viswanathan, J. W. Carey, G. Zyvoloski, and R. Pawar (2013), CO_2/brine transport into shallow aquifers along fault zones, *Environmental Science and Technology*, *47*(1), 290–297.

Kirk, K. (2011), Natural CO_2 flux literature review for the QICS project, British Geological Survey Commissioned Report, 38 pp.

Koenen, M., L. J. Wasch, M. E. van Zalinge, and S. Nelskamp (2013), Werkendam, the Dutch natural analogue for CO_2

storage – long-term mineral reactions, *Energy Procedia*, *37*, 3452–3460, doi:10.1016/j.egypro.2013.06.235.

Krooss, B. M., D. Leythaeuser, and R. G. Schaefer (1988), Light hydrocarbon diffusion in a caprock, *Chemical Geology*, *71*(1–3), 65–76.

Krooss, B. M., D. Leythaeuser, and R. G. Schaefer (1992), The quantification of diffusive hydrocarbon losses through cap rocks of natural gas reservoirs-A reevaluation, *AAPG Bulletin*, *76*(3), 403–406.

Lewicki, J. L., J. Birkholzer, and C. F. Tsang (2007), Natural and industrial analogues for leakage of CO_2 from storage reservoirs: identification of features, events, and processes and lessons learned, *Environmental Geology*, *52*(3), 457–467.

Leythaeuser, D., R. G. Schaefer, and A. Yukler (1980), Diffusion of light hydrocarbons through near-surface rocks, *Nature*, *284*(5756), 522–525.

Logan, G. A., A. T. Jones, J. M. Kennard, G. J. Ryan, and N. Rollet (2010), Australian offshore natural hydrocarbon seepage studies, a review and re-evaluation, *Marine and Petroleum Geology*, *27*(1), 26–45.

Loizzo, M., O. A. Akemu, L. Jammes, J. Desroches, S. Lombardi, and A. Annunziatellis (2011), Quantifying the Risk of CO_2 Leakage Through Wellbores, Society of Petroleum Engineers, SPE-139635-PA.

Lu, J., M. Wilkinson, R. S. Haszeldine, and A. E. Fallick (2009), Long-term performance of a mudrock seal in natural CO_2 storage, *Geology*, *37*(1), 35–38.

Lu, J., M. Wilkinson, R. S. Haszeldine, and A. Boyce (2011), Carbonate cements in miller field of the UK North Sea: a natural analog for mineral trapping in CO_2 geological storage, *Environmental Earth Sciences*, *62*(3), 507–517, doi:10.1007/s12665-010-0543-1.

Maharajh, D. M., and J. Walkley (1973), The temperature dependence of the diffusion coefficients of Ar, CO_2, CH4, CH3Cl, CH3Br, and CHCl2F in water, *Canadian Journal of Chemistry*, *51*(6), 944–952, doi:10.1139/v73-140.

Manzocchi, T., P. S. Ringrose, and J. R. Underhill (1998), Flow through fault systems in high-porosity sandstones, *Geological Society, London, Special Publications*, *127*(1), 65–82, doi:10.1144/gsl.sp.1998.127.01.06.

McGinnis, D. F., M. Schmidt, T. Delsontro, S. Themann, L. Rovelli, A. Reitz, and P. Linke (2011), Discovery of a natural CO_2 seep in the German North Sea: implications for shallow dissolved gas and seep detection, *Journal of Geophysical Research C: Oceans*, *116*(3), 1–12. doi:10.1029/2010JC006557.

McKnight, E. T. (1940), *Geology of Area Between Green and Colorado Rivers, Grand and San Juan Counties*, Utah, US Government Printing Office.

Michael, K., A. Golab, V. Shulakova, J. Ennis-King, G. Allinson, S. Sharma, and T. Aiken (2010), Geological storage of CO_2 in saline aquifers – a review of the experience from existing storage operations, *International Journal of Greenhouse Gas Control*, *4*(4), 659–667, doi:10.1016/j.ijggc.2009.12.011.

Moore, J., M. Adams, R. Allis, S. Lutz, and S. Rauzi (2005), Mineralogical and geochemical consequences of the long-term presence of CO_2 in natural reservoirs: an example from the Springerville–St. Johns field, Arizona, and New Mexico, U.S.A, *Chemical Geology*, *217*(3–4), 365–385, doi:10.1016/j.chemgeo.2004.12.019.

Mörner, N.-A., and G. Etiope (2002), Carbon degassing from the lithosphere, *Global and Planetary Change*, *33*(1–2), 185–203, doi:10.1016/S0921-8181(02)00070-X.

Pearce, J., I. Czernichowski-Lauriol, S. Lombardi, S. Brune, A. Nador, J. Baker, H. Pauwels, G. Hatziyannis, S. Beaubien, and E. Faber (2004), A review of natural CO_2 accumulations in Europe as analogues for geological sequestration, *Geological Society, London, Special Publications*, *233*(1), 29–41.

Pevear, D. R., P. J. Vrolijk, and F. J. Longstaffe (1997), Timing of Moab fault displacement and fluid movement integrated with burial history using radiogenic and stable isotopes, paper presented at Geofluids II '97: Contributions to the Second International Conference on Fluid Evolution, Migration and Interaction in Sedimentary Basins and Orogenic Belts, Belfast, The Queen's University of Belfast.

Rinaldi, A. P., P. Jeanne, J. Rutqvist, F. Cappa, and Y. Guglielmi (2014), Effects of fault-zone architecture on earthquake magnitude and gas leakage related to CO_2 injection in a multi-layered sedimentary system, *Greenhouse Gases: Science and Technology*, *4*(1), 99–120, doi:10.1002/ghg.1403.

Roberts, J. J., R. A. Wood, M. Wilkinson, and S. Haszeldine (2015), Surface controls on the characteristics of natural CO_2 seeps: implications for engineered CO_2 stores, *Geofluids*, *15*(3), 453–463, doi:10.1111/gfl.12121.

Schlömer, S., and B. M. Krooss (1997), Experimental characterisation of the hydrocarbon sealing efficiency of cap rocks, *Marine and Petroleum Geology*, *14*, 565–580.

Setzmann, U., and W. Wagner (1991), A new equation of state and tables of thermodynamic properties for methane covering the range from the melting line to 625 K at pressures up to 1000 MPa, *Journal of Physical and Chemical Reference Data*, *20*, 1061–1155.

Shipton, Z. K., J. P. Evans, D. Kirschner, P. T. Kolesar, A. P. Williams, and J. Heath (2004), *Analysis of CO_2 Leakage Through 'Low-Permeability' Faults from Natural Reservoirs in the Colorado Plateau, East-Central Utah*, Geological Society Special Publication, pp. 43–58.

Shipton, Z. K., J. P. Evans, B. Dockrill, J. Heath, A. Williams, D. Kirchner, and P. T. Kolesar (2005), Natural leaking CO_2-charged systems as analogs for failed geologic storage reservoirs, in *Carbon Dioxide Capture for Storage in Deep Geologic Formations – Results from the CO_2 Capture Project*, edited by D. Thomas, vol. 2, Elsevier Science, pp. 699–712.

Span, R., and W. Wagner (1996), A new equation of state for carbon dioxide covering the fluid region from the triple-point temperature to 1100 K at pressures up to 800 MPa, *Journal of Physical and Chemical Reference Data*, *25*(6), 1509–1596.

Tao, Q., D. A. Checkai, N. J. Huerta, and S. L. Bryant (2014), Estimating CO_2 fluxes along leaky wellbores, *SPE Journal*, *19*(2), 227–238.

Tissot, B., and D. H. Welte (1984), *Petroleum Formation and Occurrence*, Springer-Verlag, Berlin.

Tueckmantel, C., Q. J. Fisher, T. Manzocchi, S. Skachkov, and C. A. Grattoni (2012), Two-phase fluid flow properties of cataclastic fault rocks: implications for CO_2 storage in saline aquifers, *Geology*, *40*, 39–42, doi:10.1130/g32508.1.

Watson, M. N., N. Zwingmann, and N. M. Lemon (2004), The Ladbroke Grove-Katnook carbon dioxide natural laboratory: a recent CO_2 accumulation in a lithic sandstone reservoir, *Energy*, *29*(9–10), 1457–1466.

Watson, Z., W. S. Han, E. H. Keating, N.-H. Jung, and M. Lu (2014), Eruption dynamics of CO_2-driven cold-water geysers: crystal, Tenmile geysers in Utah and Chimayó geyser in New Mexico, *Earth and Planetary Science Letters*, *408*, 272–284.

Whiticar, M. J., and F. Werner (1981), Pockmarks: submarine vents of natural gas or freshwater seeps?, *Geo-Marine Letters*, *1*(3–4), 193–199.

Wilkinson, M., S. V. M. Gilfillan, R. S. Haszeldine, and C. J. Ballentine (2009), Plumbing the depths: testing natural tracers of subsurface CO_2 origin and migration, Utah, in *Carbon Dioxide Sequestration in Geological Media – State of the Science*, edited by M. Grobe, J. C. Pashin and R. L. Dodge, pp. 619–634, AAPG Studies in Geology 59, American Association of Petroleum Geologists, Tulsa, OK.

Wollenweber, J., S. Alles, A. Busch, B. M. Krooss, H. Stanjek, and R. Littke (2010), Experimental investigation of the CO_2 sealing efficiency of caprocks, *International Journal of Greenhouse Gas Control*, *4*(2), 231–241.

Worden, R. H., and S. A. Barclay (2000), Internally-sourced quartz cement due to externally-derived CO_2 in sub-arkosic sandstones, North Sea, *Journal of Geochemical Exploration*, *69–70*, 645–649.

Worden, R. H., and L. K. Smith (2004), Geological sequestration of CO 2 in the subsurface: lessons from CO_2 injection enhanced oil recovery projects in oilfields, *Geological Society*, *233*, 211–224, doi:10.1144.

Yielding, G., B. Freeman, and D. T. Needham (1997), Quantitative fault seal prediction, *AAPG Bulletin*, *81*(6), 897–917.

Yielding, G., P. Bretan, and B. Freeman (2010), Fault seal calibration: a brief review, *Geological Society, London, Special Publications*, *347*(1), 243–255, doi:10.1144/sp347.14.

Yielding, G., N. Lykakis, and J. R. Underhill (2011), The role of stratigraphic juxtaposition for seal integrity in proven CO_2 fault-bound traps of the southern North Sea, *Petroleum Geoscience*, *17*(2), 193–203, doi:10.1144/1354-0793/10-026.

Zhang, M., and S. Bachu (2011), Review of integrity of existing wells in relation to CO_2 geological storage: what do we know?, *International Journal of Greenhouse Gas Control*, *5*(4), 826–840, doi:10.1016/j.ijggc.2010.11.006.

15

A Review of Studies Examining the Potential for Groundwater Contamination From CO_2 Sequestration

Charuleka Varadharajan[1], Ruth M. Tinnacher[1,2], Robert C. Trautz[3], Liange Zheng[1], Baptiste Dafflon[1], Yuxin Wu[1], Matthew T. Reagan[1], Jens T. Birkholzer[1], and J. William Carey[4]

ABSTRACT

Geologic carbon sequestration (GCS) is the process of injecting CO_2 into deep subsurface rock formations such as into depleted oil and gas reservoirs or deep saline aquifers for long-term storage. Over the last decade, a number of field, laboratory, and modeling studies have been undertaken to assess the feasibility and safety of CO_2 geologic storage, including their potential impact on shallow groundwater aquifers overlying target CO_2 storage reservoirs. This chapter provides a review of the literature on the primary concerns to groundwater quality from GCS and practices to mitigate or avoid impacts to water. The review covers possible pathways by which CO_2 can leak from GCS reservoirs, the potential hazards associated with leakage of dissolved CO_2 into shallow groundwater, state-of-the-art monitoring strategies for CO_2 leakage, and an overview of guidelines and regulations in the United States that seek to minimize potential impact. Several knowledge gaps and science needs still remain for understanding the impacts of large-scale deployment of GCS to groundwater including likely leakage scenarios, pathways and associated fluxes, key characteristics of aquifers that could make them vulnerable to degradation upon CO_2 intrusion, and the development of reduced-order models and effective monitoring strategies for understanding and designing appropriate responses to CO_2 leakage.

15.1. INTRODUCTION

Anthropogenic carbon dioxide (CO_2) emissions have become a serious concern because of their impact on climate. As a result, carbon capture and storage (CCS), which refers to the capture of CO_2 emissions from industrial (e.g., steel and cement production) or energy-related sources (e.g., a coal or natural gas power plant or natural gas processing facility) followed by storage, is considered an important potential mitigation measure for climate change. GCS is the process of injecting CO_2 into deep subsurface rock formations such as into depleted oil and gas reservoirs or deep saline aquifers for long-term storage. Over the last decade, many studies have been undertaken to assess the feasibility and safety of CO_2 geologic storage, with emphasis primarily on hydrological, geochemical, and mechanical processes affecting deep injection and containment in storage reservoirs.

Geologic storage of CO_2 can potentially pose some environmental hazards to groundwater, air, and ecology, which must be considered when weighing the benefits of reducing greenhouse gas emissions to the atmosphere [*Lions et al.*, 2014a; *Jones et al.*, 2015; *Pawar et al.*, 2015]. Proper selection and permitting of storage reservoirs for sequestration of carbon dioxide (CO_2) should evaluate

[1] *Earth and Environmental Sciences Area, Lawrence Berkeley National Laboratory, Berkeley, CA, USA*

[2] *Department of Chemistry & Biochemistry, California State University East Bay, Hayward, CA, USA*

[3] *Electric Power Research Institute, Palo Alto, CA, USA*

[4] *Earth and Environmental Sciences, Los Alamos National Laboratory, Los Alamos, NM, USA*

Geological Carbon Storage: Subsurface Seals and Caprock Integrity, Geophysical Monograph 238, First Edition.
Edited by Stéphanie Vialle, Jonathan Ajo-Franklin, and J. William Carey.

the environmental hazards of CCS, including the possibility of CO_2 leakage into potable groundwater overlying CO_2 storage reservoirs, and the consequent impacts to water quality. Efforts to determine the hazards and risks of CCS for specific groundwater aquifer systems will first need to understand the baseline groundwater quality conditions, then predict future changes that may occur in the event of CO_2 or brine migration from depth, and finally implement long-term monitoring to detect any changes to water quality due to potential CO_2 leakage.

This chapter describes the current understanding in literature of the potential hazards to shallow aquifers that are located above target CO_2 storage reservoirs and the means by which potential impacts can be minimized. The chapter covers the following topics:

- What are the leakage pathways by which CO_2 can escape from storage reservoirs and intermediate zones into shallow groundwater?
- What would happen to groundwater quality if leakage were to occur? This includes concerns about the possibility of groundwater contamination due to interaction of acidified groundwater with aquifer rocks and the migration of brine or organic compounds from depth.
- Where, when, and how should monitoring be conducted to detect for possible leakages of CO_2?
- What are some guidelines and regulations that should be followed to avoid or mitigate impacts to shallow groundwater overlying GCS sites?
- What are some of the current knowledge gaps that need to be addressed to assess the safety of GCS operations with respect to groundwater?

15.2. POTENTIAL FOR LEAKAGE OF CO_2 FROM RESERVOIRS AND SHALLOW GROUNDWATER

Injecting CO_2 deep underground increases fluid pressure in the target formation, which can potentially push CO_2 out of the storage reservoir and into an overlying aquifer through leakage pathways, if they exist. Since CO_2 is less dense than water, it can migrate into shallow zones by buoyancy forces alone, even in the absence of a pressure gradient. The potential mechanisms by which CO_2 can leak include fast flow pathways such as wells, faults, or fractures or slow processes such as dissolution through the caprock [*Wildenborg et al.*, 2005; *Lemieux*, 2011; *Lions et al.*, 2014a]. Another concern is that CO_2 injection pressures could push naturally occurring brine, typically found in deep storage reservoirs, into groundwater through these same pathways [*Birkholzer et al.*, 2015]. Formation brines contain high concentrations of salts, trace metals, and other ions and can potentially degrade freshwater quality upon intrusion.

The three main potential leakage pathways that can result in migration of CO_2 to aquifers, that is, naturally occurring faults or fractures, operating or abandoned wells, and the pore space of caprock, are summarized in Figure 15.1 Pathways must exist, or be created, before migration can occur, and while a multitude of possible pathways exist, the likelihood of pathway occurrence and the probabilities and rates of leakage through these pathways vary considerably. Studies of leakage from both natural analogues and industrial CO_2 reservoirs found most leaks occurred via unsealed fault and fracture zones or through improperly constructed or abandoned wells, at highly variable rates [*Lewicki et al.*, 2007; *IEAGHG*, 2011a]. The presence of a confining caprock layer overlying a storage reservoir is also important to ensure that the CO_2 and brine will remain in the reservoir.

The most effective way to ensure long-term CO_2 storage is to choose sites of sufficient depth (typically deeper than 800m), with adequate capacity and an overlying sealing system, such as impermeable caprock, to ensure containment of fluids [*CCP*, 2015]. There is significant amount of literature on the prediction and monitoring of various leakage pathways out of the intended storage reservoir, including fluid flow models that have been used to evaluate the possibility and extent of leakage [e.g., *Nicot*, 2008; *Birkholzer et al.*, 2009; *Yamamoto et al.*, 2009; *IEAGHG*, 2011b; *Lemieux*, 2011]. The key issues related to man-made (wells) and naturally occurring (faults, fractures, permeable rock) leakage pathways are described in this section.

15.2.1. Wells as Leakage Pathways

In general, wells are considered to be the main hazard to geologic sequestration site integrity, as locations where the physical and chemical trapping mechanisms are disrupted [e.g., FutureGen 2007; *Newmark et al.*, 2010; *USEPA*, 2010]. Leakage can occur through operating wells (such as injection or monitoring wells) or through abandoned wells [e.g., *Nordbotten et al.*, 2009; *Humez et al.*, 2011]. Wells provide isolation through a multiple barrier system consisting of steel, cement, and packers; along the length of the well within the caprock, any small element of these can act as a barrier to isolate the CO_2 storage reservoir from the surface. Well leakage pathways involve failure of one or more of these barriers. Within operating wells, failure of packers, failure of tubular joints, and hydrostatic imbalance can result in leakage. Failure of cement plugs can lead to leakage through abandoned wells. External to the well, failure of the cement sheath or the steel casing can allow fluids to escape from the reservoir via either operating or abandoned wells. The leakage path is not necessarily a single point of failure and may involve a circuitous path through many wellbore elements or via crosswell flow.

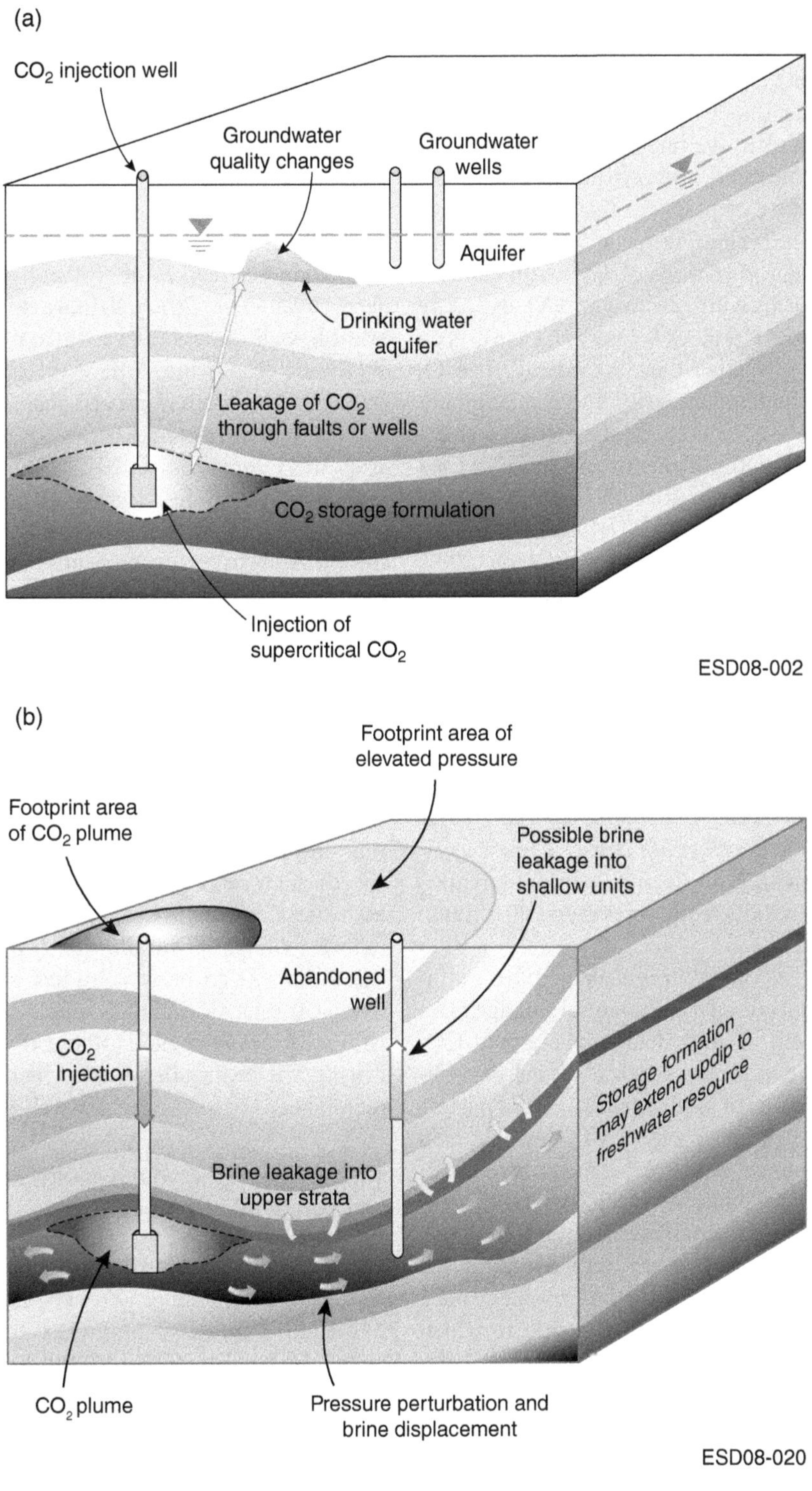

Figure 15.1 Schematics showing (a) potential groundwater leakage pathways and impact scenarios for CO_2 and (b) different regions of influence related to CO_2 storage. Reprinted from *Birkholzer et al.* [2009]. Reprinted with the permission of Elsevier.

Recent research studies on the potential for leakage of CO_2 through wellbores have been conducted toward understanding the likelihood and magnitude of CO_2 leakage from existing wells, including (i) the frequency with which wells fail and leak fluids, (ii) the rate of fluid flow in leaking wells (effective permeability and proximity to the injection well), (iii) the factors or practices that correlate with increased risk of well-integrity failure, and (iv) impacts to long-term integrity due to reactions of CO_2 with cement. Some field studies of wells with a history of CO_2 exposure in both CO_2-EOR and natural CO_2 reservoirs have been focused on determining whether there is evidence for CO_2 migration, analysis of material integrity, and measurement of effective permeability of wellbore systems [*Carey et al.*, 2007; *Crow et al.*, 2010]. Other field surveys have focused on developing an understanding of the frequency and severity of well-integrity failure events [*Watson and Bachu*, 2008, 2009; *Duncan et al.*, 2009; *Jordan and Benson*, 2009; *Jackson et al.*, 2014]. There have been a number of experimental studies on the behavior of wellbore materials on exposure to CO_2, mainly focused on carbonation of cement [*Barlet-Gouedard et al.*, 2006; *Kutchko et al.*, 2007, 2008, 2009; *Rimmelé et al.*, 2008; *Fabbri et al.*, 2009; *Duguid and Scherer*, 2010]. More recently, experimental studies have focused on CO_2-induced reactions during flow through wellbore defects [*Carey et al.*, 2010; *Huerta et al.*, 2013; *Newell and Carey*, 2013; *Walsh et al.*, 2013]. A few experimental studies of corrosion of steel in wellbore environments have also been conducted [*Carey et al.*, 2010; *Han et al.*, 2011b].

Computational studies of wellbore leakage have examined scales ranging from cumulative potential leakage from a large region of wells to the details of cement-CO_2 reactions. For example, a semi-analytical model of wellbore leakage has been applied to numerous wells as part of a hypothetical sequestration project with an assumed probability distribution of effective permeability values [*Gasda et al.*, 2004; *Nordbotten et al.*, 2005, 2009; *Celia et al.*, 2011]. The model allows calculation of the cumulative loss of CO_2 from the storage reservoir but lacks validated values for permeability. Similar calculations have been conducted within a risk assessment framework, in which multiple realizations are used to develop a probability distribution of potential leakage [e.g., *Viswanathan et al.*, 2008; *Oldenburg et al.*, 2009; *Stauffer et al.*, 2009]. Computational models for cement carbonation at the small scale have also been developed [Carey and Lichtner, 2007; *Carey et al.*, 2007; *Huet et al.*, 2010; *Gherardi et al.*, 2012]. Corrosion modeling in sequestration systems has been described by *Han et al.*, [2011a]. Relatively little work has been done on two-phase flow in the wellbore system [e.g., *Carey and Lichtner*, 2011].

Although not specific to CO_2 sequestration, the hazards of degraded abandoned wells were highlighted by a 1989 US Government Accountability Office study of Class II wells across the United States [*USGAO*, 1989] that found that one third of Class II injection-related contamination incidents were due to injection near an improperly plugged abandoned oil and gas well. Formations intended for CO_2 injection may require assessment of nearby abandoned wells before new wells are drilled and completed to guard against gas or brine migration between target formations and overlying aquifers [*Cihan et al.*, 2013]. While also not specific to sequestration, wellbore issues were also highlighted in studies of hydraulic fracturing, where isotopic studies of fugitive gas emissions from fractured shales were more likely to be traced to well-related pathways than natural or induced fractures [*Darrah et al.*, 2014].

Proper system design and operation, including the following recommended construction guidelines, should minimize leakage through wellbore pathways. Monitoring annular pressure is already an industry standard or requirement and provides continuous evaluation of potential leakage within the well. Internal mechanical integrity tests (MITs) (tests of pressure tightness) are effective at demonstrating lack of leakage within the well. Monitoring of external mechanical integrity is much less common and is not generally a requirement, except in the rules for geologic sequestration injection wells (although pressure tests on casing shoes do test the external integrity of each cemented interval). These involve the use of acoustic, temperature, or radioactive tracer logs that can detect leakage [*Thornhill and Benefield*, 1990, 1992]. There appear to be few studies that evaluate the sensitivity of external MITs or document results of external MITs [*McKinley*, 1994]. Other possible methods for monitoring the external annulus include soil-gas measurements near the well or water-well monitors near the well.

15.2.2. Leakage Through Faults, Fractures, and Permeable Caprock

Fault and fracture discontinuities are key features in geomechanical studies [*Zoback*, 2007] and can be potential fluid migration pathways to aquifers [*Oldenburg et al.*, 2009; *IEAGHG*, 2011a; *Song and Zhang*, 2013]. The act of CO_2 injection can have geomechanical impacts and has the potential to create fractures, induce fault activation, and change rock properties [*Rutqvist et al.*, 2013]. The formation of permeable pathways via injection into otherwise low-permeability formations has been examined in a limited number of simulation studies. Coupled flow-geomechanical modeling [*Kim and Moridis*, 2012; *Kim et al.*, 2014] found inherent limitations to the extent of fracture propagation during injection into low-permeability reservoirs, while the potential for fault activation during injection (as a means of permeable pathway creation or a source of seismicity) was found to be a remote possibility [*Rutqvist et al.*, 2013].

For fractured formations and caprock, model simulations indicate that migration of fluids from a fractured formation is possible for high-permeability fractures and permeable bounding formations, but on 1000-year time frames for the chosen reservoir parameters [*Gassiat et al.*, 2013]. Others conclude that upward migration through fractured or permeable bounding formations, if possible at all, is likely an even slower process operating at much longer timescales (in their estimate, ~1,000,000 years) [*Flewelling and Sharma*, 2014]. However, these, like all simulation studies, make assumptions about the size, extent, and configuration of potential pathways. In contrast, simulations of gas reservoirs connected to nearby aquifers via highly permeable pathways show the possibility of rapid gas migration under certain conditions [*Reagan et al.*, 2015]. Thus, such studies cannot rule out (or guarantee) migration through unknown preexisting pathways. In addition, experimental work suggests CO_2-acidified brines can geochemically alter the permeability of preexisting fractures [*Ellis et al.*, 2011].

Several laboratory studies have been conducted to assess the impact of CO_2 on mechanical rock properties [Rohmer *et al.*, 2016 *and references therein*]. The studies suggest that the short-term impact to mechanical properties remains limited in many sedimentary reservoirs, with the largest impact to carbonate reservoirs that exhibit a broad range of responses. The review concluded that extrapolation from laboratory to reservoir scale is difficult due to dependencies on site-specific conditions.

CO_2 migration rates due to diffusive leakage through the pore space of intact caprocks are considered to be very small [e.g., *Lewicki et al.*, 2007; *Busch et al.*, 2008; *IEAGHG*, 2011a; *Song and Zhang*, 2013], and thus, it is less important as a leakage pathway, even when considering long-term CO_2 storage over geological timescales. For example, the rates of movement through fine-grained muddy cap seal rocks are of the order of 700s of years per meter [*Lindeberg and Bergmo*, 2003] to over 33,000 years per meter of water [*Dewhurst et al.*, 1999] for the low volumes of CO_2 that could migrate through these submicron pore spaces.

Proper site characterization and monitoring for leakage pathways in overlying formations serves to identify conductive features such as fault/fracture zones or wells, helps planning of intervention strategies, and is described later in Section 15.4.1.

15.3. POTENTIAL HAZARDS AND RISKS OF CO_2 LEAKAGE INTO SHALLOW GROUNDWATER

The impact of CO_2 release into shallow freshwater aquifers is expected to be different from that in deep storage reservoirs, since temperatures, pressures, and salinities will be lower near the surface. CO_2 will also transition from a supercritical phase to a gaseous phase at shallow depths, where it will partially or fully dissolve into native waters.

Dissolution of CO_2 into groundwater can produce carbonic acid, which decreases the pH (by increasing H^+ concentrations in solution) and increases the dissolved carbonate content:

$$CO_{2(g)} + H_2O = H_2CO_{3(aq)} \tag{15.1}$$

$$H_2CO_{3(aq)} = HCO_3^- + H^+ \tag{15.2}$$

In pure water, the dissolution of CO_2 at elevated pressure yields a pH near 3 (Fig. 15.2). In natural systems, the pH drop is not as pronounced (values typically near or above 5) because of buffering by various reactions with mineral phases, natural organic matter, and so on.

The decrease in pH and increase in carbonate ligands can trigger the release of metal(loid)s (hereafter referred to as metals and designed by the letter M in the chemical equilibrium reactions below) that are naturally present in aquifer rocks and sediments [*Zheng et al.*, 2009; *Apps et al.*, 2010]. Trace metals in sediments can be found adsorbed on the surfaces of carbonates, iron (oxy)hydroxides, and silicates, substituted in clay interlayers, coprecipitated in secondary carbonates, or present in trace quantities of host minerals (e.g., galena or arsenopyrite).

Reactions that can occur in the presence of elevated levels of CO_2 include processes such as dissolution of carbonates, sulfides, or iron (oxy)hydroxides, ion exchange from mineral surfaces, and competitive desorption driven by competing carbonate ligands [*Apps et al.*, 2010]. Mineral dissolution caused by lower pH mostly affects carbonates and sulfides:

$$MCO_{3(s)} + H^+ \rightarrow M^{2+} + HCO_3^- \tag{15.3}$$

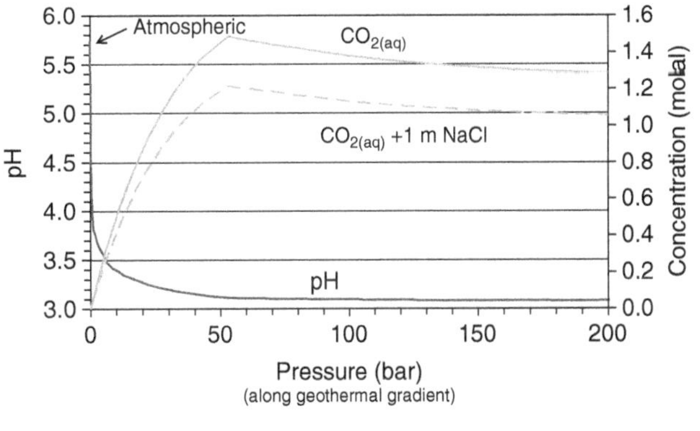

Figure 15.2 Decrease in pH expected with increase in dissolved CO_2 concentrations at elevated pressures.

$$MS_{(s)} + H^+ \rightarrow M^{2+} + HS^- \quad (15.4)$$

Adsorption-desorption reactions are also expected to play an important role in metal mobilization, as well as in buffering pH in the absence of significant amounts of fast-reacting carbonate minerals:

$$> SOM^+ + H^+ \rightarrow > SOH + M^{+2} \quad (15.5)$$

$$> SOH + H^+ \rightarrow > SOH_2^+ \quad (15.6)$$

Competitive sorption/desorption with bicarbonate ions (e.g., As, Se) could also release metal oxyanions:

$$> SMO_4^{-n} + HCO_3^- \rightarrow > SHCO_3^- + MO_4^{-n} \quad (15.7)$$

In addition, metals released by the above reactions (e.g., Ca) may trigger exchange reactions that cause the further release of other metals into solution:

$$X - M + Ca^{+2} \rightarrow X - Ca + M^{+2} \quad (15.8)$$

The kinetics of the different release mechanisms will vary, with ion exchange and desorption reactions likely occurring faster than dissolution from host minerals. It is important to add that some contaminants can be favorably attenuated either through coupled dissolution-precipitation reactions [*Ruiz-Agudo et al.*, 2014] or through adsorption/sorption of some pollutant species on minerals with high specific surface area, such as iron oxides/oxyhydroxides [*Montes-Hernandez et al.*, 2013]. Some potential elements/chemicals of concern in a CO_2 leakage scenario and their associated Environmental Protection Agency (EPA) primary and secondary drinking water standards are summarized in Table 15.1.

Table 15.1 EPA Maximum Contaminant Levels (MCLs) or Secondary Drinking Water Standards for Selected Contaminants in Drinking Water for Public Supply Systems [*USEPA*, 2015].

Contaminant	Maximum contaminant level (MCL in ug/l)	Secondary drinking water standards
Arsenic (As)	10	
Barium (Ba)	2000	
Cadmium (Cd)	5	
Chloride (Cl)		250,000
Chromium (Cr)	100	
Copper (Cu)		1000
Iron (Fe)		300
Lead (Pb)	15	
Manganese (Mn)		50
Mercury (Hg)	2	
Nitrate (NO3 as N)	10,000	
Selenium (Se)	50	
Silver (Ag)	50	
Sulfate (SO4)		250,000
Thallium (Tl)	2	
Uranium	30	
Zinc		5000
Total dissolved solids		500,000
pH		6.5–8.5 standard pH units
Benzene	5	
Ethylbenzene	700	
Benzo(a)pyrene (PAHs)	0.2	

Limits in μg/l except as indicated.

A number of field, laboratory, and modeling studies have been recently conducted to evaluate the impacts of CO_2 intrusion into shallow aquifers. These are described briefly below.

15.3.1. Controlled Release Field Experiments

Field experiments involving a controlled release of CO_2-saturated waters into shallow aquifers, simulating a leak, can help determine the changes that may occur in aquifers under realistic conditions. In particular, field tests that have been coupled with laboratory characterizations and reactive transport modeling are very useful to identify elements that can be mobilized in a potential leakage scenario, as well as to understand and predict the migration and impacts of the CO_2 plume. Several field experiments simulating CO_2 leakage into shallow aquifers have been conducted in the recent decade and have found that short-term geochemical changes, including a drop in groundwater pH and a concurrent increase in dissolved concentrations of major cations and some metals, can occur under such a scenario.

In the Zero Emission Research and Technology (ZERT) field test in Montana, USA, CO_2 was released, over a time period of a few weeks, into a perforated pipe placed just below the groundwater table at a depth of 2 m [*Kharaka et al.*, 2010; *Spangler et al.*, 2010; *Kharaka and Cole*, 2011]. Rapid changes (within days) in pH, alkalinity, and conductivity were observed, along with an initial increase in the concentrations of some major cations (e.g., Ca, Mg), trace metals (e.g., Fe, Mn), and organics (BTEX). However, the pH and metal concentrations in the groundwater returned quickly to background levels after CO_2 injection ended, within the short ~1-week duration of post-injection monitoring. Measured metal and organic concentrations were below EPA MCLs.

Another controlled release field experiment, which was carried out in conjunction with a laboratory study, involved the controlled release of CO_2 into a shallow groundwater formation in Mississippi, USA, at a depth of

~50 m (Fig. 15.3). A change in groundwater pH from ~8 to ~5 was observed in monitoring wells, and the concentrations of some major cations (Ca, Mg, Na, K) and trace metals (e.g., Ba, Sr, Fe, Mn) as well as the groundwater conductivity and alkalinity exhibited a fast initial increase, followed by a drop associated with the passing of the CO_2 front. When the system returned to natural groundwater flow conditions after ~5 months of injection, the pH of the groundwater remained low, while constituent concentrations rebounded to values lower than the initial pulse, but higher than the original background. The concentrations of several trace elements (e.g., As, Cd, Pb) remained below analytical detection limits through the course of the field test [*Trautz et al.*, 2012, 2013].

In another controlled release field test in Brandenburg, Germany, CO_2 was injected through three wells for a period of 10 days into an aquifer at a depth of 18 m [*Peter et al.*, 2012]. Water samples from downgradient monitoring wells showed elevated concentrations of some major cations (Ca, Mg, K, Al, Si) and trace metals (Fe, Mn, Cu, Ni, Ba, Zn, Cd, Pb) in comparison with baseline water. *Cahill and Jakobsen* [2013] reported a field-scale pilot test in which CO_2 gas was injected at 5–10 m depth into an unconfined aquifer in Denmark for 2 days, with monitoring of water geochemistry for more than 100 days. In addition to a one-unit drop in pH and a twofold increase in electrical conductivity (EC), increases in major and trace element concentrations WERE OBSERVED.

A few push-pull tests have also been conducted, where CO_2-saturated water is injected through a well, left in the formation for a while, and pumped back out of the same injection well. *Yang et al.* [2013] conducted a single-well push-pull field test in a Gulf Coast aquifer in Cranfield, Mississippi, at ~73 m depth. Increases in major ion concentrations (Ca, Mg, K, Si) were also observed and were attributed to the dissolution of silicates and possibly carbonate minerals. The mobilization and retardation of major and trace elements were quantified using a mixing model to evaluate the concentration trends observed during the test. *Yang et al.* [2013] also conducted a push-pull test in the Newark Basin, New Jersey, where CO_2 was injected into a fracture zone in a sandstone-siltstone-mudstone interbedded aquifer located ~360 m below the ground surface. The water recovered after a 3–6-week period of incubation exhibited a decrease in pH; increase in alkalinity, Ca, Mg, and Si concentrations; decrease in sulfate and Mo concentrations; and increase in concentrations of trace elements including Fe, Mn, Cr, Co, Ni, Cu, Zn, Rb, Sr, Ba, and U. These changes in aquifer water geochemistry were attributed to the dissolution of silicate and carbonate minerals, and elemental release was dependent on the pH, pCO_2, and the altered redox conditions in the aquifer. *Rillard et al.* [2014] conducted a push-pull test in a shallow fractured sandstone aquifer, ~56 m deep in southern France, where groundwater from the aquifer was previously pumped, saturated in CO_2, and reinjected into the aquifer. Again, rapid (within a few hours of pumping) and systematic changes in pH and alkalinity as well as an increase in the aqueous concentrations of major cations (Ca, Mg) and trace element species (Fe, Mn, Zn, As) were observed in recovered water. Thermodynamic calculations showed that production of alkalinity and release of Ca^{2+} and Mg^{2+} were related to dolomite dissolution, while iron release was related to the dissolution of hydroxide mineral (ferrihydrite) and carbonate mineral (siderite). Additional modeling showed that trace element release could be enhanced by complexation reactions of metals with bicarbonate and sulfate anions.

(a) (b)

Figure 15.3 Field setup of the controlled release experiment at Plant Daniel, Mississippi.

15.3.2. Other Field Studies

There have been large-scale CO_2 field injections for CCS coupled with enhanced oil recovery (CCS-EOR), where the aquifers overlying the target formations have been monitored continuously for changes in water quality. In SACROC, Texas, monitoring conducted for over 35 years of injection found no differences between freshwaters within the site and trends in regional groundwater chemistry outside of the area [*Romanak et al.*, 2012; *BEG*, 2015]. Similarly, no degradation of potable water quality has been observed so far at Weyburn, Canada [*Whittaker et al.*, 2011]. Since no leaks were reported at either site, these results could possibly reflect the safety of a properly sited CCS operation, rather than the lack of impact on groundwater in the event of leakage.

Studies of natural accumulations and analogues can be useful in understanding CO_2 trapping and migration mechanisms and assessing the risks associated with high levels of CO_2 dissolved in freshwaters. In addition, these can also yield insights into geochemical processes that can be used to monitor impacts. For example, a study of a natural analogue in Chimayó, New Mexico [*Keating et al.*, 2010], found no evidence of trace metal mobilization caused by the high levels of CO_2 dissolved in the shallow waters; instead, increases in As, U, and Pb were found to be associated with the CO_2-brine mixture rising from deep formations. Sr isotopes were used as tracers to distinguish between the CO_2 and CO_2-brine source terms [*Keating et al.*, 2011].

Higgs et al. [2015] studied a natural analogue in the Pretty Hill Formation of the Otway Basin, Australia, using HyLogging, a new noninvasive, nondestructive technique to quickly obtain mineral spectroscopy covering the visible-near infrared (VNIR), short-wave infrared (SWIR), and thermal infrared (TIR) wavelengths, combined with robotic sample handling and semiautomated interpretation software. Additional mineralogical information was obtained using petrographic analyses, scanning electron microscopy (SEM), QEMSCAN, X-ray diffraction (XRD), and stable isotope analysis. They found that reservoir heterogeneities could present local barriers to the flow of CO_2 and also affect the distribution of carbonate cements, with carbonate-rich zones concentrating near lithological boundaries. Their results suggest that the greatest amount of CO_2-related reaction and carbonate deposition will occur at the basal CO_2-water contact and at reservoir-baffle/seal boundaries, creating effective internal seals to the CO_2-charged fluids.

Lions et al. [2014b] studied the major and trace element geochemistry of shallow aquifers overlying a natural CO_2 reservoir in Montmiral, France, using isotope tracers potentially sensitive to leakage of CO_2 ($\delta^{13}C$ of dissolved inorganic carbon (DIC), $^{87}Sr/^{86}Sr$, and stable isotopes of water). Most observed $\delta^{13}C$ values were compatible with modeled carbonate dissolution under open or closed conditions with respect to CO_2. The isotopic signatures clearly indicated dissolution of Mg-Sr-calcite or dolomite, rather than mixing of isotopically heavy deep CO_2. The absence of any sign of CO_2 leakage into shallow groundwaters in this study is consistent with the fact that the reservoir and caprock have been trapping the CO_2 efficiently over millions of years.

A field and modeling study of a natural reservoir near Springerville, Arizona, where CO_2 has leaked to the surface along a fault zone for thousands of years, found that the fault zone likely provides hydrologic communication between the shallow aquifer and the deeper reservoir [*Keating et al.*, 2014b]. Although CO_2 leak rates have declined significantly, the shallow aquifer is still highly enriched in CO_2, which was assumed to be migrating from the deeper CO_2 reservoir. The study also collected water chemistry data and found that besides arsenic, trace metal concentrations in the aquifer were not elevated. It was suggested that the source of As was external, rather than an in situ release of As due to CO_2 reaction with shallow aquifer sediments. Observations at this site demonstrate that hydraulic communications between shallow and deep layers and upward CO_2 migration do not preclude the long-term viability of a substantial CO_2 reservoir at depth and that some degree of upward CO_2 leakage may not be necessarily incompatible with the overarching goals of sequestering CO_2 and protecting shallow groundwater.

However, reactions in natural analogues may not accurately represent the changes that could result from accidental leakage of CO_2 into a freshwater aquifer, since the sediments would have been equilibrated with CO_2-saturated waters over very long timescales.

15.3.3. Laboratory Studies

An understanding of the chemical, physical, and biological processes that control the distributions of trace elements between the aqueous and solid phases in a variety of aquifer settings is needed to assess the impacts of CO_2 leakage on water quality. Laboratory experiments are useful in this regard, particularly because field experiments simulating CO_2 intrusion into overlying aquifers are expensive and time intensive and can involve complex permitting requirements. Furthermore, laboratory experiments allow for a better control of experimental conditions and a more specific investigation of the underlying processes involved. A few laboratory studies have been carried out to study the impact of CO_2 leakage into shallow aquifer settings.

Short-duration (4-week) batch studies have observed pH drops and elevated concentrations of metals such as Ba,

Ca, Mg, Fe, Mn, and Sr when diverse aquifer rocks were reacted with CO_2-charged waters. *Little and Jackson* [2010] conducted batch experiments, exposing sediments from several locations in three aquifers to CO_2 for (>)300 days, and found a decrease in pH, accompanied by increases in concentrations of some metals such as Mn, Co, Ni, and Fe by one to two orders of magnitude. However, the experiments were carried out under oxidizing conditions, potentially altering redox reactions in the batch cells. *Humez et al.* [2013] conducted a water-mineral-CO_2 batch experiment with samples taken from the Albian aquifer in the Paris Basin (France) combined with extensive sediment characterization. They observed an increase in concentrations of Ca, Si, Na, Al, B, Co, K, Li, Mg, Mn, Ni, Pb, Sr, and Zn, a decline in Fe and Be concentrations, and no changes for Cl and SO_4 after initial CO_2 influx. *Smyth et al.* [2009] and *Lu et al.* [2010] conducted laboratory batch experiments exposing aquifer materials from the Texas Gulf Coast region to elevated levels of CO_2. Two different types of responses were observed in these studies: *type I cations* (Ca, Mg, Si, K, Sr, Mn, Ba, Co, B, Zn) had rapidly increasing concentrations at the start of CO_2 injection that became steady by the end of the experiment, whereas *type II cations* (Fe, Al, Mo, U, V, As, Cr, Cs, Rb, Ni, and Cu) showed an initial concentration increase at the start of CO_2 injection followed by a decrease to values lower than prior to injection.

Varadharajan et al. [2013] carried out laboratory experiments with extensive mineralogical characterization of samples from the Plant Daniel field site in Mississippi [*Trautz et al.*, 2012, 2013]. Sequential leaching experiments were conducted under in situ redox conditions, exposing unconsolidated sandy and organic-rich sediments from the field site to synthetic groundwater solutions saturated with CO_2 (pH ~ 5) or N_2 (pH ~ 8.5) at formation pressures. Comparable leaching experiments were carried out using CO_2-free, pH-amended solutions to identify the underlying mechanisms of metal release, particularly to distinguish between two driving forces for metal release that can result from CO_2 dissolution in groundwater, that is, the decrease in solution pH and the release of carbonate ligands in solution. It was found that some constituents can be released from sediments due to increased concentrations of dissolved CO_2, by primarily pH-driven processes such as carbonate dissolution, and subsequent ion exchange reactions. Similar to other experimental studies, elements that were quickly mobilized, primarily due to the decrease in pH, were alkali and alkaline earth metals (Ca, Mg, Ba, Sr, Na, Li, Rb), and a few other elements (Co, Fe, Ge, Mn, Ni, Si, Zn), with release concentrations significantly decreasing over time. In addition, carbonate ligands appeared to either enhance (U, Ba, As, Mo, Sr, Mn, Co, Ge, Mg) or suppress (weak trends observed for Fe and Li) the release of metals. In general, the constituents that were mobilized were As, Ba, Ca, Fe, Ge, Mg, Mn, Na, Ni, Si, Sr, and Zn, which are largely consistent with the group of type I cations reported by *Lu et al.* [2010].

Wunsch et al. [2014] reacted rock samples from three limestone aquifers with 0.01–1 bar CO_2 for up to 40 days and found increasing concentrations of Ca, Mg, Sr, Ba, Tl, U, Co, As, and Ni from the dissolution of mostly calcite and to a lesser extent pyrite. Researchers from the same group [*Kirsch et al.*, 2014] reacted sandstone samples with CO_2 under similar conditions for 27 days, showing a rapid increase in major (Ca, Mg) and trace (As, Ba, Cd, Fe, Mn, Pb, Sr, U) elements, which was attributed to the dissolution of calcite.

Yang et al. [2015a] conducted batch and continuous flow experiments using the fine fraction of sedimentary rock from the Newark Basin, in conjunction with a field experiment [*Yang et al.*, 2014]. Similar to other experiments, elevated CO_2 concentrations caused a decrease in pH; an increase in the concentrations of major ions including Ca, Mg, Si, K, and alkalinity; an enhanced dissolution of carbonate minerals; and an increase in the concentrations of trace elements including Mn, Fe, Be, Cr, Co, Cu, Zn, Rb, Zr, Cd, Sb, Ba, Pb, and U. The normalized (by reactive surface area) logarithmic elemental release rates of several elements (Ca, Mg, Si, Mn, Fe, and Zn) showed a linear decrease with increase in pH (or decrease in log (pCO_2)).

Viswanathan et al. [2012] conducted a batch experiment to specifically test the leaching of arsenic (As) from samples collected at the Chimayó, New Mexico, natural analogue site, aiming to shed light on the processes leading to high concentrations of As observed in a few wells with elevated levels of natural CO_2 [*Keating et al.*, 2010]. They observed a sharp increase in As concentrations as soon as pH dropped but then a slow decrease of concentrations although pH remained low, suggesting that the initial metal release was driven by the pH decrease, but subsequently the source of As was depleted.

Batch experiments are useful since they provide a relatively simple and inexpensive means to determine the chemicals that can be mobilized by CO_2 leakage in a variety of aquifer settings. The experiments mentioned above analyzed several types of aquifer rocks and sediments, under a variety of redox and pressure conditions, and reaction times ranging from a couple days to several months. As expected, the wide range of experimental conditions and sediment types resulted in different responses; however, one common observation found in these experiments was the rapid increase in concentrations of alkali and alkaline earth metals.

It is important to interpret laboratory results, with an understanding of the limitations of the experimental setup and conditions. Many of the studies mentioned above were conducted at atmospheric pressure (1 bar), which is much lower than typical aquifer pressures (5–15 bars). Hence, the concentrations of metals released could be underestimated, because higher hydrostatic pressures at depth would be expected to result in increased CO_2 dissolution and pH decrease, potentially further enhancing CO_2-induced mobilization of metals (e.g., at the deep Frio test site [*Kharaka et al.*, 2009]). Experiments using unconsolidated sediments also maximize the mineral surface area available for reactions and can overpredict the amount of metals released, especially in batch settings where the sediments are allowed to equilibrate with CO_2 over extended periods of time [*Gilfillan and Haszeldine*, 2011]. Hence, it is expected that the magnitude of metal release in field settings will be significantly lower than that observed in laboratory experiments.

A recent experiment investigated the potential for contaminants (Cd, As) to be attenuated upon intrusion of a CO_2-brine mixture in a shallow aquifer [*Shao et al.*, 2015]. Batch and column experiments were conducted using sediments from an unconsolidated sand and gravel aquifer in Kansas, containing 0–4 wt% carbonate with solutions spiked with cadmium and arsenic to represent formation brines. Cadmium and arsenic were found to be adsorbed to the sediments, even after the solution pH decreased following CO_2 dissolution. The results suggest that sediments could potentially mitigate the effects of CO_2 leakage and brine intrusion, although the mitigation capacity will depend on the sediment mineralogy, such as the content of constituents like carbonates and phosphates.

Another set of laboratory experiments has investigated constituents that can be leached from source rocks into supercritical CO_2-brine mixtures. For example, batch experiments exposing reservoir and caprocks (sandstone and shale) from the In Salah, Algeria, storage formation to a supercritical CO_2-brine mixture found that metals such as As, Cd, Cr, Cu, Ni, Pb, and U could be released into the formation fluids [*Carroll et al.*, 2011]. Other experiments were conducted to evaluate the potential for mobilization of organic compounds from representative reservoir materials and caprock (Fruitland coal and Gothic shale) and their transport after release through porous media (quartz sand) and sandstone. Their results indicate that lighter organic compounds (benzene, toluene) were more susceptible to mobilization by scCO2 and transport through overlying media compared to heavier compounds [*Zhong et al.*, 2014; *Cantrell et al.*, 2015].

Farquhar et al. [2015] reacted cores from five Jurassic-aged Surat Basin sandstones and siltstones of varying mineralogy in low-salinity water with supercritical CO_2 at simulated in situ reservoir conditions (P = 12 MPa and T = 60°C) for 16 days (384 h) to characterize CO_2-water-rock interactions in fresh or low-salinity potential siliciclastic CO_2 storage formations in Queensland, Australia. Detailed mineral and porosity characterization was also conducted. They observed an increase in concentrations of most major (e.g., Ca, Fe, Si, Mg, Mn) and minor (e.g., S, Sr, Ba, Zn) components during reaction with CO_2. SEM and registered 3D images from micro–computed tomography collected prior and after the experiments, combined with numerical modeling, indicate release due to initial dissolution of carbonates, chlorite, and biotite and, in the long term, due to dissolution of feldspars. Similarly, *Dawson et al.* [2015] reacted samples of Berea Sandstone, USA, with brine saturated with pure CO_2 or mixed SO_2-CO_2 gas at 50°C and 10 MPa and found increases in the concentrations of Ca, Mg, Fe, Mn, and Si throughout experiments, likely released due to carbonate and reactive silicate dissolution. SEM images confirmed dissolution of carbonates, but showed no reaction for minerals such as K-feldspar, potentially due to the short duration of the experiment (15 days). The magnitude of apparent mineral reaction was higher for the mixed gas SO_2-CO_2-brine experiment, with geochemical modeling also indicating greater dissolution of reactive silicates such as chlorite and potential precipitation of amorphous silica.

15.3.4. Modeling Studies

Modeling can play a central role in understanding the potential impact of CO_2 leakage on groundwater quality, provided that model development is constrained by experiments, characterization, and monitoring. Models can be used to predict impacts by testing hypothetical CO_2 leakage scenarios in aquifer settings [*Carroll et al.*, 2009; *Zheng et al.*, 2009; *Apps et al.*, 2010; *Humez et al.*, 2011; *Vong et al.*, 2011; *Schwartz*, 2014] and to develop screening technologies based on predictions of changes to aquifer geochemistry induced by CO_2 leakage [*Wilkin and Digiulio*, 2010]. Reactive transport models can be used to understand the key physical and chemical processes that control the response of an aquifer to CO_2 leakage such as the release of metals from aquifer rocks [*Apps et al.*, 2011; *Zheng et al.*, 2015a] or transport of organics from the storage formation to shallow aquifers, as well as to interpret data from laboratory experiments [e.g., *Viswanathan et al.*, 2012; *Kirsch et al.*, 2014; *Zheng et al.*, 2015b] and field tests [e.g., *Zheng et al.*, 2012, 2015b].

For instance, geochemical modeling based on a principal component analysis of the observations at the ZERT site suggests that the observed increases in Pb, Cu, Cd, Sr, and Zn were mostly caused by ion exchange within clays driven by Ca^{2+} from calcite dissolution and desorption reactions [*Zheng et al.*, 2012]. Model results also indicated that the increase in anions such as arsenate was a result of competitive sorption of bicarbonate ions, whose concentrations were increased as a result of CO_2

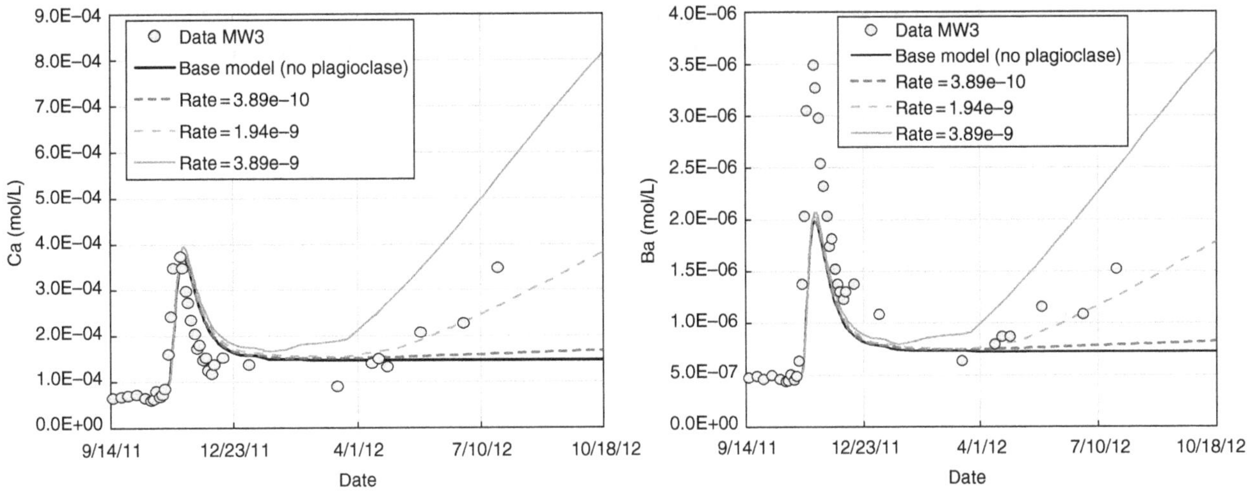

Figure 15.4 Observed (symbols) and modeled concentration time profiles (lines) for Ca (left) and Ba (right). Reprinted from *Zheng et al.* [2015b]. Reprinted with the permission from Wiley.

dissolution [*Apps et al.*, 2011; *Zheng et al.*, 2011]. The geochemical concepts developed for the reactive transport model at ZERT also applied to a different field site in Plant Daniel, Mississippi, even though the different hydrogeological and geochemical conditions, sediment types, and concentration profiles were different for the two sites. The observations at the ZERT site showed a progressive increase in metal concentrations with a pH decrease (and alkalinity increase) upon the injection of CO_2 and a quick return to background concentrations after the injection stopped, whereas the Mississippi site showed a pulse-like change in concentrations of trace and major cations during the injection period followed by slowly increasing concentration levels during the post-injection period (Fig. 15.4).

The model results from the two sites suggest that the chemical reactions controlling the release of trace metals are fairly similar for sites with similar sediment types, but the actual concentration changes in the groundwater depend on the extent to which chemical reactions can occur as well as groundwater flow and residence time. For example, at these two field sites, the amount of calcite available for dissolution led to distinct concentration profiles, as calcium driven cation exchange also controls the concentrations of alkali and alkaline earth metals such as Ba and Sr. An additional difference noticed was that concentrations dropped quickly after the injection of CO_2 stopped at the ZERT site, but slowly increased during the post-injection period at the Mississippi site. This can be explained by differences in flow rate and groundwater residence times and that some relatively slow reactions were not observed at ZERT due to the high flow rate. However, these were observed at the Mississippi site, where flow rates were considerably lower. Ultimately, the Mississippi results were modeled using (i) a fast-reacting but limited pool of reactive minerals that respond quickly to changes in pH to result in the pulse releases of ions and (ii) a slow-reacting but essentially unlimited mineral pool that yields rising concentrations upon decreased groundwater velocities after pumping and injection stopped [*Zheng et al.*, 2015b].

Modeling is also an important tool for assessing system-level risk, to update risk assessment as new data become available, and to evaluate which processes or parameters are important over space and time. These models can be used to integrate risk assessment of groundwater impacts into decision-making processes and can aid in efforts to rank aquifer vulnerability considering several configurations, including differences in hydrogeology, mineralogy, and CO_2 leakage conditions [e.g., *Siirila et al.*, 2012; *Carroll et al.*, 2014; *Dai et al.*, 2014]. An area of current research focuses on the information that is needed to inform these models and to develop reduced-order models that capture the important chemical, biological, and physical processes while using lesser computational resources [e.g., *Bianchi et al.*, 2014; *Last et al.*, 2014; *Zheng et al.*, 2014; *Bacon*, 2012]. Finally, reactive transport models should be used to predict the potential long-term changes in aquifer response to CO_2-brine leakage, to conduct uncertainty quantification to lay the foundation for risk assessment studies, and to provide guidance for risk management and mitigation [*Bacon et al.*, 2014; *Carroll et al.*, 2014; *Hou et al.*, 2014].

15.4. MONITORING FOR CO_2 LEAKAGE

In the context of geologic sequestration of CO_2, monitoring involves both the detection of injected CO_2 through remote sensing and the estimation of processes and properties modified by the injection of CO_2 (such as fluid pressure increases or displaced fluids). Monitoring can be conducted at the near surface or at depth near the storage formation. Baseline site characterization data of aquifer hydrology, mineralogy, and biogeochemistry are also needed to interpret monitoring data and for usage in models that predict changes due to potential migration of CO_2 or other fluids from depth. Extensive reviews for state of the art for sequestration monitoring including descriptions of a variety of CO_2 monitoring tools are available in published literature [*Jenkins et al.*, 2015 and references therein] and are summarized below.

15.4.1. Baseline Site Characterization

Baseline studies that capture the natural conditions, current water usage, and sampling variability are important for monitoring efforts because leak detection, and evaluation of the magnitude and risk of impact, will be made against available baselines. In addition, the hydrology, geology, and mineralogy of overlying groundwater formations need to be characterized to predict their response to potential CO_2 leakage. One example of a potential aquifer characterization method is to measure the buffering capacity of the aquifer to resist changes in pH or redox potential from CO_2 and brine leakage. Physical properties of sediments, such as grain size, cation exchange capacity, and specific surface area, are needed to provide more accurate inputs into reactive transport models. Mineral characterization through optical methods such as petrographic analysis, XRD, SEM, and micro/bulk X-ray spectroscopy can also be used to determine the dominant mineral phases, and mineral-metal associations, which provides information complementary to wet chemical analyses. Other methods include sequential/selective extractions [e.g., *Tessier et al.*, 1979], which can help to identify the associations of trace elements with different sediment phases and potentially provide an indicator for the risk of metal mobilization from sediments. Lab tests, such as acidification with hydrochloric acid to mimic a CO_2-induced drop in groundwater pH, in conjunction with routine sediment characterizations, can be cost-effective means to predict constituent release from sediments when field tests are not possible [*Varadharajan et al.*, 2013].

15.4.2. Atmospheric and Near-Surface Monitoring

A range of technologies can be used to measure CO_2 concentrations and fluxes in atmosphere and shallow subsurface, including infrared gas analyzers and chamber measurements for measurement of point CO_2 concentrations, eddy covariance to measure regional fluxes, airborne measurements, and light detection and ranging (LiDAR) to measure CO_2 concentrations over an integrated path [*Oldenburg et al.*, 2003; *Abshire et al.*, 2013; *Biraud et al.*, 2013]. However, atmospheric methods have their limitations; for example, some of these methods such as eddy covariance and LiDAR are expensive to deploy, and it is also hard to detect small amounts of CO_2 leakage with atmospheric monitoring alone, given naturally varying background concentrations and fluxes.

In the soil and vadose zone, geochemical monitoring often focuses on the direct sampling of CO_2 and its reaction products [*Litynski et al.*, 2012]. For instance, near-surface CO_2 releases may be quantified with soil-flux monitors, supplemented by the analyses for tracers that were specifically injected into the storage reservoir along with CO_2, isotopic analyses, and ecosystem stress monitoring [*Litynski et al.*, 2012].

Other indirect measurements such as monitoring of surface deformations can be used to monitor for pressure buildup in shallow zones. Fluids moving to shallow zones can significantly elevate pressures and produce distinctive surface elevation patterns that can be detected by an array of technologies, including satellite-based methods such as InSAR, GPS, and surface and downhole tilt [*Vasco and Ferretti*, 2005; *Vasco et al.*, 2010]. Gravity could also be a tool for detecting shallow gas accumulations related to preferred pathways, as could various seismic and electrical methods, described in Section 15.4.3.

15.4.3. Geophysical Monitoring

Several geophysical techniques can be used for CO_2 leakage monitoring, due to the sensitivity of some geophysical attributes (e.g., electrical resistivity, seismic velocity) to CO_2-induced changes in the physical, and sometimes geochemical, properties of the aquifers. For electrical and seismic methods, they can be used with various configurations including surface, downhole, downhole to surface, or crosshole configurations [*Litynski et al.*, 2012]. The choice and evaluation of the proper geophysical methods to use for CO_2 leakage detection depends on the characteristics of the aquifers, such as the expected size of the CO_2-induced plume, the depth of investigation, the availability of monitoring wells, the geologic conditions, and the magnitude of changes in physicochemical properties induced by CO_2 leakage. CO_2 escaping close to the storage depth can exist as supercritical, gaseous, or dissolved CO_2 (often mixed with brine) but is normally present in the dissolved phase at shallower depths [*Harvey et al.*, 2012].

The contrast in geophysical properties between leaking CO_2 and background geological and aquifer conditions is critical for evaluating whether the monitoring strategies will be effective. While electrical methods are more frequently used for shallow groundwater aquifer studies, seismic and gravity methods have been applied more frequently for CO_2 monitoring in the deep subsurface. Joint monitoring through the combination of multiple geophysical methods can often reduce monitoring uncertainty and improve accuracy. Oftentimes, other types of measurements, such as well logging, wellbore fluid analysis, surface deformation, and CO_2 flux measurements, can be used to constrain and improve geophysical data interpretation.

Due to the sensitivity of seismic velocity and attenuation to changes in reservoir pressure, CO_2, and brine saturation, time-lapse seismic methods have been applied for the monitoring of CO_2 plume migration in deep storage reservoirs [*Saito et al.*, 2006; *Chadwick et al.*, 2009; *Kim et al.*, 2011; *Ivanova et al.*, 2012; *Ajo-Franklin et al.*, 2013]. While not widely used, these methods could be applied to monitor potential CO_2 leakage in shallower subsurface, particularly when gaseous CO_2 is involved. Field-scale surface seismic measurements offer less resolution than well logging [*Xue et al.*, 2006] and crosswell seismic [*Saito et al.*, 2006; *Daley et al.*, 2007] but enable the monitoring of much larger areas. In addition to active seismic monitoring involving the use of active sources, passive seismic monitoring can detect small microseismic events associated with pressure-induced fracturing or alteration of flow paths.

Electrical methods using either galvanic contact, capacitive coupled electrodes, or electromagnetic signals are used to monitor changes in subsurface electrical properties that can be associated with CO_2 leakages. These methods can be applied in various configurations including surface, downhole, crosshole, or downhole to surface. Several studies show the potential of these electrical methods for monitoring potential CO_2 leakage into shallow aquifers. These include applications from the surface [*Strazisar et al.*, 2009; *Auken et al.*, 2014; *Doetsch et al.*, 2015] using a downhole approach [*Denchik et al.*, 2014] or crosshole tomography [*Dafflon et al.*, 2013; *Yang et al.*, 2015b]. These studies typically involved an injection of dissolved CO_2 into shallow groundwater systems and demonstrated that electrical resistivity changes associated with geochemical changes induced by CO_2 (e.g., mineral dissolution) are typically larger than the changes from the CO_2 plume alone. The use of complex resistivity, which measures changes in both the resistivity and the phase shift in electrical signals, can provide added value to differentiate various responses to dissolved CO_2 plumes [*Dafflon et al.*, 2013]. Electrical resistivity and electromagnetic methods have also been used for monitoring of the migration of supercritical CO_2 plumes in deep subsurface reservoirs [*Kiessling et al.*, 2010; *Bergmann et al.*, 2012; *Carcione et al.*, 2012; *Carrigan et al.*, 2013; *Doetsch et al.*, 2013; *Park et al.*, 2013; *Schmidt-Hattenberger et al.*, 2013].

In addition to electrical and seismic methods, high-precision gravity monitoring can be used to detect changes in density and pressure caused by CO_2 replacing formation water or oil, thus tracking the migration or potential leakage of CO_2 in the subsurface [*Alnes et al.*, 2011].

15.4.4. Groundwater Sampling and Monitoring

Special attention should be given to selecting appropriate spatial locations for monitoring groundwater quality. A recent study modeled a hypothetical CO_2 release into the shallow Edwards Aquifer (Texas) from a deep CO_2 storage reservoir [*Keating et al.*, 2014a]. The probability-based contaminant transport model simulated 128 monitoring well locations in the aquifer to detect the leak. The authors found that there was less than a 20% probability of detecting dissolved CO_2 if the monitoring well was located more than 400 m from the release location. This is due to the fact that the relatively small dissolved CO_2 plume that develops decreases rapidly with distance from the leak. Local buffering of the dissolved CO_2 plume by naturally occurring carbonates causes a rapid decrease in CO_2 concentrations, requiring a dense monitoring network for CO_2 detection. One approach would be to select and utilize groundwater monitoring intervals with low buffering capacity consisting of nonreactive, quartz-rich sands (if available) to optimize detection. In addition, locating monitor wells at or near potential high-risk leakage pathways, such as known or suspected well penetrations and faults, or upgradient from domestic or municipal water supply wells to protect potential receptors, may be a more practical and cost-effective detection strategy.

Special consideration should also be given to the selection of the monitoring depth. Installing and sampling deep monitoring wells penetrating the zone of interest (usually the injection reservoir or the overlying formation) is preferred from the standpoint of early leak detection; however, there are significant logistical drawbacks. For example, using EPA-recommended low-flow purge techniques for sampling wells greater than a few tens of meters deep is impractical because of the time required to purge and sample deep wells. An electrical submersible pump could be used to initially purge a deep well three times as recommended by EPA. However, the cost of disposing of 25–30 m^3 (several thousands of gallons) of brine per well for wells approaching the minimum depth needed for optimum CO_2 storage (i.e., deeper than 800 m) could be prohibitively expensive. Other technical challenges include separation of fluid mixtures containing CO_2, brine, and hydrocarbons in the wellbore and preservation of temperature, solubility conditions, and concentrations of dissolved gases. Hence, specialized sampling equipment including a wireline-deployed Kuster sampler or U-tube sampling device [*Freifeld et al.*, 2005] must be used to prevent rapid sample depressurization that can affect water quality results

[*Conaway et al.*, 2016]. Sample analysis can then be used to determine the size, movement, and stability of the CO_2 plume and to characterize fluid properties, as well as their potential changes, in the storage reservoir. Tracers can be injected in order to track the migration of CO_2 and additionally to assess the phase partitioning of CO_2 in the reservoir. Different tracers may be used for subsurface and near-surface monitoring in order to avoid an overlap of release signals from different geologic horizons.

Shallow groundwater monitoring has its drawbacks too. Shallow monitoring zones need to be carefully selected to ensure that natural processes like recharge [*Apps et al.*, 2011] or human activities like groundwater pumping do not disturb baseline groundwater quality conditions that could mask or be misinterpreted as a false leakage signal. In the event that a release is detected, a shallow monitoring network may not provide sufficient warning to develop an emergency response program needed to protect potable groundwater resources. Project proponents will need to develop site-specific groundwater monitoring plans that take into account the technical issues described above and address public health and safety, environmental protection, and stakeholder concerns.

For the monitoring of near-surface groundwater systems in proximity to CO_2 underground storage reservoirs, shallow groundwater monitoring wells, typically at a depth of 100 m or less, are installed for the collection of groundwater samples. Geochemical analyses of these samples commonly include measurements of pH, alkalinity, EC, and the quantification of various cations (e.g., Na^+, Ca^{2+}, Mg^{2+}, Fe^{2+}, Fe^{3+}) and anions (e.g., HCO_3^-, CO_3^{2-}, Cl^-, SO_4^{2-}) [*Litynski et al.*, 2012]. Furthermore, a characterization of isotopic C, H, and O signatures and analyses of DIC and co-injected tracer concentrations may be performed. The specific characteristics and heterogeneity of a CO_2 storage site may affect the applicability of the various analytical techniques. For instance, carbon isotope analysis may be of limited use for the identification of CO_2 sources in shallow groundwater systems [*Romanak et al.*, 2010] at very complex sites with mixing of groundwater and saline waters, temporal variations in geochemistry due to pumping or irrigation, and so on. Current research is focused on the identification of a small number of chemical parameters to clearly indicate a CO_2 release into overlying formations. For instance, the use of trace element concentration levels and pH measurements as *markers* for CO_2 leakage has been suggested [e.g., *Little and Jackson*, 2011; *Varadharajan et al.*, 2013].

15.5. REGULATIONS IN THE UNITED STATES

The protection of groundwater quality, along with the development of groundwater-specific monitoring and mitigation tools, is recognized as necessary practice for successful CO_2 storage.

The US EPA has developed regulations for CO_2 geologic sequestration projects under the authority of the Safe Drinking Water Act's Underground Injection Control (UIC) Program [*USEPA*, 2010]. These regulations, also known as the Class VI rule (for CO_2 injection wells), are designed by the EPA's Office of Water to protect underground sources of drinking water (USDW). The Class VI rule builds on existing UIC Program requirements, with extensive tailored requirements that address carbon dioxide injection for long-term storage, to ensure that wells used for geologic sequestration are appropriately sited, constructed, tested, monitored, funded, and closed. The EPA's Office of Air and Radiation also has reporting requirements under the Greenhouse Gas Reporting Program for facilities that inject CO_2 underground for geologic sequestration and all other facilities that inject CO_2 underground.

The UIC Program Class VI Well Site Characterization Guidance [*USEPA*, 2013] describes data requirements and information that are typically used to characterize the geology and geochemistry of a site and addresses the collection of background information for proposed project sites. Various aspects of site selection are covered, including the detailed geologic characterization of the proposed injection zone and confining zones, as well as development of sufficient geochemical sampling and analysis plans to establish baseline water quality. The rule indicates that the confining unit has to be free of transmissive faults/fractures and has sufficient extent to contain the injected volume of CO_2 and displaced brine within the injection zone. It also discusses testing the quality of a site's seal, via samples, logging, and pressure communication tests, and geomechanical characterization of fractures and faults. Additional testing may be required to determine whether non-transmissive faults may become transmissive due to increase pressure during injection and storage.

In the United States, current US EPA regulations do not require the use of specific monitoring tools, but recommend that appropriate monitoring tools be selected based on project- and site-specific conditions and needs [*Litynski et al.*, 2012]. Guidance documents provide technical guidelines for the monitoring, verification, and accounting of carbon dioxide stored in geologic formations [*NETL*, 2017; *Litynski et al.*, 2012].

Regulations for CO_2 injection and storage operations have also been developed in several regions, including the United States, Europe [*OSPAR*, 2007, the EU Directive on *GCS*, 2009], Canada [*Alberta Energy*, 2013], and Australia [*Ministerial Council on Mineral and Petroleum Resources*, 2005].

15.6. SCIENTIFIC KNOWLEDGE GAPS

Despite the extensive studies that have been conducted over the past decade on this topic, some knowledge gaps and science needs remain and should be addressed in future studies to properly determine the potential impacts to groundwater from large-scale GCS operations.

Initially, while selecting sites, we need a better understanding of characteristics that could cause aquifers to be more or less vulnerable to CO_2 intrusion or lead to problems with monitoring. For instance, the role of sediment buffering capacity in CO_2-driven reactions needs to be further investigated, since high buffering capacities could mitigate the impacts of CO_2 plumes but also interfere with the early detection of CO_2 leakage. We also need a better understanding of key risk-related issues including the large-scale hydrologic impacts of large-volume CO_2 injections, the driving mechanisms behind pressure buildup and brine displacement, and the roles of fractures/fault leakage, well leakage, and induced seismicity. With respect to understanding the consequences of CO_2 intrusion into groundwater, the variability of aquifer responses stemming from a range of redox conditions, pressures, spatial scales, and porosities on water quality is unknown. Furthermore, the effects of brines and organics co-migrating with CO_2 plumes, as well as microbial responses and natural attenuation processes following CO_2 leakage, need to be further evaluated.

In the context of site monitoring and mitigation, we need to determine a set of baseline screening parameters for water quality monitoring that include cost considerations, improve the resolution of geophysical methods for the detection of CO_2-brine leakage, and assess existing remediation technologies for their potential application in cases of CO_2-related contamination. Finally, current computational modeling tools are limited and need further development for large-scale deployment of GCS. Although current reactive transport simulators have the capability to include a range of chemical and physical parameters, the complex chemical systems and potential addition of microbial/ecological processes require significant computational resources. Consequently, there is a desire to reduce the chemical, biological, and physical parameter space to focus on the most important parameters that relate leakage rates to groundwater quality. It is also important to understand the uncertainty in conceptual models and the relative importance of the parameters used to describe the models. The accuracy and robustness of any predictions regarding the persistence and reversibility of potential impacts within aquifers will rely on the ability to scale chemical and transport processes to field-scale reactive transport [*Varadharajan et al.*, 2012].

15.7. SUMMARY

Carbon dioxide (CO_2) emissions from industrial sources or coal- and gas-fired power plants could be mitigated by the sequestration of CO_2 into deep underground storage systems. However, an understanding of the potential impacts on water and air quality and ecology is needed in order to minimize or mitigate potential environmental hazards [*Lions et al.*, 2014a; *Jones et al.*, 2015]. The three main potential leakage pathways that can result in migration of CO_2 to aquifers are naturally occurring faults or fractures, operating or abandoned wells, and the pore space of caprock. Over the last decade, several field, lab, and modeling studies have been conducted to assess the impacts of potential leakage of CO_2 on different types of shallow aquifers. A particular focus of many of the studies has been the potential degradation of groundwater quality due to the release of major ions and trace metals from natural sediments by various mechanisms including dissolution by acidified groundwater, desorption by carbonate ligands, and cation exchange reactions.

Hence, it is necessary to conduct thorough characterization of the geology and geochemistry of proposed CO_2 sequestration sites [*USEPA*, 2013] to make appropriate choices for site selection such as deep sites with adequate capacity and impermeable barriers to ensure fluid containment. The appropriate construction, testing, and monitoring of injection wells [*USEPA*, 2010] have to be ensured, and monitoring plans (including collection of baseline water quality data) need to be developed to detect any potential CO_2 plumes or other unacceptable changes in water quality in neighboring aquifers. Currently proposed monitoring techniques include geophysical methods, such as remote sensing, electrical resistivity, and seismic velocity measurements and high-precision gravity monitoring, as well as geochemical analysis focusing on changes in fluid composition with regard to pH, alkalinity, and cation, anion, and co-injected tracer concentrations.

Protecting groundwater resources during CCS implementation must address several different areas including (i) the initial site selection and characterization, (ii) the identification of potential CO_2 release pathways, (iii) the technical understanding of parameters and processes that can lead to a degradation of groundwater quality after a CO_2 plume has emerged, and (iv) the development of effective monitoring strategies and guidelines for appropriate responses to leak detection.

REFERENCES

Abshire, J., A. Ramanathan, H. Riris, J. Mao, G. Allan, W. Hasselbrack, C. Weaver, and E. Browell (2013), Airborne measurements of CO_2 column concentration and range using a pulsed direct-detection IPDA Lidar, *Remote Sens.*, *6*(1), 443–469, doi:10.3390/rs6010443.

Ajo-Franklin, J. B., J. Peterson, J. Doetsch, and T. M. Daley (2013), High-resolution characterization of a CO_2 plume using crosswell seismic tomography: Cranfield, MS, USA, *Int. J. Greenh. Gas Control*, *18*, 497–509, doi:10.1016/j.ijggc.2012.12.018.

Alberta Energy (2013), Carbon Capture and Storage – Summary Report of the Regulatory Framework Assessment. www.solutionsstarthere.ca

Alnes, H., O. Eiken, S. Nooner, G. Sasagawa, T. Stenvold, and M. Zumberge (2011), Results from Sleipner gravity monitoring: updated density and temperature distribution of the CO_2 plume, *Energy Procedia*, *4*, 5504–5511, doi:10.1016/j.egypro.2011.02.536.

Apps, J. A., L. Zheng, Y. Zhang, T. Xu, and J. T. Birkholzer (2010), Evaluation of potential changes in groundwater quality in response to CO_2 leakage from deep geologic storage, *Transp. Porous Media*, *82*(1), 215–246, doi:10.1007/s11242-009-9509-8.

Apps, J. A., L. Zheng, N. Spycher, J. T. Birkholzer, Y. Kharaka, J. Thordsen, E. Kakouros, and R. Trautz (2011), Transient changes in shallow groundwater chemistry during the MSU ZERT CO_2 injection experiment, *Energy Procedia*, *4*, 3231–3238, doi:10.1016/j.egypro.2011.02.241.

Auken, E., J. Doetsch, G. Fiandaca, A. V. Christiansen, A. Gazoty, A. G. Cahill, and R. Jakobsen (2014), Imaging subsurface migration of dissolved CO2 in a shallow aquifer using 3-D time-lapse electrical resistivity tomography, *J. Appl. Geophys.*, *101*, 31–41, doi:10.1016/j.jappgeo.2013.11.011.

Bacon, D. H. (2012), Reduced Order Model for the Geochemical Impacts of Carbon Dioxide, Brine and Trace Metal Leakage into an Unconfined, Oxidizing Carbonate Aquifer, Version 2.0. PNNL-22049 (Limited Distribution), Pacific Northwest National Laboratory, Richland, WA.

Bacon, D. H., Z. Dai, and L. Zheng (2014), Geochemical impacts of carbon dioxide, brine, trace metal and organic leakage into an unconfined, oxidizing limestone aquifer, *Energy Procedia*, *63*, 4684–4707, doi:10.1016/j.egypro.2014.11.502.

Barlet-Gouedard, V., G. Rimmele, B. Goffe, and O. Porcherie (2006), Mitigation strategies for the risk of CO_2 migration through wellbores, in *IADC/SPE Drilling Conference*, SPE–98924–MS, Society of Petroleum Engineers, Miami, FL, 21–23 February.

BEG (2015), SACROC groundwater study final report. http://www.beg.utexas.edu/files/gccc/research/sacroc/SACROC%20Final%20Report_v1.pdf.

Bergmann, P., C. Schmidt-Hattenberger, D. Kiessling, C. Rücker, T. Labitzke, J. Henninges, G. Baumann, and H. Schütt (2012), Surface-downhole electrical resistivity tomography applied to monitoring of CO_2 storage at Ketzin, Germany, *Geophysics*, *77*(6), B253–B267, doi:10.1190/geo2011-0515.1.

Bianchi, M., L. Zheng, N Spycher, J Birkholzer (2014), Reduced-order models for prediction of groundwater quality impacts from CO_2 and brine leakage – application to the high plains aquifer, NRAP-TRS-III-00X-2014; NRAP Technical Report Series; U.S. Department Energy, National Energy Technology Laboratory: Morgantown, WV.

Biraud, S. C., M. S. Torn, J. R. Smith, C. Sweeney, W. J. Riley, and P. P. Tans (2013), A multi-year record of airborne CO_2 observations in the US Southern Great plains, *Atmos. Meas. Tech.*, *6*(3), 751–763, doi:10.5194/amt-6-751-2013.

Birkholzer, J., Q. Zhou, and C. Tsang (2009), Large-scale impact of CO_2 storage in deep saline aquifers: a sensitivity study on pressure response in stratified systems, *Int. J. Greenh. Gas Control*, *3*(2), 181–194, doi:10.1016/j.ijggc.2008.08.002.

Birkholzer, J. T., C. M. Oldenburg, and Q. Zhou (2015), CO_2 migration and pressure evolution in deep saline aquifers, *Int. J. Greenh. Gas Control*, *40*, 203–220, doi:10.1016/j.ijggc.2015.03.022.

Busch, A., S. Alles, Y. Gensterblum, D. Prinz, D. Dewhurst, M. Raven, H. Stanjek, and B. Krooss (2008), Carbon dioxide storage potential of shales, *Int. J. Greenh. Gas Control*, *2*(3), 297–308, doi:10.1016/j.ijggc.2008.03.003.

Cahill, A. G., and R. Jakobsen (2013), Hydro-geochemical impact of CO_2 leakage from geological storage on shallow potable aquifers: a field scale pilot experiment, *Int. J. Greenh. Gas Control*, *19*, 678–688, doi:10.1016/j.ijggc.2013.03.015.

Cantrell, K. J., L. Zhong, D. A. Bryce, C. J. Thompson, D. H. Bacon, and Shao, H. (2015), Mobilization and transport of organic compounds from geologic carbon sequestration reservoirs, NRAP-TRS-III-002-2015; NRAP Technical Report Series; U.S. Department of Energy, National Energy Technology Laboratory: Morgantown, WV, p 54.

Carcione, J. M., D. Gei, S. Picotti, and A. Michelini (2012), Cross-hole electromagnetic and seismic modeling for CO_2 detection and monitoring in a saline aquifer, *J. Pet. Sci. Eng.*, *100*, 162–172, doi:10.1016/j.petrol.2012.03.018.

Carey, J. W., and P. Lichtner, (2007), Calcium silicate hydrate (C-S-H) solid solution model applied to cement degradation using the continuum reactive transport model FLOTRAN. In B. Mobasher, and J. Skalny, editors, *Transport Properties and Concrete Quality: Materials Science of Concrete, Special Volume*, pp. 73–106. American Ceramic Society; John Wiley & Sons, Inc.

Carey, B., and P. C. Lichtner (2011), Computational studies of two-phase cement/CO_2/brine interaction in wellbore environments, *Soc. Pet. Eng. J.*, *16*(4), SPE–126666–PA, doi:10.2118/126666-PA.

Carey, J. W., M. Wigand, S. J. Chipera, G. WoldeGabriel, R. Pawar, P. C. Lichtner, S. C. Wehner, M. A. Raines, and G. D. Guthrie (2007), Analysis and performance of oil well cement with 30 years of CO_2 exposure from the SACROC unit, West Texas, USA, *Int. J. Greenhouse Gas Control*, 1(*1*), 75–85, doi:10.1016/S1750-5836(06)00004-1.

Carey, J. W., R. Svec, R. Grigg, J. Zhang, and W. Crow (2010), Experimental investigation of wellbore integrity and CO_2 – brine flow along the casing – cement microannulus, *Int. J. Greenh. Gas Control*, *4*(2), 272–282, doi:10.1016/j.ijggc.2009.09.018.

Carrigan, C. R. et al. (2013), Electrical resistance tomographic monitoring of CO_2 movement in deep geologic reservoirs, *Int. J. Greenh. Gas Control*, *18*, 401–408, doi:10.1016/j.ijggc.2013.04.016.

Carroll, S., Y. Hao, and R. Aines (2009), Geochemical detection of carbon dioxide in dilute aquifers, *Geochem. Trans.*, *10*, doi:10.1186/1467-4866-10-4.

Carroll, S. A., W. W. McNab, and S. C. Torres (2011), Experimental study of cement – sandstone/shale – brine – CO_2 interactions, *Geochem. Trans.*, *12*(1), 9, doi:10.1186/1467-4866-12-9.

Carroll, S. A., E. Keating, K. Mansoor, Z. Dai, Y. Sun, W. Trainor-Guitton, C. Brown, and D. Bacon (2014), Key factors for determining groundwater impacts due to leakage from geologic carbon sequestration reservoirs, *Int. J. Greenh. Gas Control*, *29*, 153–168, doi:10.1016/j.ijggc.2014.07.007.

CCP (2015), Carbon Capture Project. www.co2captureproject.org. Accessed 10 December 2015.

Celia, M. A., J. M. Nordbotten, B. Court, M. Dobossy, and S. Bachu (2011), Field-scale application of a semi-analytical model for estimation of CO_2 and brine leakage along old wells, *Int. J. Greenh. Gas Control*, *5*(2), 257–269, doi:10.1016/j.ijggc.2010.10.005.

Chadwick, R. a., D. Noy, R. Arts, and O. Eiken (2009), Latest time-lapse seismic data from Sleipner yield new insights into CO_2 plume development, *Energy Procedia*, *1*(1), 2103–2110, doi:10.1016/j.egypro.2009.01.274.

Cihan, A., J. T. Birkholzer, and Q. Zhou (2013), Pressure buildup and brine migration during CO_2 storage in multilayered aquifers, *Ground Water*, *51*(2), 252–267, doi:10.1111/j.1745-6584.2012.00972.x.

Conaway, C. H., Thordsen J. J., Manning, M. A., Cook, P. J., Trautz, R. C., Thomas, B., and Kharaka, Y. K. (2016), Comparison of geochemical data obtained using four brine sampling methods at the SECARB Phase III Anthropogenic Test CO2 injection site, Citronelle Oil Field, Alabama, *Int. J. Coal Geol.*, *162*, 85–95.

Crow, W., J. W. Carey, S. Gasda, D. Brian Williams, and M. Celia (2010), Wellbore integrity analysis of a natural CO_2 producer, *Int. J. Greenh. Gas Control*, *4*(2), 186–197, doi:10.1016/j.ijggc.2009.10.010.

Dafflon, B., Y. Wu, S. S. Hubbard, J. T. Birkholzer, T. M. Daley, J. D. Pugh, J. E. Peterson, and R. C. Trautz (2013), Monitoring CO_2 intrusion and associated geochemical transformations in a shallow groundwater system using complex electrical methods, *Environ. Sci. Technol.*, *47*(1), 314–321, doi:10.1021/es301260e.

Dai, Z., E. Keating, D. Bacon, H. Viswanathan, P. Stauffer, A. Jordan, and R. Pawar (2014), Probabilistic evaluation of shallow groundwater resources at a hypothetical carbon sequestration site, *Sci. Rep.*, *4*, 4006, doi:10.1038/srep04006.

Daley, T. M., L. R. Myer, J. E. Peterson, E. L. Majer, and G. M. Hoversten (2007), Time-lapse crosswell seismic and VSP monitoring of injected CO_2 in a brine aquifer, *Environ. Geol.*, *54*(8), 1657–1665, doi:10.1007/s00254-007-0943-z.

Darrah, T. H., A. Vengosh, R. B. Jackson, N. R. Warner, and R. J. Poreda (2014), Noble gases identify the mechanisms of fugitive gas contamination in drinking-water wells overlying the Marcellus and Barnett Shales, *Proc. Natl. Acad. Sci.*, *111*(39), 14076–14081, doi:10.1073/pnas.1322107111.

Dawson, G. K. W., J. K. Pearce, D. Biddle, and S. D. Golding (2015), Experimental mineral dissolution in Berea Sandstone reacted with CO_2 or SO_2–CO_2 in NaCl brine under CO_2 sequestration conditions, *Chem. Geol.*, *399*, 87–97, doi:10.1016/j.chemgeo.2014.10.005.

Denchik, N., P. A. Pezard, D. Neyens, J. Lofi, F. Gal, J.-F. Girard, and A. Levannier (2014), Near-surface CO_2 leak detection monitoring from downhole electrical resistivity at the CO_2 field laboratory, Svelvik ridge (Norway), *Int. J. Greenh. Gas Control*, *28*, 275–282, doi:10.1016/j.ijggc.2014.06.033.

Dewhurst, D. N., Y. Yang, and A. C. Aplin (1999), Permeability and fluid flow in natural mudstones, *Geol. Soc. Lond. Spec. Publ.*, *158*(1), 23–43, doi:10.1144/GSL.SP.1999.158.01.03.

Doetsch, J., M. B. Kowalsky, C. Doughty, S. Finsterle, J. B. Ajo-Franklin, C. R. Carrigan, X. Yang, S. D. Hovorka, and T. M. Daley (2013), Constraining CO_2 simulations by coupled modeling and inversion of electrical resistance and gas composition data, *Int. J. Greenh. Gas Control*, *18*, 510–522, doi:10.1016/j.ijggc.2013.04.011.

Doetsch, J., G. Fiandaca, E. Auken, A. V. Christiansen, and A. G. Cahill (2015), Field-scale time-domain spectral induced polarization monitoring of geochemical changes induced by injected CO_2 in a shallow aquifer, *Geophysics*, *16*(2), 10294, doi:10.1190/geo2014-0315.1.

Duguid, A., and G. W. Scherer (2010), Degradation of oilwell cement due to exposure to carbonated brine, *Int. J. Greenh. Gas Control*, *4*(3), 546–560, doi:10.1016/j.ijggc.2009.11.001.

Duncan, I. J., J.-P. Nicot, and J.-W. Choi (2009), Risk assessment for future CO_2 sequestration projects based CO_2 enhanced oil recovery in the U.S.*Energy Procedia*, *1*(1), 2037–2042, doi:10.1016/j.egypro.2009.01.265.

EC 2009. Implementation 2009/31/EC on the geological storage of carbon dioxide – Guidance Document 3 – Criteria for Transfer of Responsibility to the Competent Authority. http://ec.europa.eu/clima/policies/lowcarbon/docs/gd3_en.pdf.

Ellis, B., C. Peters, J. Fitts, G. Bromhal, D. McIntyre, R. Warzinski, and E. Rosenbaum (2011), Deterioration of a fractured carbonate caprock exposed to CO_2 -acidified brine flow, *Greenh. Gases Sci. Technol.*, *1*(3), 248–260, doi:10.1002/ghg.25.

Fabbri, A., J. Corvisier, A. Schubnel, F. Brunet, B. Goffé, G. Rimmele, and V. Barlet-Gouédard (2009), Effect of carbonation on the hydro-mechanical properties of Portland cements, *Cem. Concr. Res.*, *39*(12), 1156–1163, doi:10.1016/j.cemconres.2009.07.028.

Farquhar, S. M., J. K. Pearce, G. K. W. Dawson, A. Golab, S. Sommacal, D. Kirste, D. Biddle, and S. D. Golding (2015), A fresh approach to investigating CO_2 storage: experimental CO_2 – water-rock interactions in a low-salinity reservoir system, *Chem. Geol.*, *399*, 98–122, doi:10.1016/j.chemgeo.2014.10.006.

Flewelling, S. A., and M. Sharma (2014), Constraints on upward migration of hydraulic fracturing fluid and brine, *Ground Water*, *52*(1), 9–19, doi:10.1111/gwat.12095.

Freifeld, B. M., R. C. Trautz, Y. K. Kharaka, T. J. Phelps, L. R. Myer, S. D. Hovorka, and D. J. Collins (2005), The U-tube: a novel system for acquiring borehole fluid samples from a deep geologic CO2 sequestration experiment, *J. Geophys. Res. Earth*, *110*(B10), doi:10.1029/2005jb003735.

FutureGen (2007), Final risk assessment report for the futuregen project environmental impact statement. Technical Report Contract No. DE-AT26-06NT42921, FutureGen Project.

Gasda, S., S. Bachu, and M. Celia (2004), Spatial characterization of the location of potentially leaky wells penetrating a deep saline aquifer in a mature sedimentary basin, *Environ. Geol.*, *46*(6–7), 707–720, doi:10.1007/s00254-004-1073-5.

Gassiat, C., T. Gleeson, R. Lefebvre, and J. McKenzie (2013), Hydraulic fracturing in faulted sedimentary basins: numerical simulation of potential contamination of shallow aquifers over long time scales, *Water Resour. Res.*, *49*(12), 8310–8327, doi:10.1002/2013WR014287.

Gherardi, F., P. Audigane, and E. C. Gaucher (2012), Predicting long-term geochemical alteration of wellbore cement in a generic geological CO_2 confinement site: tackling a difficult reactive transport modeling challenge, *J. Hydrol.*, *420–421*, 340–359, doi:10.1016/j.jhydrol.2011.12.026.

Gilfillan, S. M. V., and R. S. Haszeldine (2011), Comment on Potential impacts of leakage from deep CO_2 Geosequestration on overlying freshwater Aquifers, *Environ. Sci. Technol.*, *45*(7), 3171–3174, doi:10.1021/es104307h.

Han, J., J. W. Carey, and J. Zhang (2011a), A coupled electrochemical–geochemical model of corrosion for mild steel in high-pressure CO_2–saline environments, *Int. J. Greenh. Gas Control*, *5*(4), 777–787, doi:10.1016/j.ijggc.2011.02.005.

Han, J., J. W. Carey, and J. Zhang (2011b), Effect of sodium chloride on corrosion of mild steel in CO_2-saturated brines, *J. Appl. Electrochem.*, *41*(6), 741–749, doi:10.1007/s10800-011-0290-3.

Harvey, O. R., N. P. Qafoku, K. J. Cantrell, G. Lee, J. E. Amonette, and C. F. Brown (2012), Geochemical implications of gas leakage associated with geologic CO_2 storage-a qualitative review, *Environ. Sci. Technol.*, *47*(1), 23–36.

Higgs, K. E., R. R. Haese, S. D. Golding, U. Schacht, and M. N. Watson (2015), The Pretty Hill formation as a natural analogue for CO_2 storage: an investigation of mineralogical and isotopic changes associated with sandstones exposed to low, intermediate and high CO_2 concentrations over geological time, *Chem. Geol.*, *399*, 36–64, doi:10.1016/j.chemgeo.2014.10.019.

Hou, Z., D. H. Bacon, D. W. Engel, G. Lin, Y. Fang, H. Ren, and Z. Fang (2014), Uncertainty analyses of CO_2 plume expansion subsequent to wellbore CO_2 leakage into aquifers, *Int. J. Greenh. Gas Control*, *27*, 69–80, doi:10.1016/j.ijggc.2014.05.004.

Huerta, N. J., M. A. Hesse, S. L. Bryant, B. R. Strazisar, and C. L. Lopano (2013), Experimental evidence for self-limiting reactive flow through a fractured cement core: implications for time-dependent wellbore leakage, *Environ. Sci. Technol.*, *47*(1), 269–275, doi:10.1021/es3013003.

Huet, B. M., J. H. Prevost, and G. W. Scherer (2010), Quantitative reactive transport modeling of Portland cement in CO2-saturated water, *Int. J. Greenh. Gas Control*, *4*(3), 561–574, doi:10.1016/j.ijggc.2009.11.003.

Humez, P., P. Audigane, J. Lions, C. Chiaberge, and G. Bellenfant (2011), Modeling of CO_2 leakage up through an abandoned well from deep saline aquifer to shallow fresh Groundwaters, *Transp. Porous Media*, *90*(1), 153–181, doi:10.1007/s11242-011-9801-2.

Humez, P., V. Lagneau, J. Lions, and P. Negrel (2013), Assessing the potential consequences of CO_2 leakage to freshwater resources: a batch-reaction experiment towards an isotopic tracing tool, *Appl. Geochem.*, *30*, 178–190.

IEAGHG (2011a), Caprock Systems for CO_2 Geological Storage. http://ieaghg.org/docs/General_Docs/Reports/2011-01.pdf

IEAGHG (2011b), Potential impacts on groundwater resources of CO_2 geological storage. https://ieaghg.org/docs/General_Docs/Reports/2011-11.pdf.

Ivanova, A., A. Kashubin, N. Juhojuntti, J. Kummerow, J. Henninges, C. Juhlin, S. Lüth, and M. Ivandic (2012), Monitoring and volumetric estimation of injected CO_2 using 4D seismic, petrophysical data, core measurements and well logging: a case study at Ketzin, Germany, *Geophys. Prospect.*, *60*(5), 957–973, doi:10.1111/j.1365-2478.2012.01045.x.

Jackson, R. B., A. Vengosh, J. W. Carey, R. J. Davies, T. H. Darrah, F. O'Sullivan, and G. Pétron (2014), The environmental costs and benefits of fracking, *Annu. Rev. Environ. Resour.*, *39*(1), 327–362, doi:10.1146/annurev-environ-031113-144051.

Jenkins, C., A. Chadwick, and S. D. Hovorka (2015), The state of the art in monitoring and verification – ten years on, *Int. J. Greenh. Gas Control*, *40*, 312–349, doi:10.1016/j.ijggc.2015.05.009.

Jones, D. G., S. E. Beaubien, J. C. Blackford, E. M. Foekema, J. Lions, C. De Vittor, J. M. West, S. Widdicombe, C. Hauton, and A. M. Queirós (2015), Developments since 2005 in understanding potential environmental impacts of CO_2 leakage from geological storage, *Int. J. Greenh. Gas Control*, *40*, 350–377, doi:10.1016/j.ijggc.2015.05.032.

Jordan, P., and S. Benson (2009), Well blowout rates and consequences in California Oil and Gas District 4 from 1991 to 2005: implications for geological storage of carbon dioxide, *Environ. Geol.*, *57*(5), 1103–1123, doi:10.1007/s00254-008-1403-0.

Keating, E. H., J. Fessenden, N. Kanjorski, D. J. Koning, and R. Pawar (2010), The impact of CO(2) on shallow groundwater chemistry: observations at a natural analog site and implications for carbon sequestration, *Environ. Earth Sci.*, *60*(3), 521–536, doi:10.1007/s12665-009-0192-4.

Keating, E. H., J. A. Hakala, H. S. Viswanathan, R. Capo, B. Stewart, J. Gardiner, G. D. Guthrie, W. J. Carey, and J. Fessenden (2011), The challenge of predicting groundwater quality impacts in a CO_2 leakage scenario: results from field, laboratory, and modeling studies at a natural analog site in New Mexico, USA, *Energy Procedia*, *4*, 3239–3245, doi:10.1016/j.egypro.2011.02.242.

Keating, E., Z. Dai, D. Dempsey, and R. Pawar (2014a), Effective detection of CO_2 leakage: a comparison of groundwater sampling and pressure monitoring, *Energy Procedia*, *63*, 4163–4171, doi:10.1016/j.egypro.2014.11.448.

Keating, E., D. Newell, D. Dempsey, and R. Pawar (2014b), Insights into interconnections between the shallow and deep systems from a natural CO_2 reservoir near Springerville, Arizona, *Int. J. Greenh. Gas Control*, *25*, 162–172, doi:10.1016/j.ijggc.2014.03.009.

Kharaka, Y. K., and D. R. Cole (2011), Geochemistry of geologic sequestration of carbon dioxide, in R. S. Harmon, and A. Parker, editors, *Frontiers in Geochemistry*, pp. 133–174, John Wiley & Sons, Ltd, Chischester.

Kharaka, Y. K., J. J. Thordsen, S. D. Hovorka, H. S. Nance, D. R. Cole, T. J. Phelps, and K. G. Knauss (2009), Potential environmental issues of CO_2 storage in deep saline aquifers: geochemical results from the Frio-I brine pilot test, Texas,

USA, *Appl. Geochem.*, *24*(6), 1106–1112, doi:10.1016/j.apgeochem.2009.02.010.

Kharaka, Y. K. et al. (2010), Changes in the chemistry of shallow groundwater related to the 2008 injection of CO_2 at the ZERT field site, Bozeman, Montana, *Environ. Earth Sci.*, *60*(2), 273–284, doi:10.1007/s12665-009-0401-1.

Kiessling, D., C. Schmidt-hattenberger, H. Schuett, F. Schilling, K. Krueger, B. Schoebel, E. Danckwardt, and J. Kummerow (2010), Geoelectrical methods for monitoring geological CO2 storage: first results from cross-hole and surface–downhole measurements from the CO2SINK test site at Ketzin (Germany), *Int. J. Greenh. Gas Control*, *4*, 816–826, doi:10.1016/j.ijggc.2010.05.001.

Kim, J., and G. J. Moridis (2012), Gas Flow Tightly Coupled to Elastoplastic Geomechanics for Tight- and Shale-Gas Reservoirs: Material Failure and Enhanced Permeability, in *SPE Americas Unconventional Resources Conference*, Society of Petroleum Engineers, Pittsburgh, PA, 5–7 June 2012.

Kim, J., T. Matsuoka, and Z. Xue (2011), Monitoring and detecting CO_2 injected into water-saturated sandstone with joint seismic and resistivity measurements, *Explor. Geophys.*, *42*(1), 58, doi:10.1071/EG11002.

Kim, J., E. S. Um, and G. J. Moridis (2014), Fracture Propagation, Fluid Flow, and Geomechanics of Water-Based Hydraulic Fracturing in Shale Gas Systems and Electromagnetic Geophysical Monitoring of Fluid Migration. SPE 168578, in *SPE Hydraulic Fracturing Technology Conference*, Society of Petroleum Engineers, Woodlands, TX, 4–6 February.

Kirsch, K., A. K. Navarre-Sitchler, A. Wunsch, and J. E. McCray (2014), Metal release from sandstones under experimentally and numerically simulated CO_2 leakage conditions, *Environ. Sci. Technol.*, *48*(3), 1436–1442, doi:10.1021/es403077b.

Kutchko, B. G., B. R. Strazisar, D. A. Dzombak, G. V Lowry, and N. Thaulow (2007), Degradation of well cement by CO_2 under geologic sequestration conditions, *Environ. Sci. Technol.*, *41*(13), 4787–4792, doi:10.1021/es062828c.

Kutchko, B. G., B. R. Strazisar, G. V Lowry, D. A. Dzombak, and N. Thaulow (2008), Rate of CO_2 attack on hydrated class H well cement under geologic sequestration conditions, *Environ. Sci. Technol.*, *42*(16), 6237–6242, doi:10.1021/es800049r.

Kutchko, B. G., B. R. Strazisar, N. Huerta, G. V Lowry, D. A. Dzombak, and N. Thaulow (2009), CO_2 reaction with hydrated class H well cement under geologic sequestration conditions: effects of Flyash admixtures, *Environ. Sci. Technol.*, *43*(10), 3947–3952, doi:10.1021/es803007e.

Last, G., C. Murray, Y. Bott, and C. Brown (2014), Threshold values for identification of contamination predicted by reduced-order models, *Energy Procedia*, *63*, 3589–3597, doi:10.1016/j.egypro.2014.11.389.

Lemieux, J.-M. (2011), Review: the potential impact of underground geological storage of carbon dioxide in deep saline aquifers on shallow groundwater resources, *Hydrogeol. J.*, *19*(4), 757–778, doi:10.1007/s10040-011-0715-4.

Lewicki, J., J. Birkholzer, and C.-F. Tsang (2007), Natural and industrial analogues for leakage of CO_2 from storage reservoirs: identification of features, events, and processes and lessons learned, *Environ. Geol.*, *52*(3), 457–467, doi:10.1007/s00254-006-0479-7.

Lindeberg, E., and Bergmo, P. (2003), The long-term fate of CO_2 injected into an aquifer, in J. Gale, and Y. Kaya, editors, *Greenhouse Gas Control Technologies*, pp. 489–494, Pergamon, Amsterdam, The Netherlands.

Lions, J., N. Devau, L. de Lary, S. Dupraz, M. Parmentier, P. Gombert, and M.-C. Dictor (2014a), Potential impacts of leakage from CO_2 geological storage on geochemical processes controlling fresh groundwater quality: a review, *Int. J. Greenh. Gas Control*, *22*, 165–175, doi:10.1016/j.ijggc.2013.12.019.

Lions, J., P. Humez, H. Pauwels, W. Kloppmann, and I. Czernichowski-Lauriol (2014b), Tracking leakage from a natural CO_2 reservoir (Montmiral, France) through the chemistry and isotope signatures of shallow groundwater, *Greenh. Gases Sci. Technol.*, *4*(2), 225–243, doi:10.1002/ghg.1381.

Little, M. G., and R. B. Jackson (2010), Potential impacts of leakage from deep CO_2 Geosequestration on overlying freshwater aquifers, *Environ. Sci. Technol.*, *44*(23), 9225–9232, doi:10.1021/Es102235w.

Little, M. G., and R. B. Jackson (2011), Response to comment on "potential impacts of leakage from deep CO(2) Geosequestration on overlying freshwater aquifers", *Environ. Sci. Technol.*, *45*(7), 3175–3176, doi:10.1021/Es200421f.

Litynski, J., T. Rodosta, and B. Brown (2012), *Best Practices for Monitoring, Verification, and Accounting of CO_2 Stored in Deep Geologic Formations 2012 Update*, US Department of Energy, Pittsburgh, PA.

Lu, J., J. Partin, S. Hovorka, and C. Wong (2010), Potential risks to freshwater resources as a result of leakage from CO_2 geological storage: a batch-reaction experiment, *Environ. Earth Sci.*, *60*(2), 335–348, doi:10.1007/s12665-009-0382-0.

McKinley, R. M. (1994), Temperature, radioactive tracer, and noise logging for injection well integrity. Technical Report EPA/600/R-94/124, Environmental Protection Agency.

Ministerial Council on Mineral and Petroleum Resources (2005). Carbon Dioxide Capture and Geologic Storage Australian Regulatory Guiding Principles. https://industry.gov.au/resource/LowEmissionsFossilFuelTech/Documents/Regulatory-Guiding-Principles-CDC-GS.pdf

Montes-Hernandez, G., Renard, F., & Lafay, R. (2013). Experimental assessment of CO_2-mineral-toxic ion interactions in a simplified freshwater aquifer: implications for CO_2 leakage from deep geological storage. *Environ. Sci. Technol.*, *47*(12), 6247–6253.

NETL (2017), Best Practices for Monitoring, Verification, and Accounting of CO_2 Stored in Deep Geologic Formations 2017 Revised Edition, DOE/NETL-2017/1847, Pittsburgh, PA. https://www.netl.doe.gov/File%20Library/Research/Carbon-Storage/Project-Portfolio/BPM-MVA-2012.pdf.

Newell, D. L., and J. W. Carey (2013), Experimental evaluation of wellbore integrity along the cement-rock boundary, *Environ. Sci. Technol.*, *47*(1), 276–282, doi:10.1021/es3011404.

Newmark, R., S. Friedmann, and S. Carroll (2010), Water challenges for geologic carbon capture and sequestration, *Environ. Manag.*, *45*(4), 651–661, doi:10.1007/s00267-010-9434-1.

Nicot, J.-P. (2008), Evaluation of large-scale CO_2 storage on fresh-water sections of aquifers: an example from the Texas Gulf Coast Basin, *Int. J. Greenh. Gas Control*, *2*(4), 582–593, doi:10.1016/j.ijggc.2008.03.004.

Nordbotten, J., M. Celia, and S. Bachu (2005), Injection and storage of CO_2 in deep saline aquifers: analytical solution for CO_2 plume evolution during injection, *Transp. Porous Media*, *58*(3), 339–360, doi:10.1007/s11242-004-0670-9.

Nordbotten, J. M., D. Kavetski, M. A. Celia, and S. Bachu (2009), Model for CO_2 leakage including multiple geological layers and multiple leaky wells, *Environ. Sci. Technol.*, *43*(3), 743–749, doi:10.1021/es801135v.

Oldenburg, C. M., J. L. Lewicki, and R. P. Hepple (2003), Near-surface monitoring strategies for geologic carbon dioxide storage verification. Lawrence Berkeley National Laboratory. http://escholarship.org/uc/item/1cg241jb

Oldenburg, C. M., S. L. Bryant, and J.-P. Nicot (2009), Certification framework based on effective trapping for geologic carbon sequestration, *Int. J. Greenh. Gas Control*, *3*(4), 444–457, doi:10.1016/j.ijggc.2009.02.009.

OSPAR (2007), OSPAR Guidelines for Risk Assessment and Management of Storage of CO_2 Streams in Geological Formations, OSPAR 07/24/1-E, Annex 7.

Park, J., M. Fawad, I. Viken, E. Aker, and T. I. Bjørnarå (2013), CSEM sensitivity study for Sleipner CO_2-injection monitoring, *Energy Procedia*, *37*, 4199–4206, doi:10.1016/j.egypro.2013.06.322.

Pawar, R. J., G. S. Bromhal, J. W. Carey, W. Foxall, A. Korre, P. S. Ringrose, O. Tucker, M. N. Watson, and J. A. White (2015), Recent advances in risk assessment and risk management of geologic CO_2 storage, *Int. J. Greenh. Gas Control*, *40*, 292–311, doi:10.1016/j.ijggc.2015.06.014.

Peter, A. et al. (2012), Investigation of the geochemical impact of CO_2 on shallow groundwater: design and implementation of a CO_2 injection test in Northeast Germany, *Environ. Earth Sci.*, *67*(2), 335–349.

Reagan, M. T., G. J. Moridis, N. D. Keen, and J. N. Johnson (2015), Numerical simulation of the environmental impact of hydraulic fracturing of tight/shale gas reservoirs on near-surface groundwater: background, base cases, shallow reservoirs, short-term gas, and water transport, *Water Resour. Res.*, *51*(4), 2543–2573, doi:10.1002/2014WR016086.

Rillard, J., P. Gombert, P. Toulhoat, and P. Zuddas (2014), Geochemical assessment of CO_2 perturbation in a shallow aquifer evaluated by a push–pull field experiment, *Int. J. Greenh. Gas Control*, *21*, 23–32, doi:10.1016/j.ijggc.2013.11.019.

Rimmelé, G., V. Barlet-Gouédard, O. Porcherie, B. Goffé, and F. Brunet (2008), Heterogeneous porosity distribution in Portland cement exposed to CO_2-rich fluids, *Cem. Concr. Res.*, *38*(8–9), 1038–1048, doi:10.1016/j.cemconres.2008.03.022.

Rohmer, J., Pluymakers, A., & Renard, F. (2016). Mechano-chemical interactions in sedimentary rocks in the context of CO_2 storage: weak acid, weak effects?. *Earth Sci. Rev.*, *157*, 86–110.

Romanak, K. D., R. C. Smyth, C. Yang, and S. D. Hovorka (2010), Detection of anthropogenic CO_2 in dilute groundwater: field observations and geochemical modeling of the Dockum aquifer at the SACROC oilfield, West Texas, USA, GCCC Digital Publication Series 10–06: presented at the 9th Annual Conference on Carbon Capture and Sequestration, Pittsburgh, PA, 10–13 May 2010.

Romanak, K. D., R. C. Smyth, C. Yang, S. D. Hovorka, M. Rearick, and J. Lu (2012), Sensitivity of groundwater systems to CO_2: application of a site-specific analysis of carbonate monitoring parameters at the SACROC CO_2-enhanced oil field, *Int. J. Greenh. Gas Control*, *6*, 142–152, doi:10.1016/j.ijggc.2011.10.011.

Ruiz-Agudo, E., Putnis, C. V., & Putnis, A. (2014). Coupled dissolution and precipitation at mineral–fluid interfaces, *Chem. Geol.*, *383*, 132–146.

Rutqvist, J., A. P. Rinaldi, F. Cappa, and G. J. Moridis (2013), Modeling of fault reactivation and induced seismicity during hydraulic fracturing of shale-gas reservoirs, *J. Pet. Sci. Eng.*, *107*, 31–44, doi:10.1016/j.petrol.2013.04.023.

Saito, H., D. Nobuoka, H. Azuma, Z. Xue, and D. Tanase (2006), Time-lapse crosswell seismic tomography for monitoring injected CO_2 in an onshore aquifer, Nagaoka, Japan, *Explor. Geophys.*, *37*(1), 30, doi:10.1071/EG06030.

Schmidt-Hattenberger, C., P. Bergmann, D. Bösing, T. Labitzke, M. Möller, S. Schröder, F. Wagner, and H. Schütt (2013), Electrical resistivity tomography (ERT) for monitoring of CO_2 migration – from tool development to reservoir surveillance at the Ketzin pilot site, *Energy Procedia*, *37*, 4268–4275, doi:10.1016/j.egypro.2013.06.329.

Schwartz, M. O. (2014), Modelling leakage and groundwater pollution in a hypothetical CO_2 sequestration project, *Int. J. Greenhouse Gas Control*, *23*, 72–85.

Shao, H., N. P. Qafoku, A. R. Lawter, M. E. Bowden, and C. F. Brown (2015), Coupled geochemical impacts of leaking CO_2 and contaminants from subsurface storage reservoirs on groundwater quality, *Environ. Sci. Technol.*, *49*(13), 8202–8209, doi:10.1021/acs.est.5b01004.

Siirila, E. R., A. K. Navarre-Sitchler, R. M. Maxwell, and J. E. McCray (2012), A quantitative methodology to assess the risks to human health from CO_2 leakage into groundwater, *Adv. Water Resour.*. doi:10.1016/j.advwatres.2010.11.005.

Smyth, R. C., S. D. Hovorka, J. Lu, K. D. Romanak, J. W. Partin, C. Wong, and C. Yang (2009), Assessing risk to fresh water resources from long term CO_2 injection laboratory and field studies, *Energy Procedia*, *1*(1), 1957–1964, doi:10.1016/j.egypro.2009.01.255.

Song, J., and D. Zhang (2013), Comprehensive review of caprock-sealing mechanisms for geologic carbon sequestration, *Environ. Sci. Technol.*, *47*(1), 9–22, doi:10.1021/es301610p.

Spangler, L. et al. (2010), A shallow subsurface controlled release facility in Bozeman, Montana, USA, for testing near surface CO_2 detection techniques and transport models, *Environ. Earth Sci.*, *60*(2), 227–239.

Stauffer, P. H., H. S. Viswanathan, R. J. Pawar, and G. D. Guthrie (2009), A system model for geologic sequestration of carbon dioxide, *Environ. Sci. Technol.*, *43*(3), 565–570, doi:10.1021/es800403w.

Strazisar, B. R., A. W. Wells, J. R. Diehl, R. W. Hammack, and G. a. Veloski (2009), Near-surface monitoring for the ZERT shallow CO_2 injection project, *Int. J. Greenh. Gas Control*, *3*(6), 736–744, doi:10.1016/j.ijggc.2009.07.005.

Tessier, A., P. G. C. Campbell, and M. Bisson (1979), Sequential extraction procedure for the speciation of particulate trace-metals, *Anal. Chem.*, *51*(7), 844–851.

Thornhill, J. T. and B. G. Benefield (1990), Injection well mechanical integrity. Technical Report EPA/625/9–89/007, U.S. Environmental Protection Agency.

Thornhill, J. T. and B. G. Benefield (1992), Detecting water flow behind pipe in injection wells. Technical Report CR-815283, U.S. Environmental Protection Agency.

Trautz, R. C. et al. (2012), Evaluation of dissolved CO_2-induced metals mobilization in groundwater using a controlled release experiment, in *11th International Conference on Greenhouse Gas Technologies (GHGT)*, Kyoto, Japan, 18–22 November 2012.

Trautz, R. C. et al. (2013), Effect of dissolved CO_2 on a shallow groundwater system: a controlled release field experiment, *Environ. Sci. Technol.*, *47*(1), 298–305.

USEPA (U.S. Environmental Protection Agency) (2015), Drinking Water Contaminants – Standards and Regulations. http://www.epa.gov/dwstandardsregulations#List.

USEPA (United States Environmental Protection Agency) (2010), Federal Requirements under the Underground Injection Control (UIC) Program for Carbon Dioxide (CO_2) Geologic Sequestration (GS) Wells, Final Rule, 40 CFR Parts 124, 144, 145, 146, and 147, EPA–HQ–OW–2008–0390FRL–9232–7. http://www.gpo.gov/fdsys/pkg/FR-2010-12-10/pdf/2010-29954.pdf.

USEPA (United States Environmental Protection Agency) (2013), Underground Injection Control (UIC) Program Class VI Well Site Characterization Guidance, EPA 816-R-13-004. https://www.epa.gov/sites/production/files/2015-07/documents/epa816r13004.pdf.

USGAO (Government Accountability Office) (1989), DRINKING WATER. Safeguards are not Preventing Contamination from Injected Oil and Gas Wastes. GAO/RCED-89-97, Washington, DC. http://www.gao.gov/assets/150/147952.pdf.

Varadharajan, C., Birkholzer J.T., Kraemer S., Porse S., Carroll S., Wilkin R., Maxwell R., Bachu S., Hovorka S., Daley T., Digiulio D., Carey W., Strasizar B., Huerta N., Gasda S., and W. Crow (2012), Summary report on CO_2 geologic sequestration & water resources workshop. LBNL Report, LBNL-5346E, January 2012.

Varadharajan, C., R. M. Tinnacher, J. D. Pugh, R. C. Trautz, L. Zheng, N. F. Spycher, J. T. Birkholzer, H. Castillo-Michel, R. A. Esposito, and P. S. Nico (2013), A laboratory study of the initial effects of dissolved carbon dioxide (CO2) on metal release from shallow sediments, *Int. J. Greenh. Gas Control*, *19*, 183–211, doi:10.1016/j.ijggc.2013.08.017.

Vasco, D., and A. Ferretti (2005), On the use of quasi-static deformation to understand reservoir fluid flow, *Geophysics*, *70*(4), O13–O27, doi:10.1190/1.1993711.

Vasco, D. W., A. Rucci, A. Ferretti, F. Novali, R. C. Bissell, P. S. Ringrose, A. S. Mathieson, and I. W. Wright (2010), Satellite-based measurements of surface deformation reveal fluid flow associated with the geological storage of carbon dioxide, *Geophys. Res. Lett.*, *37*(3), doi:10.1029/2009GL041544.

Viswanathan, H. S., R. J. Pawar, P. H. Stauffer, J. P. Kaszuba, J. W. Carey, S. C. Olsen, G. N. Keating, D. Kavetski, and G. D. Guthrie (2008), Development of a hybrid process and system model for the assessment of wellbore leakage at a geologic CO(2) sequestration site, *Environ. Sci. Technol.*, *42*(19), 7280–7286, doi:10.1021/Es800417x.

Viswanathan, H., Z. Dai, C. Lopano, E. Keating, J. A. Hakala, K. G. Scheckel, L. Zheng, G. D. Guthrie, and R. Pawar (2012), Developing a robust geochemical and reactive transport model to evaluate possible sources of arsenic at the CO2 sequestration natural analog site in Chimayo, New Mexico, *Int. J. Greenh. Gas Control*, *10*, 199–214, doi:10.1016/j.ijggc.2012.06.007.

Vong, C. Q., N. Jacquemet, G. Picot-Colbeaux, J. Lions, J. Rohmer, and O. Bouc (2011), Reactive transport modeling for impact assessment of a CO_2 intrusion on trace elements mobility within fresh groundwater and its natural attenuation for potential remediation, *Energy Procedia*, *4*, 3171–3178, doi:10.1016/j.egypro.2011.02.232.

Walsh, S. C., W. Du Frane, H. Mason, and S. Carroll (2013), Permeability of wellbore-cement fractures following degradation by carbonated brine, *Rock Mech. Rock. Eng.*, *46*(3), 455–464, doi:10.1007/s00603-012-0336-9.

Watson, T. L., and S. Bachu (2008), Identification of wells with high CO_2-leakage potential in mature oil fields developed for CO_2-enhanced oil recovery, *Soc. Pet. Eng.*, doi:10.2118/112924-MS.

Watson, T. L., and S. Bachu (2009), Evaluation of the potential for gas and CO_2 leakage along wellbores, *Soc. Pet. Eng.* (*106817*), doi:10.2118/106817-PA.

Whittaker, S., B. Rostron, C. Hawkes, C. Gardner, D. White, J. Johnson, R. Chalaturnyk, and D. Seeburger (2011), A decade of CO_2 injection into depleting oil fields: monitoring and research activities of the IEA GHG Weyburn-Midale CO_2 monitoring and storage project, *Energy Procedia*, *4*, 6069–6076, doi:10.1016/j.egypro.2011.02.612.

Wildenborg, T., T. Leijnse, E. Kreft, and M. Nepveu, and A. Obdam (2005), Long-term safety assessment of CO_2 storage: the scenario approach, in E. S. Rubin, D. W. Keith, C. F. Gilboy, M. Wilson, T. Morris, J. Gale, and K. Thambimuthu, editors, *Greenhouse Gas Control Technologies 7*, Elsevier, pp. 1283–1287, doi:10.1016/B978-008044704-9/50139-7.

Wilkin, R. T., and D. C. Digiulio (2010), Geochemical impacts to groundwater from geologic carbon sequestration: controls on pH and inorganic carbon concentrations from reaction path and kinetic modeling, *Environ. Sci. Technol.*, *44*(12), 4821–4827, doi:10.1021/Es100559j.

Wunsch, A., A. K. Navarre-Sitchler, J. Moore, and J. E. McCray (2014), Metal release from limestones at high partial-pressures of CO_2, *Chem. Geol.*, *363*, 40–55, doi:10.1016/j.chemgeo.2013.10.036.

Xue, Z., D. Tanase, and J. Watanabe (2006), Estimation of CO_2 saturation from time-lapse CO_2 well logging in an onshore aquifer, Nagaoka, Japan, *Explor. Geophys.*, *37*(1), 19, doi:10.1071/EG06019.

Yamamoto, H., K. Zhang, K. Karasaki, A. Marui, H. Uehara, and N. Nishikawa (2009), Numerical investigation concerning the impact of CO_2 geologic storage on regional groundwater flow, *Int. J. Greenh. Gas Control*, *3*(5), 586–599, doi:10.1016/j.ijggc.2009.04.007.

Yang, C., P. J. Mickler, R. Reedy, B. R. Scanlon, K. D. Romanak, J.-P. Nicot, S. D. Hovorka, R. H. Trevino, and T. Larson (2013), Single-well push-pull test for assessing potential impacts of CO_2 leakage on groundwater quality in a shallow Gulf Coast aquifer in Cranfield, Mississippi, *Int. J. Greenh. Gas Control*, *18*, 375–387, doi:10.1016/j.ijggc.2012.12.030.

Yang, Q. et al. (2014), Groundwater hydrogeochemistry in injection experiments simulating CO_2 leakage from geological storage reservoir, *Int. J. Greenh. Gas Control*, *26*, 193–203, doi:10.1016/j.ijggc.2014.04.025.

Yang, Q., J. Matter, T. Takahashi, M. Stute, G. O'Mullan, K. Clauson, K. Umemoto, and D. Goldberg (2015a), Groundwater geochemistry in bench experiments simulating CO_2 leakage from geological storage in the Newark Basin, *Int. J. Greenh. Gas Control*, *42*, 98–108, doi:10.1016/j.ijggc.2015.06.024.

Yang, X., R. N. Lassen, K. H. Jensen, and M. C. Looms (2015b), Monitoring CO_2 migration in a shallow sand aquifer using 3D crosshole electrical resistivity tomography, *Int. J. Greenh. Gas Control*, *42*, 534–544, doi:10.1016/j.ijggc.2015.09.005.

Zheng, L., J. A. Apps, Y. Q. Zhang, T. F. Xu, and J. T. Birkholzer (2009), On mobilization of lead and arsenic in groundwater in response to CO_2 leakage from deep geological storage, *Chem. Geol.*, *268*(3–4), 281–297, doi:10.1016/j.chemgeo.2009.09.007.

Zheng, L., N. F. Spycher, J. T. Birkholzer, T. Xu, J. Apps, and Y. Kharaka (2011), *Modeling Studies on the Transport of Benzene and H_2S in CO_2-Water Systems*, Berkeley, CA, Lawrence Berkeley National Laboratory.

Zheng, L., J. A. Apps, N. Spycher, J. T. Birkholzer, Y. K. Kharaka, J. Thordsen, S. R. Beers, W. N. Herkelrath, E. Kakouros, and R. C. Trautz (2012), Geochemical modeling of changes in shallow groundwater chemistry observed during the MSU-ZERT CO2 injection experiment, *Int. J. Greenh. Gas Control*, *7*, 202–217.

Zheng, L., S. Carroll, M. Bianchi, K. Mansoor, Y. Sun, and J. Birkholzer (2014), Reduced order models for prediction of groundwater quality impacts from CO_2 and brine leakage, *Energy Procedia*, *63*, 4875–4883, doi:10.1016/j.egypro.2014.11.518.

Zheng, L., N. P. Qafoku, A. Lawter, G. Wang, H. Shao, and C. F. Brown (2015a), Evaluating impacts of CO_2 intrusion into an unconsolidated aquifer: II. Modeling results, *Int. J. Greenh. Gas Control*, doi:10.1016/j.ijggc.2015.07.001.

Zheng, L., N. Spycher, C. Varadharajan, R. M. Tinnacher, J. D. Pugh, M. Bianchi, J. Birkholzer, P. S. Nico, and R. C. Trautz (2015b), On the mobilization of metals by CO_2 leakage into shallow aquifers: exploring release mechanisms by modeling field and laboratory experiments, *Greenhouse Gases Sci. Technol.*, *5*(4), 403–418, doi:10.1002/ghg.1493.

Zhong, L., K. J. Cantrell, D. H. Bacon, and J. Shewell (2014), Transport of organic contaminants mobilized from coal through sandstone overlying a geological carbon sequestration reservoir, *Int. J. Greenh. Gas Control*, *21*, 158–164, doi:10.1016/j.ijggc.2013.12.014.

Zoback, M. (2007), *Reservoir Geomechanics*, Cambridge, Cambridge University Press.

16

Review of CO_2 Leakage Mitigation and Remediation Technologies

Cesar A. Castaneda-Herrera, Geoffrey W. Stevens, and Ralf R. Haese

ABSTRACT

The main purpose of CO_2 sequestration is to keep CO_2 underground in a safe and permanent environment. However, it is possible that unwanted migration of CO_2 occurs from the geological containment due to different leakage pathways. These pathways could be engineered or natural. This work reviews different possibilities of materials and techniques that may seal or avoid leakage of CO_2. These technologies have been categorized into two main groups. The first group is the high-viscosity fluid-based barrier formation, which includes cements, geopolymers, gels, and nanoparticles. These technologies are more suitable for engineered pathways. The second group is the low-viscosity fluid-based barrier formation, which are biomineralization, hydraulic barrier formation, and chemical reactive barrier formation. Each of the technologies is explained in detail showing advantages and disadvantages. At the end, a final synthesis is given, concluding that more research is needed to identify the best technology; however, the chemical reactive barrier was identified as the technology with less disadvantages.

16.1. INTRODUCTION

Global climate change driven by the emission of greenhouse gases is one of the main challenges of our time. The increase in the global average temperature is influenced by the accumulation of gases in the atmosphere, in which carbon dioxide (CO_2) plays a major role. Most of the anthropogenic CO_2 comes from the use of fossil fuels in electricity generation and industrial processes [*Le Quéré et al.*, 2015]. In order to mitigate the steep increase in CO_2 emissions, a range of technologies have been developed and deployed including carbon capture and storage (CCS) where up to 90% of CO_2 coming from industrial flue gases is captured and stored in geological formations, preventing the carbon dioxide from entering the atmosphere.

Peter Cook Centre for CCS Research, The University of Melbourne, Melbourne, Australia; Cooperative Centre for Greenhouse Gas Technologies, CO2CRC, Carlton, Victoria, Australia

One of the main purposes of CCS is to keep the CO_2 safely underground. In principle, CO_2 is stored in depleted oil and gas reservoirs, saline aquifers, or other subsurface geological formation overlain by a caprock. Reservoirs are typically high-porosity-high-permeability formations, for example, a sandstone, while the caprock is a low-permeability unit such as a shale. The combination of a reservoir and caprock is in principle suitable conditions for permanent containment of CO_2. However, in some instances, the possibility of CO_2 migration out of the permanent containment through leakage pathways to an overlying aquifer formation or the surface cannot be fully ruled out and can be a major concern to the public.

Consequences of CO_2 leakage have been identified from different risk assessments, where the main concern is the hazard of the reaction between CO_2 and CO_2-enriched brine with overlying fresh groundwater resources. According to *Lions et al.* [2014], mixing of CO_2-brine with fresh groundwater will lead to a decrease of pH, change of redox potential, change of microbial metabolism, and a possible mobilization of contaminants

Geological Carbon Storage: Subsurface Seals and Caprock Integrity, Geophysical Monograph 238, First Edition.
Edited by Stéphanie Vialle, Jonathan Ajo-Franklin, and J. William Carey.

such as trace elements. CO_2 migration into the vadose zone, accumulation in soil gas, release into the atmosphere, and accumulation in surface water have been also considered as possible hazards [*Lions et al.*, 2014].

Pawar et al. [2015] recently reviewed the development of risk management for the CCS industry over the last 10 years. At least 50 million tonnes of CO_2 was injected in different projects around the world, while risk management plans were still improved. The study states that significant progress has been made in areas such as injection performance and effectiveness of monitoring; however, there is still a need for a better assessment of risks including the risk from CO_2 leakage.

As a part of a risk assessment, procedures for CO_2 leakage mitigation and remediation options for CO_2 leakage should be identified and proposed. Risk assessments have shown that very limited knowledge and technologies are available to remediate a potential CO_2 leakage. In order to develop those, it is important to characterize different CO_2 leakage scenarios and subsequently provide solutions for the different scenarios. A number of recent studies have proposed a range of different technologies and materials to mitigate or remediate CO_2 leakage [*Brydie et al.*, 2014; *Cunningham et al.*, 2011; *Manceau et al.*, 2014; *Réveillère et al.*, 2012]. In order to be an operational technique, these procedures need to account for reservoir conditions, where the technologies need to lead to a high degree of permeability reduction and must remain stable under CO_2-enriched brine and supercritical CO_2 ($scCO_2$) conditions over a long time.

The purpose of this chapter is to provide a review of the most relevant research related to CO_2 leakage mitigation and/or remediation techniques. The following sections are divided into three parts. Firstly, possible leakage pathways are identified and explained (Section 16.2). Secondly, a summary of technologies that could be applied to prevent leakage is given in Section 16.3. The summary categorizes the technologies by the viscosity of the fluid used in the deployment of the respective technique. Finally, in the last section (Section 16.4), a synthesis of the reviewed technologies is given.

16.2. LEAKAGE PATHWAYS

A special report by the Intergovernmental Panel on Climate Change [*Metz et al.*, 2005] provided a first summary of different potential CO_2 leakage pathways. These pathways are related with the different phases (supercritical, liquid, or gas) in which CO_2 could be present. More recent studies [*Manceau et al.*, 2014; *Tongwa et al.*, 2013b] have reconsidered these different pathways and identified two main categories. Leakage associated with wells is referred to as *engineered*, while $scCO_2$ and/or CO_2-enriched brine migration through geological formations and structures, such as the caprock and geological faults, is referred to as *natural* leakage pathways.

16.2.1. Engineered Pathways

Engineered pathways are related to all possible wells involved in the process of CO_2 injection and storage as well as wells drilled for other purposes including monitoring wells and abandoned oil and gas exploration wells. Particularly preexisting plugged-and-abandoned and long-term monitoring wells require attention as part of a CO_2 site assessment. The potential problem of wells in a CCS scenario is the declination of the wellbore over time due to the contact with CO_2-enriched brine. CO_2 injected into a saline aquifer often leads to a drop in the pH and forms a corrosive environment for materials such as cement and steel. These preexisting wells were not designed for high-CO_2 environments [*Carey*, 2013].

Gasda et al. [2004] identified critical areas where CO_2 migration in non-injection wells may occur: (i) interfaces between the cement and steel casing, (ii) the cement matrix as a result of cement degradation, (iii) holes within steel casing as a result of corrosion, (iv) fractures in cement, and (v) interfaces between cement and rock. For the specific case of cement-rock interaction, reactions with CO_2-enriched water are expected, leading to the formation of calcium carbonate ($CaCO_3$). Excess of CO_2-enriched water may dissolve the precipitated $CaCO_3$, further degrading the cement [*Manceau et al.*, 2014]. Degradation rates of cements differ and field records show that actual rates are low [*Carey*, 2013].

Besides chemical degradation, poor cementing may cause mechanical deformation of well cement and casing; cement shrinkage may create important migration pathways at the interfaces between the cement and the casing or between the cement and the caprock [*Barclay et al.*, 2001].

16.2.2. Natural Pathways

Natural pathways are related to the geological conditions of the storage site such as caprock integrity and the presence of faults. Caprocks may have preexisting joints and fractures too small for seismic detection, and those may serve as migration pathways of $scCO_2$ or CO_2-enriched brine [*Tongwa et al.*, 2013b]. However, since geological carbon storage involves the injection of large volumes of CO_2, imposed geomechanical and geochemical processes may also change the geological integrity of caprocks. Of particular concern is the pressure buildup during the injection phase, potentially leading to fault reactivation and the degradation of the caprock related to mineral dissolution [*Fitts and Peters*, 2013]. Flow paths can be created when acidified brine dissolves minerals within the rock. The new flow paths will change the hydrodynamic properties of the rock, and leakage promotion can occur.

Aydin [2000] categorizes the two most common types of structural heterogeneities that facilitate hydrocarbon migration and flow as dilatant fractures (joints, veins, and dikes) and shear fractures/faults. *Fitts and Peters* [2013] explained that there is a high likelihood for caprocks to be affected by geomechanical and geochemical changes. Thus, it is critical to quantify the risk of leakage through the caprock, by means of assessing the likelihood of the occurrence, the potential for development and reactivation, and the predicted hydrodynamic properties of CO_2 flow paths through caprocks [*Fitts and Peters*, 2013].

16.3. MITIGATION AND REMEDIATION TECHNOLOGIES

The engineered pathway is more widely studied, since well leakage is not only an issue of CO_2 storage but a known problem for the oil and gas industry. Therefore, there has been considerably more research on the development of new materials and procedures to avoid leakage through wells. Meanwhile, the mitigation and remediation of natural CO_2 pathways has been considered less. At the same time, they are more difficult to identify, find, and treat [*Manceau et al.*, 2014]. Nevertheless, there are some novel techniques that focus on minimizing and inhibiting natural CO_2 pathways, which are presented here.

In this review, mitigation is the process used to avoid a hazard to occur or reduce its magnitude. Meanwhile, remediation is the restoration treatment applied to a damage that already occurred [*Manceau et al.*, 2014].

Remediation at reservoir scale implies injection of a material into a porous or fractured geological formation. One of the main factors that affect the injectivity of a material is the viscosity of it. Materials with low viscosity flow easily and can be injected at greater rate. In contrast, high-viscosity materials migrate slower and cannot be distributed widely within a formation [*Aminzadeh et al.*, 2013]. Consequently, highly viscous materials can be used for the remediation of engineered leakage where leakage occurs within or near the wellbore. In contrast, the remediation of natural CO_2 leakage pathways requires low-viscosity materials since they can be placed over a larger area on top of a caprock or storage reservoir. In this review, we categorize the different mitigation and remediation technologies in two main groups, namely, barrier formation based on high- and low-viscosity fluids.

16.3.1. High-Viscosity Fluid-Based Barrier Formation

16.3.1.1. Cements

Cements have been the best choice to plug wells in different fields, because they are excellent impervious materials. Portland cement is the most common cement used, and other cement types are derived from it [*Carey*, 2013]. This cement is hydraulic cement (it hardens when contact with water), and it is produced from the heating of limestone, clays, and quartz-bearing materials at temperatures higher than 1400°C. After this process, a powder is obtained and water is added. Thus, the hydration reaction leads to the formation of Portland cement, which has these components: 48% calcium silicate hydrate (C—S—H), 19% portlandite ($Ca(OH)_2$), 18% monosulfate ($Ca_4Al_2(OH)_{12}SO_4$ $6H_2O$), 9% ettringite ($Ca_6Al_2(SO_4)_3(OH)_{12}$ $26H_2O$), and 6% others [*Carey*, 2013].

The use of Portland cement for the construction, completion, and repair of wells is well known and commonly applied to remediate oil and gas wells. However, studies have shown that the performance of Portland cement can be affected through the exposure to CO_2-enriched brine and $scCO_2$ over long time periods [*Carey*, 2013; *Kutchko et al.*, 2007].

Kutchko et al. [2007] studied this problem, where Portland cement was exposed to brine solution saturated with CO_2 at 50°C and 30.3 MPa. Their experiments ran for 9 days under static conditions. It was found that three discrete zones were observed in the samples, which indicated partial degradation of the cement when in contact with CO_2. Figure 16.1 shows the conceptual model for a propagating reaction front penetrating into the cement with three discrete reaction zones as proposed by *Kutchko et al.* [2007].

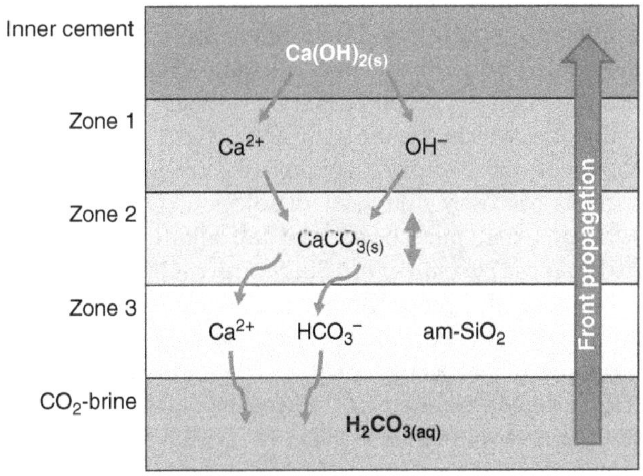

Figure 16.1 Dissolution of Portland cement in CO_2-enriched water with associated ion mobilization and formation of distinct reactions zones. Modified from *Kutchko et al.* [2007].

Kutchko et al. [2007] reported that Zone 1 develops a minor increase in porosity and a decrease in $Ca(OH)_2$ due to an acidic dissolution of portlandite by carbonic acid (Reaction 16.1). Zone 2 is characterized by reduced porosity and $CaCO_3$ formation from the reaction between leached calcium from Zone 1 and carbonated brine (Reaction 16.2). However, a total depletion of $Ca(OH)_2$ in Zone 2 generates degradation of $CaCO_3$, which decreases the pH by the formation of acid (Reaction 16.3). As a consequence of the low pH, the calcium silicate hydrate dissolves and forms amorphous silica (Reaction 16.4), as shown in Zone 3:

$$Ca(OH)_{2(s)} \rightarrow Ca^{2+}{}_{(aq)} + 2OH^{-}{}_{(aq)} \quad (16.1)$$

$$Ca^{2+}{}_{(aq)} + HCO_3^{-}{}_{(aq)} + OH^{-}{}_{(aq)} \rightarrow CaCO_{3(s)} + H_2O \quad (16.2)$$

$$H^{+}{}_{(aq)} + CaCO_{3(s)} \rightarrow Ca^{2+}{}_{(aq)} + HCO_3^{-}{}_{(aq)} \quad (16.3)$$

$$C-S-H_{(s)} \rightarrow Ca^{2+}{}_{(aq)} + OH^{-} + \text{am-}SiO_{2(s)} \quad (16.4)$$

At the end of the experiment, the outer layer of cement only consists of amorphous silica with increased permeability and porosity. It was concluded that the exposure of Portland cement to CO_2-enriched brine under reservoir conditions leads to considerable degradation.

16.3.1.2. Geopolymers and Gels

Geopolymers are aluminosilicate materials that have been proposed as a replacement for cements for construction purposes. They are formed by combining an alkaline solution with a reactive aluminosilicate solid, which results in the formation of alkali aluminosilicate gel. The two common sources of aluminosilicates are metakaolin (kaolinite clay) and fly ash (by-product of coal combustion) [*Provis and Van Deventer*, 2009]. Recently, fly ash has been considered as the better choice because of its particle shape, continuous production in coal-fired power plants, and low cost. The most common alkaline solutions used for the synthesis of geopolymers are sodium or potassium hydroxide, which are the activators of the reaction. The choice of the hydroxide solution depends on properties such as viscosity and heat dissolution. Other alkaline solutions that have been used are alkali silicates and carbonates [*Duxson et al.*, 2005; *Provis*, 2009].

In general, geopolymers are considered a good replacement of Portland cements because of improved mechanical properties such as compressive strength, reduced costs, reduced CO_2 emissions during its production, and a higher resistivity to acidic conditions. A number of studies have found geopolymers to be more resistant to strong acids such as sulfuric and nitric acid than common cement [*Provis and Van Deventer*, 2009].

The performance of geopolymers has also been studied under CO_2 storage conditions, since it has been proposed as a material to seal well leakage. *Nasvi et al.* [2012] pointed out that geopolymers have the advantage of being corrosion resistant, since formation of Na_2CO_3 or K_2CO_3 leads to a pH of 10–10.5 compared to a pH of 7–8 provided by $CaCO_3$ formation.

Nasvi and colleagues [2013a, 2013b, 2014] tested fly-ash-based geopolymers under different conditions related to CO_2 sequestration. They evaluated variables like the change in salinity and temperature when in contact with $scCO_2$ or CO_2-enriched brine. The performance of cements and geopolymers was tested at different salinity levels; a reduction in material strength was observed in both cases, but the reduction was less for geopolymers. Similarly, the permeability of geopolymers flooded with $scCO_2$ was found to be distinctively lower compared to cement. *Nasvi et al.* [2014] concluded that geopolymers are a good replacement for Portland cement since they show superior behavior under variable temperatures and salinity levels, a lower permeability, higher acid resistivity, and a more environmentally friendly production. Despite those promising results, *Zhang et al.* [2014] pointed out that the development of geopolymers is not mature and further research is required to optimize their performance. Key variables in the production of geopolymers are the composition of aluminosilicates and the curing temperature affecting the soluble alkali content and pore size. Furthermore, the application of this material has been limited to the construction industry so far, and only Nasvi and colleagues have proposed it as a material for sealing possible CO_2 leaks.

Besides geopolymers, other types of polymers in the form of gels have been studied as sealing agents for CO_2 leakage. Similar to other sealing materials, the aim of gels is to reduce permeability in existing or induced pathways. Polymer and silicate gels have been the most used gels in the oil industry, especially because of their chemical and thermal stability [*Manceau et al.*, 2014].

For the purpose of CO_2 storage applications, *Tongwa et al.* [2013a] proposed to use a gel as a plugging agent and developed a nanocomposite hydrogel. This hydrogel is a three-dimensional hydrophilic polymer, which does not dissolve but swells in water. The innovation of this hydrogel is the use of natural clays as nanoparticles and the elimination of organic cross-linkers for its synthesis. *Tongwa et al.* [2013a] synthetized this hydrogel with polyacrylamide (PAM) and Laponite (a synthetic layered silica $Na^{+}{}_{0.7}[(Si_8Mg_{5.5}Li_{0.3})O_{20}(OH)_4]^{-0.7}$). The PAM is a water-soluble, viscous, and anionic polymer, which is a product of the alkaline hydrolysis of acrylamide monomers. The negative surface charge of PAM creates electrostatic interaction with the cationic clay surface, leading to the formation of gel. The mechanical and rheological properties of the hydrogel were also tested, and it was found that the gel has resistance to high temperatures and its strength depends on the clay concentration. The main disadvantage of gels is their high viscosity limiting injectivity and deep penetration into the reservoir.

Tongwa et al. [2013b] compared the performance of cement, wax, silicate gel, and hydrogel as potential sealing agents by testing four criteria: the ability to reduce permeability, the chemical and thermal stability, the sealed-fracture integrity, and the sealed-fracture strength compared to unfractured rock. The study found that each material has the ability to seal small fractures and reduce permeability; however, only cement was able to withstand the high pressure under injection conditions. The gel showed to be effective only in fractures with a width less than 0.25 mm, since flow of $scCO_2$ developed wormholes in the gel structure within wider fractures. However, it was also found that no material could emulate the same strength of the unfractured sample [*Tongwa et al.*, 2013b]

16.3.1.3. Nanoparticle Foams

Based on previous enhanced oil recovery (EOR) research, the use of nanoparticles has also been proposed to prevent CO_2 leakage. Here chemical-coated nanoparticles adhere to the surface of CO_2-water, stabilizing the suspension with the formation of foams [*Aminzadeh et al.*, 2013]. The nanoparticle suspension is injected into the depleted oil field that contains CO_2. The contact of the nanoparticles with CO_2 will produce CO_2-charged foam [*Aminzadeh et al.*, 2013]. The formation of foam will decrease the mobility of free CO_2 and reduce CO_2 leakage. One option is mixing the nanoparticles and the CO_2-brine at surface and injecting the mixture. However, foam formation occurs quickly and may reduce injectivity of the well [*Aminzadeh et al.*, 2013; *Manceau et al.*, 2014]. Alternatively, *Aminzadeh et al.* [2013] proposed to only inject the nanoparticle suspension into the possible leakage path. If there is CO_2 present, the nanoparticles will react at the CO_2-brine interface, produce foam, and potentially seal the leaks. Based on this idea, an experiment was carried out, where high-pressure liquid CO_2 was injected into a core pretreated with a nanoparticle/brine suspension [*Aminzadeh et al.*, 2013]. The nanoparticles used in the experiment were silica nanoparticles with polyethylene glycol (PEG) as a ligand at the surface. The results of the experiments showed that the nanoparticles altered the system, but they could not completely stabilize the displacement front of the liquid CO_2. This problem may be related to the differences in viscosities of both fluids, where a complete mixing was not possible.

Previous studies [*DiCarlo et al.*, 2011] showed a high degree of mobility for coated nanoparticles in porous media and effective immobilization of CO_2 through the foam formation. However, the high viscosity of the nanoparticle-enriched injection fluid remains the main problem [*Aminzadeh et al.*, 2013].

16.3.2. Low-Viscosity Fluid-Based Barrier Formation

16.3.2.1. Biomineralization and Biofilm Formation

The use of microbiological communities to reduce leakage in wells has been proposed by different authors [*Cunningham et al.*, 2011, 2013; *Mitchell et al.*, 2009]. The underlying principle for this concept is the ability of microorganisms to induce and accelerate mineral precipitation. Microorganisms are very small in size, and they can be injected with a low-viscosity fluid penetrating deep into the reservoir or into small fractures.

Cunningham et al. [2009] proposed barrier formation through biomineralization involving ureolysis. This mechanism has been used for other purposes such as industrial plugging and immobilization of contaminants in groundwater and to strengthen cements. The principle of the mechanism is the degradation of urea through the enzyme ureases which is commonly present in a variety of microorganisms.

In this process, urea is degraded to ammonia and carbamate by the microorganism (Reaction 16.5). Then, the carbamate is hydrolyzed to form ammonia and carbonic acid (Reaction 16.6). Both of these products dissociate in water and lead to a rise in the pH (16.7). In response, the carbonate equilibrium shifts toward dominance by bicarbonate carbonate ions (Reactions 16.8 and 16.9). In the presence of ambient calcium, the carbonate ions react to precipitate the $CaCO_3$ (16.10) [*Cunningham et al.*, 2009]:

$$CO(NH_2)_2 + H_2O \rightarrow NH_2COOH + NH_3 \tag{16.5}$$

$$NH_2COOH + H_2O \rightarrow NH_3 + H_2CO_3 \tag{16.6}$$

$$2NH_3 + 2H_2O \rightarrow 2NH_4^+ + 2OH^- \tag{16.7}$$

$$H_2CO_3 + OH^- \rightarrow HCO_3^- + H_2O \tag{16.8}$$

$$HCO_3^- + OH^- \rightarrow CO_3^{2-} + H_2O \tag{16.9}$$

$$CO_3^{2-} + Ca^{2+} \rightarrow CaCO_3 \tag{16.10}$$

It has been proven in laboratory studies that precipitation of $CaCO_3$ or biomineralization enhances pore clogging and mineral trapping of CO_2 [*Cunningham et al.*, 2011; *Mitchell et al.*, 2009].

Phillips et al. [2013] carried out core sealing experiments where $CaCO_3$ precipitation was induced by biofilms in cores with and without fractures. They found that these engineered biofilms have the potential to seal and strengthen fractures as permeability was reduced. However, several conditions need to be tested and validated in order to use this technique at field scale. Live ureolytic microorganisms must resist reservoir conditions including high temperatures, high pressures, mild acidity (pH: 4–7), and high CO_2 partial pressure. Indigenous microbial communities may not have these attributes in which case cultured organisms and/or enzyme solutions must be injected into the area where biofilm formation is required. Furthermore, all the necessary reactants such as urea, calcium, and nutrients will need to be added as well [*Phillips et al.*, 2013].

Some authors [*Mu and Moreau*, 2015; *Mu et al.*, 2014] pointed out that it is important to determine the microbial community living in the CO_2 storage reservoirs before implementing biofilm technology. *Peet et al.* [2015] studied the behavior of microorganism in core samples (from different CO_2 known sites around the world) under reservoir conditions in laboratory experiments. It was found that under $scCO_2$ conditions, microbial survival and activity are possible, that in situ biomineralization at the CO_2 plume-water interface may be feasible, and that spore-forming microbes are more likely to survive after CO_2 injection.

The microbial response to high-CO_2 reservoir conditions is generally poorly understood, and lab-scale *in vitro* results need to be validated by in situ field experiments. *Mu et al.* [2014] conducted a study on changes in the microbial community before and after CO_2 injection. Their results are based on the $scCO_2$ sequestration experiment conducted at the CO2CRC Otway Project site, where they detected that there was a shift in the microbial community after CO_2 injection. However, the change in the microbial structure may be related to the disappearance of PEG, which are residual compounds from the drilling fluid used for the injection well.

Based on the rather limited number of studies directly pertaining to engineered biofilm formation and/or biomineralization, this approach is still at a conceptual to early R&D stage. However, as more information on the in situ microbial community in reservoirs with and without CO_2 become available and dedicated microbial studies become part of field experiments, engineered barrier formation using microorganisms may become feasible.

16.3.2.2. Hydraulic Barrier

Fluid management has been proposed as a way to prevent or control CO_2 leakage migration. It aims to prohibit leakage through the caprock, more specifically through faults, fractures, or high-permeability zones, and is based on an imposed hydraulic pressure gradient. *Manceau et al.* [2014] describes different cases where fluid management can be applied:

1. Temporal or permanent control of the pressure gradient within the storage reservoir
2. Imposing overpressure in the aquifer above the reservoir, leading to a *push* of CO_2 away from the leakage location
3. Enhanced nonstructural trapping mechanisms

Within these cases, *Réveillère et al.* [2012] suggested a hydraulic barrier as a fluid management technology. The concept of the hydraulic barrier is to change the driving force of the leakage, making a change to the pressure field and thereby countering the hydraulic gradient of the flow along the leakage pathway. Hydraulic barriers have been commonly used in pollution remediation technologies. One example is the use of production wells to change the hydrology of aquifers and avoid the contact between drinking and salt water [*Pareek et al.*, 2006]. This technique has been used at shallower depths, but its application for CO_2 leakage purposes has some restrictions. *Lions et al.* [2014] discussed issues relating to poor hydraulic conductivity in the overlying aquifer and the need for water management at surface.

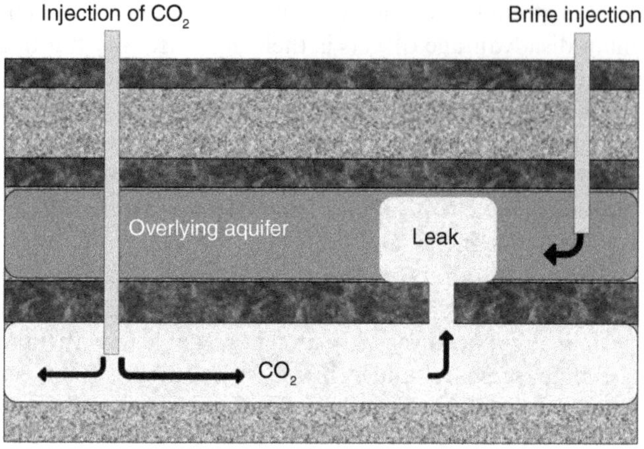

Figure 16.2 Hydraulic barrier scenario. Modified from *Réveillère et al.* [2012].

Réveillère et al. [2012] carried out numerical simulations of a leakage scenario where CO_2 was leaking through a fracture in the caprock as represented in Figure 16.2. The study found two major limitations to this approach: Firstly, a high hydraulic transmissivity of the overlying aquifer is not suitable, because the required overpressure to reverse the CO_2 leakage may be insufficient. Secondly, the distance from the brine injection well to the leak position was identified as the most important parameter that controls the effectiveness of this process. The hydraulic barrier was found to be most effective where the brine was injected in the vicinity of the leakage location. For this case, the leakage was stopped in less than 6 months under the modeled conditions. However, the injection of brine is only possible as a temporary technique. A longer injection time will make this technology unsustainable [*Réveillère et al.*, 2012].

Zahasky and Benson [2016] also modeled the implementation of a hydraulic barrier but this time under different reservoir heterogeneities with different hydraulic methods. Results show that small injection rates of water and a simultaneous brine production (called a multistage hydraulic barrier) were enough to stop CO_2 leakage, improving the efficacy of the technology. *Zahasky and*

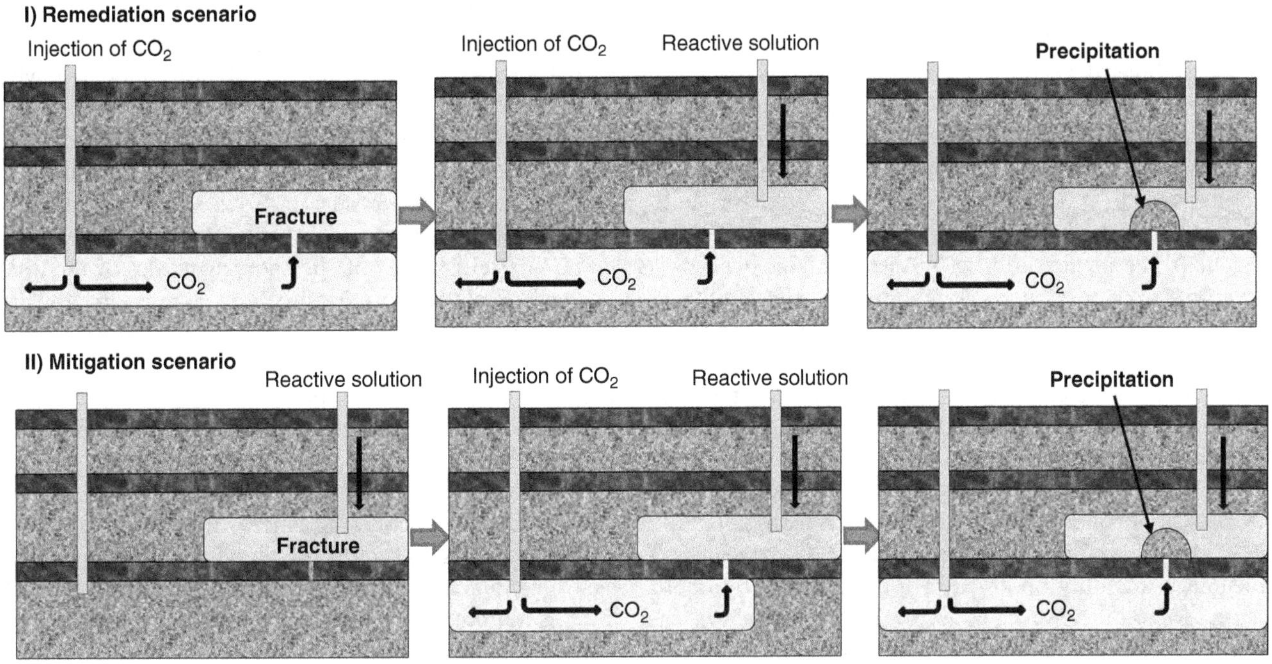

Figure 16.3 Two suggested scenarios for reactive barrier formation. Modified from *Ito et al.* [2014].

Benson [2016] also found that the presence of heterogeneous formations aids the formation of flow barriers.

In general, this technology is only applicable under certain geological conditions, and it requires a significant effort in its design and implementation, brine availability, and high operational cost, which may limit the applicability of this novel approach. However, these first findings can be used as guidance for further research into fluid management.

16.3.2.3. Chemical Reactive Barrier

The idea of a chemical reagent reacting with CO_2 to create a mineral barrier has also been proposed by different authors [*Brydie et al.*, 2014; *Ito et al.*, 2014]. The objective is a chemical reaction with CO_2 that produces a precipitate that fills the pores or open fractures of a high permeable zone and stops the leakage [*Ito et al.*, 2014]. This more recent concept evolved from the recognition that conventional materials such as cements and polymers are not suitable for sealing leaks within a reservoir, because their flow through pores and fractures is limited by their high viscosities. However, if CO_2 is used as a reactant for the grout formation, it may be possible to inject an aqueous solution with a low viscosity.

The aqueous solution would be injected on top of the caprock of the storage reservoir, where it spreads laterally and would lead to a precipitation reaction once it is in contact with $scCO_2$ or CO_2-enriched water. *Ito et al.* [2014] have suggested two different cases where this mechanism can be used. In the first case, it is considered a remediation technique; here the solution is injected on top of the caprock after CO_2 leakage is detected. In contrast, the second scenario is proposed as a mitigation technique, where the solution is emplaced on top of the caprock before the CO_2 is injected into the storage reservoir. The latter case is a precautionary measure in situations where the seal integrity is uncertain. Here the precipitation occurs in case of an eventual leak. Both scenarios are shown in Figure 16.3.

Several chemicals have been identified as suitable to carry out the desired reaction, namely, calcium hydroxide, magnesium hydroxide, and sodium silicates [*Brydie et al.*, 2014]. The calcium and magnesium hydroxide solutions are expected to precipitate carbonate minerals in contact with CO_2. However, this application has been less studied because the formation of calcite and hydromagnesite may lead to complex changes in pH and later carbonate dissolution may occur under CO_2 storage [*Ito et al.*, 2014]. The alkaline sodium silicate solution in contact with a CO_2-enriched solution leads to the precipitation of an amorphous silica phase stable under acidic and pH neutral conditions. In this case, the precipitation is triggered by a change in pH. CO_2-enriched water has a pH of approximately 4–6 under reservoir conditions depending on the alkalinity and the degree of CO_2 enrichment in the water [*Ito et al.*, 2014]. The mixing of the acidic solution with the alkaline sodium silicate solution results in a pH neutral to mildly alkaline solution, triggering the formation of amorphous silica. This behavior is illustrated in a stability diagram for sodium silicate (Fig. 16.4) under CO_2 storage conditions with a temperature of 60°C and a pressure of 140 bars.

The stability diagram shows that amorphous silica is in equilibrium with silicic acid ($H_3SiO_4^-$) between a pH of approximately 9.5 and 12. During the mixing of the two solutions, precipitation of amorphous silica starts at about a pH of 11.5 and proceeds until the mixing is completed and the final pH and silicate concentration is reached. The precipitation of amorphous silica will plug fractures and pores and is stable under high CO_2 and acidic conditions [*Ito et al.*, 2014].

Ito et al. [2014] simulated the reaction with the silicate solution using a reactive transport model. The simulation was performed with TOUGHREACT, a code that couples chemical reactions to multiphase fluid flow [*Xu et al.*, 2005]. Results of the simulation showed that precipitation of amorphous silica, permeability reduction, and CO_2 leakage mitigation were possible. A decrease in permeability was only observed in some areas of the modeling domain; however, a drop in permeability was observed in all preliminary laboratory experiments.

Brydie et al. [2014] also conducted laboratory experiments testing the above described silicate reaction in porous media under reservoir conditions. In this study, a core flood apparatus was used to induce the reaction within a sandstone core from the Berea Formation. The experiment found that the silica precipitation occurred within minutes and successfully blocked the flow through the porous medium.

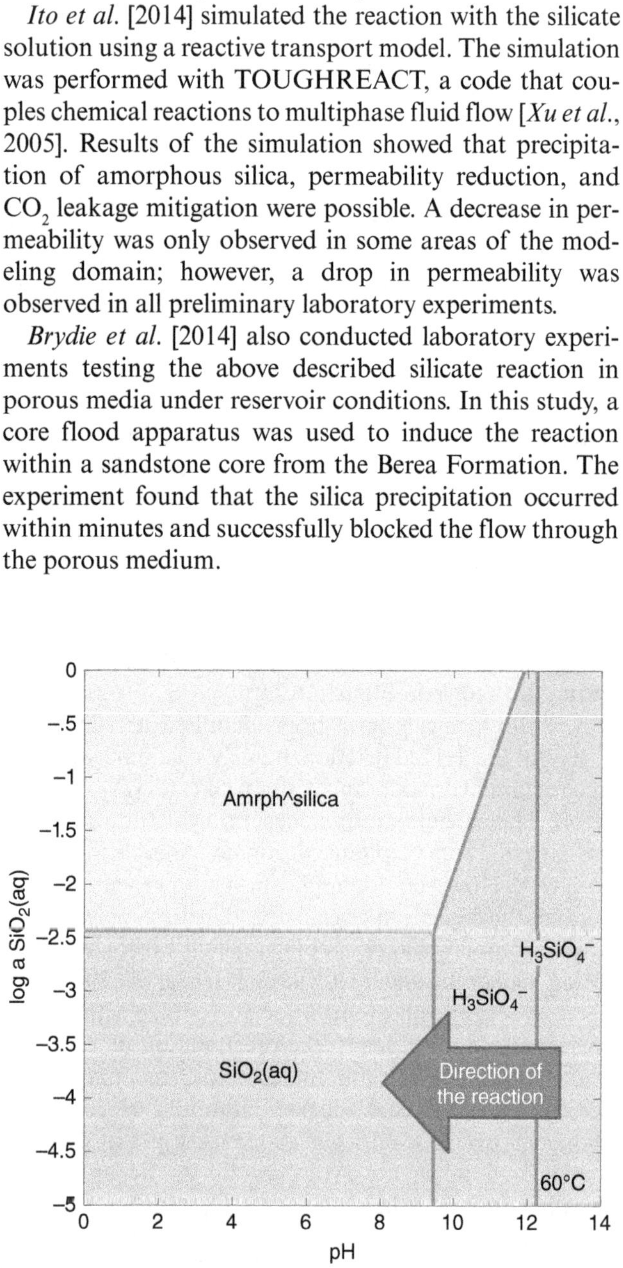

Figure 16.4 Activity diagram of aqueous silica with a representation of the proposed mixing of fluids leading to the precipitation of amorphous silica (T = 60°C, P = 140 bars).

One of the main challenges of this technology is the successful emplacement of the reactive solution near the CO_2 leakage. *Druhan et al.* [2014] also modeled with TOUGHREACT the injection of a pH-dependent sealant into a leaking zone. It was found that a well-defined mixing zone between the alkaline sealant and the CO_2-enriched water is necessary in order to have an effective seal. A correct implementation of this technology requires a good understanding of the hydrodynamics of the leak, changes in relative permeability, and changes in capillary pressure [*Vialle et al.*, 2016].

While these initial results are promising, research into chemical reactive barriers has only recently commenced. Many factors controlling the reaction rates and products such as the mixing rate of the two fluids, the reaction time, and silica concentration still need to be better understood. For field deployment, the successful detection of a leakage location is a major challenge. Further experimental research supported by geochemical modeling is required in order to optimize this technology for sealing CO_2 leakage pathways.

16.4. SYNTHESIS AND RECOMMENDATIONS

CO_2 leakage pathways can be grouped into two different categories. The first is the engineered CO_2 leakage pathway which is related to the impairment of wells, especially through the degradation of well materials such as cements and steel. The second leakage pathway relates to migration of CO_2 outside the containment through permeable zones in the seal.

In order to inhibit the migration of CO_2 through these different leakage pathways, several technologies and materials have been proposed. Dependent on the viscosity of the implemented material, the technologies have been categorized into two groups. A summary of the identified technologies can be found in Table 16.1.

The first group is the high-viscosity fluid-based barrier formation, which includes cements, geopolymers, gels, and nanoparticles. Cements and geopolymers are useful to block engineered pathways. However, settle time, chemical degradation, and injectivity have to be considered when using these materials. Cement has been the traditional sealing material of wells for a long time, especially in the oil and gas industry; however, chemical degradation of cements under CO_2 storage conditions has been demonstrated. More recently, geopolymers have been proposed as a replacement for cement because of their better chemical stability under CO_2 storage conditions. It has been found that geopolymers have a good response to low pH and high temperature conditions and high dissolved salt concentrations.

Similar to geopolymers, gels have also been proposed. They are polymer-based materials that use clay minerals

Table 16.1 Technologies for Mitigation and Remediation of CO_2 Leakage.

Barrier formation group	Material/ technology	Proposed pathway	Chemical resistance to CO_2 environment	Stage of research for CO_2 storage	Disadvantages
High-viscosity fluid based	Cements	Engineered	Low	Applied in the field	Degradation under CO_2 conditions
	Geopolymers	Engineered	High	Ongoing lab research	Manufacture needs development procedures
	Gels	Engineered Natural	Medium	Ongoing lab research	Not suitable for sealing at high pressures
	Nanoparticles	Engineered Natural	High	Ongoing lab research	Immobilization of CO_2 will not occur with difference in viscosity
Low-viscosity fluid based	Biomineralization	Natural	Medium	Ongoing lab research	Growth of microbial communities
	Hydraulic barrier	Natural	N/A	Ongoing simulation research	Requires certain geological conditions and brine availability
	Chemical reactive barrier	Natural	Medium	Ongoing lab research	Depends on precipitation rates and fluid mixing

as cross-linkers. They have the advantage of being more stable under acidic conditions than traditional cement, but the sealing capacity is limited as gels tend to be susceptible to the formation of wormholes when pressure gradients are present. Similarly, foams can form as a result of interactions between nanoparticles CO_2-enriched water, reducing the migration of CO_2.

The second group of barrier formation technologies is based on low-viscosity fluids, which are more suitable to remediate leakage through natural pathways. These technologies are biomineralization and/or biofilm formation, hydraulic barrier formation, and chemical reactive barrier formation.

Microorganisms in combination with certain reactants can be used for biomineralization. Calcite is the most common precipitate effectively sealing leaks. The application of microbes requires certain conditions such as nutrient availability for rapid microbial growth and the availability of calcium for the abundant precipitation of calcite. While it is important to further develop this technology, it is recognized that very little is known about the microbial community structure in deep reservoirs and their response to high CO_2 conditions, warranting further research as well.

Hydraulic barrier formation in the aquifer overlying the caprock can be used to change the hydrodynamics through pressure management of the site and redirect the CO_2 flow direction if the permeability is not too high. The implementation of this technique is site specific and requires economical and operational efforts.

As a last technology, we considered chemical reactive barrier formation. This technology is based on a chemical reaction between the migrating CO_2 and a reactive solution that generates a stable sealing mineral. Different reactive solutions have been proposed, but sodium silicate has been found as most suitable. One advantage of this application is the stability of amorphous silica under acidic conditions of CO_2 storage over long time. This technology may be used also as mitigation procedure. However, more studies are necessary, particularly in relation to the rate and degree of precipitation and the associated reduction in permeability within the zone of fluid mixing.

As it is shown in Table 16.1, each of the technologies proposed could be used for either of the two possible pathways; however, some of them are not resistant to a CO_2 storage environment. Cement is the least suitable material for the remediation of CO_2 leakage through engineered pathways and may well be replaced by geopolymers in the future. Regarding natural pathways, the group of low-viscosity fluid-based barrier formation technologies is most suitable since they could be emplaced easily in the subsurface. The chemical reactive barrier formation seems to be the most promising approach, since it does not need high operational costs or introduces large uncertainty, for example, relating growth conditions of microbial communities.

In conclusion, CO_2 storage risk assessments in recent years have shown that potential natural and engineered CO_2 leakage pathways represent a considerable concern. A number of new technologies and materials have been proposed to mitigate and remediate CO_2 leakage, and those have been derived from other industrial applications. All of these technologies are currently at a conceptual or evaluation stage at laboratory scale. Further process studies under well-defined laboratory conditions as well as predictive modeling studies are required to identify the most suitable and cost-effective barrier formation technology for a given leakage situation. Field validation of a particular technology will be critical to make it acceptable to the broader CCS community and ultimately contribute to CO_2 storage safety.

REFERENCES

Aminzadeh, B., Chung, D. H., Bryant, S. L., Huh, C., and DiCarlo, D. A. (2013), CO_2 leakage prevention by introducing engineered nanoparticles to the in-situ brine, *Energy Procedia*, *37*, 5290–5297.

Aydin, A. (2000), Fractures, faults, and hydrocarbon entrapment, migration and flow, *Marine and Petroleum Geology*, *17*(7), 797–814.

Barclay, I., Pellenbarg, J., Tettero, F., Pfeiffer, J., Slater, H., Staal, T., Stiles, D., Tilling, G., and Whitney, C. (2001), The beginning of the end: a review of abandonment and decommissioning practices, *Oilfield Review*, *13*(4), 28–41.

Brydie, J. R., Perkins, E. H., Fisher, D., Girard, M., Valencia, M., Olson, M., and Rattray, T. (2014), The development of a leak remediation technology for potential non-wellbore related leaks from CO_2 storage sites, *Energy Procedia*, *63*, 4601–4611.

Carey, J. W. (2013), Geochemistry of wellbore integrity in CO_2 sequestration: Portland cement-steel-brine-CO_2 interactions, in *Geochemistry of Geologic CO_2 Sequestration*, edited by D. J. DePaolo, D. R. Cole, A. Navrotsky, and I. C. Bourg, pp. 505–539, Reviews in Mineralogy and Geochemistry, Mineralogical Society of America, Washington, DC.

Cunningham, A. B., Gerlach, R., Spangler, L., and Mitchell, A. C. (2009), Microbially enhanced geologic containment of sequestered supercritical CO_2, *Energy Procedia*, *1*, 3245–3252.

Cunningham, A. B., Gerlach, R., Spangler, L., Mitchell, A. C., Parks, S., and Phillips, A. (2011), Reducing the risk of well bore leakage of CO_2 using engineered biomineralization barriers, *Energy Procedia*, *4*, 5178–5185.

Cunningham, A. B., Lauchnor, E., Eldring, J., Esposito, R., Mitchell, A. C., Ebigbo, A., Spangler, L. H., Gerlach, R., and Phillips, A. J. (2013), Abandoned well CO_2 leakage mitigation using biologically induced mineralization: current progress and future directions, *Greenhouse Gases: Science and Technology*, *3*(1), 40–49.

DiCarlo, D. A., Aminzadeh, B., Roberts, M., Chung, D. H., Bryant, S. L., and Huh, C. (2011), Mobility control through spontaneous formation of nanoparticle stabilized emulsions, *Geophysical Research Letters*, *38*(24), L24404.

Druhan, J. L., Vialle, S., Maher, K., and Benson, S. (2014), A reactive transport model for geochemical mitigation of CO_2 leaking into a confined aquifer, *Energy Procedia*, *63*, 4620–4629.

Duxson, P., Provis, J. L., Lukey, G. C., Mallicoat, S. W., Kriven, W. M., and Van Deventer, J. S. (2005), Understanding the relationship between geopolymer composition, microstructure and mechanical properties, *Colloids and Surfaces A: Physicochemical and Engineering Aspects*, *269*(1), 47–58.

Fitts, J. P., and Peters, C. A. (2013), Caprock fracture dissolution and CO_2 leakage, in *Geochemistry of Geologic CO_2 Sequestration*, edited by D. J. DePaolo, D. R. Cole, A. Navrotsky and I. C. Bourg, pp. 459–479, Reviews in Mineralogy and Geochemistry, Mineralogical Society of America, Washington, DC.

Gasda, S. E., Bachu, S., and Celia, M. A. (2004), Spatial characterization of the location of potentially leaky wells penetrating a deep saline aquifer in a mature sedimentary basin, *Environmental Geology*, *46*(6–7),707–720.

Ito, T., Xu, T., Tanaka, H., Taniuchi, Y., and Okamoto, A. (2014), Possibility to remedy CO_2 leakage from geological reservoir using CO_2 reactive grout, *International Journal of Greenhouse Gas Control*, *20*, 310–323.

Kutchko, B. G., Strazisar, B. R., Dzombak, D. A., Lowry, G. V., and Thauiow, N. (2007), Degradation of well cement by CO_2 under geologic sequestration conditions, *Environmental Science and Technology*, *41*(13), 4787–4792.

Le Quéré, C., Moriarty, R., Andrew, R. M., Peters, G. P., Ciais, P., Friedlingstein, P., Jones, S. D., Sitch, S., Tans, P., and Arneth, A. (2015), Global carbon budget 2014, *Earth System Science Data*, *7*(1), 47–85.

Lions, J., Devau, N., de Lary, L., Dupraz, S., Parmentier, M., Gombert, P., and Dictor, M.-C. (2014), Potential impacts of leakage from CO_2 geological storage on geochemical processes controlling fresh groundwater quality: a review, *International Journal of Greenhouse Gas Control*, *22*, 165–175.

Manceau, J. C., Hatzignatiou, D. G., de Lary, L., Jensen, N. B., and Réveillère, A. (2014), Mitigation and remediation technologies and practices in case of undesired migration of CO_2 from a geological storage unit-current status, *International Journal of Greenhouse Gas Control*, *22*, 272–290.

Metz, B., Davidson, O., De Coninck, H., Loos, M., and Meyer, L. (2005), IPCC, 2005: IPCC special report on carbon dioxide capture and storage. Prepared by Working Group III of the Intergovernmental Panel on Climate Change, Cambridge and New York, 442 pp.

Mitchell, A. C., Phillips, A. J., Hiebert, R., Gerlach, R., Spangler, L. H., and Cunningham, A. B. (2009), Biofilm enhanced geologic sequestration of supercritical CO_2, *International Journal of Greenhouse Gas Control*, *3*(1), 90–99.

Mu, A., and Moreau, J. W. (2015), The Geomicrobiology of CO_2 geosequestration: a focused review on prokaryotic community responses to field-scale CO_2 injection, *Frontiers in Microbiology*, *6*, 263.

Mu, A., Boreham, C., Leong, H. X., Haese, R., and Moreau, J. W. (2014), Changes in the deep subsurface microbial biosphere resulting from a field-scale CO_2 geosequestration experiment, *Frontiers in Microbiology*, *5*, 209.

Nasvi, M. M. C., Gamage, R. P., and Jay, S. (2012), Geopolymer as well cement and the variation of its mechanical behavior with curing temperature, *Greenhouse Gases: Science and Technology*, *2*(1), 46–58.

Nasvi, M. C. M., Ranjith, P. G., and Sanjayan, J. (2013a), The permeability of geopolymer at down-hole stress conditions: application for carbon dioxide sequestration wells, *Applied Energy*, *102*, 1391–1398.

Nasvi, M. C. M., Ranjith, P. G., Sanjayan, J., and Haque, A. (2013b), Sub- and super-critical carbon dioxide permeability of wellbore materials under geological sequestration conditions: an experimental study, *Energy*, *54*, 231–239.

Nasvi, M. C. M., Ranjith, P. G., Sanjayan, J., Haque, A., and Li, X. (2014), Mechanical behaviour of wellbore materials saturated in brine water with different salinity levels, *Energy*, *66*, 239–249.

Pareek, N., Jat, M. K., and Jain, S. (2006), The utilization of brackish water, protecting the quality of the upper fresh water layer in coastal aquifers, *Environmentalist*, *26*(4), 237–246.

Pawar, R. J., Bromhal, G. S., Carey, J. W., Foxall, W., Korre, A., Ringrose, P. S., Tucker, O., Watson, M. N., and White, J. A. (2015), Recent advances in risk assessment and risk management of geologic CO_2 storage, *International Journal of Greenhouse Gas Control*, *40*, 292–311.

Peet, K., Freedman, A., Hernandez, H., Britto, V., Boreham, C., Ajo-Franklin, J., and Thompson, J. (2015), Microbial growth under supercritical CO_2, *Applied and Environmental Microbiology*, *81*, 03162–03114.

Phillips, A. J., Lauchnor, E., Eldring, J., Esposito, R., Mitchell, A. C., Gerlach, R., Cunningham, A. B., and Spangler, L. H. (2013), Potential CO_2 leakage reduction through biofilm-induced calcium carbonate precipitation, *Environmental Science & Technology*, *47*(1), 142–149.

Provis, J. L. (2009), Activating solution chemistry for geopolymers, in *Geopolymers: Structures, Processing, Properties and Industrial Applications*, edited by J. L. Provis, pp. 50–71, Woodhead Publishing Limited, Cambridge.

Provis, J. L., and Van Deventer, J. S. J. (Eds.) (2009), *Geopolymers: Structures, Processing, Properties and Industrial Applications*, 454 pp., Woodhead Publishing Limited, Cambridge.

Réveillère, A., Rohmer, J., and Manceau, J. C. (2012), Hydraulic barrier design and applicability for managing the risk of CO_2 leakage from deep saline aquifers, *International Journal of Greenhouse Gas Control*, *9*, 62–71.

Tongwa, P., Nygaard, R., and Bai, B. (2013a), Evaluation of a nanocomposite hydrogel for water shut-off in enhanced oil recovery applications: design, synthesis, and characterization, *Journal of Applied Polymer Science*, *128*(1), 787–794.

Tongwa, P., Nygaard, R., Blue, A., and Bai, B. (2013b), Evaluation of potential fracture-sealing materials for remediating CO_2 leakage pathways during CO_2 sequestration, *International Journal of Greenhouse Gas Control*, *18*, 128–138.

Vialle, S., Druhan, J. L., and Maher, K. (2016), Multi-phase flow simulation of CO_2 leakage through a fractured caprock in response to mitigation strategies, *International Journal of Greenhouse Gas Control*, *44*, 11–25.

Xu, T., Apps, J. A., and Pruess, K. (2005), Mineral sequestration of carbon dioxide in a sandstone-shale system, *Chemical Geology*, *217*(3–4 SPEC. ISS.), 295–318.

Zahasky, C., and Benson, S. M. (2016), Evaluation of hydraulic controls for leakage intervention in carbon storage reservoirs, *International Journal of Greenhouse Gas Control*, *47*, 86–100.

Zhang, Z., Provis, J. L., Reid, A., and Wang, H. (2014), Geopolymer foam concrete: an emerging material for sustainable construction, *Construction and Building Materials*, *56*, 113–127.

INDEX

Geological Carbon Storage: Subsurface Seals and Caprock Integrity, Geophysical Monograph 238, First Edition.
Edited by Stéphanie Vialle, Jonathan Ajo-Franklin, and J. William Carey.

www.ingramcontent.com/pod-product-compliance
Lightning Source LLC
LaVergne TN
LVHW062144200926
848484LV00007B/1852
9781119118640